METHODS
IN AGING RESEARCH

METHODS
IN AGING RESEARCH

Edited by

Byung Pal Yu

informa
healthcare

New York   London

First published in 1999 by CRC Press LLC, 2000 Corporate Blvd., N.W., Boca Raton, FL 33431.
This edition published in 2010 by Informa Healthcare, Telephone House, 69-77 Paul Street, London EC2A 4LQ, UK.

Simultaneously published in the USA by Informa Healthcare, 52 Vanderbilt Avenue, 7th Floor, New York, NY 10017, USA.

Informa Healthcare is a trading division of Informa UK Ltd. Registered Office: 37–41 Mortimer Street, London W1T 3JH, UK. Registered in England and Wales number 1072954.

A CIP record for this book is available from the British Library.

Library of Congress Cataloging-in-Publication Data available on application

ISBN-13: 9780849331121

Orders may be sent to: Informa Healthcare, Sheepen Place, Colchester, Essex CO3 3LP, UK
Telephone: +44 (0)20 7017 5540
Email: CSDhealthcarebooks@informa.com
Website: http://informahealthcarebooks.com/

For corporate sales please contact: CorporateBooksIHC@informa.com
For foreign rights please contact: RightsIHC@informa.com
For reprint permissions please contact: PermissionsIHC@informa.com

Contents

Section D: Methods of Assessing Aging Processes

Section E: Molecular and Evolutionary Probes of Senescence
Alterations

Section F: Techniques for Exploring Age-Related
Intra- and Subcellular Changes

Section G: Techniques for Assessing Age-Related Oxidative Modification

Preface

My goal as editor of this book is to provide researchers and scientists interested in the study of aging with the technical methods necessary to investigate their specific area of concern. Naturally, the basic layout of this book is designed not to function as a review of methodologies or to provide theoretical approaches to the study of aging, but rather to serve as a practical, bench-top guide for anyone who would like to know how to plan, design, or conduct an aging research project. The methods highlighted here delineate the execution of the experimental procedures key to the investigation of biological aging phenomena.

Because biological aging is a diverse and multifaceted process, the content of this book focuses on the various technical and procedural aspects used to investigate the aging process by considering the multiple needs of scientists. To accomplish this end, the book is divided into seven sections. In Section A, authors cover general theoretical approaches to the study of aging by offering basic guidelines to follow in the development of a scientific paradigm. Topics include proven experimental designs as well as novel approaches in aging studies. In Section B and Section C, expert researchers lead readers in the selection and use of specific experimental models, ranging from simpler living systems, such as yeast and nematodes, to more complex organisms, such as the nonhuman primate. The chapters in Section D through Section G are devoted to delineating the various procedures, techniques, and probes used to assess aging processes. Experienced gerontologists share their knowledge and technical expertise on a wide range of contemporary and innovative research methods. These methods are dissected for easy comprehension with the aim, where appropriate, of guiding the reader through an in-depth molecular exploration of a subject, as in the cases of tolemere and genome investigations, for example. The chapters in these sections also describe cutting-edge approaches to aging research by introducing techniques used in the genetic modification of animals, such as those used with transgenic animal models.

As editor, I wanted to offer scientists a state-of-the-art reference that not only explores the multidisciplinary nature of modern gerontology but gives direct, clear-cut descriptions of specific methodological procedures. Close to seventy authoritative scientists from the United States and abroad have shared their knowledge and expertise on a wide range of gerontological research topics, extending well beyond any given methodology or protocol. Furthermore, I have asked these authors to critically evaluate the methods and protocols, wherever possible, by explaining the advantages and disadvantages or potential pitfalls of the technique they have described. My hope was to give potential users first-hand information well in advance of choosing the most appropriate technique.

Of course, I am honored by the eminent researchers who have agreed to share their wisdom and experience in this book. The authors were carefully selected not only for their accomplishments in the research at hand, but also for their currently active roles in cultivating and developing the present gerontological research frontier. Undoubtedly, their participation in the writing of this book is invaluable to the further advancement of gerontological research. With gratitude, I commend each of them for their thoughtful contributions.

On a personal note, I wish to thank Drs. Jeremiah Herlihy, Yuji Ikeno, and Helen Bertrand, not only for the mutual respect and intellectual enrichment we continue to share from our professional relationship, but also for their long friendship. Finally, my sincere thanks goes to Ms. Corinne Price for her editorial contributions; her abiding perseverance was exemplary and essential to the publication of this book.

The Editor

Dr. Byung Pal Yu has served on the faculty of the Department of Physiology at The University of Texas Health Science Center since 1973, where he currently holds the rank of professor. He received his undergraduate education from Central Missouri State College, Warrenburg, Missouri, where he majored in chemistry, and in 1960 entered graduate school at the University of Illinois at Urbana, earning his Doctor of Philosophy in 1965.

Dr. Yu has been actively involved in various aspects of biological aging research for more than 30 years. Devoting his most recent work to the exploration of the interaction of free radicals, lipid peroxidation, and dietary restriction on the aging process, his contributions using the proposed "Oxidative Stress Theory of Aging" have been significant to the redefining of free radical involvement in aging and age-related disease processes. Furthermore, as the originator of a proposal linking dietary restriction to the modulation of free radicals and oxidative stress, Dr. Yu has elicited great enthusiasm among biological gerontologists for dietary restriction's anti-aging action.

Among Dr. Yu's professional accomplishments is a recently completed $5 million research program project in nutritional gerontology, for which he served as director and was supported by the National Institute on Aging (NIA). He is also a past-president and current board member of the American Aging Association and has served as chair for the Biological Sciences Section of the Gerontological Society of America. Dr. Yu was also a member of the NIH Nutrition Study Section, and is the current chair of the Advisory Committee for a benchmark project of the NIA's Gerontological Research Center, Baltimore, Maryland, which is studying the effects of dietary restriction on the nonhuman primate.

Among his various editorial responsibilities, Dr. Yu has been the associated editor for the *Journal of Gerontology* for the past seven years and is an editorial member of many scientific journals, to include *AGE: Proceedings of the Society for Experimental Biology and Medicine; Journal of Nutrition, Health and Aging:* and *Aging: Clinical and Experimental Research.* Dr. Yu also has more than 235 scientific publications to his credit, to include the editing of two books published by CRC Press, *Free Radicals in Aging* (1994) and *The Modulation of the Aging Process by Dietary Restriction.*

Dr. Yu's memberships in professional societies include his role as a Gerontological Society of America Fellow and a member of the Oxygen Society and American Physiological Society.

His long-time commitment to scientific aging research has earned Dr. Yu the prestigious Ho-Am prize in medicine for 1998. The Ho-Am Award of Korea was established to honor those who have distinguished themselves with world-class achievement in academia and the arts by propagating a noble spirit of dedication to the betterment of our society. Dr. Yu is one of five laureates chosen to represent the respective fields of Medical Sciences, Basic Science, Engineering, the Arts, and Social Service.

Contributors

R. C. Allsopp Department of Pathology, Stanford University School of Medicine, Stanford, California

Robert Arking Department of Biological Sciences, Wayne State University, Detroit, Michigan

Steven N. Austad Department of Biological Sciences, University of Idaho, Moscow, Idaho

Gustavo Barja Department Animal Biology-II, Animal Physiology, Faculty of Biology, Complutense University, Madrid, Spain

Alberto Benguria Department of Biochemistry and Molecular Biology, Louisiana State University Medical Center, New Orleans, Louisiana

Helen A. Bertrand Department of Physiology, The University of Texas Health Science Center at San Antonio, San Antonio, Texas

Michael E. T. I. Boerrigter Division on Aging, Harvard Medical School and Gerontology Division, Beth Israel Deaconess Medical Center, Harvard Institutes of Medicine, Boston, Massachusetts

Ivor D. Bowen Pure and Applied Biology, University of Wales, Cardiff, Wales, United Kingdom

Palmiro Cantatore Department of Biochemistry and Molecular Biology, University of Bari and CNR Unit, Bari, Italy

James R. Carey Department of Entomology, University of California, Davis, California

Ricki J. Colman Wisconsin Regional Primate Research Center and Department of Anthropology, University of Wisconsin, Madison, Wisconsin

Vincent J. Cristofalo Center for Gerontological Research, Allegheny University of the Health Sciences, Philadelphia, Pennsylvania

Miral Dizdaroglu Chemical Science and Technology Laboratory, National Institute of Standards and Technology, Gaithersburg, Maryland

Martijn E. T. Dollé Division on Aging, Harvard Medical School and Gerontology Division, Beth Israel Deaconess Medical Center, Harvard Institutes of Medicine, Boston, Massachusetts

Gabriel Fernandes   Department of Medicine, The University of Texas Health Science Center at San Antonio, San Antonio, Texas

John F. Fogarty   Institute of Pathology, Case Western Reserve University, Cleveland, Ohio

Mary Kay Francis   Center for Gerontological Research, Allegheny University of the Health Sciences, Philadelphia, Pennsylvania

Eitan Friedman   Department of Pharmacology, Allegheny University of the Health Sciences, MCP ♦ Hahnemann School of Medicine, Philadelphia, Pennsylvania

Maria N. Gadaleta   Department of Biochemistry and Molecular Biology, University of Bari, CNR Unit, Bari, Italy

David Gershon   Department of Biology, Technion-Israel Institute of Technology, Haifa, Israel

Mitch Gore   Department of Kinesiology, University of Wisconsin-Madison, Madison, Wisconsin

Jan A. Gossen   N.V. Organon, Oss, The Netherlands

Yves Guigoz   Nestle Research Center, Nestec Ltd., Vers-Chez-Les-Blancs, Lausanne, Switzerland

C. B. Harley   Geron Corporation, Menlo Park, California

Ronald W. Hart   National Center for Toxicological Research, Jefferson, Arkansas

Karen L. Heman   Department of Pharmacology, University of Colorado Health Science Center, Denver Colorado

Jeremiah T. Herlihy   Department of Physiology, The University of Texas Health Science Center at San Antonio, San Antonio, Texas

Nikki J. Holbrook   Gene Expression and Aging Section, Laboratory of Cellular and Molecular Biology, Gerontology Research Center, National Institute on Aging, Baltimore, Maryland

Donna J. Holmes   Department of Biological Sciences, University of Idaho, Moscow, Idaho

Gene B. Hubbard   Department of Laboratory Animal Medicine, Southwest Foundation for Biomedical Research, San Antonio, Texas

Yuji Ikeno   Department of Physiology, The University of Texas Health Science Center at San Antonio, San Antonio, Texas

Donald K. Ingram   Gerontology Research Center, National Institutes of Health, Baltimore, Maryland

S. Michal Jazwinski   Department of Biochemistry and Molecular Biology, Louisiana State University Medical Center, New Orleans, Louisiana

Li Li Ji   Department of Kinesiology, University of Wisconsin-Madison, Madison, Wisconsin

Takao Kaneko   Tokyo Metropolitan Institute of Gerontology, Itabashiku, Tokyo, Japan

Joseph W. Kemnitz   Wisconsin Regional Primate Research Center, and Departments of Medicine and Institute on Aging, University of Wisconsin, Madison, Wisconsin

N. W. Kim   Geron Corporation, Menlo Park, California,

Sangkyu Kim   Department of Biochemistry and Molecular Biology, Louisiana State University Medical Center, New Orleans, Louisiana

Paul A. Kirchman   Department of Biochemistry and Molecular Biology, Louisiana State University Medical Center, New Orleans, Louisiana

Shoichi Kitano   Molecular Physiology and Genetics Section, Laboratory of Cellular and Molecular Biology, Gerontology Research Center, National Institute on Aging, Baltimore, Maryland

Gertrude C. Kokkonen   Molecular Physiology and Genetics Section, Laboratory of Cellular and Molecular Biology, Gerontology Research Center, National Institute on Aging, Baltimore, Maryland

Barbara L. Leard   The Bionetics Corporation, National Center for Toxicological Research, Jefferson, Arkansas

Sherry M. Lewis   The Bionetics Corporation, National Center for Toxicological Research, Jefferson, Arkansas

Angela M. S. Lezza   Department of Biochemistry and Molecular Biology, University of Bari, CNR Unit, Bari, Italy

Yusen Liu   Gene Expression and Aging Section, Laboratory of Cellular and Molecular Biology, Gerontology Research Center, National Institute on Aging, Baltimore, Maryland

Edward J. Masoro   Department of Physiology, University of Texas Health Science Center, San Antonio, Texas

Mitsuyoshi Matsuo   Department of Biology, Faculty of Science, Konan University, Higashinadaku, Kobe, Hyogo, Japan

Vincent M. Monnier   Institute of Pathology, Case Western Reserve University, Cleveland, Ohio

Camille S. Monnier   Institute of Pathology, Case Western Reserve University, Cleveland, Ohio

Hélène Payette   Gerontology and Geriatric Research Center, Sherbrooke Geriatric University Institute, Sherbrooke, Québec, Canada

Gowri K. Pyapali   Department of Neurosurgery, Duke University Medical Center, and Research and Surgery Services, Durham VA Medical Center, Durham North Carolina

Abraham Z. Reznick   Division of Morphological Sciences, The Bruce Rappaport Faculty of Medicine, Technion-Israel Institute of Technology, Haifa, Israel

Jay Roberts   Department of Pharmacology, Allegheny University of the Health Sciences, MCP ♦ Hahnemann School of Medicine, Philadelphia, Pennsylvania

Gregory M. Rose   Neuroscience Drug Discovery, Bristol-Myers Squibb Company, Wallingford, Connecticut

George S. Roth   Molecular Physiology and Genetics Section, Laboratory of Cellular and Molecular Biology, Gerontology Research Center, National Institute on Aging, Baltimore, Maryland

David R. Sell   Institute of Pathology, Case Western Reserve University, Cleveland, Ohio

David L. Snyder   Department of Pharmacology, Allegheny University of the Health Sciences, MCP ♦ Hahnemann School of Medicine, Philadelphia, Pennsylvania

Pamela E. Starke-Reed   Biology of Aging Program, National Institute on Aging, National Institutes of Health, Bethesda, Maryland

Maria Tresini   Center for Gerontological Research, Allegheny University of the Health Sciences, Philadelphia, Pennsylvania

Dennis A. Turner   Departments of Neurosurgery and Neurobiology, Duke University Medical Certer, and Research and Surgery Services, Durham VA Medical Center, Durham, North Carolina

Angelo Turturro   National Center for Toxicological Research, Jefferson, Arkansas

Bruno J. Vellas   Professor of Internal Medicine, CHU Purpan Casselardit, Toulouse, France

Jan Vijg   Division on Aging, Harvard Medical School and Gerontology Division, Beth Israel Deaconess Medical Center, Harvard Institutes of Medicine, Boston, Massachusetts

Craig Volker   Center for Gerontological Research, Allegheny University of the Health Sciences, Philadelphia, Pennsylvania

Christi A. Walter   Department of Cellular and Structural Biology, The University of Texas Health Science Center at San Antonio, San Antonio, Texas

S.-S. Wang   Geron Corporation, Menlo Park, California

Tarynn M. Witten Computer Medicine Group, Institute of Gerontology, University of Michigan, Ann Arbor, Michigan

R. C. Woodruff Department of Biological Sciences, Bowling Green State University, Bowling Green, Ohio

Byung P. Yu Department of Physiology, The University of Texas Health Science Center at San Antonio, San Antonio, Texas

Section A

General Theoretical Approaches to Aging Research

1

Population Study of Mortality and Longevity with Gompertzian Analysis

James R. Carey

CONTENTS

0-8493-3112-9/99/$0.00+$.50

1.1   Introduction

It is virtually impossible to separate the central premise of gerontology—that most organisms become progressively more susceptible to disease and injury and ultimately to death as they age—from its actuarial implications. The *susceptibilities* associated with the biology of aging determine the death rates and the corresponding age-specific mortality *probabilities*. These probabilities or hazard rates underlie the two most commonly used demographic summary measures in gerontology—cohort survival and life expectancy. Because of the emphasis on extended longevity and reduced mortality in gerontological research, an understanding of actuarial concepts and methods and of basic mortality models such as the Gompertz model [1] is fundamental to studies in many areas of aging research. These include not only studies concerned with changes at the level of the whole organism, but also those designed to measure the indirect effects on mortality of age-related changes in the organism's molecular, cellular, and physiological components.

Although a large literature exists on analytical methods in actuarial studies, many of these are books and articles written for professional actuaries, mathematical demographers, or statisticians rather than for research biologists [2,3]. My objectives in this chapter are to present an analytical framework for analyzing basic data on survival, mortality, and longevity in biogerontology as well as to provide a general overview of actuarial concepts. I make no attempt to be comprehensive but rather provide references for follow-up or more advanced analyses.

1.2   The Life Table

1.2.1   Concepts

The life table is one of the most important tools in biogerontological research because it serves as a framework for organizing age-specific mortality data, because it provides detailed, transparent descriptions of the actuarial properties of a cohort, and because it generates simple summary statistics such as life expectancy that are useful for comparisons. There are two general forms of the life table, both of which are relevant to biogerontology. The first is the *cohort life table*, which provides a longitudinal perspective in that it includes the mortality experience of a particular cohort from the moment of birth through consecutive ages until none remains in the original cohort. The second basic form is the *current life table*, which is cross-sectional. This table assumes a hypothetical cohort subject throughout its lifetime to the age-specific mortality rates prevailing for the actual population over a specified period and is used to construct a synthetic cohort. Cohort (longitudinal) life tables are frequently used in laboratory studies of relatively short-lived species (insects, rodents) whereas current (cross-sectional) life tables are most often used in studies of long-lived species such as humans.

1.2.2   Construction and Interpretation

The five main functions of the life table are defined and their formulae presented in Table 1.2 including l_x, the proportion of all newborn surviving to age x, p_x, the proportion of individuals alive at age x that survive to x + 1, q_x, the proportion of individuals alive at age x that die prior to x + 1, d_x, the proportion of all newborn that die in the interval x to x + 1, and e_x, the expected lifetime remaining to the average individual age x. A partial life table is presented in Table 1.1, using mortality data presented in Carey et al. on the Mediterranean fruit fly [4]. Figure 1.1 presents the complete results of the analysis through age 90. The table reveals that: (1) less than 4% of the original cohort survives to 40 days (i.e., $l_{40} = 0.03784$; (2) age-specific mortality is extremely high around 40 days ($q_{40} = 0.12125$) but decreases at older ages (e.g., $q_{120} = 0.05128$), (3) over 12% of flies die in the 3-day interval 20–22 days (i.e., $d_{20} + d_{21} + d_{22} = 0.125$); and (4) expectation of life at age 90 ($e_{90} = 27.0$ days) is greater than expectation of life at eclosion ($e_0 = 20.8$ days). This reflects the leveling off and decrease in mortality at advanced ages. More comprehensive treatments of life table concepts and techniques are included in Chiang [5] and Carey [6].

1.2.3 Additional Life Table Parameters and Relationships

1.2.3.1 Abridged Life Tables

Abridged life tables do not give the life table functions for every census period since, for practical purposes, it may not be necessary to compute expressions for every age. The concepts and notation for cohort survival, l_x, and life expectancy, e_x, in the abridged table are the same as for the complete table. However, because the functions p_x, q_x, and d_x apply to intervals, their definitions and notation must reflect changes in interval in the abridged tables. Thus, the notation and definitions for these three functions concerned with intervals are $_np_x$ = the fraction of a cohort surviving from age x to age x + n; $_nq_x$ = the fraction of a cohort alive at age x that die prior to age x + n; and $_nd_x$ = the fraction of the initial cohort dying in the interval x t x + n. For example, the probability that a 2-day-old medfly will die prior to age 10 (see Table 1.1) is given as

$$_8q_2 = 1 - \frac{1,105,164}{1,201,913} = 0.0805$$

and the probability that one of the original medflies will die in the interval 5 to 10 days is given as

$$_5d_5 = l_5 - l_{10} = 0.98320 - 0.91818 = 0.06502$$

1.2.3.2 Central Death Rate

The parameter central death rate, also known as the age-specific death rate, denoted m_x, is defined as the number of deaths occurring in a specified period in a specific age category divided by the population at risk. The formula for m_x is

$$m_x = \frac{l_x - l_{x+1}}{0.5\left(l_x + l_{x+1}\right)} \tag{1.1}$$

The central death rate is not a probability but rather an observed rate—the number of individuals that die relative to the number at risk. It is essentially a weighted average of the force of mortality between ages x and x + 1.

TABLE 1.1

Life Table Parameters for Mediterranean Fruit Fly

Age	Number Living[4]	Fraction Surviving	Age-Specific Survival	Age-Specific Mortality	Frequency of Deaths	Expectation of Life
x	N_x	l_x	p_x	q_x	d_x	e_x
(1)	(2)	(3)	(4)	(5)	(6)	(7)
0	1,203,646	1.00000	1.00000	0.00000	0.0000	20.8
1	1,203,646	1.00000	0.99856	0.00144	0.0014	19.8
2	1,201,913	0.99856	0.99599	0.00401	0.0040	18.9
3	1,197,098	0.99456	0.99492	0.00508	0.0050	17.9
4	1,191,020	0.98951	0.99362	0.00638	0.0063	17.0
5	1,183,419	0.98320	0.99247	0.00753	0.0074	16.1
6	1,174,502	0.97579	0.99023	0.00977	0.0095	15.3
7	1,163,026	0.96625	0.98768	0.01232	0.0119	14.4
8	1,148,693	0.95434	0.98358	0.01642	0.0157	13.6
9	1,129,836	0.93868	0.97816	0.02184	0.0205	12.8
10	1,105,164	0.91818	0.97018	0.02982	0.0274	12.1
20	575,420	0.47806	0.90772	0.09228	0.0441	8.3
21	522,319	0.43395	0.90320	0.09680	0.0420	8.0
22	471,756	0.39194	0.89976	0.10024	0.0393	7.9
40	45,544	0.03784	0.87875	0.12125	0.0046	7.0
41	40,022	0.03325	0.87207	0.12793	0.0043	6.9
42	34,902	0.02900	0.86986	0.13014	0.0038	6.8
90	91	0.00008	0.94505	0.05495	0.0000	27.0
91	86	0.00007	0.98837	0.01163	0.0000	27.5
92	85	0.00007	0.92941	0.07059	0.0000	26.8
120	39	0.00003	0.94872	0.05128	0.0000	18.7
121	37	0.00003	0.97297	0.02703	0.0000	18.7
122	36	0.00003	0.91667	0.08333	0.0000	18.2
164	4	0.00000	0.50000	0.50000	0.0000	4.0
165	2	0.00000	1.00000	0.00000	0.0000	6.5
166	2	0.00000	1.00000	0.00000	0.0000	5.5
172	0	0.00000	0.00000	1.00000	0.0000	0.0

From Carey et al. 1992.

TABLE 1.2

Main Life Table Functions

Function	Notation	Equation	Description
Cohort survival	l_x	$\dfrac{N_x}{N_0}$	Fraction of initial cohort surviving to age x where N_x denotes number alive at age x
Age-specific mortality	q_x	$1 - \dfrac{l_{x+1}}{l_x}$	Fraction alive at age x that die prior to x + 1
Age-specific survival	p_x	$\dfrac{l_{x+1}}{l_x}$	Fraction alive at age x that survive to x + 1
Expectation of life at age x	e_x	$\dfrac{1}{2} + \dfrac{l_{x+1} + l_{x+2} + \ldots + l_{\omega}}{l_x}$	Average days remaining to an individual age x
Frequency distribution of deaths	d_x	$l_x - l_{x+1}$	Fraction dying in interval x to x + 1

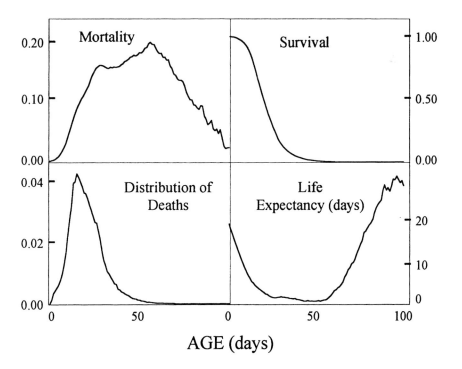

FIGURE 1.1

Example of four of the main life table functions age-specific mortality (q_x), cohort survival (l_x), distribution of deaths (d_x), and expectation of life at age x (e_x), using data on 1.2 million medflies (data from Carey et al. [4]).

The relationship between m_x and q_x is

$$m_x = \frac{q_x}{1 - \frac{1}{2}q_x} \quad \text{and} \quad q_x = \frac{m_x}{1 - \frac{1}{2}m_x} \quad (1.2)$$

For example, the central death rate at age 10 for the medfly (Table 1.1) is

$m_{10} = 0.03027$, whereas the probability of dying in the age interval 10 to 11 is $q_{10} = 0.02982$.

The parameter central death rate is used in computing another important actuarial parameter discussed in the next section—the life table aging rate.

1.2.3.3   Life Table Aging Rate

Horiuchi and Coale introduced to the demographic literature the parameter life table aging rate (LAR), denoted k_x, defined as the rate of change in age-specific mortality with age [7]. The measure is based on *relative* rather than *absolute* rate of change in mortality with age. The formula is given as:

$$k_x = \ln\left(m_{x+1}\right) - \ln\left(m_x\right) \quad (1.3)$$

where m_x denotes the central death rate discussed in the previous section. The life table aging rate is an age-specific analog of the Gompertz parameter, b, since it is a measure of the slope of mortality with respect to age. But unlike the Gompertz parameter, which assumes constancy of the mortality slope typically over a large age interval, LAR examines the change over short intervals. Example computations of k_x from the medfly data in Table 1.1 for ages 5 and 40 are $k_5 = 0.26098$ and $k_{40} = 0.05723$, which indicate that mortality is changing at over 26% per day at day 5 but less than 6% per day at day 40. Additional perspectives for LAR applied to the medfly and to the bean beetle, *Callosobruchus maculatus*, are presented in Tatar and Carey [8] and Carey and Liedo [9].

1.2.3.4   Life Table Entropy

The parameter entropy, denoted H, is a measure of heterogeneity of the distribution of deaths in a cohort.

$$H = \frac{\sum_{x=0}^{\omega} e_x d_x}{e_0} \quad (1.4)$$

If all individuals die at exactly the same age, the shape of the survival schedule, l_x, is rectangular and H = 0. If all individuals have exactly the same probability of dying at each age, the shape of the survival schedule is exponentially decreasing and H = 1.0. Values of H less than 0.5 suggest that the survival schedule is convex, and values of H greater than 0.5 suggest that the survival schedule is concave. Vaupel [10] provided three different interpretations of H: (1) the proportional increase in life expectancy at birth if every individual's first death were averted; (2) the percentage change in life expectancy produced by a reduction of 1% in the force of mortality at all ages; and (3) the number of days lost owing to death per number of days lived [11]. The H-value for the medfly life table presented in Table 1.1 is H = 0.439, indicating a slightly convex survival schedule. The sex-specific medfly H-values (data presented in Carey et al. [12]) are 0.477 and 0.393 for females and males, respectively. Thus, the entropy values indicate that the shape of the survival schedule for male medflies is more convex (rectangular) than the survival schedule for female medflies. In short, the entropy parameter provides a useful summary measure for characterizing differences in shapes of survival curves among cohorts.

1.2.3.5 Mortality Ratios and Crossovers

Manton and Stallard describe a mortality crossover as an attribute of the relative rate of change and level of age-specific mortality rates in two populations: one group is "advantaged" (i.e., lower relative mortality) and the other "disadvantaged" (i.e., higher relative mortality) [3]. The disadvantaged population must manifest age-specific mortality rates markedly higher than the advantaged population through middle age at which time the rates change. Crossovers in mortality occur due to: (1) differences in rates of aging at the individual level; and (2) demographic selection where individuals with high mortality are selected out early for one population and therefore the more robust individuals survive to the older ages. The general formula for the mortality ratio of two cohorts, A and B, is

$$R_x = \frac{\mu_x^A}{\mu_x^B} \qquad (1.5)$$

where R_x denotes the ratio at age x. Mortality is higher at age x in cohort A than in cohort B if $R_x > 1$, lower in cohort A than in cohort B if $R_x < 1$, and the same in cohort A and cohort B at age x if $R_x = 1$. An example of the use of mortality ratio to examine the relationship of male and female medfly mortality is given in Carey and co-workers [12]. For example, the male–female mortality ratios at 0, 13, 23, and 30 days were 1.00, 0.50, 1.00, and 1.29, respectively. This reveals that male mortality was only half female mortality at day 13, was equal to female mortality at day 23 (crossover age), and was nearly 1.3-fold greater at day 30. In other words, the sex mortality differentials substantially changed with age—first favoring males and then at later ages favoring females.

1.2.3.6 Average Lifetime Mortality

The inverse of life expectancy at birth, e_0, is the average mortality experienced by the cohort, denoted $\bar{\mu}$, or, more generally, the inverse of life expectancy at age x, e_x, is the average mortality experienced by the cohort beyond age x denoted $\bar{\mu}_x$. That is

$$\bar{\mu} = \frac{1}{e_0} \qquad (1.6)$$

and

$$\bar{\mu}_x = \frac{1}{e_x} \qquad (1.7)$$

Example values of $\bar{\mu}_x$ from the medfly data presented in Table 1.1 for ages 0, 50, and 100 days reveal the increase, peak, and decrease in overall patterns of age-specific mortality: $\bar{\mu}_0 = 0.048$, $\bar{\mu}_{50} = 0.150$, and $\bar{\mu}_{100} = 0.036$.

1.3   Mortality

1.3.1   Importance of the Mortality Function

The life table provides five different expressions or functions which describe the mortality and survival experience of a cohort. Because each of the functions can be independently derived from the original cohort data and all but expectation of life can be used to derive the other functions, it is often inferred that no single function has precedent. Although this is true algebraically, it is not accurate biologically, demographically, or actuarially. The age-specific mortality schedule—the series of probabilities that an individual alive at age x dies prior to age x + 1—serves as the actuarial foundation for all other functions. The basic role of mortality is evident by considering the following. *First*, death is an "event" indicating a change of state from living to dead; a failure of the system. In contrast, survival is a "nonevent" inasmuch as it is a continuation of the current state. This orientation toward events rather than nonevents is fundamental to the analysis of risk and hazard rates. *Second*, an individual can die due to a number of causes such as due to an accident or to disease. Therefore, mortality rates can be disaggregated by cause of death, thus shedding light on the biology, ecology, and epidemiology of deaths, the frequency distribution of causes, and the likelihood of dying of a particular cause by age and sex. This concept of "cause" obviously does not apply to survival. *Third*, the value of mortality rate at a specified age is independent of demographic events at other ages. In contrast, cohort survival rate (l_x) to older ages is conditional upon survival to each of the previous ages, life expectancy at age x (e_x) is a summary measure of the consequences of death rates over all ages greater than x, and the fraction of all deaths (d_x) that occur at young ages will determine how many individuals remain to die at older ages. This independence of mortality rate relative to events at other ages is important because age-specific rates can be directly compared among ages or between populations, which, in turn, may shed light on differences in relative age-specific frailty or robustness. *Fourth*, a number of different mathematical models of mortality have been developed (e.g., Gompertz [1]) which provide simple and concise means for expressing the actuarial properties of cohorts with a few parameters. Therefore, the mortality and longevity experience of different populations can be more readily compared.

1.3.2   The Force of Mortality

The force of mortality at age x, denoted $\mu(x)$, is the instantaneous mortality rate representing the limiting value of the age-specific mortality rate when the age interval to which the rate refers becomes infinitesimally short [13]. It is given as

$$\mu(x) = \frac{dl(x)}{l(x)dx} \tag{1.8}$$

and

$$l(x) = l(0) \exp\left\{-\int_0^x \mu(y)dy\right\} \tag{1.9}$$

where l(x) is the life table survival rate to age x and l(0) is the radix.

Also known as the instantaneous death rate and hazard rate, $\mu(x)$ is preferred over age-specific mortality, q_x, in gerontology, and demography because it is not bounded by unity, it is independent of the size of the census (age) intervals, and it forms the argument of numerous parametric mortality functions. Three formulae that are commonly used for computing $\mu(x)$ include:

$$\mu(x) = -\ln p_x \tag{1.10}$$

$$\mu(x) = -\frac{1}{2}\left(\ln p_{x-1} + \ln p_x\right) \tag{1.11}$$

$$\mu(x) = -\frac{1}{2n}\ln_e\left(\frac{l(x-n)}{l(x+n)}\right) \tag{1.12}$$

where n in Equation 1.12 denotes the bandwidth. For example, $\mu(x)$ is computed using 3 age classes if n = 1, 5 age classes if n = 2, and so forth. The expressions for $\mu(x)$ given in Equations (1.10–1.12) are based on mortality rates in 2 or more adjacent age classes. The relationship between $\mu(x)$ and the other measures of mortality and death is this: $\mu(x)$ denotes an instantaneous mortality rate that applies to each moment of the specified interval. In contrast, q_x denotes the probability of death over a discrete age interval and m_x denotes the death rate, which is a density function and not a probability, *per se*.

1.3.3 Smoothing Age-Specific Mortality Rates

Because of the binomial noise present in age-specific mortality schedules due to small numbers or to environmental variation, it is often useful to smooth mortality rates for plotting. A formula for computing the running geometric mean of an age-specific mortality schedule, denoted $\hat{q}_x$, is given as:

$$\hat{q}_x = 1 - \left[\prod_{y=x-n}^{x+n} p_x\right]^{-(n+1)} \tag{1.13}$$

where $p_x = 1 - q_x$ and n denotes the "width" of the running geometric average. The analytical counterpart for the running mean of force of mortality, denoted $\hat{\mu}_x$, is

$$\hat{\mu}_x = \frac{1}{n+1}\sum_{y=x-n}^{x+n}\mu_y \tag{1.14}$$

More sophisticated methods for smoothing hazard rates using locally weighted least squares techniques are described in Müller et al. [14], Wang et al. [15], and Carey et al. [16].

1.3.4   Peak-Aligned Averaging

An interesting and important characteristic of mortality in some species such as the medfly is that it both increases and decreases with age, thus creating local as well as lifetime peaks. Because the timing of these peaks may differ slightly between cohorts due to both chance and subtle differences in environmental conditions, Müller et al. [14] introduced a simple statistical technique aimed at avoiding biases associated with averaging peaks across cohorts since averaged peaks tend to "flatten out" due to the variability in the timing of peaks for individual cohorts.

The technique consists of three steps: (1) estimate the location (age) of the hazard rate peak, denoted $\hat{\theta}_j$ where j is the cohort index; (2) obtain the estimated average peak location for all cohorts, $\hat{\theta}$, by averaging all individual cohort peak locations:

$$\hat{\theta} = \frac{1}{N} \sum_{j=1}^{N} \hat{\theta}_j \qquad (1.15)$$

and (3) transform the age coordinate x for each of the N cohorts using the transformed age

$$x'_j = \frac{x\hat{\theta}}{\hat{\theta}_j} \qquad (1.16)$$

This time-scale transformation maps all individual peak location $\hat{\theta}_j$ to $\hat{\theta}$. An example application of this technique is illustrated in Müller and co-workers [14], where local maxima (i.e., "shoulders") in the age-specific mortality schedules around day 10 for 33 female medfly cohorts were peak aligned.

1.4   Mortality Models

Bowers et al. [17] states three principal justifications for postulating an analytic form for mortality and survival functions. First, many phenomena studied in the physical sciences can be explained efficiently by simple formulas. Therefore, some authors have suggested that human survival is governed by simple laws. Second, it is easier to communicate a function with a few parameters than it is to communicate a life table with several hundred parameters and probabilities. Third, a simply analytical survival function is easily estimated using a few parameters from the original determination from mortality data.

1.4.1   Common Parametric Models

Six of the most frequently used mortality models in gerontological research are presented in Table 1.3 along with the corresponding expression for age-specific survival and associated parameters. Example mortality trajectories for each model are given in Figure 1.2.

TABLE 1.3

Six of the Major Age-specific Mortality [$\mu(x)$] and Survival [$l(x)$] Models. Life Expectancy at Birth, $e(0)$, Is Computed Using the Formula $e(0) = \int_0^\omega l(x)dx$ for All Models

Model	$\mu(x)$	$l(x)$	Notes
de Moivre (1729)	$(\omega - x)^{-1}$	$1 - \dfrac{x}{\omega}$	ω = oldest age; survival can also be expressed as $l_x = a - bx$ where $a = 1.0$ (radix) and $b = 1/\omega$
Gompertz (1825)	ae^{bx}	$\exp\left[\left(\dfrac{a}{b}\right)\left(1 - e^{bx}\right)\right]$	a = Initial mortality rate; b = "Gompertz" parameter; linearized version: $\ln \mu(x) = a + bx$
Makeham (1860)	$ae^{bx} + c$	$\exp\left[\left(\dfrac{a}{b}\right)\left(1 - e^{bx}\right) - cx\right]$	c = Age-independent (accidental) mortality
Exponential	c	$\exp[-cx]$	Constant hazard rate, c
Weibull (1939)	ax^n	$\exp\left[-\left(\dfrac{a}{n + 1}\right)x^{n+1}\right]$	a = Location parameter; n = shape parameter; $n>0$; linearized version: $\ln \mu(x) = a + n \ln x$
Logistic	$\dfrac{nx^{n-1}}{g^n + x^n}$	$\left(1 + \left(\dfrac{x}{g}\right)^n\right)^{-1}$	g and n are parameters to be fitted; both parameters control shape and location

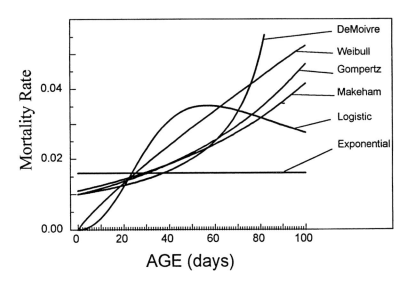

FIGURE 1.2

Example age trajectories of mortality for the six mortality models given in Table 1.2. The parameters for each model were adjusted to yield an expectation of life at age 0 to 50 days.

1.4.1.1   de Moivre Model

Mortality rate in the DeMoivre model equals the inverse of the difference between maximal and current age [18]. Thus mortality tends to unity as age approaches a putative maximum, ω. The resulting survival schedule is a linearly decreasing function of age from 1.0 at age x = 0 to zero at age $x = \omega$. The advantage of this model is essentially its simplicity. That is, the model is transparent and easily understood, requires only a single parameter (ω), and produces a linear survival function. The assumption that survival is a linear function of age is often applied over short age intervals [19].

1.4.1.2   Gompertz Model

The assumption of the Gompertz model is that mortality beyond the age of sexual maturity (or another predetermined age) is an exponentially increasing function of age [1]. The model contains two parameters—the initial mortality rate, a, which denotes mortality at the youngest age class in the specified age interval and the exponential rate of increase in death rate, b. This parameter denotes the age-specific slope of the mortality function and is often referred to as the "Gompertz" parameter. The mortality trajectory for the Gompertz model is exponential and its survival trajectory is sigmoidal. The Gompertz model provides two useful formulae (for additional details and original references, see Finch [20]). Mortality doubling time denoted MDT, defined as the time required for the mortality to increase by 2-fold where

$$\text{MDT} = \frac{\ln(2)}{b}$$

and estimated maximum life span, denoted T_{max}, and defined as the age when a population subject to Gompertzian mortality rates has diminished to one survivor where

$$T_{max} = \frac{1}{b}\ln\left\{1 + \frac{b(\ln N)}{a}\right\}$$

1.4.1.3   Makeham Model

The Makeham model is also known as the Gompertz-Makeham model since it represents an improvement in the Gompertz model rather than constitutes a separate concept [21]. Makeham found that overall mortality levels could be better represented if a constant term was added to the Gompertz formula for $\mu(x)$ to account for causes of mortality not dependent on age (i.e., accidental deaths). There is no analogous transformation for the Gompertz-Makeham equation in which linear regression can be used to estimate the parameter c. Elandt-Johnson and Johnson [2] suggest using trial and error after first estimating parameters a and b from the Gompertz regression equation and then adjusting c until the closest approach to a straight line is attained.

1.4.1.4   Exponential Model

The exponential model is effectively the Gompertz-Makeham model minus the Gompertz component. In other words, it accounts for only the accidental deaths, c, and thus is only concerned with age-independent mortality. Because the exponential mortality model assumes that mortality is constant with age, its plot is simply a horizontal line intercepting the y-axis at c and extending to the right. The survival function decreases exponentially with age.

1.4.1.5   Weibull Model

Whereas the de Moivre, Gompertz, and Makeham models were derived in an actuarial context, the Weibull Model [22] was developed in the context of reliability engineering. Lee [23] notes that the Weibull model is a generalization of the exponential model but, unlike the exponential model, does not assume a constant hazard rate and thus has broader application. The Weibull model has two parameters: the value of n determines the shape of the distribution curve and the value of a determines its scaling. Note that the Weibull hazard function increases if n > 0, decreases if n < 0, and is constant if n = 0.

1.4.1.6   Logistic Model

The logistic model was introduced to demography by Pearl [24] who used this model to estimate the ceiling or asymptote of the U.S. population. Wilson [25] showed that the logistic model provided a good fit to the medfly data presented by Carey et al. [4]. A comparable model to the logistic one is the Perks model [26] which also exhibits leveling off behavior at older ages.

1.4.2   Fitting Mortality Data To Models

Keyfitz [27] notes that fitting models to data, in this case to mortality data, is useful in several contexts including for *smoothing data*, thus making the data easier to handle by removing irregularities and inconsistencies, for *increasing precision* on the assumption that the "real" pattern underlying the observation is a smooth curve, for *aiding inferences* from incomplete data such as interpolation or extrapolation, and for *facilitating comparisons* between two cohorts using a small number of parameters.

Elandt-Johnson and Johnson [2] distinguish two broad categories of methods for fitting parametric distributions. The first group of methods is graphical and relies on plotting mortality rate against age. They note that even though graphical methods give reasonably useful results in many and perhaps most cases, "…some people are rather disturbed by their subjective nature, and prefer an analytical approach" [2]. The second broad group of methods for fitting parametric distributions is analytical and includes both ad hoc and conventional approaches. An example of the first case (ad hoc) involves the formulation and solution of two simultaneous equations. Conventional methods include maximum likelihood and least squares.

The purpose of this section is to describe several methods for fitting the Gompertz model to observed mortality data. Although I use the Gompertz as an example, the broader concepts are general and thus apply to fitting any of the conventional mortality models to data. The paper by Shouman and Witten [28] provides an important perspective on the sample sizes needed for reliable estimates of the Gompertz parameters.

1.4.2.1   Fitting the Gompertz Model

The advantage of the Gompertz model is its simplicity—it proposes a constant exponential increase of mortality rate in that it can be plotted on a semilogarithmic scale to yield a linear increase with age. Three methods are described to fit this model to mortality data:

METHOD 1: Simultaneous Equations

The underlying concept of this method for estimating the Gompertz parameters is that the observed survival rates at two different ages can be set equal to the Gompertz survival function at the respective ages giving two simultaneous equations in two unknowns (a and b). Then standard mathematical techniques can be used to solve for one parameter in terms of the other (see Method 2).

Recalling that the analytical form of the Gompertz survival function is

The parameter a can be determined by substituting the value of h into Eqn(18) and solving for g. Then determine the parameter a from the formula

$$l(x) = \exp\left[\left(\frac{a}{b}\right)\left(1 - e^{bx}\right)\right] \tag{1.17}$$

or

$$= g^{1-h^x} \tag{1.18}$$

where $g = e^{a/b}$ and $h = e^b$. Then let

$$y_1 = \ln\left[\frac{l_{x_1}}{l_{x_1+5}}\right] = h^{x_1}\left(h^5 - 1\right)\ln g \tag{1.19}$$

$$y_2 = \ln\left[\frac{l_{x_2}}{l_{x_2+5}}\right] = h^{x_1}\left(h^5 - 1\right)\ln g \tag{1.20}$$

Therefore

$$\frac{y_1}{y_2} = h^{x_1 - x_2} \tag{1.21}$$

or

$$h = \left(\frac{y_1}{y_2}\right)^{\frac{1}{(x_1 - x_2)}} \tag{1.22}$$

and

$$b = \ln h \tag{1.23}$$

The parameter a can be determined by substituting the value of h into Equation 1.18 and solving for g. Then determine the parameter a from the formula $a = \ln(g)$.

METHOD 2: Least Squares Fit

The Gompertz equation

$$\mu(x) = ae^{bx} \tag{1.24}$$

can be linearized by taking the natural logarithm of each side yielding

$$\ln \mu(x) = a + bx \tag{1.25}$$

Thus the slope, b, and the intercept, a, can be estimated by regressing the logarithms of the mortality rates on age using standard linear regression techniques. An alternative technique is simply to use a nonlinear least squares statistics program to fit the nonlinearized (original) model to the data.

METHOD 3: Maximum Likelihood

The method of least squares is designed to estimate parameter values of a model based on minimizing differences between the observed mortality rates and the rates predicted from the mortality model (i.e., hazard schedule) containing these parameters. As described in the previous section, this approach for estimating the parameters of the Gompertz model is based on age-specific mortality rates. In contrast, the method of maximum likelihood for estimating the parameters is based on the age-specific density function of deaths, the frequency distribution of deaths, d_x. The conditional probability of death on day x, q(x), is calculated as (from Fukui et al. [29]).

$$q(x) = \hat{P}(x \leq T < x + 1 \mid T > x) = 1 - \frac{\hat{S}(x + 1)}{\hat{S}(x)} \tag{1.26}$$

for x = 1, 2, …D where D is the last age of death and $\hat{S}(x)$ is the empirical (observed) survival function. The likelihood function then can be written as

$$L = \prod_{x-1}^{D} q(x)^{d_x} \left[1 - q(x)\right]^{n_x - d_x} \tag{1.27}$$

Comparison of Methods

A comparison of parameter value estimates for fitting the Gompertz from 0 to 30 and from 31 to 60 using the medfly data given in Table 1.1 is shown in Table 1.5. Although the estimates for the initial mortality rate, a, for 0 to 30 days differ by nearly 2-fold (i.e., 0.0039 for regression method vs. 0.0064 for the maximum likelihood method), the estimates for the parameter b for this interval as well as the estimates for both parameters for the 31- to 60-day interval are comparable. Each method has advantages and drawbacks. The main advantage of the method of simultaneous equations is that estimates can be made analytically. However, a drawback is that it is primarily an ad hoc approach and thus used only when first approximations of the parameter values are needed. The method has only limited value for use in comparing the Gompertz rates between experimental treatments.

The advantage of the regression method for estimating the Gompertz parameters is that it is simple and virtually all spreadsheet and statistical software packages have programs for fitting data to linear models using least squares techniques. The shortcoming of this method is that parameter estimates are less reliable when mortality rates are based on small numbers of individuals. The main reason for this is that, in regression analysis, points (i.e., mortality rates) at older ages which are based on smaller numbers of individuals are weighted equally with points at younger ages which are based on much larger numbers of individuals.

TABLE 1.4

Estimates of Gompertz Parameters, a and b, for the Medfly Mortality Data
Given in Table 1.1, Using Three Different Estimation Methods

Estimation Method	0–30 days		31–60 days	
	a	b	a	b
Method 1: Simultaneous equations	0.00406	0.16718	0.11260	0.00567
Method 2: Regression	0.00390	0.14878	0.09190	0.01024
Method 3: Maximum likelihood	0.00641	0.15540	0.10000	0.01130

The main conceptual and analytical difference between fitting the Gompertz model using the two statistical approaches is that the least squares regression method is based on fitting the distribution of age distribution of mortality rates whereas the method of maximum likelihood is based on maximizing the fit of the density function or age distribution of deaths (i.e., the d_x-schedule). Maximum likelihood is the method preferred by actuaries and statisticians for fitting models to mortality data. As Elandt-Johnson and Johnson [2] note, maximum likelihood estimators do have certain desirable properties, but (1) these properties depend on the model being a sufficiently accurate presentation of reality, and (2) even if the model is valid, the desirable properties are asymptotic—they apply when the volume of data is sufficiently large. The main drawback of the method of maximum likelihood is that most statistical programs do not have standard programs for determining maximum likelihood estimates of the Gompertz or other mortality models.

1.4.2.2   *Fitting a Two-Stage Gompertz Model*

Fukui et al. outlined methods for fitting a two-stage Gompertz model which captures the slowing of mortality at older ages by fitting two mortality curves, one for early ages and one for advanced ages. The breakpoint of the two curves is treated as an additional parameter in the model. To ensure that the two curves are continuous at the breakpoint, the model is specified with the following hazard function:

$$\mu(x) = ae^{b_1 x}I(x \leq c) + ae^{b_1 x + b_2(x-c)}I(x > c) \tag{1.28}$$

where I is the indicator function defining the breakpoint at age c. Separating this model into its component parts reveals the two stages, S_1 and S_2. That is:

$$\mu(x) = S_1 + S_2 \tag{1.29}$$

where

$$S_1 = ae^{b_1 x}I(x \leq c) \tag{1.30}$$

and

$$S_2 = ae^{b_1c + b_2(x - c)}\mathrm{I}(x > c) \tag{1.31}$$

Note that the value of I is set at either 0 or 1. For example, if the breakpoint c = 30 and x = 10, then I = 1 in the equation for S_1 and I = 0 in the equation for S_2(thus S_2= 0). However, if x = 40, then I = 0 in the expression for S_1 (thus S_1 = 0) and I = 1 in the expression for S_2. Parameter estimates of a, b_1, and b_2 using nonlinear least squares techniques are given in Table 1.5 and a plot showing the relationship between the two-stage Gompertz model and the medfly data is given in Figure 1.3.

TABLE 1.5

Nonlinear Least Squares Parameter Estimate and Standard Deviation (SD) for the Two-Stage Gompertz Fitted to the Medfly Data from 0 To 60 Days Using a Breakpoint of c = 30 days

Parameter	Estimate	SD
a	0.022846	0.002891
b_1	0.058437	0.004714
b_2	0.004090	0.001981

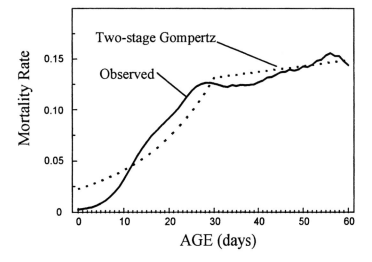

FIGURE 1.3
Relationship between actual medfly mortality data (from Table 1.1) and two-stage Gompertz model fit.

1.5   Additional Concepts

1.5.1   Demographic Selection

As populations age, they become more selected because individuals with higher death rates will die out in greater numbers than those with lower death rates, thereby transforming the population into one consisting mostly of individuals with low death rates [30]. The concept of subgroups endowed with different levels of frailty is referred to as demographic heterogeneity, and the winnowing process as the cohort ages is referred to as demographic selection. The actuarial consequence of cohorts consisting of subcohorts each of which possesses a different level of frailty is that the mortality trajectory of the whole may depart substantially from Gompertz rates even though each of the subcohorts is subject to Gompertz mortality rates.

Vaupel and Carey [31] fitted observed medfly mortality patterns with mixtures of increasing Gompertz curves using the hazard function:

$$\mu(x,z) = z\mu^0(x) \tag{1.32}$$

where $\mu(x,z)$ denotes the hazard rate of individuals with fixed frailty z at exact age x and $\mu^0(x)$ is the baseline hazard function of the form

$$\mu^0(x) = 0.003e^{0.3x} \tag{1.33}$$

Experiments with different numbers of subpopulations suggested that 12 groups were sufficient to capture the observed pattern of medfly mortality using a range of frailty values and initial proportions of subcohorts (for details see Vaupel and Carey [31]).

1.5.2   Censoring

Survival analysis deals with the fact that for some individuals brought under observation the available information on failure time may be incomplete. For example, not all individuals may have died by the close of the experiment or an individual under observation may migrate, become lost, or be accidentally killed. In each of these cases where the failure time (i.e., time of death) is incomplete, the case is said to provide censored information [32]. Censored individuals contribute days at risk but not events (deaths) to the estimation of hazard rates. Mortality rates computed from censored biogerontological data will be wrong unless corrections in the numbers at risk are considered.

1.5.3   Density Effects

Virtually by definition, all controlled studies of captive animals are subject to density effects since animals such as laboratory rodents or *Drosophila* are often co-housed, which causes an increase in the incidence of physical damage due to fighting and accidents in animals. Even for animals maintained in solitary confinement, the cage size will affect their welfare and thus longevity. A concern by some researchers (e.g., Nusbaum et al. [33]) with the results on fruit fly mortality patterns observed by Carey et al. [4] and Curtsinger et al. [34] was that the slowing of mortality at older ages was an artifact of decreasing densities in the experimental cages. However, the main conclusion from subsequent studies on the effects of density on mortality was that density affects the level but not the age pattern of mortality [35,36].

1.5.4 Threshold Mortality

A major concern of any study designed to estimate the actuarial rate of aging in a cohort (i.e., the Gompertz parameter, b) is the decrease in sample size at older ages due to attrition. However, this problem of insufficient sample size may also apply to the measurement of mortality at young ages, even when the number of individuals at risk is at or near the initial number, n. This has been referred to as the "left-hand boundary problem," [37] which occurs whenever the "actual" mortality rate is less than $1/n$. For example, a mortality rate of $\mu = 0.001$ cannot be detected with a sample size of $n = 100$, since when a single individual dies the estimate will be

$$\mu = \frac{1}{100} = 0.001$$

The main point is that even though the number of individuals at risk is highest at the youngest ages, a sample size constraint still exists inasmuch as mortality is often quite low at young ages and thus lower than $1/n$. The researcher is thus faced with either increasing sample size for measuring low mortality at young ages or using larger time intervals until a sufficient number have died. In either case, the results will have a profound effect on estimating the parameters for mortality models such as the Gompertz model.

1.5.5 Standard Deviation of Mortality

The equation for the 95% confidence interval for age-specific mortality is given as

$$CI_{95\%} \cong \hat{q}_x \pm 1.96 S_{\hat{q}_x} \tag{1.34}$$

where $S_{\hat{q}_x}$ denotes the standard deviation of the death rate at age x. The formula for $S_{\hat{q}_x}$ is

$$S_{\hat{q}_x} = \hat{q}_x \sqrt{\frac{1}{D_x}\left(1 - \hat{q}_x\right)} \tag{1.35}$$

where $\hat{q}_x$ is the age-specific mortality rate at age x and D_x is the number of deaths at age x. The upper and lower confidence intervals for the medfly age-specific mortality rates given in Table 1.1 are shown in Figure 1.4. The pattern of diverging intervals that is shown in this figure is a characteristic of all mortality schedules which are based on the absolute number dying at each age (rather than subsamples). This is because the number at risk decreases with age due to attrition and therefore the variance increases.

1.5.6 Visualizing Cohort Survival and Individual Reproduction

Understanding the relationship between mortality and reproduction is important in biogerontology studies on insects and other invertebrates because of the so-called "cost of

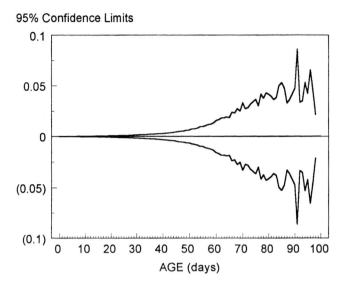

FIGURE 1.4
Plot of the 95% confidence intervals for the medfly mortality based on data given in Table 1.1.

reproduction" [38]. For example, egg laying in short-lived strains of *Drosophila* tends to occur at a younger age than egg laying in long-lived strains. Because of the importance of cost of reproduction to studies of aging and mortality, a simple graphic technique was developed [39] in which data on age-specific reproduction of individuals are portrayed using: (1) a horizontal life line, the length of which is proportional to individual longevity; (2) color-coded segments depicting the level of reproduction at each age; and (3) a cohort survival schedule created by rank-ordering individual life lines from shortest- to longest-lived. The resulting graphic, referred to as an event history diagram, portrays data at the individual level and thus allows visual comparisons of detailed life history patterns such as age of first reproduction; longevity; ages of high, medium, low, and zero reproduction; and postreproductive period.

Acknowledgments

I thank Scott Pletcher for his statistical insights and help in obtaining the maximum likelihood estimates for the Gompertz parameters given in Table 1.4. I also gratefully acknowledge the statistical and programming assistance of Shin-Min Tsai. Research was supported in part by the National Institute on Aging (grant AG08761-01).

References

1. Gompertz, B., On the nature of the function expressive of the law of human mortality, and on a new mode of determining the value of life contingencies, *Phil. Trans. R. Soc.*, London, 115, 513, 1825.

2. Elandt-Johnson, R. C. and Johnson, N. L., *Survival Models and Data Analysis*, John Wiley & Sons, New York, 1980.
3. Manton, K. and Stallard, E., *Recent Trends in Mortality Analysis*, Academic Press, Orlando, Florida, 1984, 15.
4. Carey, J. R., Liedo, P., Orozco, D., and Vaupel, J. W., Slowing of mortality rates at older ages in large medfly cohorts, *Science*, 258, 457, 1992.
5. Chiang, C. L., *The Life Table and Its Application*, Robert E. Krieger Publishing, Malabar, Florida, 1984.
6. Carey, J. R., *Applied Demography for Biologists*, Oxford University Press, New York, 1993.
7. Horiuchi, S. and Coale, A., Age patterns of mortality for older women: an analysis using the age-specific rate of mortality change with age, *Math. Pop. Stud.*, 2, 245, 1990.
8. Tatar, M. and Carey, J. R. Carey, Sex mortality differentials in the bean beetle: reframing the question, *Am. Nat.*, 144, 164, 1994.
9. Carey, J. R. and Liedo, P., Sex-specific life table aging rates in large medfly cohorts, *Exp. Gerontol.*, 30, 315, 1995.
10. Vaupel, J. W., Inherited frailty and longevity, *Demography*, 25, 277, 1988.
11. Carey, J. R., *Applied Demography for Biologists*, Oxford University Press, Oxford, 1993.
12. Carey, J. R., Liedo, P., Orozco, D., Tatar, M., and Vaupel, J. W., A male-female longevity paradox in medfly cohorts, *J. Anim. Ecol.*, 64, 107, 1995.
13. Pressat, R., *The Dictionary of Demography*, C. Wilson, Ed., Basil Blackwell, Oxford, 1985.
14. Müller, H.-G., Wang, J.-L., Capra, W. B., Liedo, P., and Carey, J. R., Early mortality surge in protein-deprived females causes reversal of male-female life expectancy relation in Mediterranean fruit flies, *Proc. Nat., Acad. Sci. U.S.A.*, 94, 2762, 1997.
15. Wang, J. L., Müller, H.-G., Capra, W. B., and Carey, J. R., Rates of mortality in populations of *Caenorhabditis elegans*, *Science*, 266, 827, 1994.
16. Carey, J. R., Liedo, P., Müller, H.-G., and Wang, J.-L., Relationship of Age Patterns of Fecundity to Mortality, Longevity, and Lifetime Reproduction in a Large Cohort of Mediterranean Fruit Fly Females, *J. Gerontol.*, in press, 1998.
17. Bowers, N. L., Gerber, H. U., Hickman, J. C., Jones, D. A., and Nesbitt, C. J., *Actuarial Mathematics*, The Society of Actuaries, Itasca, IL, 1986.
18. DeMoivre, A., Annuities on Lives: or, the Valuation of Annuities Upon Any Number of Lives, London, 1725.
19. Smith, D. and Keyfitz, N., Eds., *Mathematical Demography*, Selected Papers, Springer-Verlag, Berlin, 1977.
20. Finch, C., Longevity, in *Senescence and the Genome*, University of Chicago Press, Chicago, 1990.
21. Makeham, W. M., On the law of mortality, *J. Inst. Actuaries*, 13, 325, 1867.
22. Weibull, W., A statistical distribution of wide applicability, J. Appl. Mech., 18, 293, 1951.
23. Lee, E. T., *Statistical Methods for Survival Data Analysis*, 2nd Ed., John Wiley and Sons, New York, 1992.
24. Pearl, R. and Reed, L. J., The rate of growth of the population of the United States since 1790 and its mathematical representation, *Proc. Natl. Acad. Sci. U.S.A.*, 6, 275;, 1920.
25. Wilson, D., A comparison of methods for estimating mortality parameters from survival data, *Mech. Ageing Dev.*, 66, 269, 1993.
26. Perks, W., On some experiments in the graduation of mortality statistics, J. Inst. Actuaries, 63, 12, 1932.
27. Keyfitz, N., Choice of function for mortality analysis: effective forecasting depends on a minimum parameter representation, *Theor. Popul. Biol.*, 21, 329, 1982.
28. Shouman, R. and Witten, M., Survival estimates and sample size: what can we conclude?, *J. Gerontol.*, 50, B177, 1995.
29. Fukui, H. H., Xiu, L., and Curtsinger, J. W., Slowing of age-specific mortality rates in *Drosophila melanogaster*, *Exp. Gerontol.*, 28, 585, 1993.
30. Vaupel, J. W., Manton, K. G., and Stallard, E., The impact of heterogeneity in individual frailty on the dynamics of mortality, *Demography*, 16, 439, 1979.
31. Vaupel, J. W. and Carey, J. R., Compositional interpretations of medfly mortality, Science, 260, 1666, 1993.

32. Namboodiri, K. and Suchindran, C. M., *Life Table Techniques and Their Applications*, Academic Press, Orlando, 1987.
33. Nusbaum, T. J., Graves, J. L., Mueller, L. D., and Rose, M. R., Fruit fly aging and mortality, *Science*, 260, 1567, 1993.
34. Curtsinger, J. W., Fukui, H. H., Townsend, D. R., and Vaupel, J. W., Demography of genotypes: failure of the limited life-span paradigm in *Drosophila melanogaster*, *Science*, 258, 461, 1992.
35. Carey, J. R., Liedo, P., and Vaupel, J. W., Mortality dynamics of density in the Mediterranean fruit fly, *Exp. Gerontol.*, 30, 605, 1995.
36. Khazaeli, A. A., Xiu, L., and Curtsinger, J. W., Effect of adult cohort density on age-specific mortality in *Drosophila melanogaster*, *J. Gerontol.*, 50A, B262, 1995.
37. Promislow, D., Tatar, M., Pletcher, S., and Carey, J. R., Below-threshold mortality and its impact on studies in evolutionary ecology, unpublished manuscript.
38. Reznick, D., Costs of reproduction: an evaluation of the empirical evidence, *Oikos*, 44, 257, 1985.
39. Carey, J. R., Liedo, P., Muller, H. G., and Wang, J. L., A simple graphical technique for displaying individual fertility data and cohort survival: Case study of 1,000 Mediterranean fruit fly females, *Funct. Ecol.*, in press, 1998.

2

Design of Cross-Sectional, Longitudinal, and Sequential Studies in Gerontology

Donald K. Ingram

CONTENTS

2.1 Introduction

The study of aging is a major scientific challenge because of many conceptual and practical problems. Aging is a *theoretical construct*. It is not readily identifiable nor observable. It is inferred based on existing data which meet the concepts and definitions offered. The great majority of gerontological studies focus on age as the independent variable in order to infer that a parameter of interest is altered by a particular aging process.

The study of aging requires careful consideration in experimental design. Good experimental design involves identification of a dependent variable and an independent variable but also sources of *extraneous variation* which can affect the conclusion that manipulation of the independent variable caused a change in the dependent variable. Many sources of extraneous variation exist that offer potential confounds to conclusions regarding the degree of variation to be attributed to aging. Gerontologists have devoted much attention to this issue and have devised strategies to address it [1].

2.2   Age as an Ex Post Facto Variable

The first issue to consider regarding the design of aging studies is that age is an *ex-post facto* variable, i.e., it exists "after the fact." Indeed, it cannot be considered a true independent variable because age is not randomly assigned. Similar to other *ex-post facto* variables, like gender or race, age is inherent within the subject and thus cannot be manipulated in an experimental sense.

As an example in Figure 2.1A, if a dependent variable, such as serum cholesterol level, is being examined as a function of age, then the independent variable is age; however, this variable cannot be expressly manipulated because it is already fixed within the subjects under study. This example would be contrasted to a true experimental study in which serum cholesterol levels were being examined as a function of the percent fat in the diet as shown in Figure 2.1B. For this hypothetical study, subjects were assigned to different diets that were identical in ingredients except for the fat content. The subjects were maintained on the diet for 4 weeks at which time blood samples were drawn, which are the data presented in Figure 2.1B. In this case, the conclusion could be made with confidence that the higher fat content in the diet *caused* the increased serum cholesterol. However, confidence would be strengthened even more with the addition of a control group who were maintained on their normal diet and their cholesterol did not change or, if it did change, remained statistically significant from the experimental group. Barring the inclusion of a control group, additional confidence could be gained if the subjects had been withdrawn from the high fat diets after 4 weeks, and the data showed a return to baseline levels of serum cholesterol. It is highly reasonable to deduce in either case that manipulations of the independent variable, the fat content of the diet, was causing a change in the dependent variable, serum cholesterol. In the case of Figure 2.1A, logically it is more difficult to conclude that increased age *causes* an increase in serum cholesterol. There can be no control group for age — and thus there is no manipulation of the independent variable as applied in Figure 2.1B, in which fat content is increased or decreased accordingly. Indeed, without benefit of additional analysis, the age differences shown in Figure 2.1A could be due to differences in the fat content of the diet across these age groups rather than some age-related change in lipid metabolism.

More precisely then, age is a correlate of the dependent variable. What is being examined is the relationship between two variables, age and cholesterol level. Thus, causation is more difficult to discern. Differences in a dependent variable that exist between age groups can result from many other potential causes as will be discussed in the following sections. In summary, the design of aging studies is considered to involve *quasi-experimental* designs. Such designs are useful for collecting descriptive data that permit inferences about relationships rather than data than permit inferences about causation.

2.3   Cross-Sectional Designs

2.3.1   Major Features and Advantages

The most straightforward design in gerontological research involves the measurement of differences in a particular parameter between two or more age groups. Typically the

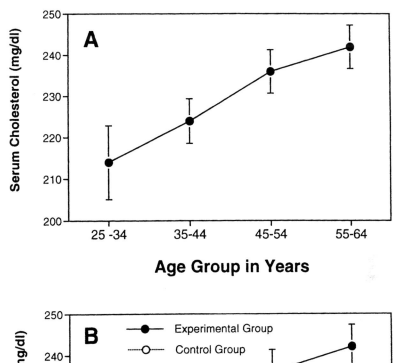

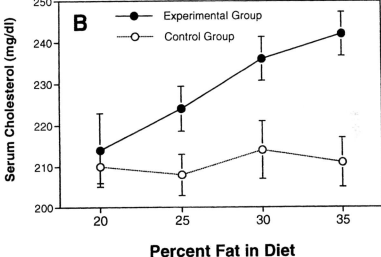

FIGURE 2.1

(A) Mean and standard error estimates of serum cholesterol in men as a function of age. (Data are derived from Hershcoph et al. [21].) (B) Mean and standard error estimates of serum cholesterol in men as a function of percent fat in the diet. Data are fictional. Control group diet is presumably maintained at 20%.

measurement is made only once and during the same time period for both groups. This design is depicted in the vertically shaded area shown in Figure 2.2. Data are collected from a *cross section* of a population; hence, the term, *cross-sectional design*.

A cross-sectional design can be appealing to an investigator because it represents a quick and efficient means of accumulating data about a parameter of interest. Returning to the example in Figure 2.1A, if the investigator were interested in testing whether aging was related to increased serum cholesterol, then this study would involve collecting serum samples from subjects representing different age groups encompassing young to elderly

Birthdates			Age		
			C.S.		
1900	70	75	80	85	90
1910	60	65	70	75	80
1920	50	55	60	65	70
1930	*Long.* 40	45	50	55	60
1940	30	35	40	45	50
1950	20	25	30	35	40
Time of Measurement	1970	1975	1980	1985	1990

FIGURE 2.2
Examples of cross-sectional (*C.S.*) and longitudinal (*Long.*) designs.

individuals. These data can be collected over a relatively short period of time. Typically the data would be represented as group means, and a statistical test would be employed to draw conclusions about the reliability of the results regarding differences between the age groups.

2.3.2   Precautions

A major consideration in the design of a cross-sectional study is the sampling procedure. Too often investigators attempt maximum economy of design by sampling only at the extremes of the age distribution. In a human study, this strategy might involve collection of data from a group of very young adults, e.g., 20 to 25 years old to compare with a group of older adults, e.g., 75 to 80 years old. This approach is viewed as economical because the investigator might conclude that if age differences do not exist between the extreme ranges of adult age, then the matter is not of sufficient interest to pursue. However, because the investigator has omitted a broad range of the age distribution, erroneous conclusions might be drawn about the relationship between age and the parameter of interest. This possibility is depicted in Figure 2.3A. The investigator would have concluded that no age difference existed between the young and old groups in the parameter if only these extremes had been sampled when in fact an age-related increase over most of the adult lifespan was observed followed by a decrease at advanced ages. Such important possibilities can exist. For example, serum cholesterol has shown a similar cross-sectional pattern as presented in Figure 2.3B.

The lesson to be gained from this example is that too often investigators assume a strict linear relationship with age. When only two age groups are employed in a cross-sectional study, the investigator will have no opportunity to examine the nonlinear aspects of the parameter under study.

In addition to these precautions, the conclusions drawn from cross-sectional studies must always be qualified in two important ways. First, it should be reemphasized that the data represent a relationship between age and cholesterol level rather than implying

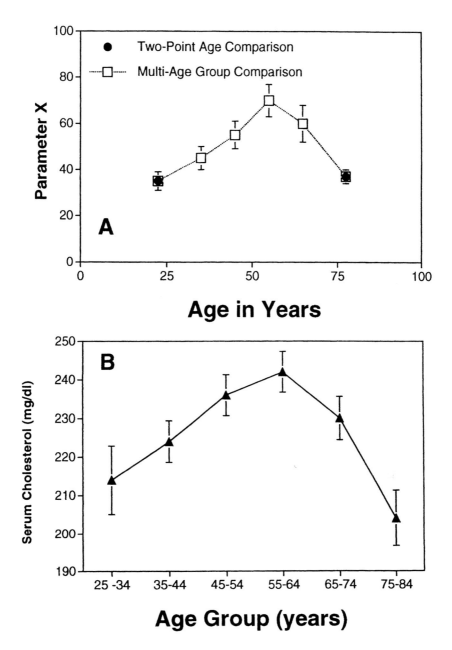

FIGURE 2.3
(A) Mean and standard error depiction of fictional data illustrating potential problem with linear assumption in gerontological studies. **(B)** Mean and standard error estimates of serum cholesterol in men as a function of age. (Data are derived from Hershcopf et al. [21].)

causation. It is accurate to refer to an age-related increase in cholesterol levels rather than to advancing age causing cholesterol to increase. Second, while the conclusion may be drawn about an age-related increase in cholesterol, it should be recognized that the cross-sectional design permits inferences about age *differences* rather than genuine age *change*. This limitation exists because the cross-sectional design collects data pertaining to *interindividual differences* rather than *intraindividual change*.

2.4   Longitudinal Designs

2.4.1   Major Features and Advantages

Only by collecting data from a group of individuals over time as they age can an investigator draw inferences about *intraindividual change*. Cross-sectional studies do not permit this inference because the interindividual differences can exist for many reasons other than age. In the case of long-lived species, like humans, it is clear that generational differences can be considerable. Thus, rather than differences due to age per se, the differences can be due to when the individuals were born, i.e., to their birth *cohort*. For example, regarding the data on serum cholesterol shown in Figure 2.1, it is highly likely that different age groups could differ in this parameter not because of age primarily but because of different dietary habits that were adopted among different cohorts. Younger generations could be watching their diets and eating foods with less cholesterol, while older generations could be sticking with the higher fat diets that they have followed throughout their lives.

Herein is the advantage of the *longitudinal* study, which is the quasi-experimental design that permits control of possible cohort differences. As depicted in the horizontally shaded area of Figure 2.2, this design in its simplest form collects data from a birth cohort on more than one measurement occasion. The scientific value of obtaining these repeated measures from one cohort can be expressed in Figure 2.4. Specifically, the conclusions about an age-related increase in serum cholesterol that might be drawn from the cross-sectional data presented in Figure 2.1A would appear compromised by the evidence from longitudinal analysis showing that age changes in cholesterol levels in the cohort born in 1941 do not match this pattern. This cohort had likely adopted different eating habits to reduce their intake of dietary cholesterol.

Longitudinal analysis of aging has provided major advances for gerontological studies in many fields. As a major example, Schaie [2] helped dramatically alter the perspective on the relationship between age and intelligence. His studies of performance in intelligence tests helped demonstrate that the marked age-related decline that had been observed in the results of many earlier cross-sectional studies were much less obvious and occurred later in life when analyzed longitudinally. His observations of this phenomenon increased awareness of the need for further longitudinal analysis of aging.

The increased complexity of the longitudinal study is manifest compared to a cross-sectional study. In the case of human studies, subjects are recruited into a study, and then contact must be kept with them throughout the duration of the study. This often means that the subjects must continue to come to a laboratory for testing. Thus, typically the cost of a longitudinal study is much higher than a cross-sectional study. Longitudinal studies of human aging might require decades to conduct. Resources are needed for a staff that keeps contact with the subjects, collects and analyzes data, and maintains records. Laboratory space and equipment must be maintained over time. Even in the case of longitudinal studies in nonhuman subjects, many of these resource issues would apply to a greater extent than in cross-sectional studies.

2.4.2   Examples of Longitudinal Studies

Because of the inherently greater resource demand in a longitudinal study, it is logical that an effort is made to collect data about many additional variables than is typically collected in a cross-sectional study. Several large-scale longitudinal studies have been initiated in the

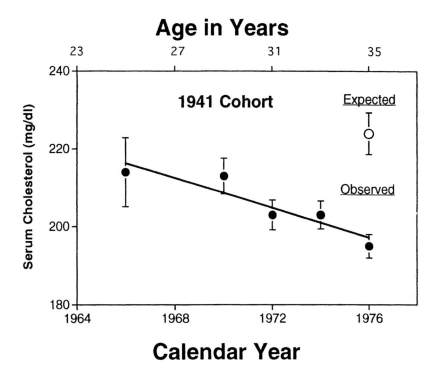

FIGURE 2.4
Mean and standard error estimates of serum cholesterol in men as a function of age of a specific cohort. (Data are derived from Hershcopf et al. [21].)

U.S. during the 20th century. Two of the oldest studies began as studies of development. Under the direction of the University of California at Berkeley, the Oakland Growth Study began in the 1920s and 1930s and recruited every fifth baby born in the city of Oakland, CA into the study. While this study was focused primarily on intellectual development, its results have provided information about a wide range of topics including aging, as many of these individuals were followed for decades and even their children became part of the study [3,4]. Another study of intellectual development begun by Louis Terman in the 1920s at Stanford University recruited children with high scores on the Stanford Binet intelligence test. This major effort known as the Terman study followed many of these individuals well into old age [5,6].

Three major longitudinal studies conducted in the U.S. that focused more on aging at their outset include the Duke Longitudinal Study, the Boston Normative Aging Study, and the Baltimore Longitudinal Study of Aging. Under the auspices of Duke University beginning in 1955, the Duke Longitudinal Study recruited community residents over the age of 60 in the area of Durham, NC [7]. Over a period of more than 20 years, these subjects (>200) underwent a battery of testing encompassing biological, psychological, and social components. Under the direction of the Boston Veterans Administration Hospital, the Boston Normative Aging Study was begun in 1962 and like the Duke Study involved a wide range of assessment for a large group of adults (>2000), many of which were followed for many years [8]. Of these three major studies, only the Baltimore Longitudinal Study of Aging (BLSA) has remained fully operational and grown under the direction of the National Institute on Aging (NIA) of the National Institutes of Health (NIH). Begun in 1958, the BLSA currently involves over 700 men and 500 women recruited at different ages across the lifespan who come to the NIA every 2 to 3 years to participate in 2 to 3 days of comprehensive testing [9]. Other large-scale longitudinal studies of aging are operational in other countries, including the Koganei Study of Aging in Japan [10], the Manitoba study of Canada [11], and, in Sweden, the Göteborg study [12] and the Swedish Adoption/Twin Study of Aging [13]. Deeg [14] can be consulted for a very comprehensive review of current and past longitudinal studies of aging.

While the aforementioned studies collect data across a wide range of biological, biomedical, and sociopsychological domains, other longitudinal studies of aging have concentrated more on a specific variable of interest. For example, the Seattle Longitudinal Study of Adult Intelligence focused on aging and changes in intellectual performance using the Thurstone Primary Mental Abilities Test [2]. Additional data were gathered about the subjects, but this was to support the major aim of the study to document and analyze age-related changes in intelligence.

While most longitudinal studies of aging in humans are prospective in design, other longitudinal studies have developed retrospectively. For example, Owens [15] was able to conduct a retrospective study of aging and intelligence. Specifically, he took advantage of locating a set of Army Alpha tests that had been taken in 1919 and then recruiting many of these individuals to retake the test in 1961. Another value of on-going longitudinal studies is that many analyses can be conducted retrospectively regarding questions that had not been important at the beginning of the study. The BLSA provides a good example. Because serum samples are banked in this study, it was possible to conduct a retrospective study of how well a new assay for prostate specific antigen (PSA) could predict the incidence of benign prostatic hypertrophy (BPH) and prostate cancer [16]. The PSA assay, of course, had not been available when the BLSA began; so, this important study emerged retrospectively.

Other longitudinal studies that do not begin with an express purpose to study aging can also eventually yield valuable gerontological information. The Oakland Growth and Terman Studies mentioned previously are good examples. The Framingham study is another. Beginning as a study in 1948 under the U.S. Public Health Service, this study has involved a large sample from Framingham, MA, and was directed toward understanding the etiology of cardiovascular disease [17]. This objective was achieved with great success, but the emerging database can also be used, primarily in a retrospective fashion, to study questions of broader gerontological interest [18].

Another example of a large-scale study that did not begin as an aging study is the National Health and Nutrition Examination Survey (NHANES) conducted during 1971 by the National Center for Health Statistics (NCHS). This was a broad-based study utilizing a national probability sample with onsite examinations. Recognizing the enormous value of the data collected, the NCHS decided to conduct a follow-up study about 10 years after the original survey [19]. This follow-up study could take advantage of the previous data collected to provide new retrospective analyses pertaining to aging and age-related disease. For example, Zonderman et al. used NHANES data to examine the relationship between self-reported symptoms of depression and risk of cancer over a 10-year period [20].

2.4.3   Disadvantages and Limitations

Although the application of longitudinal analysis is logically appealing for gerontological research, such designs are still problematic even beyond their costs and complexities. One major problem is *attrition*. It is difficult to keep track of subjects in longitudinal studies and to keep them involved. Subjects in most studies are volunteers, who can drop out of the study because they move, lose interest, get sick, or die. Thus, attrition can greatly affect the generalization of the data. Comparing data between younger ages and older ages becomes problematic because those left in the study at older ages may have generated some bias in the older sample. This problem can be alleviated to some degree by conducting longitudinal analysis only on those who remain throughout the period of analysis, but this solution again may restrict the generalization of the study to the greater population—the study would represent only those individuals who had stuck with a study for a long period of time. Clearly there is a risk of bias in this situation. However, this bias would likely depend on the type of variable being measured. For example, psychological variables might be more affected than certain biochemical variables.

Another potential problem with longitudinal studies is *progressive error*. By recruiting, then examining and testing an individual repeatedly, the investigator is likely changing that person. With exposure and practice to certain tests, the individual can become increasingly test sophisticated. The intraindividual data can thus become highly correlated, which presents problems for certain types of statistical analyses. The possibility of progressive error or practice effects can be examined by bringing in controls at various ages for initial testing and comparing these with cohorts that have been tested repeatedly. However, such controls are not used extensively because of the additional expense and effort involved.

Because longitudinal studies of long-lived species, such as humans, require extended periods for investigation, there are problems inherent in the administration and conduct of the study itself. Personnel and equipment changes can introduce bias into the test results. Data are likely to be misplaced or lost altogether. It is very difficult to imagine that test and assay conditions can remain constant over extended periods of time. Internal standards and calibrations can help determine the consistency of results; however, it is almost certain that old methods will have to be abandoned and new ones introduced. For example, an assay of cholesterol would likely change over the period of 20 years. Important reagents cannot be maintained indefinitely. The storage of tissue samples can assist in determining whether assay conditions remained unchanged by repeated testing of these. However, long-term storage of these samples may alter their biological activity in ways that would be difficult to detect.

While suitable controls may exist for the above-mentioned problems, another major conceptual problem with longitudinal studies is much more difficult to address. This is the potential confound of *time-of-measurement effects*. If the investigator is following only one or few cohorts of individuals over an extended period of time, it is highly likely that sociocultural events can intervene which can impinge upon data collection in a longitudinal study. The example of aging and serum cholesterol levels presented in Figure 2.4 further illustrates such a possibility. Following a cohort of individuals born in 1921, an investigator might discover that the dietary habits of this group could be impacted by the major health campaign that began in the 1970s and 1980s to educate the general public about the increased health risk of consuming diets high in fat or of maintaining inactive lifestyles. The fat content of their diet could be reduced substantially or their activity levels increased significantly, thereby altering the longitudinal analysis of the relationship between age and serum cholesterol. Had only this one cohort of individuals been followed, the investigator might have concluded that cholesterol levels decline as a function of age. The astute investigator would likely be aware of this possibility and even have the additional data on diet and activity to assess the impact of changes in these variables on serum cholesterol. However, it is clear that this *secular* change occurring at this period in history would have introduced a time-of-measurement effect into the longitudinal analysis. The investigator has no control over such events. Moreover, oftentimes after the fact, it is almost impossible to determine what factor contributed to this secular change.

An excellent example of such secular change can be found in the analysis by Hershcopf et al. [21] of serum cholesterol levels among males enrolled in the BLSA. These data showed that a major decrease in serum cholesterol across nearly all cohorts occurred between the 1960s and 1970s. Careful analysis of changes in obesity, dietary intake, or physical activity could not account completely for this drop of about 10 mg/dl in serum cholesterol.

2.5 Sequential Designs

2.5.1 Major Features and Advantages

The difficulty in experimental design in gerontological studies is that no design appears to avoid confounding at least one of three attributes: age, cohort, or time-of-measurement. In recognizing the challenge of this dilemma, Schaie [22] suggested the implementation of *sequential* designs that applied various combinations of cross-sectional and longitudinal analyses and could be used to analyze the contributions of these three attributes. Examples of these are depicted in Figure 2.5. The time-sequential design could be used to explore the effects of age and time-of-measurement, but confounds cohort. The cohort-sequential design could assess the effects of age and cohort but confounds time-of-measurement. The cross-sequential design could be used to investigate the effects of cohort and time-of-measurement but confounds age.

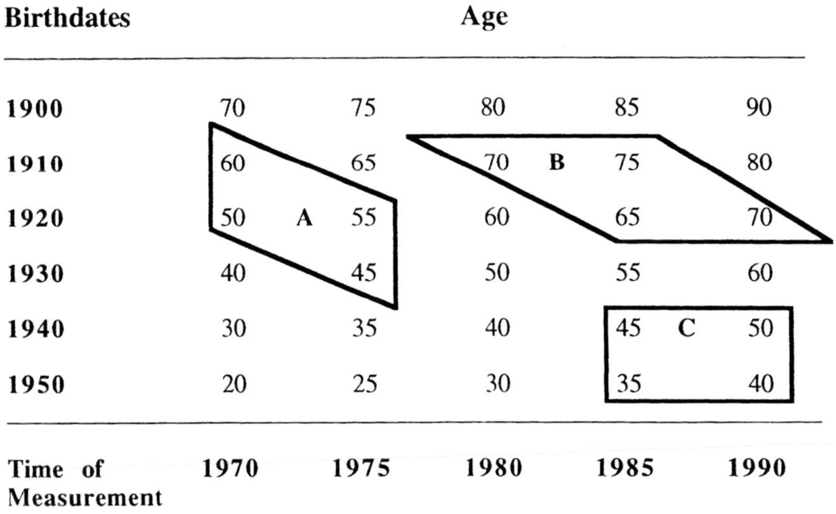

Birthdates			**Age**		
1900	70	75	80	85	90
1910	60	65	70 **B**	75	80
1920	50 **A**	55	60	65	70
1930	40	45	50	55	60
1940	30	35	40	45 **C**	50
1950	20	25	30	35	40
Time of Measurement	**1970**	**1975**	**1980**	**1985**	**1990**

FIGURE 2.5
Examples of sequential designs: **(A)** time-sequential design; **(B)** cohort-sequential design; **(C)** cross-sequential design.

A combination of these sequential designs has been described as Schaie's "Most Efficient Design," as presented in Figure 2.6. What is suggested is that four age groups (actually cohorts) of persons could be tested at the initiation of the study, for example, 1970. This, of course, would be a straightforward cross-sectional study. Then in 1980, these groups are then retested which would yield 10 years of longitudinal data. During this testing period, new subjects are recruited into the age groups of persons that were tested originally, in addition to a new cohort. This second cross-sectional study could be compared to the first. Ideally these results would match. Then in 1990 the groups recruited in 1970 would be retested providing now 20 years of longitudinal data. The subjects recruited in 1980 would yield new 10-year longitudinal data. And new subjects would be recruited in 1990 to form a third cross-sectional study.

From within this complex design, different analyses could be conducted to determine the existence of various confounds. These analyses would utilize the three sequential designs described in Figure 2.5 that could be extracted from the larger design.

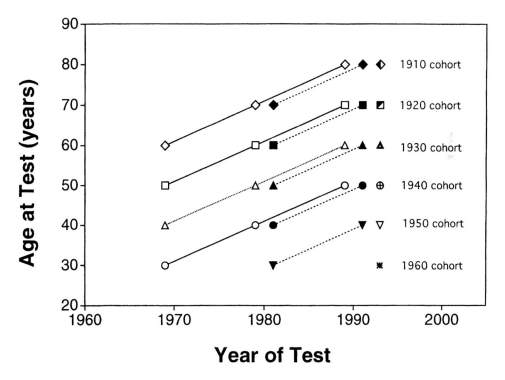

FIGURE 2.6
Example of Schaie's "Most Efficient Design." (Adapted from Schaie [22].)

2.5.2 Modified Sequential Designs and Time-Lag Studies

While the sequential design appeared to provide a very sophisticated approach for examining extraneous variation in aging studies, other gerontologists have argued that this approach is overly complex. Baltes [23] emphasized that age and cohort should be considered the primary variables, and that a modified version of the sequential design, such as shown in Figure 2.7, would suffice for most applications. Baltes and Schaie [24] eventually agreed on this matter by recognizing the strength of the formal sequential designs but also the practicality of modified versions. Formally or informally, many longitudinal studies, such as the BLSA, take advantage of such modified versions. Moreover, advances in statistical analyses of longitudinal data have provided some solutions for controlling potential sources of extraneous variation [25].

Despite the potential conceptual flaws involved in cross-sectional designs, the appeal of such studies should not be surprising given the complexities and potential biases of longitudinal studies. In cross-sectional studies, time-of-measurement effects and attrition are not problems, since all the data are collected about the same time. The latest techniques and equipment can be used. Progressive error and training effects should not be present since the subjects are only tested one time. Moreover, there are likely to be considerably less logistical support and resources required for a cross-sectional study. What remains as a major issue are cohort effects.

As a compromise on the potential confound of cohort, an efficient approach would be a modified *time-lag* study. In its simplest form, a time-lag study would focus on only one age group, for example, 65 to 75 year olds. The main questions would bear direct interest on how different cohorts had or had not changed across time. How do 65 year olds differ in 1995 than in 1975? The investigator would recruit subjects from this age group in 1975 and then follow up in 1995. However, the time-lag design can be modified to include other age groups. In effect, this would require the investigator to replicate a cross-sectional study using new samples at different times of measurements separated by a substantial period of time, e.g., 10 to 20 years for a human study. An example of this design would be represented in the two vertically shaded areas in Figure 2.7. If the cross-sectional results were equivalent across the repeated studies using new subjects on each measurement occasion, then the investigator would gain confidence in inferring the existence of a relationship between age and the parameter of interest.

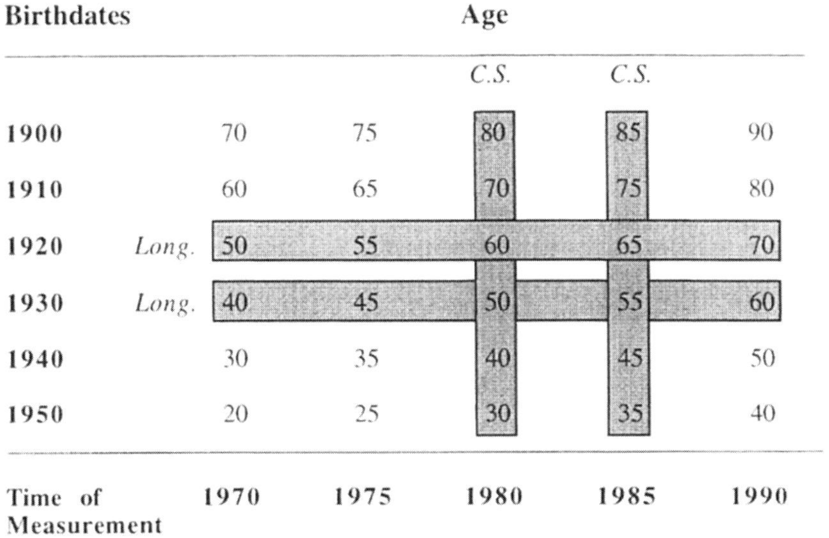

Birthdates   Age

				C.S.		C.S.		
1900		70	75	80		85		90
1910		60	65	70		75		80
1920	*Long.*	50	55	60		65		70
1930	*Long.*	40	45	50		55		60
1940		30	35	40		45		50
1950		20	25	30		35		40

Time of Measurement	1970	1975	1980	1985	1990

FIGURE 2.7
Example of a modified sequential design that involves both cross-sectional (*C.S.*) and longitudinal (*Long.*) samples.

While appealing in its simplicity, the modified time-lag study would still require a substantial time investment similar to a longitudinal study and would be risking that the results would not be repeatable due to cohort effects as well as time-of-measurement effects. Time-lag studies would probably be most efficiently conducted using a retrospective approach. Specifically, if past data were available or past tissue samples could be reanalyzed and compared to current data, an investigator would have to take less risk than proceeding with a time-lag study in a prospective manner. If the cross-sectional results were similar among different samples collected at different times and from different cohorts, then the investigator would be less concerned about cohort and time-of-measurement effects.

2.6   Sampling and Health Issues

2.6.1   Sampling Biases

Numerous difficulties encountered in designing gerontological studies in humans have been discussed thus far. Logistically the task is formidable regarding identifying, recruiting, and maintaining appropriate groups of subjects. Moreover, there are the logical problems encountered in handling cohort and time-of-measurement effects. Sampling in gerontological studies concerns other logical problems, too. Often major concerns can emerge regarding the degree of generalization of the results. Samples are meant to be representative of a larger group. What measures are taken to assure this? What shortcomings affect many gerontological studies? How generalizable are the results?

With the exception of the NHANES, no longitudinal study of aging in the U.S. has applied a true probability sample [26]. The sampling procedure for the NHANES is specifically geared to provide statistical representation of the U.S. population, exactly the same objective as opinion polls. However, nearly all cross-sectional and longitudinal studies of aging are typically regionally biased. For logistical reasons, such studies recruit persons within their locale. The NHANES is unique in that this study utilized mobile laboratories that could move from region to region based on a precise sampling strategy.

In addition to a regional bias, most studies of aging tend to recruit persons from middle to upper socioeconomic levels, which can introduce another bias into the sample. Aging studies, particularly longitudinal studies, can require a major commitment of time and resources. Participants in the BLSA provide their own resources to travel to the test site every 2 years and must take sufficient time from their jobs or other activities to make this trip. Thus, sampling in aging studies is usually biased toward upper socioeconomic groups. Even if effort is made to sample from lower socioeconomic groups, analyses of attrition in longitudinal studies have shown that that the drift toward more middle class representation would likely occur [27, 28].

2.6.2   Health Issues

In addition to the issues of regional and socioeconomic biases, sampling in aging studies also concerns bias due to health. Many early cross-sectional aging studies were heavily impacted by this bias. Young samples typically involved subjects who were very healthy, while samples of older persons were drawn from institutionalized populations. Nathan Shock, the founder of the BLSA, recognized how strongly biased such sampling could be, in effect a comparison between healthy and sick persons. Any inference about a relationship between age and a parameter of interest would be confounded. To counter this bias, the BLSA attempts to recruit healthy persons and then investigators screen the resulting data to remove subjects who later develop specific health problems that might contaminate the results [9].

While this clinical "clean-up" of the data is intended to remove the bias of disease, this objective also becomes problematic. Decisions about what conditions dictate removal from the data set can be very difficult [9]. Elimination of too many subjects might present another problem for sample bias and the generalization of the data. At the other extreme, the investigator might decide that the objective is to describe what is most "typical" during aging and decide not to screen for health status at all. A compromise solution would be to present data of interest both with and without clinical clean-up. If the findings are highly similar between these analyses, the investigator would gain confidence that the results reflect aging in the presence and absence of disease.

Returning to the issue of cholesterol, the data in Table 2.1 indicate that the cross-sectional perspective on age differences in serum cholesterol levels were little affected by an extensive clinical clean-up [21]. Fozard et al. [29] provide an excellent review of the challenge regarding the interplay between aging and disease and the techniques for deciphering it.

Another potential problem in interpreting data from aging studies is the *survivor effect*. This is akin to the issue of attrition that can bias the sample in many ways, but it is more closely aligned with health issues. For example, Metter et al. found that a group of 80 year olds recently selected for participation in the BLSA were actually healthier on many parameters compared to a group of 80 year olds who had entered the study 20 years previously. Beyond the clinical clean-up problem, it should be recognized that sampling at the end of the lifespan involves a very select group, those who have outlived most of their cohort.

TABLE 2.1

Age and Serum Cholesterol Levels in the Baltimore Longitudinal Study of Aging in Total Sample and Following Clinical Clean-up[a]

Age Group (yr)	Serum Cholesterol (mg/dl)					
	Total Sample			Clinical Clean-Up		
	N	Mean	SEM	N	Mean	SEM
20–39	249	206	2.1	245	206	2.1
40–59	373	236	1.8	306	234	1.9
60–79	319	229	2.0	204	225	2.6

[a] Data are derived from Hershcoph et al. [21].

2.7 Nonhuman Animal Studies

Many of the complexities of designing gerontological studies are presumably overcome in studies involving nonhuman studies. When laboratory animals are used, environment during aging can be controlled. Standardized strains of animals can be used. Use of inbred strains provides the investigator with genetic control. Every individual is virtually identical, and they can be aged under environmentally controlled conditions. Despite these advantages, it should be recognized that nonhuman studies can require many of the same precautions as human studies.

Because of the greater control over genetic and environmental variables as well as the short life span of most species investigated, nearly all nonhuman studies apply cross-sectional designs. Too often investigators rely upon analyzing only two age samples, comparing very young to very old groups of individuals [31]. They assume a linear relationship with age and thus run the risk of making the same mistake as depicted previously in Figure 2.3. Also many investigators will often obtain retired breeders because they are conveniently aged and made available by vendors and conduct experiments in which they compare this group to younger virgin animals [32]. This practice would obviously be ignoring a very important confounding, environmental variable.

In relying so heavily upon cross-sectional designs, investigators may inadvertently downplay the possibility of cohort effects. Investigators have found that cohorts of animals from the same colonies can differ markedly in important characteristics, such as life span [33].

Although much less applied, longitudinal designs can be used in nonhuman animal studies effectively. For example, Algeri et al. were able to document age changes in learning abilities in rats [34]. Roth et al. conducted a longitudinal analysis of wound-healing in rats [35]. These studies were conducted in short-lived animals (< 3 years); however, several research groups have begun longitudinal studies of aging in long-lived (>25 years) primate species [36–40]. Such long-term studies have begun to encounter many of the problems that impinge upon longitudinal studies in humans, but again probably have greater control over possible intervening variables [41].

Health status is also an issue that is too often ignored in nonhuman studies. Screening animals for specific pathologies that might affect the parameter of interest is commonly practiced although some simple health screens can be applied [42]. In summary, while nonhuman animal studies offer many advantages over human studies for designing gerontological research, many investigators using such models have not appeared fully aware of the potential confounds to their designs.

2.8   Summary

Because gerontology must rely upon a quasi-experimental approach, special consideration must be given to advantages and disadvantages of various study designs. Age is an *ex post facto* variable and thus not a true independent variable that limits the inferences that can be drawn. In this sense, age does not cause an effect; it is related to the effect.

The cross-sectional design permits the investigator to sample from several age groups at one point in time. This approach is efficient and cost effective. With this design, the investigator can draw inferences only about age differences and not age changes. Cohort differences can confound the inference about age differences.

The longitudinal design samples from the same cohort or cohorts over time and thus permits inferences about age changes. However, secular or time-of-measurement effects can confound inferences about age changes. Besides the increased complexity and cost of conducting longitudinal studies, other possible problems include adequate control for attrition and progressive error.

Sequential designs, including modified versions, were introduced to address the problems of confounding age, cohort, or time-of-measurement. These approaches basically use combinations of cross-sectional and longitudinal designs with new cohorts introduced at various intervals.

All studies of aging no matter how well designed must also attend to the problems of health and sampling, including the problem of interpreting biases due to socioeconomic and health status. The former problem might limit the generalization of the findings. The latter problem can be addressed by clinical clean-ups of data, but this might create another problem related to survivor effects. Specifically, older samples will include only those individuals free of disease which again might impact the generalization of the results. Methods such as examining the age-related pattern of results with and without clinical clean-up might elucidate whether survivor effects would be an issue.

While most of the caveats regarding experimental design addressed in this chapter pertain to studies of long-lived species, many can still apply to studies of laboratory animals. Cross-sectional designs are the clear design of choice because of greater control over environmental and genetic variables that presumably avoid issues of cohort and time-of-measurement effects. However, longitudinal studies can be accomplished and provide even greater control. The main issue regarding nonhuman studies is likely health status which for the typical study is not screened routinely as it is for most human studies.

Through careful consideration of the strengths and weaknesses of experimental design in gerontological research, an investigator can gain confidence with the results obtained and provide a proper interpretation for them.

References

1. Schaie, K. W., Methodological problems in descriptive developmental research on adulthood and aging, in *Life-Span Developmental Research on Adulthood and Aging*, Nesselroade, J.R. and Reese, H.W., Eds., Academic Press, New York, 1973, 253.
2. Schaie, K.W., The Seattle longitudinal study: a twenty-one year investigation of psychometric intelligence, in *Longitudinal Studies of Adult Personality Development*, Schaie, K.W., Ed., Guilford, New York, 1983, 64.
3. Bayley, N., Behavioral correlates of mental growth: birth to thirty-six years, *Am. Psychol.*, 23, 1, 1968.
4. Eichorn, D. H., The Institute of Human Development Studies: Berkeley and Oakland, in *Intellectual Functioning in Adults: Psychological and Biological Influences*, Jarvik, L. F., Eisdorfer, C., and Blum, J. E., Eds., Springer-Verlag, New York, 1973.
5. Terman, L.M. and Oden, M.H., *The Gifted Group at Midlife: Thirty-five Years Follow-Up of the Superior Child, Genetic Studies of Genius*, Stanford University Press, Stanford, CA, 1959.
6. Sears, R.R., Sources of life satisfactions of the Terman Gifted Men, *Am. Psychol.*, 32, 119, 1977.
7. Busse, E. W. and Maddox, G. L., *The Duke Longitudinal Studies of Normal Aging 1955–1980. Overview of History, Design, and Findings*, Springer-Verlag, New York, 1985.
8. Bell, B., Rose, C. L., and Damon, A., The Normative Aging Study: an interdisciplinary and longitudinal study of health and aging, *Aging Hum. Dev.*, 3, 5, 1972.
9. Shock, N. W., Greulich, R. C., Andres, R., Arenberg, D., Costa, P. T., Jr., Lakatta, E. G., and Tobin, J. D., *Normal Human Aging: The Baltimore Study of Aging*. NIH Publication No. 84-2450, U.S. Government Printing Office, Washington, D.C., 1984.
10. Shibata, H., Haga, H., Suyama, Y., Matsuzaki, T., Maeda, D., Koyano, W., and Hatano, S., A ten-year comprehensive survey of the Japanese urban elderly: the Koganei Study, *Soc. Gerontol.* (Tokyo), 27, 68, 1988.
11. Roos, L. L., Nicol, J. P., and Cageorge, S. M., Using administrative data for longitudinal research: comparisons with primary data collection, *J. Chronic Dis.*, 40, 41, 1987.
12. Svanborg, A., Cohort differences in the Göteborg studies of Swedish 70-year olds, in *Epidemiology and Aging*, Brody, J. A. and Maddox, G. L., Eds., Springer-Verlag, New York, 1988, 27.
13. McClearn, G. E., Svartengren, M., Pedersen, N. L., Heller, D. A., and Plomin, R., Genetic and environmental influences on pulmonary function in aging Swedish twins, *J. Gerontol.*, 49, 264, 1994.
14. Deeg, D. J. H., *Experiences from Longitudinal Studies of Aging*, Netherlands Institute of Gerontology, Nijmegen, The Netherlands, 49, m264, 1989.
15. Owens, W.A., Age and mental abilities: a second adult follow-up, *J. Educ. Psychol.*, 51, 311, 1966.

16. Carter, B. H., Pearson, J. D., Metter, E. J., Brant, L. J., Chan, D. W., Andres, R., Fozard, J. L., and Walsh, P. C., Longitudinal evaluation of prostate-specific antigen levels in men with and without prostate disease, *J. Am. Med. Assoc.*, 267, 2215, 1992.

17. Kannel, W.B. and Gordon, T., Evaluation of cardiovascular disease in the elderly: the Framington Study, *Bull. N.Y. Acad. Med.*, 45, 573, 1978.

18. Kannel, W.B., Nutritional considerations in occurrence and prevention of cardiovascular disease in the elderly, in *Potential for the Nutritional Modulation of Aging Process*, Ingram, D.K., Baker, G.T., III, and Shock, N.W., Eds., Food and Nutrition Press, Trumbull, CT, 1991, 17.

19. Madans, J. H., Kleinman, J. C., Cox, C. S., Barbano, H. E., Feldman, J. J., Cohen, B., Finucane, F. F., and Cornoni-Huntley, J., Ten years after NHANES I: report of initial follow-up, 1982–1984, *Public Health Rep.*, 101, 465, 1986.

20. Zonderman, A.B., Costa, P.T., Jr., and McCrae, R.R., Depression as a risk for cancer morbidity and mortality, *J. Am. Med. Assoc.*, 263, 513, 1990.

21. Hershcoph, R. J., Elahi, D., Andres, R., Baldwin, H. L., Raizes, G. S., Schocken, D. D., and Tobin, J. D., Longitudial changes in serum cholesterol in man: an epidemiologic search for an etiology, *J. Chronic Dis.*, 35, 278, 1982.

22. Schaie, K.W., A general model for the study of developmental problems, *Psychol. Bull.*, 64, 92, 1965.

23. Baltes, P. B., Longitudinal and cross-sequential sequences in the study of age and generation effects, *Hum. Dev.*, 11, 145, 1968.

24. Baltes, P. B. and Schaie, K. W., On the plasticity of intelligence in adulthood: where Horn and Donaldson fail, *Am. Psychol.*, 31, 720.

25. Brant, L. J. and Pearson, J. D., Modeling the variability in longitudinal patterns of aging, in *Biological Anthropology and Aging: Perspectives on Human Variation over the Life Span*, Crews, D. E. and Gurruto, R. M., Eds., New York, Oxford Press, 1994, 373.

26. Burt, V. L. and Harris, T., The third National Health and Nutrition and Examination Survey: contributing data on aging and health, *Gerontologist*, 34, 486, 1994.

27. Schaie, K. W., Internal validity threats in studies of adult cognitive development, in *Cognitive Development in Adulthood: Progress in Cognitive Development Research*, Howe, M. L. and Brainerd, C. J., Eds., Springer-Verlag, New York, 1988, 241.

28. Sharma, S. K., Tobin, J. D., and Brant, L. J., Factors affecting attrition in the Baltimore Longitudinal Study of Aging, *Exp. Gerontol.*, 21, 329, 1986.

29. Fozard, J. L., Metter, E. J., and Brant, L. J., Next steps in describing aging and disease in longitudinal studies, *J. Gerontol. Psychol. Sci.*, 45, P116, 1990.

30. Metter, E. J., Walega, D., Metter, E. L., Pearson, J., Brant, L. J., Hiscock, B. S., and Fozard, J. L., How comparable are healthy 60- and 80-year old men?, *J. Gerontol. Med. Sci.*, 47, M73, 1992.

31. Coleman, P, Finch, C., and Joseph, J.A., The need for multiple time points in aging studies, *Neurobiol. Aging*, 11, 1, 1990.

32. Ingram, D. K., Spangler, E. L., and Vincent, G. P., Behavioral comparison of aged virgin and retired breeder mice, *Exp. Aging Res.*, 9, 111, 1983.

33. Mos, J. and Hollander, C.F., Analysis of survival data on aging rat cohorts: pitfalls and some practical considerations, Mech. Aging and Dev., 38, 89, 1987.

34. Algeri, S., Biagini, L., Manfridi, A., and Pitsikas, N., Age-related ability of rats kept on a life-long hypocaloric diet in a spatial memory test. Longitudinal observations, *Neurobiol. Aging*, 12, 277, 1991.

35. Roth, G.S, Kowatch, M.A., Hengemihle, J., Ingram, D.K., Spangler, E.L., Johnson, L.K., and Lane, M.A., Effect of age and caloric restriction on cutaneous wound closure in rats and monkeys, *J. Gerontol. Biol. Sci.*, 52A, B98, 1997.

36. Bodkin, N.L., Ortmeyer, H.K., and Hansen, B.C., Long-term dietary restriction in older-aged rhesus monkeys: Effects on insulin resistance, *J. Gerontol. Biol. Sci.*, 50, B142, 1995.

37. Bowden, D.M., Short, R.A., and Williams, D.D., Constructing an instrument to measure the rate of aging in female pigtailed macaques (*Macaca nemestrina*), *J. Gerontol. Biol. Sci.*, 45, B59, 1990.

38. Cefalu, W.T., Wagner, J.D., Want, Z.Q., Bell-Farrow, A.D., Collins, J., Haskell, D., Bechtold, R., and Morgan, T., A study of caloric restriction and cardiovascular aging in cynomolgus monkeys (*Macaca fascicularis*): a potential model for aging research, *J. Gerontol. Biol. Sci.*, 52A, B10, 1997.

39. Ingram, D.K., Cutler, R.G., Weindruch, R., Renquist, D.M., Knapka, J.J., April, M., Belcher, C.T.,Clark, M.A., Hatcherson, C.H., Marriott, B.M., and Roth, G.S., Dietary restriction and aging: the initiation of a primate study, *J. Gerontol. Biol. Sci.*, 45, B148, 1990.

40. Kemnitz, J.W., Weindruch, R., Roecker, E.B., Crawford, K., Kaufman, P.L., and Ershler, W.B., Dietary restriction of male rhesus monkeys: design, methodology, and preliminary findings from the first year of study, *J. Gerontol., Biol. Sci.*, 48, B17, 1993.

41. Nakamura, E., Lane, M.A., Roth, G.S., Cutler, R.G., and Ingram, D.K., Evaluating measures of hematology and blood chemistry in male rhesus monkeys as biomarkers of aging, *Exp. Gerontol.*, 29, 151, 1994.

42. Spangler, E. L. and Ingram, D. K., Utilization of the rat as a model of mammalian aging: impact of pathology on behavior, *Gerontology*, 32, 707–717, 1997.

3

Mathematical and Computational Tools for Gerontological Research

Tarynn M. Witten

CONTENTS

0-8493-3112-9/99/$0.00+$.50
© 1999 by CRC Press LLC

3.1   Modeling, Simulation, and Analysis of Aging Living Systems

APOLOGETICA   This is a text on *"methods."* Typically such texts contain extensive *"how to's"* with worked examples. This chapter will be a bit different from the others in this book due to the breadth of material addressed. This chapter is designed to serve as a pointer to the variety of literature on mathematical and computational tools and their potential applications in the field of aging. Each section is, in and of itself, a complete research discipline with numerous potential citations. Page limitations forbid me from being either expository or comprehensive. Hence, it is truly not possible to go into any detail about the methodologies and how to apply them to both general and specific classes of problems arising in the various disciplines of gerontological research. My approach to this chapter has been to provide a collection of articles and books that researchers may find helpful as a starting point for obtaining a more comprehensive explanation of the various mathematical and computational methodologies commonly found in a given discipline, as well as their potential application to problems arising in the study of aging.

3.1.1   Philosophy and Phenomenology

The complexity of life is fascinating for both experimental biologists as well as applied mathematicians. Attempts by mathematical biologists over the past century to study these intricate biological systems have lead to a variety of mathematical and computational models of living systems. With the increasing cost-effectiveness of computation, both desktop as well as large-scale supercomputer, simulation modeling has become a more widely used tool for incorporating the necessary biological complexity into the original, often highly simplified, mathematical models. The advent of the new generation computer technologies greatly expands the capability of modelers and analysts to implement simulations and to construct mathematical models that are biologically realistic. Coupled with our increased understanding of the behaviors of these systems, mathematical and computer modeling can be a formidable ally. Mathematical complexity is not nearly as serious an issue, as speeds of computation are now practical enough to enable vast numbers of mathematical and simulation modeling computations to be performed in a relatively short period of time, thereby creating a *"virtual modeling/simulation laboratory"* within which gerontological research can pose numerous *"what if"* questions. The following citations will serve to open the door to the general field of modeling and simulation in the biosciences.

3.1.1.1 General Modeling and Philosophy

Citations in this section encompass general overviews of mathematical and computer simulation modeling, philosophical and methodological issues in modeling and simulation, case studies, and general introductory texts. They should be seen as gateways to the larger, more focused literature on mathematical/computer simulation modeling. Those who are novices and who are interested in becoming familiar with the general concepts involved in modeling and simulation in biological and biomedical areas should begin with the following references: [10,66,86,106,107,116,133,148,188,234,235,237,244,289,321,325,435,444,447, 450,453,456,457,461,468,471]. For those who are beginners, these references will provide a gentle introduction to the field.

3.1.1.2 Mathematical Methods in Biosciences

Citations in this section focus on books that address a particular area of mathematics (differential equations, dynamical systems theory, fractals, iterative equations, partial differential equations, finite element methods, to name a few) and its application to biological systems modeling [47,136,170,171,174,185,252,324]. These references are primarily mathematical and use the biosciences as motivation or for examples of application illustration.

3.1.1.3 Specialty Modeling

Citations in this section focus on books and articles that deal with mathematical modeling and/or computer simulation as applied to a particular subject area such as compartmental modeling, molecular kinetics, metabolic modeling, fluid dynamics, neurological models, cellular dynamics, physiological modeling, population dynamics, and numerous other areas [2,21,55,65,85,122,155,156,176,192,202,256,261,271,272,298,336,344,370,377,407, 408]. These references focus on the bioscientific problem and subsequently develop the requisite mathematics tools to solve that specific problem or problems. These citations are frequently well balanced between the mathematical and the bioscientific approaches. Those who are interested in a particular biomedical application area should examine some of the books in this section.

3.2 Data Analysis, Visualization, and Presentation

3.2.1 Visualization Philosophy

The life sciences are, by their very nature, visual/tactile sciences: (1) they handle data from images (X-ray, molecular modeling, computational chemistry, etc.) and (2) they handle complex datasets and transform them to a visual representation of the data (mathematical and computer models, enzymic reaction simulations, physiological process models, image reconstruction, medical diagnostics). Examining the various general types of medical data visualization problems, it is clear that we may categorize them in the following fashion:

- *Synthesis* — Integration of information; preferably interactively and in realtime,
- *Analysis* — Interpretation and evaluation of the data by selectively displaying experimental and/or computational results within a comprehensible framework, and
- *Communication* — Bridging the gap between science and the people, between the sciences, and between the scientists themselves.

In a sense, the computer becomes the laboratory. And, as a consequence of this new development, *what was once done in the tube is now being done on the tube.*

Perhaps the most obvious use of computers, in the visualization arena, is the handling of patient image data and/or the construction/reconstruction of medical image data for the purposes of clinical diagnosis and analysis. In the area of imaging, we recognize the following three major problem classes:

- Simple rebuilding of two-dimensional image (graphics) data into a useful/usable two-dimensional image on a screen,

- Reconstruction of three-dimensional images from two-dimensional scans, as well as direct 3-D full reconstruction/visualization, and

- Visualization of an image in a realtime, interactive mode. Given a 3-D image (say of the pelvis) can we manipulate it, in realtime, for the purposes of clinical examination? Can we slice it to ascertain a measure of osteoporotic development, rotate it to look for hidden fractures?

A further extension of medical visualization may be found in the development of computer-based, surgical planning systems. With these systems, physicians are now able to simulate a surgical procedure before it is performed on the patient, thereby minimizing potential hazard to the patient, and increasing surgical accuracy [183]. Experimental science benefits as well. These same methods, when applied to experimental biology, allow the researcher to understand biological and physiological functions of various organs and organ systems. For example, brain mapping studies have allowed us to investigate the cognitive function, as well as the physiological behavior of the brain. Recent work at the University of Michigan has applied image reconstruction techniques to PET scans of elderly non-Alzheimer and Alzheimer patients in order to ascertain differences in neural activation. Other interesting questions arise as to how we can effectively visualize the combined CAT, PET, NMR, and simulation data for a particular patient so that it is possible to visualize the bone, muscle, and metabolic data at the same time? One could imagine how such multidatatype technologies could be applied to neurological and physiological studies of biomedical aging processes. Diverse biological disciplines turn to computer graphics and advanced imaging techniques as a means of studying (visualizing) subcellular, cellular, and supercellular biological phenomena. Molecular biologists visualize the dynamics of molecules and the issue of structure/function. Virologists use computer aided design techniques to visualize viral capsid structures and the dynamics of viral infection. Neurophysiologists/neurologists map the three-dimensional behavior of the brain, in an effort to understand questions of disease, behavior, and physiology. Biomechanicists study the behavior of muscles and bones as they attempt to investigate the healing process and the potential for using the computer to aid in bone replacement surgery [36,147,342]. All of these areas can be seen as having relevance to the fields of geriatrics and gerontology.

3.2.1.1 General Visualization Methods

The following books address basic issues in data visualization with respect to graphical representation of data. The emphasis is on charting, graphing, and alternative representations for the display and presentation of data. The increasing multi-dimensional complexity of healthcare delivery data, for example, mandates that we better understand the problems, tricks, and traps of effective data visualization and presentation. It is easy to see how difficult it can be to visualize multiwave, multiform, multiquestion, multi-individual, longitudinal datasets in a fashion that most easily allows the viewer to extract the salient points of such a highly complex dataset. The references in this section will give the reader a good overview of how different types of data can be effectively (and not necessarily so effectively) presented. The first three of these references should be part of everyone's reference library [71,79,119,194,221,277,301,388–390,404].

Basic methodological approaches to analyzing data (exploratory data analysis, data modeling methods, general getting started methods for looking at datasets) are addressed in [1,38,121,146,149,179,241,335,352,391,394] and [427]. More advanced visualization methods may be found in [110,117] and [297].

The following texts address focused methodological approaches to data analysis (spatial statistics, longitudinal data analysis, categorical data analysis, principal component analysis) [11,96,141,149,241,394].

3.2.1.2   Data Animation, Sonification, and Virtual Reality

Typically, scientific visualization carries with it the implication of the *"visual."* More recently, researchers have begun to consider alternative ways to *"visualize"* or *"render"* data. One of the more recent attempts to extract information from complex datasets involves the use of sound or sonic representation of the data. Extensive discussion, along with an expansive review of the literature, can be found in [462] and [460]. Good starting points for virtual reality applications may be found in [8,45] and [46]. Currently, there is a wide variety of medical applications of VR. Among these application areas are surgical training, surgical evaluation, telemedical evaluation, and treatment planning. More recent applications of VR have occurred in psychiatry and in rehabilitative medicine [120,183]. It is easy to see how these "general" applications areas could be applied to the more topic-specific areas of gerontology. For example, consider the creation of a VR environment to help geriatric patients feel less lonely, cope with the loss of a loved one, or even envision themselves as doing something they are otherwise incapable of doing. Whether it is regular surgery or surgery arising out of geriatric complications, the VR training methodologies still apply. Animation and advanced, visualization methodologies may be found in [13,117,150,410] and [411].

3.2.1.3   Tricks and Traps in Data Visualization

Issues of data visualization and the tricks and traps associated with graphical representation can be found in the following references: [71,388,389,404]. It is important to understand that the psychology of visualization (interpreted in its more broadly used definition) and perception can profoundly affect the way in which we interpret sight and sound. Consequently, it is crucial that we understand these issues before we apply a particular visualization technique to a given dataset lest we find ourselves communicating our results incorrectly.

3.2.1.4   Visualization Software, Hardware, and Platforms

Computer graphics and scientific visualization hardware and software are available on nearly every computer platform. There are many things to think about when initiating a purchase. For example, will there be one or more users of the package (do you need a site license, a multiple user license, or a single user purchase)? Will the software be sitting on a single machine, on a network, available over a network, will it need a license manager package, or sit on a server? What operating system will you be using (UNIX, NT, DOS, W95, MAC, other)? What type of capabilities do you want in your software package? For example, do you want presentation graphics, scientific graphics, data visualization, image processing capability, statistical processing and analysis, graphical rendering capability, animation, or any other more advanced capabilities? Many of the more advanced packages require a *programming* or *nonpoint and click* environment. This must be taken into account when considering who will be using the package. It is beyond the scope of this short discussion to cover the breadth of available materials in the domain of data analysis and visualization. For a more extensive review of these and related issues, see the following citations: [452,454,455].

3.2.2   Imaging Methodologies

Image processing methodologies (image extraction, image enhancement, edge detection, contouring, etc.) can occur in a number of different areas in the gerontology domain. At the cell and molecular level, identification of bands in gels is one obvious application area. It is also easy to imagine how these methods could be used to study whole cell and organelle level changes over the cell cycle and as a cell ages. At the physiological systems level, image processing is routine in PET, NMR, and CAT scanning. Basics of image processing may be found in the following citations: [6,57,140,177,178,429]. The use of image processing as a diagnostic tool, as well as a quantitation tool, has a great deal of applications to the field of *"aging"* research.

3.2.3   Morphometrics

Morphometrics is one of the modern developments of statistical science. It is the quantitative study of biological shape variation and represents the fusion of quantitative multivariate analysis techniques and the direct visualization of shape data. It has immediate relevance to the statistical analysis of morphological changes (shape changes) with age. Primary citations and expansive bibliographies to the literature of morphometrics may be found in the following citations: [32,33,251].

3.2.4   Data Analysis

The mathematical and computational analysis needs of the gerontologic and geriatric research community encompass the full spectrum of currently available tools and methodologies.

3.2.4.1   *Biostatistical Issues*

In nearly every aspect of aging research, there is some component of statistical methodology. Attention to statistical methods cannot be over emphasized. For those interested in a basic biostatistical methodology text, consider any one of the following: [14,15,30,35,49,76, 87,99,118,253,266,368,466].

Many studies take place over a long period of time and require repeated examinations, questionnaires, or other forms of data gathering. Such studies are termed longitudinal studies. Basic longitudinal analysis texts include [73,226,247,328,399] and [400].

As more and more complex, multidimensional datasets are being acquired through highly complex clinical trials, it is becoming increasingly common to ask questions, which if the individual datasets from diverse studies had been combined, the answers to those questions might become available. However, if the datasets come from independent studies that were not designed to answer these new questions, then it is reasonable to ask under what conditions may one fuse the data from these independent clinical trials — for example — in order that one might answer the new question. Attempts to perform this fusion are called *"meta-analysis."* While there is a significant amount of dispute over the validity and value of meta-analysis, it is an often used technique [93,104,145,160,220,243, 292,326,403].

Central to the field of gerontology is the ability to observe changes in survival (lifespan) and to relate those changes to underlying biomedical processes as well as to environmental effects. In fact, the whole concept of *"biomarkers of aging"* is based upon this assumption, as is the study of the evolutionary biology of aging. For those wishing to find out more about the basic mathematical and statistical foundations of survival (reliability) theory, the following texts are worth examining: [17,198,225,231]. More biogerontological discussion may be found in the following citations (as well as the chapter in this volume) and the extensive bibliographies cited therein [101,115,131,348,465].

Many problems in aging research involve understanding differences between groups, treatments, protocols, etc. and what sets of independent variables contribute to those differences. In addition, not only would we like to ascertain the existence of potential differences, we would also like to quantify how much each variable contributes to that difference. Further complicating the process of understanding these behaviors may be the fact that there are costs to consider, the patient pool may be limited, patients may leave before completing the study, confounding factors may mitigate our understanding, the study may be longitudinal, and there may be potentially lethal side effects. All of these factors introduce statistical issues into the experimental methodology. For those interested in basic multivariate analysis texts, the following represents a reasonable list of elementary to more advanced multivariate analysis tools and techniques texts: [29,135,211,229].

Lastly, one can never be too overloaded with good handbooks of basic mathematical and statistical formulae, integrals, infinite series, and other important and useful obscure (and not so obscure) formulae and equations. Basic mathematical and statistical handbooks include [42,142,309,347] and [378].

3.2.4.2   Fitting Functions to Data

Fitting functions to data involves a variety of different tools and techniques. The most common of the statistical methodologies is called *"regression."* Classical regression models come in two flavors, linear and nonlinear. Other forms of parameter estimation include maximum likelihood estimation (MLE) [231]. MLE methods are particularly useful for parameter estimation in survival models. Various methodologies exist for performing regression and most good statistical software packages have regression sub-routines, of various types, in them. It is important to understand that for parametric estimation, the model, the methodology, and the sample size may all play an important role in the value of the parametric estimates obtained. Data fitting may involve fitting a specific model or distribution to the data or it may be "model or distribution free" [39,54,72,88,94,132,168, 175,200,232,242,274,281,287,332,378]. Regression models are commonly used to estimate survival rates in clinical trials, rates of aging in the biology of aging, as well as in numerous physiological modeling, drug trial, and healthcare delivery studies.

Engineering disciplines take a different approach to the problem of estimation in that they also use curve fitting methods (splines, b-splines, NURBS, and finite elements) to provide data estimates [22,111,302,309,402]. Methods of this type are used in some of the newer approaches to estimate mortality and hazard rates in studies of the evolutionary biology of aging.

3.2.4.3 *Dynamical Systems*

Ever since its popularization in the late 1970s, dynamical systems theory has emerged as a powerful tool for the analysis of complex systems. Increasingly, it is finding more and more application in the analysis of biological systems. Recently, Gerontological Society of America Annual Meeting 1996, dynamical systems theory had begun to be applied to problems in gerontology and the term "gerontodynamics" has been coined to indicate this application domain. From these studies has emerged the word "gerontometrics" meaning the study of the dynamics of aging.

There is a large literature in the area of dynamical systems and much of it is cited in the section below on chaos and fractals. Good starting references include [95,114,134,169,187, 189,254,255,284,291,386,419,432] and [460]. An important caveat in this area is to remember that the concepts of dynamical systems, while they may look as if they map onto gerontological problems, do not, in point of fact, necessarily do so. Among the most common of fatal mistakes is the tacit assumption that because something *"looks"* chaotic, that it (a) is chaotic and (b) that iterative or continuous dynamical systems theory can be applied to the study of that particular system. Great care must be taken in the borrowing of one discipline's terminology and methodology and the application of that terminology and methodology to another, particularly gerontology and geriatrics.

3.3   Chemistry and Pharmacology/Pharmacokinetics

In his Philosophie Positif, A. Comte stated that *"Every attempt to employ mathematical methods, in the study of chemical questions, must be considered profoundly irrational and contrary to the spirit of chemistry. If mathematical analysis should ever hold a prominent place in chemistry — an aberration which is happily almost impossible — it would occasion a rapid and widespread degeneration of that science."* Needless to say, the field of computational chemistry is a burgeoning field which is readily divided into the following major topic areas:

- *Biophysical properties* such as crystallographic reconstruction and molecular visualization,
- *Molecular Biochemistry* which encompasses such areas as structure-function studies, enzyme/substrate studies (reaction studies, pathway analysis), and protein dynamics and their properties,
- *Pharmacokinetics/Pharmacodynamics* which encompasses such areas as interactive molecular modeling, drug design/drug interactions (cancer chemotherapy, orphan drugs), binding studies (binding site properties), structure/function relationships as applied to drug effectiveness, and cell receptor structures, and
- *Nuclear Chemistry/Medicine*

All of these areas involve intensive numeric computation and subsequent realtime graphics for the visualization of the final molecular/chemical structures. Within the discipline of computational chemistry, the field of neural nets is taking hold. Neural nets are being used to learn to identify similar protein structures based upon recognition of pattern representations for the three-dimensional structure of proteins [7,28,126–129,151,152,157,166, 207,209,222,223,236,240,256,311,313,317,354,456,457,459,461,473,474].

While superworkstations and supercomputers are used for molecular visualization and dynamics, another computational issue is that of large-scale pharmacologic simulation, both at the molecular and the physiological levels. Here, we wish to assess the impact of a particular drug upon a given biological and/or physiological system or systems. Such problems involve large-scale numeric and symbolic data bases, numerous systems of non-linear differential/partial differential equations (with possible stochastic factors), and sophisticated numerical techniques for handling the numerical computations related to their solution, not to mention the associated scientific visualization problems.

The graying of America and the increased healthcare requirements accompanying that aging have clear ramifications with respect to the need for the methods discussed in this section as related to the pharmacologic needs of this increasing community.

3.4 Molecular Biology and Genetics

3.4.1 Overview

The analysis of genetic structure contains numerous interesting questions ripe for investigation. One of the more important of these questions is the issue of the evolution of the structure/complexity of the human genome. As experimental molecular biology sequences more and more of the genes in various species, we have a greater data base of information which can be used to study evolutionary biology (see later section on the matrix of biological knowledge). Evolutionary biology questions are intensely scalar computing problems, involving the use of complex tree search problems. Computational tree search algorithms, implemented on parallel computers and on networks of smaller work-stations (functioning in a parallel-like environment), can assist scientists in evaluating genetic evolutionary trees and, in doing so, help them to understand the evolution of these genomic structures. In addition, these same algorithms can be used to study the related problem of the origin of the species. Linkage analysis, a complex mathematical analysis involving tree search algorithms, is used to locate and to map gene structures, for the purposes of understanding inherited disorders [24,90,112,113,188,197,233,271,382,408,409]. What is equally fascinating is that this same set of tools can be used to look for patterns of disease and mortality in a pathology data base or patterns of healthcare delivery and utilization in a nursing home data base.

3.4.2 Sequencing

At the gene sequencing level, mathematical and computational algorithms are utilized to align gene sequences, match gene sequences, to determine similarity between sequences, to reconstruct gene sequences from sequence fragments, and to construct theoretical three-dimensional structures based upon those sequences. These hypothesized structures are subsequently compared to the experimental data and binding and/or transcription predictions can be made and/or analyzed for new insights into the biological dynamics of the gene [21,24,50,69,98,143,164,190,224,277,278,295,301,356,371,408,417].

Beyond the issue of algorithms for searching, folding, alignment, and graphics, the gene sequencing problem gives rise to important questions in data base design and analysis, particularly as related to the concept of integrated datasets. The incessant daily increase in biological information requires information systems that can handle massive amounts of data at rapid speeds. The integration of structural, sequence, biochemical, biological, genetic, and physicochemical datasets is not only increasingly important, but mandatory for the effective pursuit of the new generation of biological/medical research questions. In time, we will see artificial intelligence and neural net overlays to complex, high performance, data base systems. These systems, learning from their users, will eventually be able to hypothesize biological questions and to search the database for possible examples/counterexamples. As their usage continues, it is possible to postulate that the systems will garner enough knowledge to formulate and/or deduce rules; thereby entering the *metaquery* stage of evolution [21,24,68,74,77,79,89,127,186,222,228,268,269,308,313,351, 354,359,360,362,380,396,412,473,474]. Recent applications of this work are discussed in the chapters of Carey and Curtsinger in this volume.

3.4.3   Topological Modeling

The interplay of knot theory, differential geometry, and topology (mathematical disciplines) with computer science and molecular biology has led to a number of interesting and important results on DNA coiling, super-coiling, and winding [23–26,62,81,83,151,152,279, 303,397, 398,424,425,463]. Such questions are of great importance as we study the existence and meaning of structural changes in DNA and other "coiled" structures in a cell as it ages.

3.4.4   Molecular Bioinformatics

Molecular bioinformatics is the newly emergent discipline comprising the development and application of computational algorithms for the purpose of analysis, interpretation, and prediction of data and for the design of experiments in the biosciences. It crosses the disciplines of computer science with molecular biology to create a truly unique and new discipline [21,341].

3.5   Cell Biology, Physiology, and Metabolism

The dynamics of cell populations is of great interest to both biologists and clinicians. Understanding the cell cycle would have impact upon our understanding of how to better control the development of various forms of cancer. From a therapeutic perspective, a mathematical model or enhanced simulation of cellular processes could be used to test treatment protocols and regimens before they were actually implemented upon the patient. Accurate molecular dynamics and simulation models of membrane channel dynamics could open the way towards a deeper understanding of membrane transport mechanisms [27,82,123,161–163,230,258,259,264,327,330,331,383–385,387,392,413,414,434,448,451,458]. Question of how age-related membrane changes, changes in enzymatic levels, and potential long-term effects of free radical damage can also be examined within the context of simulation models [161–163]. Models of cellular physiology can be used to examine questions of cellular aging from a variety of perspectives such as *in vitro* clonal senescence. Such models can also be used to examine questions of how older cell populations might behave in the presence of tumors (malignant or benign). Similarly, a good model of cell population aging might be used to further dissect out potential processes involved in aging and/or to test new hypotheses of cellular aging.

3.6   Large-Scale Physiology

Human physiology attempts to explain the physical and the chemical factors that are responsible for the origin, the development, and the progression of human life. Human beings are complex machines built from equally complex systems (immune system, digestive system, nervous system, *etc.*). These systems contain multiple parts or organs (each, itself, a complex hierarchy of systems and subsystems). Realistic models of these systems can lead to deeper understanding of the basic biology of these interacting bodily systems. And, as the models become more sophisticated, they can lead to a deeper understanding of the important synergism between the systems.

3.6.1   Reproductive Biology

The ovaries are the repository of the female reproductive component, the follicles. Of the approximately 500,000 follicles present in the 2 ovaries at birth, only about 375 of these follicles will eventually develop into ova (eggs). Worldwide, it has been demonstrated that there are increasing levels of infertility in both sexes. This is particularly true in the U.S. and in Poland. It is not at all clear what is causing such an increase to occur. As a consequence of this fact, it is of no small importance that the dynamics of the reproductive cycle be studied in detail. Such models might give insight into how the environment and/or other factors might play into the level of infertility displayed in a particular country or population. In addition, such models can be used to study the dynamics of aging in the mammalian reproductive system [5,213–215,238,239,263,265,357,451]. Further, it is possible to imagine how the combination of a neurobiological and an endocrinological model might be used to examine (simulate) how homeostatic adjustments to the neuroendocrine system decline over time.

3.6.2   Cardiovascular Dynamics

Central to the cardiovascular system is the heart. Much work has been done in such areas as the mathematical modeling and computer simulation of nonlinear waves in excitable media. One particular example of an excitable medium is the heart muscle. One particular form of excitation of recent interest is reentrant tachycardia in the heart as it is related to the onset of fibrillation. Other researchers have been investigating the electrical energy of the heart and its effect upon the organ's natural rhythm. Particular interest has been expressed in how such a model could be utilized to study the interaction of cardiac function and cardiac drugs. Other researchers have been performing 2-, and now 3-dimensional modeling of the heart, including valves and ventricles, and is now involved in adding atria and other vessels. This working model beats and moves the blood through the chambers of the heart. Other questions such as the timing between the atrial and ventricular contraction; a clinically important facet of cardiac function, as sophisticated pacemakers can now separately pace the chambers can also be examined with such a model. In fact, such a model can be used to study heart disease and its effect on cardiac dynamics. Additionally, one could imagine how finite element models of arteries and veins could be used to study age-related changes in mechanical properties, the general age-related pathogenesis of artherosclerotic lesions, and the dynamics of vascular pharmaceuticals [12,248,260, 262,299,300,430,431].

3.6.3 The Nervous System

The nervous system (along with the endocrine system) provides the control functions for the human body. The nervous system is responsible for the rapid activities of the body: muscular contraction, rapidly changing visceral events, and even the rates of secretion of some of the endocrine glands. It is a unique system in that it can control and perform a vast complex of actions. The human brain is estimated to contain approximately 10^{12} neurons. Many of these neurons are connected to 10,000 other neurons. Thus in many ways, the brain is itself a sophisticated supercomputer. Models of the nervous system range from molecular and/or biochemical models, cellular and cell population models, all the way through models of the cortex, the brain, problems in brain imaging and the modeling of brain dynamics, to modeling of cognition, memory, and even consciousness. With but a bit of thought, it is easy to see how such models could be used to study senile dementia, Alzheimer's disease, the evolution of age-related changes in motor neurons and its effect on motor abilities [4,20,44,63,64,85,202,208,216–219,249,294,296,314,345,355,367,467].

3.6.4 Renal Physiology

The body fluids are extremely important to the basic physiology of the human being. The renal system, of which the kidneys are part, is intimately tied to the dynamics of the body fluids. The kidneys perform two major functions: (1) they excrete most of the end products of bodily metabolism and, (2) they control the concentrations of most of the constituents of the bodily fluids. Loss of kidney function can lead to death. It is known that all parameters of renal function are affected by the process of aging. Coupled with these changes is increased evidence that the elderly have a reduced ability to handle excretion of drugs. Common renal problems in the aged include renal failure, impaired drug excretion, urinary tract infections, hypertension, and miscellaneous disorders such as tuberculosis, nephritis, and diabetes. Many of these disorders could readily be modeled or simulated using known physiological, biochemical, and morphological data [59,180,227,307,364–366, 374,375,405,422,423].

3.6.5 Biomechanics

Mathematical and computer models of the biomechanical dynamics of humans are of great importance in the study of (1) age-related changes in biomechanical structure and function, (2) the development of methods for the prevention of injury (*i.e.,* hip fracture), (3) clinical intervention when such injuries do occur, and (4) the development of daily life tools and instruments to enhance the quality of life for the elderly in our population. One can imagine, for example, a finite element model of the hip which incorporates known changes of bone properties, as a means of understanding the effects of osteoporosis on hip fracture and fall rates in the elderly [18,36,43,65,75,97,144,147,153,245,246,260,262,273,299,318, 321–323,329,339,340,342,395,418].

3.6.6 Patient–Based Physiological Simulation

Patient-based physiological modeling has come of age. More and more computer systems are being targeted at taking patient image data and reconstructing it so that a physician can view/review, in 3-D, a patient's X-ray, CAT, PET, NMR, or other clinical image data. Computerized surgery systems are currently available in the marketplace. Virtual reality has created a synergy between mathematics, computers, and medicine. Facial reconstruction and surgical simulation are no longer a pipe dream. They are a practiced reality. The day of the computer enhanced mathematical scalpel has arrived [36,43,58,67,110,124,181–184, 312,337,361,370,372,376].

3.6.7 Project Human

Beyond the complexity of such ultralarge-scale simulations and models is the no longer unreasonable goal of an ultralarge-scale simulation of a human being. Such a simulation would rely upon the patient's image data, noninvasive measurements of his physiological functions, and assorted clinical tests. One can envision scenarios in which chemotherapy can be simulated, in a given patient, before the therapy is actually performed. Radical drug treatments can be simulated and the results can be examined and evaluated, not based upon an idealized mathematical model but, rather, as based upon an integrated model and patient system. Eventually, one can envision the possibility of actually testing newly designed drugs (now designed in computers) in computer-based large-scale simulations (*flight simulators for humans*). While it will be a long time before such a complex simulation/ modeling system can be put into place, it is no longer a pipe dream to imagine its existence.

3.7 Population Biology

The study of populations, particularly human populations, is called demography. The canonical demographic system is the McKendrick/Von Foerster system and is discussed in [9,51,100,106,167,195,258,259,315,383,413,434,436,440,443,444] and [447]. The study of such models is of great interest for a number of reasons. Phenomonologically, with the use of a simulation model, it would be possible to study effects of various interventions on the compression of morbidity and mortality in a population with particular age/health characteristics. In particular, given the increasing cost of healthcare and the associated increase of the aged component of the population, it is of great importance to understand the dynamics of the human population, in an effort to hold down the cost of healthcare. In addition, models of this type arise in the study of toxicological effects of the environment upon a population (ecotoxicology). Mathematical modeling of diseases, particularly of such diseases as AIDS and Lyme disease, requires the use of advanced mathematical and computational methods [173,436,440,443,444,447].

Epidemiology and biostatistics study population dynamics and characteristics from a probabilistic perspective. With large-scale epidemiological models, it is possible to investigate questions with respect to the epidemiology of disease, sex differentials in mortality rates, and social correlations of disability in the elderly. Clinical trials often generate large datasets. Statistical analysis of these datasets is often intense, due to the sample size and the complexity of the interactions. In addition, there are often issues of multicenter trials and the more recent problems arising in metastatistical analysis (see previous discussion on meta-analysis) [3,108,109,122,158,193,196,199,288,346,444,447,457].

3.8 Dentistry and Oral Physiology

What can be done with visualization of the bones and muscles in the torso can be done, as well, with the jaw and the facial muscles. Many dental schools are collaborating with mechanical and biomedical engineers for the purpose of finite element modeling of the jaw. The resultant models are then applied to examining the problems of orthodonture and of computer automated patient-based, dental prosthetics design. One of the greatest dental healthcare costs is TMJ (Temporal Mandibular Joint syndrome). Modeling TMJ involves not only the use of sophisticated mathematical equations to describe the dynamics of the bones and muscles of the jaw, but also patient-based data as the input for describing these same structures. Thus, the models will be based upon real patient data rather than hypothesized and/or idealized dental structures [18,34,75,153,318–320,329,418]. In the same vein as the modeling of cardiovascular disorders of aging, one could build mathematical models of swallowing and pharyngeoesophageal function. These models could be used to study the dynamics and treatment of dysphagia (difficulty in swallowing, a common complaint in the elderly) or presbyesophagus (old esophagus, the most common disorder of the esophageal motility problems with age).

3.9   Time, Space, Chaos, Fractals, and Dyamical Systems

Complex systems are ubiquitous. Living systems are complex. Therefore, it is natural to assume that such systems would be amenable to analysis using the new generation techniques arising in dynamical systems theory, fractal dynamics, and related subject areas.

3.9.1   Time

Time, one of the major means by which aging is measured and discussed, and by which dynamical systems are studied, is fundamental to any analysis of aging processes. The following citations address the concept of time from various perspectives: [37,53,70,91,102, 134,201,275,282,304–306,343,419,426,439,449] and [472].

3.9.1.1   Signal Analysis

Many biological datasets may be represented as *signals* in that they are a measure of the change of some biomedical/biophysical variable over time. The analysis of such signals, with respect to circadian behaviors, periodic behaviors, and nonlinear behaviors, falls in the domain of signal analysis. The following references provide good starting points in this area: [31,37,40,41,52,53,56,61,80,125,172,210,283,304–306,316,333,338,349,369,373,381, 401,406,416] and [421].

3.9.1.2   Dynamical Systems

Over the past 20 years, dynamical systems has emerged as one of the most powerful tools for analyzing complex temporal behaviors. The following citations provide a wide variety of levels, backgrounds, and insights into the application of dynamical systems theory: [16,48,60,84,92,95,114,134,138,139,159,165,189,191,254,255,257,275,282,284–286,290,293,304, 305,316,334,353,358,363,379,386,393,419,420] and [432].

3.9.2   Space

Dynamical systems tend to deal with temporal changes that are typified by *"biological signals."* Examples may include measurements of physiological changes over time, nerve transmission, or other neurological signals, as well as a variety of other possible temporally changing biological variables. However, it is also reasonable to assume that spatial structure can evolve over time. The following citations deal with issues of structure and time: [70,92] and [105].

3.9.2.1 Fractals

Fractal geometry represents a focused look at specific geometric structures; those that are typically called self-similar. The following citations deal with fractal dynamics from a variety of perspectives: [19,84,154,187,250,257,275,291,419] and [420].

3.9.2.2 Complexity, Structure, Networks, and Information

For the author, the interplay of aging, complexity, information, and temporal dynamics, and biological structure is a fascination. It is here, I believe, that we will find the ultimate definition for *aging*. In fact, I propose [464] that *aging may be defined as the process of structural and/or temporal transitions and/or alterations increasing the body's inability to handle information and informational changes.* The following citations deal with various aspects of complexity, structure, and information: [37,70,78,103,105,130,137,203–206,212,267,270,276,280,285,286, 304–306,310,350,415,420,428,433,437,438,441,442,445,446,469,470] and [475].

3.10 Closing Thoughts

Biocomputing and biomathematics have, very rapidly, gone from their conception to their infancy. The technological advances of recent years carry with them the necessary fuel to allow these fields to grow into adulthood. Once thought of as unnecessary, and even a hindrance, the computer has gained a strong foothold in the biosciences. As technology emerges from its annual metamorphosis, the need for biocomputing becomes stronger, the problems being tackled become more complex, and the technology more ready and able to handle both the needs and the demands of the biosciences. The fields of gerontology and geriatrics need not and should not be strangers to these technological advances.

Biological computing and high performance modeling have taught us numerous lessons as scientists. We have seen that a good model/simulation can aid us in testing hypotheses that might be extremely expensive to test in the laboratory and/or might be otherwise impossible to test. We have seen that we can ask the all important *"what if"* questions thereby allowing us to investigate what might happen under various hypothetical or unattainable experimental conditions. We have also learned to face the limitations of this approach. Good models and good simulations must depend upon *good data! We* must return to the living system and check our findings. Certain measurements must be made, experimentally, in order for the mathematical/computer model to have validity. We have seen that many forms of models can give rise to similar quantitative results. Hence, careful hypotheses must be involved in the formative modeling stages, lest ambiguity be introduced. As we increase the size and complexity of the models, we often introduce more parameters. These parameters must be known in order for the model to be an effective tool. Statisticians are fond of the warning: *"Give me enough parameters and I can model the world."* However, even with enough parameters, one must know their values if one is to study the dynamics of the world. And, while measurement techniques are getting better, there will always be parameters that cannot be known and/or easily measured. Thus, like any tool, one must take the good with the bad, refining the tool so that it can be more effective.

The future of biocomputing, computational and mathematical modeling, and computers and mathematics in the field of gerontology and geriatrics is one of great excitement. Experimentation, clinical research, and computational/mathematical modeling can work hand-in-hand to improve the quality of life—across the lifespan—for all living species.

Acknowledgments

This work has been partly supported by an NIA grant 1R01 AG0179 to T.M.W. The author would like to thank Ann Titus for typing a significant fraction of the references. The author would also like to thank Ari Gafni, Peter Macpherson, Gordon Lynch, Steven Devor, Joe Schauerte, Al Schultz, Steve Goldstein, James Sneyd, and John Aarsvold for their contributions of additional relevant reference citations. I would also like to thank the Department of Mathematics, University of Michigan, Ann Arbor, for providing additional computer support on this project.

References

1. Abramson, J.H., *Making Sense of Data*, Oxford University Press, New York, 1994.
2. Adam, J.A. and Bellomo, N., Ed., *A Survey of Models for Tumor-Immune System Dynamics*, Birkhäuser, Boston, 1997.
3. Ahlbom, A. and Norell, S., *Introduction to Modern Epidemiology*, Epidemiology Resources, Ind., Chestnut Hill, MA, 1984.
4. Aidley, D.J., *The Physiology of Excitable Cells*, Cambridge University Press, Cambridge, 1978.
5. Akin, E. and Lacker, H.M., Ovulation control: the right number or nothing, *J. Math. Biol.*, 20, 113, 1984.
6. Aldroubi, A. and Unser, M., Ed., *Wavelets in Medicine and Biology*, CRC Press, Boca Raton, FL, 1996.
7. Allison, S.A., Bacquet, R.J., and McCamon, J.A., Simulation of the diffusion controlled reaction between superoxide and superoxide dismutase. II. Detailed models, *Biopolymers*, 27, 251, 1988.
8. Ames, A.L., Nadeau, D.R., and Moreland, J.L., *VMRL 2.0 Sourcebook*, John Wiley & Sons, New York, 1997.
9. Arino, O., Axelrod, D.E., and M., Kimmel, Ed., *Mathematical Population Dynamics*, Vol. 131, *Lecture Notes in Pure and Applied Mathematics*, Marcel Dekker, New York, 1991.
10. Aris, R. and Penn, M., The mere notion of a model, *Math. Comput. Modeling*, 1, 1, 1980.
11. Arlinghaus, S.L., Ed., *Practical Handbook of Spatial Statistics*, CRC Press, Boca Raton, FL, 1996.
12. Backx, P.H., de Tombe, P.P., Van Deen, J.H.K., Mulder, B.J.M., and ter Keurs, H.E.D.J., A model of propagating calcium-induced calcium release mediated by calcium diffusion, *J. Gen. Physiol.*, 93, 963, 1989.
13. Badler, N.I., Barsky, B.A., and Zeltzer, D., Ed., *Making Them Move*, Series in Computer Graphics and Geometric Modeling, Morgan Kaufmann Publishers, San Mateo, CA, 1991.
14. Bailar, J.C. and Mosteller, F., Ed., *Medical Uses of Statistics*, New England Journal of Medicine Books, Boston, 1986.
15. Bailey. N.T.J., *Statistical Methods in Biology*, Hodder and Stoughton, London, 2nd ed., 1981.
16. Baker, G.L. and Gollub, J.P., *Chaotic Dynamics An Introduction*, Cambridge University Press, Cambridge, 1990.
17. Barlow, R.E. and Proschan, F., *Statistical Theory of Reliability and Life Testing Probability Models*, Holt, Reinhart and Winston, New York, 1975.

18. Barnes-Svarney, P., Improving durability of bone implants, *Super-comput. Rev.*, July, 24, 1991.
19. Barnsley, M.F., *Fractals Everywhere*. Academic Press, Cambridge, 2nd ed., 1993.
20. Baron, R.J., *The Cerebral Computer: An Introduction To The Computational Structure Of The Human Brain*, Lawrence Erlbaum Pub., New Jersey, 1987.
21. Barron, S., Witten, M., and Liu, G., A bibliography on computational molecular biology and genetics, *Math. Comput. Modeling*, 16(6/7), 245, 1992.
22. Barsky, B.A., *Computer Graphics and Geometric Modeling Using Beta-splines*, Springer-Verlag, Berlin, 1988.
23. Bauer, W.R., Crick, F.H.C., and White, J.H., Supercoiled DNA, *Sci. Am.*, 243, 118, 1980.
24. Bell, G.I. and Marr, T.G., Ed., *Computers and DNA*, Vol. 7, *Sante Fe Institute Studies in the Sciences of Complexity*, Addison-Wesley, Redwood City, CA, 1990.
25. Benham, C.J., Elastic model of supercoiling, *Proc. Natl. Acad. Sci. U.S.A.*, 74, 2397, 1977.
26. Benham, C.J., Theoretical analysis of conformation equilibria in superhelical DNA, *Annu. Rev. Biophys. Biophys. Chem.*, 14, 23, 1985.
27. Bertuzzi, A., Gandolfi, A., and Giovenco, M.A., Mathematical models of the cell cycle with a view to tumor studies (introductory review), *Math. BioSci.*, 53, 159, 1981.
28. Beveridge, D.L. and Jorgensen, W.L., Ed., *Computer Simulation of Chemical and Biomolecular Systems*, Vol. 482, *Annals of the New York Academy of Sciences*, New York Academy of Science Press, New York, 1986.
29. Bishop, Y.M.M., Fienberg, S.E., and Holland, P.W., *Discrete Multivariate Analysis: Theory and Practice*, MIT Press, Cambridge, MA, 1991.
30. Bland, M., *An Introduction to Medical Statistics*, 2nd ed., Oxford Medical Publications, Oxford University Press, Oxford, 1995.
31. Bloomfield, P., *Fourier Analysis of Time Series: An Introduction*, John Wiley & Sons, New York, 1976.
32. F.L., Bookstein, *Morphometric Tools For Landmark Data*, Cambridge University Press, New York, 1991.
33. Bookstein, F.L., Biometrics, biomathematics, and the morphometric synthesis, *Bull. Math. Biol.*, 58(2), 313, 1996.
34. Borchers, L. and Reichart, P., Three-dimensional stress distribution around a dental implant at different stages of interface development, *J. Dent. Res.*, 62, 155, 1983.
35. Box, G.E.P., Hunter, W.G., and Hunter, S.J., *Statistics for Experimenters*, Wiley Series in Probability and Mathematics Statistics, John Wiley & Sons, New York, 1978.
36. Bragdon, C. and Lowenstein, J., Solving the problems of hip joint replacement. *Sci. Comput. Automation*, September, 15, 20, 1993.
37. Brand, R.M., Midgley, A.R., and Williams, W.J., *Convolution: A Method for Data Analysis in Perifusion Systems*, American Physiuological Society, 1994, 759.
38. Branham, R.L., Scientific Data Analysis, Springer-Verlag, New York, 1990.
39. Breen, R., *Regression Models: Censored, Sample Selected, or Truncated Data*, Vol. 111, *Quantitative Applications in the Social Sciences*, Sage Pub., Thousand Oaks, CA, 1996.
40. Brock, M.A., Age-related changes in circannual rhythms of lymphocyte blastogenic responses in mice, *Am. J. Physiol.*, 252, R299, 1987.
41. Brock, M.A., Temporal order *vs.* variability in activation of lymphocytes from aging mice, *Mech. Aging Dev.*, 37, 197, 1987.
42. Bronshtein, I.N. and Semendyayev, K.A., *Handbook of Mathematics*, Van Nostrand Reinhold Company, New York, 1985. An excellent source of all kinds of useful mathematical formulae.
43. Buford, W.L. and Thompson, D.E., A system for three-dimensional interactive simulation of hand biomechanics, *IEEE Trans. Biomed. Eng.*, 36(6), 444, 1987.
44. Bunow, B., Segev, I., and Fleshman, J.W., Modeling the electrical properties of anatomically complex neurons using a network analysis program: excitable membrane, *Biol. Cyber.*, 53, 41, 1985.
45. Burdea, G., *Force and Touch Feedback for Virtual Reality*, John Wiley & Sons, New York, 1996.
46. Burdea, G. and Coiffet, P., *Virtual Reality Tecyhnology*, John Wiley & Sons, New York, 1994.
47. Burton, T.A., Ed., *Modeling and Differential Equations in Biology*, vol, 58, Marcel Dekker, New York, 1980.

48. Cambel, A.B., *Applied Chaos Theory A Paradigm for Complexity*, Academic Press, San Diego, 1993.
49. Campbell, R.C., *Statistics for Biologists*, 3rd ed., Cambridge University Press, Cambridge, 1989.
50. Cantor, C.R. and Smith, C.L., Mapping and sequencing the human genome, in *Molecular Genetic Approaches to Neuro-Psychiatric Diseases*, Brosius, J. and Fremeau, R.T., Eds., Academic Press, New York, 1991.
51. Carey, J.R., *Applied Demography for Biologists with Special Emphasis on Insects*, Oxford University Press, New York, 1993.
52. Carnes, M., Goodman, B., and Lent, S.J., High resolution spectral analysis of plasma ACTH reveals a multifactorial frequency structure, *Endocrinology*, 128, 902, 1991.
53. Carpenter, G.A., Ed., *Some Mathematical Questions in Biology Circadian Rhythms*, Vol. 19, American Mathematical Society, Providence, 1987.
54. Carroll, R.J., Ruppert, D., and Stefanski, L.A., *Measurement Error in Nonlinear Models*, Vol. 63, *Monographs on Statistics and Applied Probability*, Chapman and Hall, London, 1995.
55. Carson, E.R., Cobelli, C., and Finkestein, L., Mathematical Modeling of Metabolic and Endocrine Systems, John Wiley & Sons, New York, 1983.
56. Casdagli, M. and Eubank, S., Ed., *Nonlinear Modeling and Forecasting*, Vol. XII, *Santa Fe Institute Studies in The Sciences of Complexity*, Addison-Wesley, Reading, MA, 1992.
57. Castleman, K.R., *Digital Image Processing, Prentice Hall*, Englewood Cliffs, NJ, 1996.
58. Centofani, M., Pinpoint the problem: tailor the answer, *Hopkins Med. News*, 11(3),16, 1988.
59. Chandhoke, P.S. and G.M., Saidel, Mathematical model of mass transport throughout the kidney: effects of nephron heterogeneity and tubular-vascular organization, *Annu. Biomed. Engr.*, 9, 263, 1983.
60. Chandra, J., Ed., *Chaos in Nonlinear Dynamical Systems*, Society for Industrial and Applied Mathematics, Philadelphia, 1984.
61. Chatfield. C., *The Analysis of Time Series: An Introduction*, 5th ed., Texts in Statistical Science, Chapman and Hall, London, 1996.
62. Chaturvedi, M.M. and Kanungo, M.S., Analysis of chromatin of the brain of young and old rats by micrococcal nuclease and DNase I, *Biochem. Int.*, 6, 357, 1983.
63. Chay, T.R., Analyzing stochastic events in multi-channel patch-clamp data, *Biol. Cybern.*, 58, 19, 1988.
64. Chay, T.R., Kinetic modeling for the channel gating process from single patch-clamp data, *J. Theor. Biol.*, 132, 477, 1988.
65. Cheer, A.Y. and van Dam, C.P., Ed., *Fluid Dynamics in Biology*, volume 141 of *Contemporary Mathematics*. American Mathematical Society, Providence, RI, 1993.
66. Cherruault, Y., *Mathematical Modeling In Biomedicine*, D. Reidel Publishing Co., Boston, 1986.
67. Chinnock, C., Volumetric images offer interaction and group viewing, Diagnostic Imaging, September, 25, 1995.
68. Chothia, C., The classification and origins of protein folding patterns, *Annu. Rev. Biochem.*, 59, 1007, 1990.
69. Churchill, G.A., Stochastic models for heterogeneous DNA sequences, *Bull. Math. Biol.*, 51(1), 79, 1989.
70. Cladis, P.E. and Palffy-Muhoray, P., Ed., *Spatio-Temporal Patterns in Nonequilibrium Complex Systems*, vol. 21, Addison-Wesley, Reading, MA, 1995.
71. Cleveland, W.S., *The Elements of Graphing Data*, Hobart Press, Summit, NJ, 1994.
72. Cohen, A.C., *Truncated and Censored Samples: theory and Applications*, Marcel Dekker, New York, 1991.
73. Coleman, J.S., *Longitudinal Data Analysis*, Basic Books, New York, 1981.
74. Collins, J.F. and Reddaway, S.F., High-efficiency sequence database searching: Use of the distributed array processor, *Computers and DNA*, Bell, G. and Marr, T., Ed., vol. VII, *SFI Studies in The Sciences of Complexity*, Addison-Wesley, New York, 1990.
75. Cook, S.D., Klawitter, J.J., and Weinstein, A.M., The influence of implant geometry on the stress distribution around dental implants, *J. Biomed. Mater. Res.*, 16, 369, 1982.
76. Cooke, D., Crave, A.H., and Clarke, G.M., *Basic Statistical Computing*, Edward Arnold, London, 1982.
77. Coulson, A., High performance searching of biosequence databases, *Tibtech*, 12, 76, 1994.

78. Cowan, G.A., Pines, D., and Meltzer, D., Ed., *Complexity Metaphors, Models, and Reality,* Vol. 19, Addison-Wesley, Reading, MA, 1994.

79. Cowin, J.E., Jellis, C.H., and Rickwood, D., A new method of representing DNA sequences which combines ease of visual analysis with machine readability., *Nucl. Acids Res.,* 14, 509, 1986.

80. Cox, D.R., Hinkley, D.V., and Barndorff-Nielsen, O.E., Ed., *Time Series Models In Econometrics, Finance, and Other Fields,* Vol. 65, *Monographs On Statistics And Applied Probability,* Chapman and Hall, London, 1996.

81. Cozzarelli, N.R., DNA gyrase and the supercoiling of DNA, *Science,* 207, 953, 1980.

82. Creekmore, S.P., Aroesty, J., Willis, K.L., Morrison, P.F., and Lincoln, T.L., A cell kinetics model which includes heredity, differentiation and regulatory control, in *Biomathematics and Cell Kinetics,* Valleron A.J. and MacDonald, P.D.M., Ed., Elsevier/North–Holland, Amsterdam, 1978.

83. Crick, F.H.C., Linking numbers and nucleosomes, *Proc. Natl. Acad. Sci. U.S.A.,* 73, 2639, 1976.

84. Crilly, A.J., Earnshaw, R.A., and Jones, H., Ed., *Applications of Fractals and Chaos The Shape of Things,* Springer-Verlag, Heidelberg, 1993.

85. Cronin-Scanlon, J., *Mathematical Aspects Of Hodgkin-Huxley Neural Theory,* Cambridge University Press, Cambridge, 1987.

86. Cross, M. and Moscardinia, A.O., *Learning the Art of Mathematical Modeling,* John Wiley & Sons, New York, 1985.

87. Daniel, W.W., *Biostatistics: A Foundation for Analysis in the Health Sciences,* Wiley Series in Probability and Mathematical Statistics, John Wiley & Sons, New York, 1991.

88. Davidian, M. and Giltinan, D.M., *Nonlinear Models for Repeated Measurement Data,* Vol. 62, *Monographs on Statistics and Applied Probability,* Chapman and Hall, London, 1995.

89. Davison, D.B., Sequence searching on supercomputers, in *Computers and DNA,* Bell, G. and Marr, T., Ed., Vol. VII, *SFI Studies in the Sciences of Complexity,* Addison-Wesley, Reading, MA, 1990.

90. DeLisi, C., Computers in molecular biology: current applications and emerging trends, *Science,* 240, 47, 1988.

91. Denbigh, K.G., *Three Concepts of Time,* Springer-Verlag, New York, 1981.

92. Dendrinos, D.S. and Sonis, M., Chaos and Socio-Spatial Dynamics, Springer-Verlag, New York, 1990.

93. DerSimonian, R. and Laird, N., Meta-analysis in clinical trials, *Controlled Clinical Trials,* 7, 177, 1986.

94. Desu, M.M. and Raghavarao, *Sample Size Methodology,* Statistical Modeling and Decision Science, Academic Press, San Diego, CA, 1990.

95. Devaney, R.L., *An Introduction to Chaotic Dynamical Systems,* Addison-Wesley, Redwood City, CA, 1987.

96. Diggle, P.J., Liang, K-Y., and Zeger, S.L., *Analysis of Longitudinal Data,* Vol. 13, *Oxford Statistical Science Series,* Clarendon Press, Oxford, 1995.

97. Dong, C. and Skalak, R., Leukocyte deformability: finite element modeling of large viscoelastic deformation, *J. Theor. Biol.,* 158, 173, 1992.

98. Doolittle, R.F., Molecular Evolution: Computer Analysis of Protein and Nucleic Acid Sequences, Academic Press, New York, 1990.

99. Duncan, R.C., Knapp, R.G., and Miller, M.C., Ed., *Introductory Biostatistics for the Health Sciences,* Wiley Medical Publications, John Wiley & Sons, New York, 2nd Ed., 1983.

100. Dyke, B. and MacCluer, J.W., Ed., *Computer Simulation in Human Population Studies.* Studies in Anthropology, Academic Press, New York, 1984.

101. Eakin, T., Shouman, R., Qi, Y., Liu, G., and Witten, M., Estimating parametric survival model parameters in gerontological aging studies: methodological problems and insights, *J. Gerontol.,* 50A(3), B166, 1995.

102. Eakin, T. and Witten, M., A gerontological distance metric for analysis of survival dynamics, *Mech. Ageing and Dev.,* 78, 85, 1995.

103. Eakin, T. and Witten, M., How square is the survival curve of a given species?, *Exp. Gerontol.,* 30(1), 33, 1995.

104. Eddy, D.M., Hasselblad, V., and Shacter, R., *Meta-Analysis by the Confidence Profile Method,* Statistical Modeling and Decision Science, Academic Press, Boston, 1992.

105. Edelman, G.M., Code, scale, and place, in *Topobiology An Introduction to Molecular Embryology*, Basic Books, Inc., 1988, 9.
106. Edelstein-Keshet, L., *Mathematical Models in Biology*, Mathematics. Random House/Birkhauser, New York, N.Y., 1987.
107. Eisenfeld, J. and Witten, M., Ed., *Modeling of Biomedical Systems*, North-Holland, Amsterdam, 1986.
108. Elliott, P., Cuzick, J., English, D., and Stern, R., Ed., *Geographical and Environmental Epidemiology: Methods for Small Area Studies*, Oxford University Press, Oxford, 1992
109. Esteve, J., Benhamou, E., and Raymond, L., Descriptive Epidemiology, Vol. IV, *Statistical Methods in Cancer Research*, Oxford University Press, Oxford, 1994.
110. Ezquerra, N., Ed., *First Conference On Visualization In Biomedical Computing*, Ieee Computer Society Press, Los Alamitos, CA, 1990.
111. Farin, G., *Nurb Curves and Surfaces*, A.K. Peters, Wellesley, MA, 1995.
112. Felsenstein, J., Phylogenies from molecular sequences: inference and reliability, *Annu. Rev. Genet.*, 22, 521, 1988.
113. Feng, D.-F. and Doolittle, R.F., Progressive sequence alignment as a prerequisite to correct phylogenetic trees, *J. Mol. Evol.*, 25, 351, 1987.
114. Field, R.J. and Gyorgyi, L., Ed., *Chaos in Chemistry and Biochemistry*, World Scientific, Singapore, 1993.
115. Finch, C.E., *Longevity, Senescence, and the Genome*, University of Chicago Press, Chicago, 1990.
116. Finkelstein, L. and E.R., Carson, *Mathematical Modeling of Dynamic Biological Systems*, John Wiley & Sons, New York, 1986.
117. Foley, J.D. and Van Dam, A., *Fundamentals of Interactive Computer Graphics*, The Systems Programming Series, Addison-Wesley, Reading, MA, 1984.
118. Forthofer, R.N. and Lee, E.S., *Introduction to Biostatistics*, Academic Press, San Diego, CA, 1995.
119. Fortner, B., *The Data Handbook*, Spyglass, Inc., Champaign, IL, 1992.
120. Plastic Surgery Research Foundation, Ed., *Medicine Meets Virtual Reality: Discovering Applications for 3D Multi-Media Interactive Technology in the Health Sciences*, Aligned Management Associates, San Diego, CA, 1992.
121. Fox, J. and Long, J.S., *Modern Methods of Data Analysis*, Sage Pub., Newbury Park, CA, 1990.
122. Frauenthal, J.C., *Mathematical Models in Epidemiology*, Springer-Verlag, New York, 1980.
123. Frenzen, C.L. and Murray, J.D., A cell kinetics justification for Gompertz' equation, *SIAM J. Appl. Math.*, 46(4), 614, 1984.
124. Fuchs, H., Levoy, M., and Pizer, S.M., Interactive visualization of 3D medical data, *IEEE Comp.*, 22(8), 46, 1989.
125. Fuller, W.A., *Introduction to Statistical Time Series*, Wiley Series In Probability And Statistics, Wiley-Interscience, New York, 2nd ed., 1996.
126. Gafni, A., Molecular origins of the aging effects in glyceraldehyde-3-phosphate dehydrogenase, *Biochim. Biophys. Acta*, 742, 91, 1983.
127. Gafni, A., Age-related effects in enzyme metabolism and catalysis, in *Review of Biological Research in Aging*, Vol. 4, Alan R. Liss, New York, 1990.
128. Gafni, A., Altered protein metabolism in aging, in *Annual Review of Gerontology and Geriatrics*, Cristofalo, V.J. and Lawton, M.P., Ed., Springer Publishing Co., New York, 1990.
129. Galas, D., Form and function: perspectives on structural biology and resources for the future. *U.S. Department of Energy*, PUB-682/12-90, Washington, D.C., 1990.
130. Gatlin, L.L., *Information Theory and the Living System*, Columbia University Press, 1972, 1.
131. Gavrilov L.A. and Gavrilova, N.S., *The Biology of Life Span: A Quantitative Approach*, Harwood Academic Publishers, Chur, Switzerland, 1991.
132. Gifi, A., *Nonlinear Multivariate Analysis*, John Wiley & Sons, Chichester, England, 1990.
133. Giordano, F.R. and Weir, M.D., *A First Course In Mathematical Modeling*, Brooks/Cole Pub. Co., California, CA, 1985.
134. Glass, L. and Mackey, M.C., *From Clocks to Chaos*, Princeton University Press, Princeton, NJ. 1988.
135. Gnanadesikan, R., *Methods for Statistical Data Analysis of Multivariate Observations*, 2nd ed., Wiley Series in Probability and Statistics, John Wiley & Sons, New York, 1997.

136. Goel, N.S. and Richter-Dyn, N., *Stochastic Models In Biology*, Academic Press, New York, 1974.
137. Goel, N.S. and Thompson, R.L., *Organization of Biological Systems: Some Principles and Models*, Academic Press, New York, 1986, 2.
138. Goldberger, A.L., Chaos and order in the human body, *MD Computing*, 1992.
139. Goldbeter, A., *Biochemical Oscillations and Cellular Rhythms*, Cambridge University Press, Cambridge, 1996.
140. Gonzalez, R.C., *Digital Image Processing*, Addison-Wesley, Reading, MA, 1992.
141. Gower, J.C. and Hand, D.J., *Biplots*, Vol. 54, *Monographs on Statistics and Applied Probability*, Chapman and Hall, London, 1996.
142. Gradshteyn, I. and Ryzhik, I.M., *Table of Integrals, Series, and Products*, Academic Press, New York, 1980. An excellent source of information on doing difficult integrals or series summations.
143. Gray, N., A program to find regions of similarity between homologous protein sequences using dot-matrix analysis, *J. Mol. Graphics*, 8, 11, 1990.
144. Greene, A.S., Tonellato, P.J., Lui, J., Lombard, J.H., and Crowley, A.W., Microvascular rarefaction and tissue vascular resistance in hypertension, *Am. J. Physiol.*, 256, H126, 1989.
145. Greenland, S., Quantitative methods in the review of epidemiologic literature, *Epidemiol. Rev.*, 9, 1, 1987.
146. Guttorp, P., *Stochastic Modeling of Scientific Data*, Chapman and Hall, London, 1995.
147. Hademenos, G.J., The physics of cerebral aneurysms, *Physics Today*, February, 24, 1995.
148. Haefner, J.W., *Modeling Biological Systems: Principles and Applications*, Chapman and Hall, New York, 1996.
149. Hagenaars, J.A., *Categorical Longitudinal Data*, Sage Pub., Newbury Park, CA, 1990.
150. Hall, R., *Illumination and Color in Computer Generated Imagery*, Monographics in Visual Communication. Springer-Verlag, New York, 1989.
151. Hao, M-H. and Olson, W.K., Modeling DNA supercoils and knots with b-spline functions, *Biopolymers*, 28, 873, 1989.
152. Hao, W.K., and Olson, M-H., Molecular modeling and energy refinement of supercoiled DNA, *J. Biomol. Struct. Dyn.*, 7(3), 661, 1989.
153. Harriman, L.P., Snowdon, D.A., Messer, L.B., Rysavy, D.M., Ostwald, S.K., Lai, C.-H., and Soberay, A.H., Temporomandibular joint dysfunction and selected health parameters in the elderly, *Oral Surg. Oral Med. Oral Pathiol.*, 70, 406, 1990.
154. Hastings, M. and Sugihara, G., *Fractals: A User's Guide for the Natural Sciences*, Oxford University Press, New York, 1993.
155. Hayashi, K. and Sakamoto, N., *Dynamic Analysis of Enzyme Systems: An Introduction*, Japan Scientific Societies Press, Tokyo, Japan, 1986.
156. Heinmets, F., *Concepts and Models of Biomathematics*, Marcel Dekker, New York, 1969.
157. Heller, H., Grubmüller, H., and Schulten, K., Molecular simulation on a parallel computer, *Molecular Simulation*, 5, 133, 1990.
158. Hennekens, C.H., Buring, J.E., and Mayrent, S.L., *Epidemiology in Medicine*, Little, Brown, Boston, 1987.
159. Hilborn, R.C., *Chaos and Nonlinear Dyamics*, Oxford University Press, New York, 1994.
160. Hine, L.K., Laird, N.M., Hewitt, P., and Chalmers, T.C., Meta-analysis of empirical long-term antiarrhythmic therapy after myocardial infarction, *JAMA*, 262(21), 3037, 1989.
161. Hirsch, H.R., The waste-product theory of aging: Waste dilution by cell division, *Mech. Ageing Dev.*, 8, 51, 1978.
162. Hirsch, H.R., The waste-product theory of aging: Cell division rate as a function of waste volume, *Mech. Ageing Dev.*, 36, 95, 1986.
163. Hirsh, H.R., Coomes, J.A., and Witten, M., The waste-product theory of aging: Transformation to unlimited growth in cell cultures, *Exp. Gerontol.*, 24, 97, 1989.
164. Hodgman, T.C., The elucidation of protein function by sequence motif analysis, *CABIOS*, 5(1), 1, 1989.
165. Hofbauer, J. and Sigmund, K., *The Theory of Evolution and Dynamical Systems*, Vol. 7, Cambridge University Press, Cambridge, 1988.

166. Holley, L.H. and Karplus, M., Protein secondary structure prediction for a neural network, *Proc. Natl. Acad. Sci. U.S.A.*, 86, 152, 1989.

167. Hoppensteadt, F.C., *Mathematical Methods of Population Biology*, Vol. 4, *Cambridge Studies in Mathematical Biology*, Cambridge University Press, Cambridge, 1982.

168. Hoyle, R.H., Ed., *Structural Equation Modeling: Concepts, Issues, and Applications*, Sage Publications, Thousand Oaks, CA, 1995.

169. Iannaccone, P.M. and Khokha, M., Ed., *Fractal Geometry in Biological Systems*, CRC Press, Boca Raton, FL 1996.

170. Ingram, D. and Bloch, R.F., *Mathematical Methods In Medicine*, vol. 2, John Wiley & Sons, New York, 1984.

171. Ingram, D. and Bloch, R.F., *Mathematical Methods In Medicine*, vol. 1, John Wiley & Sons, New York, 1984.

172. Ingram, D.K., London, E.D., and Reynolds, M.A., Circadian rhythmicity and sleep: Effects of aging in laboratory animals, *Neurobiol. Aging*, 3, 287, 1982.

173. Isham, V. and Medley, G., Ed., *Models for Infectious Human Diseases: Their Structure and Relation to Data*, Publications of the Newton Institute, Cambridge University Press, Cambridge, 1996.

174. Iyengar, S.S., Ed., *Structuring Biological Systems: A Computer Modeling Approach*, CRC Press, Boca Raton, FL, 1992.

175. Jaccard, J. and Wan, C.K., *Lisrel Approaches to Interaction Effects in Multiple Regression*, Vol. 114, *Quantitative Applications in the Social Sciences*, Sage Pub. Co., Thousand Oaks, CA, 1996.

176. Jacquez, J.J., *Compartmental Analysis in Biology and Medicine*, University of Michigan Press, Ann Arbor, MI, 1985.

177. Jahne, B., *Digital Image Processing: Concepts, Algorithms, and Scientific Applications*, Springer-Verlag, Berlin, 1992.

178. Jain, A.K., *Fundamentals of Digital Image Processing*, Prentice-Hall, Englewood Cliffs, NJ, 1989.

179. Jambu, M., *Exploratory and Multivariate Data Analysis*, Academic Press, San Diego, CA, 1991.

180. Jen, J.F., Wang, H., Tewarson, R.P., and Stephenson, J.L., Comparison central core and radially separated models of the renal inner medulla, *Am. J. Physiol.*, 268(4), F693, 1995.

181. Johnson, T.K. and Vessella, R.L., On the application of parallel processing to the computation of dose arising from the internal deposition of radionuclides, *Comput. Phys.*, 3(3), 69, 1989.

182. Jolesz, F.A. and Kikinis, R., Interoperative imaging revolutionizes therapy, *Diagnostic Imaging*, September, 62, 1995.

183. Jolesz, F.A., Lorenson, W.E., Kikinis, R., *et al.*, Virtual endoscopy: Three dimensional rendering of cross-sectional images for endoluminal visualization, *Radiology*, 193, 469, 1994.

184. Jolesz, F.A. and Shtern, F., The operating room of the future: Report of the National Cancer Institute Workshop: "Imaging-Guided Stereotactic Tumor Diagnosis and Treatment". *Invest. Radiol.*, 27, 326, 1992.

185. Jones, D.S. and Sleeman, B.D., *Differential Equations and Mathematical Biology*. George Allen and Unwin Ltd., Boston, 1983.

186. Jones, R., Taylor, W., Zhang, X., Mesirov, J.P., and Lander, E., Protein sequence comparison on the connection machine CM-2, in *Computers and DNA*, Bell, G. and Marr, T., Ed., Vol. VII, *SFI Studies in the Sciences of Complexity*, Addison-Wesley, New York, 1990.

187. Kaandorp, J.A., *Fractal Modelling Growth and Form in Biology*, Springer-Verlag, Heidelberg, 1994.

188. Kapis, M.B. and Gad, S.C., Ed., *Non-Animal Techniques in Behavioral and Research Testing*, Lewis Publishers, Boca Raton, FL, 1993.

189. Kaplan, D. and Glass, L., *Understanding Nonlinear Dynamics*, Springer-Verlag, New York, 1995.

190. Karlin, S., Ost, F., and Blaisdell, B.E., Patterns in DNA and amino acid sequences and their statistical significance, in *Mathematical Methods For DNA Sequences*, Waterman, M., Ed., CRC Press, Boca Raton, FL, 1989.

191. Kawakami, H., *The Theory of Dynamical Systems and Its Applications to Nonlinear Problems*, World Scientific, Singapore, 1984.

192. Keener, J. and Sneyd, J., *Mathematical Physiology*, Springer-Verlag, New York.

193. Kelsey, J.L., Whittemore, A.S., Evans, A.S., and Thompson, W.D., *Methods In Observational Epidemiology*, 2nd ed., Vol. 26, *Monographs in Epidemiology and Biostatistics*, Oxford University Press, New York, 1996.

194. Kerlow, I.V. and Rosebush, J., *Computer Graphics for Designers and Artists*, Van Nostrand-Reinhold, New York, 1986.

195. Keyfitz, N., *Applied Mathematical Demography*, Wiley-Interscience, New York, 1977.

196. Khoury, M.J., Beaty, T.H., and Cohen, B.H., *Fundamentals of Genetic Epidemiology*, Vol. 22, *Monographs in Epidemiology and Biostatistics*, Oxford University Press, New York, 1993.

197. Kingsbury, D.T., Computational biology for biotechnology. I. The role of the computational infrastructure. *Trends. Biotech.*, 7, 82, 1989.

198. Klein, J.P. and Moeschberger, M.L., *Survival Analysis: Techniques for Censored and Truncated Data*, Springer-Verlag, New York, 1997.

199. Kleinbaum, D.G., Kupper, L.L., and Morgenstern, H., *Epidemiologic Research: Principles and Quantitative Methods* Van Nostrand Reinhold, New York, 1982.

200. Kleinbaum, D.G., Kupper, L.L., and Muller, K.E., *Applied Regression Analysis and Other Multivariable Methods*, 2nd ed., PWS-Kent Publishing Company, Boston, 1988.

201. Kloeden, P.E., Rössler, R., and Rössler O.E., Timekeeping in genetically programmed aging, *Exper. Gerontol.*, 28, 109, 1993.

202. Koch, C. and Segev, I., Ed., *Methods In Neuronal Modeling*, MIT Press, Cambridge, MA, 1989.

203. Kohn, M.C., *Computer Simulation of Metabolism in Palmitate-Perfused Rat Heart. Part III. Sensitivity Analysis*, Pergamon Press, Elmsford, NY, 1984, 2.

204. Kohn, M.C., and Chiang, E., *Metabolic Network Sensitivity Analysis*, Academic Press, New York, 1982, 825.1.

205. Kohn, M.C. and Chiang, E., *Sensitivity to Values of the Rate Constants in a Neurochemical Metabolic Model*, Academic Press, New York, 1983, 961.1.

206. Kohn, M.C. and Garfinkel, D., *Computer Simulation of Metabolism in Palmitate-Perfused Rat Heart. Part I. Palmitate Oxidation*, Pergamon Press, Elmsford, NY, 1984, 2.

207. Kollman, P., Molecular modeling, *Annu. Rev. Phys. Chem.*, 38, 303, 1987.

208. Koshland, D.E., Frontiers in neuroscience (editorial), *Science*, 242(4879), 641, 1988.

209. Kozack, R.E. and Subramaniam, S., Brownian dynamics simulations of molecular recognition in antibody-antigen system, *Protein Sci.*, 2, 915, 1993.

210. Krieger, D.T., *Endocrine Rhythms*, Raven Press, New York, 1979.

211. Krzanowski, *Principles of Multivariate Analysis: A User's Perspective*, Vol. 3, *Oxford Statistical Science Series*, Oxford University Press, Oxford, 1990.

212. Kursunoglu, B., Mintz, S.L., and Perlmutter, A., *Information Processing in Biological Systems*, Plenum Press, New York, 1985, 55.

213. Lacker, H.M., The regulation of ovulation number in mammals an interaction law which controls the follicle maturation, *Biophys. J.*, 35, 433, 1981.

214. Lacker, H.M. and Akin, E., How do the ovaries count?, *Math. BioSci.*, 90, 305, 1988.

215. Lacker, H.M., Seuer, M.E., and Akin, E., Cell to cell signaling through circulatory feedback, in *Cell To Cell Signalling: From Experiments To Theoretical Models*, Goldbeter, A., Ed., Academic Press, New York, 1989.

216. Lagerlund, T.D. and Low, P.A., A mathematical simulation of oxygen delivery in rat peripheral nerve, *Microvascular Res.*, 34, 211, 1987.

217. Lagerlund, T.D. and Low, P.A., Mathematical modeling of hydrogen clearance blood flow measurements in peripheral nerve, Research Report UMSI 93/16, University of Minnesota Supercomputing Center, Minneapolis, MN, 1993.

218. Lagerlund, T.D. and Sharbrough, F.W., Computer simulation of neuronal circuit models of rhythmic behavior in the electroencephalogram, *Comput. Bio. Med.*, 18(4), 267, 1988.

219. Lagerlund, T.D. and Sharbrough, F.W., Computer simulation of the generation of the electroencephalogram, *Electroencephalogr. Clin. Neurophysiol.*, 72, 31., 1989.

220. Laird, N.M. and Mosteller, F., Some statistical methods for combining experimental results, *Int. J. Technol. Assessment Health Care*, 6, 5, 1990.

221. Landsdown, J. and Earnshaw, R.A., Ed., *Computers in Art, Design, and Animation*, Springer-Verlag, New York, 1989.

222. Langone, J.J., Ed., *Molecular Design And Modeling: Concepts And Applications. Part A. Proteins, Peptides, And Enzymes*, Academic Press, New York, 1991.

223. Langone, J.J., Ed., *Molecular Design And Modeling: Concepts And Applications. Part B, Antibodies And Antigens, Nucleic Acids, Polysaccharides, And Drugs*, Academic Press, New York, 1991.

224. Lapedes, A., Barnes, C., Burkes, C., Farber, R., and Sirotkin, K., Application of neural networks and other machine learning algorithms to DNA sequence analysis, in *Computers and DNA*, Bell, G. and Marr, T., Ed., Vol. VII, *SFI Studies in the Sciences of Complexity*, Addison-Wesley, New York, 1990.

225. Lawless, J.F., *Statistical Models and Methods for Lifetime Data*, John Wiley & Sons, New York, 1982.

226. Lawton, M.P. and Herzog, A.R., Ed., *Special Research Methods for Gerontology*, Baywoord Pub. Co., Amityville, NY, 1989.

227. Layton, H.E., Distribution of Henle's loops may enhance urine concentrating capability, *Biophys. J.*, 49(5), 1033, 1986.

228. Le, S-Y., Owens, J., Nussinov, R., Chen, J-H., Shapiro, B., and Maizel, J.V., RNA secondary structures: comparison and determination of frequently recurring substructures by consensus, *CABIOS*, 5(3), 205, 1989.

229. Lebart, L., Morineau, A., and Warwick, K.M., *Multivariate Descriptive Statistical Analysis*, Wiley Series in Probability and Mathematical Statistics, John Wiley & Sons, New York, 1984.

230. Lebowitz, J.L. and Rubinow, S.I., A theory for the age-time distribution of a microbial population, *J. Math. Biol.*, 1, 17, 1974.

231. Lee, E.T., *Statistical Methods for Survival Data Analysis*, 2nd ed.,Wiley Series in Probability and Mathematical Statistics, Wiley-Interscience, New York, 1992.

232. Lemeshow, S., Hosmer, D.W., Klar, J., and Lwanga, S.K., *Adequacy of Sample Size in Health Studies*, John Wiley & Sons, Chichester, England, 1990.

233. Lesk, A.M., Ed., *Computational Molecular Biology: Sources and Methods for Sequence Analysis*, Oxford University Press, New York, 1988.

234. Levin, S. and Kingsley, D., Mathematics and biology: the interface — challenges and opportunities, *U.S. Department of Energy*, PUB 701/1992, 1992.

235. Levin, S.A., Ed., *Frontiers in Mathematical Biology*, Vol. 100, *Lecture Notes in Biomathematics*, Springer-Verlag, Berlin, 1994.

236. Levitt, M. and Warshel, A., Computer simulation of protein folding, *Nature*, 253, 694, 1975.

237. Lewis, R., Computer simulation in science teaching: pros and cons, *The Scientist*, pages 25–26, 27 November 1989.

238. Li, Y-X. and Goldbeter, G., Frequency specificity in intercellular communication, *Biophys. J.*, 55, 125, 1989.

239. Li, Y-X., Rinzel, J., Keizer, J., and Stojilković, SS., Calcium oscillation in pituitary gonadotrophs: Comparison of experiment and theory, *Proc. Natl. Acad. Sci., U.S.A.*, 91, 58, 1994.

240. Liebman, M.N., Molecular modeling of protein structure and function: a bioinformatic approach, *J. Comp.-Aided Molec. Design*, 1, 323, 1987.

241. Lindsey, J.K., *Modeling Frequency and Count Data*, Vol. 15, *Oxford Statistical Science Series*, Clarendon Press, Oxford, 1995.

242. Long, J.S., *Regression Models for Categorical and Limited Dependent Variables*, Vol. 7, *Advanced Quantitative Techniques in the Social Sciences*, Sage Pub., Thousand Oaks, CA, 1997.

243. Longnecker, M.P., Berlin, J.A., Orza, M.J., and Chalmers, T.C., A meta-analysis of alcohol consumption in relation to risk of breast cancer, *JAMA*, 260(5), 652, 1988.

244. Lotka, A.J., *Elements of Mathematical Biology*, Dover Publications, New York, 1956.

245. Lotz, J.C., Cheal, E.J., and Hayes, W.C., Fracture prediction for the proximal femur using finite element models. I. Linear analysis, *J. Biomech. Eng.*, 113, 353, 1991.

246. Lotz, J.C., Cheal, E.J., and Hayes, W.C., Fracture preduction for the proximal femur using finite element models. II. Nonlinear analysis, *J. Biomech. Eng.*, 113, 361, 1991.

247. Magnusson, D. and Bergman, L.R., Ed., *Data Quality in Longitudinal Research*, Cambridge University Press, Cambridge, 1990.

248. Mahoney, D.P., Medical researchers take simulations to heart, *Comput. Graphics World*, September, 47, 1995.

249. Maisel, M., Reconstructing neurons in the computer, 1988, Science at the San Diego Super-computing Center — 1987.

250. Mandelbrot, B.B., *The Fractal Geometry of Nature*, W.H. Freeman, New York, 1977.

251. Markus, L.F., Corti, M., Loy, A., Naylor, G., and Slice, D., Ed., *Advances in Morphometrics: Proceedings of the 1993 NATO ASI on Morphometrics*, Plenum Press, New York, 1996.
252. Marmarelis, V.Z., Ed., *Advanced Methods of Physiological System Modeling*, Biomedical Simulation Resource, University of Southern California, Los Angeles, 1987.
253. Matthews, D.E., *Using and Understanding Medical Statistics*, S. Karger, Basel, 1988.
254. May, R.M., Biological populations with nonoverlapping generations: Stable points, stable cycles, and chaos, *Science*, 186, 645, 1974.
255. May, R.M., Simple mathematical models with very complicated dynamics, *Nature*, 261, 459, 1976.
256. McCammon, J.A. and Harvey, S.C., *Dynamics of Proteins and Nucleic Acids*, Cambridge University Press, Cambridge, 1987.
257. McCauley, J.L., *Chaos, Dynamics and Fractals: An Algorithmic Approach to Deterministic Chaos*, Vol. 2, *Cambridge Nonlinear Science Series*, Cambridge University Press, Cambridge, 1993.
258. McKendrick, A.G., Applications of mathematics to medical problems, *Proc. Edinburgh Math. Soc.*, 44, 98, 1926.
259. McKendrick, A.G. and Pai, M.K., The rate of multiplication of microorganisms: a mathematical study, *Proc. R. Soc. Edinburgh*, 31, 649, 1910.
260. McQueen, D.M. and Peskin, C.S., Computer-assisted design of butterfly bileaflet valves for the mitral position, *Scand. J. Thor. Cardiovasc. Surg.*, 19, 139, 1985.
261. Medley, V.I. and Medley, G., Ed., *Models for Infectious Human Diseases: Their Structure and Relation to Data*, Cambridge University Press, Cambridge, 1996.
262. Meisner, J.S., McQueen, D.M., Ishida, Y., Vetter, H.O., Burtolotti, U., Strom, J., Frater, R.W.M., Peskin, C.S., and Yellin, E.L., Effects of timing of atrial systole on LV filling and mitral valve closure: computer and dog studies, *Am. J. Physiol.*, 249, H604, 1985.
263. Melnyk, T.W., Richardson, I.W., Simpson, A.A., and Smith, W.R., Hypothalamic regulation of pituitary secretion of luteinizing hormone, *Bull. Math. Biol.*, 38, 387, 1976.
264. Metz, J.A.J. and Diekmann, O., *The Dynamics of Physiologically Structured Populations*, Springer-Verlag, New York, 1986.
265. Meuli, L.E., Lacker, H.M., and Thau, R.B., Experimental evidence supporting a mathematical theory of the physiological mechanism regulating follicle development and ovulation number, *Biol. Reprod.*, 37, 589, 1987.
266. Miké, V. and Stanley, K.E., Ed., *Statistics in Medical Research*, Wiley Series in Probability and Mathematical Statistics, John Wiley & Sons, New York, 1982.
267. Mikhailov, A.S., *Foundations of Synergetics I*, Springer Series in Synergetics, Springer-Verlag, Heidelberg, 1990.
268. Miller, P.L., Nadkarni, P.M., and Carriero, N.M., Parallel computation and FASTA: confronting the problem of parallel database search for a fast sequence comparison algorithm, *CABIOS*, 7(1), 36, 1991.
269. Milne, R., Computer array interprets the human genome, *New Sci.*, 122(36), 36, 1989.
270. Mintz, S.L., and Perlmutter, A., Ed., *Information Processing in Biological Systems*, Plenum Press, New York, 1985, 55.
271. Miura, R.M., Ed., *Some Mathematical Questions in Biology: DNA Sequence Analysis*, Lectures on Mathematics in the Life Sciences, American Mathematical Society, Providence, RI, 1986.
272. Moller, D., Ed., *Advanced Simulation in Biomedicine*, Springer-Verlag, New York, NY, 1989.
273. Monheit, G. and Badler, N.I., A kinematic model of the human spine and torso, *IEEE Computer Graphics and Applications*, March, 29, 1991.
274. Montgomery, D.C. and Peck, E.A., *Introduction to Linear Regression Analysis*, 2nd ed., Wiley Series in Probability and Mathematical Statistics, John Wiley & Sons, New York, 1992.
275. Moon, F.C., *Chaotic and Fractal Dynamics*, John Wiley & Sons, New York, 1992.
276. Morowitz, H.J. and Singer, J.L., Ed., *The Mind, the Brain, and Complex Adaptive Systems*, vol. 22, Addison-Wesley, Reading, MA, 1995.
277. Morris, G.M., The matching of protein sequences using color intrasequence homology displays, *J. Mol. Graphics*, 6, 135, 1988.
278. Mott, R.F., Kirkwood, T.B.L., and Curnow, R.N., A test for the statistical significance of DNA sequence similarities for application in databank searches, *CABIOS*, 5(2), 123, 1989.

279. Mullinger, A.M., and Johnson, R.T., The organization of supercoiled DNA from human chromosomes, *J. Cell Sci.*, 38, 369, 1979.

280. Nadel, L. and Stein, D.L., Ed. *1993 Lectures in Complex Systems*, vol. 6, Addison-Wesley, Reading, MA, 1995.

281. Nash, J.C. and Walker-Smith, M., *Nonlinear Parameter Estimation*, Marcel Dekker, New York, 1987.

282. Nayfeh, A.H. and Holden, A.V., Ed., *Applied Nonlinear Dynamics*. John Wiley Series in Nonlinear Science, John Wiley & Sons, New York, 1995.

283. Nerlove, M., Grether, D., and Carvalho, J.L., *Analysis of Economic Time Series: A Synthesis*, rev. ed., Academic Press, San Diego, 1995.

284. Nicolis, G., *Introduction to Nonlinear Science*, Cambridge University Press, Cambridge, 1995.

285. Nicolis, J.S., *Chaos and Information Processing*, page 283. World Scientific.

286. Noest, A.J., Tuning stochastic resonance, *Nature*, 378, 341, 23 November 1995.

287. Odeh, R.E. and Fox, M., *Sample Size Choice*, 2nd ed., Marcel Dekker, New York, 1991.

288. Olsen, J. and Trichopoulos, D., Ed., *Teaching Epidemiology: What You Should Know and What You Could Do*, Oxford University Press, Oxford, 1992.

289. Othmer, H.G., Adler, F.R., Lewis, M.A., and Dallon, J.C., *Case Studies in Mathematical Modeling — Ecology, Physiology, and Cell Biology*, Prentice-Hall, Englewood Cliffs, NJ, 1997.

290. Ott, E., Chaos in Dynamical Systems, Cambridge University Press, Cambridge, 1993.

291. Ott, E., Sauer, T., and Yorke, J., Ed., *Coping with Chaos*, Wiley Series in Nonlinear Science, John Wiley & Sons, New York, 1994.

292. Parker, K.C.H., Hanson, R.K.,and Hunsley, J., MMPI, Rorschach and WAIS: A meta-analytic comparison of reliability, stability, and validity, *Psychol. Bull.*, 103(3), 367, 1988.

293. Parker, T.S. and Chua, L.O., *Practical Numerical Algorithms for Chaotic Systems*, Springer-Verlag, New York, 1989.

294. Patton, P., Thomas, E., and Wyatt, R., A comnputational model of the vertical anatomical organization of primary visual cortex, *Biol. Cyber.*, 65, 189, 1991.

295. Pearson, W.R. and Lipman, D.J., Improved tools for biological sequence comparison, *Proc. Natl. Acad. Sci. U.S.A.*, 85, 2444, 1988.

296. Pechura, C.M. and Martin, J.B., Ed., *Mapping the Brain and Its Functions*, National Academy Press, Washington, D.C., 1991.

297. Peitgen, H-O. and Saupe, D., Ed., *The Science of Fractal Images*, Springer-Verlag, Oxford, 1988.

298. Perelson, A.S. Ed., *Theoretical Immunology*, Addison-Wesley, New York, 1988.

299. Peskin, C.S., The fluid dynamics of heart valves: experimental, theoretical, and computational methods, *Annu. Rev. Fluid. Mech.*, 14, 235, 1982.

300. Peskin, C.S. and McQueen, D.M., A three dimensional computational method for blood flow in the heart: immersed elastic fibers in a viscous incompressible fluid, *J. Comput. Phys.*, 81(2), 372, 1989.

301. Pickover, C.A., DNA vectorgrams: representation of cancer genes as movements on a 2D cellular lattice, *IBM J. Res. Dev.*, 31(1), 111, 1987.

302. Piegl, L. and Tiller, W., *The Nurbs Book*, Monographs in Visual Communication, Springer-Verlag, New York, 1995.

303. Pienta, K.J. and Coffey, D.S., A structural analysis of the role of the nuclear matrix and DNA loops in the organization of the nucleus and chromosome, *J. Cell Sci. Suppl.*, 1, 123, 1984.

304. Pincus, S.M., Cummins, T.R., and Hadda, G.G., *Heart Rate Control in Normal and Aborted SIDS Infants*, American Physiological Society, 1993, 638.

305. Pincus, S.M. and Keefe, D.L., *Quantification of Hormone Pulsatility via an Approximate Entropy Algorithm*, American Physiological Society, 1992, 741.

306. Pincus, S.M. and Viscarello, R.R., *Approximate Entropy: A Regularity Measure for Fetal Heart Rate Analysis*, vol. 79, The American College of Obstetricians and Gynecologists, 1991, 249.

307. Pitman, E.B. and Layton, H.E., Tubuloglomerular feedback in a dynamic nephron, *Comm. Pure Appl. Math.*, 42, 49, 1989.

308. Powell, P.A., Remin – fast algorithms for finding the similarity of regular expression based patterns and sequences. University of Minnesota Computer Science Department Technical Report 9#0-16, 1990.

309. Press, W.H., Teukolsky, S.A., Vetterling, W.T., and Flannery, B.P., *Numerical Recipes in C*, Cambridge University Press, Cambridge, 1992. Also available in a FORTRAN version — Great source of computer programs for all kinds of scientific calculations.

310. Preuss, L., *Maximum Specific Entropy, Knowledge, Ordering and Physiucal Measurements*, Kluwer Academic Publishers, 1989.

311. Prohofsky, E., *Statistical Mechanics and Stability of Macromolecules*, Cambridge University Press, Cambridge, 1995, 405.

312. Qi, Y. and Wissler, E., A combined analytic/finite difference technique for analyzing two-dimensional heat transfer in hbuman limbs which contain major arteries and veins, Technical Report Series CHPC - TR1992-0001, University of Texas Center For High Performance Computing, Austin, TX, 1992.

313. Qian, N. and Sejnowski, T.J., Predicting the secondard structure of globular proteins using neural network models, *J. Mol. Biol.*, 202, 865, 1988.

314. Reggia, J.A., Ruppin, E., and Berndt, R.S., Ed., Neural Modeling of Brain and Cognitive Disorders, World Scientific Pub. Co., River Edge, NJ, 1996. Has a discussion of modeling memory impairment in Alzheimer's disease.

315. Renshaw, E., *Modeling Biological Populations in Space and Time*, Vol. 11, *Cambridge Studies in Mathematical Biology*, Cambridge University Press, Cambridge, 1991.

316. Rensing, L., an der Heiden, U., and Mackey, M.C., Ed., *Temporal Disorder in Human Oscillatory Systems*, Springer-Verlag, Berlin, 1986.

317. Richards, F.M. and Kundrot, C.E., Identification of structural motifs from protein coordinate data: Secondard structure and first-level supersecondary structure, *PROTEINS: Struct. Funct. Genet.*, 3, 71, 1988.

318. Rieger, M.R., Adams, W.K., and Kinzel, G.L., A finite element survey of eleven endosseous implants, *J. Prosthet. Dent.*, 63, 457, 1990.

319. Rieger, M.R., Adams, W.K., Kinzel, W.K., and Brose, M.O., Finite element analysis of bone adapted and bone bonded endosseous implants, *J. Prosthet. Dent.*, 62, 436, 1989.

320. Rieger, M.R., Mayberry, M., and Brose, M.O., Finite element analysis of six endosseous implants, *J. Prosthet. Dent.*, 63, 671, 1990

321. Roach, G.F., Ed., *Mathematics in Medicine and Biomechanics*, Birkhauser-Boston, Boston, 1984.

322. Robinovitch, S.N., Hayes, W.C., and McMahon, T.A., Prediction of femoral impact forces on falls on the hip, *J. Biomech. Eng.*, 113, 366, 1991.

323. Rohlmann, A., Cheal, E.J., Hayes, W.C., and Bergmann, G., A nonlinear finite element analysis of interface conditions in porous coated hip endoprostheses, *J. Biomech.*, 21(7), 605, 1988.

324. Rosen, R., *Dynamical Systems Theory in Biology*, John Wiley & Sons, New York, 1970.

325. Rosen, R., *Foundations of Mathematical Biology*, Academic Press, New York, 1972.

326. Rosenthal, R., *Meta-Analytic Procedures for Social Research*, Vol. 6, *Applied Social Research Mechods Series*, Sage Pub., Newbury Park, CA, 1991.

327. Rotenberg, M., Theory of distributed quiescent state in the cell cycle, *J. Theor. Biol.*, 96, 495, 1982.

328. Rovine, M.J. and von Eye, A., *Applied Computational Statistics in Longitudinal Research*, Academic Press, Boston, 1991.

329. Rubin, C., Krishnamurthy, N., Capilouto, E., and Yi, Y., Stress analysis of the human tooth using a three-dimensional finite element model, *J. Dent. Res.*, 62, 82, 1983.

330. Rubinow, S., A maturity-time representation for cell populations, *Biophys. J.*, 8, 1055, 1968.

331. Rubinow, S.I., Cell kinetics, in *Mathematical Models of Molecular Biology*, L.A., Segel, Ed., Cambridge University Press, Cambridge, 1980.

332. Ryan, T.P., *Modern Regression Methods*, Wiley Series in Probability and Statistics, John Wiley & Sons, New York, 1997.

333. Samis, H.V. and Capobianco, S., Ed., *Aging and Biological Rhythms*, Plenum Press, New York, 1978.

334. Sandefur, J.T., *Discrete Dynamical Systems Theory and Applications*, Clarendon Press, Oxford, 1990.

335. Sapsford, R. and Jupp, V., *Data Collection and Analysis*, Sage Pub., London, 1996.

336. Savageau, M.A., *Biochemical Systems Analysis*, Addison-Wesley, Reading, MA, 1976.

337. Schench, J.F. et al., Superconducting open-configuration MR imaging system for image-guided therapy, *Radiology*, 195, 805, 1995.

338. Scheving, L.E., Halberg, F., and Pauly, J.E., Ed., *Chronobiology*, Igaku Shoin Ltd., Tokyo, 1974.

339. Schultz, A.B., Mobility impairment in the elderly: challenges for biomechanics research, *J. Biomech.*, 25, 519, 1992.

340. Schultz, A.B., Alexander, N.B., and Ashton-Miller, J.A., Biomechanical analyses of rising from a chair, *J. Biomech.*, 25(2), 1383, 1992.

341. Schulze-Kremer, S., *Molecular Bioinformatics*. Walter de Gruyter, Berlin, 1995. Excellent source of Internet/Web related resources for documents, software, agents, etc.

342. Schwartz, M.H., Leo, P.H., and Lewis, J.L., A microstructural model for the elastic response of articular cartilage. Research Report UMSI 93/44, University of Minnesota Supercomputiong Center, Minneapolis, MN, 1993.

343. Scott, G.P., Ed., *Time, Rhythms, and Chaos in the New Dialogue with Nature*, Iowa State University Press, Ames, IA, 1991.

344. Segel, L.A., Ed., *Mathematical Models in Molecular and Cellular Biology*, Cambridge University Press, Cambridge, 1980.

345. Sejnowski, T., Koch, C., and Churchland, P., Computational neuroscience, *Science*, 141, 1299, 1988.

346. Selvin, S., *Statistical Analysis of Epidemiological Data*, Vol. 25, *Monographs in Epidemiology and Biostatistics*, Oxford University Press, New York, 1996.

347. Sheskin, D.J., *Handbook of Parametric and Nonparametric Statistical Procedures*, CRC Press, Boca Raton, FL, 1997.

348. Shouman, R. and Witten, M., Survival estimates and sample size: what can we conclude?, *J. Gerontol.*, 50A(3), B177, 1995.

349. Shumway, R.H., *Applied Statistical Time Series Analysis*, Prentice-Hall Series in Statistics. Prentice-Hall, Englewood Cliffs, NJ, 1988.

350. Simon, H.A., The architecture of complexity, *Proc. Am. Philos. Soc.*, 106(6), 467, December 1962.

351. Singh, A.K. and Overbeek, R., Derivation of efficient parallel programs: an example from genetic sequence analysis, ANL Mathematics and Computer Science Division, MCS-P104-0989, 1989.

352. Sivia, D.S., *Data Analysis: A Bayesian Tutorial*, Clarendon Press, Oxford, 1996.

353. Skarda, C.A. and Freeman, W.J., How brains make chaos in order to make sense of the world, *Behav. Brain Sci.*, 1987.

354. Skolnick, J., Computer simulations of globular protein folding and tertiary structure, *Annu. Rev. Phys. Chem.*, 40, 207, 1989.

355. Smith, D.O., Cellular and molecular correlates of aging in the nervous system, *Exp. Gerontol.*, 23, 399, 1988.

356. Smith, R.F. and Smith, T.F., Automatic generation of primary sequence patterns from sets of related protein sequences, *Proc. Natl. Acad. Sci. U.S.A.*, 87, 118, 1990.

357. Smith, W.R., Hypothalamic regulation of pituitary secretion of luteinizing hormone. II. Feedback control of gonadotropin, *Bull. Math. Biol.*, 42, 57, 1980.

358. Sole, R.W. and Valls, J., On Structural Stability and Chaos in Biological Systems, Academic Press, New York, 1992, 87.

359. Soll, D. and Roberts, R.J., *The Applications of Computers to Research on Nucleic Acids* II, Part 1, IRL Press, Washington, D.C., 1984.

360. Soll, D. and Roberts, R.J., *The Applications of Computers to Research on Nucleic Acids II*, Part 2, IRL Press, Washington, D.C., 1984.

361. Spitzer, V.M. and Whitlock, D.G., A 3-d database of human anatomy, *Adv. Imaging*, March, 48, 1989.

362. Staden, R. Methods for discovering novel motifs in nucleic acid sequences, *CABIOS*, 5(4), 293, 1989.

363. Stadler, P.F. and Schuster, P., *Dynamics of Small Autocatalytic Reaction Networks. I. Bifurcations, Permanence and Exclusion*, vol. 52, Pergamon Press, Oxford, 1990, 485.

364. Stephenson, J.L., Urinary concentration and dilution: Models, in *Handbook of Physiology*, Windhager, E., Ed., Oxford University Press, New York, 1992, 1349.

365. Stephenson, J.L., Jen, J.F., Wang, H., and Tewarson, R.P., Convective uphill transport of NaCl from ascending thin limb of the loop of henle. *Am. J. Physiol.*, 268(4), F680, 1995.
366. Stephenson, J.L., Wang, H., and Tewarson, R.P., Effect of vasa recta flow on concentrating ability of the renal inner medula, *Am. J. Physiol.*, 268(4), F698, 1995.
367. Stiles, J.R., Monte Carlo simulation of synaptic transmission, Forefronts: Cornell Theory Center, 1994. Cornell Theory Center Newsletter.
368. Stine, R. and Fox, J., Ed., *Statistical Computing Environments for Social Research*, Sage Pub., Thousand Oaks, CA, 1997.
369. Stupfel, M., Gourlet, V., and Court, L., Effects of aging on circadian and ultradian respiratory rhythms of rats synchronized by an LD12:12 lighting (L = 100 lx), *Gerontology*, 32, 81, 1986.
370. Swan, G., *Optimization Of Human Cancer Radiotherapy*, Springer-Verlag, New York, 1981.
371. Tavare, S. and Giddings, B.W., Some statistical aspects of the primary structure of nucleotide sequences, in *Mathematical Methods For DNA Sequences*, Waterman, M., Ed., CRC Press, Boca Raton, FL, 1989.
372. Taylor, J.M.G., Withers, H.R., and Hu, Z., A comparison of mathematical models for regeneration of acutely responding tissues, *Int. J. Radiat. Oncol. Biol. Phys.*, 15, 1389, 1988.
373. Tenover, J.S. and Bremner, W.J., Circadian rhythm of serum immunoreactive inhibin in young and elderly men. *J. Gerontol. Med. Sci.*, 46(5), M181, 1991.
374. Tewarson, R.P. and Moon, I.H., Renal concentrating mechanism: central core and vasa recta models, *Appl. Math. Lett.*, 10(2), 39, 1997.
375. Tewarson, R.P., Stephenson, J.L., and Juang, L.L., A note on solution of large sparse systems of nonlinear equations, *J. Math. Anal. Appl.*, 63, 439, 1978.
376. Thompson, D.E., Buford, W.L., Myers, L.M., and Giurintano, D.J., Advances in hand biomechanics simulation, *J. Hand Ther.*, April–June, 142, 1989.
377. Thompson, J.R. and Brown, B.W., Ed., Cancer Modeling, Marcel Dekker, New York, 1987.
378. Thompson, W.J. *Atlas for Computing Mathematical Functions*, John Wiley & Sons, New York, 1997.
379. Topsoe, F., *Spontaneous Phenomena*, Academic Press, San Diego, 1990.
380. Torney, D.C., Burks, C., Davison, D., and Sirotkin, K.M., Computation of d^2 — a measure of sequence dissimilarity, in *The Interface between Computational Science and Nucleic Acid Sequencing*, Bell, G., and Marr, T., Ed., SFI Studies in the Sciences of Complexity, Addison-Wesley, Reading, MA, 1990.
381. Touitou, Y., Reinberg, A., Bogdan, A., Auzéby, A., Beck, H., and Touitou, C., Age-related changes in both circadian and seasonal rhythms of rectal temperature with special reference to senile dementia of the Alzheimer type, *Gerontology*, 32,110, 1986.
382. Trifonov, E.N., Nucleotide sequences as a language: Morphological classes of words. *Classification and Related Methods of Data Analysis*, 1988.
383. Trucco, E., Mathematical models for cellular systems: the Von Foerster equation. I, *Bull. Math. Biophys.*, 27, 285, 1965.
384. Trucco, E., Mathematical models for cellular systems: the Von Foerster equation. II, *Bull. Math. Biophys.*, 27, 449, 1965.
385. Trucco, E., On the use of the Von Foerster equation for the solution and generalization of a problem in cellular studies, *Bull. Math. Biophys.*, 27, 39, 1965.
386. Tsonis, A.A., *Dynamical Systems as Model for Physical Processes*, John Wiley & Sons, New York, 1996, 23.
387. Tucker, S.L., and Zimmerman, S.O., A nonlinear model of population dynamics containing an arbitrary number of continuous structure variables, *SIAM J. Appl. Math.*, 48(3), 549, 1988.
388. Rufte, E.R., *The Visual Display of Quantitative Information*, Graphics Press, Cheshire, CT, 1983.
389. Tufte, E.R., *Envisioning Information*, Graphics Press, Cheshire, CT, 1990.
390. Tufte, E.R., *Visual Explanations: Images and Quantities, Evidence and Narrative*, Graphics Press, Cheshire, CT, 1997.
391. Tukey, J.W., *Exploratory Data Analysis*, Addison-Wesley, Reading, MA, 1977.
392. Tyson, J.J., and Hannsgen, K., Cell growth and division: global asymptotic stability of the size distribution in probabilistic models of the cell cycle, *J. Math. Biol.*, 23, 231, 1986.

393. Verhulst, F., *Nonlinear Differential Equations and Dynamical Systems*, Springer-Verlag, Heidelberg, 1990.
394. Viertl, R., *Statistical Methods for Non-Precise Data*, CRC Press, Boca Raton, FL, 1996.
395. Vincent, J.F.V., Ed., *Biomechanics: A Practical Approach*, Vol. 105. *The Practical Approach Series: IRL Press*, Oxford University Press, Oxford, 1992.
396. Vingron, M. and Argos, P., A fast and sensitive multiple sequence alignment algorithm, *CABIOS*, 5(2), 115, 1989.
397. Vollenweider, H.J., Fiandt, M., and Szybalsk, W., A relationship between DNA helix stability and recognition sites for RNA polymerase, *Science*, 205, 509, 1979.
398. Vologodskii, A., *Topology and Physics of Circular DNA*, CRC Press, Boca Raton, FL, 1992.
399. von Eye, A., Ed., *Statistical Methods in Longitudinal Research*, Vol. 1, Academic Press, Boston, 1990.
400. von Eye, A., Ed., *Statistical Methods in Longitudinal Research*, vol. 2, Academic Press, Boston, MA, 1990.
401. von Mayersbach, H., Ed., *The Cellular Aspects of Biorhythms*, Springer-Verlag, New York, 1967.
402. von Seggern, D.H., *CRC Handbook of Mathematical Curves and Surfaces*, CRC Press, Boca Raton, FL, 1990.
403. Wachter, K.W. and Straf, M.L., Ed., *The Future of Meta-Analysis*, Russell Sage Foundation, New York, 1990.
404. Wallgren, A., Wallgren, B., Persson, R., Jorner, U., and Haaland, J-A., *Graphing Statistics and Data*, Sage Pub., Thousand Oaks, CA, 1996.
405. H., Wang, Stephenson, J.L., Deng, Y., and Tewarson, R.P., An efficient parallel algorithm for solving n-nephron models of the renal inner medulla, *Comput. Math. Appl.*, 28(5), 1, 1994.
406. Wang, Y. and Brown, M.B., A flexible model for human circadian rhythm, *Biometrics*, 52, 588, 1996.
407. Wasserman, P.D., *Neural Computing: Theory and Practice*, Van Nostrand Reinhold, New York, 1989.
408. Waterman, M., *Introduction to Computational Biology: Maps, sequences and genomes*, Chapman and Hall, London, 1995.
409. Waterman, M.S., *Mathematical Methods for DNA Sequences*, CRC Press, Boca Raton, FL, 1989.
410. Watt, A., *Fundamentals of Three-Dimensional Computer Graphics*, Addison-Wesley, Wokingham, England, 1989.
411. Watt, A. and Watt, M., *Advanced Animation and Rendering Techniques*, ACM Press/Addison-Wesley, Wokingham, England, 1992.
412. Watts, S., Parallel thinking takes on the human genome, *New Sci.*, 120, 34, 1988.
413. Webb, G.F., *Theory Of Non-Linear Age-Dependent Population Dynamics*, Marcel Dekker, New York, 1985.
414. Webb, G.F., A model of proliferating cell populations with inherited cycle length, *J. Math. Biol.*, 23, 269, 1986.
415. Weber, B.H., *Entropy, Information and Evolution*, MIT Press, Cambridge, 1987, 74.
416. Weigend, A.S. and Gershenfeld, N.A., Ed., *Time Series Prediction: Forecasting the Future and Understanding the Past*, Vol. XV, *Santa Fe Institute Studies In The Sciences of Complexity*, Addison-Wesley, Reading, MA, 1994.
417. Weir, B.S., Statistical analysis of DNA sequences, *J. Natl. Cancer Inst.*, 80, 495, 1988.
418. Weiss, R., High-tech tooth repair, *Sci. News*, 134, 376, 1988.
419. West, B.J., *Fractal Physiology and Chaos in Medicine*, vol. 1., World Scientific, Singapore, 1990.
420. West, B.J., Ed., *Patterns, Information and Chaos in Neuronal Systems*, Vol. 2, *Studies of Nonlinear Phenomena in Life Science*, World Scientific, Singapore, 1993.
421. Wever, R., Die Bedeutung der circadianen Periodik für den alternden Menschen, *Verh. Dtsch. Ges. Path.*, 59, 160, (in German).
422. Wexler, A.S., Kalaba, R.E., and Marsh, D.H., Three-dimensional anatomy and renal concentrating mechanism. I, Modeling results, *Am. J. Physiol. (Renal Fluid Electrolyte Physiol. 29)*, 260, F368, 1991.

423. Wexler, A.S., Kalaba, R.E., and Marsh, D.H., Three-dimensional anatomy and renal concentrating mechanism. II, Sensitivity results, *Am. J. Physiol. (Renal Fluid Electrolyte Physiol. 29)*, 260, F384, 1991.

424. White, J.H., Self-linking and the Gauss integral in higher dimensions. *Am. J. Math.*, 91, 693, 1969.

425. White, J.H., Introduction to the geometry and topology of DNA structure, in *Mathematical Methods for DNA Sequences*, Waterman, M., Ed., CRC Press, Boca Raton, FL, 1989.

426. Whitrow, G.J., Ed., *The Natural Philosophy of Time*, Thomas Nelson and Sons, London, 1963.

427. Whittaker, J., *Graphical Models in Applied Multivariate Statistics*, Series in Probability and Multivariate Statistics, John Wiley & Sons, Chichester, 1990.

428. Whyte, L.L., Wilson, A.G., and Wilson, D., Ed., *Hierarchical Structures*, Huntington Beach, 1969. Douglas Advanced Research Laboratories, American Elsevier Publishing Company.

429. Wickerhauser, M.V., *Adapted Wavelet Analysis from Theory to Software*, A.K. Peters, Wellesley, MA, 1994.

430. Winfree, A.T., *The Geometry of Biological Time*, Springer-Verlag, New York, 1980.

431. Winfree, A.T., *When Time Breaks Down*, Princeton University Press, Princeton, NJ, 1987.

432. Witten, M., Some generalized conjugacy theorems and the concepts of fitness and survival in logistic growth models, *Bull. Math. Biol.*, 42, 507, 1980.

433. Witten, M., Ed., *Investigating the aging mammalian system: Cellular levels and beyond*, volume Proceedings of the 25th Annual Meeting, Washington, DC, 1981. Society for General Systems Research.

434. Witten, M., Modeling cellular systems and aging processes, I. Mathematics of cell system models — a review, *Mech. Ageing Dev.*, 17, 53, 1981.

435. Witten, M., Thoughts about the essentials of simulation modeling. *Math. Comput. Modeling*, 2, 393, 1981.

436. Witten, M., Ed., *Hyperbolic Partial Differential Equations*, Vol. 1, *Advances In Hyperbolic Partial Differential Equations*, Pergamon Press, Oxford, 1983.

437. Witten, M., A return to time, cells, systems, and aging. I. Rethinking the concepts of senescence in mammalian systems, *Mech. Ageing Dev.*, 21, 69, 1983.

438. Witten, M., A return to time, cells, systems, and aging: II. Relational and reliability theoretic aspects of senescence in mammalian systems, *Mech. Ageing Dev.*, 27, 323, 1984.

439. Witten, M., Time abberation in living organisms: stochastic effects, *Math. Modeling*, 5, 97, 1984.

440. Witten, M., Ed., *Hyperbolic Partial Differential Equations*, Vol. 2, *Advances in Hyperbolic Partial Differential Equations*, Pergamon Press, Oxford, 1985.

441. Witten, M., Reliability theoretic methods and aging: Critical elements, hierarchies, and longevity—interpreting survival curves, in *The Molecular Biology Of Aging*, Woodhead, A., Blackett, A., and Setlow, R., Ed., Plenum Press, New York, 1985.

442. Witten, M., A return to time, cells, systems and aging: III. Critical elements, hierarchies, and Gompertzian dynamics, *Mech. Ageing Dev.*, 32, 141, 1985.

443. Witten, M., Ed., *Hyperbolic Partial Differential Equations*, Vol. 3, *Advances in Hyperbolic Partial Differential Equations*, Pergamon Press, Oxford, 1986.

444. Witten, M., *Advances in Mathematics and Computers*, vol. 1, Pergamon Press, New York, 1987.

445. Witten, M., Information content of biological survival curves arising in aging experiments: some further thoughts, in *Evolution Of Aging Processes In Animals*, Woodhead, A., and Thompson, K.H., Ed., Plenum Press, New York, 1987.

446. Witten, M., A return to time, cells, systems, and aging. V. Further thoughts on Gompertzian survival dynamics — the geriatric years, *Mech. Ageing Dev.*, 46, 175, 1988.

447. Witten, M., *Advances in Mathematics and Computers*, vol. 2, Pergamon Press, New York, 1989.

448. Witten, M., Modeling of the aging-cancer interface: some thoughts on a complex biological dynamics, *J. Gerontol.*, 44(6), 72, 1989.

449. Witten, M., Quantifying the concepts of rate and acceleration/deceleration of aging, *Growth Dev. Ageing*, 53, 7, 1989.

450. Witten, M., Peering inside living systems: physiology in a supercompter, *Supercomputing Rev.*, December 1990, 34.

451. Witten, M., Mathematical modeling of cell population dynamics as applied to the study of cellular aging, in *Mathematical Population Dynamics*, Arino, O., Axelrod, D.E., and Kimmel, M., Ed., Marcel Dekker, New York, 1991, 127.

452. Witten, M., Adding it up, *SunWorld*, 5(10), 61, 1992.

453. Witten, M., The Frankenstein project: building a man in the machine and the arrival of the computational physician, *Int. J. Super. Appl.*, 6(2), 127, 1992.

454. Witten, M., Visualization software, *SunWorld*, 5(1), 79, 1992.

455. Witten, M., Numbers to pictures, *Adv. Sys.*, November 1994, 34.

456. Witten, M., Ed., *Building a Man in the Machine*, vol. 1, World Scientific Press, Singapore, 1995.

457. Witten, M., Ed., *Building a Man in the Machine*, vol. 3, World Scientific Press, Singapore, 1995.

458. Witten, M., Cell cycle models of aging and apoptosis: towards a model of stage-specific cell cycle dynamics, in *Building A Man In The Machine: Computational Medicine, Publixc Health, and Biotechnology*, Vol. 1, *Series in Mathematical Biology and Medicine — Volume 5*, World Scientific, Singapore, 1995. 442.

459. Witten, M., Proceedings of the 1993 workshop on computational issues in neuroscience, in *Mathematics and Computers in Simulation*, Vol. 40, Witten, M., Ed., North Holland, Amsterdam, 1995.

460. Witten, M., The sounds of science. II. Listening to dynamical systems — towards a musical exploration of complexity, *Comput. Math. Appl.*, 32(1), 145, 1996.

461. Witten, M. and Vincent, D.J., Ed., *Building a Man in the Machine*, Vol. 2, World Scientific Press, Singapore, 1995.

462. Witten, M. and Wyatt, R.E., Increasing our understanding of biological models through visual and sonic representations: a cortical case study, *Int. J. Supercomput. Appl.*, 6(3), 257, 1992.

463. Witten, T.M., *Aging By The Numbers*, The Johns Hopkins University Press, Baltimore, Maryland, 1997. In preparation - chapters available from author upon request.

464. Witten, T.M., A return to time, cells, systems and aging. VI. An exploration into the interplay of structure and time as a mechanism for understanding the processes associated with aging, *Mech. Aging Dev.*, in preparation, 1997.

465. Witten, T.M. and Eakin, T., Multiphasic models of survival analysis of mortality rate change regions and the issue of finite species lifespan, *Exp. Gerontol.*, 32(3), 259, 1997.

466. Woolson, R.F., *Statistical Methods for the Analysis of Biomedical Data*, Wiley Series in Probability and Mathematical Statistics, John Wiley & Sons, New York, 1987.

467. Wyatt, R.E. and Driver, J.W., Computational brain dynamics: visualization of activity flow in the cerebral cortex, Cray Channels, 1991.

468. Yates, F.E., Good manners in good modeling: mathematical models and computer simulations of physiological systems, *Am. J. Physiol.*, 3, R159, 1978.

469. Yates, F.E., *Quantumstuff and Biostuff: A View of Patterns of Convergence in Contemporary Science*, Plenum Press, 1987, 617.

470. Yates, F.E., Ed., *Self-Organizing Systems: The Emergence of Order*, Life Science Monographs, Plenum Press, New York, 1987.

471. Yeargers, E.K., Shonkwiler, R.W., and Herod, J.V., *An Introduction to the Mathematics of Biology with Computer Algebra Models*, Birkhauser, Boston, 1996.

472. Yvette, G.S., AIP conference proceedings 226, in *The Living Cell in Four Dimensions*, Paillotin, G., Ed., New York, 1991, 3–586, Societe Francaise de Biophysique, American Institute of Physics.

473. Zuker, M., The use of dynamic programming algorithms in RNA secondary structure prediction, *Mathematical Methods For DNA Sequences*, Waterman, M., Ed., CRC Press, Boca Raton, FL, 1989.

474. Zuker, M. and Somorjai, R.L., The alignment of protein structures in 3 dimensions, *Bull. Math. Biol.*, 51(1), 55, 1989.

475. Zurek, W.H., Ed., *Complexity, Entropy, and the Physics of Information*, Vol. 8, Addison-Wesley, Redwood City, CA, 1990.

Section B

Experimental *In Vitro* Models in Aging Research

4

Use of the Fibroblast Model

Vincent J. Cristofalo, Maria Tresini, Craig Volker, and Mary Kay Francis

CONTENTS

0-8493-3112-9/99/$0.00+$.50
© 1999 by CRC Press LLC

4.1   Introduction

Normal human diploid fibroblasts (HDF) have a limited proliferative potential and lose their ability to divide after a finite number of population doublings [78]. After initial explantation, cells proliferate rapidly and can be subcultured frequently. This rapid growth is followed by a period of declining proliferation during which time numerous morphological and physiological changes occur until the cultures are no longer capable of proliferating. These progressive events have been defined as cellular aging or senescence (for reviews, see References 13, 40, 41).

In this chapter, we present an historical perspective of the development of this cell system as an *in vitro* model to study cellular aging, the strengths and limitations of the system, and the standard techniques for maintaining a defined and manipulable environment for a single cell type. Adherence to these culture procedures provides a well-controlled and reproducible cell system in which to study the mechanisms that determine the limited life span and regulate the deteriorative changes associated with aging.

4.2   Historical Development of Cell Cultures

In *Metazoa*, following fertilization of the egg and initial cell divisions, the cells of the embryo differentiate into a germ cell lineage and a somatic cell lineage. The germ cell lineage is potentially immortal, in the sense that the complement of genes can be passed on indefinitely. In contrast, cells of somatic lineage are destined for further differentiation and, ultimately, for aging and death. A question of major interest is whether biological aging is a cellular phenomenon or whether it derives from failures in integrative function that occur at the supracellular level. Thus, in addition to the organismic "clock(s)" that determine maximum life span potential, are there cellular clocks which can operate in the organism and independently outside the organism (*in vitro*) to determine cellular life span?

In vitro studies by Carrel and co-workers (16–19) suggested that individual cells, when separated from the organism, were potentially immortal in the same way that bacteria and most protozoa are considered immortal. This philosophy was generally accepted until the 1960s when the work of Swim and Parker [163], Hayflick and Moorhead [78], and others established that populations of normal human diploid fibroblasts can proliferate in culture for only finite periods of time. These authors examined a variety of human fetal, neonatal, and adult tissues and showed that they were incapable of unlimited proliferation.

The inability of cell cultures to proliferate indefinitely was initially ascribed to various technical difficulties, such as inadequate nutrition, pH variation, toxic metabolic products, and microcontaminants. Hayflick and Moorhead [78], however, showed that when mixtures of young and old populations, distinguishable by karyotypic markers, were grown in the same pool of medium, the older population phased out (completed replicative life span) after it had undergone a total of approximately 50 population doublings while the younger population continued to proliferate until the 50 or so doublings expected for this population had been completed. These results seem to rule out that cell degeneration is due to any direct effect of media composition, the presence of contaminating microorganisms, or toxic end products of metabolism. Hayflick and Moorhead [78] concluded that the limited life span phenomenon could be programmed and/or that genetic damage may accumulate, and they interpreted their observation as a cellular expression of senescence.

Over the last 35 years, in a large number of laboratories throughout the world, limited proliferative life span has been noted for chick and human fibroblasts, as well as for cells from other mammalian species. Moreover, limited *in vitro* life span has been described not only for fibroblasts, but for a variety of cell types including glial cells [132], keratinocytes [138], vascular smooth muscle cells [5], lens cells [165], endothelial cells [108], lymphocytes [166], and melanocytes [103]. Based on these findings, and those of Hayflick and coworkers [76,78], the notion that isolated animal cells are capable of unlimited proliferation in culture, as proposed by Carrel [16–19] and Ebeling [55,56], seems not to be the case.

4.3   Importance to Current Research

The use of cells and tissues in culture has made vast contributions to the understanding of the biology of growth, metabolism, differentiation, development, and numerous disease states. Therefore, it is not unreasonable to extend the cell culture system to study the biology of aging. In fact, cell culture has been used sporadically as an experimental tool to answer aging-related questions for more than 70 years. However, its generalized use began only more recently, in the 1960s.

While little is known about the fundamental process of aging, the literature suggests that it is a complex phenomenon that is likely to have both environmental and genetically programmed components. Furthermore, aging does not occur in exactly the same manner in all species or in all organisms of the same species. In fact, senescent changes in an organism are reflective of those changes that occur in different cell and tissue types that are likely to be initiated by different mechanisms and proceed at different rates. For example, while some tissues may lose functional capacity precipitously, others may exhibit a slower decline in functional activity. To understand and determine the mechanistic foundation of the aging process, we need to dissect the process of aging in multiple species, tissues, and cells. At the most basic level, the use of cell culture has the advantage of providing a controlled environment to study genetic and stochastic contributions to cellular senescence. However, there are also limitations inherent in the use of cell culture, specifically that of isolating a type from normal control elements that might be provided by other cell types.

4.3.1   Cell Cultures As Models

In considering the relevance of using cell culture as a model to study *in vivo* aging, one must make several assumptions. First, one must assume that aging has a cellular basis. This is a reasonable assumption since all processes of the organism are based upon cellular activity. Second, one must show that cells in culture show a deterioration in structure and function that leads to a decline in homeostasis. These deteriorative changes are often accompanied by alterations in gene expression which can be primary events or secondary changes. By these criteria, serial subcultivation of fibroblasts show aging characteristics and provide a model to study changes associated with the aging process.

Studies that examine the aging process in cell culture have used two models. In the predominant one, cell lines, such as WI-38 human diploid fibroblast-like cells, are serially subcultured to replicative senescence. These lines are well-characterized and with serial subculture show deteriorative changes, some of which parallel changes observed in *in vivo* aging. The other related type of cell culture model for aging studies is to use cells, such as skin fibroblasts, derived from donors of different ages. These are typically used for only a few passages to define characteristics of that cell type defined by the age of the donor. In either case, changes in the individual cell lines that occur during the life history or replicative lifespan form the basis of the model.

Models in biology are based on homology and analogy and cells in culture can be models of organismic aging by both criteria. While cell types within an organism differ from one another primarily due to different patterns of gene expression, every human somatic cell contains the same genetic information and can therefore be a model to another cell by homology. This comparison also extends to somatic cells in culture. Cells in culture are also potential models by analogy due to the various morphologic and metabolic changes which are similar to those *in vivo* (as discussed below). However, findings in these models may or may not be related in any simple direct way to the regulatory mechanisms that limit *in vivo* life span (32).

4.3.2   Strengths: *In Vivo/In Vitro* Parallels

There has been a long-standing debate as to whether *in vitro* studies using isolated populations of cells in culture, detached from the effect of neighboring cells *in situ*, accurately reflect the *in vivo* physiological processes of an individual cell in an organism. Opposing arguments have been put forth in mechanistic studies of the development of various disease states and neoplasia, as well as aging. Based on several lines of correlative evidence, as described below, we suggest that the *in vitro* culturing of HDF cells provides an excellent model for studying the aging process. The procedure which we present provides a controlled environment in which cells can be stimulated to proliferate by serial subcultivation and can in turn be used to decipher the fundamental mechanisms that drive replicative senescence.

The relationship of a limited replicative life span *in vitro* to organismic aging springs from evidence presented over the last 35 years in volumes of work and recently reviewed [13,40,41]. A brief summary of that information is presented here. First, any direct relationship that exists between *in vivo* and *in vitro* aging, as described for fibroblasts, is likely to involve cells that are capable of continued proliferation *in vivo*. As initially described by Leblond [85], the proliferation rate of some cells *in vivo* gradually slows down after an initially high rate of cell doubling. Some cells of the gastrointestinal tract, however, appear to be an exception in that they show an increased proliferative rate [3]. In a review of early works, Buetow [12] shows a tabulation of age-associated changes in cellular proliferation rates *in vivo*. In general in a wide variety of human and rodent tissues, there is a decline in mitotic events which leads to a reduced proliferative capacity. Therefore, the ability to study the biochemical and morphological parameters that accompany the decline in proliferative capacity may be of considerable importance to the understanding of the mechanism of senescence and the control of proliferation *in vivo*. The controlled environment of the cell culture system provides a model for these studies.

Second, perhaps one of the strongest lines of evidence for the relationship between *in vitro* replicative senescence and organismic aging is the relationship between the age of a cell donor and the proliferative lifespan of the derived cells. In general, cells from old donors tend to undergo fewer population doublings than cells from young donors. This has been found in humans and several other species [5,11,86,100,127,147]. The work of Martin et al. [100] on human subjects (ages 0 to 90) showed a regression of approximately 0.20 cell doublings per year. Although the variance was large, the regression was highly significant ($p < 0.001$). These findings suggest that the replicative life span of cells in renewable tissues is gradually reduced during the chronological life of an organism. However, in recent results from our laboratory, we have found for 42 *healthy* individuals from the Baltimore Longitudinal Study on Aging no correlation with donor age (Cristofalo et al., unpublished). This suggests that perhaps some of the previously reported differences in replicative life span may be in part due to disease.

Third, similar types of studies examined and compared *in vitro* proliferative life spans of short-lived and long-lived species. On the average, cells from short-lived species underwent senescence much quicker than comparable cells from a longer-lived species [60,77,141]. For example, fibroblasts from a Galapagos tortoise (maximum lifespan of more than 100 years) exhibit >100 population doublings, whereas those from a mouse (3-year maximum life span) achieved on average 10 to 15 population doublings before senescence. Thus, the maximum life span potential of a given species is reflected in the proliferative ability of their derived fibroblasts in culture. Furthermore, these results suggest that the controlling elements for the chronological lifespan of the organism may partially overlap the control mechanisms during *in vitro* replicative senescence.

Fourth, correlated with the above findings are several pieces of evidence that show age-related characteristics in cellular physiology. For example, lysosomal enzyme activities in adult lung derived cultures after 12 to 14 passages are similar to those of fetal cells after 35 to 50 passages [38]. Additionally, early passage human adult skin cells show cell cycle properties that are analogous to late passage human embryonic fibroblasts [96,97]. Furthermore, cells from individuals who have been diagnosed with hereditary premature aging disorders (Hutchinson-Guilford syndrome or Werner syndrome) often exhibit a reduced proliferative capacity as compared to age-matched controls [10,59,99,145]. Decreased mitotic activity, DNA synthesis, and cloning efficiency are often observed in early passage cells from affected patients [44,112]. These age-related physiologies are further evidence in support of similar or overlapping control mechanisms of organismal aging and *in vitro* replicative senescence.

In addition to the life span studies and age-related physiologies just described, numerous biochemical, genetic, and morphological changes occur at the cellular level during *in vitro* replicative senescence which parallel *in vivo* changes. Some of these are tabulated in Table 4.1 with selected references. These changes include, but are not limited to, decreased mitotic activity, saturation density, mitogenic response to growth factors, and colony size distribution. These and certain genetic alterations are described in more detail in Section 4.4, "Cellular and Molecular Markers of *In Vitro* Senescence." Overall, it is clear that characteristic changes of *in vivo*/organismal aging are expressed to varying degrees in cell culture models. Finally, a detailed analysis of these *in vivo* changes shows some parallels to *in vitro* aging that are apparently tightly coupled to the proliferative capacity of a cell.

TABLE 4.1

Parallel Changes with Cellular Aging *In Vitro* and *In Vivo*

	References
Decreases:	
Proliferative capacity	76, 100, 147
Saturation density	31, 147
Mitogenic response to serum/growth factors	130, 154
Colony size distribution	156
Response to calcium ion	136
Expression of TIMP-1 (tissue inhibitors of metalloproteinases)	105
Expression of IGFBP-3 (insulin-like growth factor binding protein-3)	61
Increases:	
Ploidy	45, 56
Cell size	153, 162
Collagenase expression	106, 157
Stromelysin expression	105
Changes in cellular morphology	153, 162

4.3.3   Limitations/Considerations/Caveats of the System

While cell cultures provide an easily manipulated system to monitor proliferative changes throughout the replicative life span, the biggest drawback of the system is that one is working with a single population of cells. In other words, any contributing effects from interactions with neighboring cell types are absent.

Several other criticisms have been directed at the use of cell cultures as models to address aging-related questions. For example, if the limited life span observed *in vitro* is truly relevant to organismal aging, then it should be characteristic of cultures derived from all vertebrate species. *In vitro* replicative senescence has been well-documented in human [75,76,78] and avian [74] cell cultures. Alternatively, there are numerous reports of rodent cell cultures [82,120,167,182], as well as others [137,159], that seem to proliferate indefinitely. However, when these cultures are followed more closely, many of these cultures actually exhibit a period of declining proliferation prior to the development of the immortal phenotype [127,141,177]. In one respect these cultures exhibit a limited period of senescence.

A second area of past controversy focused on the interpretation of data that describe the population dynamics of diploid cultures. This has recently been reviewed [13]. Initially, it was not clear whether it was due to a uniform increase in cell division time in the culture or whether the population was becoming heterogeneous with an increasing percentage of cells that were incapable of division, or both. In general, it appears that the decline in proliferative capacity in diploid cells develops as a result of an increased heterogeneity in the population. The changes include both an increased cell division time and an increased fraction of nondividing cells [43]. Furthermore, similar changes were observed by Soukupova et al. [158] when they examined cellular outgrowth characteristics of explanted tissue from young and old donors.

A third issue involves the effects of environmental factors on the lifespan of diploid cells. Numerous examples have been cited (for a review, see Reference 41) in which components of medium have been tested and determined to extend the life span of human and chicken cells. Additionally, chemically defined substances, such as hydrocortisone and cortisone, have been used to extend the proliferative life span of fibroblast-like cells in culture [29,55,68,95,143,155]. In all of these experimental descriptions, however, replicative senescence is only postponed, not eliminated.

These examples and others support the view that environmental conditions can indeed alter the proliferative capacity of cells in culture. More importantly, however, it is not whether environmental conditions can alter the proliferative life span of cells in culture, but whether there is a uniform "program" that underlies these environmental effects. In support of this possibility is the observation by numerous investigators that cultures of a single stock, such as the WI-38 human diploid fibroblast cell line, exhibit a similar number of divisions before undergoing a proliferative decline to senescence. This is quite striking and points to the reproducibility of this culture model since the cells have been grown in numerous laboratories with various commercially available media.

One final point in considering the cell culture model focuses on what actually determines the end point of life. Does *in vitro* replicative senescence reflect a genetically determined number of population doublings or is it merely a function of the time spent in culture? Initial evidence using chick cells seemed to support the latter [73,74,173]. Cultures died out within a defined calendar time, irrespective of the treatments which altered the number of divisions achieved. In contrast to these studies, many other studies have demonstrated that cultures can be maintained in a stationary phase for long periods of time and then subcultured for approximately the same number of population doublings as the continuously subcultured controls [37,46,62,104,111]. These reports suggest that the replicative life span is determined primarily by cumulative population doublings and that chronological time in culture is less important.

4.4 Cellular and Molecular Markers of *In Vitro* Senescence

As human diploid fibroblast-like cells approach the end of their replicative life span, they often exhibit differences in a number of cellular and molecular characteristics (for recent reviews, see References 13, 40, 41). We have included a brief review of the more salient attributes, broken down into several subsections. Furthermore, a number of these changes are enumerated in Table 4.2 for easy access.

4.4.1   Changes in Phenotype and Growth Characteristics

Although on average when cells approach the end of their replicative life span, they exhibit a shift toward larger cell size with an increase in both nuclear and nucleolar size, the population on the whole shows much greater heterogeneity [4,36,66,107]. In addition, morphological characteristics, such as large lysosomal bodies, vacuoles in the cytoplasm, increased microfilaments, prominent Golgi, and an evacuated endoplasmic reticulum have been observed in senescent cells [7,26,88]. There is also an increase in the number of multinucleated cells, tetraploid cells, and cells of higher ploidy in senescent cultures [45,56,101,183].

As the cells begin to show morphologic and phenotypic changes, they also show growth changes, which are reflected in a loss of replicative potential. Cultures of senescent cells display reduced harvest densities and lowered saturation densities at the plateau phase of growth [31,98]. At the end of their *in vitro* life span, substantial cell death occurs; however, a stable population emerges which can exist in a viable, although nonproliferating state for many months [102]. This stable, metabolically active population is capable of maintaining only extremely low saturation density representing less than five percent of that achieved by early passage cultures. The observed decrease in saturation densities most likely reflects an increased sensitivity to intercellular contact, since the forced contact of senescent cells, produced by overseeding, promotes the loss of cells from the culture until a density is achieved that is comparable to the density normally reached in the absence of such forced contact [129]. Thus the original diploid cell life history formulation of Hayflick and Moorhead [78], which included a primary outgrowth phase, a period of vigorous growth, and a period of declining proliferation, must be amended to include a phase of cell death and the emergence of a long-lived postmitotic population.

4.4.2   Changes in Macromolecular Synthesis and Content

In addition to the alterations in morphologic and growth parameters as cells approach the end of their proliferative life span, the synthetic rate of macromolecules decreases, while the cellular content of macromolecules, except DNA, increases. In senescent cells, it appears that DNA synthesis becomes uncoupled from the synthesis of macromolecules such as RNA, protein, glycogen, and lipid. In general, this leads to a further dysregulation of coordinated processes in the cell.

In senescent populations of WI-38 fibroblasts, most of the nondividing cells have a 2C DNA content [184]. There is, however, an increase in cells with a 4C DNA content which appear to be G_1 tetraploids rather than G_2 diploids [183]. Thus, senescent WI-38 cells remain essentially diploid to very near the end of their *in vitro* life span when they exhibit structural chromosomal changes and increasing levels of tetraploidy and polyploidy. Similar chromosomal alterations seem to occur with *in vivo* cell aging [45,56]. At least one study reported a small decrease in DNA content in human fibroblasts with *in vitro* age [146]. It should be noted, however, that there is an inherent bias in all of the studies designed to evaluate chromosome changes with aging *in vitro*, in that cytogenetic analyses can only be carried out with proliferating cells, that is, the least senescent cells in the aged culture.

TABLE 4.2

Changes in Gene Expression and Proteins in Human Diploid Fibroblast-Like Cells with Senescence

Gene/Protein	Description	Change with Senescence	References
Calmodulin	Ca^{2+}-binding/regulatory protein	Overexpressed protein[a]; uncoupled	9
EGF-R	Growth factor receptor	Detergent-solubilized form loses ability to autophosphorylate[a]	15
IGF-I	Insulin-like growth factor-I	Underexpressed	54
IFN-β	Interferon-β	Overexpressed	164
IGFBP-3	IGF-I binding protein-3	Overexpressed	109
IL-1 α and β	Interleukin-1 α and β cytokines	Overexpressed	83
IL-6	Interleukin-6 cytokine	Underexpressed	63
WS3-10	Ca^{2+}-binding protein	Overexpressed	109
E2F-1	Component of E2F transcription factor	Underexpressed	49
c-fos	Component of AP-1 transcription factor	Underexpressed	139, 151
Id1, Id2	Inhibitors of bHLH transcription factors	Underexpressed	71
h-Twist	bHLH transcription factor	Underexpressed	52, 172
CR	Cortisone reductase	Underexpressed protein[a]	2
2-5AS	2′, 5′-Oligoadenylate synthetase	Overexpressed	164
6-16	Interferon-inducible gene	Overexpressed	164
DHFR	Dihydrofolate reductase	Underexpressed	116
EF-Iα	Protein synthesis elongation factor-Iα	Underexpressed protein[a]; uncoupled	58
H1, H2a, H2b, H3, H4	Replication-dependent histones	Underexpressed	116, 151, 185
HLA B7	Histocompatibility antigen class I protein	Overexpressed	164
L5, L7, P1, S3, S6, S10	Ribosomal proteins	Underexpressed	152
ODC	Ornithine decarboxylase	Underexpressed protein[a]; similar mRNA expression	21, 22, 151
PCNA	Proliferating cell nuclear antigen/DNA polymerase δ co-factor	Underexpressed	20
polα	DNA polymerase α	Underexpressed protein[a]	117
RNR	Ribonucleotide reductase	Underexpressed	116
TK	Thymidine kinase	Underexpressed	21
TS	Thymidylate synthetase	Underexpressed	116
cdc2 (p34)	Cell division control	Underexpressed	161
cdk2	Cyclin-dependent protein kinase-2	Underexpressed; reduced phosphorylation[a]	1, 53
cdk4	Cyclin-dependent protein kinase-4	Underexpressed	94

TABLE 4.2   (continued)
Changes in Gene Expression and Proteins in Human Diploid Fibroblast-Like Cells
with Senescence

Gene/Protein	Description	Change with Senescence	References
cyc A, cyc B	Regulatory components of cyclin-dependent protein kinases	Underexpressed	161
cyc D1, cyc E	Regulatory components of cyclin-dependent protein kinases	Overexpressed; reduced serum-stimulated mRNA induction	53, 94, 179
cyc D3	Regulatory component of cyclin-dependent protein kinases	Reduced serum-stimulated mRNA induction	179
gas1, gas6	Growth arrest specific genes	Underexpressed	28
p16 (CDKN2, INK4A, CDK4I, MTS1)	Cyclin-dependent kinase inhibitor	Overexpressed mRNA and protein[a]	69
p21 (SDI-1, CIP1, WAF1)	Cyclin-dependent kinase inhibitor	Overexpressed	113
Prohibitin	Growth inhibitor	Loss of postsynthetic modification[a]	60
Rb protein	Tumor suppressor, growth inhibitor	Lack of phosphorylation[a]	160
Statin	Nuclear protein; associates with Rb	Overexpressed protein[a]	170
α1(I) Collagen	Extracellular matrix components	Overexpressed	109
Collagenase	Protease; extracellular matrix remodeling	Overexpressed mRNA and protein[a]	105, 175
Fibronectin	Extracellular matrix component	Overexpressed; uniquely antigenic	109, 133
mig-5	Tissue inhibitor of metalloproteinase	Underexpressed	176
t-PA	Tissue-type plasminogen activator	Overexpressed	174
u-PA	Urokinase-type plasminogen activator	Overexpressed	174
PAI-1, PAI-2	Plasminogen activator inhibitors-1 and 2	Overexpressed	83, 174
Stromelysin	Protease; extracellular matrix remodeling	Overexpressed mRNA and protein[a]	105
TIMP-1	Tissue inhibitor of metalloproteinase	Underexpressed	175
Catalase	Antioxidant enzyme	Overexpressed protein[a]	Cristofalo et al., unpublished
MnSOD	Superoxide dismutase	Overexpressed	84
hsp70, 90	Heat shock proteins	Underexpressed mRNA and protein[a]	23, 91, 92
Cathepsin B	Protease	Two-chain form is overexpressed; one-chain form is underexpressed[a]	51
cyt b	Cytochrome b	Overexpressed	52

TABLE 4.2   (continued)
Changes in Gene Expression and Proteins in Human Diploid Fibroblast-Like Cells
with Senescence

Gene/Protein	Description	Change with Senescence	References
EPC-1	Serine protease inhibitor homolog	Underexpressed mRNA and protein[a]	52, 128
LPC-1	p63; Endoplasmic reticulum/Golgi protein transport	Expression is uncoupled from the cell cycle	52
ND4	NADH dehydrogenase subunit 4	Overexpressed	52
SAG	Senescence-associated gene	Overexpressed	178
Terminin	Lysosomal protein	Protein is cleaved to smaller size[a]	171
WS9-14	Clone from Werner's syndrome cDNA library	Overexpressed	109

[a] Protein differences associated with senescence.

Schneider and Shorr [149] reported the amount of all species of cellular RNA (ribosomal, transfer, and messenger) to be elevated in late passage cultures with the ratio of mRNA to total cellular RNA remaining constant in cells before and after senescence. However, Johnson et al. [80] showed that the ratio of mRNA to rRNA was about 30% lower in senescent cells than in young cells. Protein content also increased with increasing *in vitro* age at least in part due to decreased protein turnover [35,48].

DNA, RNA, and protein synthetic rates decreased in late passage cultures and may be related to the altered template activity in senescent cells. The loss of DNA synthetic capacity and arrest in a state with characteristics of the late G_1 phase of the cell cycle (see below) would limit the absolute number of transcriptional events (and subsequently the number of potential translations) of genes normally induced during S phase, G_2, and mitosis. Changes in chromatin structure with senescence could also result in a net decrease of transcription events due to an inaccessibility of local template regions. Ryan and Cristofalo [144], for example, showed an age-associated reduction in chromatin template activity in WI-38 cells, consistent with a decrease in the number of net transcriptional events.

In senescent cells there was a marked reduction in the rate of RNA synthesis [98,144,148] and in transcription levels in response to serum stimulation [131]. Bowman et al. [18], however, suggested that the *in vitro* age-related decline in RNA synthesis may be due almost entirely to the decreased synthesis of nucleolar RNA.

The rate of protein synthesis in senescent cells is diminished relative to proliferating early-passage cells [42] and is similar to that of quiescent young cells [152]. It is unclear if these changes have to do with impaired translational capacity, translational control, or mRNA template availability. Despite their similar rates of protein synthesis, the mRNAs of six ribosomal proteins (L5, L7, P1, S3, S6, and S10) are downregulated specifically in senescent, but not quiescent young cells [152]. This downregulation can not be due simply to growth arrest or a general reduction in protein synthesis and points to a clear distinction between senescence and quiescence.

With senescence there is a change in the expression of many genes, including many associated with cell cycle progression. Yet, how these changes are orchestrated so that specific cell populations undergo the same characteristic pattern of senescence within a narrow window of number of cell divisions is at present unknown. It has been suggested that the shortening of telomeres during cell division may function as a mitotic clock to trigger at least some of these changes. As the physical ends of chromosomes, telomeres are not fully replicated during DNA replication. Consequently, telomeres shorten with each successive cell division. Shortened telomeres may provide signals for the expression of specific gene products. Furthermore, shortened telomeres may result in chromosome instability, which in turn may activate a DNA-damage pathway that produces slowed growth, eventual cell cycle arrest, and other processes observed in cellular senescence (for recent reviews, see References 27, 67, 72).

4.4.3 Cell Cycle Position of Arrested Cells

The hallmark of senescence in culture is the inability of cells to replicate their DNA. It is possible that replicative enzymes themselves and/or replication-associated processes, such as control of DNA hierarchical structural orders, are reduced or altered. There is, for example, a direct relationship between the concentration of DNA polymerase α and the rate of entry into S phase [117], suggesting that the local availability of a replicative enzyme may be rate limiting. This observation is supported by the failure of senescent fibroblasts to complement a temperature-sensitive DNA polymerase α mutant [114]. Similarly, senescent cells are unable to express the proliferating cell nuclear antigen (PCNA), a cofactor of DNA polymerase δ, presumably due to a posttranscriptional block [20]. In senescent cells, the replication-dependent histones are also repressed and a variant histone polyadenylated RNA is uniquely expressed [151]. Although these findings suggest gross changes in the DNA synthetic machinery of senescent cells, the observation that the SV40 virus can initiate an additional round of semiconservative DNA replication in senescent human fibroblasts [64] indicates that this machinery is still intact.

Although the machinery for DNA replication appears to be intact, senescent cells do not enter S phase of the cell cycle. In fact, there are several lines of evidence, as described in this section, that suggest that senescent cells are blocked in a state with characteristics of the late G_1 phase of the cell cycle. One of the first indications was that thymidine kinase activity in old, slowly proliferating cultures of WI-38 cells was found to be similar to that of young, rapidly dividing populations [30,115]. Since thymidine kinase is cell cycle-regulated and appears at the G_1/S boundary, this apparent anomaly first raised the possibility that senescent cells retained the ability to phosphorylate thymidine and thus may be arrested in late G_1. Subsequently, Olashaw et al. [115] found that thymidine triphosphate synthesis, which normally occurs in late G_1, is not impaired in senescent cells. Furthermore, the nuclear fluorescence pattern of senescent cells stained with quinacrine dihydrochloride is typical of cells blocked in late G_1 or at the G_1/S boundary [65]. Nucleolar association is also consistent with the hypothesis that senescent cells arrest in late G_1 (Cristofalo, V.J., Martin, B., Razi, B., Anderson, M., and Pignolo, R.J., unpublished results).

Rittling et al. [140] demonstrated that 11 cell cycle-dependent genes, whose time of increased expression ranges from early G_1 through the G_1/S boundary, are mitogen-inducible in both young and senescent cells during G_1. Yet, senescent cells lose the ability to complete the mitogen-initiated cascade of signal transduction pathways and synthesize DNA. Although the genes that encode ornithine decarboxylase and thymidine kinase remain inducible in senescent cells, the activities of their gene products have been reported to be diminished in senescent cells [21]. Furthermore, the c-*fos* gene, which encodes a component of the AP-1 transcription factor, and the *ld1* and *ld2* genes, which encode negative regulators of basic helix-loop-helix transcription factors, apparently do not respond to mitogens in senescent cells [71,151]. In addition, *cdc2*, *cycA*, and *cycB*, growth-regulated genes that are expressed in G_1, are also repressed in senescent cells [161]. The above findings suggest the possibility that senescent cells are irreversibly arrested in a unique state different from the normal cell cycle stages. However, the fact that some G_1 growth-regulated genes are not repressed in cells that fail to replicate suggests that only particular pathways are affected.

4.4.4 Signal Transduction Pathways

Signal transduction pathways play significant roles in the regulation of cell proliferation and adaptive responses and defects may contribute to the decline of these responses in replicative senescence. Human diploid fibroblasts, at or near the beginning of their *in vitro* replicative life span, vigorously respond to serum or a defined combination of growth factors by DNA synthesis and mitosis. As these cells approach the end of their proliferative potential in culture, they become increasingly refractory to mitogenic signals [34,79,123]. A decreased mitogenic response to growth factors *in vitro* has also been observed for *in vivo*-aged cells [130,154]. The basis for this loss of responsiveness cannot be attributed to any dramatic reductions in the number of cell surface growth-factor receptors, nor in the affinities with which these receptors bind ligands. In WI-38 cells, for example, both the number of receptors for EGF, PDGF, and IGF-I (per unit cell surface area) and receptor affinity for their ligands remain essentially unchanged with *in vitro* age [57,124,125]. Interestingly, senescent WI-38 cells produce neither the mRNA nor the protein for IGF-I [54]. The IGF-I receptor, however, is equivalently made in young and senescent WI-38 cells [150]. Moreover, the tyrosine kinase activity of the IGF-I receptor, as measured by the autophosphorylation of the receptor upon ligand binding, is unchanged in senescent cells [150]. Similar results have been obtained for the PDGF receptor and for the EGF receptor *in situ* [8,25]. However, the EGF receptor from detergent-solubilized, immunoprecipitated WI-38 senescent cell preparations is not autophosphorylated either in the presence or absence of EGF stimulation [15]. This ability appears to be dependent upon the nature of the extraction used to isolate the EGF receptor and indicates the existence of a more labile form of the EGF receptor or a closely associated moiety, in senescent cells. Indeed, Carlin et al. [14] have shown that an altered form of the EGF receptor is present in detergent extracts from senescent, but not young cells. This altered form is the cleavage product of an endogenous, leupeptin-sensitive proteolytic activity present in senescent cells.

Mitogenic signals are normally transmitted via the cytoplasm into the nucleus through a series of complicated events which include phospholipid turnover, calcium mobilization, and activation of protein kinases. Alterations in signal transduction pathways have been documented for arachidonic acid and prostaglandin (PG) metabolism from phospholipids in senescent cells [39]. Substantial increases in both arachidonate and in the subsequent production of prostaglandins can be seen in late-passage human fibroblasts [39,110]. Given the ability of exogenously added PG_{E2} to inhibit the proliferation of replication-competent cells, the above results indicate a possible role for prostaglandins as negative regulators of growth. Although the release of arachidonate from membrane phospholipids is mediated by phospholipase A_2, whose activity appears to be intact in senescent WI-38 cells, phospholipid turnover by phospholipase $C_\gamma 1$ in senescent IMR-90 cells is impaired [24].

Calcium ion (Ca^{2+}) is a potent modulator of cell proliferation for fibroblasts [135] and at supraphysiological concentrations can completely replace serum or growth factors as a mitogen [134]. Both proliferatively senescent and *in vivo*-aged fibroblasts have a reduced ability to respond to extracellular calcium as measured by increases in saturation density [136]. Young and senescent WI-38 fibroblasts have similar baseline levels of intracellular Ca^{2+} and exhibit similar changes in cytosolic Ca^{2+} fluxes following growth factor stimulation [9].

Skin fibroblasts from aged donors have slightly reduced free cytosolic Ca^{2+} levels [119] and slightly elevated amounts of bound Ca^{2+} [118]. The mRNA of the calcium-binding/regulatory protein, calmodulin, is expressed similarly in young and senescent cells in terms of timing and steady-state levels; however, expression of the protein, as detected by radioimmune assay, is uncoupled from the cell cycle and exists in variable amounts in senescent WI-38 cells [9]. The calmodulin-associated phosphodiesterase activity also appears to be diminished in late-passage cells (Cristofalo et al., unpublished results). Thus, an immunologically active, but functionally inactive pool of calmodulin may be present in senescent cells.

There is evidence that demonstrates that the protein abundance of protein kinase A (PKA) and various isoforms of protein kinase C (PKC) are not decreased in senescent fibroblasts [6,81]. These data, however, do not exclude the possibility that the activity of these enzymes is impaired in late passage cells. Indeed it has been shown that PKC translocation from the cytoplasm to the plasma membrane, an event that occurs upon PKC activation, is impaired in both WI-38 and IMR-90 senescent fibroblasts [47,168]. Similarly, the protein abundance of several members of the MAPK family, such as ERK1, ERK2, and ERK3 and the upstream activator of these kinases, raf-1, is at similar levels in young and senescent fibroblasts [81]. Senescent cells, however, seem to be impaired in their ability to activate the ERK kinases in response to serum stimulation (Cristofalo et al., unpublished observation). Taken together, the above results suggest that late-passage fibroblasts do not respond to mitogenic stimuli in the same manner as early-passage fibroblasts, in part, because of faulty or unregulated signalling pathways.

As mentioned above, a number of early and late cell cycle-regulated genes are expressed in senescent WI-38 fibroblasts, including c-*myc*, H-*ras*, and thymidine kinase [140]. The immediate-early response gene c-*fos*, however, is repressed in late-passage cells [151] under certain conditions. Furthermore, the ability of these cells to form active AP-1 complexes is also severely reduced [139]. As a result of repressed c-*fos* expression in senescent cells, there is a tendency toward Jun:Jun homodimers in contrast to the Jun:Fos heterodimers seen in young cells (139). Since Jun:Jun homodimers are not as stable as Jun:Fos heterodimers, fewer DNA binding events occur, ultimately resulting in decreased transcriptional activation through the AP-1 site. The forced expression of an inducible c-*fos* construct by transient transfection of senescent cells results in as much as a six-fold increase in the number of cells capable of initiating DNA synthesis [126]. In contrast to these findings, the induction of c-*fos* in senescent fibroblasts by microinjection of oncogenic c-*Ha-ras* protein was not sufficient to initiate DNA synthesis [142].

While the repression of c-*fos* transcription provides evidence for an early block in one or more pathways potentially required for DNA synthesis, later events may also be required. For example, the serum-stimulated mRNA induction of cyclins is diminished in senescent fibroblasts [53,94,161,179]. The expression of two cyclin-dependent kinase genes, *cdk2* and *cdk4*, is also diminished in senescent cells [1,94]. The retinoblastoma (Rb) gene product fails to be phosphorylated in senescent cells [160], most likely due to the absence of the cdc2 protein and associated cyclins [161]. The hypothesis that unregulated tumor suppressor genes may in some way prevent senescent cells from entering S phase is supported by studies in which antisense oligomers to Rb and p53 can extend the *in vitro* life span of human diploid fibroblasts by about 30% [70].

As cells senesce, there is a progressive reduction in their ability to respond to environmental stresses (for a review, see Reference 90). For example, in response to hyperthermic exposure, senescent cells give a reduced heat shock gene expression [89,92]. Similarly, in response to heavy metals, such as cadmium, induction of metallothioneins is reduced in senescent cells, and the cells are more sensitive to heavy metal toxicity [93]. In addition, there is limited evidence that the transcript for SOD-2, the manganese-containing form of superoxide dismutase, is increased in senescent cells [84]. Furthermore, we have observed an increase in the expression of catalase protein in senescent WI-38 cells (Cristofalo et al., unpublished). Perhaps, to help compensate for their reduced ability to adapt to environmental stresses, the basal levels of protective enzymes, such as heat shock proteins, metallothioneins, and antioxidant enzymes, are elevated in senescent cells.

4.4.5 Biomarkers of Cellular Senescence

At early passages (low cumulative population doubling levels (CPDLs)), WI-38 cells proliferate rapidly. With continued subcultivation, the rate of cellular proliferation declines until the cultures can no longer be propagated. Generally, early passage cells are defined as those that have progressed through <50% of their replicative life span, whereas senescent cells are defined as those that have completed >95% of their replicative life span. In our laboratory, the CPDL at "phase out," i.e., the end of the replicative life span, for 36 separate sublines of WI-38 cells was determined to be 66 ± 6 (mean ± standard deviation) [31].

As the rate of cell proliferation decreases when cultures approach "phase out," DNA synthesis also declines. Thus, the percent life span completed, as well as the CPDL at "phase out," may be estimated from the observed changes in cell number at subcultivation, as well as by the ability of the culture to incorporate [^{3}H-methyl]-thymidine [43]. We find that the percentage of nuclei labeled in actively growing cells in the presence of [[^{3}H-methyl]-thymidine for 30 h by our standard method decreases from approximately 90% for cells that have progressed through 50% of their replicative life span to approximately 50% for cells that have progressed through 90% of their replicative life span.

Another parameter typically used as a biomarker of cellular senescence is saturation density, the density that cells achieve when they are allowed to become confluent and quiescent. For cultures that have progressed through approximately 50% of their proliferative life span, cell density increases from the original plating density of 1×10^4 cells per cm^2 to approximately 1.5×10^5 cells per cm^2 at confluence. Saturation density decreases with increased population doubling level and reaches only ~2.0×10^4 cells per cm^2 in senescent cultures [31].

A β-galactosidase activity has recently been identified and reported to be associated only with senescent cells [50]. This activity may be detected in single cells that have been fixed and treated with X-gal (5-bromo-4-chloro-3-indolyl-β-D-galactoside), a compound that gives a blue precipitate when cleaved by β-galactosidase. Cells that have β-galactosidase activity stain blue and can easily be visualized by light microscopy. Thus, this activity can be a useful biomarker for cellular senescence. We find, however, that this activity is also associated with growth state and may be readily detected not only in senescent cells, but in quiescent, early passage cells, as well. Similarly, damaged and degenerating cells show β-galactosidase staining (Cristofalo et al., unpublished).

4.5 Propagation of Human Fibroblasts: Materials and Methods

The procedures presented here are the standard techniques employed to study the senescence of human diploid fibroblast-like cells, such as WI-38 cells, in culture. These procedures have been adapted primarily from previous publications and include methods to determine replicative life span, saturation density, and thymidine incorporation [31,33,43,122]. These methods have been well tested over many years and provide highly reproducible results.

The serum-supplemented and serum-free growth factor-supplemented formulations presented each give optimal growth of human diploid fibroblast-like cells. Cells in culture require carbon dioxide for growth and survival, and we have found that carbon dioxide/bicarbonate buffered media gives superior growth compared with media buffered with synthetic buffers, such as HEPES. Serial propagation is generally performed in serum-supplemented media; yet serum is a complex fluid with numerous known and unknown bioactive components. Combinations of growth factors have been identified that stimulate cell proliferation to the same extent as serum [121, 122]. Thus, for many studies, it is often desirable to use a serum-free growth medium of defined composition.

All reagents and materials for cell culture must be sterile, and all manipulations are performed in a laminar flow hood. Cell cultures are routinely examined microscopically for contamination and tested for mycoplasma every five weeks [87]. Cell cultures are subcultivated weekly. Multiple identical sister flasks are prepared at subcultivation, as a hedge against potential contamination or other anomalies. As a substantial fraction of the cells do not survive subcultivation, the number of cells does not generally increase above the seeded cell number until approximately 24 h after subcultivation.

4.5.1   Materials

The items required for the growth and biomarker measurements of human diploid-like fibroblast cells are listed in Tables 4.3 through 4.8.

TABLE 4.3

Laboratory Equipment

Equipment	Supplier	Model No.
Laminar flow hood	Sterilgard	VBM 400
Water bath	Precision Scientific	183
Incubator	Forma Scientific	3172
Inverted phase microscope, minimum 400× magnification	Nikon	MS
Micrometer, eyepiece with 10 × 10 grid	Swift	MA6659
Micrometer, stage	Swift	MA663
Centrifuge, refrigerated	International Equipment	PR-J
Coulter counter	Coulter	ZM

4.5.2   Preparation of Solutions and Other Items Required for Procedures

4.5.2.1   *Solutions for Cell Propagation in Serum-Supplemented Medium*

Suppliers and more detailed information on the items required for the preparation of serum-supplemented media are listed in Table 4.5.

4.5.2.1.1   *100X Basal Medium Eagle Vitamins*

Filter-sterilized 100X basal medium Eagle vitamins are purchased in 100-ml bottles and stored at –20°C. When first thawed, the vitamin solution is aliquoted (10-ml portions) in a laminar flow hood using sterile procedures into sterile, 15-ml centrifuge tubes and stored at –20°C until use.

TABLE 4.4

General Laboratory Supplies

Equipment	Supplier	Model No.
Cell culture flask, 25 cm²	Becton Dickinson	3014
Cell culture flask, 75 cm²	Becton Dickinson	3024
Cell culture flask, 175 cm²	Becton Dickinson	3028
Cell culture plate, 24-well	Becton Dickinson	3047
Cell culture plate, 6-well	Becton Dickinson	3046
Cell culture dish, 35 mm	Becton Dickinson	3001
Cell culture dish, 60 mm	Becton Dickinson	3002
Cell culture dish, 100 mm	Becton Dickinson	3003
Centrifuge tube, 15 ml	Becton Dickinson	2097
Centrifuge tube, 50 ml	Becton Dickinson	2070
Filter, bottle-top, 0.2 µm, 100 ml	Corning Costar	8310
Filter, bottle-top, 0.2 µm, 500 ml	Corning Costar	8330
Isoton II	Coulter	8546719
Coulter Clenz	Coulter	8546931
Cryogenic vial, 1.0 ml	Nalge	5000-1012
Pipette, plastic, 1 ml	Sarstedt	86.1252.001
Pipette, plastic, 5 ml	Sarstedt	86.1253.001
Pipette, plastic, 10 ml	Sarstedt	86.1254.001

TABLE 4.5

Components of Standard Growth Medium

Component	Supplier	Cat. No.
Auto-Pow™, autoclavable powder minimal essential medium Eagle modified with Earle's salts without glutamine and without sodium bicarbonate	ICN	11-100-22
100X Basal Medium Eagle Vitamins	ICN	16-004-49
L-Glutamine, 200 mM	Sigma	G3126
Sodium Bicarbonate	Sigma	S5761

Note: Fetal Bovine Serum (FBS) is from a variety of suppliers and tested on a lot-by-lot basis.

4.5.2.1.2 Incomplete Eagle's Modified Minimum Essential Medium

To prepare 1 l of incomplete Eagle's modified minimum essential medium, Auto-Pow™ powder (9.4 g) and 100X basal medium Eagle vitamins (10 ml) are added to 854 ml deionized, distilled water with magnetic stirring. It is important that the highest quality deionized, distilled water be used to prepare growth medium and all other reagents used for cell culture.

The incomplete medium is split into equal portions (432 ml each), each of which is placed into a 1-l bottle. The caps are screwed on loosely, autoclave tape is applied, and the bottles

TABLE 4.6

Components of Serum-Free Growth Medium

Component	Supplier	Cat. No.
MCDB-104, a modified basal medium with L-glutamine, without $CaCl_2$, without Na_2HPO_4, without $NaHCO_3$, and without HEPES, where sodium pantothenate is substituted for calcium pantothenate	Gibco BRL	82-5006EA
Sodium phosphate, dibasic	Sigma	S5136
Sodium chloride	Sigma	S5886
Calcium chloride dihydrate	Sigma	C7902
Sodium bicarbonate	Sigma	S5761
N-2-Hydroxyethylpiperazine-N′-2-ethanesulfonic acid (HEPES)	Sigma	H9136
Sodium hydroxide, 1 M	Sigma	S2770
Epidermal growth factor (EGF), human recombinant	Gibco BRL	13247-010
Insulin-like growth factor-I (IGF-I), human recombinant	Gibco BRL	13245-014
Insulin	Sigma	I6634
Ferrous sulfate heptahydrate	Sigma	F8633
Hydrochloric acid, 1 M	Sigma	H9892
Dexamethasone	Sigma	D4902
Ethanol, 95% (not denatured)	Pharmco	111000-190CSGL

TABLE 4.7

Components of Trypsinization Solution

Component	Supplier	Cat. No.
Sodium chloride	Sigma	S5886
Potassium chloride	Sigma	P5405
Sodium phosphate monohydrate, monobasic	Sigma	S5655
Glucose	Sigma	G6152
50× MEM amino acids, without glutamine	Gibco BRL	11130-051
100× basal medium Eagle vitamins	ICN	16-004-49
Phenol red, 0.5%	ICN	16-900-49
Sodium bicarbonate	Sigma	S5761
Trypsin, 2.5% in Hanks' balanced salt solution	Sigma	T4674
Soybean trypsin inhibitor, type I-S	Sigma	T6522

are autoclaved for 15 min at 121°C. When the sterilization cycle is finished, the bottles are quickly removed from the autoclave, as prolonged heat destroys some medium components. With the caps still loose, the bottles are placed in a laminar flow hood and allowed to cool to room temperature. When cooling is complete, the caps are tightened, and the incomplete medium is stored at 4°C in the dark.

TABLE 4.8

Items for Thymidine Incorporation

Item	Supplier	Cat. No.
[³H-methyl]-thymidine, 2 Ci/mmol; 1 mCi/ml	Dupont NEN	NET-027A
Cover slip, no. 1, 22 mm × 22 mm	Thomas	6662-F55
Cover slip rack, ceramic	Thomas	8542-E30
Cover slip rack, glass	Fisher	08-812
Chloroform	Sigma	C5312
Ethanol (95%), not denatured	Pharmco	111000190CSGL
Sulfuric acid, 95%	Sigma	S1526
Nitric acid, 70%	Sigma	25,811-3
Sodium hydroxide	Sigma	S5881
Petri dish, glass, 100 mm	Thomas	3483-K33
Emulsion, NTB-2	Eastman Kodak	165 4433
Developer, D-19	Eastman Kodak	146 4593
Acid fixer	Eastman Kodak	197 1746
Hematoxylin, Harris modified	Fisher	SH30-500D
Permount	Fisher	SP15-100
Microscope slide, 3 in × 1 in	Thomas	6684-H61
Lab-Tek® Chamberslide™, 2-chamber	Nalge Nunc	177380
Lab-Tek® Chamberslide™, 4-chamber	Nalge Nunc	177437
Lab-Tek® Chamberslide™, 8-chamber	Nalge Nunc	177445
Sodium phosphate, dibasic	Sigma	S5136
Potassium phosphate, monobasic	Sigma	P5655
Methanol	Fisher	A408-1
Slide mailer, polypropylene	Thomas	6707-M27
Slide box, polypropylene	Thomas	6708-G08

4.5.2.1.3 *Sodium Bicarbonate (7.5% w/v)*

Sodium bicarbonate (37.5 g) is added to 500 ml of deionized, distilled water with magnetic stirring. This solution is then sterilized by filtration into a sterile bottle through a 0.2-μm pore size bottle-top filter in a laminar flow hood, and stored at 4°C.

4.5.2.1.4 *Fetal Bovine Serum (FBS)*

Prior to purchase, various lots of fetal bovine serum (FBS) are tested by determination of the rate of WI-38 cell proliferation and saturation density for three consecutive weeks. The best available serum lot is chosen, and serum is purchased in quantities that will last about one year. The serum is stored at –20°C until use. Once thawed, the serum that remains in the bottle is not refrozen, but is stored at 4°C for subsequent use.

4.5.2.1.5 *L-Glutamine (200 mM)*

L-Glutamine (14.6 g) is added to 500 ml of deionized, distilled water with magnetic stirring. This solution is then sterilized by filtration into a sterile bottle through a 0.2-μm pore size bottle-top filter in a laminar flow hood. Aliquots (50 ml) are added to sterile 100-ml bottles that are then capped and stored at –20°C until use. When thawed for use, the remainder of the glutamine solution is aliquoted (5 ml portions) into 15-ml centrifuge tubes using sterile procedures in a laminar flow hood. These smaller aliquots are also stored at –20°C until use.

4.5.2.1.6 Standard Serum-Supplemented Growth Medium (Complete Medium with 10% v/v Fetal Bovine Serum)

To prepare the standard serum-supplemented growth medium (complete medium with 10% v/v FBS), to each 1-l bottle that contains 432 ml of sterile, incomplete Eagle's modified minimum essential medium (Section 4.5.2.1.2), filter-sterilized 7.5% (w/v) sodium bicarbonate (13 ml) and sterile FBS (50 ml) are added in a laminar flow hood. Just before use the medium is prewarmed to 37°C in a warm water bath, then transferred to a laminar flow hood where filter-sterilized 200 mM L-glutamine (5 ml) is added. Complete medium is generally prepared fresh for each use, but any remaining medium may be stored at 4°C for subsequent use. If this medium is stored longer than one week at 4°C, additional L-glutamine (1 ml per 100 ml of complete medium) is added just before use.

4.5.2.2 Solutions for Cell Propagation in Serum-Free Medium

Suppliers and more detailed information on the items required for the preparation of serum-free media are listed in Table 4.6.

4.5.2.2.1 Bicarbonate-Buffered Incomplete Serum-Free Growth Medium

To prepare 1 l of incomplete medium, one packet of powdered MCDB-104 medium (with L-glutamine, without $CaCl_2$, without Na_2HPO_4, without $NaHCO_3$, and without HEPES, where sodium pantothenate is substituted for calcium pantothenate) is added to 700 ml of deionized, distilled water with magnetic stirring. The packet is rinsed several times with deionized, distilled water and the rinsings are added, as well. The following additional components are then added in the order listed: 0.426 g Na_2HPO_4, 1.754 g NaCl, 1.0 ml of a 1 M $CaCl_2$ solution, and 1.176 g $NaHCO_3$. HEPES is not to be used in the cell culture growth medium. The final volume is brought to 1 l with deionized, distilled water. Incomplete medium is sterilized by filtration through a 0.2-μm bottle-top filter into sterile glass bottles. Using sterile procedures in a laminar flow hood, a 5% CO_2 to 95% air mixture is passed through a sterile, cotton-filled $CaCl_2$ drying tube, through a sterile pipette and bubbled into the medium. As the medium is gassed, it changes in color from pink to peach. The final pH is 7.3–7.5. Incomplete medium is generally prepared fresh for each use, but any remaining medium may be stored for subsequent use for up to 3 weeks at 4°C. If the unused medium is stored longer than 1 week, additional L-glutamine (1 ml per 100 ml of complete medium) is added just before use.

4.5.2.2.2 HEPES-Buffered Incomplete Medium for Soybean Trypsin Inhibitor and Concentrated Growth Factor Stock Solutions

The pH of carbon dioxide/bicarbonate-buffered MCDB-104 solutions rise during freezing, resulting in precipitate formation. Thus, growth factor and soybean trypsin inhibitor solutions that are stored frozen are prepared in HEPES-buffered solutions. To prepare 1 l of HEPES-buffered incomplete medium, one packet of powdered MCDB-104 medium (with L-glutamine, without $CaCl_2$, without Na_2HPO_4, without $NaHCO_3$, and without HEPES, where sodium pantothenate is substituted for calcium pantothenate) is added to 700 ml of deionized, distilled water with magnetic stirring. The packet is rinsed several times with deionized, distilled water and the rinsings are added, as well. The volume is brought to 900 ml with additional deionized, distilled water; then the following are added in order: 0.426 g Na_2HPO_4, 11.9 g HEPES free acid, 1.0 ml of 1 M $CaCl_2$, and 25.0 ml of 1 M NaOH. The pH of the solution is adjusted to 7.5 by titration with additional 1 M NaOH and the volume is brought to 1 l with deionized, distilled water. The medium is sterilized by filtration in a laminar flow hood through a 0.2-μm bottle-top filter into sterile glass bottles. The HEPES-buffered incomplete medium may be stored at –20°C until needed.

4.5.2.2.3 *Concentrated Growth Factor Stock Solutions*

Using sterile plastic pipettes in a laminar flow hood, 100X growth factor stock solutions in HEPES-buffered incomplete medium are prepared at the following concentrations: EGF (2.5 μg/ml) and either IGF-I (10 μg/ml) or insulin (500 μg/ml). Insulin and IGF-I both stimulate growth through the IGF-I receptor, although insulin has lower affinity for the IGF-I receptor and 50-fold higher concentrations are required to achieve comparable growth. Insulin is less expensive than IGF-I, and despite the reduced specificity, insulin is completely satisfactory for most experiments. All stock solutions are dispensed with sterile plastic pipettes into sterile 1.0-ml cryogenic vials. The stock solutions may be stored at –20°C for short periods (up to 4 weeks) or at –70°C for longer periods (3 to 4 months).

A solution of dexamethasone (5 mg/ml) in 95% ethanol (not denatured) is prepared. This solution is then diluted into HEPES-buffered incomplete medium to give a 100X stock solution (5.5 μg/ml). Dexamethasone is stored in sterile, siliconized test tubes.

Ferrous sulfate is prepared fresh, just before use. A 10-ml 100X solution of ferrous sulfate (0.5 mM) is prepared to which 5 μl of 1 M hydrochloric acid is added. This solution is sterilized by filtration through a 0.2-μm filter.

4.5.2.2.4 *Complete Serum-Free Growth Medium*

For 100 ml of complete serum-free growth medium, 1 ml of each of the 100X growth factor stock solutions (Section 4.5.2.2.3) are added to 96 ml of incomplete medium (MCDB-104). The resultant concentrations in the serum-free medium are: EGF, 25 ng/ml; IGF-I, 100 ng/ml or insulin 5 μg/ml; dexamethasone, 55 ng/ml; ferrous sulfate, 5 μM.

4.5.2.2.5 *Soybean Trypsin Inhibitor Solution for Serum-Free Propagation*

Soybean trypsin inhibitor (100 mg) is added to 100 ml of HEPES-buffered incomplete medium (Section 4.5.2.2.2). This solution is sterilized by filtration through a 0.2-μm bottle-top filter into a sterile bottle. The sterile solution is then aliquoted into sterile 15-ml centrifuge tubes in 7-ml portions and stored at –20°C. When needed, the solution is thawed and diluted 1:1 with bicarbonate-buffered incomplete medium (Section 4.5.2.2.1).

4.5.2.3 **Trypsinization Solution**

Suppliers and more detailed information on the items required for the preparation of trypsinization solution are listed in Table 4.7.

4.5.2.3.1 *Ca^{2+}/Mg^{2+}-Free Medium*

Trypsin-treated cells tend to aggregate less readily in medium that is low in Ca^{2+} and Mg^{2+}. To prepare 1 l of Ca^{2+}/Mg^{2+}-free medium, the following ingredients are added in order to 900 ml of deionized, distilled water with magnetic stirring: 6.8 g NaCl, 0.4 g KCl, 0.14 g $NaH_2PO_4 \cdot H_2O$, 1 g glucose, 20 ml of 50X MEM amino acids without glutamine, 10 ml of 100X basal medium Eagle vitamins, and 10 ml of a 0.5% (w/v) solution of phenol red. The solution is then diluted to 1 l with deionized, distilled water and sterilized by filtration in a laminar flow hood through a 0.2-μm pore size bottle-top filter into sterile bottles. The Ca^{2+}/Mg^{2+}-free medium is stored at 4°C until use.

4.5.2.3.2 Trypsin Stock Solution (2.5%)

Filter-sterilized trypsin (2.5%) in Hanks' buffered salts solution is purchased in 100-ml bottles and stored at –20°C. When first thawed, the trypsin solution is aliquoted (5-ml portions) into 15-ml centrifuge tubes in a laminar flow hood. Aliquots are stored at –20°C until use.

4.5.2.3.3 Trypsin Solution (0.25%)

To 40 ml of ice-cold Ca^{2+}/Mg^{2+}-free medium, sterile 7.5% sodium bicarbonate (5 ml) is added, followed by the 2.5% trypsin stock solution (5 ml). This solution is prepared just before the cells are treated and is kept on ice.

4.5.2.4   Preparation of Items for Thymidine Incorporation

Suppliers and more detailed information on the items required for measurement of thymidine incorporation are listed in Table 4.8.

4.5.2.4.1 Preparation of Cover Slips

  A. Cover slips are placed in a clean, glass rack using forceps, and remain in the rack until step H.
  B. The cover slips are lowered into a solution of chloroform and 95% ethanol (1:1) and allowed to soak for 30 min.
  C. The cover slips are rinsed in deionized water.
  D. The cover slips are lowered into a 95:5 solution of concentrated sulfuric acid (95%) and concentrated nitric acid (70%), previously prepared in a fume hood and allowed to cool to room temperature. The cover slips are soaked in this solution for 30 min.
  E. The cover slips are rinsed in deionized water.
  F. The cover slips are lowered into a solution 0.2 M NaOH and allowed to soak for 30 min.
  G. The cover slips are removed from the NaOH solution and rinsed three times in deionized water.
  H. The cover slips are removed from the rack and allowed to air dry on lint-free disposable wipers.
  I. When completely dry, the cover slips are baked for 3 h at 180°C for sterilization.

4.5.2.4.2 Preparation of [³H-methyl]-Thymidine Stock Solution

Under sterile conditions, [³H-methyl]-thymidine (2 Ci/mmol, 1 mCi/ml) is diluted to a concentration of 5 µCi/ml in sterile medium. This stock solution is aliquoted (5-ml portions) in a laminar flow hood using sterile procedures into sterile, 15-ml centrifuge tubes and stored at –20°C until use.

4.5.2.4.3 Preparation of Phosphate-Buffered Saline Solution

Phosphate-buffered saline (PBS) is prepared as follows. In a total volume of 1 l, 8 g of NaCl, 0.2 g of KCl, 1.44 g of Na_2HPO_4, and 0.24 g of KH_2PO_4 are dissolved with magnetic stirring. The pH is adjusted to 7.4 with HCl, and the solution is autoclaved for 20 min at 121°C.

4.5.2.4.4 Emulsion

Kodak NTB-2 emulsion is stored at 4°C. The emulsion must not be stored near high energy sources of radioactivity.

4.5.2.4.5 Developer and Fixer

A. Kodak D-19 developer is purchased in packets that make 1 gal when reconstituted. The entire packet is used at one time and the solution is stored in a brown bottle in the dark. The developer remains useable for 1 to 3 months. When the developer turns yellow, it is discarded.

B. Acid fixer is made and stored in the same manner as the D-19 developer.

4.5.3 Cell Culture Procedures

4.5.3.1 Cell Propagation in Serum-Supplemented Medium

To propagate adherent cells, such as WI-38 cells:

A. Fresh trypsin solution (0.25%) is prepared, placed on ice, and fresh growth medium is warmed to 37°C.

B. In a laminar flow hood, using sterile procedures, the spent growth medium is removed from the culture vessel. For flasks and bottles, the medium is removed by aspiration or decantation from the side opposite the cell growth surface. For cell culture plates and dishes, the medium is removed by aspiration from the edge of the growth surface.

C. The adherent cells are gently washed twice (4 ml[1]) with trypsin solution (0.25%).

D. Residual trypsin solution is removed by aspiration from the side opposite the cell growth surface (flasks) or from the edge of the growth surface (plates, dishes, and slides) as appropriate.

E. Enough trypsin solution (0.25%) to wet the entire cell sheet (2 ml) is added.

F. The culture vessel is tightly capped for sterility and placed at 37°C.

G. The cells will round up as they lift off the growth surface. Using the microscope, the cells are periodically checked to monitor detachment. Generally detachment is complete within 15 min. The trypsinization process may be speeded up by gently tapping the sides of the flask. Care should be taken to not splatter the cells against the top and sides of the flask, as this will lead to errors in the determination of the number of cells in the flask.

H. When all of the cells have detached from the growth surface, as determined by inspection under the microscope, the flask is returned to the laminar flow hood. Complete medium with 10% v/v FBS is carefully pipetted down the growth surface of the vessel to stop the trypsinization process and to wash down and collect the cells. For a T-75 flask, 8 ml of complete medium is used for a final harvest volume of 10 ml.

---

[1] Cells may be grown in a variety of culture vessels. Amounts given are for a T-75 flask. Proportional amounts are used for other size vessels, i.e., for a T-25 flask, one third of all of the amounts given is used.

I. Before counting the cell harvest, cell clumps are dispersed by drawing the entire suspension into a 10-ml pipette and then allowing it to flow out gently against the wall of the vessel. The process is repeated at least three times. The procedure is then repeated with a 5-ml pipette to obtain a single-cell suspension. Until the procedure becomes routine, a sample is withdrawn and examined under the microscope to ensure that a single-cell suspension has been achieved. During this process, the cells are kept on ice to inhibit cell aggregation and reattachment.

J. Using sterile procedures, an aliquot is removed from the cell suspension, then diluted into Isoton II in a counting vial. Typically, 0.5 ml of the cell suspension is diluted into 19.5 ml of Isoton II.

K. The sample is then counted with a Coulter Counter.

L. The number of cells in the harvest is calculated, and the volume of cell suspension and complete medium to add to new cell culture growth vessels are determined. (Cells normally are seeded at a density of 1×10^4 cells per cm² of cell growth surface, and the total volume of cell suspension plus complete medium added to the culture vessels is 0.53 ml per cm² of cell growth surface.)

M. In the laminar flow hood, the calculated amounts of complete medium are added to the new culture vessels.

N. Other than serum, dissolved CO_2 in equilibrium with HCO_3^- is the principal buffer system of the medium. Since CO_2 is volatile, the gas phase in the flasks are adjusted to the proper pCO_2 to maintain the pH of the medium at 7.4. Using sterile procedures in a laminar flow hood, a 5% CO_2 to 95% air mixture is passed through a sterile, cotton-filled $CaCl_2$ drying tube, through a sterile pipet, and into the gas phase of the cell culture flask with the growth surface down. As the mixture is flushed over the medium surface, the color of the medium will change from a dark red toward a red-orange. The flask is flushed until the medium no longer changes color. At this point, the gas above the medium is 5% CO_2 and the pH of the medium is 7.4. The flask is then tightly capped to prevent gas exchange with the outside environment. Cells grown in culture plates, dishes, and Lab-Tek® slides, which are not gas-tight, are not gassed, but must be grown in incubators that provide a humidified, 5% CO_2 atmosphere.

O. The cell harvest is resuspended with 10-ml and 5 ml pipettes, as before. Then, each culture vessel is inoculated to give a density of 1×10^4 cells per cm² of growth surface.

P. The culture vessel is then flushed as before for several seconds with the 5% CO_2 to 95% air mixture to replace the CO_2 lost when the vessel was opened for inoculation. The flask is then tightly capped and incubated at 37°C. Periodically, the color of the medium is examined to ensure that the seal is gas tight.

Q. The cumulative population doubling level (CPDL) at each subcultivation is calculated directly from the cell count.

Example:

A week after seeding a T-75 flask with the standard inoculum of 7.5×10^5 cells at a CPDL of 37.2, the cells are harvested. One doubling would give $2 \times 7.5 \times 10^5 = 1.5 \times 10^6$ cells. Two doublings would give $2 \times 2 \times 7.5 \times 10^5 = 3.0 \times 10^6$ cells. Three doublings would give $2 \times 2 \times 2 \times 7.5 \times 10^5 = 6.0 \times 10^6$ cells, etc. Thus, the population doubling increase is calculated by the formula: $N_H = 2^X \times N_I$ (where N_H = number of cells harvested, N_I = number of cells inoculated, and X = the population doubling increase). If for example, 9.1×10^6 cells are

harvested, then the population doubling increase may be calculated from the expression:
9.1×10^6 cells $= 2^X \times 7.5 \times 10^5$ cells.

Thus, $X \log_{10} 2 = \log_{10} (9.1 \times 10^6) - \log_{10} (7.5 \times 10^5)$,
and $X = 3.6$

The population doubling increase is added to the previous CPDL to give the new CPDL of the cell population when seeded. Thus, for this example, the new CPDL is $37.2 + 3.6 = 40.8$.

4.5.3.2 Cell Propagation in Serum-Free Medium

A. As undefined mitogens and inhibitors present in serum complicate the interpretation of cell growth response results, soybean trypsin inhibition solution (Section 4.5.2.2.5) is used instead of complete medium with 10% v/v FBS to wash and collect the cells from the growth surface of the flask. Otherwise, cells are treated exactly as described in Section 4.5.3.1, "Cell Propagation in Serum-Supplemented Medium," parts A through L.

B. The cell harvest is then washed to remove mitogens, rather than used directly to inoculate the culture flasks:

1. Under sterile conditions, the cell harvest is subject to centrifugation at $75 \times g$ for 5 min at 4°C.

2. The centrifuge tubes are placed in ice, transferred to a laminar flow hood, the supernatant is removed, and the cells are resuspended in 10 ml of incomplete serum-free growth medium (Section 4.5.2.2.1).

3. Under sterile conditions, the cells are again subject to centrifugation, the supernatant is removed, and the cells are resuspended in 10 ml of complete serum-free growth medium (Section 4.5.2.2.4).

C. An aliquot of the cell suspension (0.5 ml) is removed and the number of cells is determined with the Coulter Counter as before.

D. Cells are then treated exactly as described in Section 4.5.3.1, "Cell Propagation in Serum-Supplemented Medium," parts M through R, except that the cell growth medium used is serum-free (Section 4.5.2.2.4)

4.5.4 Replicative Life Span

Cells are considered to have reached the end of their proliferative life span when the cell number has not doubled after 4 weeks of growth and 3 weekly refeedings. The maximum proliferative capacity of the cells is determined as follows:

When cell cultures are near the end of their proliferative life span, at least four identical sister flasks are prepared at subcultivation. One flask is harvested each week. If the number of cells harvested is at least double the number inoculated, cells are subcultivated as usual. One of the sister flasks may also need to be harvested to provide enough cells for subcultivation into four flasks. If the number of cells harvested is not at least double the number inoculated, all of the sister flasks are refed by replacement of the spent medium with fresh complete medium and gassing with 5% CO_2. The following week, a second flask is harvested. If the number of cells harvested is again not at least double the number inoculated, cultures of the remaining sister flasks are refed, as before. The following week, a third flask is harvested. If the number of cells harvested is again not at least double the number inoculated, the remaining sister flask is refed, as before. The following week, a fourth flask is harvested. If the number of cells harvested is again not at least double the number inoculated, the cell culture is considered to be "phased out."

4.5.5   Saturation Density

Cultures are grown until the cells are densely packed and no mitotic figures are apparent. This ranges from 7–10 days after seeding for early passage cells, and greater than 9 days for higher passage cells. To estimate the saturation density, these confluent and quiescent cells are then harvested and counted as described in Section 4.5.3.

4.5.6   Thymidine Incorporation

4.5.6.1   *Preparation of Cell Slides*

   A. In a laminar flow hood, under sterile conditions, cells are harvested and counted in the usual manner.
   B. Cells are seeded at a density of 1×10^4 cells per cm^2 on Lab-Tek® slides or in cell culture dishes that contain cover slips that have been placed on the bottom using clean, sterile forceps and arranged so that they do not overlap one another.
   C. Immediately after seeding, the slides and dishes are placed in an incubator at 37°C in a humidified atmosphere of 5% CO_2 and 95% air.
   D. Twenty-four hours after the cells are seeded, the stock solution of [^{3}H-methyl]-thymidine (Specific activity 2 Ci/mmol; section 4.5.2.4.2) is added to the cultures to a final concentration of 0.1 mCi/ml.
   E. At appropriate time points (30 h is sufficient to achieve maximum labeling of WI-38 cells; Reference 43), the labeling medium is removed, cells are immediately washed twice with PBS (Section 4.5.2.4.3), fixed in 100% methanol for 15 min, then allowed to dry. If cells are grown on cover slips, the cover slips are removed from the dishes and placed in a clean, ceramic or glass rack using forceps prior to the wash and fix steps. If a Lab-Tek® slide is used, the plastic container and gasket are removed prior to the wash and fix steps. These procedures are done rapidly to limit damage to the cells. The cells must not be permitted to dry before they are fixed.
   F. If cover slips are used, then each cover slip is mounted with the cell surface up, on a separate microscope slide, using mounting resin. The resin is allowed to dry overnight.

4.5.6.2   *Autoradiography*

   A. The box of Kodak NTB-2 emulsion is removed from storage at 4°C and placed in a warm room at 37°C. The emulsion will liquefy in 3–4 h. The emulsion may also be melted by placing the bottle of emulsion in a 40°C water bath. About 1–1.5 h are required for this method. The bottle should not be shaken, as the resultant bubbles may cause irregularities in the emulsion thickness.

B. In the dark, the desired amount of emulsion is gently, but thoroughly mixed in a 1:1 ratio with deionized, distilled water.

C. The slide mailer previously set up in a 40°C water bath in the dark is filled (approximately 15–20 ml) with the 1:1 emulsion to water solution.

D. Each slide is dipped individually into the slide mailer. One dip is sufficient to coat the slide.

E. After dipping, each slide is placed in a standing (vertical) position in a wire test tube rack in the dark to drain off excess emulsion. The slides are allowed to dry for 30 min.

F. The dried slides are placed into a slide box with a desiccant. The box is covered and sealed with electrical tape. The box is placed inside a second light-tight container that also contains a desiccant and is sealed with electrical tape.

G. The container is placed at 4°C for 4 days.

4.5.6.3   Development of Cell Slides

A. Kodak D-19 developer and acid fixer are poured into separate large glass dishes.

B. In the dark, the slide containers are opened, and the slides are removed and placed in racks.

C. The slides are placed in developer for 5 min.

D. The slides are placed in fixer for 5 min.

E. At this point, the room light may be turned on, if desired. The slides are then gently rinsed for 15 min in cold running water.

4.5.6.4   Staining Slides

A. Harris modified hematoxylin stain (a stain that enhances nuclear visualization) is poured into glass staining dishes.

B. Slides are placed in the hematoxylin to achieve a light staining (5–10 min).

C. Slides in slide racks are placed in a Pyrex® dish containing water.

D. Slides are gently rinsed repeatedly under running water until the excess purple stain is gone.

E. Excess emulsion is rubbed from the back of each slide while the slides are damp.

F. Slides are allowed to air dry.

4.5.6.5   Counting Labeled Nuclei on Slides

A. For ease in identifying the limits of individual chambers under the microscope, if Lab-Tek® slides are used, the stain between the individual chambers is removed by scratching with the end of a paper clip or push pin.

B. There will be silver grains over nuclei where [³H-methyl]-thymidine has been incorporated into the DNA that will be readily visible under 400× magnification. Any nuclei over which there are five or more grains are considered labeled.

C. To determine the percentage of labeled nuclei, at least 400 cells are counted per cover slip or chamber using random fields. Typically, determinations are done in duplicate.

4.5.7 Microscopic Estimate of Cell Density

Often, it is desirable to obtain an estimate of cell density without harvesting the cells. The stage micrometer is used to calibrate the eyepiece micrometer and determine the diameter of the field of view for each objective and occular lens used. The area of the field of view is $\pi (D^2/4)$, where D is the diameter of the field of view.

The sample is scanned to ensure that the cells are uniformly distributed. Then, at least 400 cells are counted using random fields. Both the number of cells and the number of fields counted are recorded. The cell density is then calculated as follows:

cell density = (# of cells counted) / ((# of fields counted) · (area per field))

4.6 Summary and Conclusions

The modern era of aging research in tissue culture began over 35 years ago with the recognition by Hayflick and Moorhead that the phenomenon of *in vitro* senescence paralleled, at least in part, cell aging *in vivo*. It was the work of these investigators that led to the development of human diploid fibroblast-like cells as a model to study the mechanisms that underlie cellular senescence. In order to maintain a defined and consistent cell culture environment in which to obtain reproducible results, we present the standard techniques for subcultivation. Adherence to these methods will provide the appropriate culture conditions in which to pursue studies that examine mechanisms that regulate *in vitro* replicative senescence.

The early studies with this system began with a detailed characterization and comparison of young vs. senescent cells at a physiological and morphological level. This work revealed that the life history of fibroblasts and fibroblast-like cells includes an initial stage of outgrowth and establishment in culture, a period of vigorous proliferation that has a variable length (depending on the tissue of origin, age of donor, etc.), a period of declining proliferative vigor that includes cell death, and finally the emergence of an apparently long-lived population that is unable to proliferate in response to mitogens. A comparable history of limited proliferative life span has been described for a number of other cell types, which include glial cells, keratinocytes, vascular smooth muscle cells, endothelial cells, melanocytes, and lymphocytes.

The period in the life history which is usually referred to as the "senescent phase" is probably more complicated than was originally thought since studies with life span modulators suggest that there is a "conditionally" senescent state from which cells can be "rescued" for one or more additional rounds of DNA synthesis. Ultimately, the cells enter an "obligatory" arrested state in which only SV40 infection can reverse the block to DNA synthesis but not the block to mitosis.

Many of these studies provided the basis for a wide variety of subsequent studies which addressed possible mechanisms underlying cell senescence. These included studies on DNA repair, protein synthetic errors, chromatin structure and function, and mechanisms for modulating replicative life span. Many of the alterations which accompany senescence *in vitro* have also been demonstrated to occur *in vivo*.

There have been two major directions of research in the area of cellular senescence. The first began as studies of cell cycle kinetics, which evolved into studies of differential responses to growth factors, signal transduction mechanisms, and differential gene expression. The results show that as cells in culture age there is an increase in cell cycle time, due primarily to an increase in the G_1 phase. Cells are arrested in senescence primarily with a 2C DNA content, suggesting a block prior to DNA replication or S phase.

Young cells that are deprived of serum/growth factors respond to the addition of fresh serum (growth factors) by traversing G_0/G_1, synthesizing DNA, and dividing. Senescent cells handled in precisely the same way bind and respond (at least partially) to growth factors. They carry out many G_0/G_1 events, but do not synthesize DNA. Two noteworthy differences in the cascade of reactions that follow growth factor addition to senescent cells are an attenuated expression of the early response gene c-*fos* and a substantial reduction in the ability to translocate protein kinase C to the cell membrane. The significance of either of these changes is not understood at this writing.

Although the hallmark of senescent cells is the inability to replicate their DNA, SV40 T-antigen initiates another round of DNA replication, without completion of mitosis. Other studies have taken advantage of T-antigen as a probe of the mechanisms of proliferative arrest. These studies suggest that overcoming senescence involves overcoming two different blocks, M1, a block to DNA synthesis and M2, a block to post-DNA synthetic events [180,181].

In parallel with these studies, a second approach through genetic analysis has been used. Studies began with the observation that in heterokaryons senescence was dominant, which suggested that an inhibitor was expressed in senescent cells that prevented them from replicating. Later studies in which preparations of senescent (and quiescent young) cell mRNA were microinjected into young cells further suggested that senescence was controlled by negative regulators. The next major thrust in the evolution of these studies has been based on the search for differentially expressed genes, especially the putative antiproliferative gene(s) that may be characteristic of senescence. These studies have been performed by using various methodologies including the construction of selective subtractive libraries. To date, although several "antiproliferative" genes have been identified, these have all been associated with nongrowing cells whether nongrowth is the result of contact inhibition or serum deprivation in young cells or senescence. In contrast, at least one gene, *EPC-1*, has been shown to be expressed in young, serum-starved or contact inhibited cells (quiescent), but not senescent cells treated in the same way.

As a result of all of this research, several overriding concepts have emerged. (1) The senescent state is not merely a case of irreversible quiescence, except in a very limited operational sense. (2) There is no clear evidence that the process by which normal cells regulate proliferation through the expression of proliferation suppressors is related to the process by which senescence is regulated. Furthermore, senescent cells appear to be in a unique state which is not identical to any of the defined segments of the cell replication cycle. Senescence is characterized by a general loss of integrative function which results in cellular dysfunction leading to either cell death or acquisition of a strikingly different phenotype that is irreversibly postreplicative. Perhaps senescence in this postreplicative cell type is a process by which cell populations are prevented from unlimited proliferation and thus would have an adaptive value *in vivo*. (3) Finally, specific genes which are differentially expressed in senescent cells may be regulators of the process or, alternatively, may be the downstream targets of a higher order change which then results in a new but dysfunctional phenotype. Thus far there is no information which will allow us to interpret the meaning and role of each of these in the senescent phenotype.

References

1. Afshari, C. A., Vojta, P. J., Annab, L. A., Futreal, P. A., Willard, T. B., and Barrett, J. C., Investigation of the role of G_1/S cell cycle mediators in cellular senescence, *Exp. Cell Res.*, 209, 231, 1993.
2. Aronson, J. F., McClaskey, J. W., and Cristofalo, V. J., Human fetal lung fibroblasts: observations on origin and stability in culture, *Mech. Ageing Dev.*, 21, 229, 1983.
3. Atillasoy, E. and Holt, P. R., Gastrointestinal proliferation and aging, *J. Gerontol.*, 48, B43, 1993.
4. Bemiller, P. M. and Miller, J. E., Cytological changes in senescing WI-38 cells: a statistical analysis, *Mech. Ageing Dev.*, 10, 1, 1979.
5. Bierman, E. L., The effect of donor age on the *in vitro* lifespan of cultured human arterial smooth-muscle cells, *In Vitro Cell. Dev. Biol.*, 14, 951, 1978.
6. Blumenthal, E. J., Miller, A. C. K., Stein, G. H., and Malkinson, A. M., Serine/threonine protein kinases and calcium-dependent protease in senescent IMR-90 fibroblasts, *Mech. Ageing Dev.*, 72, 13, 1993.
7. Brandes, D., Murphy, D. G., Anton, E. B., and Barnard, S., Ultrastructural and cytochemical changes in cultured human lung cells, *J. Ultrastruct. Res.*, 39, 465, 1972.
8. Brooks, K. M., Phillips, P. D., Carlin, C. R., Knowles, B., and Cristofalo, V. J., EGF-dependent phosphorylation of the EGF receptor in plasma membranes isolated from young and senescent WI-38 cells, *J. Cell. Physiol.*, 133, 523, 1987.
9. Brooks-Frederich, K. M., Cianciarulo, F. L., Rittling, S. M., and Cristofalo, V. J., Cell cycle dependent regulation of Ca^{++} in young and senescent WI-38 cells, *Exp. Cell Res.*, 205, 412, 1993.
10. Brown, W. T., Genetic diseases of premature aging as models of senescence, *Annu. Rev. Gerontol. Geriatrics*, 10, 23, 1990.
11. Bruce, S. A., Deamond, S. F., and T'so, P. O., *In vitro* senescence of Syrian hamster mesenchymal cells of fetal to aged adult origin. Inverse relationship between *in vivo* donor age and *in vitro* proliferative capacity, *Mech. Ageing Dev.*, 34, 151, 1986.
12. Buetow, D. E., Cell numbers vs. age in mammalian tissues and organs, in, *CRC Handbook of Cell Biology of Aging*, Cristofalo, V. J., Ed., CRC Press, Boca Raton, FL, 1985, 1.
13. Campisi, J., Dimri, G., and Hara, E., Control of Replicative Senescence, in *Handbook of the Biology of Aging*, Schneider, E. L. and Rowe, J. W., Eds., Academic Press, San Diego, 1996, 121.
14. Carlin, C. R., Phillips, P. D., Brooks-Frederich, K. M., Miller, S., Knowles, B. B., and Cristofalo, V. J., Cleavage of the EGF receptor by an endogenous leupeptin-sensitive protease active in detergent extracts of senescent but not young human diploid fibroblasts, *J. Cell. Physiol.*, 160, 427, 1994.
15. Carlin, C. R., Phillips, P. D., Knowles, B. B., and Cristofalo, V. J., Diminished *in vitro* tyrosine kinase activity of the EGF receptor of senescent human fibroblasts, *Nature*, 306, 617, 1983.
16. Carrel, A. and Burrows, M. T., Cultivation of adult tissues and organs outside of the body, *JAMA*, 55, 1379, 1910.
17. Carrel, A. and Burrows, M. T., Cultivation of sarcoma outside of the body, *JAMA*, 55, 1554, 1910.
18. Carrel, A. and Burrows, M. T., On the physiochemical regulation of the growth of tissues, *J. Exp. Med.*, 13, 562, 1911.
19. Carrel, A. and Ebling, A. H., Age and multiplication of fibroblasts, *J. Exp. Med.*, 34, 599, 1921.
20. Chang, C., Phillips, P. D., Lipson, K. E., Cristofalo, V. J., and Baserga, R., Senescent human fibroblasts have a post-transcriptional block in the expression of the proliferating cell nuclear antigen, *J. Biol. Chem.*, 266, 8663, 1991.
21. Chang, Z. F. and Chen, K. Y., Regulation of ornithine decarboxylase and other cell cycle-regulated genes during senescence of IMR-90 human diploid fibroblasts, *J. Biol. Chem.*, 263, 11431, 1988.
22. Chen, K. Y., Chang, Z. F., and Liu, A. Y. C., Changes of serum-induced ornithine decarboxylase activity and putrescine content during aging of IMR-90 human diploid fibroblasts, *J. Cell. Physiol.*, 129, 142, 1986.

23. Choi, H. S., Lin, Z., Li, B. S., and Liu, A. Y. C., Age-dependent decrease in the heat-inducible, DNA sequence-specific binding activity in human diploid fibroblasts, *J. Biol. Chem.*, 265, 18005, 1990.

24. Choudhury, G. G., Sylvia, V. L., and Sakaguchi, A. Y., Decline of signal transduction by phospholipase $C_\gamma 1$ in IMR-90 human diploid fibroblasts at high population doubling levels, *FEBS Lett.*, 293, 211, 1991.

25. Chua, C. C., Geiman, D. E., and Ladda, R. L., Receptor for epidermal growth factor retains normal structure and function in aging cells, *Mech. Ageing Dev.*, 34, 35, 1986.

26. Comings, D. E. and Okada, T. A., Electron microscopy of human fibroblasts in tissue culture during logarithmic and confluent stages of growth, *Exp. Cell Res.*, 61, 295, 1970.

27. Counter, C. M., The roles of telomeres and telomerase in cell life span, *Mutation Res.*, 366, 45, 1996.

28. Cowled, P. A., Ciccarelli, C., Coccia, E., Philipson, L., and Sorrentino, V., Expression of growth arrest specific (*gas*) genes in senescent murine cells, *Exp. Cell Res.*, 211, 197, 1994.

29. Cristofalo, V. J., Metabolic aspects of aging in diploid human cells, in *Aging in Cell and Tissue Culture*, Holeckova, E. and Cristofalo, V. J., Eds., Plenum Press, New York, 1970, 83.

30. Cristofalo, V. J., Cellular senescence, factors modulating cell proliferation *in vitro*, in *INSERM*, Bourliere, F., Courtois, Y., Macieira-Coelho, A., and Robert, L., Eds., INSERM, Paris, 1973, 65.

31. Cristofalo, V. J., Cellular biomarkers of aging, *Exp. Gerontol.*, 23, 297, 1988.

32. Cristofalo, V. J. On models and the study of senescence, reflections on the state of biogerontology and a farewell, *J. Gerontol.*, 46, B207, 1991.

33. Cristofalo, V. J. and Charpentier, R., A standard procedure for cultivating human diploid fibroblast-like cells to study cellular aging, *J. Tissue Cult. Methods*, 6, 117, 1980.

34. Cristofalo, V. J., Doggett, D. L., Brooks-Frederich, K. M., and Phillips, P. D., Growth factors as probes of cell aging, *Exp. Gerontol.*, 24, 367, 1989.

35. Cristofalo, V. J., Howard, B. V., and Kritchevsky, D., The biochemistry of human cells in culture, in *Research Progress in Organic, Biological, and Medicinal Chemistry*, Gallo, U. and Santamaria, L., Eds., North-Holland, Amsterdam, 1970, 95.

36. Cristofalo, V. J. and Kritchevsky, D., Cell size and nucleic acid content in the human diploid cell line WI-38 during aging, *Med. Exp. Int. J. Exp. Med.*, 19, 313, 1969.

37. Cristofalo, V. J., Palazzo, R., and Charpentier, R. L., Limited lifespan of human fibroblasts *in vitro*, metabolic time or replications?, in *Neural Regulatory Mechanisms During Aging*, Adelman, R. C., Roberts, J., Baker, G. T., Baskin, S. I., and Cristofalo, V. J., Eds., Alan R.Liss, New York, 1980, 203.

38. Cristofalo, V. J., Parris, N., and Kritchevsky, D., Enzyme activity during the growth and aging of human cells *in vitro*, *J. Cell. Physiol.*, 69, 263, 1967.

39. Cristofalo, V. J., Phillips, P. D., Sorger, T., and Gerhard, G., Alterations in the responsiveness of senescent cells to growth factors, *J. Gerontol.*, 44, 55, 1989.

40. Cristofalo, V. J. and Pignolo, R. P., Replicative senescence of human fibroblast-like cells in culture, *Physiol. Rev.*, 73, 617, 1993.

41. Cristofalo, V. J. and Pignolo, R. P., Cell culture as a model, in *Handbook of the Physiology of Aging*, Masoro, E. J., Ed., Oxford University Press, Oxford, 1995, 53.

42. Cristofalo, V. J. and Stanulis-Praeger, B. M., Cellular senescence *in vitro*, in *Advances in Cell Culture*, Mararmorosch, K., Ed., Academic Press, New York, 1982, 1.

43. Cristofalo, V. J. and Sharf, B. B., Cellular senescence and DNA synthesis, thymidine incorporation as a measure of population age in human diploid cells, *Exp. Cell Res.*, 76, 419, 1973.

44. Danes, B. S., Progeria: a cell culture study on aging, *J. Clin. Invest.*, 50, 2000, 1971.

45. Deknudt, G. and Leonard, A., Aging and radiosensitivity of human somatic chromosomes, *Exp. Gerontol.*, 12, 237, 1977.

46. Dell'Orco, R. T., Mertens, G. B., and Kruse, P. F., Doubling potential and calendar time of human diploid cells in culture, *Exp. Cell Res.*, 84, 363, 1974.

47. De Tata, V., Ptasznik, A., and Crisofalo, V. J., Effects of the tumor promoting agent phorbol 12-myristate 13-acetate (PMA) on the proliferation of young and senescent WI-38 human diploid fibroblasts, *Exp. Cell Res.*, 205, 261, 1993.

48. Dice, J. F., Altered intracellular protein degradation in aging, a possible cause of proliferative arrest, *Exp. Gerontol.*, 24, 451, 1989.

49. Dimri, G. P., Hara, E., and Campisi, J., Regulation of two E2F-related genes in presenescent and senescent human fibroblasts, *J. Biol. Chem.*, 269, 16180, 1994.

50. Dimri, G. P., Lee, X., Basile, G., Acosta, M., Scott, G., Roskelley, C., Medrano, E. E., Linsken, M., Rubel, I., Pereira-Smith, O., Peacocke, M., and Campisi, J., A biomarker that identifies senescent human cells in culture and in aging skin *in vivo*, *Proc. Natl. Acad. Sci. U.S.A.*, 92, 9363, 1995.

51. Di Paolo, B. R., Pignolo, R. P., and Cristofalo, V. J., Overexpression of the two-chain form of cathepsin B in senescent WI-38 cells, *Exp. Cell Res.*, 201, 500, 1992.

52. Doggett, D. L., Rotenberg, M. O., Pignolo, R. J., Phillips, P. D., and Cristofalo, V. J., Differential gene expression between young and senescent quiescent WI-38 cells, *Mech. Ageing Dev.*, 65, 239, 1992.

53. Dulic, V., Kaufmann, W. K., Wilson, S. J., Tisty, T. D., Lees, E., Harper, J. W., Elledge, S. J., and Reed, S. I., p53-Dependent inhibition of cyclin-dependent kinase activities in human fibroblasts during radiation-induced G1 arrest, Cell 76, 1013, 1994.

54. Ferber, A., Chang, C. D., Sell, C., Ptasznik, A., Cristofalo, V. J., Ozer, H., Leroith, D., and Baserga, R., Failure of senescent human fibroblasts to express the IGF-I gene, *J. Biol. Chem.*, 268, 17883, 1993.

55. Fulder, S. J., The growth of cultured human fibroblasts treated with hydrocortisone and extracts of the medicinal plant *Panax ginseng*, *Exp. Gerontol.*, 12, 125, 1977.

56. Galloway, S. M. and Buckton, K. E., Aneuploidy and ageing, chromosome studies on a random sample of the population using G-banding, *Cytogenet. Cell Genet.*, 20, 78, 1978.

57. Gerhard, G. S., Phillips, P. D., and Cristofalo, V. J., EGF and PDGF-stimulated phosphorylation in young and senescent WI-38 cells, *Exp. Cell Res.*, 193, 87, 1991.

58. Giordano, T., Kleinsek, D., and Foster, D. N., Increase in abundance of a transcript hybridizing to elongation factor I alpha during cellular senescence and quiescence, *Exp. Gerontol.*, 24, 501, 1989.

59. Goldstein, S., Human genetic disorders that feature premature onset and accelerated progression of biological aging, in *The Genetics of Aging*, Schneider, E. L., Ed., Plenum Press, New York, 1978, 171.

60. Goldstein, S., Replicative senescence, the human fibroblast comes of age, *Science*, 249, 1129, 1990.

61. Goldstein, S., Moerman, E. J., and Baxter, R. C., Accumulation of insulin-like growth factor binding protein-3 in conditioned medium of human fibroblasts increases with chronologic age of donor and senescence *in vitro*, *J. Cell. Physiol.*, 156, 294, 1993.

62. Goldstein, S. and Singal, D. P., Senescence of cultured human fibroblasts, mitotic vs. metabolic time, *Exp. Cell Res.*, 88, 359, 1974.

63. Goodman, L. and Stein, G. H., Basal and induced amounts of interleukin-6 mRNA decline progressively with age in human fibroblasts, *J. Biol. Chem.*, 269, 19250, 1994.

64. Gorman, S. D. and Cristofalo, V. J., Reinitiation of cellular DNA synthesis in BrdU-selected nondividing senescent cells by simian virus 40 infection, *J. Cell. Physiol.*, 125, 122, 1985.

65. Gorman, S. D. and Cristofalo, V. J., Analysis of the G_1 arrest position of senescent WI-38 cells by quinacrine dihydrochloride nuclear fluorescence-evidence for a late G_1 arrest, *Exp. Cell Res.*, 167, 87, 1986.

66. Greenberg, S. B., Grove, G. L., and Cristofalo, V. J., Cell size in aging monolayer cultures, *In Vitro Cell. Dev. Biol.*, 13, 297, 1977.

67. Greider, C. W., Telomere length regulation, *Annu. Rev. Biochem.*, 65, 337, 1996.

68. Grove, G. L. and Cristofalo, V. J., Characterization of the cell cycle of cultured human diploid cells: Effects of aging and hydrocortisone, *J. Cell. Physiol.*, 90, 415, 1977.

69. Hara, E., Smith, R., Parry, D., Tahara, H., Stone, S., and Peters, G., Regulation of p16^{CDKN2} expression and its implications for cell immortalization and senescence, *Mol. Cell Biol.*, 16, 859, 1996.

70. Hara, E., Tsurui, H., Shinozaki, A., Nakada, S., and Oda, K., Cooperative effect of antisense-Rb and antisense-p53 oligomers on the extension of lifespan in human diploid fibroblasts, TIG-1, *Biochem. Biophys. Res. Commun.*, 179, 528, 1991.

71. Hara, E., Yamaguchi, T., Nojima, H., Ide, T., Campisi, J., Okayama, H., and Oda, K., Id-related genes encoding helix-loop-helix proteins are required for G_1 progression and are repressed in senescent human fibroblasts, *J. Biol. Chem.*, 269, 2139, 1994.

72. Harley, C. B. and Kim, N. W., Telomerase and cancer, *Imp. Adv. Oncol.*, 1, 57, 1996.

73. Hay, R. J., Menzies, R. A., Morgan, H. P., and Strehler, B. L., The division potential of cells in continuous growth as compared to cells subcultivated after maintenance in stationary phase, *Exp. Gerontol.*, 3, 35, 1968.

74. Hay, R. J. and Strehler, B. L., The limited growth span of cell strains isolated from the chick embryo, *Exp. Gerontol.*, 2, 123, 1967.

75. Hayakawa, M., Progressive changes of the growth characteristics of the human diploid cells in serial cultivation *in vitro. Tohoku J. Exp. Med.*, 98, 171, 1969.

76. Hayflick, L., The limited *in vitro* lifetime of human diploid cell strains, *Exp. Cell Res.*, 37, 614, 1965.

77. Hayflick, L., The cellular basis for biological aging, in *Handbook of the Biology of Aging*, Finch, C. E. and Hayflick, L., Eds., van Nostrand Reinhold Co., New York, 1977, 159.

78. Hayflick, L. and Moorhead, P. S., The serial cultivation of human diploid cell strains, *Exp. Cell Res.*, 25, 585, 1961.

79. Hosokawa, M., Phillips, P. D., and Cristofalo, V. J., The effect of dexamethasone on epidermal growth factor binding and stimulation of proliferation in young and senescent WI-38 cells, *Exp. Cell Res.*, 164, 408, 1986.

80. Johnson, L. F., Abelson, H. T., Penman, S., and Green, H., The relative amounts of the cytoplasmic RNA species in normal, transformed, and senescent cultured cell lines, *J. Cell. Physiol.*, 90, 465, 1976.

81. Keogh, B. P., Tresini, M., Cristofalo, V. J., and Allen, R. G., Effects of cellular aging on the induction of *c-fos* by antioxidant treatments, *Mech. Ageing Dev.*, 86, 151, 1996.

82. Krooth, R. S., Shaw, M. W., and Campbell, B. K., A persistent strain of diploid fibroblasts, *J. Natl. Cancer Inst.*, 32, 1031, 1964.

83. Kumar, S., Millis, A. J., and Baglioni, C., Expression of interleukin 1-inducible genes and production of interleukin 1 by aging human fibroblasts, *Proc. Natl. Acad. Sci. U.S.A.*, 89, 4683, 1992.

84. Kumar, S., Vinci, J. M., Millis, A. J., and Baglioni, C., Expression of interleukin-1 alpha and beta in early passage fibroblasts from aging individuals, *Exp. Gerontol.*, 28, 505, 1993.

85. Leblond, C. P., Classification of cell populations on the basis of their proliferative behavior, *Natl. Cancer Inst. Monogr.*, 14, 119, 1964.

86. Le Guilly, Y., Simon, M., Lenoir, P., and Bourel, M., Long term culture of human adult liver cells: Morphological changes related to *in vitro* senescence and effect of donor's age on growth potential, *Gerontologia*, 19, 303, 1973.

87. Levine, E. M., Mycoplasma contamination of animal cell cultures, a simple, rapid detection method, *Exp. Cell Res.*, 74, 99, 1972.

88. Lipetz, J. and Cristofalo, V. J., Ultrastructural changes accompanying the aging of human diploid cells in culture, *J. Ultrastruct. Res.*, 39, 43, 1972.

89. Liu, A. Y. C., Choi, H. S., Lee, Y. K., and Chen, K. Y., Molecular events involved in transcriptional activation of heat shock genes become progressively refractory to heat stimulation during aging of human diploid fibroblasts, *J. Cell. Physiol.*, 149, 560, 1991.

90. Liu, A. Y. C., Lee, Y. K., Manalo, D., and Huang, L. E., Attenuated heat shock transcriptional response in aging: molecular mechanism and implication in the biology of aging, *EXS* 77, 393, 1996.

91. Liu, A. Y. C., Lin, Z., Choi, H. S., Sorhage, F., and Li, B., Attenuated induction of heat shock gene expression in aging diploid fibroblasts, *J. Biol. Chem.*, 264, 12037, 1989.

92. Luce, M. C. and Cristofalo, V. J., Reduction in heat shock gene expression correlates with increased thermosensitivity in senescent human fibroblasts, *Exp. Cell Res.*, 202, 9, 1992.

93. Luce, M. C., Schyberg, J. P., and Bunn, C. L., Metallothionein expression and stress responses in aging human diploid fibroblasts, *Exp. Gerontol.*, 28, 17, 1993.

94. Lucibello, F. C., Sewing, A., Brusselbach, S., Burger, C., and Muller, R., Deregulation of cyclins D1 and E and suppression of *cdk2* and *cdk4* in senescent human fibroblasts, *J. Cell Sci.*, 105, 123, 1993.

95. Macieira-Coelho, A., Action of cortisone on human fibroblasts *in vitro*. *Experientia*, 22, 390, 1966.

96. Macieira-Coelho, A., The decreased growth potential *in vitro* of human fibroblasts of adult origin, in *Aging in Cell and Tissue Culture*, Holeckova, E. and Cristofalo, V. J., Eds., Plenum Press, New York, 1970, 83.

97. Macieira-Coelho, A. and Ponten, J., Analogy in growth between late passage human embryonic and early passage human adult fibroblasts, *J. Cell Biol.*, 43, 374, 1969.

98. Macieira-Coelho, A., Ponten, J., and Phillipson, L., The division cycle and RNA synthesis in diploid human cells at different passage levels *in vitro*, *Exp. Cell Res.*, 42, 673, 1966.

99. Martin, G. M., Genetic syndromes in man with potential relevance to pathobiology of aging, in *Genetic Effects on Aging*, Bergsma, D. and Harrison, D. E., Eds., Alan R. Liss, New York, 1978, 5.

100. Martin, G. M., Sprague, C. A., and Epstein, C. J., Replicative lifespan of cultivated human cells. Effects of donor age, tissue, and genotype, *Lab. Invest.*, 23, 86, 1970.

101. Matsumura, T., Multinucleation and polyploidization of aging human cells in culture, *Adv. Exp. Med. Biol.*, 129, 31, 1980.

102. Matsumura, T., Zerrudo, Z., and Hayflick, L., Senescent human diploid cells in culture: survival, DNA synthesis and morphology, *J. Gerontol.*, 34, 328, 1979.

103. Medrano, E. E., Yang, F., Boissy, R., Farooqui, J., Shah, V., Matsumoto, K., Nordlund, J. J., and Park, H. Y., Terminal differentiation and senescence in the human melanocyte: repression of tyrosine-phosphorylation of the extracellular signal-regulated kinase 2 selectively defines the two phenotypes, *Mol. Biol. Cell*, 5, 497, 1994.

104. Mellgren, J., Effects of the number of cell divisions and of added isologous nucleic acids on aging of normal human fibroblasts *in vitro*, *Pathol. Eur.*, 10, 215, 1975.

105. Millis, A. J., Hoyle, M., McCue, H. M., and Martini, H., Differential expression of metalloproteinase and tissue inhibitor of metalloproteinase genes in aged human fibroblasts, *Exp. Cell Res.*, 201, 373, 1992.

106. Millis, A. J., Sottile, J., Hoyle, M., Mann, D. M., and Diemer, V., Collagenase production by early and late passage cultures of human fibroblasts, *Exp. Gerontol.*, 24, 559, 1989.

107. Mitsui, Y. and Schneider, E. L., Relationship between cell replication and volume in senescent human diploid fibroblasts, *Mech. Ageing Dev.*, 5, 45, 1976.

108. Mueller, S. N., Rosen, E. M., and Levine, E. M., Cellular senescence in a cloned strain of bovine fetal aortic endothelial cells, *Science*, 207, 889, 1980.

109. Murano, S., Thweatt, R., Shmookler-Reis, R. J., Jones, R. A., Moerman, E. J., and Goldstein, S., Diverse gene sequences are overexpressed in Werner syndrome fibroblasts undergoing premature replicative senescence, *Mol. Cell Biol.*, 11, 3905, 1991.

110. Murato, S.-I., Mitsui, Y., and Kawamura, M., Effect of *in vitro* aging on 6-ketoprostaglandin $F^{1\alpha}$-producing activity in cultured human diploid lung fibroblasts, *Biochim. Biophys. Acta*, 574, 351, 1979.

111. Nielsen, P. J. and Ryan, J. M., Cumulative population doublings as the determinant of chick cell lifespan *in vitro*, *J. Cell. Physiol.*, 107, 371, 1981.

112. Nienhaus, A. J., De Jong, B., Ten Kate, L. P., and Oswald, F. H., Fibroblast culture in Werner's syndrome, *Humangenetik*, 13, 244, 1971.

113. Noda, A., Ning, Y., Venable, S. F., Pereira-Smith, O. M., and Smith, J. R., Cloning of senescent cell derived inhibitors of DNA synthesis using an expression screen, *Exp. Cell Res.*, 211, 90, 1994.

114. Norwood, T. H., Saulewicz, A., Hanaoka, F., and Pendergrass, W. R., Failure by senescent fibroblasts to complement a DNA polymerase α mutant (Abstract) Ann. Meet. Gerontol. Soc. Am. 44th San Francisco, CA, 1991, p. 354. Published in *The Gerontologist* by the Gerontological Society of America (GSA).

115. Olashaw, N. E., Kress, E. D., and Cristofalo, V. J., Thymidine triphosphate synthesis in senescent WI-38 cells. Relationship to loss of replicative capacity, *Exp. Cell Res.*, 149, 547, 1983.

116. Pang, J. H. and Chen, K. Y., Global change of gene expression at late G_1/S boundary may occur in human IMR-90 diploid fibroblasts during senescence, *J. Cell. Physiol.*, 160, 531, 1994.

117. Pendergrass, W. R., Angello, J. C., Kirschner, M. D., and Norwood, T. H., The relationship between the rate of entry into S phase, concentration of DNA polymerase α, and cell volume in human diploid fibroblast-like monokaryon cells, *Exp. Cell Res.*, 192, 418, 1991.

118. Peterson, C. and Goldman, J. E., Alterations in calcium content and biochemical processes in cultured skin fibroblasts from aged and Alzheimer donors, *Proc. Natl. Acad. Sci. U.S.A.*, 83, 2758, 1986.

119. Peterson, C., Ratan, R. R., Shelanski, M. L., and Goldman, J. E., Cytosolic free calcium and cell spreading decrease in fibroblasts from aged and Alzheimer donors, *Proc. Natl. Acad. Sci. U.S.A.*, 83, 7999, 1986.

120. Peterson, G., Coughlin, J. G., and Meylan, C., Long-term cultivation of diploid rat cells, *Exp. Cell Res.*, 33, 60, 1964.

121. Phillips, P. D. and Cristofalo, V. J., Classification system based on the functional equivalency of mitogens that regulate WI-38 cell proliferation, *Exp. Cell Res.*, 175, 396, 1988.

122. Phillips, P. D. and Cristofalo, V. J., Cell culture of human diploid fibroblasts in serum-containing and serum-free media, in *Cell Growth and Division: A Practical Approach*, Baserga, R., Ed., IRL Press, Washington, D.C., 1989, 121.

123. Phillips, P. D., Kaji, K., and Cristofalo, V. J., Progressive loss of the response of senescing WI-38 cells to platelet-derived growth factor, epidermal growth factor, insulin, transferrin, and dexamethasone, *J. Gerontol.*, 39, 11, 1984.

124. Phillips, P. D., Kuhnle, E., and Cristofalo, V. J., [^{125}I] EGF binding ability is stable throughout the replicative life span of WI-38 cells, *J. Cell. Physiol.*, 114, 311, 1983.

125. Phillips, P. D., Pignolo, R. J., and Cristofalo, V. J., lnsulin-like growth factor-1: specific binding to high and low affinity sites and mitogenic action throughout the life span of WI-38 cells, *J. Cell. Physiol.*, 133, 135, 1987.

126. Phillips, P. D., Pignolo, R. J., Nishikura, K., and Cristofalo, V. J., Renewed DNA synthesis in senescent WI-38 cells by expression of an inducible chimeric *c-fos* construct, *J. Cell. Physiol.*, 151, 206, 1992.

127. Pignolo, R. J., Masoro, E. J., Nichols, W. W., Bradt, C. I., and Cristofalo, V. J., Skin fibroblasts from Fischer 344 rats undergo similar changes in replicative life span but not immortalization with caloric restriction of donors, *Exp. Cell Res.*, 201, 16, 1992.

128. Pignolo, R. J., Rotenberg, M. O., and Cristofalo, V. J., Senescent WI-38 cells fail to express *EPC-1*, a gene induced in young cells upon entry into the G_0 state, *J. Biol. Chem.*, 268, 8949, 1993.

129. Pignolo, R. J., Rotenberg, M. O., and Cristofalo, V. J., Alterations in contact and density-dependent arrest state in senescent WI-38 cells, *In Vitro Cell Dev. Biol.*, 30, 471, 1994.

130. Plisko, A. and Gilchrest, B. A., Growth factor responsiveness of cultured human fibroblasts declines with age, *J. Gerontol.*, 38, 513, 1983.

131. Pochron, S. F., O'Meara, A. R., and Kurtz, M. J., Control of transcription in aging WI-38 cells stimulated by serum to divide, *Exp. Cell Res.*, 116, 63, 1978.

132. Ponten, J. Aging properties of glia, in *INSERM*, Bourliere, F., Courtois, Y., Macieira-Coelho, A., and Robert, L., Eds., INSERM, Paris, 1973, 53.

133. Porter, M. B., Pereira-Smith, O. M., and Smith, J. R., Novel monoclonal antibodies identify antigenic determinants unique to cellular senescence, *J. Cell. Physiol.*, 142, 425, 1990.

134. Praeger, F. C. and Cristofalo, V. J., The growth of WI-38 cells in a serum-free, growth factor-free, medium with elevated calcium concentrations, *In Vitro Cell. Dev. Biol.*, 22, 355, 1986.

135. Praeger, F. C. and Cristofalo, V. J., Modulation of WI-38 cell proliferation by elevated levels of $CaCl_2$, *J. Cell. Physiol.*, 129, 27, 1986.

136. Praeger, F. C. and Gilchrest, B. A., Influence of increased extracellular calcium concentration and donor age on density dependent growth inhibition of human fibroblasts, *Proc. Soc. Exp. Biol. Med.*, 182, 315, 1986.

137. Regan, J. D., Sigel, M. M., Lee, W. H., Llamas, K. A., and Beasley, A. R., Chromosomal alterations in marine fish cells *in vitro, Can. J. Genet. Cytol.*, 10, 448, 1968.

138. Rheinwald, J. G. and Green, H., Serial cultivation of strains of human epidermal keratinocytes, the formation of keratinizing colonies from single cells, *Cell*, 6, 331, 1975.

139. Riabowol, K., Schiff, J., and Gilman, M. Z., Transcription factor AP-1 activity is required for initiation of DNA synthesis and is lost during cellular aging, *Proc. Natl. Acad. Sci. U.S.A.*, 89, 157, 1992.

140. Rittling, S. R., Brooks, K. M., Cristofalo, V. J., and Baserga, R., Expression of cell cycle-dependent genes in young and senescent WI-38 fibroblasts, *Proc. Natl. Acad. Sci. U.S.A.*, 83, 3316, 1986.

141. Röhme, D., Evidence for a relationship between longevity of mammalian species and life-spans of normal fibroblasts *in vitro* and erythrocytes *in vivo, Proc. Natl. Acad. Sci. U.S.A.*, 78, 5009, 1981.

142. Rose, D. W., G. McCabe, J. R. Feramisco, and M. Adler. Expression of c-fos and AP-1 activity in senescent human fibroblasts is not sufficient for DNA synthesis. J. Cell Biol. 119, 1406-1411, 1992.

143. Rowe, D. W., Starman, B. J., Fujimoto, W. Y., and Williams, R. H., Differences in growth response to hydrocortisone and ascorbic acid by human diploid fibroblasts, *In Vitro Cell. Dev. Biol.*, 13, 824, 1977.

144. Ryan, J. M. and Cristofalo, V. J., Chromatin template activity during aging in WI-38 cells, *Exp. Cell Res.*, 90, 456, 1975.

145. Salk, D., Bryant, E., Hoehn, H., Johnson, P., and Martin, G. M., Growth characteristics of Werner syndrome cells *in vitro, Adv. Exp. Med. Biol.*, 190, 305, 1985.

146. Schneider, E. L. and Fowlkes, B. J., Measurement of DNA content and cell volume in senescent human fibroblasts utilizing flow multiparameter single cell analysis, *Exp. Cell Res.*, 98, 298, 1976.

147. Schneider, E. L. and Mitsui, Y., The relationship between *in vitro* cellular aging and *in vivo* human age, *Proc. Natl. Acad. Sci. U.S.A.*, 73, 3548, 1976.

148. Schneider, E. L., Mitsui, Y., Tice, R., Shorr, S. S., and Braunschweiger, K., Alteration in cellular RNAs during the *in vitro* lifespan of cultured human diploid fibroblasts. II. Synthesis and processing of RNA, *Mech. Ageing Dev.*, 4, 449, 1975.

149. Schneider, E. L. and Shorr, S. S., Alteration in cellular RNAs during the *in vitro* lifespan of cultured human diploid fibroblasts, *Cell*, 6, 179, 1975.

150. Sell, C., Ptasznik, A., Chang, C. D., Swantek, J., Cristofalo, V. J., and Baserga, R., IGF-I receptor levels and the proliferation of young and senescent human fibroblasts, *Biochem. Biophys. Res. Commun.*, 194, 259, 1993.

151. Seshadri, T. and Campisi, J., Repression of *c-fos* transcription and an altered genetic program in senescent human fibroblasts, *Science*, 247, 205, 1990.

152. Seshadri, T., Uzman, J. A., Oshima, J., and Campisi, J., Identification of a transcript that is down-regulated in senescent human fibroblasts, *J. Biol. Chem.*, 268, 18474, 1993.

153. Simons, J. W. I. M. and van den Broek, C., Comparison of ageing *in vitro* and ageing *in vivo* by means of size analysis using a Coulter counter, *Gerontologia*, 16, 340, 1970.

154. Slayback, J. R. B., Cheung, L. W. Y., and Geger, R. P., Comparative effects of human platelet growth factor on the growth and morphology of human fetal and adult diploid fibroblasts, *Exp. Cell Res.*, 110, 462, 1977.

155. Smith, B. T., Torday, J. S., and Giroud, C. J. P., The growth promoting effect of cortisol in human fetal lung cells, *Steroids*, 22, 515, 1973.

156. Smith, J. R., Pereira-Smith, O., and Schneider, E. L., Colony size distribution as a measure of *in vivo* and *in vitro* aging, *Proc. Natl. Acad. Sci. U.S.A.*, 75, 1353, 1978.

157. Sottile, J., Mann, D. M., Diemer, V., and Millis, A. J., Regulation of collagenase and collagenase mRNA production in early- and late-passage human diploid fibroblasts, *J. Cell. Physiol.*, 138, 281, 1989.

158. Soukupova, M., Holeckova, E., and Hnevkovsky, P., Changes of the latent period of explanted tissues during ontogenesis, in *Aging in Cell and Tissue Culture*, Holeckova, E. and Cristofalo, V. J., Eds., Plenum Press, New York, 1970, 41.
159. Stanley, J., Pye, D., and Macgregor, A., Comparison of doubling numbers attained by cultured animal cells with life span of species, *Nature*, 255, 158, 1975.
160. Stein, G. H., Beeson, M., and Gordon, L., Failure to phosphorylate retinoblastoma gene product in senescent human fibroblasts, *Science*, 249, 666, 1990.
161. Stein, G. H., Drullinger, L. F., Robetorye, R. S., Pereira-Smith, O. M., and Smith, J. R., Senescent cells fail to express *cdc2*, *cyc A*, and *cyc B* in response to mitogen stimulation, *Proc. Natl. Acad. Sci. U.S.A.*, 88, 11012, 1991.
162. Steinhagen, M., Effect of donor age on clonal differentiation of human skin fibroblasts *in vitro*, *Gerontology*, 31, 27, 1985.
163. Swim, H. E. and Parker, R. F., Culture characteristics of human fibroblasts propagated serially, *Am. J. Hyg.*, 66, 235, 1957.
164. Tahara, H., Kamada, K., Sato, E., Tsuyama, N., Kim, J. K., Hara, E., Oda, K., and Ide, T., Increase in expression levels of interferon-inducible genes in senescent human diploid fibroblasts and in SV40-transformed human fibroblasts with extended lifespan, *Oncogene*, 11, 1125, 1995.
165. Tassin, J., Malaise, E., and Courtois, Y., Human lens cells have an *in vitro* proliferative capacity inversely proportional to the donor age, *Exp. Cell Res.*, 123, 388, 1979.
166. Tice, R. R., Schneider, E. L., Kram, D., and Thorne, P., Cytokinetic analysis of impaired proliferative response of peripheral lymphocytes from aged humans to phytohemagglutinin, *J. Exp. Med.*, 149, 1029, 1979.
167. Valenti, C. and Friedman, E. A., Long-term cultivation of diploid rabbit skin cells, *Tex. Rep. Biol. Med.*, 26, 363, 1968.
168. Venable, M. E., Blobe, G. C., and Obeid, L. M., Identification of a defect in the phospholipase D/diacylglycerol pathway in cellular senescence, *J. Biol. Chem.*, 269, 26040, 1994.
169. Wadhwa, R., Kaul, S. C., Sugimoto, Y., and Mitsui, Y., Induction of cellular senescence by transfection of cytosolic mortalin cDNA in NIH 3T3 cells, *J. Biol. Chem.*, 268, 22239, 1993.
170. Wang, E., A 57,000-mol-wt. protein uniquely present in nonproliferating cells and senescent human fibroblasts, *J. Cell Biol.*, 100, 545, 1985.
171. Wang, E. and Tomaszewski, G., Granular presence of terminin is the marker to distinguish between the senescent and quiescent states, *J. Cell. Physiol.*, 147, 514, 1991.
172. Wang, S. M., Phillips, P. D., Sierra, F., and Cristofalo, V. J., Altered expression of the *twist* gene in young versus senescent human diploid fibroblasts, *Exp. Cell Res.*, 228, 138, 1996.
173. Weissman-Shomer, P. and Fry. M., Chick embryo fibroblast senescence *in vitro*: pattern of cell division and life span as a function of cell density, *Mech. Ageing Dev.*, 4, 159, 1975.
174. West, M. D., Shay, J. W., Wright, W. E., and Linskens, M. H. K., Altered expression of plasminogen activator and plasminogen activator inhibitor during cellular senescence, Exp. Gerontol., 31, 175, 1996.
175. West, M. D., Pereira-Smith, O. M., and Smith, J. R., Replicative senescence of human skin fibroblasts correlates with a loss of regulation and overexpression of collagenase activity, *Exp. Cell Res.*, 184, 138, 1989.
176. Wick, M., Burger, C., Brusselbach, S., Lucibello, F. C., and Muller, R., A novel member of human tissue inhibitor of metalloproteinases (TIMP) gene family is regulated during G_1 progression, mitogenic stimulation, differentiation and senescence, *J. Biol. Chem.*, 269, 18953, 1994.
177. Williams, J. R. and Dearfield, K. L., Nonhuman fibroblast-like cells in culture, in *CRC Handbook of Cell Biology of Aging*, Cristofalo, V. J., Ed., CRC Press, Boca Raton, FL, 1985, 433.
178. Wistrom, C. and Villeponteau, B., Cloning and expression of SAG, a novel marker of cellular senescence, *Exp. Cell Res.*, 199, 355, 1992.
179. Won, K.-A., Xiong, Y., Beach, D., and Gilman, M. Z., Growth-regulated expression of D-type cyclin genes in human diploid fibroblasts, *Proc. Natl. Acad. Sci. U.S.A.*, 89, 9910, 1992.

180. Wright, W. E., Pereira-Smith, O. M., and Shay, J. W., Reversible cellular senescence: implications for a two-stage model for the immortalization of normal human diploid fibroblasts, *Mol. Cell. Biol.*, 9, 3088, 1989.
181. Wright, W. E. and Shay, J. W., The two-stage mechanism controlling cellular senescence and immortalization, *Exp. Gerontol.*, 27, 383, 1992.
182. Yaffe, D., Retention of differentiation potentialities during prolonged cultivation of myogenic cells, *Proc. Natl. Acad. Sci. U.S.A.*, 61, 477, 1968.
183. Yanishevsky, R. and Carrano, A. V., Prematurely condensed chromosomes of dividing and non-dividing cells in aging human cell cultures, *Exp. Cell Res.*, 90, 169, 1975.
184. Yanishevsky, R., Mendelsohn, M. L., Mayall, B. H., and Cristofalo, V. J., Proliferative capacity and DNA content of aging human diploid cells in culture, a cytophotometric and autoradiographic analysis, *J. Cell. Physiol.*, 84, 165, 1974.
185. Zambetti, G., Dell'Orco, R., Stein, G., and Stein, J., Histone gene expression remains coupled to DNA synthesis during *in vitro* cellular senescence, *Exp. Cell Res.*, 172, 397, 1987.

5

Electrophysiological Assessment of the Aged Hippocampus In Vivo and In Vitro

Gregory M. Rose, Karen L. Heman, Gowri K. Pyapali, and Dennis A. Turner

CONTENTS

5.1   Introduction

5.1.1   Characteristics of CNS and Hippocampal Aging

Aging is a process that causes a variety of biochemical, cellular, and immunological changes in an organism, partially in response to lifelong exposure to environmental insults, stress, genetic mutations, and various other factors [1–6]. Among the effects of aging are a number of changes in the central nervous system (CNS), which can be variable from individual to individual, but which often result in impaired behavioral function. Age-related changes in the CNS have been reported to include regional loss of neurons or their connections, structural modifications in dendrites, alterations in neuronal excitability and synaptic function, and differences in the intracellular handling of calcium by neurons (reviewed in [7–13]).

One of the consequences of aging is a highly variable degree of cognitive deficits in older individuals. The appearance of such cognitive problems, particularly in the areas of learning and memory, has focused research upon age-related changes in the function of the hippocampus, a brain structure that plays a critical role in encoding new information into memory [14, 15]. A large number of morphological, biochemical, and electrophysiological differences between the hippocampus of young adults and aged individuals have been reported. (See [16–23] for recent reviews.)

5.1.2   Role of Electrophysiological Studies In Aging Research

Electrophysiological techniques can provide useful information at a variety of levels, from the individual neuron to overall circuit functioning. Studies of cell firing patterns reveal when brain structures are active during particular behaviors, and characterizing the behavioral correlates of individual neurons can give insights as to what kind of information is being processed (e.g., hippocampal "place" cells [24]). Extracellular field potential recordings provide information about synaptic responses in groups of neurons. Extracellular recording methods can be used to examine plasticity mechanisms, and, in combination with drug application, to evaluate the status and function of neurotransmitter receptors. Such studies are often the groundwork for intracellular and whole cell recording methods, which are used to investigate single cell biophysical properties, provide a window into the functioning of various membrane channels, and furnish information about the structural organization of a neuron or neuronal circuit.

A specific rationale for using electrophysiological techniques for evaluating age-related changes in neuronal circuitry is that, by their very nature, these methods select for neurons that are alive and are therefore still contributing to the function of a particular brain region. Information gathered from functioning neurons is more likely to provide insights into the status of neuronal circuits within the whole animal than will data obtained using less discriminating or selective neuroanatomical or biochemical techniques. In addition, studying functioning neurons in the aged brain is valuable in itself, in that it may reveal characteristics or mechanisms by which these cells have maintained their viability in spite of the aging process. However, it is important to note that electrophysiological studies of the aged nervous system are difficult, in that the neurons are more fragile. This property increases the likelihood that the invasive techniques used to obtain electrophysiological data may, in themselves, alter the characteristics of the neurons or circuits being studied.

In our opinion, the most powerful application of electrophysiological techniques to the study of aging involves their use in combination with other methods that can provide information about the functional consequences of age-related changes in electrophysiological parameters. For example, age-related studies evaluating both behavioral function and cellular characteristics in the same animals can provide critical information about brain–behavior relationships in aging [25–28]. Intracellular injections of substances, like HRP, biocytin, or neurobiotin, through recording electrodes allow visualization of the structure of electrophysiologically characterized neurons [29–32] and can furnish insights into the relationship between age-related structural alterations and electrophysiological parameters [28,33]. Thus, a fruitful approach with aging has been to, first, measure some aspect of behavior relevant to the region under study to determine how well a particular brain region or circuit is working in the intact brain. Studies then can be performed in the behaving animal that provide direct correlative evidence between electrophysiological parameters and behavioral function. Subsequent examination of the brain region *in vitro* can reinforce conclusions derived from intact animal work and can go further to reveal cellular or subcellular mechanisms of age-related dysfunction.

5.1.3   Advantages of Electrophysiology for Assessment of Age-Related Changes in the Hippocampus

It has been difficult to relate the diverse changes observed in the aged hippocampus to each other, much less to the behavioral dysfunction that the studies seek to ultimately explain. It is unlikely that any single approach will provide the answer to how age-related changes in the hippocampus cause learning and memory impairments. However, electrophysiological techniques provide a relatively high level of assessment of hippocampal function and have the particular advantage that they may be employed in both reduced preparations and, in some forms, in behaving animals. Thus, the breadth of electrophysiological methods available makes it possible to directly correlate age-related changes in neuronal firing or circuit interactions to behavior and then to identify the mechanisms involved in the changes at the cellular or subcellular (e.g., ionic) level.

5.2   Background

5.2.1   Brief History of Hippocampal Electrophysiology

Electrophysiological methods have been applied to the central nervous system since 1875, when Robert Caton first reported electroencephalographic activity in animals [34]. However, general interest in the electroencephalogram did not increase until the seminal studies of Hans Berger were published, beginning in 1929 [35]. Although recordings were made directly from the hippocampus as early as 1940 [36], special features of hippocampal electrophysiology (notably hippocampal theta rhythm) were not identified until the decade of the 1950s [37]. Later work [38–40], particularly the contributions of Andersen [41] and Spencer and colleagues [42], provided a thorough electrophysiological characterization of hippocampal neurons and the actions of their intrinsic and afferent pathways. While this initial work was performed in anesthetized animals, later technical modifications permitted data to be obtained from the hippocampus of unanesthetized, behaving subjects [43-45].

Understanding of the biophysical properties of hippocampal neurons and intrinsic hippocampal circuitry has benefited enormously from the development of the *in vitro* hippocampal slice preparation [46,47]. This method has gained wide popularity and is now the predominant preparation for studying hippocampal electrophysiology. However, since *in vivo* and *in vitro* electrophysiological studies continue to contribute valuable information about the aged hippocampus, methods for both techniques are presented here.

5.2.2   Advantages and Disadvantages of *In Vivo* and *In Vitro* Recording Techniques

In vivo recordings offer the only possibility to directly observe the relationship between neuronal firing and behavior. In addition, *in vivo* recordings are necessary to understand the relationships between distant brain structures and have provided valuable information about the function of connections between regions. Although very difficult to perform, intracellular recordings made *in vivo* have proven invaluable to visualizing complete dendrite morphology, axonal arborization, and the projections of individual neurons in the brain [29,30,32]. A major disadvantage of *in vivo* recordings is that they are done "blind," i.e., recording and stimulation sites cannot be verified until the experiment is over. Another disadvantage is that pharmacological studies performed *in vivo* are complicated by the inability to precisely control drug concentrations at the recording site, or even to know what they are. Further, with *in vivo* recordings it is possible to observe electrophysiological effects of a systemically administered drug for which there is no local or direct site of action, because of the agent's influence on a different structure that provides an input to the region under study.

In vitro brain slice preparations are commonly used for experiments that are difficult or impossible to perform *in vivo*. Slices are more accessible, in both physical and biochemical senses, than the intact brain: cell layers can be visualized with minimal magnification, and electrodes can be precisely placed in the axonal tracts, synaptic termination regions, or in cell body layers. Since slice tissue has a minimal diffusion barrier, and no blood–brain barrier, it is possible to control the composition of the extracellular fluid and to make precise pharmacological evaluations merely by adding drugs to the support medium in known concentrations. It is also feasible to influence the intracellular compartment of neurons by using membrane-permeable compounds or by direct intracellular injection. Slices are freed from the pulsation of tissue that exists *in vivo*, which facilitates obtaining stable intracellular recordings. However, this enhanced accessibility and control comes at the cost of decreased synaptic connections and the loss of extrinsic inputs. Fortunately, intrinsic connections between various regions of the hippocampus are relatively well preserved and, along with fibers from some extrinsic afferents, can be discretely activated with electrical stimulation.

5.2.3   Choice of Experimental Subject

Rats and mice are both commonly used for aging research. In the U.S., aged animals of well-defined strains of both species are maintained by specific vendors via a contractual arrangement with the National Institute on Aging. For electrophysiological studies, the choice of subject is usually dictated by technical constraints. For example, the size of the rat brain makes precise stereotaxic placement of electrodes easier, and the thickness of the rat skull provides a firmer foundation for anchoring a headpiece for chronic recording. In addition, complex behavioral studies are more easily performed with rats. On the other hand, the large number of inbred mouse strains available makes genetically oriented studies of aging processes possible, and the mouse is the animal of choice for approaches involving molecular biological manipulations such as "knockouts" or gene transfer [48–51]. In addition, mice respond much more dramatically to treatments that prolong the life span, such as dietary restriction [52,53].

In our experience, mice make poor subjects for *in vivo* electrophysiological studies of the hippocampus for the following reasons: (1) stable anesthesia, particularly necessary for acute recording studies, is difficult to maintain; (2) consistent placement of electrodes for stimulating hippocampal afferent pathways requires great precision; (3) small skull size complicates the geometry of multiple electrode placements; (4) the skull itself is thin and fragile, making it difficult to permanently attach headpieces for chronic recording; and (5) the small size of the animal makes it more likely that hardware attached for chronic recording will interfere with normal behavior. In spite of these problems, some investigators successfully use mice for their *in vivo* electrophysiological work [54–56]. However, we suggest that electrophysiological studies of the mouse hippocampus can be better performed *in vitro*. The results of such studies can be validly correlated with behavioral results previously obtained on the same animals [25,26]. In addition, the small size of the mouse brain offers the possibility to study connected brain regions in a single slice.

5.3 General Description

5.3.1 *In Vivo* Recording

In vivo recordings are primarily used to obtain two types of data: (1) information about hippocampal neuron firing; or (2) information about the status of afferent inputs or intrinsic hippocampal connections. Both these types of data can be acquired using extracellular recording techniques. Intracellular techniques are necessary to make true biophysical measurements or to correlate such information with the morphology of individual hippocampal neurons and are more easily accomplished using the *in vitro* hippocampal slice preparation. This section will focus on the tools and techniques used for making extracellular recordings from the hippocampus of either anesthetized or unanesthetized behaving rats.

5.3.1.1 *Recording Electrodes*

Many different materials have been used to fabricate probes for recording from the hippocampus, including glass capillaries, carbon fibers, and wire. The type of probe used depends to a certain extent upon the type of information sought. For recording hippocampal EEG or field potentials, any probe with a diameter of less than 100 μm will provide an adequate signal and spatial resolution. Since micropipettes made from glass capillaries are usually filled with concentrated solutions of electrolytes (e.g., NaCl), it is important that the tip diameter not be larger than a few microns since diffusion gradients will cause rapid efflux of ions into the brain. For this reason, as well as because of friability and evaporative problems, fluid-filled electrodes are not well suited for chronic recordings. Newly available small (20–100 μm) diameter carbon fibers make stable recording probes, but insulating them and providing connections to recording equipment can be troublesome. Thus, our feeling is that small diameter wires offer the greatest flexibility and ease of use for making extracellular recordings from the hippocampus.

There are two options for fabricating recording electrodes from metal wire. First, a relatively large (100–250 μm) diameter rod may be etched to a fine (1–5 μm) tip, then insulated.

The electrode material is usually tungsten or stainless steel. Flame-etching using an oxy-acetylene torch is by far the fastest and most convenient technique for preparing fine-tipped electrodes [57]. Tungsten is superior in terms of stiffness (allowing finer diameters to be used) and recording characteristics (lower noise), but stainless steel offers the advantage of being able to easily mark the location of the tip of the recording electrode. This is accomplished by passing a small (50 μA) anodal current through the electrode to deposit ferrous ions at the recording site, which are later visualized in histological sections using the Prussian Blue reaction [58]. The presence of a blue-green spot in the brain makes identifying the approximate recording site much easier. The insulating material for these electrodes is usually varnish (e.g., Epoxylite) or glass, although other materials are used in some commercially available electrodes. Our experience is that Epoxylite is the best material for electrodes manufactured in the laboratory, because it is stable, tough, and easy to use. With the proper equipment, glass insulation is faster to apply, but adjustment of impedance (a function of the size of the recording surface at the electrode tip) is more difficult, and the impedance of electrodes insulated with glass tends to continuously lower due to creep of the glass back from the electrode tip.

A second option for metal recording electrodes is to use previously insulated, fine-diameter wire directly from the spool. Again, although a choice of materials is available, stainless steel is recommended because of its adequate stiffness, ability to mark the electrode tip position, and reasonable cost. Of all insulating materials, Teflon is the best in terms of its reliability. The potential disadvantage of Teflon insulation is its thickness, but this usually does not present a practical problem. Electrodes made from fine wire are prepared by simply cutting the wire with a good pair of scissors. The electrode tip is blunt, the insulation leaving only the cross section of the wire exposed as the recording surface. An advantage of this type of electrode is that it is very easy to make, and the recording characteristics are very consistent. A disadvantage of this type of electrode is that the only way to alter the recording characteristics is to change the diameter or type of wire used. For recordings from the hippocampus, this problem is relevant primarily for studies of the activity of single neurons. The characteristics of the probe will determine the cell type most often recorded. In our experience, dentate granule cells are hard to find with cut wire electrodes.

Is there a best recording electrode for a given situation? For acute studies in anesthetized animals, electrolyte-filled glass micropipettes are usually the probe of choice. Control of tip characteristics (e.g., shape and resistance) is relatively easy with modern microelectrode pullers. An additional advantage of glass microelectrodes is that the resistance can be reduced to the desired value by, after filling, gently bumping the tip against a solid surface to slightly enlarge its diameter. (This operation is best performed using a microscope for visual guidance.) A major advantage of glass micropipettes is that they easily penetrate brain tissue, so they neither compress the tissue as they are lowered nor drag excessively as they are raised. The major disadvantage of these electrodes is that they can become plugged with blood or tissue as they are lowered into the brain, changing their resistance and thereby altering their recording characteristics. For chronic studies in unanesthetized animals, cut wire electrodes are recommended (with due consideration given to the caveats mentioned in the preceding paragraph). Cut wire electrodes have the advantage that they have very uniform impedances, which facilitates comparisons of data obtained from different recording sites, either within a given animal or across animals. The disadvantage of cut wire electrodes is that they compress the brain as they are being lowered, so they tend to settle in a location different from their initial placement. In addition, these electrodes drag substantially when they are being raised, so that it is impossible to accurately replace an electrode at a given recording site based upon depth criteria alone.

5.3.1.2 Stimulating Electrodes

The hippocampus is well suited for extracellular studies of intrinsic and extrinsic inputs using electrical stimulation. Afferents to either the dentate gyrus or hippocampus proper can be selectively studied if stimulating and recording electrodes are appropriately placed. Stimulation is accomplished by completing a circuit, from an active electrode to an indifferent electrode, such that the current path activates the input of interest. For extracellular studies, it is usually desirable to activate a relatively large fiber bundle to generate population field EPSPs. Stimulating electrodes made from metal wire are well suited for this purpose. We recommend using cut wire electrodes made of Teflon-insulated stainless steel because they are easy to make and allow the opportunity to mark the electrode tip position.

The diameter of the wire depends upon whether so-called "monopolar" or "bipolar" electrodes are used (shown in Figure 5.1). Twisting the active and indifferent electrode (usually 75 μm diameter) wires together produces a bipolar electrode. An advantage of bipolar electrodes is that they are very rigid, and, since the current path between the two electrodes is physically short, the brain area stimulated is small. Usually bipolar electrodes are trimmed to length after the wires are twisted together, so that the tips are even. When the electrodes are prepared this way it is very important to check (using an ohmmeter) that the leads are not accidentally making electrical connection at the tips. To increase the distance between the electrodes, thereby extending the current path, one lead can be rotated past the other until the desired tip separation is achieved. Another advantage of bipolar electrodes is that the direction of current flow (i.e., the polarity of the electrodes) can be switched to optimize the efficacy of stimulation and minimize the stimulus artifact. The disadvantages of bipolar electrodes are that their larger size and uneven surface tends to drag and compress brain tissue, so that the electrode takes some time to settle and is difficult to retract if moved beyond the optimum site, and that these electrodes do enough damage that it is usually not possible to make more than a single penetration into the region of interest.

For monopolar stimulation, the active electrode is made from a single strand of cut wire that must be thick enough to stay straight while being lowered into the brain. While 75-μm diameter Teflon-insulated stainless wire is adequate for this purpose, we have found that 125-μm wire is generally a better choice. When using the monopolar stimulation technique, the current path is completed through an indifferent electrode that is a wire placed in a hole drilled in the skull at a distant site. (A screw with a lead attached to it that is screwed into the skull can also serve as the indifferent electrode; see Figure 5.1.) Placement of the indifferent electrode is somewhat a matter of convenience, but, to prevent unacceptably large stimulus artifacts, it is important that the current path for stimulation not directly overlap with the path between the recording electrode and its indifferent (or ground). For monopolar stimulation the active electrode must be the cathode (negative) in order for the stimulation to be effective. An advantage of monopolar stimulation is that the single electrode penetrates the brain more easily and may be replaced several times, if necessary, without creating unacceptable damage. A potential disadvantage of monopolar stimulation is that the longer current path can produce larger artifacts and must necessarily result in less discrete stimulation (including activation of head musculature at very high currents). In addition, it is not possible to compensate for electrode drift by reversing stimulation polarity as with bipolar electrodes.

5.3.1.3 Microdrives

Proper placement of the recording electrode is usually the most important element of the recording process. For acute (nonsurvival) preparations, accurate electrode placement is

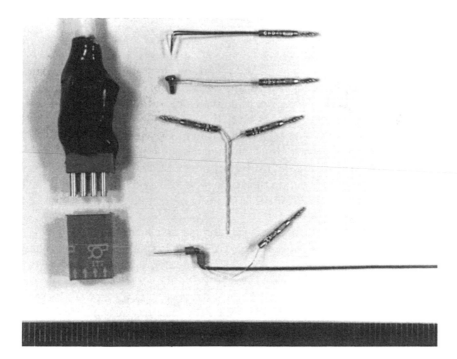

FIGURE 5.1
Hardware and electrodes for *in vivo* recording. On the left side of the figure is shown the end of the recording cable containing the preamplifier (top) that connects to the headpiece (bottom) which has been cemented to the rat's skull. A picture of the cable in use is shown in Figure 5.3. The connectors are made from Cannon Centiloc strips that are designed to hold gold-plated male and female pins. Female pins are used on the recording cable. Male pins are crimped to the electrodes and inserted into the headpiece during the implantation surgery. On the right side of the figure are shown (top to bottom): a ground electrode with two leads, a screw for brain surface recording or to serve as a stimulation indifferent, a twisted bipolar stimulator and a monopolar stimulator. The monopolar stimulation electrode is attached to a wire armature using small pieces of heat shrink tubing. The Z-bend in the armature provides surface for the dental cement, and is necessary to prevent the electrode from being pulled out of position when the lead pin is inserted into the headpiece. The portion of the armature above the electrode lead (right side) is cut off after the electrode is cemented in place. The scale at the bottom of the figure is in millimeters.

accomplished using a stereotaxic apparatus, frequently supplemented with an hydraulic drive apparatus to provide finer control of movement. This equipment is also used for placing electrodes in the hippocampus during the surgery done to prepare for subsequent recording in the unanesthetized, behaving state. Unfortunately, recording electrodes drift from their original locations during the postsurgical recovery period. This means that an electrode placed in the CA1 pyramidal cell layer during surgery will have moved up to several hundred microns (as indicated by stimulus-evoked field potentials) by the next day. For recordings of hippocampal EEG, this movement may not be important. Similarly, under certain conditions where field potentials having a broad distribution are being studied, such as the negative dendritic potentials in stratum radiatum evoked by commissural stimulation (e.g., [59]), or the positive somatic potentials in the dentate hilus evoked by perforant path stimulation (e.g., [60]), some electrode movement will not significantly affect the quality of the response. However, for recording from single neurons in the hippocampal pyramidal or dentate granule cell layers, or from narrow fields, such as the dendritic zone of input from the medial entorhinal cortex in the dentate molecular layer, it is necessary to relocate the recording electrode after surgery to optimize the response. A miniature head-mounted microdrive is used for this purpose.

Designs for a number of microdrives suitable for use with rats have been published [61–78]. The best of these devices incorporate the following elements: (1) they are small and lightweight, so that they do not interfere with normal behavior; (2) the recording electrode can be moved without rotating it, to minimize tissue damage; and (3) they maintain the recording electrode in a stable position once it has been moved to the site of interest. Other features that distinguish various types of microdrives are the number and type of electrodes they carry and whether they may be removed to change electrodes.

We have found that implanting a microdrive to record from the hippocampus of behaving rats is advantageous for every kind of preparation and is a necessity for recording cellular activity or field EPSPs from the hippocampal pyramidal or dentate granule cell layers. We use two different types of microdrives, depending upon the recording situation (shown in Figure 5.2). One type of microdrive is removable and can be used with either glass or metal microelectrodes [61]. The advantage of this microdrive is that the electrode can be changed as often as necessary to maintain optimal recording characteristics. This is important because the impedance of Expoylite-insulated metal microelectrodes with finely etched tips tends to drop after several days in the brain. In addition, we have found that relatively small differences in impedance (as little as 0.3 MΩ) are important in determining whether hippocampal pyramidal cells or dentate granule cells are isolated. Another advantage of the removable microdrive is that it can be used with glass electrodes filled with

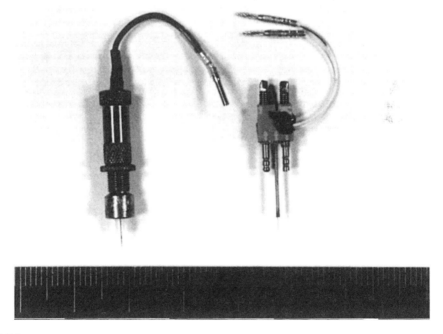

FIGURE 5.2

Two types of microdrives for *in vivo* recording. The microdrive on the left, more fully described by Deadwyler et al. [61] contains a single replaceable microelectrode. The microdrive on the right, patterned after the design of McNaughton et al. [72], contains a pair of 50-μm diameter, Teflon-insulated, stainless steel wires. The tips of the two wires are spaced approximately 275 μm apart to allow simultaneous recording from the CA1 pyramidal cell layer and the stratum radiatum apical dendritic zone. The microdrive on the left is screwed onto a base that has been cemented to the top of the rat's skull, while legs of the microdrive on the right are directly embedded in the dental acrylic applied to the top of the skull during the implant. To adjust electrode position, the cylinder on the left microdrive is turned by hand, while the legs of the microdrive on the right are turned with a slotted tool that fits over the flanges ground on the tops of the legs. The scale at the bottom of the figure is in millimeters.

solutions to mark recording sites or label the neuron being recorded. However, because of differences in osmolarity between the filling solution and the brain, such electrodes must be removed at the end of the daily recording session or the brain tissue will be irreversibly damaged. Finally, a removable microdrive should be used for *in vivo* electrochemical studies because the probes used for this work deteriorate after prolonged contact with brain tissue [75,79]. The disadvantages of using a removable microdrive are that: (1) significant time and effort must be expended to repeatedly place the recording electrode at the appropriate site; (2) there is the possibility that repeated penetrations into the same region will damage the tissue, thereby altering the electrophysiology; (3) extra handling of the animal may induce stress, influencing subsequent behavior or electrophysiology; and (4) in our experience, recordings made using a removable microdrive are not as stable as when the microdrive has been fixed to the skull. For example, we have found that it is rare to be able to record from a single hippocampal pyramidal neuron for more than a one day when using the removable drive, whereas we and others [72,80] have had stable recordings for many days using a fixed microdrive.

Thus, the use of a fixed microdrive is recommended for long-term recordings of neuronal activity or stimulus-evoked potentials. In both types of experiments, it is often advantageous to use more than a single recording electrode. Our standard fixed microdrive for extracellular field potential recording in hippocampal area CA1 contains 2 electrodes made from 50 μm diameter, Teflon-insulated stainless steel wire. The tips of the electrodes are spaced approximately 275 μm apart to permit simultaneous recording of somatic and dendritic responses to commissural stimulation. This microdrive, shown in Figure 5.2, is a modified version of the design described by McNaughton et al.[72]. The electrodes are protected outside the brain by a stainless steel guide cannula. Two legs, made from 1/72-threaded, 1/8-inch diameter stainless steel rod, are turned to adjust the position of the recording electrodes. (This process is further described below.) Detailed instructions for making this microdrive will not be presented here, but are available from the authors (G.M.R. and K.L.H.). An advantage of this microdrive is that it is small enough to fit on the skull of a rat and still have room to accommodate stimulating electrodes and/or a cannula for intracerebroventricular (i.c.v.) drug infusion.

5.3.1.4   *Choice of Anesthetic*

Many studies of age-related alterations in hippocampal electrophysiology have been performed in anesthetized animals. The primary advantage of this preparation is that it is easier to work with an immobile subject in which potential variables associated with different behavioral states have been eliminated. This advantage is also a disadvantage if the goal of the work is to relate some aspect of hippocampal electrophysiology to behavioral function. Other potential disadvantages are that: (1) the measure of interest may be directly or indirectly affected by the particular drug used as the anesthetic; (2) level of anesthesia may vary during the course of the experiment, perhaps influencing the data; and (3) differential sensitivity to anesthesia in aged animals may result in different drug levels between individual animals or across age groups, thereby causing what only appear to be age-related differences in electrophysiological measurements. For these reasons, it is clearly preferable to work in an unanesthetized preparation. The use of paralytic agents offers one alternative, but these agents are not sanctioned, or even recommended, because of the significant stress to the subject due to the lack of analgesia [81]. The remaining alternative is to record from the hippocampus of behaving animals. The advantages of this preparation are (1) the lack of potential confound of the presence of anesthesia; (2) the ability to directly correlate electrophysiological variables with ongoing behavior; and (3) the ability to make repeated measurements in the same animal.

For acutely anesthetized preparations, achieving and maintaining a stable plane of anesthesia is the goal. For survival surgical procedures, a well tolerated anesthetic with minimal aftereffects must be used. In both cases, volatile inhalant anesthetics (e.g., methoxyflurane) are probably the best choice, because the level of anesthesia is very easily controlled. However, inhalants cannot be properly administered without special equipment, and care must be taken to exhaust the gas to a safe place so that the investigator is not exposed. For these reasons, inhalant anesthetics often are not a practical choice. Urethane (1.0–1.5 mg/kg, i.p., for rats) is a good choice for nonsurvival surgery because the duration of anesthesia is very long and so does not require supplementation. (Urethane should not be used for survival surgeries because it is carcinogenic.) However, it is important to keep in mind that urethane causes a profound elevation of plasma corticosterone levels, with attendant effects on hippocampal plasticity mechanisms [82]. Urethane solutions should be made at concentrations of 20–25% in physiological saline. The total dose must be divided into two or more injections, because giving the entire dose in a single bolus will have lethal consequences.

Chloral hydrate (350 mg/kg, i.p., for rats) is also used as an anesthetic for rodent surgery. We have found two drawbacks to this drug. First, the plane of anesthesia cannot be reliably assessed using traditional measures (e.g., respiratory rate, response to toe or tail pinch). This often translates into the animal suddenly "waking up" and struggling violently in the stereotaxic, usually disturbing electrode placements and ruining the experiment. Second, chloral hydrate frequently causes profound gastroenteritis when it is used for survival surgeries. Although this problem can be reduced by using very dilute solutions of the drug (100 mg/ml), it is very frustrating to have an animal develop a swollen abdomen, and sicken and die 3–7 days after successful implant surgery.

Barbiturate anesthetics are very commonly used for both acute and survival surgeries. The dose range for rats is 50–60 mg/kg, i.p. Stable anesthesia is achieved by administering supplements of approximately 2.5 mg of drug every 15 min after the initial dose. The supplemental dose can be adjusted somewhat up or down based upon the signs of anesthesia present (respiratory rate, response to toe or tail pinch), but it is important to give regular injections to maintain pharmacodynamic equilibrium. Although sodium pentobarbital is the most commonly used barbiturate for anesthesia, for survival surgeries we have had better results with sodium secobarbital. These two agents are equipotent and have similar durations of action, but secobarbital has a higher therapeutic index. This translates into an important margin of safety for survival surgeries. We have found that rats anesthetized with pentobarbital usually do not survive if the duration of anesthesia is longer than about 2.5 h. By contrast, rats given secobarbital nearly always survive surgeries lasting more than twice this long. Since the time required to implant stimulating and recording electrodes in chronic preparations is frequently more than 3 h, we feel that secobarbital is the anesthetic of choice for survival surgeries, particularly for aged animals.

5.3.2 *In Vitro* Recording

The *in vitro* hippocampal slice offers the advantages of relatively fast preparation, the maintenance of a healthy state for up to 6–8 h, the ability to manipulate the extracellular environment and temperature, direct viewing of electrode location, absence of brain pulsation and other movement artifacts, easy biochemical accessibility, and direct imaging capability. However, the *in vitro* preparation also has its drawbacks. Slice tissue has lost its inputs from the rest of the brain, and the time during which the tissue may be examined in a healthy state is relatively short. Important aspects of slice preparation and maintenance that can affect the quality and responsiveness of the tissue include the method of slicing, the composition of the support medium, and the type of recording chamber used. In general, the viability of hippocampal slices made from the brains of aged rats is somewhat less good than for slices prepared from the brains of very young (21 to 30-day-old) rats.

5.3.2.1 General Aspects of Slice Preparation

Anesthetics, particularly barbiturates, ketamine, or volatile anesthetics such as halothane may reduce the traumatic effects of the slicing process [83]. A potential disadvantage of using anesthetics is that most will remain present to some degree in the slice tissue, and could thereby influence experimental results. In this regard, volatile anesthetics are cleared quickly and are least likely to have residual effects. Thus, their use offers a more humane method of euthanasia for the donor animal and is likely to improve the viability of the slices through direct neuroprotection of the tissue.

Slice preparation involves acute euthanasia of the subject animal, rapid removal of the brain, selection of the region of interest by blocking the brain or dissection, and cutting the desired area into sections 100–700 μm (most commonly 400–500 μm) thick. This process exposes the tissue to an unavoidable period of hypoxia that should be minimized. It is particularly important to reduce the temperature of the brain as rapidly as possible after it has been removed from the skull. This is accomplished by a placing the brain in a volume of artificial cerebrospinal fluid (ACSF, described in Section 5.3.2.2.) chilled to 0°C prior to dissection, and then performing the dissection over ice. For the dissection procedure, some investigators replace much of the NaCl in the ACSF with sucrose, reduce the Ca^{2+} or elevate the Mg^{2+}, or add dextran or other agents to the ACSF to better preserve the tissue [83,84].

The method of cutting the slices also seems to influence viability [83]. Some investigators slice the tissue by hand, using a razor blade with a cutting guide [85]. A far more common procedure is to use some kind of tissue chopper [86–88]. Older mechanical choppers (such as the McIlwain tissue chopper [89]) tend to be somewhat rough on tissue. At the other extreme, a Vibratome sectioner, which uses an oscillating movement to gently cut the tissue while it is suspended in cold ACSF, seems to be the most gentle. Although electrophysiologically viable slices of hippocampus may be obtained using any cutting method, more fragile tissue, such as cerebellum, requires the use of a Vibratome. Whatever the method, the entire slice preparation procedure should be done quickly to insure good viability and electrophysiological responses [83,90] (but see [91]).

5.3.2.2 Perfusion Media

Artificial cerebrospinal fluid (ACSF) is used for the dissection procedures and as a support medium for the slices. Concentrated stock solutions of the following individual components can be prepared and stored in the refrigerator: sodium chloride (NaCl), 1 M; potassium chloride (KCl) 1 M; sodium bicarbonate ($NaHCO_3$), 0.5M; sodium dihydrogen phosphate (NaH_2PO_4), 1 M; magnesium sulfate ($MgSO_4$), 1 M; and calcium chloride ($CaCl_2$), 1 M. Final concentrations for a standard ACSF recipe are (in mM): NaCl, 124; KCl, 3.25; $NaHCO_3$, 26; NaH_2PO_4, 1.25; $MgSO_4$, 2.0; $CaCl_2$, 2.4, and glucose, 10. The concentrations of some ions, particularly potassium, calcium, and magnesium, can be increased or decreased depending on the experiment and vary in the ACSF used by different investigators [85,92,93]. However, changing ACSF constituents can affect hippocampal slice electrophysiology. For example, elevating potassium levels (particularly above 6 mM) may lead to enhanced epileptogenicity; potassium concentrations above 8 mM can cause spontaneous field bursts indicative of spontaneous seizures in the slice [94,95]. Changing magnesium or calcium levels will alter synaptic potentials, particularly glutamatergic NMDA-receptor mediated responses, because magnesium ions normally block these channels and calcium fluxes through them [96,97]. Extracellular calcium concentrations also critically regulate the induction hippocampal long-term potentiation [98]. Thus, it is important to recognize that the ease of manipulating the constituents of ACSF can lead to using non-physiological solutions that may even have pathological effects.

5.3.2.3 Types of Recording Chambers

Many types of chambers for *in vitro* recording have been described [47,85,92,99-104]. Recording chambers can be loosely divided into two categories, interface chambers and submerged tissue chambers, although most of them can be used either way. Whether the slices are maintained at interface or are submerged in ACSF is based on the experimental requirement. However, the consensus is that interface slices have better electrophysiological properties [93]. In the interface chamber, the bottom surface of the slice is bathed in ACSF, while the top surface is exposed to moist superoxygenated (95% O_2 to 5% CO_2) atmosphere. The ACSF surrounding the slices can either be static or constantly replenished through perfusion. Because slices in interface chambers are constantly exposed to saturating concentrations of oxygen, the tissue is able to tolerate higher temperatures (35–37°C) and can remain healthy for long intervals. However, equilibration of perfused pharmacological agents is usually not as rapid in interface chambers, and drug washout can take very long times, particularly for lipophilic agents.

In the submerged chamber, slices are completely immersed in continuously flowing, freshly oxygenated ACSF. Unfortunately, insufficient oxygen can be dissolved in ACSF to allow submerged slices to remain viable at normal mammalian temperature. This problem can be somewhat mitigated by using fast perfusion rates (up to 10 ml/min), which is also advantageous for pharmacological studies involving drug application via the ACSF, but maintaining the slices at lowered temperature (usually 30–33°C) is also required [93,105]. The lowered temperatures do not affect some electrophysiological properties of the tissue including, surprisingly, synaptic conductances through ligand-gated channels [106]. However, other electrophysiological characteristics are altered, including voltage-based conductances [107] that affect responses to electrical stimulation [92,108–111]; in addition, responses to experimental manipulations (e.g., hypoxia) are also affected [112–115]. These alterations may actually be advantageous for studying certain processes electrophysiologically (e.g., single channel conductances), but it is important to remember that the results of such studies do not necessarily directly apply to the normally functioning hippocampus [106,116,117].

5.3.2.4 Stimulation and Recording Electrodes

The most commonly used electrodes for stimulating afferents in hippocampal slices are either stainless steel bipolar electrodes, similar to those described above for *in vivo* recording, or monopolar tungsten electrodes with a tip diameter of 10–20 µm which are commercially available. Bipolar wire electrodes are easy to prepare using insulated stainless steel, nichrome, or platinum wire, as described above (Section 5.3.1.2.). For discrete stimulation, wire diameters should be less than 50 µm. For very discrete stimulation protocols, where only a few afferent fibers are to be activated (e.g., [33]), a small tungsten wire (similar to that used for single unit recordings) or a glass pipette is used for stimulation in a monopolar mode. Responses evoked by small stimulating electrodes tend to be more stable over time than those evoked using larger bipolar electrodes, because their position is less affected by minor movement or swelling of the slice tissue.

Recording electrodes are most often micropipettes made from glass capillary tubing using a pipette puller. The tubing has a small glass filament fused to the inner wall to facilitate filling of the electrode in a reverse manner, from the shank into the tip. For extracellular recording, the micropipette should have a low resistance (less than 10 MΩ) after filling with an extracellular electrolyte (1–3 M NaCl). The tip diameter (1–3 μm) must be small enough to prevent efflux of large amounts of the electrolyte, since this will damage the tissue; alternatively, the electrode should be filled with agar containing the electrolyte, to prevent leakage. For intracellular recordings, a sharp microelectrode filled with intracellular electrolyte, such as KCl or K-acetate (1–3 M), is used. Electrodes with a resistance of 90–150 MΩ are optimal for impaling hippocampal neurons for so-called "sharp electrode" recordings. We have found that glass pipettes made from tubing with an outer diameter (o.d.) of 1.0 mm and an inner diameter (i.d.) of 0.75 mm are ideal for making recording electrodes for intracellular recording from hippocampal pyramidal neurons and for iontophoresis of dyes into the cells. However, for recording from the smaller dentate granule cells, glass tubing with the smaller ID of 0.5 mm is better because it creates an electrode with a finer tip. The electrodes may be filled with a solution containing 2–4% biocytin, neurobiotin, horseradish peroxidase, or a fluorescent marker dissolved in 1 M potassium acetate buffer (pH 7.4) for recording from a neuron and then filling it for later visualization.

Another commonly used technique, whole cell patch recording, uses an electrode that forms a tight seal with the neuronal membrane [118]. Advantages of this technique are that the recording electrode has low access resistance (less than 15 MΩ), producing better recording characteristics, and that the tight seal decreases the leak around the electrode (reflected in reduced membrane resistance) thought to occur with sharp electrodes. However, since patch electrodes have large openings, the solution in the electrode rapidly exchanges with intracellular contents. This perfusion effect necessitates that the patch pipette filling solution closely match the intracellular milieu, particularly with respect to the presence of ATP, potassium, and other ions, to avoid relatively rapid degradation of recorded signals and neuronal responsiveness [84,107]. A current methodological limitation to using whole cell patch recordings for aging research is that the ability to form a tight seal between the patch electrode and cell becomes very difficult if the slices are made from rats that are more than 21 days old. In addition, with patch recording there is the concern that abnormalities present in an aged cell (e.g., changes in calcium levels or fluxes) might be obscured by the intracellular perfusion from the patch electrode, whereas this process is minimal when sharp electrodes are used.

High quality micromanipulators are necessary for stable placement of electrodes for *in vitro* recording, especially when intracellular techniques are used. The best equipment for this task is remote controlled, either by a stepper motor mounted on the manipulator itself, or via hydraulic tubes from a remote device. Newer manipulator designs allow micropositioning in three dimensions, which is advantageous for both sharp electrode and patch electrode recordings. Optical indicators of the electrode location are helpful, including infrared camera control of the electrode position, and phase contrast or other forms of contrast enhancement to be able to visualize when the electrode is close to a cell. Optical aids are particularly useful for patch clamp recording, since the electrode can then be moved to the cell membrane under direct visual guidance.

5.3.2.5   *Optical Imaging*

Optical properties of CNS tissue have been described over many years, beginning with light transmittance through large axons [119] (reviewed by Salzberg [120]). As action potentials pass through axons the optical properties of the tissue change at both short and long periods, with the fast changes likely reflecting directly ion fluxes and the slower changes compensatory water and ion movements that occur following activation. Optical imaging is a technique that can be used to detect these neuronal signals over the surface of a brain tissue slice [121–123]. The addition of dyes permits direct imaging of membrane voltage changes (using fast voltage-sensitive dyes), calcium, or other ions, so that these measures can be directly related to cell activity [124–130]. There is considerable rationale for using optical imaging of aged CNS tissue to compare the spatial patterns of synaptic activation, to measure responses to hypoxia, to analyze calcium fluxes during synaptic activation and action potential generation, and, eventually, to decipher intracellular signals such as those leading to long-term potentiation.

The procedures involved with optical imaging include the physiological recording techniques described above in combination with imaging hardware. The basic principle of optical imaging is to subtract a high resolution control image from a high resolution signal image, such as that obtained after synaptic or pathological (e.g., from spreading depression or after osmotic changes) stimulation of the tissue. After subtraction, a difference as small as 1–2% can be visualized, depending on the background noise and the capabilities of the imaging system. Synaptic stimulation commonly evokes a signal that differs by 5–8%, whereas pathological changes may induce signals that are different by up to 25–30%, from the control image. Imaging can be done with a variety of devices, including a video camera (linked to standard video frame rate of 30 per second and with video resolution of 640 × 480 pixels and depth of 7–8 bits), a confocal microscope (using a photomultiplier tube and a mirror which is scanned at video rates), a scientific (often cooled) CCD camera which is not linked to video frame rates or resolution (ranging from 512 × 512 to 3000 × 2000 pixel resolution, at 12–14 bits depth), or a frame array of photodiodes (often 12 × 12 elements). There are advantages and disadvantages of each, depending on the resolution needed (both planar and depth in terms of gray levels), the time course of the signal, the light level of the signal and whether or not the signal is a fluorescent emission. For example, fast (< 1 msec) action potential detection requires voltage sensitive dyes, a fluorescent light source, and a photodiode array in which each pixel is a real-time stream of analog levels representing light intensity. Slower intrinsic signals may be detected by video or cooled CCD cameras. Calcium signals have also been detected by both types of cameras, using a fluorescent light source or a laser as the excitation source.

5.4 Procedures

5.4.1 Hippocampal Electrophysiology *In Vivo*

The procedure for preparing a rat for recording in the unanesthetized state will be detailed below. The same protocols apply for placing electrodes in the hippocampus for acute studies.

5.4.1.1 *Electrode Implantation*

The rat should be water deprived the night before surgery. First, the animal is injected with atropine methyl nitrate (0.02 mg/kg, i.p.) to reduce salivation and excessive body fluid secretion during the surgery. Twenty minutes later, the rat is anesthetized with secobarbital (50mg/kg, i.p.). A surgical plane of anesthesia is indicated by loss of corneal blink reflex and the presence of only minor response (e.g., elevation of respiration) to strong tail pinch.

After the rat is anesthetized, its head is secured in a stereotaxic apparatus with the top of the skull oriented in the horizontal plane (i.e., parallel to the base of the stereotaxic).

A heating pad must be used to maintain the animal's body temperature at a physiological level (36–37°C). An injection butterfly, attached to a 1-cc syringe filled with secobarbital solution (25 mg/ml in sterile 0.9% NaCl), is inserted through the abdomen into the peritoneum to provide anesthetic supplements. Level of anesthesia is examined every 15 min. Supplemental doses range from 0.05–0.10 ml, determined by the rapidity and location of respiration (thoracic vs. abdominal) or the degree of response to tail pinch.

The top of the head is shaved and swabbed with Betadyne (an iodine-containing antiseptic). A midline incision is made, and the skin is retracted to expose the skull. Connective tissue is removed and the skull is cleaned and dried thoroughly. Bleeding can be controlled with pressure, application of a 1% solution of norepinephrine, or by using a small electrocautery unit. It is very important that the skull be clean and completely dry, both to facilitate locating skull landmarks (e.g., bregma or lambda) and for good adhesion of the dental acrylic used in survival surgeries.

Holes to allow placement of electrodes in the brain, as well as support screws in the case of survival surgeries, are then drilled through the skull. The diameter of the drill bit should match the diameter of the support screws. Holes for support screws are drilled at the following locations (bilaterally, from bregma): A.P. + 4.0 mm, M.L. ± 2.0 mm; A.P. – 6.0 mm, M.L. ± 3.5 mm. An additional screw is placed behind lambda, along midline. The screws should not be inserted deeper than about one millimeter, the approximate thickness of the skull; penetration of the dura or brain must be avoided. Holes for the stimulating and recording electrodes need to be large enough (1–2 mm diameter) to allow for the possibility that an electrode will need to be moved from the first placement to improve the electrophysiological response. However, care should be taken to leave as much skull as possible intact if the electrodes are to be cemented in place.

Coordinates for electrode placements in either the hippocampus or the dentate gyrus are given in Table 5.1. However, only the procedure for locating recording electrode(s) in area CA1 and a stimulating electrode in the hippocampal commissure will be described in the text. Procedural aspects of electrode placement are very similar for all sites, with only stereotaxic coordinates and final electrode depths being different. The first step for any recording is to place the ground wires (shown in Figure 5.1) in the brain; in a survival surgery, they are cemented to the anterior support screws and skull with dental acrylic. Next, the dura is slit using the tip of a 25-gauge needle and the stimulating electrode is lowered 1 mm into the brain. If a two-electrode microdrive is being implanted, its legs are screwed to approximately midposition (Figure 5.2, right). The lower portions of the microdrive legs and the electrode guide cannula (but *not* the electrodes themselves) are lightly coated with petroleum jelly so that dental acrylic will not adhere. This step is critical for free movement of the microdrive after surgery. The microdrive is then placed in a clamping device so that recording electrode movements can be controlled by the stereotaxic electrode carrier during the implant surgery.

The recording electrode is now slowly lowered into the dorsal hippocampus. Robust complex-spike discharges, indicative of CA1 pyramidal cells, are usually found at a site located between 1.9 and 2.4 mm below brain surface. Once the CA1 pyramidal cell layer has been reached, the recording electrode is left to settle for 5 min. When implanting the 2-electrode microdrive, after the pyramidal cell layer is reached with the longer electrode, the entire assembly is lowered an additional 250 μm. The cell layer response will now be recorded on the shorter electrode, while the dendritic response will be recorded on the longer electrode. Wait 5 min to allow the recording electrodes to settle into position. The commissural stimulating electrode is then slowly lowered until an optimal placement is found, usually between 2.6 and 3.8 mm below brain surface. During stimulator placement the stimulation current is initially set at 200 μA (monophasic pulse width 150 μsec) and stimuli are delivered at 20- or 30-s intervals. Initially, the responses evoked by the

TABLE 5.1

STEREOTAXIC COORDINATES FOR ELECTRODE PLACEMENTS[A]

Site	Electrode Type	Coordinates
CA1	Recording	A.P. –4.0, M.L. 2.3
		D.V. 2.1 (pyramidal cell layer)
		D.V. 2.4 (stratum radiatum)
	Commissural stimulator	A.P. –1.8 to –2.5[b], M.L. 1.0
		D.V. 2.6–3.8
Dentate Gyrus	Recording	A.P. –4.0, M.L. 2.3
		D.V. 2.8 (middle molecular layer)
		D.V. 3.0 (granule cell layer)
		D.V. 3.6 (hilus)
	Entorhinal stimulator	A.P. –8.1, M.L. 3.0
		D.V. 2.8–3.8
		(electrode is angled 10° away from midline)
Ground	Ground	A.P. +2.0, M.L. ±1.0[c]

[a] All coordinates are in millimeters from bregma, with the top of the skull oriented in a flat horizontal plane (parallel to the stereotaxic base). Dorso-ventral (D.V.) coordinates are ± 0.5 mm from the surface of the brain.

[b] A.P. placement of the commissural stimulating electrode is influenced by rat strain and age. Start with the most caudal coordinate, as is discussed in the text.

[c] Ground electrodes are implanted bilaterally. See Figure 5.1 for illustration of electrode.

stimulation will probably be small or absent. The stimulating electrode is lowered in 100 μm steps, with each movement being performed after a stimulus has been delivered. At some point the size of the evoked responses will begin to grow rapidly, and the potentials will assume their characteristic shapes (positive in the pyramidal cell layer, negative in the stratum radiatum dendrites). The stimulating electrode is then advanced further in 50 μm steps, and is moved no more frequently than once per minute. When the optimal stimulation site is reached, a good response (e.g., population spike amplitude > 2 mV) will be evoked by a current of less than 100 μA. If this criterion is not met, it is necessary to move the stimulating electrode. Usually the new placement should be approximately 0.50 mm anterior to the first penetration; a long latency (>5 msec) response is a good indicator that the stimulating electrode will need to be moved to a more anterior site. In our experience, the monopolar stimulating electrode can be replaced up to three times, if necessary, without a detrimental effect on the preparation.

Since the brain has been compressed by placing the electrodes, allow 15 min after final placements have been made for the response to stabilize. Occasionally the electrodes will require minor adjustment. For acute preparations, data acquisition can then begin. For survival surgeries, the electrodes and microdrive are now cemented in place. All electrodes and the microdrive legs are initially stabilized by applying a thin layer of very fluid dental acrylic to the top of the skull. If necessary, carefully reclean the skull and wait for it to dry before applying the acrylic. Once the first layer of dental acrylic has hardened (test with the tip of a 25-gauge needle), add additional layers of acrylic to firmly cement the electrodes and microdrive legs in place. During this process the wire armature supporting the stimulating electrode (shown in Figure 5.1) is cut away with pliers at the top of the Z-bend and the site is completely covered with acrylic. Connecting pins on the electrode wires are then inserted into the headpiece connector (see Figure 5.1), which is then cemented to the top of the skull. Additional layers of dental acrylic should be applied to form the smooth surface necessary to

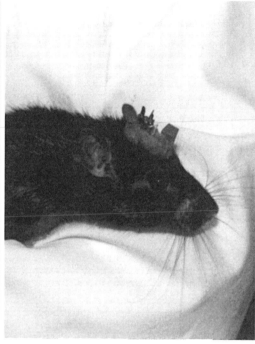

FIGURE 5.3
A rat prepared for recording in the unanesthetized state. The figure on the left shows the animal in the chamber with the recording cable plugged into the headpiece. The figure on the right shows details of the implant. The headpiece is located at the front, just above the rat's eyes. The microdrive is located behind the headpiece; because dental acrylic has been built up around the edges of the implant to form a protective wall, only the top of the microdrive and the legs are visible. The rat is a Norway Brown, one of the inbred strains maintained by the National Institute on Aging contract colonies.

prevent postoperative skin irritation and to protect the microdrive. Finally, suture if necessary and apply antibiotic powder to edges of the skin incision. The rat is then removed from the stereotaxic and placed on a heating pad until the effects of anesthesia have worn off. The animal is then returned to long-term (single) housing for a period of 3–10 days, until wound healing is complete. An example of a rat implanted with a microdrive is shown in Figure 5.3.

5.4.1.2 Microdrive Adjustment

Even when the implant procedure is done very slowly and carefully, by the time the rat has recovered from surgery it is unusual to have the recording electrodes in the same locations they were during implantation. Our experience has been that the electrodes continue to settle in the brain, ending up several hundred microns ventral to the initial placement. To reoptimize the electrode positions, the animal is placed in the recording environment and the recording cable is attached to the headpiece so that the electrophysiological responses can be observed (Figure 5.3). The legs of the microdrive are alternately rotated in increments of one-eighth turn (approximately 50 μm; clockwise to raise, counterclockwise to lower). After both legs have been rotated, a 15-min period is allowed to elapse before evaluating the change in the evoked response. Performing this adjustment process slowly is the key to having the electrode positions remain stable for multiday periods after final placements have been made. Illustrations of responses recorded at different steps of the microdrive adjustment process are shown in Figure 5.4.

5.4.1.3 Data Acquisition and Analysis

The most practical means of data acquisition and storage is via one of the many commercially available software packages designed for use with a microcomputer. With such software, both on-line (during the experiment) and off-line analysis of EEG, extracellular field potentials evoked by electrical or sensory stimulation, or the firing patterns of single neurons can be performed. Some software also provides for video input to facilitate making correlations between hippocampal electrophysiology and behavior. A computer-based acquisition system is particularly useful when signals from multiple electrodes need to be monitored and related to each other. In addition, some software allows discrimination and tracking of the activity of single neurons in a multiunit record. Off-line analysis may include many different forms, including Fourier analysis of EEG, averaging of evoked potentials, and generating peri-event histograms for examining the relationship of neuronal firing to sensory stimuli or behavioral responses.

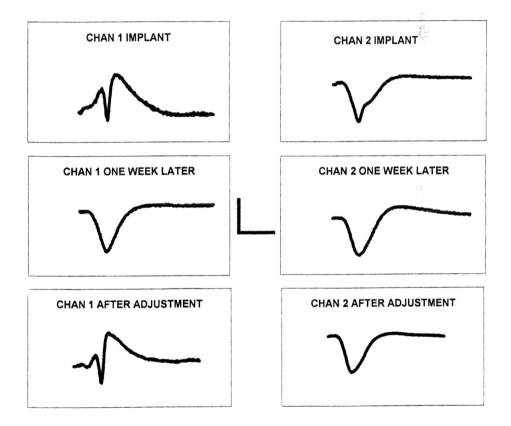

FIGURE 5.4

Examples of extracellular field EPSPs recorded in hippocampal area CA1 in response to stimulation of commissural afferents. The top row shows the potentials recorded using the two-electrode microdrive (Figure 5.2) in the pyramidal cell layer (left) and in stratum radiatum (right) at the conclusion of the implant surgery. The shape of the potentials in the middle row indicates that, by one week after surgery, the recording electrodes have drifted deeper into the hippocampus. After turning the microdrive legs to raise the electrodes approximately 300 μm, the dorsal electrode (left) has been relocated in the pyramidal cell layer. If the adjustment process is done slowly, as is described in the text, the electrode positions will now remain in the proper positions for many days. Scale bar: 2 mV, 10 msec (channel 1); 4 mV, 10 msec (channel 2).

5.4.2   Hippocampal Electrophysiology *In Vitro*

A general schematic that illustrates the important elements of an *in vitro* recording setup is shown in Figure 5.5. A widely used method for preparation of hippocampal slices will be described below. Tools for extraction and dissection of the brain include: scalpel, large scissors, rongeur, a paint brush with small bristles, a plastic spoon, and a plastic spatula. Good tools are essential for the rapid dissection necessary for healthy slices. Most investigators individualize their tools considerably; so, these suggestions are only a starting point. In general, slices from young and aged animals are prepared the same way, except that the skull of older animals is thicker so more effort is required to extract the brain in a rapid, yet gentle, manner.

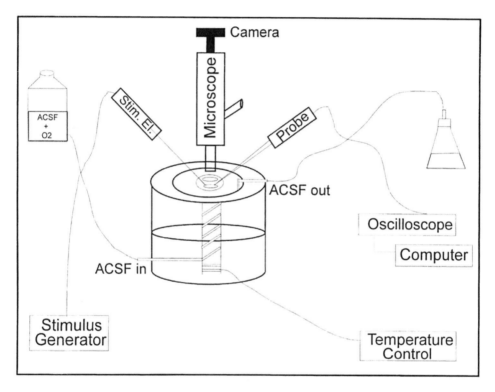

FIGURE 5.5
Schematic of an *in vitro* recording setup. The slices are maintained in a chamber, where they are perfused with warmed and oxygenated ACSF. A microscope is used to visualize electrode positions or for optical imaging. Details of the stimulation and recording protocols depend upon the type of experiment being performed and are further described in the text.

5.4.2.1   *Slice Preparation*

Two 100 to 150-ml beakers, one containing chilled ACSF (0°C) for cooling the dissected brain and the other containing ACSF at room temperature for maintaining the slices (referred to as the holding chamber), are required. ACSF in both the beakers is bubbled with a 95% O_2 to 5% CO_2 gas mixture for about 15 min prior to the dissection.

The rat is first anesthetized with halothane. A suitably sized closed container (jar or rectangular box) made of either glass or polyurethane can be used for this purpose. Surgical gauze or cotton soaked with halothane is placed in the chamber to saturate the air with halothane vapor. The rat is placed into the container until fully anesthetized (about 90 s). At this point, response to toe or tail pinch is checked to make sure that the rat is unconscious.

Decapitate the animal using a guillotine or large scissors. Cut through the skin over the top of the skull with a scalpel from front (nose) to back (neck). With the rongeur, remove the roof of the skull to expose the brain. Using a plastic spoon or a small plastic spatula, gently lift the brain out of the skull and quickly place it into the beaker of ice-cold ACSF. Allow the brain to cool in this solution for 30–60 s, which somewhat stiffens the soft tissue and makes slicing easier. The cooling at this stage also enhances the brain's tolerance to the ischemia and so is highly protective.

After cooling, place the brain on a piece of filter paper that has been moistened with ACSF. Cut the brain down the midline into two halves, and return one half to the cold ACSF. Dissect the hippocampus out of the other half. To do this, gently hold the cortex in place and ease the midbrain away from it, exposing the ventral surface of the hippocampal formation. Free the hippocampus at its septal and temporal ends using a plastic dissecting tool (such as a small plastic knife filed down to a sharp edge at one end) and flip the hippocampus out from the ventricular surface, separating it from the neocortex. Once free, remove the entorhinal cortex and trim to remove excess cortex and white matter. A detailed illustration of the dissection procedure has been published by Teyler [92].

To prepare the slices, use a small plastic spatula to place the hippocampus on ACSF-moistened filter paper on the stage of a tissue chopper and cut the slices at desired thickness (usually 400–450 µm). Gently remove each slice from the surface of the cutting blade using a fine tip paint brush and place it in the beaker containing room temperature oxygenated ACSF (the holding chamber). Up to 12 slices can be obtained from each hippocampus. Alternatively, a vibratome can be used to section the hippocampus. Vibratome sectioning is very useful for preparing complex sections, such as slices that include both the hippocampus and the entorhinal cortex. In addition, it has been suggested that vibratome sectioning is superior because it does not stretch the hippocampus during the sectioning procedure [83]. To obtain the vibratome slices, follow the same steps for removing the brain that were described above. After placing the brain on ACSF-moistened filter paper, make coronal cuts with a scalpel or razor blade to trim off the anterior and posterior ends. Glue the tissue block onto the vibratome stage using cyanoacrylate (super glue), posterior end down. Once the brain is firmly glued, divide the two hemispheres with a scalpel cut. Fix the vibratome stage into the vibratome, and fill the reservoir surrounding the stage with cold, oxygenated ACSF. Cut slices at the desired thickness; unwanted tissue is then carefully cut away. It is important to keep the brain tissue moistened with cold, oxygenated ACSF throughout the dissection and slicing procedure. Absolute requirements for obtaining healthy slices include: (1) slicing should be accomplished quickly, in approximately 5 (best slices) to 10 (good slices) min; (2) cuts through the tissue during slicing must be clean, made using a sharp blade at a high speed; and (3) slices should receive minimal handling to prevent stretching, distortion, or other injury.

Maintain the slices in the holding chamber at room temperature by bubbling the ACSF with 95% O_2 to 5% CO_2 to maintain the pH at 7.4. The slices are allowed to stabilize for 90–120 min prior to performing *in vitro* recordings. For this, slices are gently transferred from the holding chamber to the recording chamber using a dropper. It is critical to select only those slices that do not have any visible damage and in which all subregions of the hippocampal formation are intact. This selection procedure maximizes the chances for recording from healthy slices.

5.4.2.2　Extracellular Assessment

Excitatory postsynaptic potentials (EPSP), representing the postsynaptic response to activation of presynaptic fibers, provide a general measure of neuronal health and connectivity.

These responses can be assessed using either extracellular or intracellular recording. Monopolar or bipolar stimulating electrodes (described in Section 5.3.2.4) are placed in the path of the afferent fibers that terminate on cell population of interest. For extracellular examination, the recording electrode is placed in the dendritic region or the cell body layer of the postsynaptic neurons to measure synaptic EPSPs or population EPSPs, respectively. Maximum stimulation rate should be no greater than 1 pulse every 10 s to prevent inadvertent potentiation of evoked responses. Paired-pulse paradigms, employing either one or two stimulating electrodes, can be used to assess the status of feed-forward and feed-back inhibitory circuitry.

5.4.2.3   Intracellular Assessment

To perform intracellular recordings, slices are placed in the chamber and the stimulating electrode is placed in the afferent pathway of the region of interest. A high impedance microelectrode (described in Section 5.3.2.4.) is placed above the cell body layer and then lowered into the slice. The electrode is slowly advanced using a microdrive until a DC shift is observed, and usually accompanied by spontaneous action potential firing. The microelectrode tip can then be gently "buzzed" (oscillated using a switch on the amplifier) to penetrate the cell. Once intracellular, hyperpolarizing current is injected until the membrane resistance recovers to approximately –65 mV, indicating that the cell membrane has resealed around the recording electrode. Characteristics for a good impalement include a minimal resting potential of –55 mV and repetitive firing in response to intracellular depolarizing current pulses. The neuronal input resistance (R_N) is estimated from the potential changes observed following the injection of hyperpolarizing current pulses, (0.1–0.5 nA, 100 msec duration). The response to orthodromic stimulation in the stratum radiatum is then assessed over a series of graded stimulus intensities, and other physiological manipulations (e.g., induction of long-term plasticity) are performed if desired.

Following electrophysiological assessment, cells can be filled with dyes (neurobiotin or biocytin) using intracellular current stimulation (4 Hz depolarizing pulses, 150 msec in duration, 2–5 nA), with superimposed hyperpolarization (–0.2 nA) to prevent electrode blocking. Neurons located near the middle of the slice (Z axis, or depth) are the best candidates for this procedure, because of the increased chance that all their processes will be contained within the slice [28].

5.4.2.4   Cell Visualization and Morphometry

After cells are labeled, the slice is fixed and sectioned, and the intracellular dye is chemically processed (if necessary) for visualization. Various quantitative measures can then be made, including simple measures such as somatic or dendritic length or volume, and more complicated measures of dendritic branching. Dendrograms can also be constructed to allow comparisons of basic dendritic structure. Estimates of dendritic function, such as electrotonic length, may be modeled based upon the morphological data. Details of the application of these procedures for comparisons of young and aged hippocampal neurons are given in Pyapali and Turner [28].

5.4.2.5   Whole Cell Patch Recording

Whole cell clamping involves forming very high resistance (gigaohm) seals on cells. Clean cell membranes are required for this process. Patching cell membranes in brain slices is particularly difficult because the surface of the brain slices consists mostly of cellular debris that will prevent the formation of good seals. Various approaches used to circumvent this problem are described below.

First, the hippocampal slice may be physically torn at the cell layer to expose pyramidal or granule cell somata. The surface of the slice is treated with proteolytic enzymes for a short period and then pulled apart with fine forceps. The membranes of cell bodies exposed at the tear are clean enough to permit the patch electrode to form an adequate seal. The disadvantage of this technique is that dendritic processes are often damaged, so that synaptic inputs cannot be studied. A second approach involves cleaning the debris from the surface of the slice by directing a stream of ACSF onto the surface through a large-bore pipette. As the debris is loosened, it is suctioned out. When performed carefully, this procedure does not damage the underlying neurons or their dendrites, and is therefore suitable for synaptic studies. A third technique involves viewing the slice under high magnification to locate exposed but healthy cells. For this procedure, thinner slices (usually less than 200 μm thickness) are necessary to improve the optical properties of the slice so that individual cells can be visualized. Preparing the slices with a Vibratome allows thinner slices with cleaner surfaces to be obtained. An upright microscope with water immersion objectives is required to use this technique, and the addition of Nomarski optics greatly enhances the ability to visualize individual cells in the slice. A fourth technique for whole cell patching in hippocampal slices uses no specialized equipment and produces good results. This method involves simply advancing the patch electrode through the middle of the cell layer, applying suction until a seal is formed. The advantage of this technique is that it works well with standard slices and requires no preparation (e.g., cleaning or proteolysis) of the tissue prior to recording. The disadvantage of this technique is that one cannot patch onto a specific cell or process.

5.4.2.6 *Optical Imaging*

Once the image detector is chosen for the particular experiment according to the temporal and spatial resolution required (as discussed in Section 5.3.2.5.), it must be linked to a suitable source of magnification, usually either a compound microscope or a dissecting microscope, and an appropriate light source. Many signals are detected as epifluorescent emission signals; so, a fluorescent microscope with high quality lenses may be critical since the objective functions as the condenser for the light source. The preparation must thus be suitably illuminated (so as not to bleach the dye), proper filters installed to detect the emission, and the image detector appropriately positioned. We use either a Nikon UM-2 upright compound metallurgical microscope or a Nikon SMZ-U high-quality dissecting microscope to image the hippocampal slice in a slice chamber, with either a laser light source or an epifluorescent light source. The transmitted and/or fluorescent emission is then routed through the optics to a Photometrics cooled CCD camera (Star-1). The image is then directly digitized at 12 bits resolution (1 in 4000) and transferred to a host computer for processing. Since most of the imaging is for differences with respect to a control image, linear subtractions can be performed in relation to the control image (Δ intensity/intensity) and then played back as a movie. In this way, perturbations are clearly apparent, and the original data from the camera can be retrieved for numerical estimates of changes in regions of interest. For example, individual images are taken during the induction of spreading depression at one frame every 2–3 s. The spreading front of the depression can then be mapped out on the images and correlated with physiological measurement of the spreading depression [122]. Careful attention must be given to keeping the preparation motionless, because movements will decrease the effectiveness of subtraction and thus reduce the resolution of the technique.

5.5 Conclusions

The hippocampus is an ideal brain region for using electrophysiological tools to examine the effects of aging. The intrinsic circuitry of the hippocampus is fairly simple, and afferent pathways are organized such that they can be easily activated using electrical stimulation. Many basic properties of the hippocampus are similar *in vivo* and *in vitro*, which allows a wide variety of electrophysiological techniques to be employed to examine characteristics ranging from morphology and fundamental biophysical properties to circuit interactions and correlates of neuronal firing with behavior. Beyond its utility as an experimental preparation, the hippocampus plays a central role in processes of memory formation that are altered during the aging process. Electrophysiological studies of the hippocampus offer the opportunity for a detailed understanding of how age-related changes in the elements of a neuronal circuit affect cognitive behavior. Such information is vital to developing treatments for the learning and memory problems that so commonly appear with aging.

Acknowledgments

This review was in part supported by grants from the National Institute on Aging (AG-10755 to G.M.R. and AG-13165 to D.A.T.) and Veterans Affairs Merit Review Awards to G.M.R. and D.A.T.

References

1. Miller, R. A., The aging immune system: primer and prospectus, *Science*, 273, 70, 1996.
2. Jazwinski, S. M., Longevity, genes, and aging, *Science*, 273, 54, 1996.
3. Ames, B. N., Shigenaga, M. K., and Hagen, T. M., Mitochondrial decay in aging, *Biochim. Biophys. Acta*, 1271, 165, 1995.
4. Sohal, R. S. and Weindruch, R., Oxidative stress, caloric restriction, and aging, *Science*, 273, 59, 1996.
5. Reiter, R. J., Oxidative processes and antioxidative defense mechanisms in the aging brain, *FASEB J.*, 9, 526, 1995.
6. Evans, D. A., Burbach, J. P., and van Leeuwen, F. W., Somatic mutations in the brain: relationship to aging?, *Mutat. Res.*, 338, 173, 1995.
7. Gareri, P., Mattace, R., Nava, F., and De Sarro, G., Role of calcium in brain aging, *Gen. Pharm.*, 26, 1651, 1995.
8. Agnati, L. F., Benfenati, F., Solfrini, V., Biagini, G., Fuxe, K., Guidolin, D., Carani, C., and Zini, I., Brain aging and neuronal plasticity, *Ann. N. Y. Acad. Sci.*, 673, 180, 1992.
9. Agnati, L. F., Zoli, M., Grimaldi, R., Fuxe, K., Toffano, G., and Zini, I., Cellular and synaptic alterations in the aging brain, *Aging*, 2, 5, 1990.
10. Bertoni-Freddari, C., Fattoretti, P., Paoloni, R., Caselli, U., Galeazzi, L., and Meier-Ruge, W., Synaptic structural dynamics and aging, Gerontology, 42, 170, 1996.
11. Biessels, G. and Gispen, W. H., The calcium hypothesis of brain aging and neurodegenerative disorders: Significance in diabetic neuropathy, *Life Sci.*, 59, 379, 1996.
12. Goldman, J. E., Calingasan, N. Y., and Gibson, G. E., Aging and the brain, *Curr. Opin. Neurol.*, 7, 287, 1994.

13. Landfield, P. W. and Eldridge, J. C., Evolving aspects of the glucocorticoid hypothesis of brain aging: hormonal modulation of neuronal calcium homeostasis, *Neurobiol. Aging*, 15, 579, 1994.

14. Rosenzweig, M. R., Aspects of the search for neural mechanisms of memory, *Annu. Rev. Psychol.*, 47, 1, 1996.

15. Scoville, W. B. and Milner, B., Loss of recent memory after bilateral hippocampal lesions, *J. Neurol. Neurosurg. Psychiatry*, 20, 11, 1957.

16. Disterhoft, J. F., Thompson, L. T., Moyer, J. R., Jr., and Mogul, D. J., Calcium-dependent afterhyperpolarization and learning in young and aging hippocampus, *Life Sci.*, 59, 413, 1996.

17. Landfield, P. W., Aging-related increase in hippocampal calcium channels, *Life Sci.*, 59, 399, 1996.

18. Hof, P. R. and Morrison, J. H., Hippocampal and neocortical involvement in normal brain aging and dementia: morphological and neurochemical profile of the vulnerable circuits, *J. Am. Geriat. Soc.*, 44, 857, 1996.

19. Meaney, M. J., D, O. D., Rowe, W., Tannenbaum, B., Steverman, A., Walker, M., Nair, N. P., and Lupien, S., Individual differences in hypothalamic-pituitary-adrenal activity in later life and hippocampal aging, *Exp. Gerontol.*, 30, 229, 1995.

20. Geinisman, Y., Detoledo-Morrell, L., Morrell, F., and Heller, R. E., Hippocampal markers of age-related memory dysfunction: behavioral, electrophysiological and morphological perspectives, *Prog. Neurobiol.*, 45, 223, 1995.

21. Barnes, C. A., Normal aging: regionally specific changes in hippocampal synaptic transmission, *Trends Neurosci.*, 17, 13, 1994.

22. Gallagher, M. and Nicolle, M. M., Animal models of normal aging: relationship between cognitive decline and markers in hippocampal circuitry, *Behav. Brain Res.*, 57, 155, 1993.

23. Lamour, Y., Bassant, M. H., Potier, B., Billard, J. M., and Dutar, P., Aging of memory mechanisms, *C. R. Seances Soc. Biol.*, 188, 469, 1994.

24. O'Keefe, J. and Nadel, L.,*The Hippocampus as a Cognitive Map*, Clarendon Press, 1978,

25. Deupree, D. L., Turner, D. A., and Watters, C. L., Spatial performance correlates with *in vitro* potentiation in young and aged Fischer 344 rats, *Brain Res.*, 554, 1, 1991.

26. Deupree, D. L., Bradley, J., and Turner, D. A., Age-related alterations in potentiation in the CA1 region in F344 rats, *Neurobiol. Aging*, 14, 249, 1993.

27. Engstrom, D. A., Bickford, P., De La Garza, R., Young, D., and Rose, G. M., Increased responsiveness of hippocampal pyramidal neurons to nicotine in aged, learning-impaired rats, *Neurobiol. Aging*, 14, 259, 1993.

28. Pyapali, G. K. and Turner, D. A., Increased dendritic extent in hippocampal CA1 neurons from aged Fisher 344 rats, *Neurobiol. Aging*, 17, 601, 1996.

29. Buckmaster, P. S. and Schwartzkroin, P. A., Physiological and morphological heterogeneity of dentate gyrus-hilus interneurons in the gerbil hippocampus *in vivo*, *Eur. J. Neurosci.*, 7, 1393, 1995.

30. Buckmaster, P. S., Wenzel, H. J., Kunkel, D. D., and Schwartzkroin, P. A., Axon arbors and synaptic connections of hippocampal mossy cells in the rat *in vivo*, *J. Comp. Neurol.*, 366, 271, 1996.

31. Turner, D. A. and Schwartzkroin, P. A., Steady-state electrotonic analysis of intracellularly stained hippocampal neurons, *J. Neurophysiol.*, 44(1), 184, 1980.

32. Turner, D. A., Li, X. G., Pyapali, G. K., Ylinen, A., and Buzsaki, G., Morphometric and electrical properties of reconstructed hippocampal CA3 neurons recorded *in vivo*, *J. Comp. Neurol.*, 356, 580, 1995.

33. Turner, D. A. and Deupree, D. L., Functional elongation of CA1 hippocampal neurons with aging in Fischer 344 rats, *Neurobiol. Aging*, 12, 201, 1991.

34. Caton, R., The electrical currents of the brain, *Br. Med. J.*, 2, 278, 1875.

35. Berger, H., Ueber das Elektrencephalogramm des Menchen, *Arch. Psychiatr. Nervenkr.*, 87, 527, 1929.

36. Renshaw, B., Forbes, A., and Morison, B. R., Activity of isocortex and hippocampus: electrical studies with microelectrodes, *J. Neurophysiol.*, 3, 74, 1940.

37. Green, J. D. and Arduini, A. A., Hippocampal electrical activity in arousal, *J. Neurophysiol.*, 17, 533, 1954.

38. Green, J. D. and Adey, W. R., Electrophysiological studies of hippocampal connections and excitability, *Electroencephalogr. Clin. Neurophysiol.*, 8, 245, 1956.
39. Cragg, B. G. and Hamlyn, L. H., Action potentials of the pyramidal neurons in the hippocampus of the rabbit, *J. Physiol. (London)*, 129, 608, 1955.
40. Green, J. D. and Machne, X., Unit activity of rat hippocampus, *Am. J. Physiol.*, 181, 219, 1955.
41. Andersen, P., Interhippocampal impulses. I. Origin, course and distribution in cat, rabbit and rat, *Acta Physiol. Scand.*, 47, 63, 1959.
42. Kandel, E. R., Spencer, W. A., and Brinkley, F. J., Electrophysiology of hippocampal neurons. I. Sequential invasion and synaptic organization, *J. Neurophysiol.*, 24, 225, 1964.
43. Bliss, T. V. P. and Gardner-Medwin, A. R., Long-lasting potentiation of synaptic transmission in the dentate area of the unanaesthetized rabbit following stimulation of the perforant path, *J. Physiol. (London)*, 232, 357, 1973.
44. Ranck, Jr., J. B., Studies on single neurons in dorsal hippocampal formation and septum in unrestrained rats. I. Behavioral correlates and firing repertoires, *Exp. Neurol.*, 41, 462, 1973.
45. Olds, J. and Hirano, T., Conditioned responses of hippocampal and other neurons, *Electroencephalogr. Clin. Neurophysiol.*, 26, 159, 1969.
46. Skrede, K. K. and Westgaard, R. H., The transverse hippocampal slice: a well-defined cortical structure maintained *in vitro*, *Brain Res.*, 35, 589, 1971.
47. Schwartzkroin, P. A., Characteristics of CA1 neurons recorded intracellularly in the hippocampal *in vitro* slice preparation, *Brain Res.*, 85, 423, 1975.
48. Tonegawa, S., Li, Y., Erzurumlu, R. S., Jhaveri, S., Chen, C., Goda, Y., Paylor, R., Silva, A. J., Kim, J. J., Wehner, J. M. et al., The gene knockout technology for the analysis of learning and memory, and neural development, *Prog. Brain Res.*, 105, 3, 1995.
49. Wehner, J. M., Bowers, B. J., and Paylor, R., The use of null mutant mice to study complex learning and memory processes, *Behav. Genet.*, 26, 301, 1996.
50. Bach, M. E., Hawkins, R. D., Osman, M., Kandel, E. R., and Mayford, M., Impairment of spatial but not contextual memory in CaMKII mutant mice with a selective loss of hippocampal LTP in the range of the theta frequency, *Cell*, 81, 905, 1995.
51. Bellinger, F. P., Madamba, S. G., Campbell, I. L., and Siggins, G. R., Reduced long-term potentiation in the dentate gyrus of transgenic mice with cerebral overexpression of interleukin-6, *Neurosci. Lett.*, 198, 95, 1995.
52. Masoro, E. J., Dietary restriction and aging, *J. Am. Geriat. Soc.*, 41, 994, 1993.
53. Masoro, E. J., Dietary restriction, *Exp. Gerontol.*, 30, 291, 1995.
54. Namgung, U., Valcourt, E., and Routtenberg, A., Long-term potentiation *in vivo* in the intact mouse hippocampus, *Brain Res.*, 689, 85, 1995.
55. Jaffard, R. and Jeantet, Y., Posttraining changes in excitability of the commissural path-CA1 pyramidal cell synapse in the hippocampus of mice, *Brain Res.*, 220, 167, 1981.
56. Galey, D., Destrade, C., and Jaffard, R., Relationships between septo-hippocampal cholinergic activation and the improvement of long-term retention produced by medial septal electrical stimulation in two inbred strains of mice, *Behav. Brain Res.*, 60, 183, 1994.
57. Braga, P. C., Dall'oglio, G., and Fraschini, F., Microelectrode tip in five seconds. A new simple, rapid, inexpensive method, *Electroencephalogr. Clin. Neurophysiol.*, 42, 840, 1977.
58. Gomori, G., Microtechnical demonstration of iron, *Am. J. Pathol.*, 12, 655, 1936.
59. Staubli, U. and Lynch, G., Stable hippocampal long-term potentiation elicited by 'theta' pattern stimulation, *Brain Res.*, 435, 227, 1987.
60. McNaughton, B. L., Evidence for two physiologically distinct perforant pathways to the fascia dentata, *Brain Res.*, 199, 1, 1980.
61. Deadwyler, S. A., Biela, J., Rose, G., West, M., and Lynch, G., A microdrive for use with glass or metal microelectrodes in recording from freely-moving rats, *Electroencephalogr. Clin. Neurophysiol.*, 47, 752, 1979.
62. Brozek, G., Buresova, O., and Bures, J., Electrophysiological analysis of retrieval of conditioned taste aversion. Physiological techniques and data processing, *Physiol. Bohemoslov.*, 28, 537, 1979.
63. Sinnamon, H. M. and Woodward, D. J., Microdrive and method for single unit recording in the active rat, *Physiol. Behav.*, 19, 451, 1977.

64. Ainsworth, A. and J, O. K., A lightweight microdrive for the simultaneous recording of several units in the awake, freely moving rat, *J. Physiol. (London)*, 269, 8P, 1977.
65. Zhang, J. X. and Harper, R. M., A new microdrive for extracellular recording of single neurons using fine wires, *Electroencephalogr. Clin. Neurophysiol.*, 57, 392, 1984.
66. Veregge, S. and Frost, Jr., J. D., A simple inexpensive, hydraulic microdrive for recording neocortical unit activity in the unanesthetized rat, *Electroencephalogr. Clin. Neurophysiol.*, 61, 94, 1985.
67. Pager, J., A removable head-mounted microdrive for unit recording in the free-behaving rat, *Physiol. Behav.*, 33, 843, 1984.
68. Sainsbury, R. S., Montoya, C. P., Westra, I., and Mani, T., An inexpensive direct-line, non-rotating, chronically implanted microdrive for small rodents, *Physiol. Behav.*, 31, 729, 1983.
69. Bland, B. H., Colom, L. V., and Mani, T. E., An improved version of a direct-drive, nonrotating manual microdrive, *Brain Res. Bull.*, 25, 441, 1990.
70. Chang, F. C., Scott, T. R., and Harper, R. M., Methods of single unit recording from medullary neural substrates in awake, behaving guinea pigs, *Brain Res. Bull.*, 21, 749, 1988.
71. Smith, S. S. and Chapin, J. K., A paradigm for determination of direct drug-induced modulatory alterations in Purkinje cell activity during treadmill locomotion, *J. Neurosci. Meth.*, 21, 335, 1987.
72. McNaughton, B. L., O'Keefe, J., and Barnes, C. A., The stereotrode: a new technique for simultaneous isolation of several single units in the central nervous system from multiple unit records, *J. Neurosci. Meth.*, 8, 391, 1983.
73. Korshunov, V. A., Miniature microdrive for extracellular recording of neuronal activity in freely moving animals, *J. Neurosci. Meth.*, 57, 77, 1995.
74. Goldberg, E., Minerbo, G., and Smock, T., An inexpensive microdrive for chronic single-unit recording, *Brain Res. Bull.*, 32, 321, 1993.
75. Rebec, G. V., Langley, P. E., Pierce, R. C., Wang, Z., and Heidenreich, B. A., A simple micro-manipulator for multiple uses in freely moving rats: electrophysiology, voltammetry, and simultaneous intracerebral infusions, *J. Neurosci. Meth.*, 47, 53, 1993.
76. Wilson, M. A. and McNaughton, B. L., Dynamics of the hippocampal ensemble code for space, *Science*, 261, 1055, 1993.
77. Winson, J., A compact micro-electrode assembly for recording from the freely-moving rat, *Electroencephalogr. Clin. Neurophysiol.*, 35, 215, 1973.
78. Fox, S. E. and Ranck, Jr., J. B., Hippocampal field potentials evoked by stimulation of multiple limbic structures in freely moving rats, *Neuroscience*, 4, 1467, 1979.
79. Rose, G. M. and Gratton, A., Eating behavior is accompanied by dopamine release in rat medial striatum, *Neurosci. Abstr.*, 16, 437, 1990.
80. Kubie, J. L., A driveable bundle of microwires for collecting single-unit data from freely-moving rats, *Physiol. Behav.*, 32, 115, 1984.
81. Khan, I. M., Taylor, P., and Yaksh, T. L., Cardiovascular and behavioral responses to nicotinic agents administered intrathecally, *J. Pharmacol. Exp. Ther.*, 270, 150, 1994.
82. Diamond, D. M., Bennett, M. C., Fleshner, M., and Rose, G. M., Inverted-U relationship between the level of peripheral corticosterone and the magnitude of hippocampal primed burst potentiation, *Hippocampus*, 2, 421, 1992.
83. Lipton, P., Aitken, P. G., Dudek, F. E., Eskessen, K., Espanol, M. T., Ferchmin, P. A., Kelly, J. B., Kreisman, N. R., Landfield, P. W., Larkman, P. M., et al., Making the best of brain slices: comparing preparative methods, *J. Neurosci. Meth.*, 59, 151, 1995.
84. Alreja, M. and Agajanian, G. K., Intracellular application of macromolecules through patch pipettes in brain slices, in *Brain Slices in Basic and Clinical Research*, Schurr, A. and Rigor, B.M., Eds., CRC Press, Boca Raton, FL, 1995, 117.
85. Alger, B. E., Dhanjal, S. S., Dingeldine, R., Garthwaite, J., Henderson, G., King, G. L., Lipton, P., North, A., Schwartzkroin, P. A., Sears, T. A., Segal, M., Whittingham, T. S., and Williams, J., Brain slice methods, in *Brain Slices*, Dingledine, R., Eds., Plenum Press, New York, 1984, 381.
86. Duffy, C. J. and Teyler, T. J., A simple tissue slicer, *Physiol. Behav.*, 14, 525, 1975.
87. Bennett, G. W., Sharp, T., Marsden, C. A., and Parker, T. L., A manually-operated brain tissue slicer suitable for neurotransmitter release studies, *J. Neurosci. Meth.*, 7, 107, 1983.

88. Magnuson, D. S., Johnson, R., Peet, M. J., Curry, K., and McLennan, H., A novel spinal cord slice preparation from the rat, *J. Neurosci. Meth.*, 19, 141, 1987.

89. Yamamoto, C. and McIlwain, H., Electrical activities in thin sections from the mammalian brain maintained in chemically-defined media *in vitro*, *J. Neurochem.*, 13, 1333, 1966.

90. Dunwiddie, T. V., Age-related differences in the *in vitro* rat hippocampus. Development of inhibition and the effects of hypoxia, *Dev. Neurosci.*, 4, 165, 1981.

91. Leonard, B. W., Barnes, C. A., Rao, G., Heissenbuttel, T., and McNaughton, B. L., The influence of postmortem delay on evoked hippocampal field potentials in the *in vitro* slice preparation, *Exp. Neurol.*, 113, 373, 1991.

92. Teyler, T. J., Brain slice preparation: hippocampus, *Brain Res. Bull.*, 5, 391, 1980.

93. Aitken, P. G., Breese, G. R., Dudek, F. E., Edwards, F., Espanol, M. T., Larkman, P. M., Lipton, P., Newman, G. C., Nowak, Jr., T. S., Panizzon, K. L., et al., Preparative methods for brain slices: a discussion, *J. Neurosci. Meth.*, 59, 139, 1995.

94. Tancredi, V. and Avoli, M., Control of spontaneous epileptiform discharges by extracellular potassium: an "*in vitro*" study in the CA1 subfield of the hippocampal slice, *Exp. Brain Res.*, 67, 363, 1987.

95. Leschinger, A., Stabel, J., Igelmund, P., and Heinemann, U., Pharmacological and electrographic properties of epileptiform activity induced by elevated K+ and lowered Ca2+ and Mg2+ concentration in rat hippocampal slices, *Exp. Brain Res.*, 96, 230, 1993.

96. Lobner, D. and Lipton, P., Intracellular calcium levels and calcium fluxes in the CA1 region of the rat hippocampal slice during *in vitro* ischemia: relationship to electrophysiological cell damage, *J. Neurosci.*, 13, 4861, 1993.

97. Collingridge, G. L., Herron, C. E., and Lester, R. A., Synaptic activation of N-methyl-D-aspartate receptors in the Schaffer collateral-commissural pathway of rat hippocampus, *J. Physiol. (London)*, 399, 283, 1988.

98. Dunwiddie, T. V. and Lynch, G., The relationship between extracellular calcium concentrations and the induction of hippocampal long-term potentiation, *Brain Res.*, 169, 103, 1979.

99. Spencer, H. J., Gribkoff, V. K., Cotman, C. W., and Lynch, G. S., GDEE antagonism of iontophoretic amino acid excitations in the intact hippocampus and in the hippocampal slice preparation, *Brain Res.*, 105, 471, 1976.

100. White, W. F., Nadler, J. V., and Cotman, C. W., A perfusion chamber for the study of CNS physiology and pharmacology *in vitro*, *Brain Res.*, 152, 591, 1978.

101. Haas, H. L., Schaerer, B., and Vosmansky, M., A simple perfusion chamber for the study of nervous tissue slices *in vitro*, *J. Neurosci. Meth.*, 1, 323, 1979.

102. Corrigall, W. A. and Lucato, R. M., A simple modification to permit fast-flow perfusion of brain slices, *Brain Res. Bull.*, 5, 481, 1980.

103. Kelso, S. R., Nelson, D. O., Silva, N. L., and Boulant, J. A., A slice chamber for intracellular and extracellular recording during continuous perfusion, *Brain Res. Bull.*, 10, 853, 1983.

104. Palovcik, R. A. and Phillips, M. I., A constant perfusion slice chamber for stable recording during the addition of drugs, *J. Neurosci. Meth.*, 17, 129, 1986.

105. Bingmann, D. and Kolde, G., PO2-profiles in hippocampal slices of the guinea pig, *Exp. Brain Res.*, 48, 89, 1982.

106. Kang, H. and Schuman, E. M., Long-lasting neurotrophin-induced enhancement of synaptic transmission in the adult hippocampus, *Science*, 267, 1658, 1995.

107. McAllister-Williams, R. H. and Kelly, J. S., The importance of temperature control when investigating high threshold calcium currents in mammalian neurones, in *Brain Slices in Basic and Clinical Research*, Schurr, A. and Rigor, B.M., Eds., CRC Press, Boca Raton, FL, 1995, 149.

108. Schiff, S. J. and Somjen, G. G., The effects of temperature on synaptic transmission in hippocampal tissue slices, *Brain Res.*, 345, 279, 1985.

109. Thompson, S. M., Masukawa, L. M., and Prince, D. A., Temperature dependence of intrinsic membrane properties and synaptic potentials in hippocampal CA1 neurons *in vitro*, *J. Neurosci.*, 5, 817, 1985.

110. Shen, K. F. and Schwartzkroin, P. A., Effects of temperature alterations on population and cellular activities in hippocampal slices from mature and immature rabbit, *Brain Res.*, 475, 305, 1988.

111. Berg-Johnsen, J. and Langmoen, I. A., Temperature sensitivity of thin unmyelinated fibers in rat hippocampal cortex, *Brain Res.*, 576, 319, 1992.

112. Mitani, A., Kadoya, F., and Kataoka, K., Temperature dependence of hypoxia-induced calcium accumulation in gerbil hippocampal slices, *Brain Res.*, 562, 159, 1991.

113. Morris, M. E., Leblond, J., Agopyan, N., and Krnjevic, K., Temperature dependence of extracellular ionic changes evoked by anoxia in hippocampal slices, *J. Neurophysiol.*, 65, 157, 1991.

114. Taylor, C. P. and Weber, M. L., Effect of temperature on synaptic function after reduced oxygen and glucose in hippocampal slices, *Neuroscience*, 52, 555, 1993.

115. Yamashima, T., Takita, M., Akaike, S., Hirano, M., Miyakawa, A., Miyazawa, A., Kudo, Y., and Yoshioka, T., Temperature-dependent Ca2+ mobilization induced by hypoxia-hypoglycemia in the monkey hippocampal slices, *Biochem. Biophys. Res. Commun.*, 205, 1843, 1994.

116. Williams, J. H., Li, Y. G., Nayak, A., Errington, M. L., Murphy, K. P., and Bliss, T. V., The suppression of long-term potentiation in rat hippocampus by inhibitors of nitric oxide synthase is temperature and age dependent, *Neuron*, 11, 877, 1993.

117. McNaughton, B. L., Shen, J., Rao, G., Foster, T. C., and Barnes, C. A., Persistent increase of hippocampal presynaptic axon excitability after repetitive electrical stimulation: dependence on N-methyl-D-aspartate receptor activity, nitric-oxide synthase, and temperature, *Proc. Natl. Acad. Sci. U.S.A.*, 91, 4830, 1994.

118. Edwards, F. A., Patch-clamp recording in brain slices, in *Brain Slices in Basic and Clinical Research*, Schurr, A. and Rigor, B.M., Eds., CRC Press, Boca Raton, FL, 1995, 99.

119. Hill, D. K. and Keynes, R. D., Opacity changes in stimulated nerve, *J. Neurophysiol.*, 108, 278, 1949.

120. Salzberg, B. M., Optical recording of voltage changes in nerve terminals and in fine neuronal processes, *Annu. Rev. Physiol.*, 51, 507, 1989.

121. Sick, T. J. and LaManna, J. C., Intrinsic optical properties of brain slices: Useful indices of electrophysiology and metabolism, in *Brain Slices in Basic and Clinical Research*, Schurr, A. and Rigor, B.M., Eds., CRC Press, Boca Raton, FL, 1995, 47.

122. Turner, D. A., Aitken, P. G., and Somjen, G. G., Optical mapping of translucence changes in rat hippocampal slices during hypoxia, *Neurosci. Lett.*, 195, 209, 1995.

123. Iijima, T., Witter, M. P., Ichikawa, M., Tominaga, T., Kajiwara, R., and Matsumoto, G., Entorhinal-hippocampal interactions revealed by real-time imaging, *Science*, 272, 1176, 1996.

124. Segal, M. and Auerbach, J. M., Imaging of intracellular calcium in hippocampal slices: methods, limitations, and achievements, in *Brain Slices in Basic and Clinical Research*, Schurr, A. and Rigor, B.M., Eds., CRC Press, Boca Raton, FL, 1995, 89.

125. Schwartz, R. D. and Yu, X., Optical imaging of intracellular chloride in living brain slices, *J. Neurosci. Meth.*, 62, 185, 1995.

126. Ebner, T. J. and Chen, G., Use of voltage-sensitive dyes and optical recordings in the central nervous system, *Prog. Neurobiol.*, 46, 463, 1995.

127. Plenz, D. and Aertsen, A., Current source density profiles of optical recording maps: A new approach to the analysis of spatio-temporal neural activity patterns, *Eur. J. Neurosci.*, 5, 437, 1993.

128. Wu, L. G. and Saggau, P., Presynaptic calcium is increased during normal synaptic transmission and paired-pulse facilitation, but not in long-term potentiation in area CA1 of hippocampus, *J. Neurosci.*, 14, 645, 1994.

129. Hess, G. and Kuhnt, U., Presynaptic calcium transients evoked by paired-pulse stimulation in the hippocampal slice, *Neuroreport*, 3, 361, 1992.

130. Saggau, P., Galvan, M., and ten Bruggencate, G., Long-term potentiation in guinea pig hippocampal slices monitored by optical recording of neuronal activity, *Neurosci. Lett.*, 69, 53, 1986.

6

Using Drosophila in Experimental Aging Research

Robert Arking and R. C. Woodruff

CONTENTS

6.1   Introduction

The little fruit fly, *Drosophila melanogaster*, has been used profitably in aging research since 1913 [1]. Its continued use over the better part of a century suggests that using this organism as a model system with which to study certain aspects of the aging process must confer certain advantages when compared to other organisms. Certainly, the disadvantages of using *Drosophila* in experimental aging research are obvious. It is a small animal, difficult to analyze at the tissue level using many standard physiological assays. As an invertebrate, the experimental results are difficult to translate directly to mammals. They escape. There are certain offsetting advantages, of course. First, even long-lived strains of the fruit fly are

145

relatively short lived, making possible not only the contemplation of long-term multigeneration experiments, but their probable completion before the grant runs out or the investigator retires. Second, they are comparatively inexpensive to raise and assay, requiring excessive amounts of neither money nor space. Third, the fact that the cells of the adult insect are, with the exception of the germ cells and some specialized blood cells, entirely postmitotic allows one to examine senescent processes uncomplicated by the effects of cell division and replacement. The fly can be viewed as an organism composed of synchronously aging differentiated cells. And fourth, many of the cellular level phenomema that occur in the insect are similar to those known to occur in vertebrates, suggesting that the similarity in fundamental aging mechanisms crosses species boundaries and that fundamental findings obtained with the fruit fly may yield insights into mammalian aging processes [2].

But their prime attribute is their unparalled utility as a genetic organism. Nearly a century of research has yielded a treasure of genetic reagents, ranging from a multitude of special chromosomes to thousands of specific single gene mutants to a variety of specifically designed genetic probes. The *Drosophila* Genome Project will soon complete the sequencing of this model genome [3–5]. Equally important is the fact that almost the entire genome is contained within cosmids and/or P1 clones, and that these clones are keyed to the salivary chromosome bands via FlyBase [6]. The latter is the computerized database in which all of this wealth of genetic, developmental, physiological, and ecological data is being integrated (Flybase, 1995). This means that, if we can somehow mutationally identify a key gene in some gerontological process and localize it by *in situ* hybridization to the salivary gland chromosomes, we can then skip almost immediately to a sequenced clone containing the gene. This eliminates much technically demanding but noninformative work and markedly speeds up the pace of the work. It seems to us that the one step limiting our ability to make this a routine process is the current difficulty in identifying discrete genes importantly involved in aging and longevity. We also believe that this difficulty is transient. As the aging process becomes defined less in terms of longevity and more in terms of fundamental biological mechanisms, then the easier it will be to devise experimental procedures with which to detect the genes involved. Any question which can be phrased in genetical terms can be profitably studied in *Drosophila*. Of course, that question could also be studied in other well characterized genetic organisms such as yeast, nematodes, and mice, which are described elsewhere in this volume. But none of these combine the practical advantages enumerated above with the power of genetics and the advantage of being a moderately complex multicellular organism capable of existing in epigenetic states as do mammals.

A certain healthy skepticism regarding the utility of using *Drosophila* in aging research has been expressed by Johnson et al. [7], based on a concern about the failure to identify candidate genes, the assumed involvement of large numbers of genes with heterotic effects, the large difference in life histories between mammals and insects, and the inability to ascertain causality of extended longevity in *Drosophila*. We believe that recent findings have adequately answered these concerns. Genes conferring resistance to various stresses, such as the antioxidant defense system [8] or the heat shock genes [9,10], are known to increase longevity. These candidate genes are viewed as comprising the proximal longevity assurance mechanisms. Single gene mutational techniques are being used to identify the specific genes involved [11,12]. These same experiments are designed to provide evidence for a causal relationship between the candidate genes and the extended longevity phenotype, as Mackay et al. [13] have demonstrated for other complex phenotypes of the fly. Although theory suggests that many genes could well be involved in the expression of the extended longevity phenotype, the empirical data suggest that a much smaller number may actually play key roles. Orr and Sohal [14] were able to significantly increase longevity in some (but not all) of their transgenic lines by introducing only two antioxidant defense system genes. Changes in the expression of these same genes are thought to be responsible for the extended longevity characteristic of our selected strains [8]. There are no heterotic effects noted in our strains, the genes involved being normal recessives [15]. There is undoubtedly a large difference in life history strategies between mammals and dipteran insects; yet the disposable soma theory of aging [16] suggests that longevity is the outcome of a series of trade-offs between reproduction and repair/defense processes. Given the fundamental similarity of basic biological processes, then it is not surprising that enhanced antioxidant defenses, for example, have been implicated in the extended longevity mechanisms probably operative in fungi, nematodes, flies, and mammals [17,18]. We suggest that appropriate investigations into the extended longevity of *Drosophila* may provide insights useful for analysis of similar mechanisms in mammals, including humans.

There are too many specialized procedures available for the genetic analysis of complex processes in *Drosophila* for this article to list and describe each one. Instead, our goal in writing this chapter was to present an overview of the genetic strategies available now and describe those few which seemed to us to present the greatest opportunity for future understanding while adequately referencing the more standard techniques.

6.2   Background

6.2.1   Earlier Genetical Analysis in Biogerontology

A detailed review of the earlier history of genetic investigations into the aging processes of *Drosophila* has been presented elsewhere [2,19,19a] and needs not be duplicated here. Between the two of them, these two papers contain more than 500 references to this earlier work. Suffice it to say that our overview of this earlier work concluded that the analysis of single-gene, short-lived mutants (i.e., the bulk of the then-existing mutants) was not a promising approach because there was no reliable method to decide which mutants were affecting the aging process and which were affecting unrelated but vital functions. Leffelaar and Grigliatti [20] showed us one way of resolving this puzzle by using behavioral landmarks as biomarkers of aging, specifically employing them to sort out adult temperature-sensitive lethal mutations causing premature death due to accelerated aging from those causing premature death due to a nonaging-related cause. This earlier work also made it clear that the comparative analysis of existing ordinary laboratory strains with different longevities was not useful due to the inability to decide whether any given biochemical feature bore a causal or a correlative relationship to the difference in strain-specific life spans (i.e., see Ganetsky and Flanagan [21]). The wisdom of using artificial selection to construct long-lived and short-lived strain specifically designed for aging studies was pioneered by Rose and Charlesworth [22] and verified by the reports of Rose, Luckinbill et al., Partridge and Fowler, and Zwaan et al. [23–26]. We found that the use of biomarkers was very useful with these selected strains as well, for their use allowed us to confirm that our selected long-lived strain lived long due to a delayed onset of senescence rather than because of some effect unrelated to aging [27].

Most of the more recent work has been done with either selected strains or else with specific mutants which theoretical considerations had suggested would be involved in the aging processes.

6.2.2 Genetic Strategies

There are two classic genetic approaches to solving complex phenotypes: the "top-down" approach and the "bottom-up" approach [28,29]. In the former, one starts with with a longevity phenotype altered by means of either a single gene mutation or by multiple genes (i.e., as a result of a mutagenesis or selection experiment) and attempts to construct a series of causal events linking the altered phenotype to the altered genotype. This causality may take the form of a physiological explanation in which a series of plausible biological events are deduced. Subsequent experimental steps would likely attempt to critically test these deductions. Alternatively, the causal connection may take the form of a statistical linkage between the phenotype and certain loci statistically associated with a particular longevity phenotype (i.e., the QTL approach).

In the "bottom-up" approach, one starts with a specific gene sequence / gene product and attempts to causally link these altered gene products with the altered phenotype. There are several possible methods of identifying the genes. One might search for genes which have age-specific changes in their expression patterns and then attempt to relate this alteration to the longevity phenotype being examined. Alternatively, the specific genes may have been initially unknown and their loci identified at first solely by statistical means, such as the QTL procedures. Finally, genes thought likely to play a role in a plausible biological process may be identified and assayed to determine whether they are involved in the production of the altered phenotype (i.e., the "candidate gene" approach). One might do cross-species searches by using the expressed sequence tag (EST) techniques to search for *Drosophila* homologues to interesting genes known in other organisms [30].

Modern molecular biological techniques have allowed us the experimental freedom to shift back and forth between these two approaches as best suits the particular experimental scheme. One hypothetical example of this new approach would be the use of artificial selection to create strains with an altered longevity, the use of genetic techniques (i.e., mutation) or molecular techniques (i.e., subtractive cDNA libraries or differential display; Liang and Pardee, [31]) to identify genes associated with the phenotypic alteration, and the cloning of each such gene product and the determination of its cytogenetic location, followed by its sequencing and comparison with the genome data in FlyBase in order to confirm the map location and identify the function of the gene. Once the gene has been identified, more traditional mutational analysis can be undertaken in order to characterize the gene's functions, or newer techniques such as the FLP-FRT technique [32–34] may be used in order to directly screen for new mutations that affect the function of the identified gene. In principle, the techniques of classical and reverse genetics now available make it possible to construct as complete an explanation of the relationship between genotype and phenotype as resources permit.

6.2.3 Recent Examples of the Genetical Approach to Biogerontology

6.2.3.1 *Evolutionary Analysis of the Phenotype: Strategies*

The basic evidence supporting the evolutionary theory of aging is well summarized in Rose's book [35]. In brief, aging arises because the force of natural selection declines with age. As a result, there is strong selection for high early fitness but weak or no selection for late fitness. Two different mechanisms have been proposed to explain this situation: mutation accumulation [36] and antagonistic pleiotropy [37]. Much effort has been expended in the effort to experimentally test these two hypotheses, but they have proved to be difficult to evaluate. Curtsinger et al. [28] suggests that the reasons for this lack of progress involve both tactical and strategic errors. For example, the search for trade-offs between early and late life history traits, as is postulated by the antagonistic pleiotropy hypothesis, usually involves the measurement of genetic correlations which are "...strain specific, ...difficult to estimate, subject to large statistical error, and dependent upon gene frequencies" [28]. An alternative approach to testing robustly this question might be to use QTL mapping (see below) to map the genes responsible for early and for late fitness components [28]. Those that map to the same locus are likely to be pleiotropic traits and could be interpreted as constituting positive evidence for the validity of the hypothesis, provided that other processes that might logically yield the same empirical result are experimentally eliminated. The point is that the experimental tool employed must be robust enough to provide a positive answer without being misled by the noise inherent in the system. In an ideal experimental system, even the negative evidence would be strongly informative. In addition, demographic information can often be used profitably to complement the genetic information and suggest new experimental strategies.

6.2.3.2 Mechanistic Analysis of the Phenotype: Strategies

6.2.3.2.1 The Candidate Gene Approach

The candidate gene approach is based on identifying plausible biological mechanisms involved in the aging process, followed by testing genes known to be involved in that process in order to determine if they are differentially expressed in long-lived vs. normal-lived animals. It was this approach which underlaid the decision of several labs to investigate the expression patterns of known antioxidant defense genes and elucidate their role in the extended longevity phenotype [8,14]. The data suggesting that free radical-caused oxidative damage might be playing a significant role in the aging process of *Drosophila* was persuasive, albeit indirect and correlative [2]; thus, choosing this system for a detailed analysis was, to some extent, a well-educated guess. Not all such guesses are correct. Elongation factor 1α ($EF1\alpha$) is essential for protein synthesis and its expression shows a sharp age-related decline consistent with its playing an important role in bringing about the onset of senescence [38,39]. Subsequent work showed that, despite the plausibility of the supporting evidence, this gene appears to play no important regulatory role in the aging of *Drosophila* [40–43]. Thus one drawback of the candidate gene approach is that one may invest substantial resources only to determine that the existence of plausible evidence need not indicate the existence of a causal relationship between that gene and the longevity phenotype in question. In addition, the candidate gene approach is limited to known genes, which is a definite drawback. However, one could use the EST techniques described above to expand the set of known genes from those known only in *Drosophila* to those known in any organism [44,45]. And finally, the experimental determination of causality will probably require mutational tests. Nonetheless, the candidate gene approach has provided an important and useful means of inserting an experimental wedge into a difficult analytical problem.

6.2.3.2.2 Differential Gene Expression

Differential gene expression provides another means of identifying genes which may be involved in the expression of the aging phenotype. In this case, one searches for age-related changes in specific gene expression, usually at the mRNA level. Once identified, such genes could then serve as candidate genes, as described above. Such procedures allowed Fleming et al. [46] to identify the flies' response to heat shock, and the subsequent expression of ubiquitin [47], as an interesting candidate gene system. Tower and his colleagues [9] have more specifically identified *hsp22, hsp23*, and *hsp70* as candidate genes, while Khazaeli et al. [10] have shown that thermal stress can increase longevity presumably because of heat shock gene activity.

Direct visualization of differential gene expression has been developed. Using an approach in which an enhancer trap was used to drive a β-galactosidase gene, Helfand and his colleagues have characterized the expression patterns of a number of different marked genes in the adult antennae [48]. It remains to be seen whether this approach can be used to detect mutated genes. Calleja et al. have also been able to visualize gene expression in living adult flies, using a *Gal4* containing P-element which interacts with a *lacZ* marked upstream activating site (UAS), to yield the tissue- and time-specific patterns of the genes [49]. In both cases, gene expression patterns were found to be altered as the background genotype was altered.

6.2.3.2.3   *Statistical Identification of Genes*

Statistical identification of genes significantly associated with long life provides a useful means for identifying presently unknown genes. This approach requires the use of two strains which differ in some aspect of their longevity phenotype, as well as a number of genetic markers. By controlled crosses, one can create backcross progeny and determine whether a statistical association exists between a particular phenotypic trait and some subset of the genetic markers. If such an association is uncovered, then one has identified a chromosomal region containing one or more loci which may be viewed as candidate genes. Curtsinger's group has reported one such candidate locus on the third chromosome of a selected long lived strain of *Drosophila* [50]. Subsequent tests should allow one to distinguish the marker gene from the actual candidate loci. Quantitative trait loci (QTL) mapping is particulary well adapted to the identification of unknown genes involved in the expression of a complex quantitative phenotype without the necessity of making guesses about the nature of the biological mechanisms operating in the organism. In this sense, it is the most direct approach.

6.2.3.2.4   *Mutational Identification*

Mutational identification of genes associated with long life has, until recently, not been very useful. In large part, this is due to the fact that extended longevity is not due to one gene and it is very unlikely that a mutagenesis treatment would induce the multiple gain-of-function genes necessary to convert a normal animal into a long-lived mutant. Loss-of-function alleles, on the other hand, are easy to obtain but are likely to shorten longevity due to effects not specific to the aging process. This logical impasse is most unfortunate since the advances in *Drosophila* developmental genetics since 1980 well illustrate the power of this classical genetic approach. A possible solution to this problem demonstrates that one could use loss-of-function mutants to identify genes involved in extended longevity *if* an appropriately specialized system was first constructed. Prior work on this system involved the construction via artificial selection of an extended longevity strain [51], the selection of the antioxidant defense system as a likely biological mechanism involved in the phenotype [52], the identification of the *CuZnSOD* (Cu-Zn superoxide dismutase) and *Cat* (catalase) loci as candidate genes [8], and the development of an efficient mutant screen [53]. This was then followed by the mutational identification of loci coordinately regulating the expression of both these candidate genes and the longevity phenotype [12]. In this system, the mutant phenotype is that of a normal-lived animal. The loss of function mutants can be identified as longevity assurance genes whose continued expression is essential to the continued expression of the extended longevity phenotype. Proof of their causal involvement, along with identification of the genes, should lead to an understanding of the mechanisms regulating the expression of the antioxidant candidate genes. Classical mutational analysis should become increasingly more useful as other specialized systems for the analysis of longevity are devised. One virtue of the mutational approach is that it allows for the identification of multiple pathways affecting the longevity phenotype and for the relative ordering of the different loci within those pathways. The medium term goal of *Drosophila* geneticists should be to attain this level of understanding of the genetic circuitry regulating longevity, senescence, and aging.

6.2.3.2.5 Clonal Analysis

Once the analysis gets down to a single gene level, then it should be possible and feasible to incorporate some of the more advanced techniques of *Drosophila* genetics into our inventory of tricks. Clonal analysis, for example, will allow us to determine the tissue specificity and the autonomy or nonautonomy of these candidate genes in exerting their effect on the longevity phenotype. At the moment, most of our biochemical analyses are mired at the level of whole-body assays. It is most unlikely that important aging mechanisms operate at that level; indeed, one expects that important regulatory events will have a tissue- and time-specific pattern of candidate gene expression. The ability to use mosaic techniques, such as the gynandromorph [54] or X-ray induced somatic recombination (the *Minute* technique; see [54a]) procedures, to create a chimeric animal composed of distinct mutant and nonmutant tissues should prove to be most valuable in delineating these regulatory tissues. Indeed, Trout and Kaplan [55] used the gynandromorph technique to determine that the *Hyperkinetic*[1] (*Hk*[1]) and *Shaker*[5] (*Sh*[5]) mutant alleles shortened the life span due to their deleterious effects in the anterior thoracic ganglion. More recently, Golic and Lindquist [32] developed a more efficient technique for production of mosaics. The "FLP-FRT" procedure can be used with all chromosomes, its initiation is under the control of the investigator, it has no lethality or other untoward effects, and it can be used to screen for new mutations that directly affect the functioning of specific tissues [56,57]. This last point means that it might be possible to use it to efficiently isolate mutants in first-generation screens. John Tower has used the FLP-FRT technique in his analysis of the effect of overexpressing the *CuZnSOD* and *Cat* genes on the animal's longevity (personal communication). In principle, we should be able to determine whether the same sets of genes would have different epistatic patterns if expressed in different tissues and/or ages, or against different genetic backgrounds [58]. This information may be important in detecting tissues with different age-related failure thresholds. In addition, novel protein interactions potentially characteristic of tissue-specific age-related failures may be detected *in vitro* using the yeast interaction trap system [59].

Future work will undoubtedly identify genes regulating important aging processes, but if their regulatory function is altered significantly when they are placed in different backgrounds, then this may suggest that the new genome may have revealed some upstream regulatory complex. An understanding of genome-specific alteration of gene function may help us to understand the genesis of strain-specific aging patterns.

6.3 General Description of Genetic Tools and Techniques Needed

There is a large amount of information regarding *Drosophila* which is readily accessible. *Drosophila* is easy to maintain and handle, and information on its husbandry is available [61–67a,69,74a]. There is extensive information on its biology [61,62,68–83]. There is a wealth of genetic markers (both genic and chromosomal) that are useful in aging studies [84].

The FlyBase data base (http://cbbridges.harvard.edu.7081) contains information on the genetics, molecular, and cell biology of *Drosophila melanogaster* and other *Drosophila* species [6]. As of January 1997, FlyBase contains information on 38,000 alleles of more than 11,000 genes (more than twice the number of bands in the salivary gland polytene chromosomes); 13,300 chromosomal rearrangements; *Drosophila* genetic maps; functions of gene products; lists of wild-type strains and chromosomes; lists of over 12,200 nucleic acid and over 3900 protein sequence accession numbers; lists of 7000 genomic clones; lists of 1000 transposable DNA elements and P-DNA insertion mutations; a bibliography of more than 81,000 *Drosophila* citations (from 1684 to the present); the addresses of over 5600 *Drosophila* researchers; allied data bases; the bionet *Drosophila* archives; Berkeley and European Genome Projects data; lists of *D. melanogaster* stocks in the Bloomington *Drosophila* Stock Center (Indiana University; e-mail: matthewk@indiana.edu), European *Drosophila* Stock Center (Umeą, Sweden; e-mail: karin.ekstrom@genetik.ume.se), and the Mid-America *Drosophila melanogaster* Stock Center (Bowling Green State University, Bowling Green Ohio; e-mail: Dmelano@opie.bgsu.edu); list of *Drosophila* species in the National *Drosophila* Species Resource Center (Bowling Green State University; e-mail: kayoon@bgnet.bgsu.edu); and lists of stocks in private laboratories. FlyBase also allows for searchers for genes, gene symbols, alleles, aberrations, stocks, species, transposons, clones (cosmid, P1, and YAC), DNA vectors, enhancer-trap expression patterns, people, addresses, and citations.

There are also sites on the World Wide Web that are useful for *Drosophila* researchers, including BCM Gene Finder (http://dot.imgen.bcm.tmc.edu:9331/gene-finder/gf.html) (splice sites, protein coding exons and gene model construction, promoter and poly-A region recognition); *Drosophila* Virtual Library (heep://www-leland.stanford.edu/~ger/*drosophila*.html)(internet resources for research on *Drosophila melanogaster*); Flybrain (http://flybrain.neurobio.arizona.edu/) (an on-line atlas and data base of the *Drosophila* nervous system); FlyView (http://flyview.uni-muenster.de/) (a *Drosophila* image database); Gifts (http://www-biol.univ-mrs.fr/~lgpd/GIFTS_home_page.html) (gene interactions in the Fly TranWorldServer); and The Interactive Fly (http://cbbridges.harvard.edu:7081/allied-data/lk/interactive-fly/aiman/laahome.htm) (a cyberspace guide to *Drosophila* genes and their roles in development).

Genetic stocks of *D. melanogaster* and other species can be obtained free of charge from the Mid-America and European stock centers and by a small fee from the other centers. Unfortunately, the Mid-America *Drosophila melanogaster* Stock Center, which has been funded by the National Science Foundation for over 30 years, will be terminated in 1998; part of the 4300 stocks in the Mid-America Center will be transferred to the Bloomington center. Stocks and species can also be ordered directly from the Bloomington, European, Mid-America, and Species centers from FlyBase.

Stock lists, notes on mutation, techniques and research, and e-mail addresses of researchers are also published in the *Drosophila Information Service* (or DIS) (Editor, J. N. Thompson, Jr., Department of Zoology, University of Oklahoma). A cumulative subject index of all *Drosophila Information Service* volumes from 1934 to 1992 has been published [85]. Gartner (1986) has also published a bibliography on aging in *Drosophila* [86].

For individuals interested in beginning research with *Drosophila*, no amount of reading can replace the knowledge gained from direct experience in a "fly lab." One should also attend the Annual *Drosophila* Research Conferences in the U.S. or Europe, where over 1000 scientists present research, techniques, and teaching talks and posters, and demonstrations are given on how to use FlyBase. At the recent fly meeting in Atlanta, one of us overheard locals in an elevator saying: "Not only are they studying flies, but there are over 1000 of them." In addition, *Drosophila* supplies for preparing food and rearing flies are advertised at these meetings. There are also regional meetings (see FlyBase and the *Drosophila*

Information Service for organizers, time, and meeting locations), and *Drosophila* courses are also offered each year at the Cold Spring Harbor Laboratory. Finally, some of the best information and help on *Drosophila* stocks, husbandry, supplies, etc. can be obtained from the stock centers.

In addition, there is a large amount of information available regarding aging in general. Finch's book is probably the single indispensable reference source for any one interested in the intersection between genetics and aging [87]. Post (1996) has compiled a list of internet addresses of aging resources, some of which are biologically oriented [88]. She writes an occasional column on this topic (with internet addresses) for *The Gerontologist*. Given the hyperlinks between sites, entering via one of the more biologically oriented addresses should allow one to fully sample the available information.

With the abundance of information and extensive genetics of *Drosophila*, and the well-deserved reputation for cooperation among *Drosophila* workers, it is possible to set up and begin running a *Drosophila* laboratory to study aging without a large budget and in a short period of time.

6.4 Selected Procedures in Detail

6.4.1 Selection Experiments

Different laboratories using different starting strains have successfully generated a variety of long- and short-lived strains, the assertions of Baret and Lints [89] that such strains are artifactual being contradicted by the facts [90,91]. It does not seem to matter whether one uses an indirect selection for longevity by actually selecting for age at reproduction [23–25,51] or whether one uses a direct selection for longevity [26]. It has long been an article of faith that direct selection for extended longevity was impractical. Zwaan's overturning of this "dogma" came about because he established a set of selection lines for which he kept the potential parents at 15°C (where the mean longevity >120 days) and measured the longevity of their sibs at 29°C (where maximum longevity <60 days). This represents a creative application of a well-known fact of *Drosophila* life history to a otherwise insoluble problem. Other dual-environment solutions may be feasible for other experimental regimes. The techniques involved in both types of experiments are well described in these papers and the reader should refer to them for details. There are five technical issues we will deal with here.

First is the nature of the intial strain. It is generally better to use an outbred rather than an inbred strain, simply because the success of any selection experiment depends upon the presence in the starting population of alleles capable of yielding the desired phenotype. Rare alleles may not be present in isogenic lines nor in highly inbred lines.

Second, it is important to retain the initial strain as an unselected baseline control for future comparisons with the selected strain. This requires an active procedure for counteracting the effects of inadvertent selection (see below). Comparing a selected long-lived strain with a selected short-lived strain is logically permissible only if it is known that the processes leading to the two phenotypes are the equal and opposite of one another. This condition is not met in the case of the selected long-lived L_A and short-lived $2E_A$ strains [92]; there is no evidence that would allow one to *assume* that it is operative in other strains. Unless explicitly supported by reported empirical data, this type of comparison should not be used.

Third, is the problem of avoiding the effects of unintended directional selection acting on the baseline strain. If this occurs, then one's comparisons might well yield inaccurate data. In our case, in each generation we reproduce the R control strains at a randomly chosen age between 1 and 31 days, while the selected strains are reproduced at a changing age which is operationally equal to that age at which about 75% of the cohort have died [51]. We know that a constant reproduction at 15 days (the mean of the 1–31 day interval) yielded a set of strains some of which were significantly shorter-lived than the R strains; this result showed that the precaution was necessary. The extra trouble taken with the controls enabled us to avoid the effects of inadvertent directional selection associated with our indirect approach.

Fourth, it is useful to fully identify which strains/sublines are being compared and their identifying characteristics when reporting your data. Failure to do this might lead to peer confusion regarding your data and conclusions, and may limit the usefulness of the report. This is particulary important in light of the next statement.

Fifth, it is generally accepted that selection experiments require the simultaneous selection of multiple replicate or sister lines, and that outcome measurements should be averaged across these replicate lines in order to demonstrate statistical validity [93]. Failure to do this has been viewed as an experimental shortcoming. The unspoken assumption here is that the application of a uniform selection pressure should bring about the same type of changes in all replicates. But this is true only of the phenotypic levels under the direct control of the investigator, i.e, longevity. The hidden cellular and subcellular levels of the phenotype are free to vary within this overall context. If this should happen, then two sister lines may achieve the same longevity phenotype via the use of quite different physiological mechanisms. We know that this has in fact occurred in our La and Lb strain, both of which use enhanced antioxidant resistance to achieve an extended longevity but both of which show dramatically different patterns of antioxidant gene expression and of oxidative damage [94]. Averaging these molecular values across the two lines yields a mean value actually descriptive of neither strain. They cannot be viewed as true replicates, and averaging across them is an experimental error. The worst consequence of this error may be that the false average values obscure the real mechanisms from your view.

6.4.2 Quantitative Trait Loci (QTL) Mapping

The recent identification of polymorphic DNA sequences (RFLP, VNTR, microsatellites, transposable DNA elements, etc.) that can be used as molecular markers has led to renewed attempts to identify QTL (polygenes) in plants and animals [95–98]. As with many areas of genetics, *Drosophila* has played an important role in the development of QTL mapping, mainly by the use of the bristle-number model system [13,99–104]. For example, using six visible markers Shrimpton and Robertson identified 17 QTL on the third chromosome of *D. melanogaster* that influenced sternopleural bristle numbers [102]. With the identification of molecular markers that can be localized to precise cytological sites, and the availability of selection lines that have increased (or decreased) lifespan in *Drosophila*, there is now a golden opportunity to identify QTL that influence aging in a metazoan.

The objective of identifying QTL is to map and ultimately clone these genes. This objective is attained by first identifying small segments of chromosomes that affect quantitative traits and are closely linked to molecular markers of known location; second, determining if a QTL containing region also contains a "candidate" gene that is known to alter the trait in question; and third, cloning and sequencing the DNA segment that is altered. The identified QTL could then be used to understand better quantitative genetic theory and to possibly improve selection experiments and treatments of multifactorial diseases [98].

In *Drosophila*, QTL that influence aging may be alleles of candidate loci, such as superoxide dismutase [8,14,103], catalase [8,14,105], xanthine dehydrogenase [8,106], and heat shock proteins [9], or, more likely, alleles of unknown genes (see discussion of candidate genes in [28,107]). The completion of the mapping and sequencing of the *D. melanogaster* genome will be of immense help in identifying aging QTL [3–5].

By use of chromosome substitution experiments and crosses between inbred lines, genes on the X, second, and third chromosomes have been observed to be responsible for changes in lifespan in long-lived *D. melanogaster* lines (the small fourth chromosome is seldom tested since it comprises only 2% of the genome but its analysis would triple the workload). However, aging QTL may be different in independently derived lines, and these aging polygenes may interact in a positive or negative manner [107]. For example, Buck et al. [15] observed that genes on the third chromosome are mainly responsible for extended longevity in their line, whereas Yonemura et al. [108] observed that genes on the X and an autosome caused an increase in lifespan. Furthermore, Hutchinson and Rose [109], Hutchinson et al., [110], and Luckinbill et al. [111] have reported that extended life span involves additive polygenes located on all chromosomes (see Rose, [35] for a discussion of this topic).

Although the more abundant molecular markers are just now beginning to be used in *Drosophila* to identify aging QTL, molecular markers already have been used successfully to identify QTL for bristle numbers. For example, Long et al. [104] used *roo* and *B104* transposable DNA elements as markers to identify bristle-number QTL. These transposable elements are useful molecular markers because they are neutral in relation to bristle number and fitness, transpose at a low rate, are scattered throughout the genome, can be localized to polytene chromosomal sites by *in situ* hybridization, and are localized at different sites in independently derived selection lines (see Dixon, [95] for a discussion of molecular markers for QTL experiments in *Drosophila*). Long et al. identified seven QTL for abdominal bristle numbers to the X and third chromosomes. Others have observed that some of the bristle-number QTL are presumed alleles for candidate genes that have known effects on bristle morphology, including *achaete-scute* complex, *bobbed, daughterless, Delta, Enhancer of split, extramacrochaetae, hairy, malformed abdomen, Notch, and scabrous* [98,102,103,112] (see Lindsley and Zimm, [84] or FlyBase [6] for descriptions of mutant alleles).

Similar techniques to those used to identify QTL for bristle numbers [104], or techniques that use other molecular markers such as microsatellites [113-117], could also be employed to identify QTL that affect aging in *D. melanogaster*. Curtsinger has reported the isolation of a QTL that accounts for ca. 30% of the difference in longevity between a long-lived and control strain [50]; thus, the utility of the procedure is proved in principle and in practice. The reader should see [95,104,118,119] for discussions and references on the theory, experimental design, and analyses of QTL mapping data and on the subsequent isolation of aging QTL.

6.4.3 Multiple Gene Mutagenesis

Since aging is due to some number of different genes, it might be thought that one way to identify the several genes involved would be to simultaneously mutate a number of genes. This is not a good idea. Multiple mutations are often the inadvertent result when one uses chemical or radiation mutagenesis protocols, there being no way to effectively limit the number of hits to one per genome. For the analysis of nonaging, single-gene phenotypes, it is customary that the resulting mutation must be recombinationally separated from all other mutants on the affected chromosome; if not, then one cannot deal effectively with the objection that the observed phenotype is the result of multiple genes. Yet failure to clean up the chromosomes can leave one open to the objection that the aging phenotype being studied, particulary if it involves a shortened longevity, is simply the nonspecific outcome of multiple mutants.

On the other hand, screening for second-site suppressors or enhancers represents a much more feasible approach to identifying the multiple genes involved in aging.

6.4.4   Single Gene Mutagenesis

As we use selected strains, transgenic animals, and/or multiple mutant lines to acquire more detailed information on aging mechanisms and to construct hypotheses regarding their operation, it will become increasingly important to be able to critically test our conclusions. One way to do this will be to phrase our conclusions in terms of the (unknown) genes or alleles hypothesized to act in some particular system, and to then use mutagenesis procedures to induce new mutants affecting the hypothesized genes. Mutational identification of putative candidate genes is a time-honored and powerful technique, since a positive result not only confirms the hypothesis but also provides the investigator with a powerful lever with which to pry open the system in question. A negative result, however, is not informative since it might mean either that the hypothesis is wrong or else that the screen used to identify the desired mutants was somehow defective and so the hypothesis was not adequately tested. It is obviously very important then to construct an efficient mutant screen, since this is the key to success. Ideally, such a screen will let only the desired mutants through and exclude all others. Knowing that we will fall short of the ideal suggests that it is even more important that the screen lets the desired mutants through even if it cannot exclude all others. In practical terms, the use of an inherently inefficient screen means that one usually needs to run putative mutants through a series of screening steps, each designed to identify flies with certain partial attributes of the desired phenotype and to exclude flies lacking such characters. In principle, only those mutants whose phenotype matches all of the tested attributes will make it through this multiple screen. In reality, errors occur and so some sort of confirmatory tests are always useful.

There is no generic screen that one can use for all phenotypes. Rather, each investigator needs to draw out from a common tool kit those tested devices that will isolate mutants with the desired phenotypes. Leffellaar and Grigliatti [20] give an early example of one such screen designed to detect sex-linked accelerated aging mutants; a summary and critique of that screen is presented in Table 6.1 and should make clear the multistep nature of mutant screens in general. Of course, that screen was designed to isolate aging mutants in one step, without any requirement that they should interact in a defined way with other genes known to be involved in the aging processes. As our knowledge progresses, it is more likely that we would want to mutationally identify such interactive loci. In addition, the efficient use of the *Drosophila* data bases and preexisting knowledge requires that we design our screens so as to be able to quickly and efficiently move from a mutant phenotype to a cloned mutant gene. An example of a P-transposable DNA element screen we recently used to isolate second chromosome genes regulating the expression of third chromosome antioxidant genes is summarized in Table 6.2. The use of P-element insertional mutagenesis has several major advantages over chemical or radiation mutagenesis:

1. It allows one to map the locus of the mutated gene by performing *in situ* hybridization with a nonradioactive labeled probe to the particular P-element used [5].
2. It permits the relatively easy physical recovery of the mutated gene especially if the P-element has both a plasmid origin of replication and an Amp^R gene.

Together, these permit the rapid isolation of the inserted DNA followed by sequencing of the *Drosophila* DNA flanking the P-element [119a].

3. Knowing the map location and some part of the DNA sequence gives one sufficient information to localize the mutated gene to the appropriate P1 clone [3]. These are keyed to the salivary chromosome locations and can be ordered from the various stock centers (see above).

With the wild-type version of the gene of interest in hand, the investigator does not have to spend much time isolating the gene but can now move into the more interesting aspects of the science. Of course, screens designed to identify genes involved in other aging processes must be custom designed to take advantage of the unique properties of the particular experimental system involved; yet they can and should use the generic balancer chromosomes and mutagenesis available to all.

However, P-element mutagenesis carries with it certain disadvantages. One disadvantage is that P-elements insert and disrupt only about 50% of the genes [5]. Hence, achieving effective saturation of mutants may require the use of other transposable elements with different distribution kinetics such as Hobo and/or Minos [120,121]. Another major disadvantage involves the possibility of wrongly identifying some locus as being importantly involved in some particular system due to the extensive pleiotropy observed in the system. This possible confounding effect was illustrated by Clark et al. [122]. In their study, they showed that single P-element insertions into the second or third chromosome had large pleiotropic effects such that each insertion had a high probability of affecting significantly several different metabolic enzyme activities, suggesting that the mutational target for each is quite large and involves many other genes. The genome, in other words, is highly integrated. "Whenever genes share a step in a regulatory hierarchy, such as a transcription factor, it is plausible that mutations in that regulatory chain will have pleiotropic effects on expression of genes downstream from the mutation. The extensive pleiotropy that was observed is therefore not entirely unexpected..." (Clark et al. [122], p. 347). When Lyman et al. [123] analyzed the effects of single P-element mutants on bristle number in *Drosophila*,

TABLE 6.1

Procedure	Comment
Oregon-R strain	General lab strain
Chemical mutagenesis of males using ethylmethanesulfonate (EMS)	Easy method of inducing base substitution mutations; no control over number of hits/chromosome, thus phenotype may be due to multiple genes. Difficult to isolate mutant clone at later time
Mate mutagenized males x marker female virgins and allow to lay eggs in two vials.	Standard protocol for isolating sex-linked mutants; convenient but assays only 20% of the genome
Raise one vial at 22°C and the other at 29°C; select lines which are both viable and long-lived at the permissive temperature, but which die prematurely when raised at the restrictive temperature	Standard protocol for isolating temperature-sensitive mutants. Advantage is that one can control time of expression of mutant phenotype by controlling timing of restrictive temperature (29°C); disadvantage is that only some small fraction of potential mutations affect protein conformation in such a manner so as to yield a temperature-sensitive phenotype
Use biomarkers to decide which mutants die prematurely because they are aging faster and which die early due to non-aging causes	Essential step for all protocols involving short-lived mutants

Adapted from Lefflelaar and Grigliatti [20].

TABLE 6.2

Procedure	Comment
Use the La strain	This selected outbred strain was developed specifically for use in aging studies and is known to live long due to enhanced antioxidant activities
Use single-shot P-element mutagenesis protocol and arrange so that mutants are induced in second chromosome	This insertional mutagenesis protocol requires several generations of specific pair matings and so is labor intensive. It also requires the transient use of balancer chromosomes. But it controls the number of hits/chromosome, the target chromosome can be specified, and the nature of the P-element can be manipulated to best serve one's experimental needs
Screen for mutants significantly resistant or sensitive to effects of exogenous paraquat	It is technically easier to select through some trait causally linked to longevity. There is much prior data showing that paraquat is toxic due to free radical production, that the La strain is significantly resistant to paraquat relative to appropriate control strain, and that this resistance is dependent on enhanced antioxidant activities. This step of the screen isolates loss-of-function or gain-of-function mutants
Test selected mutants for antioxidant gene expression levels	This step is necessary to confirm the putative designation of each mutant
Test selected mutants for effects on longevity	If hypothesis is correct, then loss-of-function mutations decreasing antioxidant activities should simultaneously decrease both paraquat resistance and longevity. Gain of function mutants should increase both traits. Separation of the two traits is interesting but unpredicted
Do appropriate biochemical and/or molecular tests to confirm specificity of mutants'effect	This step is necessary to confirm that the mutant's effect is restricted to the desired subset of genes
Map mutants; isolate P-element insert and sequence enough flanking DNA so as to conclusively identify gene if it is in data base. Use map location to locate and order P1 clones containing DNA region of interest	There is no scientific reason to waste time on chromosome walking and gene isolation when determining the cytogenetic locus can key you into the appropriate clones and a partial sequence might enable you to identify the gene in question

Adapted from Arking et al., in preparation.

they agreed with Clark et al [122] that the variance and effects of such mutants as a group were larger than those inferred from spontaneous mutants. However, they pointed out that most of the variance was accounted for by a few inserts with large effect. They also noted that "Mutations with large effects, such as P-element insertions, define the locus as one at which spontaneous mutations affecting the trait could potentially arise. The question is then whether or not allelic variation at the locus contributes to genetic variation for the trait in natural populations." (Lyman et al., [123], p.291). Since two of the mutants identified by their P-element insertions corresponded to known bristle mutants, then it is reasonable to conclude that P-element mutagensis is, when appropriately analyzed, a powerful method for identifying important quantitative trait loci.

As a final caveat on P-element insertional mutagenesis, and one which supports the somatic mutation theory of aging, it has been observed that P-element and *mariner* element movement in somatic cells reduces *Drosophila* life span [124–126]. Hence, if one uses the P[ry+ delta2-3](99B) element as a transposase source in P-element mutagenesis, then one should use a dominantly marked delta2-3 chromosome, or homologue, and remove the delta2-3 bearing chromosome from subsequent progeny.

6.4.5 Transgenic Animals

Transgenic animals offer a technically demanding but conceptually attractive method of testing whether the overexpression of a particular gene(s) can alter the phenotype. The least of the difficulties by now are the construction of the transgene vector and its injection into the host animal; these have been explained elsewhere [14,39,128,130] and the results of these experiments have been ably reviewed by Tower [107]. However, it has been recognized from the first that certain difficult experimental problems plague the interpretation of the results observed in any transgene experiment. The most important is that of position effect: the integration of the transgene into a portion of the genome that does not permit the normal qualitative or quantitative expression of the transgene. Not every transgene that successfully integrates into the genome will be expressed. In the successful experiment of Orr and Sohal (1994), they generated 15 different transgene lines containing both the *CuZn-SOD* and catalase genes [14]. The goal was to increase longevity by increasing the expression of these antioxidant defense products of these genes. Of the 15 lines, 8 had an increased life span, 1 had a decreased life span, and 6 were unchanged. It was presumed, but not proven, that the six lines with an unchanged life span did not express their transgenes. It was also presumed that all of the eight lines with an increased life span were expressing their transgenes; confirmatory data were acquired for only three lines. The decreased life span in the one line might present an interesting exception to the expected results had it been shown that the transgene was actually being expressed. The difficulties attendant upon assuming the transgene to be active are well illustrated by the "cautionary tale" involving the elongation factor EF-1α. It was first assumed that the EF-1α transgenic flies expressed the transgene [39] and an extensive analysis was done [43,131]. Only later was it found that the animals did not overexpress EF-1α mRNA, and that the observed increase in life span of the transgenic animals could be attributed solely to the position of the inactive transgene, the size of the inactive transgene and/or the genetic background [132]. It seems clear that transgene experiments will require the actual measurement of the gene product in question in every line produced, and at a number of appropriate ages as well. The question of genetic background is more difficult to solve unless one constructs animals isogenic for all four chromsomes. A better suggestion offered by Tower (1996) involves constructing inducible expression systems so that the effect of having a transgene expressed or not expressed could be analyzed in identical genetic backgrounds [106].

6.4.6 Manipulation of Environmental Variables

Answering some questions may require the manipulation of environmental variables. Some of these variables (i.e., ambient temperature, dessication) are experimentally quite straightforward. Others, such as dietary restriction, are technically involved. Dietary (or caloric) restriction clearly increases life span in rodents, and *Drosophila* has been used as a model system to test diet restriction and aging in invertebrates [87]. The results vary, with the majority of studies showing no effect of diet restriction on life span of *D. melanogaster* [133,134], or a small increase in adult life span in one experiment with underfed male larvae [135]. Chippindale et al. [136] reported that life span in females and males is increased by a low yeast diet, but Partridge et al. [137] previously found that the observed reduction in life span was actually caused by a reduction in egg laying. However, Arking et al. [138] found that paraquat resistance and longevity covaried significantly as a result of the amount of yeast allowed per animal during the larval period, although the differences were not large. No measurement was actually made of the amount of yeast consumed. Riha and Luckinbill actually showed that long-lived stocks have a lowered feeding rate, an altered use of nutrients, and an altered metabolic flux relative to the normal-lived control [139]. Although it seems that additional experiments are needed to determine if and how diet restriction does increase life span in *Drosophila*, the above studies give another example of the role of the environment in aging. In particular, they suggest that the manipulation of diet in the developmental phase of the life span may have important effects on the eventual expression of the genes involved in life span regulation.

Acknowledgments

The writing of this article was supported in part by NIA Rol-08834 to R. A.

References

1. Lints, F.A. and Soliman, M.H., Eds., *Drosophila as a Model Organism for Ageing Studies*, Blackie, Glasgow, 1988.
2. Arking, R. and Dudas, S.D., Review of genetic investigations into the aging processes of *Drosophila*, *J. Am. Geriatr. Soc.*, 37, 757, 1989.
3. Kimmerly, W., Stultz, K., Lewis, S., Lewis, K., Lustre, V., Romero, R., Benke, J., Sun, D., Shirley, G., Martin, C., and Palazzolo, M., A P1-based physical map of the *Drosophila* euchromatic genome, *Genome Res.*, 6, 414, 1996.
4. Rubin, G.M. Around the genome: the *Drosophila* genome project, *Genome Res.*, 6, 71, 1996.
5. Spradling, A.C., Stern, D.M., Kiss, I., Roote, R., Laverty T., and Rubin, G.M., Gene disruptions using P transposable elements: an integral component of the *Drosophila* genome project, *Proc. Natl. Acad. Sci. U.S.A.*, 92, 10824, 1995.
6. FlyBase Consortium, FlyBase: the *Drosophila* database, *Nucleic Acids Res.*, 24, 53, 1996.
7. Johnson, T., Tedesco, P.M., and Lithgow, G.J., 1994. Comparing mutants, selective breeding, and transgenics in the dissection of aging processes of *Caenorhabditis elegans*, in *Genetics and Evolution of Aging*, Rose, M.R. and Finch, C.E., Eds., Kluwer Academic Publishers, Dordrecht, The Netherlands, 1994, 83.
8. Dudas, S.D. and Arking, R., A coordinate upregulation of antioxidant gene activities is associated with the delayed onset of senescence in al long-lived strain of *Drosophila*, *J. Gerontol. Biol. Sci.*, 50A, B117, 1995.
9. Wheeler, J.C., Bleschke E.T., and Tower, J., Muscle-specific expression of *Drosophila hsp70* in response to aging and oxidative stress, *Proc. Natl. Acad. Sci. U.S.A.*, 10408, 1995.
10. Khazaeli, A.A., Tatar, M., Pletcher, S.D., and Curtsinger, J.W., Heat induced longevity extension in *Drosophila*. I. Heat treatment, mortality, and thermotolerance, *J. Gerontol. Biol. Sci.*, 52A, B48, 1997.
11. Phillips, J.P. and Hilliker, A.J., 1990. Genetic analysis of oxygen defense mechanisms in *Drosophila melanogaster*, *Adv. Genetics*
12. Arking, R., unpublished data.
13. Mackay, T.F.C., The genetic basis of quantitative variation: numbers of sensory bristles of *Drosophila melanogaster* as a model system, *Trends in Genet.*, 11, 464, 1995.
14. Orr, W. and Sohal, R.S., Extension of life-span by overexpression of superoxide dismutase and catalase in *Drosophila melanogaster*, *Science*, 263, 1128, 1994.
15. Buck, S., Wells, R.A., Dudas, S.P., Baker, G.T., III, and Arking, R., Chromosomal localization and regulation of the longevity determinant genes in a selected strain of *Drosophila melanogaster*, *Heredity*, 71, 11, 1993.
16. Kirkwood, T., The disposable soma theory of aging, in *Genetic Effects on Aging II*, Harrison, D.E., Ed., Telford Press, Caldwell, NJ, 1990, 9.

17. Kowald, A. and Kirkwood, T.B.L., Towards a network theory of aging: a model combining the free radical theory and the protein error theory, *J. Theoret. Biol.*, 168, 75, 1994.
18. Martin, G.M., Austad, S.N., and Johnson, T.E., Genetic analysis of ageing: role of oxidative damage and environmental stresses, Nature Genet., 13, 25, 1996.
19. Baker, G.T., III, Jacobson, M., and Mokrynski, G., Aging in *Drosophila*, in *Handbook of Cell Biology of Aging*, Cristafalo, V., Ed., CRC Press, Boca Raton, FL, 1985, 511.
19a. Mayer, P.J. and Baker, G.T., III, Genetic aspects of *Drosophila* as a model system of human aging, *Int. Rev. Cytol.*, 95, 61, 1985.
20. Leffelaar, D. and Grigliatti, T.A., A mutation in *Drosophila* that appears to accelerate aging, *Dev. Genet.*, 4, 199, 1984.
21. Ganetsky, B. and Flanagan, J.R., On the relationship between senescence and gae-related changes in two wild-type strains of *Drosophila melanogaser*, *Exp. Gerontol.*, 13, 189, 1976.
22. Rose, M.R. and Charlesworth, B., Genetics of life-history in *Drosophila melanogaster*. II. Exploratory selection experiments, *Genetics*, 97, 187, 1981.
23. Rose, M.R., Laboratory evolution of postponed senescence in *Drosophila melanogaster*, *Evolution*, 38, 1004, 1984.
24. Luckinbill, L.S., Arking, R., Clare, M., Cirocco, W., and Buck, S., Selection for delayed senescence in *Drosophila melanogaster*, *Evolution*, 38, 996, 1984.
25. Partridge, L. and Fowler, K., Direct and correlated responses to selection on age at reproduction in *Drosophila melanogaster*, *Evolution*, 46, 76, 1993.
26. Zwaan, B., Bijlsma, R., and Hoekstra, R.F., Direct selection on lifespan in *Drosophila melanogaster*, *Evolution*, 49, 649, 1995.
27. Arking, R. and Wells, R.A., Genetic alteration of normal aging processes is responsible for extended longevity in *Drosophila*, *Dev. Genet.*, 11, 141, 1990.
28. Curtsinger, J.W., Fukui, H.H., Khazaeli, A.A., Kirscher, A., Pletcher, S.D., Promisolow, D.E.L., and Tatar, M., Genetic variation and aging, *Annu. Rev. Genet.*, 29, 553, 1995.
29. Wilkins, M., *Genetic Analysis of Animal Development*, 2nd ed., Wiley-Liss, New York, 1993.
30. Banfi, S., Borsani, G., Rossi, E., Bernard, L., Guffanti, A., Rubboli, F., Marchitiello, A., Giglio, S., Coluccia, E., Zollo, M., Zuffardi, O., and Ballabio, A., Identification and mapping of human cDNAs homologous to *Drosophila* mutant genes through EST database searching, *Nature Genet.*, 13, 167, 1996.
31. Liang, P., Bauer, D., Averboukh, L., Warthoe, P., Rohrwild, M., Muler, H., Strauss, M., and Pardee, A.B., Analysis of altered gene expression by differential display, *Methods Enzymol.*, 254, 304, 1995.
32. Golic, K.G. and Lindquist, S., The FLP recombinase of yeast catalyzes site-specific recombination in the *Drosophila* genome, *Cell*, 59, 499, 1989.
33. Golic, K.G., Site-specific recombination between homologous chromosomes in *Drosophila*, *Science*, 252, 958, 1991.
34. Xu, T. and Harrison, S.D., Mosaic analysis using FLP recombinase, in *Methods in Cell Biology*, Vol. 44, *Drosophila melanogaster: Practical Uses in Cell and molecular Biology*, Goldstein, S.B. and Fryberg, E.A., Eds., Academic Press, New York, 1994, 655.
35. Rose, M.R., *Evolutionary Biology of Aging*, Oxford University Press, New York, 1991.
36. Medawar, P.B., *An Unsolved Problem of Biology*, H.K. Lewis, London, 1952.
37. Williams, G.C., Pleiotropy, natural selection, and the evoluton of senescence, *Evolution*, 11, 398, 1957.
38. Webster, G.C., Effect of aging on the components of the protein synthesis system, in *Insect Aging: Strategies and Mechanism*, Collatz K.G. and Sohal, R., Eds., Springer-Verlag, Heidelberg, 1986, 207.
39. Shephard, J.W.C., Walldorf, U., Hug, P., and Gehring, W.J., Fruit flies with additional expression of the elongation factor EF-1α live longer, *Proc. Natl. Acad. Sci. U.S.A.*, 86, 7520, 1989.
40. Dudas, S.P. and Arking, R., The expression of the EF1α genes of *Drosophila* is not associated with the extended longevity phenotype in a selected long-lived strain, *Exp. Gerontol.*, 29, 645, 1994.
41. Shikama, N., Ackermann R., and Brack, C., Protein synthesis elongation factor EF-1α expression and longevity in *Drosophila melanogaster*, *Proc. Natl. Acad. Sci. U.S.A.*, 91, 4199, 1994.

42. Shikama, N. and Brack, C., Changes in the expression of genes involved in protein synthesis during *Drosophila* aging, *Gerontology*, 42, 123, 1996.

43. Kaiser, M., Gaser, M., Ackermann, R., and Stearns, S.C., P-element inserts in transgenic flies: a cautionary tale, *Heredity*, 78, 1, 1997.

44. Gerhold, D. and Caskey, C.T., It's the genes: EST access to human genome content, *BioEssays*, 18, 973, 1996.

45. Hartley, D., *Drosophila* inherit diseases (News & Views), *Nature Genet.*, 13, 133, 1996.

46. Fleming, J.E., Walton, J.K., Dubitsky, R., and Bebsch, K.G., Aging results in an unusual expression of *Drosophila* heat shock proteins, *Proc. Natl. Acad. Sci. U.S.A.*, 85, 4099, 1988.

47. Niedwieck, A. and Fleming, J.E., Heat shock induces changes in the expression and binding of ubiquitin in senescent *Drosophila melanogaster*, *Dev. Genet.*, 14, 78, 1993.

48. Helfand, S.L., Blake, K.J., Rogina, B., Stracks, M.S., Centurion, A., and Naprta, B., Temporal patterns of gene expression in the antenna of the adult *Drosophila melanogaster*, *Genetics*, 140, 549, 1995.

49. Calleja, M., Moreno, E., Pelaz, S., and Morata, G., Visualization of gene expression in living adult *Drosophila*, *Science*, 274, 252, 1996.

50. Curtsinger, J.W., Talk presented at annual meeting of Gerontology Society of America, November 1996.

51. Arking, R., Successful selection for increased longevity in *Drosophila*: analysis of the survival data and presentation of a hypothesis on the genetic regulation of longevity, *Exp. Gerontol.*, 22, 199, 1987.

52. Force, A.G., Staples, T., Soliman, S., and Arking, R., Comparative biochemical and stress analysis of genetically selected *Drosophila* strains with different longevities, *Dev. Genet.*, 17, 340, 1995.

53. Arking, R., Buck, S., Berrios, A., Dwyer, S., and Baker, G.T., III, Elevated paraquat resistance can be used as a bioassay for longevity in a genetically based long-lived strain of *Drosophila*, *Dev. Genet.*, 12, 362, 1991.

54. Hinton, C.W., The behavior of an unstable ring chromosome of *Drosophila melanogaster*, *Genetics*, 40, 951, 1955.

54b. Morata, G. and Ripoll, P., Minutes: mutants of *Drosophila* autonomously affecting cell division rate, *Dev. Biol.*, 42, 211, 1975.

55. Trout, W.E. and Kaplan, W.D., Mosaic mapping of foci associated with longevity in the neurological mutants Hk1 and Sh5 of *Drosophila melanogaster*, *Exp. Gerontol.*, 16, 461, 1981.

56. Simpson, P., Flipping fruit flies: a powerful new technique for generating *Drosophila* mosaics, *Trends Genet.*, 9, 227, 1993.

57. Xu, T. and Rubin, G.S., Analysis of genetic mosaics in developing and adult *Drosophila* tissues, *Development*, 117, 1223, 1993.

58. Rorth, P., A modular misexpression screen in *Drosophila* detecting tissue-specific phenotypes, *Proc. Natl. Acad. Sci. U.S.A.*, 93, 12418, 1996.

59. Kunst, C.B, Mezey, E., Brownstein, M.J., and Patterson, D., Mutations in SOD1 associated with amyotrophic lateral sclerosis cause novel protein interactions, *Nature Genet.*, 15, 91, 1997.

61. Ashburner, M., *Drosophila: A Laboratory Handbook*, Cold Spring Harbor Laboratory Press, Cold Spring Harbor, NY, 1989.

62. Ashburner, M., *Drosophila: A Laboratory Manual*, Cold Spring Harbor Laboratory Press, Cold Spring Harbor, NY, 1989.

63. Matthews, K.A., Care and feeding of *Drosophila melanogaster*, in *Methods in Cell Biology*, Vol. 44, *Drosophila melanogaster: Practical Uses in Cell and Molecular Biology*, Goldstein, L.S.B. and Fyrberg, E.A., Eds., Academic Press, New York, 1994, 13.

64. Hamilton, B.A. and Zinn, K., From clone to mutant gene, in *Methods in Cell Biology*, Vol. 44, *Drosophila melanogaster: Practical Uses in Cell and Molecular Biology*, Goldstein, L.S.B. and Fyrberg, E.A., Eds., Academic Press, New York, 1994, 81.

65. Shaffner, C.D., Wuller, J.M., and Elgin, S.C.R., Raising large quantities of *Drosophila* for biochemical experiments, in *Methods in Cell Biology*, Vol. 44, *Drosophila melanogaster: Practical Uses in Cell and Molecular Biology*, Goldstein, L.S.B. and Fyrberg, E.A., Eds., Academic Press, New York, 1994, 33.

66. Wolfner, M.F. and Goldberg, M.L., Harnessing the power of *Drosophila* genetics, in *Methods in Cell Biology*, Vol. 44, *Drosophila melanogaster:Practical Uses in Cell and Molecular Biology*, Goldstein, L.S.B. and Fyrberg, E.A., Eds., Academic Press, New York, 1994, 33.

67. Greenspan, R.J., *Fly Pushing: The Theory and Practice of Drosophila Genetics*, Cold Spring Harbor Laboratory Press, Cold Spring Harbor, NY, 1997.

67a. Goldstein, L.S.B., Sources of information about the fly: where to look it up, in *Methods in Cell Biology*, Vol. 44, *Drosophila melanogaster: Practical Uses in Cell and Molecular Biology*, Goldstein L.S.B. and Fyrberg, E.A., Eds., Academic Press, New York, 1994, 3.

68. Demerec, M., *Biology of Drosophila*, Hafner Pub. Co., New York, 1950.

69. Roberts, D.B., Basic *Drosophila* care and techniques, in *Drosophila, A Practical Approach*, Roberts, D.B., Ed., IRL Press, Oxford, 1986, 1.

69b. Okada, T., *Systematic Study of the Early Stages of Drosophilidae*, Bunka Augeisha Co., Tokyo, 1968.

70. King, R.C., *Ovarian Development in Drosophila melanogaster*, Academic Press, New York, 1970.

71. Ashburner, M., Novitski, E., Wright, T.R.F., Carson, H.L., and Thompson, Jr., J.N., Eds., *The Genetics and Biology of Drosophila*, Vol 1a–c, 2a–d, 3a–e, Academic Press, London, 1976–86.

72. Siddiqi, O., Babu, P., Hall, L.M., and Hall, J.C., *Development and Neurobiology of Drosophila*, Plenum Press, New York, 1980.

73. Roberts, D.B., *Drosophila: A Practical Approach*, IRL Press, Oxford, 1986.

74. Ransom, R., *A Handbook of Drosophila Development*, Elsevier Biomedical Press, Amsterdam, 1992.

74a. Ransom, R., Techniques, in *A Handbook of Drosophila Development*. Ransom, R., Ed., Elsevier Biomedical Press, Amsterdam, 1992, 1.

75. Campos-Ortega, J.A. and Hartenstein, V., *The Embryonic Development of Drosophila melanogaster*, Springer-Verlag, Berlin, 1985.

76. Lawrence, P.A., *The Making of a Fly*, Blackwell Scientific Publications, Oxford, 1992.

77. Hartenstein, V., *Atlas of Drosophila Development*, Cold Spring Harbor Laboratory Press, Cold Spring Harbor, NY, 1993.

78. Bate, M. and Martinez Arias, A., *The Development of Drosophila melanogaster*, Cold Spring Harbor Laboratory Press, Cold Spring Harbor, NY, 1993.

79. Maroni, G., *An Atlas of Drosophila Genes: Sequences and Molecular Features*, Oxford University Press, Oxford, 1993.

80. Goldstein, L.S.B. and Fryberg, E.A., *Methods in Cell Biology*, Vol. 44, *Drosophila melanogaster: Practical Uses in Cell and Molecular Biology*, Academic Press, New York, 1994.

81. Hartl, D.L. and Lozovskaya, E.R., *The Drosophila Genome Map: A Practical Guide*, R.G. Landes Co., Austin, 1995.

82. Lasko, P.F., *Molecular Genetics of Drosophila Oogenesis*, R.G. Landes Co., Austin, 1994.

83. Arkhipova, I.R., Lyubomirskaya, N.V., and Ilyin, Y.V., *Drosophila Retrotransponsons*, R.G. Landes Co., Austin, 1995.

84. Lindsley, D.L. and Zimm, G.G., *The Genome of Drosophila melanogaster*, Academic Press, New York, 1992.

85. Flybase, The Bibliography of *Drosophila*: 1982–1993. DIS 74

86. Gartner, L.P., Aging in *Drosophila*: A selectively annotated bibliography, Jen House Pub. Co. Baltimore, MD. 1986, 1986.

87. Finch, C.E., *Longevity, Senescence, and the Genome*, University of Chicago Press, Chicago, 1991, 1991.

88. Post, J.A., Internet resources on aging: research, *Gerontologist*, 36, 277, 1996.

89. Baret, P. and Lints, F.A., Selection for increased longevity in *Drosophila* melanogaster: a new interpretation, *Gerontology*, 39, 252, 1993.

90. Fukui, H.H., Pletcher, S.D., and Curtsinger, J.W., Selection for increased longevity in *Drosophila melanogaster*: a response to Baret and Lints, *Gerontology*, 41, 65, 1995.

91. Arking, R. and Buck, S., Selection for increased longevity in *Drosophila melanogaster*: a reply to Lints, *Gerontology*, 4169, 1996.

92. Arking, R., Antioxidant genes and other mechanisms involved in the extended longevity of *Drosophila*, in *Oxidative Stress and Aging*, Cutler, R.G., Packer, L., Bertram, J., and Mori, A., Eds., Birkhauser Verlag, Basel, 1995, 123.

93. Rose, M.R., Genetics of increased life span in *Drosophila*, *BioEssays*, 11, 132, 1989.

94. Arking, R., Graves, K., and Ward, K., Differential patterns of antioxidant gene expression in long-, normal- and short-lived strains of *Drosophila*, *Gerontologist*, 36, 295 (abstr.), 1996.
95. Dixon, L.K., Use of recombinant inbred strains to map genes of aging, *Genetica*, 91, 151, 1993.
96. Tanksley, S.D., Mapping polygenes, *Annu. Rev. Genet.*, 27, 205, 1993.
97. Routman, E. and Cheverud, J.M., Individual genes underlying quantitative traits: Molecular and analytical methods, in Schierwater, B., Street, B., Wagner, G.P., and DeSalle, R., Eds., *Molecular Ecology and Evolution: Approaches and Applications*, Birkhauser Verlag, Basel, 1994, 593.
98. Falconer, D.S. and Mackay, T.F.C., *Introduction to Quantitative Genetics*, Longman Group Limited, Essex, 1996.
99. Thoday, J.M., Location of polygenes, *Nature*, 191, 368, 1961.
100. Thoday, J.M., Polygene mapping: uses and limitations, in *Quantitative Genetic Variation*, Thompson, Jr., J.N. and Thoday, J.M., Eds., Academic Press, New York, 1979, 220.
101. Shrimpton, A.E. and Robertson, A., The isolation of polygenic factors controlling bristle score in *Drosophila melanogaster*, I. Allocation of third chromosome bristle effects to chromosome sections, *Genetics*, 118, 437, 1988a.
102. Shrimpton, A.E. and Robertson, A., The isolation of polygenic factors controlling bristle score in *Drosophila melanogaster*. II. Distribution of third chromosome bristle effects within chromosome sections, *Genetics*, 118, 445, 1988b.
103. Lai, C., Lyman, R.F., Long, A.D., Langley C.H., and Mackay, T.F.C., Naturally occurring variation in bristle number associated with DNA sequence polymorphisms at the scabrous locus in *Drosophila melanogaster*, *Science*, 266, 1697, 1994.
103a. Phillips, J.P., Campbell, S.D., Michaud, D., Charbonneau, M., and Hilliker, A.J., Null mutations of copper/zinc superoxide dismutase in *Drosophila* confers hypersensitivity to paraquat and reduced longevity, *Proc. Natl. Acad. Sci. U.S.A.*, 86, 2761–2765, 1989.
104. Long, A.D., Mullaney, S.L., Reid, L.A., Fry, J.D., Langley C.H., and Langley, T.F.C., High resolution mapping of genetic factors affecting abdominal bristle number in *Drosophila melanogaster. Genetics*, 139, 1273, 1995.
105. Bewley, G.C. and Mackay, W.J., Development of a genetic model for acatalasemia: testing the oxygen free radical theory of aging, in *Genetic Effects on Aging II*, Harrison, D.E., Ed., Telford Press, Caldwell, NJ, 1990, 359.
106. Hilliker, A., Duyf, J., Evans, B., and Phillips, J.P., Urate-null rosy mutants of *Drosophila melanogaster* are hypersensitive to oxygen stress, *Proc. Natl. Acad. Sci. U.S.A.*, 89, 4343, 1992.
107. Tower, J., Aging mechanisms in fruit flies, *BioEssays*, 18, 799, 1996.
108. Yonemura, I., Motoyama, T., and Hasekura, H., Mode of inheritance of major genes controlling the life span differences between two inbred strains of *Drosophila melanogaster*, *Hereditas*, 111, 207, 1989.
109. Hutchinson, E.W. and Rose, M.R., Quantitative genetics of postponed ageing in *Drosophila melanogaster*, I. Analysis of outbred populations, *Genetics*, 127, 719, 1991.
110. Hutchinson, E.W., Shaw, A.J., and Rose, M.R., Quantitative genetics of postponed ageing in *Drosophila melanogaster*. II. Analysis of selected lines, *Genetics*, 127, 729, 1991.
111. Luckinbill, L.S., Graves, J.L., Reed, A.H., and Koetsawag, S., Localizing genes that defer senescence in *Drosophila melanogaster*, *Heredity*, 60, 367, 1988.
112. Mackay, T.F.C. and Langley, C.H., Molecular and phenotypic variance in the achaete-scute region of *Drosophila melanogaster*, *Nature*, 348, 64, 1990.
113. Goldstein, D.B. and Clark, A.G., Microsatellite variation in North American populations of *Drosophila melanogaster*, *Nucl. Acids Res.*, 23, 3882, 1995.
114. England, P.R., Briscoe, D.A., and Frankham, R., Microsatellite polymorphisms in a wild population of *Drosophila melanogaster*, *Genet. Res. Camb.*, 67, 285, 1996.
115. Michalakis, Y. and Veville, M., Length variation of CAG/CAA trinucleotide repeats in natural populations of *Drosophila melanogaster* and its relation to the recombination rate, *Genetics*, 143, 1713, 1996.
116. Schug, M.D., Mackay, T.F.C., and Aquadro, C.F., Low mutation rates of microsatellite loci in *Drosophila melanogaster*, *Nature Genet.*, 15, 99, 1997a.
117. Schug, M.S., Wetterstrand, K.A., Gaudette, M.S., Lim, R.H., Hutter, C.M., and Aquando, C.F., The distribution and frequency of microsatellite loci in *Drosophila melanogaster*.

118. Haley, C.S. and Knott, S.A., A simple regression method for mapping quantitative loci in line crosses using flanking markers, *Heredity*, 69, 315, 1992.

119. Haley, C.S., Knott, S.A., and Elsen, J., Mapping quantitative loci in crosses between outbred lines using least squares, *Genetics*, 136, 1195, 1994.

119a. Bier, E., Vaessin, H., Shepherd, S., Lee, K., McCall, K., Barbel, S., Ackerman, L., Carretto, R., Vemura, T., Grell, E., Jan L.Y., and Jan, Y.N., Searching for pattern and mutation in the *Drosophila* genome with a P-lacZ vector, *Genes Dev.*, 3, 1273, 1989.

120. Smith, D., Wohlgemuth, J., Calvi, B.R., Franklin, I., and Gelbart, W.M., Hobo enhancer trapping mutagenesis in *Drosophila* reveals an insertion specificity different from P elements, *Genetics*, 135, 1063, 1993.

121. Loukeris, T.G., Arca, B., Livadaras, I., Dialektaki, G., and Savakis, C., Introduction of the transposable element Minos into the germ line of *Drosophila melanogaster*, *Proc. Natl. Acad. Sci. U.S.A.*, 92, 9485, 1994.

122. Clark, A.G., Wang, L., and Hulleberg, T., P-element-induced variation in metabolic regulation in *Drosophila*, *Genetics*, 139, 337, 1995.

123. Lyman, R.F., Lawrence, F., Nuzhdin, S.V., and Mackay, T.F.C., Effects of single P-element insertions on bristle number and viability in *Drosophila melanogaster*, *Genetics*, 143, 277, 1996.

124. Driver, C.J.I. and McKechnie, S.W., Transposable elements as a factor in the aging of *Drosophila melanogaster*, *Ann. N.Y. Acad. Sci.*, 673, 83, 1992.

125. Woodruff, R.C., Transposable DNA elements and life history traits. I. Transposition of P-DNA elements in somatic cells reduces the lifespan of *Drosophila melanogaster*, *Genetica*, 86, 143, 1992.

126. Woodruff, R.C. and Nikitin, A.G., P DNA element movement in somatic cells reduces lifespan in *Drosophila melanogaster*: Evidence in support of the somatic mutation theory of aging, *Mutat. Res.*, 338, 35, 1995.

127. Nikitin, A.G. and Woodruff, R.C., Somatic movement of the mariner transposable element and lifespan of *Drosophila* species, *Mutat. Res.*, 338, 43, 1995. Woodruff, R.C. 1992.

128. Orr, W.C. and Sohal, R., The effects of catalase gene overexpression on life span and resistance to oxidative stress in transgenic *Drosophila melanogaster*, *Arch. Biochem. Biophys.*, 297, 35, 1992.

129. Reveillaud, J., Niedzwiecki, A., Bensch K.G., and Fleming, J.E., Expression of bovine super-oxide dismutase in *Drosophila melanogaster* augments resistance to oxidative stress, *Mol. Cell. Biol.*, 11, 632, 1991.

130. Seto, N.O.L., Hayashi, S., and Tener, G.M., *Proc. Natl. Acad. Sci. U.S.A.*, 87, 4270, 1990.

131. Stearns, S.C. and Kaiser, M., The effects of enhanced expression of elongation factor EF1-α on lifespan in *Drosophila melanogaster*, *Genetica*, 91, 167, 1993.

132. Kaiser, M., Gasser, M., Ackermann, R., and Stearns, S.C., P-element inserts in transgenic flies: a cautionary tale, *Heredity*, 78, 1, 1997.

133. David, J., Van Herrewege, J., and Fouillet, P., Quantitative under-feeding of *Drosophila*: Effects on adult longevity and fecundity, *Exp. Gerontol.*, 6, 249, 1971.

134. Le Bourg, E. and Minois, N., Failure to confirm increased longevity in *Drosophila melanogaster* submitted to a food restriction procedure, *J. Gerontol. Biol. Sci.*, 51A, B280, 1996.

135. Alpatov, W.W., Experimental studies on the duration of life. XIII. The influence of different feeding during the larval and imaginal stages on the duration of life of the imago of *Drosophila melanogaster*, *Am. Nat.*, 64, 37, 1930.

136. Chippindale, A.K., Leroi, A.M., Kim, S.B., and Rose, M.R., Phenotypic plasticity and selection in *Drosophila* life-history evolution. I. Nutrition and the cost of reproduction, *J. Evol. Biol.*, 6, 171, 1993.

137. Partridge, L., Green, A., and Fowler, K., Effects of egg-production and of exposure to males on female survival in *Drosophila melanogaster*, *J. Insect Physiol.*, 33, 745, 1987.

138. Arking, R., Force, A.G., Dudas, S.P., Buck, S., and Baker, III, G.T., Factors contributing to the plasticity of the extended longevity phenotypes of *Drosophila*, *Exp. Gerontol.*, 31, 623, 1996.

139. Riha, V.F. and Luckinbill, L.S., Selection for longevity favors stringent metabolic control in *Drosophila melanogaster*, *J. Gerontol. Biol. Sci.*, 51A, B284, 1996.

140. Sohal, R.S. and Weindruch, R., Oxidative stress, caloric restriction, and aging, *Science*, 273, 59, 1996.

7

Experimentation with Nematodes

Abraham Z. Reznick and David Gershon

CONTENTS

0-8493-3112-9/99/$0.00+$.50
© 1999 by CRC Press LLC

7.1   Introduction: Advantages and Disadvantages of the Nematode as a Model for Aging Research

In this review we do not intend to give a very extensive compilation of the many papers that describe experimental work dealing with nematode aging in general and with *Caenorhabditis elegans*, in particular. We refer the reader to three excellent books which deal with the biology of the free-living nematodes and the cellular, genetic, and molecular biological methods used in experimentation with these organisms: (1) *The Nematode Caenorhabditis elegans*, edited by W. B. Wood (Cold Spring Harbor Laboratory, 1988); (2) *Methods in Cell Biology*, Vol. 48, *Caenorhabditis elegans: Modern Biological Analysis of an Organism*, edited by H. F. Epstein and D. C. Shakes (Academic Press, 1995); and (3) *Caenorhabditis elegans II*, edited by D. L. Riddle, T. Blumenthal, B. J. Meyer, and J. R. Priess (Cold Spring Harbor Laboratory, 1997). All three reviews on aging in nematodes by Johnson [1], Lithgow [2], and Kenyon [3] should be consulted by the readers. Johnson's review [1] is particularly thoughtful, comprehensive, and still very relevant in discussing the attributes (and predicaments) of the nematode as an experimental aging system. We have the advantage of having the experience and knowledge of the system from working on it via a slightly different approach and outlook (as will become obvious below) from those used in the current mainstream of aging research on nematodes. Also, we introduced the nematode as a model for aging research in the late 1960s and at that time had pondered extensively the essential criteria that should be fulfilled in a paradigm suitable for aging research. As will be seen below, we still feel that the nematode is an attractive model for aging research. However, its limitations must be taken into account when its contributions to general aging research are evaluated. The reader will have to draw conclusions on how and for what problems this system can be best used.

7.1.1   General Background

In the late 1960s one of us (D.G.) set about establishing a laboratory to study aging. It was obvious that one of the major models to be studied should be a short-lived mammal such as a rodent which has obvious "relevance" to human senescence. However, it was also apparent that the mammalian system is very complex, with a generation time incompatible with the urgency of answering some of the major questions that we were considering in our thoughts on aging. It was, therefore, decided to search for a paradigm which could be used in parallel to a rodent system and which would be found suitable (and agreeable to the community of researchers on aging) for gerontological inquiry and experimentation at a relatively rapid pace. It was decided that one should set strict criteria to aid in making a successful new experimental paradigm for the study of basic mechanisms of aging. The model organism was expected to have the following attributes: it should (1) have a **short** life cycle; (2) have short mean and maximal life spans; (3) be multicellular and yet of simple morphology with a limited number of fully differentiated cell types that should be similar to most of the cellular systems of adult mammals, which are essentially postmitotic; and (4) be relatively easy to culture, with high fecundity, under well-defined and easily controlled nutritional conditions.

The search for the chosen animal system finally narrowed down to rotifers and nematodes. Both of these groups fulfilled most of the criteria listed above. Rotifers were discarded despite Lansing's interesting and enticing experiments on the effect of maternal age on the longevity of their offspring [4]. The decisive factor against the rotifers was that they had to be fed with ciliates and axenic cultures could not be obtained;. The presence of a second organism in the cultures would complicate biochemical analysis of age-associated changes in our experimental organism. Free-living nematode cultures had been maintained axenically in Rothstein's laboratory for the purpose of studies on intermediary metabolism [5]. Nematodes were obtained from Rothstein and evaluated for their suitability for laboratory studies on aging. Two species were chosen for study: *Caenorhabditis briggsae* and *Turbatrix aceti*. All the work of the 1970s from our laboratory and subsequently from that of Rothstein used these two species.

C. briggsae, a soil-dwelling, free-living nematode, is very closely related to the organism of choice of more recent work on aging in nematodes, *C. elegans*. As will be described below, both *Caenorhabditis* species are self-fertilizing hermaphrodites which produce a small proportion of males that allow outbreeding. The males, therefore, perform a very significant genetic function in natural populations of these species and have been used very elegantly and cleverly for genetic analysis of development and architectural body construction of these organisms by Brenner [6], his disciples, and a host of other investigators in the past two decades. *T. aceti*, the vinegar eel, belongs to the same family of Rhabditidae. It is very similar in anatomy and life cycle to *C. elegans* and *C. briggsae*, but differs from them in being bisexual and, therefore, outbred by nature. The latter characteristic makes it more "pertinent" for comparison with mammalian systems, on the one hand, and less amenable (although by no means unfit) to genetic analysis than the *Caenorhabditis* species. It has other attributes that make it favorable for some aspects of aging research, as will be argued in later sections of this review.

7.1.2 General Description

C. elegans and *C. briggsae* are ubiquitous, diminutive (1–1.5 mm), poikilothermic, free-living soil nematodes. They are essentially self-fertilizing hermaphrodites. An attractive feature, for those intending to run molecular genetic analysis of development, is that they possess a very small genome of about 100 megabase (as compared to the 3000 megabase of the mammalian genome) [6]. They feed mainly, but not solely, on bacteria with life cycles of 3–5 days *under constant laboratory conditions and can be raised under axenic conditions*. This fact must be emphasized, as in their natural habitat they are subject to substantial temperature fluctuations, irregular nutrition, variable humidity, and other environmental adversities. Like most of the other members of the phylum Nematoda, they are highly adaptable and have evolved to become very durable and successful organisms. For instance, when extreme conditions prevail, they are able to differentiate into very resistant, **Dauer larvae**, which maintain very low metabolic activity for periods of time much longer than the observed life span under permissive laboratory conditions.

High durability and potentially longer life spans under natural conditions pose a problem in evaluating experimental results on extension of the life span by laboratory manipulations such as specific gene mutations and various nutritional conditions. It must be emphasized that *this difficulty is not unique to nematodes* and must be addressed in studies of most aging paradigms, including the mammalian ones. However, as will be mentioned later, the problem of interpretation of life span results is more serious with nematodes, due to their unique resilience.

T. aceti is very similar to the other two species, except that its natural habitat is in environments in which fermentation takes place (as the name *vinegar eel* indicates). It is also bisexual, with high fecundity in liquid cultures at pH 3.0, which facilitates its maintenance under axenic conditions.

7.1.3   The Life Cycle and Morphological Features

Each adult is composed of a fixed number of cells [7]. In *C. elegans*, the hermaphroditic form produces both oocytes and sperm and usually reproduces by self-fertilization. Males arise spontaneously at low frequency (~0.1%) and can fertilize the hermaphrodites. Hermaphrodites consist of only 959 cells while the male has 1031 cells. Larvae hatch and develop through four stages, without obvious metamorphosis, with each culminating in a molt. The mature adult starts reproducing between days 3 and 5. On average, a hermaphrodite produces 300 eggs. The rate of reproduction usually diminishes after day 9 (this depends on the nutritional and temperature conditions).

The **body plan** is typical of all nematodes (see an excellent and comprehensive treatise on the anatomy of *C. elegans* by White [8]). There is an outer tube which consists of cuticle, hypodermis, neuronal network, and muscle. This tube surrounds a pseudocoelomic space that contains the intestine and the reproductive system. The shape of the worm is maintained by internal turgor which is controlled by osmoregulation.

The **cuticle** is an extracellular three-layered structure composed mainly of collagen secreted by the cells of the underlying hypodermis. In the adult, the hypodermis is essentially a syncytium formed by cell fusion. The cuticle is covered by a thin layer of phospholipids, which can be removed by detergents. The adult cuticle cannot be removed by gentle methods, such as digestion with phospholipases and collagenase, for the purpose of obtaining the intact cells of the organism. We have resolutely attempted, albeit without success, a large number of techniques for cuticle peeling (mainly of *T. aceti*), in order to separate the various cell types for individual biochemical analysis (D. Gershon, unpublished results). As discussed later in this review, the failure to obtain preparations of individual cell types due to the tough cuticle is a problem that is only very partially circumvented by various studies using *in situ* hybridization [9], immunofluorescence microscopy [10], and electron microscopy [11].

The **musculature** [12] is composed of a fixed number of cells with a basic structure that differs from striated mammalian skeletal muscle. There are body-wall muscle cells that are grouped into four strips on the dorsal and ventral sides running the length of the worm. There are also pharyngeal, vulval, anal, and intestinal muscle complexes. In general, the major muscle components are similar to those of other animals [12]. Extensive genetic, molecular, biological, biochemical, and anatomical studies have been carried out which make the functions of this system well understood.

The **nervous system** of these nematodes has been extensively studied, by a combination of microscopic, molecular, and genetic methods, because of its relative simplicity and its involvement in easily discernible behavioral patterns of the animals (see, e.g., References [13] and [14]). The knowledge accumulated on this relatively simple system is awesome; suffice it to refer to the "neuronal connectivity diagrams" that were already available in

1988 to appreciate this fact [15]. Most of the nerve cells are localized about the pharynx, the ventral midline, and the tail. Processes from these neurons form a ring about the pharynx and also send appendages in bundles along the long axis of the body. In contrast to most other animals, innervation of muscle cells is attained by processes produced and sent by the muscle cells to motor neurons. This is one of several features that signify differences between nematodes and higher eukaryotes.

The **digestive system** (see Reference [8] for a comprehensive description) is composed of the pharynx, intestine, and rectum, which together consist of a relatively small number of cells involved in digestion and absorption. Most of these cells have microvilli. The pharynx has a dual function: it serves as a pump to send the food to the intestine, and it also crushes the food on its way to the rest of the alimentary tract. The intestine is composed of two rows of cells in association with an anterior ring of cells forming a tube which ends in an intestinal-rectal valve. The rectum ends in an anus that is located near the tail. Small numbers of muscle cells are connected with the various parts of the digestive system; they are coupled by gap junctions and are innervated by a very small number of nerve cells. This is an example of the very economical design of the body structure of these nematodes which enables the construction of a whole multicellular organism with a complex differentiation pattern using a minimal cell number.

The **excretory system** is very simple. It consists of four cells: a gland cell, a large excretory cell, a duct cell, and a pore cell that connects the duct cell to the hypodermis (the external epithelium). The function of this system is only partially understood. Laser ablation of the duct cell indicates that it is involved in osmoregulation. We have observed a hint of "hormonal" activity in extracts of *C. briggsae* which affected the reproduction of hermaphrodites exposed to it (J. Epstein and D. Gershon, unpublished results). This "hormonal" activity could be due to the coelomytes, which have the structure of gland cells, although there is evidence they can be involved in scavenging activity [8] (primitive macrophages?).

The **reproductive system** [8] in the hermaphrodite includes a bilobed gonad. One lobe extends anteriorly from the middle part of the body, and the other extends posteriorly. Each lobe contains an ovary and uterus and also a male gonad. The ovaries are syncytial. The distal nuclei are mitotic and move proximally while entering the first meiotic division. They reach diakinesis in the oviduct prior to fertilization. At that stage they become enclosed in cytoplasm and membranes to form complete oocytes. The oocytes enlarge and mature as they move in the oviduct. The oviduct terminates with the spermatheca, which contain ameboid sperm cells that fertilize the mature oocytes as they pass through into the uterus [16]. Embryogenesis takes place in the eggs in the uterus. The eggs are laid through a vulva (which is also a syncytial structure [17]) which is situated on the ventral side in the middle of the animal. The male gonad is a single-lobed, U-shaped tubular structure which terminates in a seminal vesicle. At the distal end it contains mitotic germline cells. Cells in progressive stages of meiosis are arranged sequentially in the gonad starting at the distal end. Spermatids accumulate in the vesicle and are released during copulation with the hermaphrodite.

7.1.4 Developmental Biology

A comprehensive description of germline development and embryogenesis is provided by Kimble and Ward [18] and Wood [19], respectively.

In this review we stress the developmental aspects of these organisms, because it is our conviction that senescence is closely related to development and to a large extent is the result of terminal differentiation of most cell types in multicellular organisms. Terminal differentiation entails the acquisition of very specialized functions at the expense of the capacity to perform other functions that pluripotent embryonic cells or omnipotent stem cells are endowed with. Stem cells are absent in tissues of adult nematodes which are, therefore, devoid of the capacity to regenerate and replace damaged cells. This factor is most probably a major contributor to the relatively short life span of these organisms. Moreover, as will be discussed in a later section of the review, an additional major factor is the very limited capacity of protein turnover in the fully differentiated cells that comprise the mature adults as demonstrated in *T. aceti* [20,21].

The nematode *C. elegans* has become a major model for the study of animal development, due to the pioneering and visionary work of Brenner [6,22]. He chose this organism based on the seminal work of Pai [7] in the 1920s and others, who described the anatomy and embryology by painstaking microscopy. Brenner combined his insight into molecular biology and genetics and his appreciation of the role of the nervous system in behavior [23] with the modern tools of Nomarski differential interference contrast microscopy [24], electron microscopy, and molecular genetic methods already available in the 1970s and 1980s [25]. By the late 1980s, his disciples had established the entire cell lineage of *C. elegans*. Thus, the entire developmental history of each adult cell has been traced back to the zygote [26–29].

Analysis of the cell lineage in *C. elegans* has helped to elucidate some of the major problems of development: the temporal and spatial control of the formation of the final organismal architecture; the interrelationship between genes which decides the activation and maintenance of the developmental blueprint; the nature of the factors which lead to determination of developmental fate in specializing cell types; the nature of cell–cell interactions which aid in shaping the structure of tissues and organs; the function of programmed cell death in normal development; and the morphogenetic movement of cells during the embryonic and larval stages of the life span.

In *C. elegans* all of this has been made possible by the use of various methods, especially that of Nomarski microscopy (which is currently being supplemented and supplanted by confocal microscopy). This microscopy allows high-level resolution in relatively thick specimens. It visualizes differences in refractivity and, thus, differentiates nuclei from the more refractive cytoplasm and uses them as cell markers (see, e.g., Reference [28]). When properly mounted, live animals can be microscopically observed while they continue to feed and develop [29]. During the period of direct observation, nuclei are photographed at various time intervals in order to follow the spatial position and dimensions of the cells. Differential focusing (which is better achieved by the currently available confocal microscopy) allows for the tracing of the fate of cells at different structural levels of the organism. (For further details, see, e.g., Reference [28].)

A detailed picture of the emergence of specialized cell structures and cell–cell interactions during development has also been attained by electron microscopy [29]. We have also successfully employed electron microscopy to study age-associated changes in *C. briggsae* [30,31] (see below).

Laser ablation of individual cells has also been very useful in studying cell lineage at various stages of development. The design of the apparatus for this purpose, which requires training in a laboratory where it is routinely performed, was first described by White and Horvitz [32,33].

Genetic analysis of developmental processes has been very fruitful in *C. elegans* [25]. Random mutagenesis and selection for interruption in the array of developmental steps of the organism has been a productive approach. Mapping and isolation of the responsible gene(s) has been performed using available molecular and classical methods (33 and chapters 2, 3, 6, and 7 in the same book). Cloning, ectopic expression, and characterization of the protein products of the genes have been accomplished. In this way, it can be determined which genes are involved in developmental decisions, i.e., control genes, which direct or control the temporal and spatial expression of other genes (e.g., [34,35]). With the imminent completion of the sequencing of the whole *C. elegans* genome and the virtual completion of its physical mapping [36], the task of identifying and characterizing the genes involved in development and their function will be facilitated (although it will still require a monumental effort).

The process of **programmed cell death** has been superficially, and to a certain extent erroneously, related to the phenomenon of senescence. To determine its relevance to senescence, one should critically analyze the features of this fascinating process, which has aroused great interest in recent years. In *C. elegans* this process has been examined extensively [37,38]. Of the 1090 cells that are generated in the hermaphrodite, 131 undergo programmed cell death. That the process is programmed has been shown by genetic and morphological methods since it always involves the activity of the same genes and follows identical morphological alterations in the dying cells. Very significantly, cell lineage analysis reveals that the same cells always die with the same timing and in the same location in the embryo. Mutant analysis has revealed that elimination of the specific 131 cells is part of a developmental program. In certain mutants, the cells that normally die do survive, and on certain occasions, but not always, this has deleterious effects on the development of specific regions in the developing nematode [38]. No such program has been observed in tissues of senescent nematodes or other multicellular organisms. Rather, random loss of cells occurs which occasionally, but not always, follows an apoptotic pattern in tissues of senescent higher eukaryotes. Systematic work on programmed cell death in the nematode, and in other organisms, has helped set strict criteria for the definition of the phenomenon. No equivalent criteria apply in the phenomenon of senescence.

7.1.5 Genetics and Genetic Manipulation

Mutant analysis in *C. elegans* by a variety of methods has proven to be extremely useful for the elaboration of the complex processes of development and behavior. The genome is relatively small for eukaryotes, 100 megabases, which is half the size of the *Drosophila* genome, 1/30 that of the human and 20 times the size of the *E. coli* genome. Sequencing of the entire genome is expected to be completed in 1998. Moreover, virtually the total physical map of this genome is already available [39]. The genome consists of six chromosomes, one of which is the X chromosome [40]. The number of genes is estimated to be between 12,000–14,000. Genome analysis has revealed minimal repetitive sequences (17%). Unlike higher eukaryotes, it does not contain methylated cytosine, a fact that should be taken into account when the nematode paradigm is evaluated for its reflection on aging in mammals.

For those interested in isolating and characterizing new genes there is no necessity to form new libraries as virtually the whole genome is available in yeast artificial chromosomes and cosmids. Databases for physical map construction and editing are also available [39]. The methodology required for the physical mapping of the *C. elegans* genome is well detailed [36,39].

7.2 Evolution of the Nematode Paradigm in Aging Research

7.2.1 The Use of *Turbatrix Aceti* in the 1970s: Advantages of the System

In choosing the nematode *T. aceti*, it was possible to show that this organism indeed significantly fulfills most of the criteria mentioned in Section 7.1.1. (1) Its life span is relatively short with a mean life span of about 25–30 days and a maximum life span of 35–40 days [40]. (2) It has a small and fixed number of postmitotic cells which, with the exception of the reproductive organs, lack proliferative capacity. (3) All of the above postmitotic cells exhibit full differentiation at the time of hatching. Therefore, using *T. aceti* it was possible to show that one could use fluorodeoxyuridine (FUDR) to arrest reproduction while not affecting life span [40]. Synchronized populations were obtained by the use of FUDR whose growth curves were identical to those of untreated animals [40]. (4) These animals can be raised under axenic conditions in quantities suitable for biochemical studies [41].

Following the establishment of *T. aceti* as a suitable system for aging studies, it was shown that the activities of several enzymes representative of various biochemical pathways tended to decline in aging nematodes [42]. Thus, the activities of acetylcholinesterase (AChE), representing the nervous system; α-amylase, for the digestive system; malic dehydrogenase, for the respiratory cycle; and ribonuclease I and acid phosphatase for lysosomal activity declined consistently as the nematodes grew older. However, some exceptions to this general phenomenon were observed for the lysosomal enzymes AChE and RNase, the activities of which were elevated considerably in very old (35-day-old) nematodes [42].

In later work on *T. aceti*, methods for extraction and labeling of RNA and protein with radioactive precursors were attempted [41]. The best precursors were [^{14}C]formate for RNA and [^{35}S]methionine for proteins. Subsequently, methods for extraction of ribosomes, polysomes, and ribosomal subunits were described [43], which will be elaborated on in Chapter 3. Indeed, it was found that the relative distribution of monosomes to polysomes was altered with age, and there was a distinct age-dependent accumulation of 60S particles in aging nematodes [44].

7.2.2 Emergence of the Concept of Accumulation of Altered Enzyme Molecules with Age

The fact that in the nematode *T. aceti* several enzymes have reduced activity with age [42] did not explain whether this decline in activity was the result of a reduction in the number of enzyme molecules per cell or of the progressive formation and/or accumulation of inactive enzyme molecules. Harriet and David Gershon [45] used a novel immunological approach to ascertain this question. Using isocitrate lyase, it was found by immunotitration that indeed enzyme-specific activity decreased considerably in the aging nematode. The results clearly demonstrated that in homogenates of old animals a greater amount of antibody was needed to precipitate a comparable amount of enzyme activity [45]. This provided strong evidence for the existence of a partially or totally inactive form of isocitrate lyase molecules in the form of immunologically cross-reactive material in the old nematodes.

Further investigations performed by mixing homogenates of young and old animals showed that older *T. aceti* nematodes contain two populations of isocitrate lyase molecules—one active and one totally inactive [45]. In subsequent studies, using antiserum prepared against crude homogenates of young adult *T. aceti*, faulty enzyme molecules of fructose-1,6-diphosphate aldolase were also observed [20,46]. Study of the biochemical properties of the enzymes from young versus old revealed no change in electrophoretic mobility nor changes in K_m and thermal denaturation characteristics [20]. This latter research was performed measuring aldolase properties in crude homogenates [20]. Thus, it was of great interest to isolate this enzyme to maximal purity from both young and old nematodes. This was accomplished, by using several biochemical and biophysical procedures. It was found that nematode aldolase was different from all three known mammalian aldolases (aldolases A, B, and C), thus implicating nematode aldolase as a product of an ancient gene from which mammalian aldolases probably evolved [47]. The kinetic and biochemical characteristics of the purified aldolase from young and old nematodes revealed a distinct reduction of about 45% in specific activity, accumulation with age of antigenically cross-reacting material, and changes in heat stability. The enzyme from old animals turned out to be more stable than that from young ones [48]. There was, however, complete immunological identity between the purified enzymes as determined by immunoprecipitation studies. Also, there were no changes in enzyme molecular weight nor in subunit molecular weight. Finally, no significant changes were found in charges as determined by electrophoretic mobility nor in the K_m values of the purified enzymes from young and old worms [47]. The above results suggested that the differences between enzymes from young and old worms were not due to the appearance of new isoenzymes in the old organisms. Therefore, it was postulated that the population of enzyme molecules in old worms contained a proportion of molecules which were either devoid of activity or perhaps possessed partial activity, while other molecules were as active as the enzyme molecules from young worms.

Two general mechanisms might cause the accumulation of inactive enzyme molecules as a function of age: (1) increasing infidelity of the protein synthesizing machinery during protein synthesis, with resultant amino acid substitutions, might lead to formation of inactive enzyme molecules. This possibility had already been hypothesized in the early 1960s by Leslie Orgel [49]. (2) The other possible mechanism could be the occurrence of posttranslational protein modifications with aging.

The data on aldolase as well as many other enzymes [50,51] do not support Orgel's theory that random substitutions of amino acids occur in enzymes of old animals. On the other hand, the various findings with aldolase and other enzymes strongly suggest that the observed alterations in properties of old enzymes are due primarily to postsynthetic modifications [48].

7.2.3 Discovery of the Decline in Function of the Protein Degradation System

The etiology of the accumulation of altered enzyme molecules with aging was a subject of extensive research in the mid-1970s. As early as 1973, the first report appeared that the half-life of aldolase is increased about 5-fold in 30-day-old nematodes when compared to 8-day-old worms [20]. This was measured in crude homogenates using cycloheximide (500 μg/ml) inhibition of protein synthesis and measurement of the disappearance of specific activity of labeled aldolase.

Taking the degradation studies a step further, Reznick and Gershon [21] showed that amino acid analogs incorporated into nematode proteins caused the accumulation of inactive, rapidly degraded, aldolase molecules. Concomitant with the appearance of inactive enzyme molecules, a rapid rate of mortality was observed [48]. Mortality was delayed with the disappearance of the inactive molecules. Using the $NaH^{14}CO_3$ method to estimate the half-life of total proteins, it was found that proteins in old and intermediate-age nematodes had half-lives 4 and 2.5 times longer, respectively, than those of young worms [21]. These results suggest that the pronounced slowdown in protein degradation in old nematodes may contribute to the appearance of a greater proportion of altered enzyme molecules. Similar studies in *T. aceti* demonstrated that another enzyme, enolase, had a slower turnover in aging animals [52].

The question of what is responsible for the increase in half-lives of proteins in aging animals, the faulty proteins as substrates for degradation or the degradation system itself, was answered in experiments using [^{14}C]puromycin in a mammalian system [53]. In that study, radioactive faulty puromycinyl peptides were generated in young and old mice. The rate of protein degradation in the liver was examined. Liver of old mice degraded these abnormal peptides at approximately one tenth the rate of livers of young animals. This study clearly showed that part of the reason for the accumulation of faulty proteins with age might be due to the decline in function of the protein degradation system [53].

7.2.4   *Caenorhabditis Briggsae*: The Use of Antioxidants, Reduction in Oxidative Damage, and Prolongation of the Life Span

The free-living nematode *Caenorhabditis briggsae* has been used as another model to study old nematodes. In very old animals, electron microscopical studies revealed severe damage to nematode tissues and progressive age-related accumulation of lipid granules containing a high level of acid phosphatase, especially in the intestinal epithelium. These granules were suggested to be age pigment, which is the product of peroxidation processes of cellular components [30]. The pigmented granules seemed to occupy most of the cytoplasmic volume of the aged nematode, leading to eventual tissue breakdown.

In a subsequent paper, *C. briggsae* nematodes were incubated in the presence of the antioxidant α-tocopherol quinone (α-TQ) [31]. In animals administered α-TQ, the 50% survival period was extended to 46 ± 2 days as compared to 35 ± 2 days for untreated controls. α-TQ caused a marked delay in the appearance of age pigment, especially in intestinal tissue. A strong correlation between the delay in appearance of age pigment and the prolongation of life in the α-TQ-treated animals was observed, suggesting that oxidative damage may be a crucial event leading to senescence and death in these nematodes [31].

All the subsequent studies on aging were carried out in chronological order with *C. elegans* by the laboratories of D. Hirsh, M. Klass, T.E. Johnson, and C. Kenyon among others as mentioned in various sections of this review.

7.3   Methods Used in Aging Research in Nematodes

7.3.1   Choice of Species and Strains, and of Either Axenic or Monoxenic Culture Systems

C. elegans is usually raised monoxenically on a lawn of *E. Coli* grown on agar plates as outlined in Reference 29 (which is also a very good general source for various methods of genetic analysis, microscopy, and some biochemistry, but not of axenic culture conditions) and Reference 1. Culture conditions are very important and must be carefully chosen according to the nature of the research problem which is being pursued. There are considerations of the ease and convenience of handling procedures on the one hand, and suitability and relevance of the conditions to the goals of the investigation, on the other. These are not always easily reconciled.

This lack of reconciliation in the case of aging research is apparent in the case of the mammalian systems. It is relatively easy to grow cells in culture and to describe cell senescence under culture conditions. These are quite stressful (e.g., trypsinization of membrane and extracellular proteins, and lack of communication with other cell systems) and, therefore, do not always credibly represent *in vivo* senescence. This is most apparent in the development of the concept that cells that cease dividing are essentially dead because eventually they diminish in number in the culture dishes. This erroneous concept does not take into account that *in vivo* most tissues consist of postmitotic cells which survive and function well for tens of years. Moreover, the tissue culture results suggest that diploid cells have a limited division potential, a phenomenon that implies a biological clock of aging. Careful examination of actively dividing cell systems *in vivo* does not reveal a seriously diminished ability to divide and maintain a constancy of cell numbers in the tissues throughout the life span of the individual (e.g., see review by Daniel [54]). It is, thus, absolutely necessary, although much less convenient and much more expensive, to conduct aging studies on whole mammals.

Studies on whole animals such as rodents generally employ inbred strains or strains that have been maintained in genetic isolation for a long time. These have at times developed idiosyncratic phenotypes such as propensity to develop specific malignancies at high frequency, or other age-associated maladies, which are not common to all strains and probably not prevalent in natural populations. This fact must be considered seriously when the results of studies on individual strains are evaluated. Another major concern is the fact that these animals are usually fed *ad libitum* and are maintained under constant optimal conditions of temperature, light, and either pathogen-free conditions or, at least, strict hygienic conditions. All of the above conditions are justified in order to reduce the number of parameters that have to be taken into account in the analysis of experimental results. Nevertheless, this convenience inevitably introduces some serious questions regarding the complete validity of the results as related to the aging process in outbred animals in their natural habitats. Laboratory mouse strains would have difficulty surviving in the habitats of their feral counterparts. In other words we have created laboratory "freaks" which are highly adapted to artificial conditions. The same and other issues have got to be addressed when one considers using the nematode as a paradigm of an aging system.

There are two strains of *C. elegans* which have been used in laboratory experiments. The Bergerac strain which was isolated from the soil in France in 1949 [55] and the Bristol strain which was isolated from mushroom compost in Great Britain in 1959 [56]. Dougherty [57] propagated some of the nematode species in his laboratory under monoxenic on agar slants inoculated with *E. coli*. He also maintained them quite successfully under axenic conditions, initially on undefined media containing yeast extract, tryptone, and other components and later on more defined media containing salts, amino acids, vitamins, and other components [57]. What is more relevant here is that both the monoxenic and axenic cultures were distributed to many investigators who have studied the nutrition, development, genetics, and aging [40] of the nematodes (we obtained such populations in 1967 from Rothstein who had obtained them from Dougherty and still worked on *C. briggsae* and *T. aceti* throughout the 1970s). Most of the later enormous volume of work on the above mentioned aspects of nematode biology has been conducted on the N2 Bristol strain of *C. elegans* which originated from Brenner's laboratory. This was primarily due to the fact that the Bergerac strain exhibited a high rate of spontaneous mutations due to the high copy number of the Tc1 transposon in its genome. Also, the males of this strain are essentially infertile and thus are difficult to use for genetic crosses [58]. The N2 strain has been adapted to monoxenic *ad libitum* laboratory conditions. There is good evidence that this strain is quite unique; it does not produce progeny in crosses with different isolates of *C. elegans* from many sites around the world [58]. This is of some concern as N2 was defined as the wild-type strain of *C. elegans*. Regarding aging research, the concern is that one may be dealing with a wild-type strain that possesses a genetic makeup which is quite different from that of natural populations. The strain has been in the laboratory for over 30 years and has been selected for the "easy life" of *ad libitum* feeding on *E. coli* (which is not usually encountered in the soil), constant optimal temperatures, lighting (to which this soil nematode is probably rarely exposed in its natural habitat), etc. It is, thus, very likely that N2 may have lost some of its capacity to cope with stress. The attenuation of the capacity to respond to stress is a major characteristic of biological aging. It is very possible that the life span of this strain under laboratory conditions is very different than that of the natural strains. It is, therefore, not as surprising as it seemed initially that mutations in single genes can extend the life span, e.g., [2,3,59,60,61], which is a complex characteristic presumably governed by a multigenic system. These mutations were obtained on the background of genotypes which developed by selection under laboratory conditions for many hundreds of generations (with life cycles of 3–5 days over a period of 30 years). Gene mutations usually exert deleterious effects on an organism, particularly one that has been subject to selective forces since the early stages of metazoan evolution [62]. Obviously, the selection under laboratory conditions for the last almost 40 years was directed at obtaining high fecundity and short generation time under fixed, artificial environmental conditions. During this process no particular attention was given to the life span of the evolving laboratory strain N2. This discussion is not meant to criticize all the interesting research on *C. elegans* aging, but to introduce a precautionary note when we consider the nematode paradigm in aging research. We suggest that *C. elegans* strains isolated from natural populations be examined genetically and biochemically for age-associated changes and life spans and compared to N2. Also, the nematodes used in the study of aging should be raised under conditions that are closer to those prevailing in their natural habitats: nutritional scarcity, feeding with soil-dwelling bacterial species rather than *E. coli* (which may be deleterious to *C. elegans* at certain population densities [1]), temperature fluctuations, and low light intensities. Also, related outbred bisexual species such as *T. aceti* or *Panagrellus redivivus* should be included in the studies. Inbreeding of hermaphrodites has certain advantages for genetic studies and some disadvantages when one looks for universal features of the aging phenomenon which bear some relevance to human aging. In this context, for instance, the low heterosis for life span found in *C. elegans* [63] should be regarded with some concern.

The nutritional model to be used in studies on nematode aging should also be considered. Growing an experimental animal monoxenically in the constant presence of another organism poses some problems. The most serious problem is encountered in biochemical studies, such as enzyme analysis, where extra caution must be taken not to mistake the bacterial components for those of the nematode. Also, the relative multiplicity in the culture of the number of bacteria to nematodes is a variable that must be contended with. In his comprehensive review, Johnson [1] dealt with this problem and aptly presented data (Table 1 of that review) which demonstrated that nematodes grown with 10^8 bacteria per culture had a mean life span of 25.9 ± 2.4 days with egg production of 63 per animal. At 5×10^7 bacteria per culture the mean life span was 15.1 ± 1.5 days and egg production of 14 per animal. At 10^9 bacteria the mean life span was 19.4 ± 1.2 days and egg production of 206 per animal. At 10^{10} bacteria the mean life span was 15.0 ± 1.0 days and egg production was only 26 per animal. These data indicate that quite small changes in multiplicity of bacteria to nematodes exert a large effect on both the life span and fecundity of the animals. The control of this relatively simple parameter in monoxenic cultures is problematic and, therefore, the consistency of the data obtained for large nematode populations must be of some concern to the investigator.

Axenic conditions, with the obvious major advantage of the avoidance of having another organism in the culture, allow for a good control of the nutrients available to the animals as it is possible to grow them in defined sterile media. We do not consider working under sterile conditions a major technical difficulty as suggested by Johnson [1]. However, a serious difficulty with axenic cultures of *C. elegans* is the fact that many animals are killed due to failure to lay eggs before they hatch. It is likely that axenic conditions can be found that prevent this problem. Axenic conditions do not, however, seem to affect *T. aceti* and *P. redivivus* in the same way (Gershon, unpublished results). Also, it is a commonly held postulate that these conditions are more removed from the natural nutritional conditions than the monoxenic cultures because in the soil these animals feed on bacteria. There is evidence that in the soil these animals feed on bacteria and also on an additional wide spectrum of food [16,58]. Therefore, in addition to what was mentioned earlier in this chapter, we are not entirely convinced that monoxenic cultures are closer to the authentic mode of feeding than are axenic cultures. They are technically slightly more convenient than axenic cultures, but their conditions are less well controlled than those in axenic cultures. It seems to us that if the problem of premature hatching of the eggs is solved, or other species are used, axenic cultures provide advantages, at least for biochemical studies. The choice of either mode of culture will then depend on the type of experiments that are planned. We find a slight edge in favor of axenic conditions particularly because they provide a more restricted diet, which is closer to the less than abundant diet of these nematodes in their natural habitats. This mode of culturing entails longer larval stages and, therefore, longer generation time and certainly longer life spans of the nematodes (without need of mutations in "gerontogenes").

7.3.2 Synchronization and Maintenance of Cultures of Organisms of Uniform Age

One of the major technical problems of working with large populations of nematodes is the creation of age-synchronized cultures and prevention of contamination with progeny. This is aptly covered in Reference [1]. Obtaining and maintaining small, age-synchronized populations is achieved without great difficulty, as follows. A few adults are transferred to fresh agar plates (or liquid medium plates). Follow by a few rounds of isolation of fourth stage larvae and transfer to new plates. The larvae of the last isolation are allowed to mature and lay eggs for a few hours. The adults are then removed and the eggs allowed to hatch. This provides for the establishment of a population of animals of identical age who are the progeny of comparably aged young hermaphrodites. This isolation avoids possible maternal age effects. Now the challenge is to keep the population progeny-free in order to maintain age uniformity. For small populations of hermaphrodites the preferred method is to manually separate the adults from the eggs before they hatch (e.g., [1,30]). This is done daily until it is observed that egg laying ceases which occurs sometime after the middle of the life span, usually 9–15 days after hatching. For small populations of bisexual species (e.g., *T. aceti*) it is possible to separate the males from the females and develop progeny-free cultures of either sex. Cultures must be scrutinized daily for possible eggs or larvae which have escaped the separation procedure.

Synchronization of large populations, which are required for large-scale preparations of nucleic acids, enzymes, or other components, is much more problematic [1]. Since manual separation is impractical on a large scale there are three alternative ways of obtaining large synchronized populations: (1) screening; (2) sterilization of adults with 5-fluorodeoxyuridine (FUDR); and (3) use of temperature-sensitive lethals.

1. **Screening** is achieved by passing the population through meshes that are made of either stainless steel or nylon. The diameter of the pores should obviously be smaller than that of the adult animals and larger than that of the larvae. The

screening is performed frequently enough so that larvae are eliminated before their maturation. This procedure has proven to be slightly leaky and, therefore, results in obvious difficulties in getting complete age synchrony and accurate determination of mean life spans at the population level.

2. **FUDR sterilization of the adults.** The use of inhibitors of DNA synthesis for this purpose is based on the fact that in the adult the only cells which synthesize DNA are germline cells. Inhibitors of DNA synthesis should, therefore, effectively prevent the production of gametes and consequently of progeny. Using *T. aceti*, we tested the efficacy of several inhibitors of DNA synthesis that do not function through incorporation into nucleic acids but through the inhibition of enzymes involved in DNA synthesis [40] (and unpublished results of extensive experiments in the late 1960s). FUDR proved to be most efficacious. The concentrations of the drug required for complete sterilization vary with the species and the larval stage of the population, as was shown in later studies by others [64,65]. In our hands, at the appropriate concentrations and time of application the life span of *T. aceti* was not affected. Moreover, our extensive studies showed no effect on enzyme levels and size of the adults [20,40,42,45]. This was further corroborated by Rothstein and Sharma [66]. There has been some controversy on the effect of FUDR on *C. elegans* that is mostly due to differences of concentrations and experimental conditions [1], but it still remains the best means of keeping large, age-synchronized nematode cultures. The regimen of FUDR application under the experimental conditions used should be very carefully tested. We contend that it is possible to find the correct treatment conditions with no apparent effects on the synchronized population.

3. **Temperature-sensitive (ts) mutants** are available that do not produce progeny when the organisms are transferred to nonpermissive temperatures at an appropriate stage of their life. Although there has not been any change in life span in some studies of ts mutants [1], we find this to be a risky mode of synchronization. In general, mutants which arise by mutagen treatment of the wild-type animals can carry mutations in genes other than the desirable one. The phenotype of these genes while not quite obvious may carry a subtle yet significant effect on the development or metabolism of the organism. Also, the ts mutation may carry pleiotropic effects that can create an undesirable paradigm for aging research. This mode of synchronization must, therefore, require very extensive observation of the ts mutant including, for instance, reversibility of the effect of the nonpermissive temperature, behavior characteristics, and possible developmental alterations.

7.3.3 Biochemical Studies: Enzymes, Nucleic Acids, Protein Labeling, and Degradation Studies

7.3.3.1 *Enzymes*

The activities of a number of glycolytic and nonglycolytic enzymes have been examined in homogenates of nematodes as a function of age, especially in *T. aceti*. Among these are aldolase [47,48] and enolase [51], of the glycolytic pathway, and nonglycolytic enzymes such as isocitrate lyase [45], amylase, AChE, malic dehydrogenase, acid phosphatase, and RNase [42]. The specific activities of these enzymes are referenced either to the amount of soluble proteins [42,52] or, in some cases, to the total amount of DNA present in the homogenates of these nematodes. With most of the enzymes studied so far, a consistent decline in specific activities was observed in aging organisms.

7.3.3.1.1 Preparation of Nematode Homogenates for Enzyme Studies: Cleaning the Nematodes

Usually, nematodes from age-synchronized cultures are harvested by centrifugation in the cold at 3000 × *g* for 10 min. The nematodes are then separated from the insoluble constituents of the medium by adding a cold 35% (wt/wt) sucrose solution to the pellet and exposing it to 5 min centrifugation at 1500 × *g* in a swinging bucket rotor. In this procedure, the live nematodes usually float on the surface, while the dead animals and the insoluble residues settle in the pellet. The washed nematodes are then rinsed thoroughly with water and suspended in a homogenizing buffer.

7.3.3.1.2 Homogenization

Attempts to homogenize the nematodes by Potter-Elvehjem or Dounce homogenizers have proven to be ineffective, most probably due to the tough cuticle of these animals [41]. After harvesting, the worms are usually freeze-dried, and the lyophilized animals are kept at −20°C. Before homogenization, batches of animals are thawed and suspended in 1:10 (wt/vol) of homogenization buffer, which differs with the different enzymes or components to be tested.

The suspension is applied to a French Press cell. Homogenization of the worms is carried out using a load of 2000 kg in the cold (4°C). The homogenates are then centrifuged at 20,000 × *g* for 60 min at 4°C, and the supernatants are collected for various enzyme studies. Specific assays for the different enzymes studied in nematodes have been described in the relevant publications.

7.3.3.2 Nucleic Acid Studies

7.3.3.2.1 Extraction of RNA

Several methods for extraction of RNA from nematodes have been attempted in the past. Initially, the phenol-cresol method of Kirby [67] resulted in the highest yields of pure undegraded RNA. However, new techniques and kits have been developed more recently [68].

7.3.3.2.2 Labeling of Nematode RNA with Various Radioactive Precursors

Incorporation of various precursors into RNA has been attempted. However, the most efficient labeling procedure using [14C]formate as a precursor for RNA synthesis gives the highest specific activity in RNA [41]. RNA labeling is analyzed on a sucrose gradient using absorption of 260 nm and 14C counting. Usually, efficient labeling is achieved by using concentration of 15 μCi label per milliliter of culture solution, where concentrations of the worms are about 100,000 animals per milliliter. Incorporation is linear for up to 6 h after addition of the [14C]precursors [41]. However, all attempts to chase the [14C]formate with a 1000-fold concentration of cold precursor have proved unsuccessful. This is probably due to the very large pool of nucleotides in these animals [41].

7.3.3.2.3 DNA Extraction

A description of DNA extraction from the nematode *C. elegans* is found in Reference [29]. In essence, *C. elegans* are grown in liquid cultures, collected, and frozen in 1g liquid nitrogen and stored dried in aliquots. Homogenization is carried out by grinding the dried powder in a cooled mortar in liquid nitrogen and then adding a solution containing 30 ml of 100 m*M* EDTA (pH 8.0), 0.5% sodium dodecyl sulfate, and 50:g/ml proteinase K. The solution is then incubated at 50°C for 2 h and cooled in ice, followed by extraction in concentrated phenol at 4°C for 15 min with slow shaking. The aqueous layer is then separated by centrifugation and collected by careful pipetting. The aqueous portion is then precipitated with two volumes of ethanol. The pellet is treated with 80% ethanol and allowed to dissolve in 10 mM Tris-HCl/0.5 mM EDTA buffer (pH 8.0).

7.3.3.3   Protein Labeling and Degradation Studies

To study protein turnover and degradation, it was important to establish the best methods of radioactive labeling of nematode proteins. The best labeling efficiency was achieved using [^{35}S]methionine and a synthetic medium devoid of methionine to assure good incorporation [41].

About 100,000 nematodes are incubated for 2–3.5 h in the presence of 15 µCi/ml [^{35}S]methionine over which the labeling is generally linear. After labeling, the animals are washed well with 10% trichloroacetic acid (TCA) and homogenized in a French Press. The homogenate is then precipitated by centrifugation at 10,000 × g, the precipitate is suspended in 5% TCA, recentrifuged, and dissolved in 0.1 M NaOH. Samples of 0.1–0.5 ml are removed for determination of protein concentration and radioactivity is estimated in a scintillation counter [41].

7.3.3.4   Protein Turnover Studies

The first report of enzyme half-life studies in nematodes was published in 1973 [20]. Aldolase half-life was calculated by addition of 500 µg/ml of cycloheximide, which caused 95% inhibition of protein synthesis as judged by incorporation of [^{35}S]methionine. This inhibition was verified in all ages [20]. Maximal inhibition is obtained within 30 min of exposure to cycloheximide. For calculation of aldolase half-life, the specific activity of the enzyme per milligram protein is determined at different times after blocking protein synthesis by cycloheximide. The half-life is calculated from the slope of the logarithm of specific activity plotted as a function of time after complete blocking of protein synthesis. At the age of 7 days, aldolase half-life was calculated as 40 h, and it increased progressively to 200 h in 28-day-old worms [20].

Determination of half-lives of total proteins in *T. aceti* has also been reported. Using [^{14}C]bicarbonates as a precursor for labeling of proteins, it has been shown that this procedure minimizes reutilization of the labeled precursor [69]. ^{14}C-labeled sodium bicarbonate was purchased from Amersham Radiochemical Centre (Amersham, Bucks, U.K.). Then 10mCi of NaH^{14}CO$_3$(60.5 mCi/mmol) were added per milliliter of nematode culture medium containing 50,000 worms per ml in a 200-ml culture bottle. Ten-milliliter samples of nematode culture were withdrawn at various times after administration of label. Each sample was then washed with 50 mM Tris-EDTA buffer (pH 7.4), resuspended in 2 ml cold Tris buffer, and subjected to French Press [41]. Fifty percent TCA was added to the chilled homogenates to a final concentration of ten percent. Precipitated proteins were collected by centrifugation at 25,000 × g for 10 min. Subsequently, these proteins were washed with 10% TCA, and the final precipitate was dissolved in 0.5 N NaOH solution. Aliquots were counted in a scintillation counter.

7.3.4   Developmental Studies and Dauer Larva Development

The nematode is very suitable for a detailed analysis of developmental processes. Indeed, the knowledge accumulated on the patterns of development at the cellular and organismal levels and their genetic control is astounding. Suffice it to read the recent reviews on various aspects of nematode development to be highly impressed [70–75]. The immense body of data availableaccumulated points to the fact that the basic features and the genes that are involved in the process are either homologous or very similar to those of higher eukaryotes. Each gene that has been identified in *C. elegans* has been shown to have its homologs in mammals. At times the function of the genes is somewhat different from that of the mammalian counterparts. This can be expected from our understanding of the more complex pathways, cellular composition, and body architecture of higher forms. Still, the degree of resemblance of the basic control mechanisms in nematode development and higher forms is very astonishing and exciting. The very extensive information of the developmental genetics of the nematode is a special attribute of this system also for aging research. The mouse is the equivalent among the mammals, but due to its complexity and the much longer life cycle, the information on the developmental genetics, though impressive, is lagsging behind the nematode. As a matter of fact, information from the nematode and *Drosophila* systems serves as a guidance to research on the mammalian system. Comprehensive characterization of the structure and function of the genes involved in development will provide insight into the possible involvement of these genes in the ultimate determination of the life span. This is because during the evolution of multicellular organisms gene systems were developed to create the most sophisticated mechanisms of differentiation. The main purpose was to form cell types with varied specialization which carry out certain functions most efficiently at the expense of the capacity to perform some basic functions which are carried out by embryonic cells and stem cells (in higher eukaryotes). This high degree of cell specialization, which often entails loss of the capacity to divide and thus to replace damaged cells at the tissue level, renders the multicellular organism progressively less responsive to environmental challenges as a function of time. This, in our view, is the major characteristic of senescence. Therefore, the nematode system provides us with the opportunity to study the genes which cause cell (and tissue) specialization which inevitably lead to the decline, as a function of time, in the capacity to cope with environmental stresses.

The Dauer larva is a special nematode form which underscores their exceptional ability to adapt to adverse environmental conditions. This is a characteristic that may be ostensibly likened to hibernation of higher phylogenetic forms, including some mammals. The Dauer stage of *C. elegans* develops in response to conditions that are prohibitive of reproduction such as insufficient nutrition and high temperatures. When such conditions are encountered, development is arrested in the third larval stage and the very specialized Dauer is formed. Dauer larvae do not feed and yet can survive for very long periods of time (several months as compared to 2–3 weeks for the normal life span). Dauer larvae can be easily distinguished morphologically, because they become thin and dense (see Reference [76] for a very comprehensive review of the Dauer larva) in culture after the animals are subjected to starvation. They lie motionless, presumably in order to conserve energy, but move vigorously when touched. They do not feed and maintain survival by reducing mitochondrial TCA cycle activity, but maintain a high capacity for glycogen and lipid utilization as energy sources. When conditions become favorable the Dauer larvae start feeding again and resume development to the adult stage. There is a fascinating system of "sensors" that indicate to the animal the existence of unfavorable conditions which will not allow normal reproduction. These involve a specific signal transduction pathway which also includes neurotransmitters that integrate the environmental signals and a hormonal system that activates special genes in a tissue specific manner [3,76]. The fascinating gene network that controls the entrance into the Dauer state and the exit from it upon environmental changes has been studied extensively. These genes, designated daf genes, are responsible for varied functions such as the development of cells that process the environmental signals and for the establishment of the specialized signal transduction pathways. Other Daf genes encode proteins which control transcription of genes necessary for Dauer formation. Mutations in many of these genes result in constitutive formation of Dauer larvae. Mutations in other genes block the ability to form Dauers. Of particular interest to the gerontologist is the fact that mutations in several of these genes involve life span extension that is independent of environmental stress and formation of Dauer larvae. Mutations in daf-2 and daf-23 are best known. Certain mutations within daf-2 can lead to doubling of the life span [60]. daf-23 is an allelic form of the well-known age-1 first described by Friedman and Johnson [77]. This gene was recently cloned [78] and was found to encode a phosphatidylinositol-3-OH kinase. This indicates that a phosphatidylinositol signaling pathway is involved in the formation of Dauer larvae and is associated with life span extension of the mutants. Two more genes, daf-16 and daf-18, are found downstream to daf-2 and age-1 in the Dauer pathway and are also involved in "life span control" [3]. daf-2 and age-1 "shorten the life span" [3] in the wild type by downregulating daf-16 and daf -18 which "extend the life span" [3] (quotation marks are ours, since we have some reservations about these terms as will be discussed below). It is an interesting pathway, but the whole phenomenon of "suspended animation" is quite unique to the nematodes and other lower eukaryotic forms. A search for the homologous genes in mammals and the characterization of their functions and the pathways to which they belong will tell if the Dauer genes are of general significance to the understanding of the process of aging.

We also caution the reader not to make a final conclusion about the "life span genes" since genetically identical nematodes exhibit extremely varied life spans at various temperatures (because it is a poikilothermic organism) and at varied nutritional conditions. For instance, at 14°C *C. elegans* lives roughly twice as long as at 25.5°C [79]. As mentioned previously the nutritional conditions as determined by the concentration of bacteria in monoxenic cultures affect significantly the mean life span and fecundity of *C. elegans* [1]. In our laboratory, *C. briggsae* showed a significantly longer life span under axenic conditions than when cultured monoxenically (D. Gershon, unpublished results). Thus, one can manipulate the life span of the nematode by changing the environmental conditions without any genetic changes. These extreme changes in life span are quite unique to the nematode particularly if one includes the developmental pathway of Dauer formation.

7.4 Conclusions and Perspectives

7.4.1 Impact and Relevance of Nematode Studies on the Field of Aging Research

Initially it was necessary to examine the main features of aging in the nematode in an attempt to see if they parallel those of other organisms, particularly mammals. This was established by the following criteria: (1) nematode survival curves followed the same patterns as found for other organisms [40] and the death rate increases exponentially with age in a manner that conforms with the Gompertz function [1]; (2) with age nematodes accumulated products of oxidation of various cellular components including lipofuscin [30,31]; (3) nematodes accumulated inactive enzyme molecules [20,45] in a very similar manner to that found in mammals [80–82]; (4) as in mammals [53,83], the accumulation of inactive enzyme molecules could be largely attributed to the decline in the capacity of the protein degradation system to dispose of altered molecules in senescent nematodes [20,21]; (5) a considerable proportion of the enzymes tested showed decline in specific activity with age [84,85]; (6) the motor activity of older nematodes declines quite dramatically with age. Thus, the nematode has been established as a most suitable and valuable model system for the study of the basic underlying mechanisms of aging. The attractiveness of the nematode system derives from the understanding of the specific attributes that are specified in Sections 7.1.1 and 7.4.2.

The major contributions that the nematode system has made to the general field of aging research in the past 27 years (since our first papers of 1970 [40,42,45]) are (1) it established unequivocally that lower eukaryotes can serve as most suitable paradigms for the study of basic mechanisms which underlie the aging process; and (2) it showed that genetic analysis of aging is feasible and may yield very significant insight into the most basic processes of aging; particularly, in our opinion, it will reveal pleiotropic effects of genes which act early in the life cycle of the organism and also have a late effect in the postreproductive stage of life. This aspect will be very much strengthened in the near future following the completion of the sequencing of the total *C. elegans* genome and facilitation of the identification of new genes and characterization of their functions.

7.4.2 Evaluation of the Main Merits and Disadvantages of the Nematode Model

As mentioned throughout this review the nematode system (particularly if several species are studied) has many important attributes which make it highly suitable for aging research. However, like any other system it also has some limitations that must be considered.

7.4.2.1 Merits

1. Nematodes have a short generation time of a few days and a short life span of 2–5 weeks, depending on the species and growth conditions. They can be raised cheaply under relatively simple conditions in large numbers either monoxenically or axenically. They, thus, provide the opportunity to conduct the most intensive *in vivo* experimentation with a eukaryote.

2. The genome of the nematode is small (100 megabase) and contains only about 12–14,000 genes as compared to 50–100,000 in the mammalian genome. The physical map of the entire genome is already available. Genetic methodology is very well worked out for *C. elegans*.

3. The nematode is a relatively simple eukaryote consisting of a limited number of cells and cell types. Yet, it possesses highly differentiated cell types which can be discerned by Nomarsky and confocal microscopy. The developmental lineage of each of the somatic cells has been worked out.

4. The developmental genetics of free-living nematodes has been extensively studied. A relatively large body of information on genes and development is already available. There are many excellent laboratories which are engaged in studying this aspect, thus assuring a rapid flow of new information in the next few years. We are convinced that there is a very close bond between development and senescence, because during differentiation a restriction is imposed on the repertoire of genes which are expressed in the cell. This restriction inevitably results

in a decline in the capacity of cells to perform all the functions that are necessary for coping efficiently with certain forms of stress. The outcome of this process is increased vulnerability and morbidity. One of the major results of terminal differentiation of most cell types is the loss of capacity to divide which renders the tissue lacking in regenerative capacity. This is true provided that the tissue is devoid of stem cells which can supply replacement for damaged cells. In the nematode there is no demonstrable capacity for cell replacement in the adult (see also discussion on limitations below). It, therefore, provides an excellent model system for studying the interrelationship between differentiation and aging.

7.4.2.2 *Limitations*

As has been discussed in several instances above, the nematode model possesses a few intrinsic characteristics which must be taken into account when one seeks to generalize the results in the field of aging research. We can broadly (and for the sake of convenience somewhat artificially) categorize these into genetic and biochemical characteristics.

Biochemical problems that we encountered, at least with *T. Aceti*, were (1) the pools of nucleotides and amino acids were very large; it was thus impossible to conduct basic pulse-chase experiments [41] that are required for studying rates of synthesis and degradation of proteins and RNA. We suggest that these large pools allow these organisms to cope with periods of starvation encountered in their natural habitats. Animals could be cultured in distilled water for one week without showing considerable reduction in their nucleotide and amino acid pool sizes. It is suspected that they could efficiently replenish these pools by degrading nonessential proteins and RNA. This high degree of adaptability to adverse conditions should be investigated also in *C. elegans* in both the laboratory strains and in fresh isolates from nature, in order to investigate the degree of changes in metabolism that have taken place during many hundreds of generations of *ad libitum* feeding in the laboratory (see also discussion of genetic changes below). (2) The removal of the cuticle without damaging the cells has so far proven impossible. This constitutes ais significant hindrance forin biochemical analysis of individual cell types in the aging adult. One is, therefore, forced to carry out biochemical studies on homogenates of whole animals. This is obviously a disadvantageous situation, as one needs to assess the extent, rate, and time-dependent accumulation of alterations in a variety of biochemical parameters in individual cell systems of the adult and senescent animal.

Genetic problems have been discussed in previous chapters. Essentially, our contention is that due to the very long exposure to artificial conditions, the laboratory strains, particularly of *C. elegans*, are far removed from the natural strains as far as their genetic makeup is concerned. Particularly altered must be those parts of the genome that are involved in the control and determination of the various phases of the life cycle and the ability to respond to stress. It is, therefore, not too surprising that single gene mutations in age-1 and some of the daf genes "lengthen the life span." What one really observes in these cases is changes in the genotypes which cause the laboratory strain to revert more closely to the "real" life span of the original isolates from nature. This is a fascinating study in evolutionary biology, but its relevance to the understanding of the basic underlying mechanisms of aging is still to be rigorously elucidated. We suggest that the allelic makeup of those "life span extending" genes (and indeed the actual life spans) be investigated in fresh isolates from natural habitats. We cited studies above that showed reproductive incompatibility between such isolates and the laboratory strains of *C. elegans*. This is an indication that the results of the genetic analysis of aging in the present laboratory strains should be interpreted with caution.

As also discussed above, *C. elegans* is essentially a homozygous, self-fertilizing hermaphrodite which is highly suitable for genetic analysis, but is it an appropriately analogous system to higher eukaryotes that are bisexual? We are certain it is up to a point and would suggest that comparative work on bisexual nematodes such as *T. aceti* should complement that on the hermaphrodites.

Finally, we also need to test critically the process of senescence under conditions that are closer to those the nematode encounters in nature. For this we need a very innovative system in which are included combinations of temperature fluctuations, varied availability of nutrition, a diet of soil-dwelling microorganisms rather than *E. coli*, various population densities, and, of course, different nematode strains (not only the descendants of the well-known N2 but also freshly isolated from the soil) and species. All of these will substantially increase our understanding of the aging process.

References

1. Johnson, T. E., Analysis of the biological basis of aging in the nematode, with special emphasis on *Caenorhabditis elegans*, in *Invertebrate Models in Aging Research*, Mitchell, D. H. and Johnson, T. E., Eds., CRC Press, Boca Raton, FL, 1984, 59.
2. Lithgow, G. J., Molecular genetics of *C. elegans* aging, in *Handbook of the Biology of Aging*, 4th ed., Schneider, E. L. and Rowe, J. W., Eds., Academic Press, New York, 1996, 55.
3. Kenyon, C., Environmental factors and gene activities that influence life span, in *Caenorhabditis elegans II*, Riddle, D. L., Blumenthal, T., Meyer, B. J., and Priess, J. R., Eds., Cold Spring Harbor Laboratory, Cold Spring Harbor, NY, 1997, 791.
4. Lansing, A. I., A transmissible cumulative and reversible factor in aging, *J. Gerontol.*, 2, 228, 1947.
5. Rothstein, M. and Cook, E., Nematode biochemistry. VI. Conditions for axenic culture of *Turbatrix aceti, Panagrellus redivivus, Rhabditis anomala* and *Caenorhabditis briggsae*, *Comp. Biochem. Physiol.*, 17, 683, 1966.
6. Brenner, S., The genetics of *Caenorhabditis elegans*, *Genetics*, 77, 71, 1974.
7. Pai, S., Die phasen des lebencyclus der *Anguillula aceti* und ihre experimentell-morphologisch beeinflussung, *Z. Wiss. Zool.*, 131, 293, 1928.
8. White, J., The anatomy, in *The Nematode Caenorhabditis elegans*, Wood, W. B., Ed., Cold Spring Harbor Laboratory, Cold Spring Harbor, NY, 1988, 81.
9. Albertson, D. G., Fishpool, R. M., and Birchall, P. S., Fluorescence *in situ* hybridization for the detection of DNA and RNA, in *Methods in Cell Biology*, Vol. 48, *Caenorhabditis elegans: Modern Biological Analysis of an Organism*, Epstein, H. F. and Shakes, D. C., Eds., Academic Press, San Diego, 1995, 339.
10. Miller, D. M. and Shakes, D. C., Immunofluorescence microscopy, in *Methods in Cell Biology*, Vol. 48, *Caenorhabditis elegans: Modern Biological Analysis of an Organism*, Epstein, H. F. and Shakes, D. C., Eds., Academic Press, San Diego, 1995, 365.
11. Hall, D. H., Electron microscopy and three-dimensional image reconstruction, in *Methods in Cell Biology*, Vol. 48, *Caenorhabditis elegans: Modern Biological Analysis of an Organism*, Epstein, H. F. and Shakes, D. C., Eds., Academic Press, San Diego, 1995, 395.
12. Waterston, R. H., Muscle, in *The Nematode Caenorhabditis elegans*, Wood, W. B., Ed., Cold Spring Harbor Laboratory, Cold Spring Harbor, NY, 1988, 281.
13. Chalfie, M. and White, J., The nervous system, in *The Nematode Caenorhabditis elegans*, Wood, W. B., Ed., Cold Spring Harbor Laboratory, Cold Spring Harbor, NY, 1988, 327.
14. Gannon, T. N. and Rankin, C. H., Methods of studying behavioral plasticity in *Caenorhabditis elegans*, in *Methods in Cell Biology*, Vol. 48, *Caenorhabditis elegans: Modern Biological Analysis of an Organism*, Epstein, H. F. and Shakes, D. C., Eds., Academic Press, San Diego, 1995, 205.
15. Appendix 2, Neuroanatomy (Part B: Neuronal connectivity diagrams), in *The Nematode Caenorhabditis elegans*, Wood, W. B., Ed., Cold Spring Harbor Laboratory, Cold Spring Harbor, NY, 1988, 449.

16. Chitwood, B. G. and Chitwood, M. B., *Introduction to Nematology*, University Park Press, Baltimore, 1974.

17. Podbilewicz, B., ADM-1, a protein with metalloprotease- and disintegrin-like domains, is expressed in syncytial organs and sheath cells of sensory organs of *Caenorhabditis elegans*, *Mol. Biol. Cell*, 7, 1877, 1996.

18. Kimble, J. and Ward, S., Germ-line development and fertilization, in *The Nematode Caenorhabditis elegans*, Wood, W. B., Ed., Cold Spring Harbor Laboratory, Cold Spring Harbor, NY, 1988, 191.

19. Wood, W. B., Embryology, in *The Nematode Caenorhabditis elegans*, Wood, W. B., Ed., Cold Spring Harbor Laboratory, Cold Spring Harbor, NY, 1988, 215

20. Zeelon, P. E., Gershon, H. E., and Gershon, D., Inactive enzyme molecules in aging organisms: nematode fructose-1,6-diphosphate aldolase, *Biochemistry*, 12, 1743, 1973.

21. Reznick, A. Z. and Gershon, D., The effect of age on the protein degradation system in the nematode *Turbatrix aceti*, *Mech. Ageing Dev.*, 11, 403, 1979.

22. Brenner, S., Forward, in *The Nematode Caenorhabditis elegans*, Wood, W. B., Ed., Cold Spring Harbor Laboratory, Cold Spring Harbor, NY, 1988, ix.

23. Brenner, S., The genetics of behaviour, *Br. Med. Bull.*, 29, 269, 1973.

24. Nomarski, G., MicrointerfJromJtre diffJrentiel B ondes polarisJes, *J. Phys. Radium*, 16, 9, 1955.

25. Brenner, S., Nematode research, *Trends Biochem. Sci.*, 9, 172, 1984.

26. Sulston, J., Cell lineage, in *The Nematode Caenorhabditis elegans*, Wood, W. B., Ed., Cold Spring Harbor Laboratory, Cold Spring Harbor, NY., 1988, 123.

27. Horvitz, H. R., Genetics of cell lineage, in *The Nematode Caenorhabditis elegans*, Wood, W. B., Ed., Cold Spring Harbor Laboratory, Cold Spring Harbor, NY., 1988, 157.

28. Sulston, J. E. and Horvitz, H. R., Post-embryonic cell lineages of the nematode, *Caenorhabditis elegans*, *Dev. Biol.*, 56, 110, 1977.

29. Sulston, J. and Hodgkin, J., Methods, in *The Nematode Caenorhabditis elegans*, Wood, W. B., Ed., Cold Spring Harbor Laboratory, Cold Spring Harbor, NY, 1988, 587.

30. Epstein, J., Himmelhoch, S., and Gershon, D., Studies on ageing in nematodes. III. Electron-microscopical studies on age-associated cellular damage, *Mech. Ageing Dev.*, 1, 245, 1972.

31. Epstein, J. and Gershon, D., Studies on ageing in nematodes. IV. The effect of antioxidants on cellular damage and life span, *Mech. Ageing Dev.*, 1, 257, 1972.

32. White, J. G. and Horvitz, H. R., Laser microbeam techniques in biological research, *Electro-Opt. Sys. Des.*, 11 [August], 1979.

33. Bargmann, C. I. and Avery, L., Laser killing of cells in *Caenorhabditis elegans*, in *Methods in Cell Biology*, Vol. 48, *Caenorhabditis elegans: Modern Biological Analysis of an Organism*, Epstein, H. F. and Shakes, D. C., Eds., Academic Press, San Diego, 1995, 225.

34. Huang, L. S. and Sternberg, P. W., Genetic dissection of developmental pathways, in *Methods in Cell Biology*, Vol. 48, *Caenorhabditis elegans: Modern Biological Analysis of an Organism*, Epstein, H. F. and Shakes, D. C., Eds., Academic Press, San Diego, 1995, 97.

35. Grant, B. and Greenwald, I., Structure, function, and expression of SEL-1, a negative regulator of LIN-12 and GLP-1 in *C. elegans*, *Development*, 124, 637, 1997.

36. Baumeister, R., Liu, Y., and Ruvkun, G., Lineage-specific regulators couple cell lineage asymmetry to the transcription of the *Caenorhabditis elegans* POU gene *unc-86* during neurogenesis, *Genes Dev.*, 10, 1395, 1996.

37. Hengartner, M. O., Cell death, in *Caenorhabditis elegans II*, Riddle, D. L., Blumenthal, T., Meyer, B. J., and Priess, J. R., Eds., Cold Spring Harbor Laboratory Press, Cold Spring Harbor, NY, 1997, 383.

38. Ellis, H. M. and Horvitz, H. R., Genetic control of programmed cell death in the nematode *C. elegans*, *Cell*, 44, 817, 1986.

39. Coulson, A., Huynh, C., Kozono, Y., and Shownkeen, R., The physical map of the *Caenorhabditis elegans* genome, in *Methods in Cell Biology*, Vol. 48, *Caenorhabditis elegans: Modern Biological Analysis of an Organism*, Epstein, H. F. and Shakes, D. C., Eds., Academic Press, San Diego, 1995, 533.

40. Gershon, D., Studies on aging in nematodes. I. The nematode as a model organism for aging research, *Exp. Gerontol.*, 5, 7, 1970.

41. Zeelon, P. E. and Gershon, D., Biochemical studies of the free-living nematode *Turbatrix aceti*. I. Methods of handling; isolation of RNA and labelling of RNA and proteins with radioactive precursors, *Comp. Biochem. Physiol. B.*, 46, 321, 1973

42. Erlanger, M. and Gershon, D., Studies on aging in nematodes. II. Studies of the activities of several enzymes as a function of age, *Exp. Gerontol.*, 5, 13, 1970.

43. Wallach, Z., Zeelon, P. E., and Gershon, D., Biochemical studies of the free-living nematode *Turbatrix aceti*. II. Isolation and partial characterization of polysomes, ribosomes and ribosomal subunits, *Comp. Biochem. Physiol. B.*, 46, 337, 1973.

44. Wallach, Z. and Gershon, D., Altered ribosomal particles in senescent nematodes, *Mech. Ageing Dev.*, 3, 225, 1974.

45. Gershon, H. and Gershon, D., Detection of inactive enzyme molecules in ageing organisms, *Nature*, 227, 1214, 1970

46. Gershon, H., Zeelon, P., and Gershon, D., Faulty proteins: altered gene products in senescent cells and organisms, in *Control of Gene Expression*, Kohn, A. and Shatkay, A., Eds., Plenum Press, New York, 1973, 255.

47. Reznick, A. Z. and Gershon, D., Purification of fructose-1,6-diphosphate aldolase from the free-living nematode *Turbatrix aceti*: Comparison of properties with those of other class I aldolases, *Int. J. Biochem.*, 8, 53, 1977.

48. Reznick, A. Z. and Gershon, D., Age related alterations in purified fructose-1,6-diphosphate aldolase from the nematode *Turbatrix aceti*, *Mech. Ageing Dev.*, 6, 345, 1977.

49. Orgel, L. E., The maintenance of the accuracy of protein synthesis and its relevance to ageing, *Proc. Natl. Acad. Sci. U.S.A.*, 49, 517, 1963.

50. Reiss, U. and Rothstein, M., Age-related changes in isocitrate lyase from the free living nematode, *Turbatrix aceti*, *J. Biol. Chem.*, 250, 826, 1975

51. Sharma, H. K., Gupta, S. K., and Rothstein, M., Age-related alteration of enolase in the free-living nematode, *Turbatrix aceti*, *Arch. Biochem. Biophys.*, 174, 324, 1976.

52. Sharma, H. K., Prasanna, H. R., Lane, R. S., and Rothstein, M., The effect of age on enolase turnover in the free-living nematode, *Turbatrix aceti*, *Arch. Biochem. Biophys.*, 194, 275, 1979.

53. Lavie, L., Reznick, A. Z., and Gershon, D., Decreased protein and puromycinyl-peptide degradation in livers of senescent mice, *Biochem. J.*, 202, 47, 1982.

54. Daniel, C. W., Cell longevity *in vivo*, in *Handbook of the Biology of Aging*, 1st ed., Finch, C. E. and Schneider, E. L., Eds., Academic Press, New York, 1977, 122.

55. Nigon, V. and Dougherty, E. C., Reproductive patterns and attempts at reciprocal crossing of *Rhabditis elegans Maupas*, 1900, and *Rhabditis briggsae*, *J. Exp. Zool.*, 112, 485, 1949.

56. Nicholas, W. L., Dougherty, E. C., and Hansen, E. L., Axenic cultivation of *Caenorhabditis briggsae* with chemically undefined supplements: Comparative studies with related nematodes, *Ann. N.Y. Acad. Sci.*, 77, 218, 1959

57. Dougherty, E. C., Hansen, E. L., Nicholas, W. L., Mollett, J. A., and Yarwood, E. A, Axenic cultivation of *Caenorhabditis briggsae* with unsupplemented and supplemented chemically defined media, *Ann. N.Y. Acad. Sci.*, 77, 176, 1959.

58. Riddle, D. L., Blumenthal, T., Meyer, B., and Priess, J. R., Introduction to *Caenorhabditis elegans*, in *Caenorhabditis elegans II*, Riddle, D. L., Blumenthal, T., Meyer, B. J., and Priess, J. R., Eds., Cold Spring Harbor Laboratory Press, Cold Spring Harbor, NY, 1997, 1.

59. Johnson, T. E., Increased life span of *age-1* mutants in *Caenorhabditis elegans* and lower Gompertz rate of aging, *Science*, 249, 908, 1990.49.

60. Kenyon, C., Chang, J., Gensch, E., Rudner, A., and Tabtiang, R. A., *Caenorhabditis elegans* mutant that lives twice as long as wild type, *Nature*, 366, 461, 1993

61. Lakowski, B. and Hekimi, S., Determination of life span in *Caenorhabditis elegans* by four clock genes, *Science*, 272, 1010, 1996.

62. Fitch, D.H.A., and Thomas, W.K., Evolution in *Caenorhabditis elegans II*, Riddle, D. L., Blumenthal, T., Meyer, B. J., and Priess, J. R., Eds., Cold Spring Harbor Laboratory Press, Cold Spring Harbor, NY, 1997, 815.

63. Johnson, T. E. and Hutchinson, E. W., Absence of strong heterosis for life span in the nematode *Caenorhabiditis elegans*, *Genetics*, 134, 465, 1993.

64. Mitchell, D. H., Stiles, J. W., Santelli, J., and Sanadi, D. R., Synchronous growth and aging of *Caenorhabditis elegans* in the presence of fluorodeoxyuridine, *J. Gerontol.*, 34, 28, 1979.
65. Gandhi, S.. Santelli, J., Mitchell, D. H., Stiles, J. W., and Sanadi, D. R., A simple method for maintaining large, aging populations of *Caenorhabditis elegans*, *Mech. Ageing Dev.*, 12, 137, 1980.
66. Rothstein, M. and Sharma, H. K., Altered enzymes in the free-living nematode, *Turbatrix aceti*, aged in the absence of fluorodeoxyuridine, *Mech. Ageing Dev.*, 8, 175, 1978.
67. Kirby, K. S., Isolation of nucleic acids with phenolic solvents, in *Methods in Enzymology*, Vol. 12, Part B, Grossman, L. and Moldave, K., Eds., Academic Press, New York, 1968, 87.
68. Krause, M., Techniques for analyzing transcription and translation, in *Methods in Cell Biology*, Vol. 48, *Caenorhabditis elegans: Modern Biological Analysis of an Organism*, Epstein, H. F. and Shakes, D. C., Eds., Academic Press, San Diego, 1995, 513
69. Swick, R. W. and Ip, M. M., Measurement of protein turnover in rat liver with [^{14}C]carbonate: Protein turnover during liver regeneration, *J. Biol. Chem.*, 249, 6836, 1974
70. Schedl, T., Developmental genetics of the germ line, in *Caenorhabditis elegans II*, Riddle, D. L., Blumenthal, T., Meyer, B. J., and Priess, J. R., Eds., Cold Spring Harbor Laboratory Press, Cold Spring Harbor, NY, 1997, 241.
71. Emmons, S. W. and Sternberg, P.W., Male development and mating behavior, in *Caenorhabditis elegans II*, Riddle, D. L., Blumenthal, T., Meyer, B. J., and Priess, J. R., Eds., Cold Spring Harbor Laboratory Press, Cold Spring Harbor, NY, 1997, 295.
72. Kemphues, K. J. and Strome, S., Fertilization and establishment of polarity in the embryo, in *Caenorhabditis elegans II*, Riddle, D. L., Blumenthal, T., Meyer, B. J., and Priess, J. R., Eds., Cold Spring Harbor Laboratory Press, Cold Spring Harbor, NY, 1997, 335.
73. Schnabel, R. and Priess, J. R., Specification of cell fates in early embryo, in *Caenorhabditis elegans II*, Riddle, D. L., Blumenthal, T., Meyer, B. J., and Priess, J. R., Eds., Cold Spring Harbor Laboratory Press, Cold Spring Harbor, NY, 1997, 361.
74. Greenwald, J., Development of the vulva, in *Caenorhabditis elegans II*, Riddle, D. L., Blumenthal, T., Meyer, B. J., and Priess, J. R., Eds., Cold Spring Harbor Laboratory Press, Cold Spring Harbor, NY, 1997, 519.
75. Ruvkun, G., Patterning the nervous system, in *Caenorhabditis elegans II*, Riddle, D. L., Blumenthal, T., Meyer, B. J., and Priess, J. R., Eds., Cold Spring Harbor Laboratory Press, Cold Spring Harbor, NY, 1997, 543.
76. Riddle, D. L., and Albert, P. S., Genetic and environmental regulation of Dauer larva development, in *Caenorhabditis elegans II*, Riddle, D. L., Blumenthal, T., Meyer, B. J., and Priess, J. R., Eds., Cold Spring Harbor Laboratory Press, Cold Spring Harbor, NY, 1997, 739.
77. Friedman, D. B. and Johnson, T. E., A mutation in the *age-1* gene in *Caenorhabditis elegans* lengthens life and reduces hermaphrodite fertility, *Genetics*, 118, 75, 1988.
78. Morris, J. Z., Tissenbaum, H. A., and Ruvkun, G., A phosphatidylinositol-3-OH kinase family member regulating longevity and diapause in *Caenorhabditis elegans*, *Nature*, 292, 301, 1996.
79. Klass, M. R., Aging in the nematode *Caenorhabditis elegans*: major biological and environmental factors influencing the life span, *Mech. Ageing Dev.*, 6, 413, 1977.
80. Gershon, H. and Gershon, D., Inactive enzyme molecules in aging mice: Liver aldolase, *Proc. Natl. Acad. Sci. U.S.A.*, 70, 909, 1973.
81. Reiss, U. and Gershon, D., Rat liver superoxide dismutase: Purification and age-related alterations, *Eur. J. Biochem.*, 63, 617, 1976.
82. Reznick, A. Z., Rosenfelder, L., Shpund, S., and Gershon, D., Identification of intracellular degradation intermediates of aldolase B by antiserum to the denatured enzyme, *Proc. Natl. Acad. Sci. U.S.A.*, 82, 6114, 1985.
83. Reznick, A. Z., Lavie, L., Gershon, H.E., and Gershon, D., Age-associated accumulation of altered FDP aldolase B in mice: Conditions of detection and determination of aldolase half life in young and old animals, *FEBS Lett.*, 128, 221, 1981.
84. Gershon, D., Current status of age-altered enzymes: alternative mechanisms, *Mech. Ageing Dev.*, 9, 189, 1979.
85. Rothstein, M., The formation of altered enzym es in aging organisms, *Mech. Ageing Dev.*, 9, 197, 1979.

8

Experimentation With The Yeast Model

Sangkyu Kim, Paul A. Kirchman, Alberto Benguria, and S. Michal Jazwinski

CONTENTS

0-8493-3112-9/99/$0.00+$.50

8.1 Introduction

In the past several years, *Saccharomyces cerevisiae* (baker's yeast) has emerged as an important model system for the study of aging [1]. This development follows on the heels of the application of genetic and molecular tools in yeast to the elucidation of several fundamental cellular phenomena, such as cell cycle control and secretion. It is now easy to encounter laboratories that go back and forth between yeast and mammalian systems to analyze complex cellular and organismal phenomena. This change in approach has been cemented by the realization that human disease genes very often have yeast counterparts [2]. Indeed, this fact has made possible, in some cases, the identification of the relevant human gene.

Yeast is a unicellular eukaryote, and this contributes to its status as a genetic powerhouse. Furthermore, the facility with which homologous recombination occurs in this organism has brokered a perfect marriage of classical and molecular genetics that is found nowhere else. The completion of the sequence of the yeast genome in 1996 [3] has propelled yeast into the postgenome era of functional genomics. All of these factors certify this organism as a genetic model for the analysis of the mechanisms of aging. There is yet another reason why yeast is an excellent model for the study of aging. Yeasts present a simple, stripped-down version of aging. Extracellular factors, such as hormones, and interactions with other cells can largely be ignored in analyzing the intrinsic features of aging. Cellular and organismal aging are one. Indeed, the germ line and the soma are contained in the individual cell. Yeasts can be likened to stem cells, if the experimental bent is cellular aging. In studying yeast aging, the experimenter is not misled by age changes that do not contribute to mortality; there is no accumulation of gray hair with age.

The life spans of individual yeasts can be measured accurately, because they divide asymmetrically by budding, allowing discrimination between the mother cell and its daughter. Thus, life span is measured by the number of cell divisions or daughters produced, and it is generally expressed in generations [4,5] and often called replicative life span. Time can also be used as the metric, but this is less accurate, given the fluctuations that environmental factors such as temperature changes exert, and more tedious. It follows that lifetime metabolic activity could also be used as the metric, with the same restrictions. The asymmetric division of yeasts reconciles the immortality of the population or culture with the mortality of individuals. It also creates a major difficulty in the preparation of age-synchronized populations of yeasts, because the culture always contains approximately 50% virgin cells that have never divided and the frequency of cells decreases exponentially as a function of their age.

As yeasts progress through their replicative life span, they undergo a wide array of morphological and physiological changes [6]. Some of these changes are decremental; thus, it is reasonable to speak of an aging process. Yeast aging has been reviewed extensively [1,6–11]. A contemporary laboratory guide to yeast genetics and molecular biology is available [12]. The techniques peculiar to the study of yeast aging are described in this chapter. There is no better predictor of mortality than life span itself, certainly in yeast. Therefore, the genetic analysis of longevity in this organism has been a search for and analysis of genes that determine longevity. Some dozen longevity genes have been found in yeast thus far [1]. The genetic analysis of aging in this organism points to three factors that contribute to aging: metabolic capacity, efficiency of stress responses, gene dysregulation, and genetic instability [1].

8.2 Life Span Determination

8.2.1 Microscope and Micromanipulator

Life span analyses can be performed with a standard tetrad dissection setup [13]. A micromanipulator mounted to the base of the microscope and not the stage allows the needle to remain under the objective and in focus while the stage with the specimen is moved to locate cells. This eliminates the need to "find" the needle every time the stage is moved. However, a fixed-stage microscope is necessary, in which focus is obtained by adjusting the optics. We prefer to have the joystick mounted on the right hand side of the microscope and the stage controls on the left (left-handed individuals may prefer the opposite orientation). This can usually be requested at the time of ordering. Finally some microscopes used for tetrad dissection have "stops" in the movement of the stage. When working in a position near one of these stops, the stage can sometimes move from its position into the stop, which can be at least very annoying and at worst cause the loss of cells being manipulated. These stops are not necessary for life span analyses, and microscopes with no stops can be requested when ordering.

Microscope objectives that come with a dissection microscope are usually long working–distance objectives. A good level of magnification for working with yeast can range from 150× to 300×. This can be achieved with a combination of 10 or 15× eyepieces and a 10 or 20× objective. We use microscopes with 15× eyepieces and 10×, 20×, and 2× objectives. The 2× objective is very helpful in installing a new needle. The top lens of the condenser is removed to facilitate micromanipulation under long working–distance objectives. A blue daylight filter provides a pleasing illumination, and an infrared filter prevents specimens from overheating.

There are many choices of microscopes and micromanipulators on the market. These are fully discussed elsewhere [13]. A complete and affordable tetrad dissection microscope that includes the micromanipulator (Figure 8.1) can be purchased from Micro Video Instruments, Inc. (Avon, MA). An inexpensive joystick micromanipulator that can be mounted on any microscope with assistance from a local shop is distributed by Allen Benjamin, Inc. (Tempe, AZ).

8.2.2 Microneedles

The most consistent way to make glass microneedles suitable for micromanipulation is by generating glass fibers which are then glued to a glass mounting rod. Glass mounting rods can be made from 5-μl disposable pipettes (Baxter cat. #P4518-5X). The pipette is heated

FIGURE 8.1
A tetrad dissection microscope used for life span analysis. The microscope stage and the micromanipulator are highlighted.

over a flame and bent to a right angle approximately 5 mm from one end. Needles are made by heating a 2-mm diameter glass rod until the glass begins to melt and then rapidly pulling apart the two ends as the rod is removed from the flame. Use a razor blade to cut the thin glass fiber perpendicular to its length at approximately 1-cm intervals. Do the cutting on a solid surface such as a microscope slide to generate a flat tip. A flat tip that contacts the agar smoothly will be best for moving cells and will not gouge the agar. A needle 20–40 μm in diameter is desired. Large needles are better for picking up cells, but harder to use for single cell separations. A needle that is too thin will bend and be very difficult to use for moving cells. Needles can be examined under a low power dissecting microscope to aid in choosing the proper diameter and tip.

Once a needle is chosen, it is then glued to the mounting rod. Place the needle on a microscope slide so that the best end is on the slide and the other end extends over the edge. A small amount of cyanoacrylate glue (Super glue or Crazy glue) is applied to the mounting rod by touching the short upturned end of the rod to a drop of glue. The sticky end is then used to pick up the needle (Figure 8.2). Glue the needle so that it extends in the same direction as the upturned portion of the mounting rod and a couple of millimeters beyond the end. If the needle is not parallel to the upturned shaft of the mounting rod, it must be quickly adjusted before the glue sets. Once the glue has set test the needle for its ability to pick up, release, and separate cells. Much time will be saved later if a good needle is obtained.

8.2.3 Choice of Strain

The strain chosen as the "wild-type" control may depend upon the experiments you wish to perform; however, several things should be considered. One consideration that is very important in choosing a strain is the ease with which the buds can be removed from individual cells. Some strains seem to form elongated cells, particularly during later generations, which makes bud removal very difficult. The life span of your control strain is also an important consideration. A strain with a life span that is too short may make it hard to examine variables that shorten life span, while a strain with a life span that is too long may make it difficult to detect extensions and takes too long to assay. We have chosen a strain with a mean life span of about 19 generations and a maximum of around 30 generations, for many of our studies. As with any strain that you may want to use for genetic manipulations, it is a good idea to choose a strain containing a variety of nonreverting selectable markers.

8.2.4 Maintenance of Strains

Maintaining a clonal (isogenic) population is important with any genetic analysis; however, extra care needs to be taken when dealing with a "nonselectable," polygenic

FIGURE 8.2
Orientation of the microneedle and the mounting rod. The microneedle is glued so that it extends several millimeters beyond the end of the mounting rod.

phenotype such as life span. Life span mutants of yeast can easily arise in a strain if it is not treated with care. Starvation on plates has actually been used as a selection for strains with extended life span [14]. Several other observations support this point. We have obtained the same yeast strain from another laboratory on two different occasions separated by an interval of three years. These two "isogenic" strains had significantly different life spans. We also obtained a diploid strain from another laboratory that was generated by mating two "isogenic" strains that differed only at the mating type locus. We sporulated the strain and examined the life spans of the haploids obtained. Life span segregated as a trait determined by more than two genes. We have found that there is no difference in life span between coisogenic strains of opposite mating type, that is congenic *MATa* and *MATα* strains. We serially restreaked a clonal isolate over a period of a few months. Although the strain retained all of its known genetic markers, it was now composed of many clones, each of which differed in its replicative life span.

We routinely streak out all clones as soon as they are obtained and never let cells remain on plates longer than it takes them to grow into colonies. For newly isolated strains, single colonies are picked and allowed to grow in 5 ml of liquid medium overnight. Frozen stocks should be made of strains immediately by addition of glycerol to a final concentration of 15% and freezing at –70°C, even before they are further analyzed for mutations or life span. A fresh inoculum from this stock should be used for each assay.

8.2.5   Transformations

Analysis of molecular mechanisms involved in determining longevity is greatly facilitated in yeast by the ease with which genes can be deleted, mutated, and overexpressed. However, maintenance of the genetic background is extremely important. Typical yeast lithium acetate transformations [15] involve the use of carrier DNA to increase efficiency. Because of the possibility of integration by nonhomologous recombination of the carrier DNA into the yeast chromosome, we typically eliminate it from transformations. Strains containing unintegrated replicative plasmids are easily obtained in this way and dozens of colonies containing deletions can also be obtained, more than enough for life span analyses. If for some reason transformation efficiency is too low to obtain the desired clones, carrier RNA may be used [16].

8.2.6   Procedure

Life span analysis is usually carried out on a rich medium (1% yeast extract, 2% peptone, 2% glucose). Solid media contain 2% agar, and the pH is adjusted to 6.5–6.8. However, alternate carbon sources, such as glycerol or oleic acid, may also be used depending on the experiment. If it is necessary to maintain selection for a plasmid, synthetic medium [12] lacking a required nutrient to maintain selective pressure may also be used. Sometimes it is necessary to transfer cells from one type of medium to another at some point in their life span. This can be accomplished most easily by performing the analysis on agar slabs. Slabs can be cut from two different types of media with a sterile scalpel and transferred to an empty petri plate. The plates from which the slabs are taken should be the same thickness (i.e., use a measured amount of medium to fill each plate). Once the slabs are on the plate, cells can be transferred between the two types of media with ease.

Regardless of the type of medium used, the basic procedure for determining life span of a strain [17] is as follows. Streak the strains to be assayed onto the appropriate agar plates. Colonies from this plate are then used to start a liquid culture the night before beginning the assay. Starting liquid cultures directly from freezer stocks causes problems with some strains. Spot 1 µl of an exponentially growing culture onto the appropriate medium. Plates should be fresh since the assay can take up to 2 weeks to complete. If multiple strains are being assayed, place the spots in a line near the middle of the plate. Allow the excess liquid to be absorbed. Invert the plate under the microscope, being careful not to break the needle with the edge of the plate. This is one of the many ways the needle can be broken, maybe the most frequent. Rotate the plate so that the spots of cells are aligned parallel to the needle. Locate and focus on the cells and then slowly move the needle up until it comes into view. Use the joystick to raise the needle further. Once the needle contacts the agar, make any adjustments necessary to position the joystick to a comfortable position. You may be at the microscope a while, and this will increase speed and decrease hand cramps.

To obtain virgin cells, pull cells from the spot either singly or in groups and spread unbudded cells out away from the original spot. Cells should be placed far enough away so that they are not overgrown by the large spot of cells, but close enough so that they can be reached without having to move the field of view. Once enough individual cells are separated from the original spot of cells (about twice the final number to be used in the assay), lower the needle and remove the plate. Cover the plate and wrap the circumference with parafilm. Incubate the plate at the appropriate temperature to allow buds to grow (about 2 h for healthy strains on rich medium at 30°C). Remove virgin cells (buds that have never produced daughters) from their mothers and place them in a row parallel to the needle. It is not necessary to move these cells too far, but do not place the virgin cells between the original spot and the mother cells; place them beyond the mother cells. Once you have your starting population, again incubate the wrapped plate and allow the cells to bud.

After incubating, remove the buds from the cells of your population and discard them in the direction of the original spot of cells. Each bud removed is recorded as one generation for that particular cell. Occasionally, you will need to move your cells away from the encroaching colonies formed by the discarded cells. The first generation removed from your population may be tricky since the mother and its daughter are of similar size. If there is some confusion, remember that the mother will begin to bud again before the daughter buds for the first time. This is not necessarily true for older mothers, but by then the size difference makes it apparent which cell is the mother.

One mistake often made by those trying this assay for the first time is to try to remove all of the buds every time they examine the plate. This can become quite time consuming and frustrating. If a bud cannot be removed with a few strokes and pokes with the needle it is better to move on and get it the next time around. Of course, buds are more easily removed from some strains than others, which is an important consideration in choosing an initial "control" strain. Once the mother cells are bigger, the plates can be incubated for longer periods, allowing up to two generations of growth before examination. Too much growth becomes very confusing, although you may be able to recognize the original cell, the number of times it has budded may be obscured by all of the daughters, granddaughters, etc.

Setting up this experiment is the most time-consuming step. Once all of the starting buds are lined up, it is simply a matter of checking the cells, removing the buds and recording the results. An experienced investigator can remove buds and record the results from over 200 cells in about 30 min.

A nonstop life span determination would exhaust any investigator. To provide respite, the plate with the cells can be stored overnight at a lower temperature. At 4–5°C, overnight growth is prevented. However, time is not lost by incubating the cells at 12–14°C, which allows about one cell division to be completed overnight with most strains. It has been demonstrated that repeated incubations at low temperature do not alter the life span [5].

8.2.7 Data Recording and Analysis

Data are usually recorded on a sheet with a numbered column on the left side. Each row corresponds to a single cell. The date and time of each observation/micromanipulation is entered in the top row. The entries in the second and subsequent columns indicate the number of buds produced or generations completed. For example:

Cell no.	Date/time	Date/time	Date/time	Date/time	Date/time	Date/time
1	v	1	2	4	5	...
2	v	1	3	4	6	...

Note: v indicates a virgin cell.

Variations on this that can be helpful or necessary, depending on the experiment, can also be employed. One modification is recording which cells have a nonremovable bud at the time of observation by adding a b (for a small bud) or a B (for a large bud) to the data. This may be necessary if a record of the generation time of individual cells is desired. Another notation that is helpful, particularly with strains in which bud removal is difficult, is to simply add a tick mark onto the number in the direction the bud is oriented (e.g., 6- if the bud is on the right side of the cell). This is helpful in determining which cell is the bud upon the next examination when the bud is closer in size to the mother. Columns should remain aligned and a recording should be made on every observation, even if there is no change. Simply using hash marks to indicate generations does not allow complete analysis of the data. The point at which a cell dies is not always obvious. Usually they lyse or lose refractility; however, if no bud is produced after an entire day of incubation, the cell can also be considered dead. Such cells rarely, if ever, bud again.

Once all of the data are gathered (i.e., all the cells are dead), it can be analyzed in several ways. We always plot percent survival as a function of age (generations) as shown in Figure 8.3A. Plots of age specific mortality rates can also be helpful in analyzing the data (as discussed in Chapter 1 of this volume) and show that the extension in life span seen in Figure 8.3A is due to a decrease in mortality rate throughout the life span (Figure 8.3B). Mean life span data should also be analyzed statistically to determine differences between strains. A *t*-test can be used to compare means, but a nonparametric test such as the Mann-Whitney or Wilcoxon signed rank is more appropriate to the data. The maximum life span is not statistically significant. However, the 90th or 95th percentiles of populations can be meaningfully compared [17]. There are many statistical packages on the market that can be used.

8.3 Preparation of Age-Synchronized Cells

8.3.1 Separation of Age-Synchronized Cells by Rate-Zonal Sedimentation in Sucrose Density Gradients

8.3.1.1 *Principles*

We have utilized two properties of yeasts in one protocol for preparing age-synchronized cell populations by rate–zonal sedimentation [18]. The first property involves uniform cell-cycle arrest in the nondividing, stationary phase. Yeast cells arrest in the stationary (G_0/G_1)

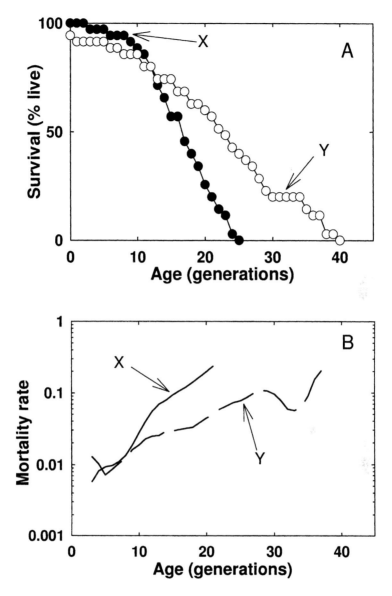

FIGURE 8.3
Analysis of life span data. **(A)** Data on life span is plotted as percent survival at a given age. **(B)** Mortality rates as a function of age. The data are smoothed over an 8-generation window. The program MORTAL 1.0 was used to calculate mortality and smoothed mortality rates (MORTAL 1.0 designed and provided by J. W. Curtsinger).

phase without losing viability significantly [19]. This stationary-phase arrest is utilized to synchronize the cells in the cell division cycle and to limit the number of generations cells go through in each step of the preparation. Haploid cells can also be arrested in the G_1 phase of the cell cycle, before the initiation of DNA synthesis, by treating them with the mating pheromone (α-factor) secreted by the opposite mating type [20]. Typically, α-factor is used to arrest **a** mating type cells. The second property involves isolation of cells of a desired age from a stationary-phase culture. This procedure utilizes the observation that old cells are generally larger in size than younger cells [21,22]. Given that young and old cells are noticeably different in size, they can be easily separated by their different sedimentation rates in density gradients.

8.3.1.2 Materials and Equipment

8.3.1.2.1 Yeast Strains

The following procedures have been developed for *Saccharomyces cerevisiae* X2180-1A (*MATa SUC2 mal mel gal2 CUP1*), obtained from the Yeast Genetic Stock Center, Berkeley, CA. For other strains, it may be necessary to modify the details of some steps, although the principles mentioned above do not change.

8.3.1.2.2 Media

YPD (2% peptone, 1% yeast extract, 2% glucose) and YPDG (2% peptone, 1% yeast extract, 0.04% glucose, 1.6% glycerol) are used. Solid media contain 2% agar.

8.3.1.2.3 Other Reagents

Sterile water or a saline solution containing 0.85% NaCl (both are equally effective in washing and resuspending cells without affecting the outcome of the preparation), sterile 10% w/v and 30% w/v sucrose solutions, Cellufluor (1 mg/ml in water made fresh daily, Polysciences, Warrington, PA), kanamycin (50 mg/ml stock in H_2O) or tetracycline (10mg/ml stock in 50% ethanol), and 1 mM α-factor stock must be prepared. To make 1 mM α-factor stock, add 2.745 ml of the α-factor buffer to one 5-mg ampule of the pheromone (Sigma). It is stored as 100-μl aliquots at –20°C. The 5× concentrated α-factor buffer is prepared by mixing 8.7 μl of concentrated HCl, 100 μl of 7.5 mg/ml EDTA, and 3.5 μl of 2-mercaptoethanol, and adding sterile water up to 10 ml.

8.3.1.2.4 Equipment

Cole-Parmer ultrasonic homogenizer (4710 series), an IEC Centra-8R centrifuge with model 216 swinging-bucket rotor, thin-wall Nalgene polyallomer centrifuge tubes (50 ml, cat.# 3117-0500), a Buchler Auto Densi Flow IIC gradient fractionator, and an epifluorescence microspcope with a UV filter cube (400 nm dichroic mirror, 330 to 380-nm excitation fiter, 420 nm barrier filter) are used.

8.3.1.3 Step-by-Step Procedures

8.3.1.3.1 Preparation of a Stationary-Phase Culture

Step 1: Inoculate 100 ml of YPD in a one liter flask with 5~10 colonies of X2180-1A freshly grown on a YPD plate. The volume of the culture flask should be at least ten times the YPD liquid volume to allow sufficient aeration.

Step 2: Incubate the culture at 30°C on a shaker at 250~300 rpm until the culture reaches and stays in the stationary phase, taking care not to lose cell viability. With X2180-1A, this usually takes 3 to 4 days and the saturated cell concentration reaches more than 5×10^8 cells per ml, determined as described below. It is important that the proportion of unbudded cells should be more than 95% of the cells examined. Although the number of budded cells usually decreases with longer incubation, it should be noted that prolonged incubation results in a sharp decrease in cell viability. Therefore, it is necessary to monitor cell viability periodically during the incubation and find the best incubation time to maximize both cell viability and the proportion of unbudded cells.

Step 3: When the culture is ready, store it at 4°C and prepare a sufficient number of linear 10–30% sucrose gradients in thin-wall Nalgene polyallomer centrifuge tubes (50 ml). The gradients are usually poured in a UV-sterilized hood to minimize bacterial contamination. They are then chilled at 4°C overnight before use to prevent cells from fermenting the sucrose during subsequent centrifugation.

8.3.1.3.2 Isolation of Virgin (g0) Cells

Day 1

Step 1: Divide the 100-ml culture into two 50-ml conical screw cap tubes. Centrifuge the tubes at 3000 rpm for 5 min at 4°C in an IEC Centra-8R centrifuge with model 216 swinging-bucket rotor.

Step 2: Transfer the pelleted cells in one tube to the other while resuspending the cells in a total volume of 50-ml sterile water. Vortex the tube and set aside a small sample (10 μl) of the cell suspension. Centrifuge the tube again. Meanwhile, dilute the sample 1/100 in water, sonicate briefly (~15 s) in a cup horn, and determine the cell concentration on a hemacytometer. Count each discernible cell even if it is attached to another cell. The probe of the sonicator can be sterilized by washing with 70% ethanol, followed by a rinse with sterile deionized water.

Step 3: Resuspend the pelleted cells in sterile water at 1×10^9 cells per ml, and transfer the cell suspension to a 250-ml Nalgene beaker (cat. #1201-0250). Put the beaker on ice. Sonicate the cell suspension for 60 sec at ~ 40 W using a 1/2" standard horn to obtain more than 95% single cells. Examine cells microscopically to check if they are well separated.

Step 4: Using a 1-ml pipette, gently layer the cell suspension (1 ml/gradient) to the top of each gradient, four at a time. Increasing the volume causes streaming of cells along the inside wall of the tubes and pelleting of unseparated cells at the bottom of the centrifuge tubes. Amounts lower than 1×10^9 cells per gradient produce better separation but this requires an impractically large number of gradients in large-scale preparations.

Step 5: Centrifuge the gradients at 4°C for 4 min at 1400 rpm with the brake off. Two distinct bands of cells should be visible when viewed with backlighting. The upper band consists of virgin cells that have never budded. Avoiding cells between the bands, collect virgin cells in the upper band in 50-ml conical screw cap tubes by fractionating the gradients. To do this, place the centrifuged gradient in the holder. Turn the function switch to DOWN. The probe will slowly descend without disturbing the bands formed. When the tip of the probe reaches just below the desired band, turn the function switch to STANDBY to stop the probe movement. The cells in the band are removed into empty 50-ml screw cap tubes through the tubing attached to the top of the probe by turning the pump to REMOVE. When finished, turn off the pump and raise the probe by turning the function switch to UP. The collected fractions are kept on ice to prevent growth. Repeat Steps 4 and 5 until all the virgin cells are collected. Pool the virgin cells.

Step 6: Centrifuge the virgin cells in the sucrose solution for 10 min at 3000 rpm with the brake on. Resuspend and pool the cells in 100 ml of sterile water divided into two 50-ml conical screw cap tubes. Centrifuge the tubes for 5 min at 3000 rpm.

Step 7:   Resuspend and pool the cells in 50 ml of sterile water in one 50-ml conical screw cap tube. Vortex the tube and remove 10 μl of the cell suspension and mix it with 90 μl of sterile water containing 0.1 mg/ml Cellufluor. Examine the cells under the fluorescence microscope for the absence of bud scars and count the cell number to determine the cell concentration in 50-ml water. A 100-ml stationary phase culture of X2180-1A normally yields 6 to 8×10^9 virgin cells. The cells are small and highly uniform in size. Although contamination by large virgin cells and 1 or 2-generation old cells are not uncommon, such contamination should not exceed 5%.

8.3.1.3.3   *Synchronization of Virgin Cells*

The virgin cells obtained from a stationary phase culture are uniformly arrested in the G_1 phase and then allowed to bud as long as synchrony lasts.

Step 1:   Resuspend the virgin cells in YPD at 8×10^7 cells per ml. Add α-factor to the YPD culture to the final concentration of 400 nM (0.4 μl of the 1 mM stock per 1 ml of the culture). Incubate the culture for 40 min at 30°C.

Step 2:   Repeat this addition 2 more times, incubating the culture for 40 min each time. It is important not to let the arresting cells "shmoo" (technical term for cell elongation) due to overdoses of the α-factor or due to unnecessarily long incubation. It is beneficial to monitor cell size and shape periodically by removing small aliquots of cells and examining them microscopically. At the end of this treatment, the cells will uniformly arrest in the G_1 phase and almost double in size. Upon examination of Cellufluor-stained cells, a brightly stained spot is usually observed on the cell surface due to a slight accumulation of chitin [23].

Step 3:   Wash the G_1-arrested cells twice in sterile water and resuspend them in YPDG at 4×10^7 cells per ml. Store the cell suspension overnight at 4°C.

8.3.1.3.4   *Isolation of Two-Generation-Old (g2) Cells*

Day 2

Step 1:   Incubate the G_1-arrested virgin cells in YPDG at 30°C on a shaker at 275 rpm to triple the cell count (i.e., up to ~ 1.2×10^8 cells per ml). This corresponds to 2 cell divisions and usually takes 8 to 10 h. Although the cell cycle synchrony can continue up to three cell divisions, the virgin cells are allowed to bud only twice because the size difference between the mother cells and their daughters is not large enough to prevent the first daughter from approaching the mother in size in three divisions. The YPDG medium, in which the amount of glucose is 50 times less than in the regular YPD medium, is used to slow down cell growth without any significant change in life span. Slower growth in media with limiting carbon source enhances the size asymmetry between the mother and daughter cells [21,24].

Step 2:   While incubating the culture, prepare linear 10–30% sucrose gradients as before and store them at 4°C. When the culture is ready, store it at 4°C.

Day 3

Step 3:   Distribute the culture into 50-ml conical screw cap tubes. Centrifuge the tubes for 5 min at 3000 rpm at 4°C.

Step 4:   Resuspend the cells in 50-ml sterile water. Count the cell number in a 1/10 dilution. Centrifuge the tube as before.

Step 5: While resuspending the pelleted cells in sterile water at 1×10^9 cells per ml, transfer the cell suspension to a Nalgene beaker. Sonicate the cell suspension on ice for 1 min at 40 W.

Step 6: Layer 1 ml of the cell suspension to each of 4 gradients, and centrifuge them 4 min at 1400 rpm with the brake off. Either two or three bands are formed (Figure 8.4A). The top band is diffuse and contains newly generated virgin cells. The middle band, if visible, is close to the top band and contains one-generation-old cells. The bottom band contains two-generation-old cells. Fractionate the g2 cells in the bottom band. Keep the fractionated cells on ice. Repeat this step until all the g2 cells are fractionated.

Step 7: The collected g2 cells in sucrose solution are washed twice in water as before and resuspended in YPDG at 2.5×10^7 cells per ml. Remove a small aliquot of the cell suspension, stain the cells with Cellufluor as before, and examine them microscopically. Store the YPDG culture overnight at 4°C.

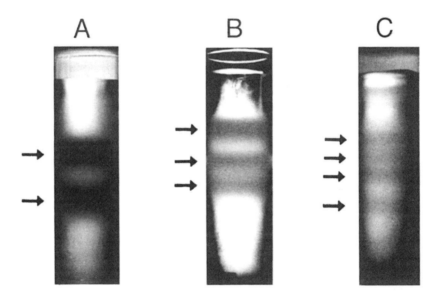

FIGURE 8.4
Separation of age-synchronized cells on sucrose gradients. Photographs of density gradients from various steps of the age-synchronized cell preparation were taken after centrifugation. The arrows indicate the bands of cells in the gradients. **(A)** Separation of two-generation-old (g2) cells (lower band) from their two daughters (upper band). **(B)** Separation of g5 cells (lowest band) from their first daughters (middle band), and from their second and third daughters and the daughters produced by the first daughters (top band). **(C)** Separation of g14 cells (lowest band) from their first daughters (second band from the bottom), their second daughters (third band from the bottom), and third daughters and the daughters derived from the first daughters (top band). (From Egilmez, N. K., Chen, J. B., and Jazwinski, S. M., *J. Gerontol.*, 45, B9, 1990. With permission.)

8.3.1.3.5 Preparation of Older Cells

Since the size difference between mother and daughter cells becomes larger starting with the g2 cells, cells are allowed to bud 3 times in YPDG per each round to obtain g5, g8, g11, g14, g17, and sometimes g20. As a result, either three or four bands are visible. Early in the preparation, usually prior to ten generations, three buddings usually result in three well-separated bands in the gradients (Figure 8.4B). However, later in the procedure, four bands are usually visible (Figure 8.4C). The lowest band corresponds to the old mother cells, the next one up to the first daughters, the third from the bottom to the second daughters, and the top band to both the third daughters and the buds of the first daughters. The two upper bands overlap, yielding a single band early in the preparation.

As preparations of older cells are continued, cultures become more susceptible to bacterial contamination. If contamination occurs, it can be removed in the following way. Cells are pelleted by centrifugation at 3000 rpm. They are suspended in 45 ml of sterile water by vortexing vigorously and recentrifuging. This is repeated four to five times. The final pellet is resuspended in water at 1×10^9 cell per ml and separated on gradients as described before. The bacteria remain at the top of the gradient and are discarded. If contamination is detected early during growth in YPDG, tetracycline at a concentration of 1 μg/ml or kanamycin at 25 μg/ml can be added to the culture. It should also be noted that yields of older cells decrease because of higher mortality rates of older cells. If the yield is poor, collect all the remaining cells after fractionation by centrifugation for 10 min at 3000 rpm. After washing them twice in water, repeat Steps 5 and 6 above (Section 8.3.1.3.4) again. The bottom band will still be visible, though fainter than during the first separation.

Day 4

Step 1: Incubate g2 cells in YPDG at 30°C for 3 generations so that the cell number increases 5-fold (up to 1.25×10^8 cells per ml). With X2180, this usually takes 8–10 h. When the culture is ready, store it at 4°C. The synchronous cell cycles can be monitored microscopically.

Step 2: Prepare sucrose gradients as before and store them at 4°C.

Day 5

Step 3: Transfer the culture to a Nalgene beaker. Sonicate the cells as before. Wash them once in water and resuspend the pelleted cells at 1×10^9 cells per ml. Layer the cell suspension onto gradients and centrifuge them as before. Fractionate the cells from the bottom band as before. If the yield matters, do the second round of centrifugation and fractionation with the remaining cells as described before. Pool all the fractionated cells together by centrifugation and wash them twice in sterile water. Determine the cell number and resuspend the g5 cells in YPDG at 2.5×10^7 cells per ml. Monitor the purity by fluorescence microscopy.

Days 6 and 7

Repeat Steps 1 to 3 above to obtain g8 cells.

Days 8 and 9

Repeat Steps 1 to 3 above to obtain g11 cells.

Days 10 and 11

Repeat Steps 1 to 3 above to obtain g14 cells.

Days 12 and 13

Repeat Steps 1 to 3 above to obtain g17 cells.

Days 14 and 15

Repeat Steps 1 to 3 above to obtain g20 cells.

8.3.1.4 Determination of the Age of Each Population of Cells Prepared

The replicative age of collected cells in each population prepared can be confirmed by the methods described below (Section 8.3.2.3.4).

8.3.1.5 Advantages and Disadvantages

One advantage of this method is that it yields sufficient quantities of cells (up to 4×10^8 cells per 100 ml of starting stationary phase culture) for molecular studies and other studies requiring large quantities of cells. A preparation of 1×10^8 17-generation-old cells can generate 60 to 80 µg of total RNA. More importantly, the cells are highly pure and well synchronized by age. In addition, the procedure does not have any known detrimental effects on the cells as evidenced by the measurement of several physiological parameters [18]. One disadvantage of this method is that it requires preparaton of relatively large amounts of sucrose gradients, which is laborious.

8.3.2. Sampling of Cells from Mixed Populations

8.3.2.1 Principles

This method utilizes the observation that cell size and the number of bud scars, which incrementally mark the mother cell at each division, increase as yeast cells progress through their replicative life span [18]. Accordingly, cells from stationary cultures are stained with Cellufluor, which specifically intercalates into the chitin of the bud scar [25]. Then, using a fluorescence-activated cell sorter (FACS), the stained cells are sorted into the desired number of age groups according to their size and fluorescence.

8.3.2.2 Materials and Equipment

8.3.2.2.1 Yeast Strains

Although any yeast strain can be used, avoid a strain in which cells are clumpy and tend to aggregate even after sonication.

8.3.2.2.2 Media

YPD (2% peptone, 1% yeast extract, 2% glucose), YPDG (2% peptone, 1% yeast extract, 0.04% glucose, 1.6% glycerol), and YP (2% peptone, 1% yeast extract, 0.01% glucose) containing 15 µg/ml of gentamycin are used. Solid media contain 2% agar.

8.3.2.2.3 Other Reagents

Cellufluor (1 mg/ml in water, Polysciences, Warrington, PA), gentamycin (10 mg/ml in water), and phosphate-buffered saline solution (137 mM NaCl, 2.7 mM KCl, 4.3 mM Na_2HPO_4 – $7H_2O$, 1.4 mM KH_2PO_4, pH ~ 7.3) containing 15 µg/ml of gentamycin are prepared.

8.3.2.2.4 Equipment

A flourescence-activated cell sorter, such as FACS Vantage (Becton Dickinson), an ultrasonic homogenizer, and a fluorescence microscope with a UV filter cube, as described above, are needed. A photometer attachment for the microscope is useful.

8.3.2.3 Step-by-Step Procedures

8.3.2.3.1 Preparation of a Stationary-Phase Culture

Step 1:   Inoculate 5 ml of YPD (or any synthetic medium) in a 50-ml erlenmyer flask with a single colony freshly grown on a YPD plate. The volume of the culture flask should be at least ten times the YPD liquid volume to allow sufficient aeration.

Step 2:   Incubate the culture at 30°C on a shaker at 250 ~ 300 rpm until the culture reaches and stays in stationary phase, without losing cell viability. It is important that the proportion of unbudded cells should be more than 95% of the cells examined. It is necessary to monitor cell viability periodically during the incubation and find the best incubation time to maximize both cell viability and the proportion of unbudded cells.

8.3.2.3.2   Staining Cells with Cellufluor

Prior to actual experimentation, it is beneficial to determine the optimal Cellufluor concentration to stain cells for each yeast strain. The concentration should be high enough for the FACS machine to detect differences in fluorescence intensity proportional to the number of bud scars. On the other hand, higher concentrations will cause higher background levels of fluorescence because this reagent is known to bind to other cell membrane components, such as other polysaccharides [25]. Moreover, higher concentrations can affect cell viability.

Step 1:   Transfer 1 ml of the stationary-phase culture to a 1.5-ml microcentrifuge tube. Sonicate cells at about 40 W for 30 s, using the microtip of the sonicator. Make sure microscopically that cells are well separated.

Step 2:   Transfer 0.5 ml of the sonicated culture into another 1.5-ml microcentrifuge tube. Add 50 µl of the Cellufluor stock solution (1 mg/ml in water) to the tube and incubate it at room temperature for 20 min in a dark place (or wrap the rack containing the microcentrifuge tube with aluminum foil). The remaining culture without Cellufluor is a negative control.

Step 3:   Under dimmed light, centrifuge the tubes (both stained and unstained) briefly (5 sec) at the maximum speed. Remove the supernatant and resuspend the pelleted cells in 0.5 ml sterile water.

Step 4:   Repeat Step 3 two more times. After the final centrifugation, resuspend cells in phosphate-buffered saline containing 15 µg/ml of gentamycin at 1×10^6 cells per ml. Examine the stained and unstained cells by fluorescence microscopy to see if the cells are properly stained.

8.3.2.3.3   Cell Sorting

Step 1:   Unstained cells are first loaded into the FACS according to the manufacturer's operational directions. With FACS Vantage (Becton-Dickinson), the UV source is a Coherent Enterprise Laser with 251–364 nm emission and 6.0 mW output. The aperture is set at 70 µm and equipped with a filter with 424/44 band pass. A typical example of unstained cell distribution is shown in Figure 8.5A. Set the vertical line to mark the background fluorescence level and the horizontal line to cut off small-sized nonyeast particles.

A

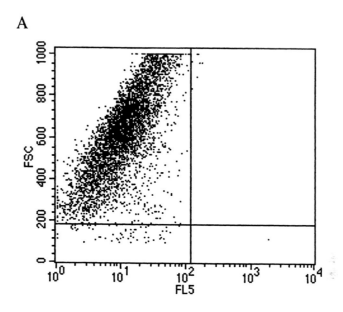

B

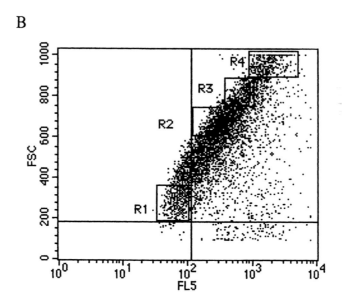

FIGURE 8.5
Distribution of individual unstained **(A)** and stained **(B)** cells of *S. cerevisiae* strain UCC519 (provided by Dan Gottschling, Univ. of Chicago) displayed by the FACS according to their size (y-axis, linear scale) and fluorescence intensity (x-axis, logarithmic scale). Approximately 5×10^4 cells were collected from the areas enclosed by each of the boxes labeled R1, R2, R3, and R4. Most of the R1 cells were virgin, the R2 and R3 cells were on average 3 and 8 generations old, respectively, and the average age of the R4 cells was 14, as determined by the two age determination methods described in the text (see also Figure 8.6). (From Kim, S., Villeponteau, B., and Jazwinski, S. M., *Biochem. Biophys. Res. Commun.*, 219, 370, 1996. With permission.)

Step 2: Load stained cells of the same strain into the FACS. A typical example of stained cell distribution is shown in Figure 8.5B. Note that the overall cell distribution is shifted to the right of the vertical line without change in cell size distribution.

Step 3: Enclose the regions from which cells are to be collected with a rectangle using the computer program provided with the FACS, as shown in Figure 8.5B (R1, R2, etc.).

Step 4: Collect cells in 5-ml snap cap tubes. The flow rate and the number of cells to be collected are adjustable. With the FACS Vantage, it usually takes 30–60 min to collect 5000 cells. Store them on ice or 4°C.

Step 5: Stain a small portion of the collected cells in each age group with Cellufluor and examine them by fluorescence microscopy to take note of the differences in cell size and bud scar number.

Step 6: Transfer the cells in phosphate-buffered saline to microcentrifuge tubes. Centrifuge the tubes 1 min in a microcentrifuge. Resuspend the pelleted cells in a desired volume of YP containing 15 µg/ml of gentamycin. In YP, which contains 0.01% glucose, cells can be store at 4°C for up to 2 weeks without noticeable viability loss [19,26].

8.3.2.3.4   *Age Determination of Collected Cells*

The average age of sorted cells in each group can be estimated by restaining them with Cellufluor and examining them under a fluorescence microscope to count the number of bud scars on individual cells. This method applies well to the virgin and young cells with fewer than five bud scars. However, the accuracy of counting bud scars on cells decreases as the number of bud scars increases [18]. Therefore, a linear regression equation relating the number of bud scars and the intensity of fluorescence is derived for each strain using unsorted stained cells with fewer than six bud scars, and the bud scar numbers are estimated from this equation [18,26]. The average age of sorted cells can also be estimated by determining their remaining life spans as described elsewhere in this chapter under life span determination. Briefly, remaining life spans are determined microscopically on 35 sorted cells from each age group by counting the number of daughters that individual cells produce (Figure 8.6). The representative replicative age of each age group, expressed in generations, is obtained by subtracting the median life span of the respective preparation from that of the virgin cells. The replicative ages of the old cells obtained using this method are in close agreement with those determined by the first procedure [18,26]. The following is the step-by-step procedure to estimate the number of bud scars in old cells.

Step 1: Cells are stained with 0.1 mg/ml Cellufluor for 20 min, washed twice with water, and mounted on a Nikon Microphot FX epifluorescence microscope with the Nikon UV-2A filter cube.

Step 2: Fluorescence readings are performed on individual cells with zero to five bud scars, using the Nikon P1 photometer. At least ten different cells of the same age are measured.

Step 3: To estimate the number of bud scars on sorted cells, a linear regression equation relating fluorescence and the number of bud scars is obtained from unsorted stationary phase cells.

Step 4: Fluorescence intensity of individual sorted cells is measured as described in Step 2.

Step 5: These measurements are converted to the estimated bud scar numbers using the linear regression equation obtained from Step 3.

8.3.2.4 Advantages and Disadvantages

One advantage of this method is that it is relatively simple and easy; old cells can be obtained without much effort and time spent. However, there are several drawbacks of this method. First, because of lengthy times for the FACS to sort cells, the yield is not sufficient for molecular studies or any other studies requiring large quantities of cells. Second, cells, other than virgin cells, collected in an age group form a continuum with cells of other nominal ages. Therefore, it is particularly important to estimate average ages of cells in different age groups by using either of the two age determination methods mentioned above. Although the cells in a fraction are not of exactly the same nominal age, their age distribution is as narrow as the investigator specifies during cell sorting. Third, there may be some detrimental effects of staining cells with Cellufluor and exposing them to UV during the FACS seperation, although such effects have not been observed thus far.

8.3.3 Other Methods

Three other methods have been described for preparing old yeast cells. Although none of them has been used in our laboratory, they could be alternatives to the methods described above. Here, we will discuss certain aspects of these procedures that we believe decrease their utility in the batch preparation of cells for aging studies.

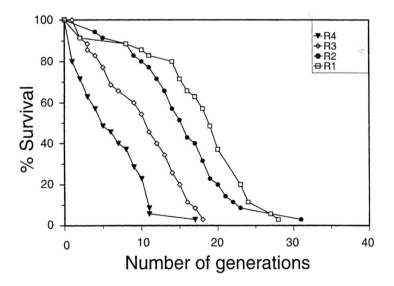

FIGURE 8.6
Survival curves of the cells in the four age groups (R1, R2, R3, and R4, as shown in Figure 8.5) of sorted UCC519 cells. The number of buds that individual cells generated during their remaining life spans was determined. The open squares are for R1 (virgin) cells (median life span, 18), closed circles for R2 (median remaining life span, 15), open diamonds for R3 (median remaining life span, 10), and closed triangles for R4 (median remaining life span, 4). The remaining life spans correspond to replicative ages of 3, 8, and 14 generations for R2, R3, and R4. (From Kim, S., Villeponteau, B., and Jazwinski, S. M., *Biochem. Biophys. Res. Commun.*, 219, 370, 1996. With permission.)

The first of the alternative methods that could theoretically yield yeasts of different ages is the so-called "baby machine" culture technique [27]. In this method, a growing population of cells is attached to a filter membrane using a specific adhesive. Each time a cell divides the mother is retained on the membrane and the daughter is released to the medium. This technique was first used for studies of the cell cycle of *Escherichia coli* strain B/r and later on *Salmonella typhimurium, E.coli* strain K-12, and *Bacillus subtilis* [28-31].

The use of this approach with *Saccharomyces cerevisiae* was reported by C.E. Helmstetter [27]. In this paper, time-lapse video recording showed that yeast cells attached to a coverslip, using adhesives such as polylysine, Concanavalin A, or Cell Tak, could undergo division. The daughter cell was released, while the mother remained firmly attached to the surface of the coverslip. This observation shows that yeasts can be used in an actual baby machine for separation of daughter cell populations from their parents. The device consists basically of a nitrocellulose membrane supported by a stainless steel screen and clamped between a lower Lucite funnel and an upper Lucite ring. First, the filter is coated with the cell adhesive, and, after washing it with distilled water, 100 ml of exponentially growing yeast culture is poured on the top of the nitrocellulose membrane. The culture is filtered by vacuum and the filter holder is then inverted and placed on the top of another Lucite funnel. Then the upper funnel is filled with medium and the whole device is connected to a peristaltic pump allowing drops of medium to pass through the filter and the hole of the lower funnel. This apparatus is held on a rack of a shaker and incubated at 30°C for 2, 15, or 25 h. Samples are collected every 10 min from the effluent and the concentration of cells is calculated in each of the eluted fractions collected.

It was shown that this device can be used for generating large numbers of daughter cells and is suitable for studying processes such as the cell cycle or the segregation of components between the mother and the daughter yeast. The author also claims that this method could be suitable for preparing old yeast cells for aging experiments just by detaching the remaining old cells from the nitrocellulose membrane following a period of incubation sufficient to achieve the desired number of cell divisions. However, no procedure for detaching cells is presented, nor is the effect of detachment of cells on viability discussed. No data on the nominal age of cells that remain attached to the membrane, as determined by the staining of bud scars, are found in this paper. It is possible that many daughter cells stick to the membrane contaminating the old cell preparation or that many parent cells detach from the membrane during the long period of time needed for obtaining 15-generation-old cells, resulting in low yields. As described, the method would only yield one age group of cells. It is possible to conceive of modifications to the procedure that would allow cells of several age groups to be prepared. However, these fractionated populations of cells would be rather heterogeneous in age, due to the lack of cell cycle synchrony. More studies on the reliability of this technique are necessary for its use in the preparation of age-fractionated yeast cells.

The second alternative method uses a centrifugal elutriator in order to separate the parent cells from their daughters [32]. The centrifugal force makes the mother cells remain in the chamber of the rotor while the daughters are eluted in the medium continuously flowing through the elutriator. The separation between the mother and daughter cells occurs when the difference in their sizes is large. This is achieved by culturing the yeast in the presence of ethanol as a carbon source which results in a big difference in volume between the parent (33 to 34 μm^3) and the daughter cell (17 μm^3). First, an exponentially growing yeast culture is loaded in both chambers of the rotor and allowed to grow for one mass doubling time. Small cells are collected (daughters cells) at a constant flow rate and centrifugal force. These cells are collected by centrifugation at 4°C and resuspended in 5 ml of medium. They are used for obtaining populations of daughter cells as well as populations of old cells by loading them again in one chamber of the rotor and allowing them to grow at 30°C for 26 h (15 generations). New daughter cells are continously collected at a flow rate of 12.5–16 ml/min at 2500 rpm. The parent (old) cell population that is retained in the rotor is finally eluted after cooling the elutriator to 4°C and increasing the medium flow rate through the rotor. Studies were first done on the daughter cells eluted, showing an increase in cell volume during the first five to seven generations. Interestingly, it was observed that parent cells maintain a constant cell cycle time throughout the incubation period spanning 15 generations, and even beyond. However, daughter cells from 15-generation-old parents showed a shorter cell cycle time than the daughters of 1-generation-old parents. Since the authors clearly show and point to a loss of synchrony after a nominal ten generations of elution, it is difficult to assign much significance to the rather small changes they observed in the cell parameters measured or to the lack of changes, for that matter.

Although this method was proposed to be suitable for obtaining large quantities of old yeast cells (up to 29 generations), bud scar staining analysis of the cell population remaining in the rotor (parent cells) after 15 generations of growth showed marked heterogeneity. Of the cells 71% contained 10 or more scars, while there was a large fraction (17%) of virgin cells that could not be removed from the parent cells by elutriation at low temperature. The rest (12%) contained cells ranging from 1 to 9 scars. Thus, the loss of synchrony indicated above is not surprising. In a 29 generation experiment, several fractions of parent cells were obtained. In the first fraction, 50% contained cells with 15 or more scars, while as high as 25% were virgin cells. This low purity in old cells renders this technique not feasible for obtaining high-purity old cell populations needed for analysis of gene expression in aging yeasts or other aging-related experiments.

The last alternative method for preparing old yeast cells [33] is based on the fact that daughter cells of *Saccharomyces cerevisiae* have a wall that is newly formed, or nearly so, and do not have any detectable remnants of the mother. So it is possible to derivatize the surface of yeast cells, let them grow for a defined number of generations, and separate specifically the derivatized cells from the daughters they produced.

The cells are reacted with an activated derivative of biotin (*N*-hydroxysuccinamide-LC-biotin) which covalently binds to primary amines such as the ε-amino group of lysine. The aliphatic chain (LC) prevents steric hinderance allowing free access to the biotin. The coating is assessed by fluorescence microscopy or by flow cytometry, using fluorescein-conjugated streptavidin which tightly binds to the biotin moiety. Biotinylation seems to be detrimental to yeasts, because life span analysis shows that as many as 20% of the cells die very young.

For preparing old cells, biotin-coated yeasts are grown for 12–14 generations in YPD and then mixed with streptavidin-coated magnetic beads. After an incubation time of 2 h at 4°C, the mixture is placed on a magnetic sorter that retains the magnetic beads having the biotin-coated cells associated with them. The sorted cells are grown for 12–14 generations and are again magnetically sorted. Calcofluor staining analysis shows that the first sorting yields mostly cells with more than ten bud scars, although there is some contamination with cells with very few scars and also with virgin cells. In the second sort, most of the cells contain more than 19 bud scars, many of them having 28–30 scars. This method seems to produce old cells after two rounds of magnetic sorting. However, these preparations have a very low plating efficency (less than 1% on YPD), indicating that instead of senescent cells this technique yields mostly dead cells. Clearly, detachment of the sorted cells from the beads is problematic. Cells that are biotinylated and immediately sorted appear to have a colony-forming ability of as little as 50%. Because of the lack of cell cycle synchrony, this method will produce preparations that are heterogeneous with respect to replicative age.

Acknowledgment

The research in the authors' laboratory is supported by grants to S.M.J from the National Institute on Aging (NIH) and from the Glenn Foundation for Medical Research. P.A.K. is the recipient of a postdoctoral fellowship from the National Institute on Aging. A.B. held a fellowship from the Ministry of Education and Culture of Spain.

References

1. Jazwinski, S. M., Longevity, genes, and aging, *Science*, 273, 54, 1996.
2. Bassett, D. E., Boguski, M. S., and Hieter, P., Yeast genes and human disease, *Nature*, 379, 589, 1996.
3. Goffeau, A., Barrell, B. G., Bussey, H., Davis, R. W., Dujon, B., Feldmann, H., Galibert, F., Hoheisel, J. D., Jacq, C., Johnston, M., Louis, E. J., Mewes, H. W., Murakami, Y., Phillippsen, P., Tettelin, H., and Oliver, S. G., Life with 6000 genes, *Science*, 274, 546, 1996.
4. Mortimer, R. K. and Johnston, J. R., Life span of individual yeast cells, *Nature*, 183, 1751, 1959.
5. Muller, I., Zimmermann, M., Becker, D., and Flomer, M., Calendar life span versus budding life span of *Saccharomyces cerevisiae*, *Mech. Ageing Dev.*, 12, 47, 1980.
6. Jazwinski, S. M., Aging and senescence of the budding yeast *Saccharomyces cerevisiae*, *Molec. Microbiol.*, 4, 337, 1990.
7. Jazwinski, S. M., An experimental system for the molecular analysis of the aging process: The budding yeast *Saccharomyces cerevisiae*, *J. Gerontol.*, 45, B68, 1990.
8. Jazwinski, S. M., The genetics of aging in the yeast *Saccharomyces cerevisiae*, *Genetica*, 91, 35, 1994.
9. Jazwinski, S. M., Longevity-assurance genes and mitochondrial DNA alterations: Yeast and filamentous fungi, in *Handbook of the Biology of Aging*, 4th ed., Schneider, E. L. and Rowe, J. W., Eds., Academic Press, San Diego, 1996, chap. 3.
10. Jazwinski, S. M., Models of aging: Invertebrates, filamentous fungi, and yeasts, in *Encyclopedia of Gerontology*, Vol. 2, Birren, J. E., Ed., Academic Press, San Diego, 1996, 151.
11. Kennedy, B. K. and Guarente, L., Genetic analysis of aging in *Saccharomyces cerevisiae*, *Trends Genet.*, 12, 355, 1996.
12. Guthrie, C. and Fink, G. R., Eds., *Guide to Yeast Genetics and Molecular Biology*, Academic Press, San Diego, 1991.
13. Sherman, F. and Hicks, J., Micromanipulation and dissection of asci, in *Guide to Yeast Genetics and Molecular Biology*, Guthrie, C. and Fink, G. R., Eds., Academic Press, San Diego, 1991, chap 2.
14. Kennedy, B. K., Austriaco, N. R., Zhang, J., and Guarente, L., Mutation in the silencing gene *SIR4* can delay aging in *S. cerevisiae*, *Cell*, 80, 485, 1995.
15. Gietz, R. D. and Woods, R. A., High efficiency transformation in yeast, in *Molecular Genetics of Yeast: Practical Approaches*, Johnston, J. A., Ed., Oxford University Press, New York, 1994, 121.
16. Schiestl, R. H. and Gietz, R. D., High efficiency transformation of intact yeast cells using single stranded nucleic acid as carrier, *Curr. Genetics*, 16, 339, 1989.
17. Egilmez, N. K. and Jazwinski, S. M., Evidence for the involvement of a cytoplasmic factor in the aging of the yeast *Saccharomyces cerevisiae*, *J. Bacteriol.*, 171, 37, 1989.
18. Egilmez, N. K., Chen, J. B., and Jazwinski, S. M., Preparation and partial characterization of old yeast cells, *J. Gerontol.*, 45, B9, 1990.
19. Kaibuchi, K., Miyajima, A., Arai, K., and Matsumoto, K., Possible involvement of *RAS*-encoded proteins in glucose-induced inositol phospholipid turnover in *Saccharomyces cerevisiae*, *Proc. Natl. Acad. Sci. U.S.A.*, 83, 8172, 1986.

20. Bücking-Throm, E., Duntze, W., Hartwell, L. H., and Manney, T. R., Reversible arrest of haploid cells at the initiation of DNA synthesis by a diffusible sex factor, *Exp. Cell. Res.*, 76, 99, 1973.

21. Hartwell, L. H. and Unger, M. W., Unequal cell division in *Saccharomuces cerevisiae* and its implications for the control of cell division, *J. Cell Biol.*, 75, 422, 1977.

22. Johnston, G. C., Pringle, J. R., and Hartwell, L. H., Coordination of growth with cell division in the yeast *Saccharomyces cerevisiae*, *Exp. Cell. Res.*, 105, 79, 1977.

23. Schekman, R. and Brawley, V., Localized deposition of chitin on the yeast cell surface in response to mating pheromone, *Proc. Natl. Acad. Sci. U.S.A.*, 76, 645, 1979.

24. Tyson, C. B., Lord, P. G., and Wheals, A. E., Dependency of size of *Saccharomyces cerevisiae* cells on growth rate, *J. Bacteriol.*, 138, 92, 1979.

25. Sherman, F., Fink, G. R., and Hicks, J. B., *Methods in Yeast Genetics*, Cold Spring Harbor Laboratory, Cold Spring Harbor, New York, 1986, 159.

26. Kim, S., Villeponteau, B., and Jazwinski, S. M., Effect of replicative age on transcriptional silencing near telomeres in *Saccharomyces cerevisiae*, *Biochem. Biophys. Res. Commun.*, 219, 370, 1996.

27. Helmstetter, C. E., Description of a baby machine for *Saccharomyces cerevisiae*, *The New Biol.*, 3, 1089, 1991.

28. Helmstetter, C. E., An improved method for the selection of bacterial cells at division, *Biochim. Biophys. Acta*, 82, 608, 1964.

29. Cooper S. and Ruettinger, T., Replication of deoxyribonucleic acid during the division cycle of *Salmonella typhimurium*, *J. Bacteriol.*, 114, 966, 1973.

30. Cummings, D. J., Synchronization of *E. coli* K12 by membrane selection, *Biochem. Biophys. Res. Commun.*, 41, 471, 1970.

31. Holmes, M., Rickert, M., and Pierucci, O., Cell division cycle of *Bacillus subtilis:* Evidence of variability in period D, *J. Bacteriol.*, 142, 254, 1980.

32. Woldringh, C. L., Fluiter, K., and Huls, P. G., Production of senescent cells of *Saccharomyces cerevisiae* by centrifugal elutriation, *Yeast*, 11, 361, 1995.

33. Smeal, T., Claus, J., Kennedy, B., Cole, F., and Guarente, L., Loss of transcriptional silencing causes sterility in old mother cells of *S. cerevisiae*, *Cell*, 84, 633, 1996.

Section C

Vertebrate Models in Aging Research

9

Long-Term Housing of Rodents under Specific Pathogen-Free Barrier Conditions

Sherry M. Lewis, Barbara L. Leard, Angelo Turturro, and Ronald W. Hart

CONTENTS

9.1 Introduction

Central to the success of long-term animal studies is the rearing of healthy rodent models. Good survival is dependent upon husbandry, nutrition, and genetic factors, each of which affect a broad range of physiological and biological parameters.

Over time, rodent species and strains have experienced increases in physiological variability, reduced survival rates, and increases in the rates of occurrence of a number of degenerative diseases [1–5]. With greater variability and lessened reproducibility of experimental findings, data have become less valid and more difficult to interpret. Improving the quality of carcinogenic, environmental, or toxicological research is linked to the control, or understanding, of nontreatment-related effects in animal models.

Until recently, rodent bioassay protocols frequently used the *ad libitum* (AL) feeding method for chronic studies. It has been suggested that AL feeding may, in part, be responsible for some of the previously cited changes in disease occurrence [6,7]. One way to address the impact of AL feeding on toxicologic parameters is to control dietary intake. One paradigm using controlled dietary intake is dietary restriction (DR) which reduces intake without inducing malnutrition. DR results in increased survival and retards the onset of degenerative disease and the rate of tumor progression [8,9]. DR has provided demonstrably healthy, uniform animal models for testing the challenges of toxicological, carcinogenic, pharmacological, and geriatric research. Animals studies have clearly shown that the level of caloric intake directly alters a number of physiological, cellular, metabolic, biochemical, and molecular functions important to survival [8].

DR has been proposed as a standard management regimen [10]. In addition to control of dietary intake, Specific Pathogen-Free (SPF) status is the other strategically critical method that serves to maximize life span and control deleterious environmental variables. Both these variables, and their interaction, have been examined in studies done at the National Center for Toxicological Research (NCTR). Long-term survival, dietary intake control, and health status are synergistically related when addressing the specialized needs of aged rodents.

9.1.1   Interagency NIA/NCTR Study

The Food and Drug Administration's NCTR and the National Institute on Aging (NIA) began collaborations in 1984 on studies of nutrition and aging. These studies were the first large-scale experiments to use multiple *ad libitum*-fed and diet controlled rodent model paradigms. The experimental design included the selection of four mouse and three rat genotypes to represent the range of genomic diversity within each species. The study included males and females of each genotype, with the animals housed singly and maintained according to requirements of the *Guide for the Care and Use of Laboratory Animals* [11,12]. A primary goal of the study was to provide diet-controlled, aged rodents to competitive investigators within the NIA Biomarkers of Aging program.

Rodent models selected for the NIA/NCTR aging studies included: (1) Fischer 344 (F344NNia) rat, a toxicology/carcinogenicity model used by the National Toxicology Program; (2) Brown Norway (BN/RijNia), an inbred feral rat strain; (3) F344 × BNF$_1$Nia hybrid rat, developed for heterosis and disease resistance; (4) C57BL6/NNia mouse, a NIA model; (5) B6D2F$_1$Nia hybrid mouse (C57BL6 × DBA); (6) DBA/2JNia mouse, a short-lived strain; and (7) B6C3F$_1$ hybrid mouse (C57BL/6N × C3H/HeN) used in toxicological studies. All animals were born and reared under SPF-barrier conditions. Breeding and pedigree documentation conformed to standard industry practices [13]. Intake levels were established from AL-intake data acquired prior to implementation of DR. At weaning, animals were AL-fed one of three diets: (1) NIH-31 open formula, natural ingredient diet; (2) EM 911-A, a semipurified, natural ingredient-based diet; and (3) 5770C-1, a purified diet. From the start of the DR period, restricted-fed diets included: NIH-31, EM911-A, and 5770C-1 supplemented with vitamins to allow equal intake per animal among DR- and AL-fed animals. Restriction was increased incrementally from 10% to 25%, and finally to 40% of AL intake at 14, 15, and 16 weeks of age, respectively [12]. After adaptation to DR, and consequent lower body weights, dietary macro- and micronutrient concentrations were adequate to support long-term survival and were similar to AL-fed rodents, on a per gram body weight basis. Vitamin fortification, however, provided additional vitamins to the DR rodents on a per gram body weight basis.

These models have demonstrated the increased survival percentage at all ages, as well as prolonged life span overall, when associated with the DR feeding paradigm [8,9,14]. Age in weeks at median, maximal (average age at death of the oldest 10% of each cohort), and maximum survival for each strain of rat and mouse used in the nutrition study are reported in Table 9.1. The values reported represent the week that the event occurred. At each interval, survivability for each group was increased by DR, regardless of type of diet fed.

In addition to the NIA/NCTR collaboration studies, chronic bioassay studies were performed at NCTR. Chronic risk studies are concluded at 104 weeks of testing with animals placed on trial at 6-8 weeks of age. For comparison purposes, survival percentages of animals at 50 and 104 weeks of tests are reported in Table 9.2.

Survivability among the Fischer 344 (F344NNia) rats maintained under SPF-barrier conditions, and *ad libitum*-fed the same NIH-31 rodent diet used in the NIA/NCTR study, but during different years, suggests a time effect in survivability characteristics. Male rats in a study reported in 1993 [15], at 104 weeks on study, had a similar survivability as did F344 rats at 107 weeks of age in the NIA/NCTR study, 52 vs. 51% (data not shown), respectively. The age adjustment accounted for a 3-week age difference in allocation between studies. However, 107-week-old female F344 rats of the NIA/NCTR study demonstrated an 18% increase in survivability, 77 vs. 65%, (data not shown) respectively, over the 104-week females of the chronic study. Both the male and female NIA/NCTR study rats, *ad libitum*-fed the NIH-31 diet, demonstrated 50 and 77% survivabilities (data not shown) at 107 weeks of age. These rats outlived rats of a chronic study reported in 1991 [16] at 104 weeks on study, 40 and 58% survivabilities for the male and females, respectively (Table 9.2). Although representative numbers were small, the persistent increase in survival associated with DR feeding regimens is also demonstrated among male and female F344 rats when caged singly in conventional housing and fed the NIH-31 diets (Table 9.2).

Differences in dietary composition may have accounted, in part, for the differences in survivability noted for NIA/NCTR study F344 rats as compared to chronic study F344 rats [17,18]. Body weight as affected by dietary intake, and, therefore diet composition, has been reported to affect survivability among C57BL6 mice [19].

Among SPF-barrier reared NIA/NCTR study B6C3F$_1$ mice, that were *ad libitum*-fed the NIH-31 diet, survivability at 107 weeks of age was 87% for both males and females (data not shown). These data are comparable to survivabilities reported for 104-week chronic study mice housed in the SPF-barrier and fed the NIH-31 diet [16,20], and the 5010 diet [21,22]. The 107-week survivability values were also similar to the 104-week chronic study B6C3F$_1$ mice that were conventionally housed and *ad libitum*-fed the NIH-07 diet [23]. Diet and/or barrier effects on survivability are evident when 107-week-old mice, *ad libitum*-fed the NIH-31 natural ingredient diet in the barrier, are compared with 104-week chronic study mice that were maintained in conventional housing and *ad libitum*-fed the AIN-76A purified diet (Table 9.2). Survivability was 1.74-fold greater, 87 vs. 50%, respectively, when the natural ingredient vs. the purified diet was fed. These data would suggest that diet type, level of intake, and/or SPF health status maintenance are important in the successful rearing of aged animals. The procedures used to maintain the NIA/NCTR SPF-barrier colonies are detailed below.

TABLE 9.1

Age in Weeks at Median, Maximal, and Maximum Survival among 40% Diet-Restricted Rodent
Models, Bred and Maintained under SPF-Barrier Conditions

		Survival, Weeks			
	n[a]	Median	Maximal	Maximum	Diet[b]
Rat					
F344NNia					
AL male	50	105	126	129	A
AL female	54	117	150	155	A
DR male	54	126	161	176	B
DR female	54	134	169	186	B
BN/RijNia					
AL male	47	130	159	161	A
AL female	47	134	163	172	A
DR male	48	151	191	196	B
DR female	48	169	195	206	B
F344 × BNF$_1$Nia					
AL male	46	147	176	181	A
AL female	48	138	170	172	A
DR male	46	177	214	224	B
DR female	44	187	229	233	B
F344NNia					
AL male	54	107	127	136	C
AL female	53	116	143	154	C
DR male	52	129	167	171	D
DR female	52	137	170	176	D
Mouse					
B6C3F$_1$Nia					
AL male	56	141	183	186	A
AL female	56	133	163	176	A
DR male	56	192	222	225	B
DR female	56	181	210	214	B
C57/BL/6NNia					
AL male	50	121	150	155	A
AL female	38	118	149	159	A
DR male	56	139	179	181	B
DR female	56	147	176	177	B
C57/BL/6NNia					
AL male	40	110	137	148	E
AL female	30	99	124	130	E
DR male	56	143	181	188	F
DR female	56	145	178	195	F
DBA/2JNia					
AL male	56	89	131	135	A
AL female	56	78	134	146	A
DR male	56	105	147	153	B
DR female	56	119	150	153	B

TABLE 9.1 (continued)
Age in Weeks at Median, Maximal, and Maximum Survival among 40% Diet-Restricted Rodent Models, Bred and Maintained under SPF-Barrier Conditions

		Survival, Weeks			
	n^a	Median	Maximal	Maximum	Diet[b]
B6D2F₁Nia					
AL male	56	139	180	191	A
AL female	55	128	162	182	A
DR male	56	189	213	227	B
DR female	56	170	208	210	B

ᵃ Animal numbers assigned to a lifetime, longevity cohort.
ᵇ Diet descriptions: A, B (NIH-31 standard, NIH-31 vitamin fortified); C, D (5770C-1 standard, 5770C-1 vitamin fortified; E, F (EM911A standard, EM911A vitamin fortified).

9.2 Background

9.2.1 History

In 1869, germ-free animals were first used to study the impact of gut microflora upon the physiology of the gut [24]. Derivation of research models was begun in the 1920s at the Universities of Notre Dame in the U.S. and Lund in Sweden; animals were generally available to the research community by the 1940s [25].

Husbandry methods improved survivability largely due to the enactment of the Animal Welfare Act in 1966. Originally written to ensure the humane care of domestic and laboratory animals, it has been amended several times (1970, 1976, 1979, and 1985) to strengthen guidelines [13]. Laboratory animals achieved an extended life span, and aging-related disease began to replace infectious disease as a major limiting factor, largely as a result of improved hygienic requirements [26]. *The Public Health Service Policy on Humane Care and Use of Laboratory Animals* [27] governs research use of vertebrate animals and stresses institutional self-regulation and responsibility for oversight via an Institutional Animal Care and Use Committee (IACUC). Animal use programs must follow recommendations in the *Guide for the Care and Use of Laboratory Animals* [11] to achieve accreditation by the Association for Assessment and Accreditation of Laboratory Animal Care (AAALAC) International, a globally recognized institution. These collective rulings have altered the course of biomedical research and strengthened the standards for husbandry, care, treatment, and transportation.

Microbiological classification of laboratory animals has painstakingly evolved, often with confusion [13,28]. The current classification scheme is generally regarded as internationally accepted: (1) germ-free (axenic), animals derived by cesarean section or embryo transfer, reared and maintained under aseptic isolator techniques, and demonstrably free of microbial, viral, protozoan, and parasitic organisms; (2) gnotobiotic (defined flora), axenic animals which have been purposefully inoculated with a known consortia (generally <8, nonpathogenic) of microorganisms, and isolator maintained; (3) pathogen-free, animals free

TABLE 9.2

Percent Survival at 50 and 104 Weeks among *Ad Libitum*-Fed, SPF-Barrier Bred[1] Control Group Rodents That Were Assigned to Toxicology Studies, and Subsequently Maintained in SPF-Barrier or Conventional Rodent Housing

Rat	n	Survival, Percent[2] 50 wk	Survival, Percent[2] 104 wk	SPF-Barrier/ Conventional	Diet[3]	Source
F344NNia, M	112	100	63	B	5010	Littlefield, N.A. et al., *Food Chem. Toxicol.*, 28, 157, 1990.
F344NNia, Fe	113	100	59	B	5010	
F344NNia, M	48	98	52	B	NIH-31	Jackson, C.D. and Blackwell, B., *J. Am. Coll. Toxicol.*, 12, 1, 1993.
F344NNia, Fe	48	100	65	B	NIH-31	
F344NNia, M	210	ND[4]	67	B	5010	Littlefield, N.A. et al., *Food Chem. Toxicol.*, 27, 239, 1989.
F344NNia, Fe	210	ND	67	B	5010	
F344NNia, M	48	98	40	B	NIH-31	Greenman, D.L. and Allaben, W., NCTR Technical Report, 1, 1991.
F344NNia, Fe	48	98	58	B	NIH-31	
F344NNia						
AL male	5	100	0	C	NIH-31	
AL female	5	100	80	40 (112)[5] C	NIH-31	
DR male	5	100	80	80 (124) C	NIH-31 + vitamins	
DR female	5	100	100	60 (124) C	NIH-31 + vitamins	

Mouse

Strain					Class	Diet	Reference
B6C3F$_1$, M	240	98	91		B	5010	Littlefield, N.A. et al., *Food Chem. Toxicol.*, 27, 455, 1989.
B6C3F$_1$, Fe	240	98	91		B	5010	
B6C3F$_1$, M	48	100	88		B	NIH-31	Jackson, C.D. and Sheldon, W., *J. Am. Coll. Toxicol.*, 12, 311, 1993.
B6C3F$_1$, Fe	48	99	85		B	NIH-31	
B6C3F$_1$, M	288	99	87		B	5010	Littlefield, N.A. et al., *Fundam. Appl. Toxicol.*, 5, 902, 1985.
B6C3F$_1$, Fe	288	100	87		B	5010	
B6C3F$_1$, M[1]	96	ND	88		C	NIH-07	Fullerton, D.L. and Greenman, D.L., NCTR Technical Report, 346, 1, 1989.
B6C3F$_1$, Fe	120	ND	90		C	NIH-07	
B6C3F$_1$, M	96	ND	50		C	AIN-76A	
B6C3F$_1$, Fe	96	ND	50		C	AIN-76A	
BALB/c, M	96	ND	59		C	NIH-07	Fullerton, D.L. and Greenman, D.L., NCTR Technical Report, 346, 1, 1989.
BALB/c, Fe	96	ND	50		C	NIH-07	
BALB/c, M	96	ND	50		C	AIN-76A	
B6C3F$_1$, M	48	100	90		B	NIH-31	Greenman, D.L. and Allaben, W., NCTR Technical Report, 1, 1991.
B6C3F$_1$, Fe	48	98	81		B	NIH-31	
B6C3F$_1$Nia							
AL male	5	100	100	60 (124)	C	NIH-31	
AL female	5	100	80	80 (124)	C	NIH-31	
DR male	5	100	100	100 (124)	C	NIH-31 + vitamins	
DR female	5	100	100	100 (124)	C	NIH-31 + vitamins	

1 NCTR, SPF-breeding colony.

2 Percentage of animals surviving at either 50 or 104 weeks on test.

3 Diets: 5010, NIH-31 (standard and vitamin fortified), and NIH-07 diets are natural ingredient, cereal-based, autoclavable diets; AIN-76A is a purified, nonautoclavable diet.

4 Value not determined.

5 Value in parenthesis, week of observed survival percentage.

of demonstrable pathogens, an ambiguous term since agreement on pathogen identification
and determinative testing varies; (4) specific pathogen-free, animals demonstrating absence
(usually by serologic, culture, or histopathologic evaluations) of a predetermined consortia
of pathogenic organisms, maintained within a barrier system; (5) virus antibody-free, animals free of antibodies to viral pathogens, identification supported by a battery of serolog-
ical tests to demonstrate absence of viral pathogens [28]; and (6) conventional, animals in
which the endogenous microbial flora and fauna are unknown, uncontrolled, or both.

9.2.2 Current Methods for Rearing Aging Rodents

Multiple studies employing DR from 60–90% of AL intake, provided to either SPF or con-
ventionally-housed animals, have demonstrated that diet restriction improved survival
rates and decreased tumorigenesis among rodents. Animals subsequently maintained
or reared on diet restricted intakes were more suitable models for studies of tumorigenesis than obese, AL-fed animals [8,10,29,30]. *Ad libitum* feeding, while an unnatural
practice, ameliorated increased labor costs required to feed animals measured amounts
of food daily.

Dietary interventions which have served to increase life span include direct restriction of
daily intake [31], alternate day feeding and periodic fasting [32], formulation of diets to
decrease caloric density [33–35], and multiple housing of animals [35]. Animals assigned
to DR protocols at NCTR were fed discrete amounts of feed daily, dependent upon restriction level, species, sex, and age [12,31].

9.2.2.1 Advantages to DR and SPF-Barrier Maintenance of Aging Animals

Clear advantages to long-term rearing of aging rodent models, accomplished via DR,
include:

1. Economic use of resources.
2. Maximization of data.
3. S longitudinal perspective on physiological effects and identification of parameters
   associated with healthy, aged animals.
4. The subsequent genetic selection and characterization of rodent models appropriate for geriatric studies.
5. The validation of methods to examine toxicity risk arising from chemical and
   environmental sources throughout sequential life phases.
6. Definition of age-related events that may have an impact on the quality of
   human life.

9.2.2.2 Disadvantages to SPF Rearing

Perceived disadvantages to rearing aging animals in SPF-barrier systems include con-
tainment-related issues, housing logistics, and administrative burdens. However,
these issues differ little from the need to develop new approaches to appropriately
breed, house, and feed a growing inventory of transgenic, specific-purpose bred
animals, and aging models of a variety of species. Nutrient requirements of novel
genotypes, and a definition of inherent metabolic changes among these models, have
yet to be defined.

9.3 General SPF-Barrier Design and Management

9.3.1 Facilities, Design, and Maintenance

Isolator and barrier engineering has been perfected and has evolved from a bulky stainless steel design or sterile glass components to lightweight plastics and flexible film which can be constructed to fit most available spaces, from complete rooms, racks within rooms, or individually protected cages [25]. The term "barrier" is currently used to describe any group of physical arrangements, procedures, and routines set up in an animal facility with the intent of minimizing the likelihood of direct or indirect animal contamination, or the escape of organisms from infected animals.

The sophistication of the barrier depends upon the category of animal being housed, research goals, and length of housing time required. Common barriers, or barrier strategies, include:

1. Physical separation of buildings, or locations away from potential hazards such as feral or infected animal populations.
2. One-way traffic flow to maximize isolation between clean and dirty operations, and the organization of work flow to minimize microbial infection or cross-contamination.
3. A separately housed unit for routine feed, water, bedding, cage, and supply sterilization.
4. Personnel restrictions with routine personnel microbial screening to identify disease carriers, and enforcement of decontamination routines.
5. Use of air pressure differentials to control airflow at entry or exit from animal areas.
6. Routine animal quarantine procedures.
7. Provision of insect and vermin barriers at entrances or exits [36,37].

The housing environment is composed of two distinct parts: a microenvironment, the immediate physical environment surrounding the animal, and a macroenvironment, the physical conditions of the enclosure or animal room containing microenvironments. Both environments are critical to the health status of the animal and, while linked, have different characteristics [13]. A laboratory animal's environment should be controlled for temperature, humidity, ventilation, illumination, noise, and microbial status. Adequate control of the physical environment, along with health surveillance monitoring, lend to the support of aging rodents in a barrier system [28].

The SPF-barrier, NIA/NCTR colony environment at NCTR contains 18 animal rooms where rodents are housed under SPF conditions. Facilities and equipment for all animal care functions, except diet preparation and feed storage, are contained within the 35,000 ft^2 area. Air supply is 100% fresh, HEPA filtered, conditioned to $22 \pm 1°$C, with relative humidity controlled between $50 \pm 10\%$. Air static pressure differentials are such that the highest pressure is maintained in the cleanest areas, and a minimum 10 changes of air per hour (range 12–15 changes per hour) provide room air ventilation. A 12-h light/dark cycle is maintained. Closed-circuit television monitoring and telephone systems provide visual and verbal communication between barrier and nonbarrier rooms [36].

9.3.2 Microbiological Support

Owing to a wide scope of maintenance conditions, two criteria of animal quality must be defined: (1) genetic quality and (2) hygienic quality. The mouse, in particular, has been well characterized genetically and has undergone genetic manipulations to produce animals with uniformly heritable phenotypes [13]. Hygienic quality includes the standardization of microflora, viruses, bacteria, and parasites associated with the defined strain.

Animal quality, in terms of defined health status, is best supported through institutionally applied microbiology, pathology and clinical controls, and persistent personnel attention. These measures serve research goals by identifying controllable experimental variables.

The NCTR maintains an institutional diagnostic laboratory where a microbiology maintenance surveillance program has been designed to be both comprehensive and quantitative (Table 9.3). In addition to the rat and mouse pathogens listed, Cilia Associated Respiratory Bacillus, as determined by Enzyme Linked Immunosorbent Assay (ELISA), is currently monitored within the colonies. Nasopharyngeal cultures for *Mycoplasma pulmonis* and *M. arthritidis* are performed quarterly in addition to antibody screening. Routine surveillance animals are screened for endoparasites; examination of the pelt and skin is conducted in order to determine the presence of ectoparasites and fungal infection.

Annual mouth rinse samples from each employee with SPF-barrier clearance are screened for select zoonotic bacterial pathogens (e.g., *Klebsiella pneumoniae, Streptococcus pneumoniae,* and *Pseudomonas aeruginosa*).

9.3.3 Chemistry Support

The NCTR Division of Chemistry provides routine surveillance on heavy metal contamination of potable water. Arsenic, cadmium, lead, mercury, selenium, iron, nickel, beryllium, chromium, and copper concentrations are monitored, as are pesticides, chloroform, and total dichlorodiphenyltrichloroethane (DDT) concentrations in water on a quarterly basis.

Sentinel nutrients (i.e., crude protein and fat, lipid soluble vitamins A and E, and thiamine) and moisture are analyzed from each lot of diet received for formulation assurance. While postautoclave analyses are not routine, steam autoclaved diet is monitored by special request as a check for effects of autoclaving on vitamin concentration.

Maximum acceptable limits for heavy metal contaminants in diets have been established: lead (1.5 ppm), arsenic (1.0 ppm), cadmium (0.25 ppm), mercury (0.1 ppm), and selenium (0.65 ppm). Maximum limits for pesticide residues have also been established: dieldrin (20 ppb), heptachlor (20 ppb), malathion (5 ppm), and total DDT (100 ppb). Maximum acceptable mycotoxin contaminants are also established: aflatoxins B_1, B_2, G_1 are set at 5 ppb, whereas G_2 is 2.5 ppb; fumonisin B_1, 100 ppb; and total fumonisins, 200 ppb. Maximum acceptable PCB concentration is 50 ppb. Contaminants and residues are monitored on each fifth lot of diet received.

Each lot of animal bedding is monitored for percent moisture, total DDT, polychlorinated biphenyl (PCB), and pentachlorophenol (PCP) levels, and is evaluated for dust fines, and sized between numbers 08 and 20 sieves.

9.3.4 Pathology and Histopathology Support

As with comprehensive health surveillance programs, histopathologic examination of major organs by qualified pathologists is standard for discrete cohorts of aging animals at

TABLE 9.3

Routine Microbiological Surveillance Parameters for SPF-Barrier, SPF-Breeding Colony, and Diet Preparation Barrier

I. Survey of Potential Pathogens of Rats and Mice

Bacteria	Endoparasites	Ectoparasites	Fungi
Bordetella bronchiseptica	*Aspiculuris tetrapetra*	*Ctenocephalides* sp. (rats)	*Aspergillus* sp.
Citrobacter freundii (mice)	*Balantidium coli* (rats)	*Dermanyssus gallinea* (rats)	*Microsporum* sp.
Corynebacterium kutscheri	*Capillaria* sp.	*Dermacentor* sp.	*Sporothrix* sp.
Erysipelothrix rhusiopathiae (mice)	*Eimeria* sp.	*Leishmania* sp.	*Trichophyton* sp.
Klebsiella oxytoca	*Entamoeba histolytica* (rats)	*Myobia musculi* (mice)	
Klebsiella pneumoniae	*Giardia muris*	*Myocoptes musculinus* (mice)	
Listeria monocytogenes	*Hexamita muris*	*Notoedres* sp.	
Mycoplasma sp.	*Hymenolepis* sp.	*Ornithonyssus bacoti*	
Pasteurella multocida	*Strongyloides* sp. (rats)	*Polypax* sp.	
Pasteurella pneumotropica	*Syphacia* sp.	*Psorergates simplex* (mice)	
Pseudomonas aeruginosa	*Taenia* sp.	*Radfordia* sp.	
Salmonella sp.	*Toxoplasma gondii*		
Streptococcus pneumoniae	*Trichomonas* sp. (rats)		
	Trichosomoides crassicauda (rats)		
	Trichostrongylus sp.		
	Trichuris sp.		

II. Serology Tests for Viral and Mycoplasma Anitbodies (ELISA)

Mice	Rats
Ectromelia Virus	Kilham Rat Virus Toolan H-1 (KRV-H1)
Lymphocytic Choriomeningitis Virus (LCM)	Pneumonia Virus of Mice (PVM)
Minute Virus of Mice (MVM)	Rat Coronavirus/Sialodacryoadenitis Virus (RCV/SDA)
Mouse Hepatitis Virus (MHV)	Sendai Virus
Pneumonia Virus of Mice (PVM)	*Mycoplasma pulmonis*
Reo Virus Type 3 (REO-3)	*Mycoplasma arthritidis*
Sendai Virus	
Theiler's Murine Encephalomyelitis Virus (TMEV)	
Mycoplasma pulmonis	
Mycoplasma arthritidis	

III. Nonanimal Microbial Surveillance

Sample	Test	Frequency
Cage water bottle	*Pseudomonas aeruginosa*	2 rack/room/month (mice); 3 (rats)
Water, processed	Bacteria, pathogens	Daily
Bedding, shipment	Bacteria, coliform, mold, pathogens	5 bags/5th shipment
Bedding, processed	Bacteria, mold, pathogens	Weekly

TABLE 9.3 (continued)

Routine Microbiological Surveillance Parameters for SPF-Barrier, SPF-Breeding Colony, and Diet Preparation Barrier

III. Nonanimal Microbial Surveillance (continued)

Sample	Test	Frequency
Biological indicators	Microbial growth (autoclave, EtO)	Daily
Surface swabs	Bacteria, mold, pathogens	Quarterly
Environmental air	Bacteria, mold, pathogens	1 SPF-barrier room/week
Personnel mouth rinse	Bacterial pathogens	Yearly

IV. Diet Preparation Barrier Assurance

Sample	Test	Frequency
Diet, shipment received	Bacteria, coliform, mold, pathogens	1/10 total bags/5th shipment
Diet, autoclaved	Bacteria, mold, pathogens	Daily
Biological indicators	Microbial growth (autoclave)	Daily
Surface swabs	Bacteria, mold, pathogens	Quarterly
Environmental air	Bacteria, mold, pathogens	Quarterly
Personnel mouth rinse	Bacterial pathogens	Yearly

NCTR. In addition, clinical biochemistry profiles have paralleled pathological changes with age for male and female DR- and AL-fed C57BL/6 mice receiving the NIH-31 or EM911A control and vitamin fortified diets, and DR- and AL-fed B6C3F$_1$ and B6D2F$_1$ mice receiving the NIH-31 diets [38].

To establish the database of developmental, age-specific pathology profiles among DR- and AL-fed rodents, 800 animals per genotype of rats and mice were evaluated for pathologic change. Groups of AL and DR animals, with both sexes represented, were randomly assigned to: (1) a lifetime group, to provide longevity data; and (2) a scheduled sacrifice group, to provide cross-sectional evaluations [14,39]. Animals were removed from the scheduled sacrifice group beginning at 12 months of study, and at 6-month intervals thereafter. Animals were necropsied, and approximately 45 tissues or organs, as well as all gross lesions, were collected for microscopic examination. Throughout their respective lifetimes, rat and mouse groups were monitored and characterized for disease incidence and changes associated with aging. Many histopathologic changes were diagnostic of specific disease, while others provided biomarkers of disease progression within the aging colony for further investigation [40,41]. Disease incidence within NCTR colonies is low, and colony health is generally regarded as excellent (Bucci, T., Pathology Associates International, Director, Pathology Services at NCTR, personal communication, 1997).

9.3.5 Animal Care Support

NCTR is unique in that it is host to two SPF-barrier animal facilities and a diet preparation barrier. Within the NCTR SPF-barrier that houses the NIA/NCTR rodent colonies, there are approximately 820 pairs of breeding stock (3 rat strains; 2 mouse strains) that produce an average of 4125 newborn pups per month. Pups from these discrete breeding colonies are allocated to treatment groups in this barrier.

A second, independent SPF-barrier system houses the remainder of the rodent breeding colonies. This breeding facility houses the nucleus stocks of breeding rodents, and the breeding stocks that are housed within the NIA/NCTR colonies SPF-barrier originate from this resource. Parent stocks are maintained in separate SPF-barrier rooms designated for mouse or rat breeding stocks. There are approximately 210 mouse breeding pairs in this barrier that produce an average 1148 newborn pups per month. The pups born in the breeding colony SPF-barrier must be transported between facilities in isolators.

These barriers meet all requirements established for barrier maintenance [28]. Written standard operating procedures (SOP's) [42] direct each operation conducted within the multiple barrier system. Equipment maintenance and personnel activities are governed by SOP compliance.

Within the 18 animal rooms in the NIA/NCTR colonies SPF-barrier, animals are individually housed, and mice and rats are housed in separate rooms. An average of 12 animal racks per room hold 84 mouse (7 shelf racks) or 72 rat (6 shelf racks) cages. Cages are handled individually in the animal rooms during husbandry procedures.

At NCTR, hardwood bedding chips are autoclaved at 121°C. Water in the SPF-breeding colony barrier is autoclaved; water in the SPF-barrier housing the NIA/NCTR colonies is sterilized by passage through bacteriological filters.

The SPF-barrier system at NCTR provides for free movement in all clean areas [36]. Barrier entry and exit procedures are enforced to ensure that personnel maintain appropriate hygienic measures upon clean corridor entry into, and subsequent exit from, the respective barriers. An air-lock pressurized entry system ensures airflow from the cleanest area; SPF-barriers are exited through return corridors to nonbarrier areas. Sterile clothing and personal protective equipment are provided within all barrier employee dressing areas. SPF-barriers are exited through return corridors to nonbarrier areas.

Equipment and supplies are initially received into clean areas of each respective barrier. These materials flow from clean areas into animal rooms for use and then are returned to processing areas for sanitation, and finally to clean areas for redistribution. All animal and support areas are maintained daily by housekeeping SOP's.

9.3.6 Diet Preparation Support

The diet preparation facility at NCTR is a self-contained, barrier system that houses equipment required for processing and handling of diet for SPF-barrier and conventionally housed rodents.

As the presence of microbial agents in diets presents a risk for potential barrier contamination, diets for barrier- and isolator-maintained animals must be sterilized, or decontaminated, to remove pathogenic bacteria, molds, viruses, and insect pests [13]. Autoclavable diets are vitamin fortified. After steam autoclave or pasteurization treatment of the diets used in the NIA/NCTR study, the diet quality is routinely sampled for adequate vitamin concentrations. Diets are stored in temperature-controlled environments before and after processing.

While steam autoclaving at 121°C for 15–20 min is frequently recommended for diet sterilization, some diet formulations (i.e., purified, high casein, high sugars) are less successfully autoclaved at these temperatures due to heat effects (e.g., Maillard reaction formation and pellet hardening). Pasteurization at 100°C for 5 min is successful in significantly reducing microbial contamination, but does not equal high temperature autoclaving in elimination of bacterial bioload. Pasteurization at 80°C for 5–10 min will remove vegetative forms of mold, but not spores [43]; therefore, pasteurized diets are not sterile, but microbial contaminants are reduced. At NCTR, semipurified diets are pasteurized at 105°C for 6 min; vitamin fortified, autoclavable natural-ingredient diets are steam autoclaved at 135°C for 5 min to ensure reliable microbial contaminant control. When diets were steam autoclaved, all microbial contaminants were effectively removed from both the pelleted and meal forms of the diet (Table 9.4). However, when diets were pasteurized at a lower temperature for a shorter interval, 92% of the microbial contaminants were eliminated from the pelleted diet, but only 14% of the total bacteria were eliminated from the meal form of the diet. Gram-negative bacteria, mold, coliforms, and pathogens were effectively removed from the diets tested when pasteurization decontamination was used. In all cases, ground meal diet was higher in numbers of microbial contaminants.

TABLE 9.4
Effect of Sterilization Method on Diet Microbial Contamination

	Treatment	Total Bacteria, CFU/g[a]	Gram-Negative Bacteria, CFU/g	Mold, CFU/g	Coliform	Pathogen[b]
NIH-31[c]	S[d]	30,000	0	0	2	0
	A	0	0	0	0	0
NIH-31, fort.	S	52,867	2	0	7	0
	A	0	0	0	0	0
EM911-A	S	6,600	60	0	0	0
	P	540	0	0	0	0
EM911-A, fort.	S	56,000	2,040	7,400	93	+
	P	48,000	0	0	0	0

[a] Colony forming unit per gram of diet.
[b] *Klebsiella oxytoca.*
[c] NIH-31 and EM911A standard formulation, pelleted; NIH-31 and EM911A vitamin fortified, meal.
[d] S = shipment diet; A = autoclaved (135°C, 5 min); P = pasteurized (105°C, 6 min).

Laboratory animal diets may also be decontaminated by ethylene oxide fumigation [13] and ionizing radiation [43]. There were no differences in intake, nitrogen retention, or dry matter, protein, and fat digestibilities among male germ-free Wistar rats fed autoclaved or irradiated diets [44].

To address costs associated with daily feeding of DR animals, a mechanized pellet sorting system was designed to meet production requirements within the diet preparation barrier. Pellets are sorted into discrete weight ranges (e.g., 2.5, 3.0, 3.5, 4.0, 4.5, and 5.0 ± 0.2 g), packaged individually by weight, and stored within the diet preparation barrier until delivered to the SPF-barriers in sealed containers [45]. In the animal room, technicians feed the daily intake requirement, by pellet weight array, based upon a computer-generated feeding schedule. The mechanized system of pellet sorting allowed for more economic delivery and animal room feeding of discrete intake levels to the DR rodents.

9.3.7   Computer Support

A network of computer support systems is in place to facilitate overall management of animals maintained within the SPF-barrier and conventional housing areas [46]. A data collection and reporting matrix serve to automate many of the tasks involved in tracking experimental data. The system provides accounting of individual animal historical data (i.e., birth date, litter identification, breeding and health status, and date of death), daily diet and water consumption, body weight measurements, and total animal census by housing area. Animal care activities are processed and record keeping transactions are updated daily.

Animal assignment to protocols is accomplished via randomized computer allocation by room, rack, and caging configuration. The network facilitates direct access to interdepartmental surveillance reports from the divisions of microbiology, chemistry, pathology, breeding colony, and general animal care. Independently tailored *ad hoc* summary reports to management personnel and investigators for protocol updates are also provided.

9.3.8 Safety Staff Support

Safety staff provide monitoring of all barrier-related activities at NCTR. In support of SPF-barrier maintenance, post-sterilization ethylene oxide air samples are tested for residual gas, and training on the use of ethylene oxide is provided. Respirator training for animal care and diet preparation personnel is routinely provided to barrier and nonbarrier personnel. At special request, safety staff take chemical or microbial swabs to determine the presence of potential biohazards. HEPA-filtered air quality is monitored on a routine maintenance schedule.

9.4 Specialized Procedures

9.4.1 Isolators

While animal quality is increasing due to sophisticated husbandry and health screening protocols, the choice of an appropriate environment for managing animal models is largely dependent upon the numbers of animals required to meet research needs. Economical use of available space is critical when complete barrier systems and isolator maintenance are integral to research programs.

A barrier system with unrestrained personnel movement is most convenient when large numbers of animals and many procedures involving the animals are required. Other types of barriers are better suited to specific purposes, such as maintaining reserve numbers of valuable breeding stocks. Smaller barrier, or isolator, units are likely more versatile, economical, and easier to maintain free of contamination. Among the isolator styles are microisolator cages with filter top or bonnet, microisolator cubicle units, isolation chambers, and plastic bubble clean rooms. As with larger barrier maintenance, written SOP's that detail isolator care are essential.

Isolators have several unique features in common, regardless of construction materials: an air circulation system to maintain ventilation, in- and outlet air filters, double doors with a sterilizable sealed passage for transfer of animals and materials in or out of the isolator, or for direct, sterile transfer from one isolator to another.

Isolators are used at NCTR for animal transport between the SPF-breeding and the NIA/NCTR SPF-barriers. Extensive SOP's designate routine inspection of isolators before, during, and after completion of transfers. Inspection records kept with each isolator unit include date, inspector identification, and equipment information. SOP's also cover cage and isolator interior cleanup, transfer of sterilized supplies and diet, animal transfers, and preparation of sterilizing agent (peracetic acid). Strict adherence to SOP protocols are critical to ensure successful animal transfers between SPF-barrier sites.

9.4.2   SPF Rodent Transport

At NCTR, the SPF-barrier reared, aged rats and mice are periodically distributed to participating researchers throughout the U.S. and Canada. To achieve maximal protection of aging SPF-maintained rodents, several types of commercial filter-protected shipping containers were subjected to aerosolized microbial contamination, and airline handling and stacking procedures. These testing criteria, under practical shipment conditions, allowed the selection of an appropriate container. In addition to container selection, a sterile agar cube was developed to provide a water source for all aged rat and mouse air freight shipments [47].

9.4.3   Compliance

An on-site Quality Assurance staff provides oversight for continuous monitoring and development of SOP's in both animal care SPF-barriers and the diet preparation barrier. Contract Nutrition and Quality Control/Safety staff provide additional support to the maintenance of the barrier systems. NCTR staff and contract veterinarians provide additional barrier surveillance, monitoring capabilities, and establish guidelines for corrective measures.

Compliance with Good Laboratory Practices (GLP's) in all animal care and diet preparation activities at the NCTR warranted the development of an automated inventory tracking system to facilitate the support of the multispecies aging studies. The objectives were to: (1) switch from manual to automated data handling, (2) simplify data retrieval and reporting capabilities, (3) coordinate internal inventory control, and (4) maintain GLP accountability. With programmed safety features, user identification and authorizations, and verified accountability, the automated system meets GLP requirements for raw data collection and reporting [48].

9.5   Conclusions

As pathogens are known to effect phenotypic expression of many disease conditions, influence the outcome of experimental events, and eradicate colonies of immunocompromised rodent models, the various types of barrier systems provide maximal protection for research efforts. Deleterious effects of unprotected environmental exposures may be most critical for aging or senescent models, particularly those assigned to protocols requiring infectious agent, drug, carcinogen, or xenotoxin challenge.

While the selection of appropriate stocks and strains of rodents is an essential element of any research plan, perhaps for aging, or long-term rearing, that selectivity becomes more critical. At minimum, an understanding of lifetime parameters and the responsiveness of those parameters to controllable and predictable modifications such as diet selection or diet restriction is essential. Diet selection and level of intake are apparently as important as maintenance of SPF health status in assuring long-term survivability among rodent models.

System controls that are required for effective and successful SPF-barrier maintenance likely provide the best insurance for research integrity. As with any system, the successful maintenance of a SPF-barrier environment is dependent upon a critical level of responsibility and dedication assumed by all involved personnel.

Acknowledgment

The authors would like to acknowledge Mr. Warren Campbell and Mr. Harold Thompson for their contributions in the development of this manuscript.

References

1. Lang, P. L., Changes in life span of research animals leading to questions about validity of toxicologic studies, *Chem. Reg. Rep.*, 14, 1518, 1991.
2. Turturro, A., Duffy, P., and Hart, R., Modulation of toxicity by diet and dietary macronutrient restriction, *Mutat. Res.*, 295, 151, 1993.
3. Turturro, A., Blank, K., Murasko, D., and Hart, R., Mechanisms of caloric restriction effecting aging and disease, *Ann. Acad. Sci.*, 719, 159, 1994.
4. Turturro, A., Duffy, P., and Hart, R., The effect of caloric modulation on toxicity studies, in *Dietary Restriction: Implications for the Design and Interpretation of Toxicity and Carcinogenic Studies*, Hart, R. W., Neumann, D. A., and Robertson, R. T., Eds., ILSI Press, Washington, D.C., 1995, 79.
5. Turturro, A., Duffy, P., Hart, R., and Allaben, W., Rationale for the use of dietary control in toxicity studies, *Toxicol. Pathol.*, 24, 769, 1996.
6. Keenan, K., Smith, P., Ballam, G., Soper, K., and Bokelman, D., The effect of diet and dietary optimization (caloric restriction) on survival in carcinogenicity studies: An industry viewpoint, in The *Carcinogenicity Debate*, McAuslane, J., Lumley, C., and Walker, S., Eds., Quay Publishing, London, 1992, 77.
7. Abelson, P. H., Flaws in risk assessments, *Science*, 270, 215, 1995.
8. Weindruch, R. and Walford, R. L., *The Retardation of Aging and Disease by Dietary Restriction*, Charles C. Thomas, Springfield, IL, 1, 1988.
9. Masoro, E. J., Food restriction in rodents: An evaluation of its role in the study of aging, *J. Gerontol.*, 43, B59, 1988.
10. Masoro, E. J., FRAR course on laboratory approaches to aging: Nutrition, including diet restriction, in mammals, *Aging*, 5, 269, 1993.
11. NRC (National Research Council) Institute of Laboratory Animal Resources, Committee to Revise the Guide for the Care and Use of Laboratory Animals, *Guide for the Care and Use of Laboratory Animals* 7th ed., National Academy Press, Washington, D.C., 1996.
12. Witt, W., Brand, C. D., Attwood, V., and Soave, O., A nationally supported study on caloric restriction of rodents, Lab. Anim., 18(3), 37, 1989.
13. NRC (National Research Council) Institute of Laboratory Animal Resources, Committee on Rodents, *Laboratory Animal Management*, National Academy Press, Washington, D.C., 1996.
14. Sheldon, W. G., Bucci, T. J., Blackwell, B. N., and Turturro, A., Effect of *ad libitum* feeding and 40% feed restriction in body weight, longevity, and neoplasms in B6C3F$_1$, C57BL6, and B6D2F$_1$ mice, in *Pathology of the Aging Mouse*, Vol. 2, Mohr, Dungworth, Capen, Carlton, Sundberg, and Word, Eds., ILSI Press, Washington, D.C., 1996, 21.

15. Jackson, C. D. and Blackwell, B. N., Two-year toxicity study of doxylamine succinate in the Fischer 344 rat, *J. Am. Coll. Toxicol.*, 12(1), 1, 1993.

16. Greenman, D. L. and Allaben, W., Triprolidine 104 week chronic dose study in rats and mice, NCTR Technical Report for NTP Experiments 05055-05/05055-06, 1991.

17. Littlefield, N. A., Sheldon, W. G., Allen, R. R., and Gaylor, D. W., Chronic toxicity/carcinogenicity studies of sulfamethazine in Fischer 344/N rats: Two-generation exposure, *Food Chem. Toxic.*, 28(3), 157, 1990.

18. Littlefield, N. A., Gaylor, D. W., Blackwell, B. N., and Allen, R. R., Chronic toxicity/carcinogenicity studies of gentian violet in Fischer 344 rats: Two-generation exposure, *Food Chem. Toxic.*, 27(4), 239, 1989.

19. Turturro, A. and Hart, R. W., Dietary alteration in the rate of cancer and aging, *Exp. Gerontol.*, 27, 583, 1992.

20. Jackson, C. D. and Sheldon, W. G., Two-year toxicity study of doxylamine succinate in B6C3F$_1$ mice, *J. Am. Coll. Toxicol.*, 12(4), 311, 1993.

21. Littlefield, N. A., Blackwell, B. N., Hewitt, C. C., and Gaylor, D. W., Chronic toxicity and carcinogenicity studies of gentian violet in mice, *Fundam. Appl. Toxicol.*, 5, 902, 1985.

22. Littlefield, N. A., Gaylor, D. W., Blackwell, B. N., and Allen, A. A., Chronic toxicity/carcinogenicity studies of sulfamethazine in B6C3F1 mice, *Food Chem. Toxic.*, 27(7), 455, 1989.

23. Fullerton, F. L. and Greenman, D. L., Comparison of the effects of two diets on spontaneous and 2-AAF induced carcinogenic lesions in two stocks of mice — a chronic study, NCTR Technical Report for Experiment 346, 1989.

24. Heneghan, J. B., Imbalance of the normal microbial flora, the germ-free alimentary tract, *Am. J. Dig. Dis.*, 10(10), 864, 1965.

25. Gordon, H. A., Germ-free animals in research: An extension of the pure culture concept, *Triangle*, 7(3), 108, 1965.

26. Heine, W., How to define SPF?, *Z. Versuchstierkd.*, 22, 262, 1980.

27. PHS (Public Health Service), Public Health Service Policy on Humane Care and Use of Laboratory Animals, U.S. Department of Health and Human Services, National Institutes of Health, Washington, DC., 1986.

28. NRC (National Research Council) Institute of Laboratory Animal Resources, Committee on Infectious Diseases of Mice and Rats, *Infectious Diseases of Mice and Rats*, National Academy Press, Washington, D.C., 1991.

29. Tannenbaum, A., The genesis and growth of tumors. II. Effects of caloric restriction *per se*, *Cancer Res.*, 2, 460, 1942.

30. Everett, R., Factors affecting spontaneous tumor incidence rates in mice: A literature review, in *CRC Critical Reviews in Toxicology*, CRC Press, Boca Raton, FL, 13(3), 235, 1984.

31. Feuers, R. J., Duffy, P. H., Leakey, J. A., Turturro, A., Mittelstaedt, R. A., and Hart, R. W., Effect of chronic caloric restriction on hepatic enzymes of intermediary metabolism in the male Fischer 344 rat, *Mech. Ageing Dev.*, 48, 179, 1989.

32. McCarter, R. J. M., Role of caloric restriction in the prolongation of life, in *Clinics in Geriatric Medicine*, W.B. Saunders, Philadelphia, 1995, 11(4).

33. Reeves, P. G., Nielsen, F. H., and Fahey, Jr., G. C., AIN-93 Purified diets for laboratory rodents: final report of the American Institute of Nutrition *ad hoc* writing committee on the reformulation of the AIN-76A rodent diet, *J. Nutr.*, 123, 1939, 1993.

34. Rao, G. N., New diet (NTP-2000) for rats in the National Toxicology Program toxicity and carcinogenicity studies, *Fundam. Appl. Toxicol.*, 32, 102, 1996.

35. Roe, F. J. C., Lee, P. N., Conybeare, G., Kelly, D., Matter, B., Prentice, D., and Tobin, G., The Biosure Study: Influence of composition of diet and food consumption on longevity, degenerative diseases and neoplasia in Wistar rats studied for up to 30 months post weaning, *Food Chem. Toxic.*, 33 (Suppl. 1), 1S, 1995.

36. Littlefield, N., Harmon, J., and Campbell, H., Barrier animal facility operations for a large-scale chronic study (ED01 study), *J. Environ. Path. Toxicol.*, 3(3), 187, 1979.

37. Clough, G., The Animal House: Design, equipment and environmental control, in *The UFAW Handbook on the Care and Management of Laboratory Animals*, 6th ed., Longman Scientific and Technical, Churchill Livingstone, New York, 1987, chap. 8.

38. Loeb, W. F., Das, S. R., Harbour, L. S., Turturro, A., Bucci, T. J., and Clifford, C. B., Clinical biochemistry, in *Pathology of the Aging Mouse*, Vol. 2, Mohr, Dungworth, Capen, Carlton, Sundberg, and Word, Eds., ILSI Press, Washington, D.C., 1996, 3.

39. Sheldon, W. G., Bucci, T. J., Hart, R. W., and Turturro, A., Age-related neoplasia in a lifetime study of *ad libitum*-fed and food-restricted B6C3F$_1$ mice, *Toxicol. Pathol.*, 23(4), 458, 1995.

40. Thurman, J. D., Bucci, T. J., Hart, R. W., and Turturro, A., Survival, body weight, and spontaneous neoplasms in *ad libitum*-fed and food-restricted Fischer 344 rats, *Toxicol. Pathol.*, 22(1), 1, 1994.

41. Bronson, R. T., Rate of occurrence of lesions in 20 inbred and hybrid genotypes of rats and mice sacrificed at 6 month intervals during the first years of life, in *Genetic Effects of Aging*, Part II, Harrison, D. E., Ed., Telford Press, Caldwell, NJ, 1990, 279.

42. Standard Operating Procedures, Animal care/diet preparation services standard operating procedures, The Bionetics Corporation, National Center for Toxicological Research, Jefferson, AR, 1996.

43. Coates, M. E., The nutrition of laboratory animals, in *The UFAW Handbook on the Care and Management of Laboratory Animals*, 5th ed., Churchill Livingstone Press, London, 1976, Chapter 3.

44. Yamanaka, M., Muneo, S., and Nomura, T., A comparison of the nutritional evaluation of irradiated and autoclaved diets in germfree rats, *Exp. Anim.*, 30(3), 299, 1981.

45. Hart, R. W., Hass, B. S., Turturro, A., and Lewis, S. M., The use of nutrition to implement refinement in animal bioassays, *Lab. Anim.*, 25(3), 20, 1996.

46. Information Technology Staff (ITS), Inlife data collection system report formats and descriptions, ITS Document #003-INLF-1.0, National Center for Toxicological Research, Jefferson, AR, 1993.

47. Lewis, S. M., Martin, J. L., Moyer, J., and Brand, C. D., Development of a novel, sterilizable, polycarbonate cup to provide water during air transport of specific pathogen free rodents, *Lab. Anim. Sci.*, 40 (5), 557, 1990.

48. Lewis, S. M., Leard, B. L., Martin, J. L., and Martin, S. A., An automated feed inventory tracking system for an animal facility, *Lab. Anim.*, 24(8), 37, 1995.

10

Choice of Rodent Model for Aging Research

Edward J. Masoro

CONTENTS

0-8493-3112-9/99/$0.00+$.50
© 1999 by CRC Press LLC

10.1 Introduction: Rodents, The Most Used Animal Models

Rodents have been by far the most widely used animal models for aging research. A survey by Weindruch of four of the leading journals in biological gerontology showed that from 1972 to 1992 rats and mice served as the animal model in 76% of the studies in which animal models were employed [1].

10.1.1 Reasons for Wide Use

There are many reasons for the wide use of rodent models in aging research. A major one is that rats and mice have been extensively used in all fields of biomedical and behavioral research; as a result, investigators have had a great deal of experience in the care and maintenance of these species and much is known about their biology and behavior. In addition, as will be discussed presently, the longevity characteristics of rodent species make them desirable for aging research. Also, the extensive data base that has been developed on their gerontologic characteristics facilitates the design of further gerontologic research.

10.1.1.1 *Longevity Characteristics*

In selecting a model for aging research, it is desirable that the longevity be less than 5 years. When the longevity exceeds 5 years, it is difficult, if not impossible, for an investigator to carry out life span longitudinal studies or to be fully in control of the life span experiences of the animals used in cross-sectional studies. Although the life span of many invertebrate species is much less than 5 years, that of most mammalian species is well in excess of 5 years [2]. In contrast, most rodents have life spans of less than 5 years, making them uniquely suited for aging research on mammalian species [3].

10.1.1.2 *Extensive Data Bases On Gerontologic Characteristics*

Since so much aging research has been done using rats and mice, there has been a positive feedback on their further use because the availability of extensive data bases on their gerontologic characteristics greatly aids in the design of aging research projects. Specifically, there is a sizable literature on the longevity of many strains of rats and mice and on the lifelong nutritional needs, disease characteristics, and physiological and behavioral characteristics of these animals [2]. Such information is scant or lacking for most mammalian species. For example, there is uncertainty about the longevity characteristics of rabbits [4]. Thus while rabbits have been a popular model for the study of atherosclerosis, the conclusions drawn on the influence of aging on atherosclerosis will remain doubtful until the longevity characteristics and the age-associated pathology of this species are known. Although some data are available on the age changes in the physiology and the age-associated diseases of dogs, cats, and primates, the amount of information is scant compared to that on mice and rats [2].

10.1.2 Rodents Used

The order Rodentia is the largest order of mammals. This order contains about 350 genera and exceeds all other mammalian orders in numbers of species and individuals [5]. Of these, only the rat, mouse, hamster, gerbil, and *peromyscus* have had significant use in aging research.

10.1.2.1 Rat

Many inbred strains, F_1 hybrids of inbred strains, and outbred stocks of rats (*Rattus norvegicus*) have been used in aging research [6]. Of these, the most widely used include: inbred F344 strain, outbred Wistar stock, outbred Sprague-Dawley stock, outbred Long-Evans stock, inbred Brown-Norway (BN) strain, inbred WAG/Rij strain, (F344 × BN) F_1 hybrid, and (WAG × BN) F_1 hybrid [2]. All of the above strains, stocks, and F_1 hybrids have been studied for longevity characteristics and age-associated pathology as well as a variety of age-associated physiological and behavioral changes. Currently the National Institute on Aging, a component of the National Institutes of Health of the U.S., is making available rats of the following strains: the F344 inbred strain, the BN inbred strain, and the (F344 × BN) F_1 hybrid [7].

10.1.2.2 Mouse

The following mouse (*Mus domesticus*) genotypes have frequently been used for aging research: the C57BL/6, DBA/2, CBA/Ca, and BALB/c inbred strains, the (C57BL/6 × CH3), (C56BL/6 × DBA/2), (BALB/c × C57BL/6) F_1 hybrids, and the Swiss Webster outbred stock [1]. Animals of various ages of all of these are available through the National Institute on Aging [7]. In the case of each of these mouse genotypes, research has been carried out on their longevity characteristics, age-associated pathology, and many other age-associated changes.

10.1.2.3 Hamster

Three species of hamsters have been used in aging studies: the Syrian hamster (*Mesocricetus auratus*), the Chinese hamster (*Cricetulus griseus*), and the Turkish hamster (*Mesocricetus brandti*). Longevity studies have been carried out with outbred stocks and inbred strains of Syrian hamsters and the Chinese hamster [8,9]. The age-associated pathology of the Syrian hamster and Chinese hamsters has also been investigated [9,10]. The longevity characteristics of the Turkish hamster, but not the age-associated pathology, have been studied [11].

10.1.2.4 Gerbil

The use of gerbils for aging research includes several studies on longevity characteristics [12]. Unfortunately almost nothing has been done on the age-associated pathology of gerbils.

10.1.2.5 Peromyscus

Several species of the genus *Peromyscus* are similar in size and appearance to *Mus domesticus* but are not close relatives [13]. Of these, *Peromyscus maniculatis* (deer mouse) and *Peromyscus leucopus* (white-footed mouse) have been used in aging studies [2]. The striking feature of these two species is that they have a much greater longevity than *Mus domesticus* [14]. Only limited research has been done on the age-associated pathology of these species. Defined genotypes of *Peromyscus* are in the process of being established [15].

10.2 Background of Use of Rodents for Aging Research

Animal models have been widely used to gain insights into both human and general biology [16]. Laboratory rodents came into favor because of the small size of these species of mammals, the ease of handling them within the laboratory setting, and their high rates of reproduction.

10.2.1 Historical Evolution

The use of rodents for aging research essentially started with McCay and his colleagues in the 1930s [17]. Although an occasional aging study using rodents had been reported prior to the 1930s, McCay and colleagues were the first to use a rodent species to intensively pursue a gerontologic phenomenon, specifically the retardation of aging in rats by restriction of caloric intake. Other than stating that white rats were used, these investigators provided no other information in regard to their rat model. Ross and his colleagues continued exploring the influence of nutrition on aging in rats with a series of exceedingly comprehensive and detailed studies carried out in the 1960s and 1970s [17]. They used a stock of Sprague-Dawley descendants from the Charles River Breeding Laboratories as their rat model, fully described the diets used, and carefully assessed the disease status of the animals. In 1972, Russell brought into focus the importance of genetic-environmental interactions in aging research and the need for the genetic characteristics of the rodent model to be defined [19]. In that article she described the genetically defined mouse strains then available for aging research. Since the 1970s, the rat (*Rattus norvegicus*) and the mouse (*Mus domesticus*) have been the most commonly used models for aging research; much of this research has been carried out on a small set of defined genotypes.

10.2.2 Positive Impact of the Extensive Use of Rodents

This extensive use of rodents, more specifically of the rat and mouse, has had a positive impact on the quality of aging research. Because of the intense focus in recent years on the use of particular rat and mouse strains for aging research, procedures have been developed to provide optimum husbandry for the life span maintenance of these particular genotypes; in addition, detailed data have accumulated on the age-associated disease processes and other biological characteristics of these genotypes. Thus investigators have been able to obtain and maintain animals that have reasonably standardized and known characteristics, which, in turn, has resulted in research at different laboratories being carried out on comparable animal models. This has been a boon to experimental gerontology, a field long plagued by the lack of reproducibility of findings which was due, in part, to the differences among animal models used in various laboratories.

10.2.3 Negative Impact of the Extensive Use of Rodents

The negative impact on gerontology of the extensive use of rats and mice stems from focusing on just two species, which makes it difficult to assess the generality of the findings, i.e., it is not possible to ascertain whether the findings on the aging of rats and mice apply to other species and in particular to humans. Moreover, an assumption underlying this extensive use of rats and mice is that the aging processes in these rapidly aging mammals are similar to those in slowly aging mammals such as the human. Although this assumption is rarely explicitly stated, it is implied and it certainly remains to be validated. Thus it is imperative that a variety of different species of well-characterized animal models be developed in order to assess generality.

10.3 Issues in Choice of Rodent for Aging Research

When choosing an animal model for aging research, two issues must be addressed. One is the assessment of the extent to which the model meets the characteristics necessary for carrying out a successful aging study. The other is the assessment of the usefulness of the model for the exploration of a specific biological or behavioral question.

10.3.1 Requirements Common to All Gerontologic Studies

There are several general criteria that investigators should consider when choosing an animal model for an aging study. Of course, it is not likely that a model will be found that ideally meets all these criteria as well as those relevant to the specific question under investigation. Thus, in practice, compromises are inevitable. Nevertheless, when considering a particular animal model, the extent to which it meets each of the following criteria should be carefully evaluated.

10.3.1.1 Availability

Ideally, standardized animals of a wide range of ages, including the oldest old, should be available. Such is the case for those strains and stocks of rats and mice available from the National Institute on Aging. A few others are available from commercial sources, for example, the Sprague-Dawley rat stock. However, for most other strains and stocks of mice and rats and for hamsters, gerbils, and other rodent species, it is necessary for the investigator either to maintain a colony of animals that is allowed to reach a variety of ages or to make arrangements for a commercial or other source to do so. Colonies that have been maintained at universities and research institutes are one source of such animals; however, to obtain such animals, it is often necessary to arrange to use them in a collaborative fashion rather than simply to purchase them. However, most rodents are relatively easy to maintain; therefore, if the investigator has the necessary animal-hold facilities, this may be the preferred route because by doing so the investigator has first-hand knowledge of the history of the animals being used. Species of *Peromyscus* present a formidable challenge in this regard because of their long life span; indeed, only recently have efforts been made to establish defined laboratory strains [15].

10.3.1.2 Cost

Because of the length of time animals must be maintained for the carrying out of aging studies, their cost is great whether the animals are purchased from a commercial source or allowed to age in the facilities of the investigator. This cost is reduced somewhat in the case of the rat and mouse strains and stocks available from the National Institute on Aging because this source of animals is subsidized by the National Institutes of Health. It is important to note that the National Institute on Aging has a program that makes available without charge a limited number of old rats and mice to investigators who are new to the field of aging research. The high cost of animals is clearly a hindrance to the rapid development of experimental gerontology and is an issue that urgently needs to be addressed by the gerontologic community, particularly by those agencies that support aging research.

10.3.1.3 Genetic Characteristics

The genetic characteristics of the rodents used in aging research must be rigorously defined to permit valid comparisons of studies done at different times and laboratories. Outbred stocks are not genetically defined and will vary from supplier to supplier and — with time — even when obtained from the same supplier. Inbred strains, which have undergone at least 20 or more consecutive generations of brother × sister mating, are considered genetically identical. These inbred animals are homozygous at virtually all loci, and colonies of them exhibit long-term genetic stability [20]. Thus inbred strains enable research carried out in different laboratories or at different times to be comparable. F_1 hybrids between inbred strains also have defined and reproducible genetic characteristics and, according to Phelan and Austad, may be a more desirable model because of less phenotypic variability than inbred strains [21]. The genetics of the animal model being used requires monitoring to guard against errors by the breeder. There are many such monitoring methods such as biochemical markers and immunological markers; DNA-based methods have many advantages and are likely to be the methods of choice in the future [20].

10.3.1.4 Nutritional Factors

Nutrition can influence both aging and the occurrence and progression of age-associated diseases [22]. Therefore, it is imperative that (1) the composition of the diet fed rodents in aging studies should be adequate for the life span maintenance of the animals; and (2) the composition should be known and consistent. Two types of diets have been used to address these issues: nonpurified diets composed of natural ingredients and defined purified (semisynthetic) diets.

The nonpurified diets are commercially available and there are two types: (1) open formula diets, which have a defined and fixed natural ingredient composition made known to the investigator; and (2) closed formula diets, for which the manufacturer does not provide the exact composition. In principle, the closed formula diet is unacceptable because the investigator is deprived of a full knowledge of the experimental system. Even the open formula diet is subject to variation in composition during the course of an aging study because the composition of natural ingredients can vary considerably from batch to batch [23]. For example, wheat may vary in protein content from 8–14%; fiber is usually not defined in regard to its nature; and contamination with pesticides and other toxic substances is not an infrequent occurrence.

The purified diets, which are prepared by or for the investigator, are composed of refined proteins, carbohydrates, and fats with vitamin and mineral mixes added. The advantage of purified diets is that the composition of the diet can be closely controlled and contaminants eliminated. However, such diets are expensive.

If an open fixed formula diet is an economic necessity, the following can help minimize problems, include a large number of ingredients to minimize the effect of variations or contaminants of one dietary component; have each batch of diet analyzed so that informed decisions can be made regarding the acceptability of the batch; and, if possible, reserve a sufficient amount of each component, carefully stored, to last through the entire study.

10.3.1.5 Environmental Factors

Rodent housing and handling procedures can markedly influence an aging study. A factor that must be avoided is infectious disease. The occurrence of infectious disease is a particularly serious problem in long-term aging studies because it can mean having to abort the study in its late stages, which represents a catastrophic loss of investigator time and other resources. An important step in avoiding infectious disease is the use of a colony of animals that has originated from full-term fetuses aseptically removed from the uterus [24]. Such a colony should be maintained in barrier facilities for the duration of the study. These barrier facilities, which vary greatly in complexity, are designed and operated to exclude the entry of pathogenic microorganisms; rodents successfully derived and maintained in this way are designated Specific Pathogen-Free (SPF). To be certain that the SPF status is being maintained, it is necessary to continuously monitor the colony for the presence of pathogenic organisms, and detailed procedures for doing so have been published [25,26].

Many other environmental factors can influence aging studies in which rodents are used as the animal model. These include: noise, number of animals per cage, size of cage, intensity of light, periodicity of the light-dark cycle, ambient temperature, humidity, presence of adventitious chemicals due to bedding material, or the use of disinfectants or insecticides [3]. These factors, which should be defined and controlled, should be reported when the findings of a study are published

10.3.1.6 Disease Characteristics

Although infectious disease should be eliminated as a factor when carrying out aging research, age-associated disease is an inevitable aspect of all aging studies. When choosing an animal model, one criterion to be considered is the availability of a literature on the age-associated pathology of that model. Lack of such information makes it difficult to intelligently design a study. Indeed, when designing a study, the investigator should make full use of the published information on the age-associated pathology of the model. In addition, both gross and histopathologic analyses should be made on the rodents sacrificed during a study. In the case of life span longitudinal studies, such an analysis should be made following the spontaneous death of each rodent; this requires frequent inspection of the animals so that the pathologic analysis is done before postmortem autolysis obscures the findings or makes it difficult to interpret them definitively.

10.3.1.7 Longevity Characteristics

In designing an aging study, the longevity characteristics of the animal model must be known. Ideally, detailed life table data should be available; minimally, one must know the mean and median length of life to be expected for a population of the model. For many stocks and strains of rats, mice, hamsters, and gerbils, a considerable amount of such data is in the literature. However, if such information is not available for the model of choice, it must be generated during the course of the study for the findings to be interpretable within a gerontologic context. Indeed, such information should be generated during the study even for those models on which published data are available because longevity characteristics are influenced by the environmental circumstances imposed by the study.

10.3.1.8 Statistical Considerations

The cost of maintaining animal models for aging research is so great that every effort must be made to use the smallest number of animals possible to answer an experimental question. Therefore, the investigator must aim for the least possible individual variability, thereby minimizing the necessary sample size and maximizing the power to detect effects of experimental treatments. Phenotypic variance, i.e., the total variation observed among individuals in an animal model population, is due to genetic variance and environmental variance and the interaction between them [21]. Genetic variance is essentially eliminated by using inbred rodent strains, which is also the case with F_1 hybrids. Much of the material presented in this chapter has focused on minimizing environmental variance. Phelan and Austad present a strong argument that the variance due to the interaction between genetic variance and environmental variance is less in F_1 hybrids than in inbred strains, and if such is the case the F_1 hybrid would be the model of choice [21].

10.3.2   Requirements Unique to a Specific Research Question

Often a research question will address a specific age-associated problem that afflicts humans. Finding a rodent model that permits the experimental exploration of this problem is desirable because, as discussed earlier, of the advantages of using rodents over other mammalian species for aging studies. A rodent model in which the age-associated problem occurs spontaneously can often be found. For example, neural presbycusis and metabolic presbycusis are types of hearing loss that often occur with age in humans. Neural presbycusis is observed with increasing age in the F344 strain of rats [27]. Metabolic presbycusis occurs naturally in gerbils [28]. Transgenic rodents provide what is likely to be an increasingly important way to explore problems of human aging with rodent models. For example, Games and associates have produced transgenic mice that express a human mutation of Amyloid Precursor Protein; and these mice show the morphologic hallmarks of Alzheimer's disease, thus enabling the study of this disease process in a rodent model [29].

Another kind of specific question often asked in aging research involves the testing of theories of aging. Transgenic rodents that overexpress or underexpress gene products have great potential for use in such studies. Another useful tool for this task is the dietary restricted rat and mouse models in which a broad spectrum of aging processes is slowed [30]. The Senescence-Accelerated Mouse strains discovered by Takeda and associates have also been used for this purpose [31]. However, Harrison properly cautions that the findings from studies involving accelerated aging are fraught with interpretational pitfalls [32].

10.4   Procedures for Choosing a Rodent Model

The issues that must be considered when choosing a rodent model for aging research have been presented. The discussion now turns to procedures for making use of this information.

10.4.1   Detailed Definition of the Study Design

The first step is to develop a detailed design of the contemplated study. This step should focus not only on the specific biological or behavioral question to be addressed but should also consider that question within a gerontologic context. Based on this information, the next steps are to define the characteristics of the animal model needed and to develop a protocol for the procurement and maintenance of the animal model. Aging studies utilizing animal models are likely to require a large investment in resources for successful execution. Substantial long-term operating funds are usually needed and, in addition, specialized facilities are often required for animal maintenance. Finally, the investment of a great amount of time by the investigators is essential. It is imperative to make certain that all these resource needs can be realistically met before beginning this kind of study.

10.4.2   Meeting the Requirements of Specific Research Question

Clearly, the specific event or process that an aging research project addresses must occur in the animal model chosen for the study. Fortunately, many aging events or processes occur in almost all mammalian species, e.g., a decline in reproductive function, a decrease in the ability to respond to noxious agents, and an increase in the frequency of neoplasia. If the phenomenon to be studied commonly occurs in rodents, the choice of a particular species and strain of rodent should be based on cost and availability, ease of executing the study, and the extent to which the general requirements of a gerontologic study are satisfied.

However, the occurrence of the same aging event in two different species or strains of a species does not necessarily mean that the mechanism underlying the event is the same in both species or strains. Thus when seeking the mechanism underlying a particular aging event, it is desirable to use more than one animal model in the investigation. The ultimate extension of the research should include nonrodent species if at all possible.

Often it is an aging process or event that occurs in humans that is the aim of the research. For such a study, what is needed is a species or strain of a species of rodent that exhibits the phenomenon of interest. A search of the literature often reveals at least one rodent model with the spontaneous occurrence of the process or event; e.g., as discussed above regarding two different types of presbycusis common to humans, there are rodent models with each type of hearing problem. Based on a careful survey of the literature, an investigator can often find a rodent model that is suited for the study of the event or process of interest. When the literature search does not uncover a rodent model in which the event or process spontaneously occurs, then the possibility that transgenic rodent technology can create such a model should be investigated. Indeed, such models have been created and are often made available to others by the laboratories of origin. For example, as previously discussed, a transgenic mouse has been produced with brain morphology comparable to that of humans with Alzheimer's disease.

In summary, this stage of the procedure of choosing a rodent model requires a thorough review of the literature for a rodent exhibiting the particular event or process of interest. If such a search fails to uncover such an animal, modern rodent transgenic technology may have produced such a model or may make it possible to create one.

10.4.3   Meeting the Gerontologic Requirements

After zeroing in on a particular model, it is then necessary to assess the extent to which the model meets the requirements for successful execution of an aging study. The extent to which the genetics of the model is defined must be considered. It should be recognized that outbred stocks of rodents that have been widely used in biological and behavioral research are not genetically well defined; and thus even if the investigator has had experience using these models, they may be inappropriate for an aging study. The effects of diet and environment over the life span have been studied for only a few strains of rats and mice, and even less is known in this regard for other rodent species. Thus when choosing a model for which such information is lacking, the investigator must be prepared to evaluate the animals during the course of the study for possible problems stemming from such factors. In any aging study, it is important to do a pathologic analysis of the animals used in the study, and this is particularly true when using models for which there is little or no published information on age-associated pathology. In such cases, the protocol for obtaining data on age-associated pathology should be a component of the study design. To intelligently design an aging study, the longevity characteristics of the model are needed. Unfortunately, this information is available for only a few genotypes, even in the case of mice and rats. The lack of such information presents an investigator with a serious challenge which can not be fully addressed during the design stage of the study, but rather it requires obtaining longevity information during the course of the study. As this information is generated, it may well be necessary to modify the study design during the course of the research.

Of course, many of these gerontologic requirements are met when the mouse and rat genotypes subsidized by the National Institute on Aging are used; in addition, the cost of these animals is less, and availability is more certain than is likely to be the case for other genotypes. Thus whenever a National Institute on Aging genotype is suitable, the investigator is well advised to choose it. However, if the specific research question to be addressed makes none of these models desirable, then the issue of cost and availability must be carefully evaluated by the investigator.

10.5 Summary

Species of the order Rodentia are the most commonly used species for aging research. Of the many species in this order, the rat (*Rattus norvegicus*) and the mouse (*Mus domesticus*) are by far the species most studied, but there has also been significant use of gerbils, three species of hamsters, and two species of the genus *Peromyscus*. The major reasons for the extensive use of mice and rats are (1) they are mammals; (2) they have a relatively short life span and small size, compared to other mammals; (3) their upkeep requires relatively low amounts of resources; and (4) there is a large body of literature on their biological, behavioral, and gerontologic characteristics. This extensive use of rodents for aging research has had the positive effect of generating genetically well-defined strains and developing excellent life span husbandry, as well as producing large data bases on age-associated pathology and other age-associated biological and behavioral characteristics. The negative side of this gerontologic focus on mouse and rat models stems from the difficulty of knowing if the gerontologic data being collected permits generality or is unique to rats and mice.

In choosing a rodent model for an aging study, the following issues must be considered: availability and cost of animals of a spectrum of ages; a defined and stable genotype; knowledge of husbandry procedures needed for successful life span maintenance of the animals; and knowledge of the age-associated disease and longevity characteristics of the model. In addition, the model must satisfy the needs of the specific research question being addressed.

References

1. Weindruch, R., Animal models, in *Handbook of Physiology*, Section 11, Masoro, E. J., Ed., Oxford University Press, New York, 1995, chap. 3.
2. Masoro, E. J., Animal models in aging research, in *Handbook of the Biology of Aging*, 3rd ed., Schneider, E. L. and Rowe, J. W., Eds., Academic Press, San Diego, 1990, chap. 5.

3. Masoro, E. J., Use of rodents as models for the study of "normal aging": conceptual and practical issues, *Neurobiol. Aging*, 12, 639, 1991.

4. Weisbroth, S. H., Neoplastic disease, in *The Biology of the Laboratory Rabbit*, Weisbroth, S. H., Flatt, R. E., and Kraus, A, L., Eds., Academic Press, New York, 1974, chap. 14.

5. Committee on Animal Models for Research on Aging, *Mammalian Models for Research on Aging*, National Academy Press, Washington, D. C., 1981, 23.

6. Hazzard, D.W. G., Bronson, R. T., McClearn, G. E., and Strong, R., Selection of an appropriate animal model to study aging processes with special emphasis on the use of rat strains, *J. Gerontol. Biol. Sci.*, 47, B63, 1992.

7. Sprott, R. L. and Austad, S. N., Animal models in aging research, in *Handbook of the Biology of Aging*, 4th ed., Schneider, E. L. and Rowe, J. W., Eds., Academic Press, San Diego, 1996, chap. 1.

8. Haverland, L. H., Yoon, C.H., and Homburger, F., Studies on the aging of inbred Syrian Golden hamsters. Effect of age on organ weight, *Prog. Exp. Tumor Res.*, 16, 120, 1972.

9. Ladiges, W. C., Diseases, in *Laboratory Hamsters*, Van Hoosier, Jr., G. L. and McPherson, C, W., Eds., Academic Press, Orlando, 1987, chap. 18.

10. Hubbard, G. B. and Schmidt, R. E., Noninfectious disease, in *Laboratory Hamsters*, Van Hoosier, Jr., G. L. and McPherson, C. W., Eds., Academic Press, Orlando, 1987, chap. 10.

11. Lyman, C. P., O'Brien, R.C., Greene, G. C., and Papafrangos, E. D., Hibernation and longevity in the Turkish hamster *Mesocricetus brandti, Science*, 212, 668, 1981.

12. Cheal, M., The gerbil: A unique model for research in aging, *Exp. Aging Res.*, 12, 3, 1986.

13. Brownell, E., DNA/DNA hybridization studies of muroid rodents: Symmetry and rates of evolution, *Evolution*, 37, 1034, 1983.

14. Sacher, G. A. and Hart, R. W., Longevity, aging, and comparative cellular and molecular biology of the house mouse, *Mus musculus*, and the white-footed mouse, *Peromyscus leucopus*, in *Genetic Effects on Aging*, Bergsma, D. and Harrison, D. E., Eds., Alan R. Liss, New York, 1978, 71.

15. Smith, G. S., Crew, M. D., and Walford, R. L., *Peromyscus* as a gerontologic animal: Aging and the MHC, in *Genetic Effects on Aging II*, Harrison, D. E., Ed., Telford Press, Caldwell, NJ, 1990, 457.

16. Davidson, M. K., Lindsey, J. R., and Davis, J. K., Requirements and selection of an animal model, *Is. J. Med. Sci.*, 23, 551, 1987.

17. McCay, C. M., Crowell, M. F., and Maynard, L. A., The effects of retarded growth upon the length of life span and upon the ultimate body size, *J. Nutr.*, 10, 63, 1935.

18. Ross, M. H., Nutritional regulation of longevity, in *The Biology of Aging*, Behnke, A., Finch, C. E., and Moment, G. B., Eds., Plenum Press, New York, 1978, chap. 10.

19. Russell, E. S., Genetic considerations in the selection of rodent species and strains for research in aging, in *Development of the Rodent as a Model System of Aging*, Gibson, D. C., Ed, DHEW Publication No. 72-121, Washington, D.C., 1972, 33

20. Festing, M. F. W., Genetic quality control of laboratory animals used in aging studies, *Neurobiol. Aging*, 12, 673, 1991.

21. Phelan, J.P. and Austad, S. N., Selecting animal models of human aging: Inbred strains often exhibit less biological uniformity than F_1 hybrids, *J. Gerontol. Biol. Sci.*, 49, B1, 1994.

22. Munro, H. N., The challenges of research into nutrition and aging: Introduction to a multi-faceted problem, in *Nutrition, Aging, and the Elderly*, Munro, H.N. and Danford, D. E., Eds., Plenum Press, New York, 1989, chap. 1.

23. Coates, M. E., Nutritional considerations in the production of rodents for aging studies, *Neurobiol. Aging*, 12, 679, 1992.

24. Weisbroth, S. H., Pathogen-free substrates for gerontologic research: Review, sources and comparison of barrier-sustained versus conventional laboratory rats, *Exp. Gerontol.*, 7, 417, 1972.

25. Sebesteny, A., Necessity of a more standardized microbiological characterization of rodents for aging studies, *Neurobiol. Aging*, 12, 663, 1991.

26. van der Logt, J. T. M., Necessity of a more standardized virological characterization of rodents for aging studies, *Neurobiol. Aging*, 12, 669, 1991.

27. Keithley, E. M., Ryan, A. F., and Feldman, M. L., Cochlear degeneration in aged rats of four strains, *Hear. Res.*, 59, 171, 1992.

28. Gratton, M. A., Schmeidt, R. A., and Schulte, B., Age-related decreases in endocochlear potential are associated with vascular abnormalities in the stria vascularis, *Hear. Res.*, 94, 116, 1996.

29. Games, D., Adams, D., Allessandrini, R., Barbour, R., Berthelette, P., Blackwell, C., Carr, T., Clemens, J., Donaldson, T. Gillespie, F., Guido, T., Hagoplan, S., Johnson-Wood, K., Khan, K., Lee, M., Leibowitz, P., Lieberburg, I., Little, S., Masilah, E., McConlogue, L., Montoya-Zavala, M., Mucke, L., Paganini, L., Pennlman, E., Power, M., Schenk, D., Seubert, P., Snyder, B. Sorlano, F., Tan, H., Vitale, J., Wadsworth, S., Wolozin, B., and Zhao, J., Alzheimer-type neuropathology in transgenic mice overexpressing V717F beta-amyloid precursor protein, *Nature*, 373, 523, 1995.

30. Masoro, E. J., Food restriction in rodents: An evaluation of its role in the study of aging, *J. Gerontol. Biol. Sci.*, 43, B59, 1988.

31. Takeda, T., Hosokawa, M., and Higuchi, K., Senescence-accelerated mouse (SAM): a novel murine model of accelerated senescence, *J. Am. Geriatr. Soc.*, 39, 911, 1991.

32. Harrison,D.E., Potential misinterpretations using models of accelerated aging, *J. Gerontol. Biol. Sci.*, 49, B245, 1994.

11

Aging Experiments Using Nonhuman Primates

Ricki J. Colman and Joseph W. Kemnitz

CONTENTS

11.1 Introduction

The average life span is longer today than ever before. This has led to an overall "aging" of the population with a shift towards greater numbers of older people [1,2]. This, in turn, has translated into an increased need for better understanding of phenomena of later life. Much of this investigation involves elucidation of disease etiology and progression, treatment regimens, and characterization of normative aging. Legal, ethical, and practical restraints limit the extent to which human subjects can be used for this type of research. There is, therefore, a need for appropriate ways in which to model human aging and age-associated diseases. While study of many species can offer insight into these types of questions, non-human primates offer the unique opportunity to examine these issues in humans' closest phylogenetic relative.

Nonhuman primates have been used in many areas of aging research; more information about aging nonhuman primates is accumulating rapidly. Centers have been established around the U.S. specifically including studies of aging nonhuman primates (see Section 11.4.1.3 of this chapter). There have been many studies into age-related disease processes in nonhuman primates and there are currently three major studies examining the effect of nutritional intervention on the aging process. With the ever-growing quantity of information about aging in nonhuman primates, they are becoming a more valuable model for the study of human aging.

11.2 Need for Nonhuman Primate Models

11.2.1 Phylogeny

The marked similarities between human and nonhuman primates make nonhuman primates uniquely useful for providing insight into our own species. Nonhuman primates share approximately 90% of the human genome [3,4]. They are similar to humans in almost all aspects of their anatomy, physiology, and behavior [5,6]. Nonhuman primates also develop in similar ways to humans and undergo many of the same changes in anatomy, physiology, and behavior during maturity and later life. This makes them an often vital link between basic research and clinical application to human health issues.

11.2.2 Longevity

In choosing a model for aging research, one must consider several issues. The selected species should progress through similar stages of life (growth and maturation, reproduction, and postreproduction) as humans do, thus exhibiting a realistic aging course. In addition, the ideal model will exhibit a fair degree of time compression, i.e., it should progress through these stages at a more rapid pace than is seen in humans. For this reason among others, it is not always most appropriate to use humans' closest phylogenetic relatives, the great apes, in aging research. While the maximal life span of the great apes (~60 years) is shorter than that of humans, it is not sufficiently compressed to allow for expeditious studies of aging. The optimal model for aging research will exhibit a combination of a realistic course of aging compressed into a reasonable time frame for study.

Macaques are often used in gerontological investigations. These species exhibit both a realistic aging course, progressing through stages of life complementary to the human life cycle with considerable time compression. It is commonly thought that macaques exhibit an aging rate 2 1/2 to 3 times faster than that of humans [3,7]. In the most general terms, this conversion factor appears to be accurate, but it may not be appropriate to apply this conversion factor across all phases of the life span. The maximum life span of a rhesus macaque in captivity is approximately 40 years and the maximum achieved life span in humans is approximately 120 years. This would support a conversion factor of 3. Very small numbers of monkeys, however, have been maintained in the protective and supportive laboratory setting throughout their life spans; so a strong statistical estimate of maximum life span is not yet available.

When examined more closely, the scaling factor for aging may not be uniform across every stage of life. For example, female rhesus macaques undergo menarche at 2.5 to 3.5 years of age [8, 9], they reach adult stature at approximately 8 years of age [10], and they experience menopause between 26 and 28 years [11–13]. This relates to approximately four times the rate of aging from birth until sexual maturity, three times the rate of aging during young adulthood, and twice the rate of aging before menopause for rhesus monkeys compared to humans. It is important to consider this sliding scale of converting ages when comparing nonhuman primates with humans when exploring the use of macaques and other nonhuman primates in research on aging.

11.3 Taxonomy and Implications for Research on Aging

11.3.1 Comparative Studies in General

There are approximately 200 identified species of nonhuman primates living today [14, 15]. Of these, approximately 30 different species are used in biomedical and behavioral research [15], with Old World monkeys being the most common. In research on aging, macaques (*Macaca mulatta, M. nemestrina, M. fuscata, M. fascicularis*) and baboons (*Papio* sp.) predominate.

There is still incomplete information about maximum life span in nonhuman primates. In general, New World species have shorter life spans than Old World species. The most studied species, the rhesus macaque, may have a maximum life expectancy of 40–45 years. In the Wisconsin Regional Primate Research Center colony, the oldest rhesus macaques have lived into their 40th year. As more species of nonhuman primates spend their entire lives in captivity where they are provided with good nutrition and prompt veterinary care, and kept in an environment free from predation and with reduced likelihood of severe aggression, we will be able to obtain more accurate data on maximum life spans.

11.3.2 Primates

11.3.2.1 Old World Species

Among the Old World species of nonhuman primates are those with the longest life spans, viz., the apes and the monkeys used most frequently in research, macaques and baboons. Aside from these species, very little is known about potential maximum life span in Old World monkeys and apes.

The most life expectancy information on apes is for chimpanzees, due to long-term study of the populations in Tanzania at the Gombe National Park and in the Mahale mountains, and several captive research groups. In the wild, chimpanzees rarely survive past 35 years of age, but in captivity they may live well into their 6th decade of life [15], with persistent reproductive cycling in females until death [16]. In a study of behavioral correlates of aging in chimpanzees in the Mahale mountains, Tanzania, an "old age" group, was defined as having individuals greater than 41 years old, and it contained only 1 male and 2 females [17]. At Gombe National Park in Tanzania, the estimated maximum age at death is 33 years for males and 37 years for females [18]. In the wild, however, so few individuals survive to old age, that little information is available on potential maximum life span and behavior and physiology of older feral chimpanzees [19,20].

Among the Old World monkeys, the macaques and baboons have been used for research on aging because of relatively abundant background information and because they adapt well in captivity. The maximum life span for rhesus macaques has been estimated to be 30–40 years [3, 7, 21, 5, unpublished data from the Wisconsin Regional Primate Research Center], much longer than for macaques living in the wild [22,23]. Among over 1000 free-ranging, provisioned rhesus macaques of Cayo Santiago, the oldest animal lived to 26 years [24].

Like macaques, baboons (*Papio* sp.) experience a naturally occurring menopause at about 25 years of age [25] and are considered elderly after about 18 years. Captive baboons at the Southwest Foundation for Biomedical Research, San Antonio, TX have lived into their fourth decade [26].

11.3.2.2 New World Species

The New World monkeys, i.e., those native to Central and South America, have been less frequently used for research; therefore, less information is available on their life spans. The majority of the New World species have not been studied long enough or continuously enough for an accurate life span determination to be made.

The New World species used most often in research related to aging is the squirrel monkey (*Saimiri* sp.), mainly because these animals are susceptible to both spontaneous and diet-induced atherosclerosis [27]. The maximum life span for squirrel monkeys is estimated to be 20 years [28]. Wild marmosets and tamarins generally do not live to old age; in captivity they are estimated to have maximum life spans of 15–16 and 20–22 years, respectively [29].

11.4 Special Considerations for Using Nonhuman Primate Models

11.4.1 Availability

11.4.1.1 Feral vs. Captive-Born

A major impediment to the gerontological study in nonhuman primates has been the availability of suitable subjects. When studying aging, it is vital to have reliable information on the age of the animals being used. Unfortunately, there is no reliable way to precisely determine the individual age of a primate without knowing its birth date. In feral populations, this information is available only in situations where the animals have been under continuous observation since their births. Even then, fixing precise ages may not be easy. Individual animals must be reliably identified by those observing them or birth dates may be inaccurately assigned. Due to these and other limitations, there have not been many feral situations in which precise aging of the populations has been possible.

An alternative to the use of feral populations or captured animals is to use animals that have been born in captivity. Under these conditions, record keeping obviously is much easier and animals are generally marked to prevent misidentification. In addition, the development of captive programs allows for the purposeful breeding and maintenance of animals of the appropriate age and condition to be used in aging research. However, one must remember that captive animals may not behave or develop in the same way as their feral counterparts.

11.4.1.2 *Pedigree Information*

Another advantage of using captive-bred nonhuman primates or well-documented feral populations is the availability of pedigree information for the study animals. Information on an animal's pedigree is potentially invaluable for determining the genetic influences on aging or the genetic component of a disease process. In addition, family histories can be traced, purposeful and planned breeding can be carried out to minimize deleterious inbreeding, or to allow selective breeding for interesting or unique traits.

11.4.1.3 *Regional Primate Research Center Program*

Research is limited by the availability of sufficient numbers of appropriate animals. Relatively long periods of gestation and generation time slow establishment of captive colonies. In addition, with the exception of the marmosets and some prosimians, nonhuman primates as a rule give birth to singletons, further limiting animal availability. In the wild, even abundant species have suffered dramatically from the effects of habitat destruction and human encroachment.

In 1961, the U.S. Congress appropriated funds for the National Institutes of Health to establish a network of regional and national primate centers designed to solve human health problems through nonhuman primate research. Today seven facilities exist as the Regional Primate Research Centers (RPRCs) and provide a major scientific resource to the advancement of biomedical and behavioral research. The centers and their host institutions are the California RPRC (University of California at Davis), the New England RPRC (Harvard Medical School), the Oregon RPRC (Oregon Health Sciences University), the Tulane RPRC (Tulane University), the Washington RPRC (University of Washington), the Wisconsin RPRC (University of Wisconsin, Madison), and the Yerkes RPRC (Emory University).

Each of the seven Regional Primate Centers includes a resident staff of scientists, as well as technical and administrative personnel, and offers the opportunity for collaborating and visiting scientists. The primary thrust of these centers is the use of nonhuman primates for biomedical and basic biological and behavioral research. However, some species are maintained because of their unique status in the wild, or, in some cases, because they are threatened or endangered [30].

11.4.1.3.1 *Species*

The 7 RPRCs maintain colonies for research and breeding purposes of over 15,000 primates, representing 32 species and including representatives of the New World monkeys, Old World monkeys, lesser apes, and great apes. The rhesus monkey predominates in several of these centers due to its versatility as a research animal.

11.4.1.3.2   Aging-Related Programs

The seven Regional Primate Research Centers encompass a broad range of interests and many areas of specialty. In 1966, the need for long-range studies of the aging process was realized. In 1978 the National Institute on Aging and the Division of Research Resources considered the joint development of an aged nonhuman primate resource. At this time four primate centers, the Wisconsin RPRC, the Oregon RPRC, the California RPRC, and the Washington RPRC had substantial colonies of older macaques [30], and in 1980 these four centers began receiving support from the NIA for the long-term noninvasive study of the aging process. These aging colonies are set aside as a resource for biological, behavioral, social, and clinical studies on the processes, conditions, and characteristics relevant to the aging process. The Yerkes Center now also has an appreciable number of rhesus monkeys older than 20 years. Ongoing research utilizing these colonies is centered primarily around noninvasive assessment of age-related change in reproduction, vision, cognitive function, carbohydrate and lipid metabolism, immunology, and other aspects of normative aging.

At the Wisconsin RPRC the practice has been to assign healthy, intact rhesus monkeys 15 years of age or greater to this set-aside colony to maintain a pool of approximately 50 animals. To date, 149 monkeys have been so assigned. An analysis conducted on this selected sample revealed 50% survival to 27 years, 10% to 35 years, and the oldest to 39.8 years of age.

11.4.2   Housing Conditions

Primates are studied in environments from free-ranging to some degree of a more controlled condition. Controlled environments can range from geographically limited, free-ranging, provisioned locales (e.g., on islands), and large enclosed social settings (e.g., corrals) to highly controlled, individually caged conditions. Not all housing situations are appropriate for all animals and all types of research; certain types of research tend to require specific housing conditions. As a general principle, however, primates should be kept in social groups or at least have abundant opportunity for visual and auditory communication.

Carrying out aging research in a controlled environment has many obvious advantages. In feral conditions, many factors influence life span. Older animals may not get enough nutritious food or potable water due to problems with general availability, competition for access, or physical limitation of accessibility to the food resource. For example in the free-ranging rhesus macaques of Cayo Santiago, osteoarthritis and osteoporosis increase in frequency with age, presenting increased locomotor difficulties and a general decrease in arboreal activities, thereby limiting access to food resources [31]. In socially living primates, older animals may lose access to other vital resources, such as sleeping sites. In general, captive nonhuman primates live longer than wild or semifree populations [32].

11.4.3   Diet and Nutritional Needs

Diet, both immediate and lifelong, can influence many biological variables in ways that can be mistaken for an influence of aging, making nutrition an important variable in aging research. A major difficulty in using humans for this type of research is the difficulty of controlling for the quantity and type of food eaten. Self-reporting of dietary habits is a notoriously unreliable method of assessing intake.

Most captive colonies of monkeys are fed an open-formula commercial diet prepared as extruded biscuits. Such diets have guaranteed minimum values of macronutrients derived from variable grain and animal sources with vitamin and mineral supplements. Defined diets, i.e., similar to open-formula but with documented food values, are also used but less frequently because of the additional expense. Purified diets provide precisely controlled nutritional formulation, but they are seldom used except for specific experiments because of their considerable expense. Purified diets are generally produced in pelleted or liquid form with much less roughage than biscuit formulations or a natural diet.

Nutritional requirements for nonhuman primates are not fully understood. The National Research Council guidelines were revised in 1978 [33] and are based on studies of growing and young-adult individuals. Updating these standards is needed and inclusion of information for postmature monkeys would be helpful.

11.5 Overview of Current Knowledge

11.5.1 Nervous System and Behavior

In humans, there is an age-related decline in several aspects of cognitive function [34,35]. Obvious limitations on the use of humans in these types of studies have slowed investigation into the neurobiological factors underlying age-related cognitive decline. The development of animal models for the study of age-related cognitive decline has helped to circumvent some of these limitations and to provide a model for therapeutic intervention studies.

Nonhuman primates are excellent models for cognitive studies for several reasons. Primarily, of all the possible laboratory models, the nonhuman primate brain most closely resembles the human brain [32,36]. In addition, nonhuman primates have a diverse and well-studied behavioral repertoire and they have the ability to perform cognitive tasks that have been adapted from human neuropsychological tests [37,38,39].

Like humans, nonhuman primates experience an age-related decline in several domains of cognitive function [32,40,41]. Also, like aged humans, old macaques take longer to learn something new, but do not appear to forget what they learn any sooner [32]. Aged monkeys may be more sensitive to tasks stressing temporal and spatial memory and those characterized by a high level of stimulus interference [32].

In humans, memory loss associated with Alzheimer's disease is characterized by the formation of senile amyloid-containing plaques and the presence of neurofibrillary tangles. Macaques, orangutans, chimpanzees, squirrel monkeys, and lemurs have been shown to form amyloid plaques [42,43], but, unlike humans, they do not develop tangles or show evidence of the associated cortical neuron loss [44–53]. Importantly, when plaque density is compared to behavioral dysfunction, there is no correlation, confirming that plaque density does not account for the cognitive decline observed in aged monkeys [54,55]. In macaques, plaque development may generally begin at about 26 years of age, but in some cases it begins as early as 16 years of age [54,56]; between 25 and 30 years of age, approximately 50% of rhesus monkeys will have plaques and for animals over 31 years of age, the incidence of amyloid plaques is 71% [57].

11.5.1.1 Vision

Of all the possible animal models for aging systems, nonhuman primates are particularly appropriate for the study of vision and its alteration with age. This is because high-acuity, stereoscopic vision is an adaptation shared only among the primates [58]. Two common impairments of the visual system exhibited with advancing age in humans are presbyopia, or the pronounced loss of near vision, and the development of cataracts, or opacities of the lens. Both of these deficits are also experienced by nonhuman primates [59,60].

The lens of the rhesus macaque eye continues to grow throughout adulthood. Continual growth of the lens is considered to be one and possibly the primary contributing factor in the development of human presbyopia, making the rhesus macaque an ideal model for the study of this affliction [61]. Among the free-ranging macaques of Cayo Santiago, the incidence of presbyopia accompanied by decreased accommodative range is known to increase with advancing age. This reduction in visual acuity has behavioral correlates including slower locomotion, reduced arboreal activity, and altered grooming orientations [61]. Diminished visual ability could contribute to decreased performance with aging on visually-dependent psychological tests. Such tests can be sensitive to vision problems as opposed to motivational or cognitive issues.

Cataract development is a common complaint of aging humans. In a study examining cataract development in captive and free-ranging rhesus macaques at the Wisconsin Regional Primate Research Center and the Cayo Santiago facility of the Caribbean Primate Center, all captive monkeys that reached 28 years of age showed lenticular opacities indistinguishable from human senile cataracts [60]. The observed level of cataract development would be expected to impair visual acuity. The free-ranging macaques at Cayo Santiago did not exhibit frank senile cataracts, but the animals were not yet of the age in which severe cataracts were regularly seen in the Wisconsin population. Precataractus changes were detected in the free-ranging Cayo Santiago macaques that were similar to those seen in the age-matched group of the Wisconsin population [60]. The Wisconsin studies showed that cataract development had begun in 20% of rhesus monkeys at 20–22 years of age, and the rate of cataract development increased significantly after 26 years of age [62] with severe cataracts being detected by 26 years of age [43].

11.5.1.2 Behavior

There is little information available about the behavior of aging nonhuman primates. This is due mainly to the fact that there are very few free-ranging populations for which enough demographic data is available to make these types of studies possible. Also, few captive colonies have been in existence long enough to accumulate this type of data on animals of known age. There are several populations that do meet these criteria to some degree; the chimpanzees of the Mahale mountains and the Arashiyama macaques are two examples.

In a study of the effects of aging on social behavior within a group of feral chimps of the Mahale mountains, Tanzania, old adults [17] (41 years and older, n = 1 male, 2 females) showed a reduction in overall activity. This decrease was most evident in activities related to traveling and grooming, with only the decrease in grooming behavior showing statistical significance. Older animals had a reduced number of grooming partners, rarely engaged in mutual grooming and were groomed much more frequently and for longer periods than they groomed their partners. In addition, the older group appeared to have less contact with the core group of animals than their younger counterparts.

In general, studies of the behavioral correlates of aging in macaques have shown an overall reduction in activity with aging. The rhesus macaques of Cayo Santiago show a reduction in arboreal activities with increasing age. This is possibly caused by an increase in the incidence of osteoarthritis and osteoporosis in the older animals [31]. The Japanese macaques (*Macaca fuscata*) of Arashiyama are divided into two groups, the East group located in Kyoto, Japan and the West group that has been transplanted to Texas. Among the West group, 9 females, 16–29 years of age, were found to spend less time in general contact with and grooming other group members. In the East group, 14 females aged 11–29 years spent more time resting and engaged in fewer social interactions than younger members of the group [63].

Another study involving the West Arashiyama female Japanese macaques (n = 10, 18–28 years old) and female stumptail macaques (*Macaca arctoides*) housed in outdoor enclosures at the University of California, Riverside (n = 4, ≥18 years old) found that old females were generally less active and involved in fewer social interactions than young adult females. The older females typically avoided or kept enough distance from other individuals to decrease the possibility of interaction, but were neither excluded from social interactions nor outcompeted in rank-related situations. Old females selectively withdrew from social interactions and maintained their rank over younger members of the group [64].

Older socially housed captive rhesus monkeys (n = 6, ≥13 years of age) show enhanced sociopositive and reduced agonistic activities, groom more often, both active and passive, and are more rarely threatened, harassed, attacked, or spatially displaced. They are seldom involved in aggressive physical contact, but threaten and chase more often. Locomotor behavior decreased up to 10 years of age and then remained constant [65]. However, with the knowledge we have about maximum life span in rhesus monkeys, we know that the animals used in this study can not truly be considered old. A small number of truly old rhesus monkeys were extensively studied by Davis and colleagues [66].

Captive monkeys at the Wisconsin Regional Primate Research Center also exhibit decreased activity with increased age. In a study of male and female rhesus macaques of three age groups, 7–9 years (n = 6 males, 6 females), 13–17 years (n = 6 males, 6 females), and greater than 20 years (n = 6 males, 6 females), time spent resting was shown to increase with age, and time spent moving was lowest for the oldest group [67].

11.5.2 Endocrine System

11.5.2.1 *Reproduction*

Reproductive aging has been examined in several Old World primate species. Macaques, baboons, and chimpanzees show increased incidence of irregular menstrual cycles with advancing age and show endocrine and ovarian changes similar to human menopause [5,62,68–77] In addition, menopausal and older male nonhuman primates show a reduction in sexual behaviors [75,78]. In female free-ranging rhesus macaques of Raccoon Key and Key Lois, 2 islands located in the Florida Keys, live birth rates declined with age in females over 8 years of age [79].

The relationship between reproduction and aging is less well defined in New World primates. Signs of reduced fertility have been observed in tamarins over age 12 [80] and, as in Old World primates, tamarins show a reduction in the number of primary ovarian follicles with advancing age [81]. However, an examination of hormonal values indicates that tamarins can continue to ovulate up to at least age 17 [82].

Insulin resistance and diabetes mellitus frequently develop during middle-age in rhesus monkeys and it is typically accompanied by obesity [83–85]. Restriction of caloric intake and prevention of obesity protect against this endocrinopathy [86, 87].

11.5.3   Musculoskeletal System

Osteoarthritis, or degenerative joint disease (DJD), is the most common disease of joints, most often seen at weight-bearing joints. It is a painful noninflammatory disorder of movable joints leading to the degeneration of articular cartilage and the growth of osteophytes. In the U.S. alone, 16 million people exhibit symptoms severe enough to receive medical treatment and 40 million adults have positive radiographs of the hands and feet [88]. Aging is one of the main etiological factors commonly evoked for this disease.

Degenerative joint disease has been observed in most mammals including many nonhuman primate species. In a study of 55 adult rhesus macaques, DeRousseau measured the occurrence and severity of DJD at the hip and spine. The results indicated a significant and positive relationship between DJD and age in these animals [89]. Similar results were obtained in a study comparing DJD in rhesus monkeys and gibbons (*Hylobates lar*). In both species, DJD increased with increasing age; however, this trend was far more apparent in the macaque [90]. A study of 58 cynomolgus macaques revealed a high prevalence of osteoarthritic lesions morphologically similar to those seen in humans [91], leading the authors to propose this species as a good model for studying the human disease.

Osteopenia, or a net loss of bone, is a well-established characteristic of human aging, found in all human populations despite nutritional and biomechanical differences [92]. Data on gross skeletal changes with aging have demonstrated substantial modification of the skeleton in late adulthood, characterized by loss of skeletal mass throughout the body [93–95]. Age-related osteopenia has been reported in numerous macaque species [92,96–103], baboons [76], chimpanzees [104], and saddleback tamarins [105]. Macaque and baboon data are particularly informative and are often used for human postmenopausal osteoporosis studies. Both human and nonhuman primate females undergo menopause [11], and both show rapid and substantial bone loss following bilateral ovariectomy [76,102].

Human studies suggest there is a 30–40% decrease in muscle mass between the 3rd and 8th decade of life [106]. This loss of net muscle mass, or sarcopenia, is the major factor in the age-related decline in muscle strength rather than a deterioration in the intrinsic capacity of the muscle cells to generate force [107].

Unfortunately, there is little direct information on muscle-specific changes related to aging in nonhuman primates. The rhesus monkey would make a good candidate for this type of study due to its phylogenetic proximity to humans. In a comparative study of captive male and female rhesus macaques of 3 age groups (6–9 years, 15–19 years, and 26–30 years), total body lean tissue mass was markedly lower in the oldest group [108]. The measurement of total body lean tissue mass includes skeletal muscle mass, but other factors are involved as well (e.g., abdominal and thoracic organ). A detailed analysis of fiber types from several musles has been reported for rhesus monkeys from 4–31 years of age, but very few animals were studied [109]. To answer questions of age-related muscle loss, more specific measurements of skeletal muscle are needed in more monkeys.

11.5.4   Immune System

In all mammals thus far studied, there is a decline in immune function with age [110,111]. In humans and rodents, decline in immune function correlates with the involution of the thymus gland and the concomitant decrease in T cell function [110]. This decline is often considered a prerequisite for the increase in infections and tumors seen in the geriatric human population.

Very little information is available on immune function across the life span in a nonhuman primate model. In rhesus macaques, data supports a reduction with age in specific antibody response to vaccines, most likely indicative of diminished T-helper cell function, and a reduction in natural killer cell activity [112]. Other research has corroborated this T cell impairment with age in the rhesus macaque [113].

11.6 Ongoing Interventionistic Studies

11.6.1 Dietary Restriction

Dietary restriction (DR), or undernutrition without malnutrition, is the only intervention which has repeatedly and strikingly increased the maximum life span and retarded the rate of aging in laboratory rodents. The ability of DR to increase life span has also been shown in fish, spiders, water fleas, and other lower animals [114,115].

Study into dietary restriction dates to the early 1900s. The first studies reported reduced tumor growth under conditions of severe DR [114]. These were followed by numerous studies documenting DR's effectiveness in reducing cancer and increasing life span in mice and rats [116–118]. An ongoing phase of research, begun in 1970, shows that rodents on DR stay younger longer [114,119]. Another phase of DR research concerns the mechanisms by which DR retards disease and aging in rodents. Finally, the study of DR in primates has recently begun with ongoing projects at the National Institute on Aging (NIA) [120] and at the Wisconsin Regional Primate Research Center [121].

Approximately 165 rhesus monkeys, both males and females, are being used in the two primate DR studies. To date, results from these two studies are remarkably similar even given differences in experimental design. Two major differences between these studies are in the age of the animals at the onset of DR and the way in which the reduced diet was calculated. The NIA study began with animals in three age groups; juvenile, adult, and old. The juveniles were all clearly still growing and it is likely that the "adult" males (3–5 years old at onset of DR) were still growing as well. The Wisconsin study began with all fully adult animals (8–14 years of age at onset of DR).

In the NIA study food allotments were based upon National Research Council Guidelines [120], while for the Wisconsin study, individual baseline intakes were determined for each animal [121]. Both studies then imposed a 30% restriction on the subjects.

As predicted from rodent studies, monkeys on DR show smaller body size and altered body composition. Significant reductions are seen rapidly in body weight, total body fat mass, and total body percent fat [122–124]. Reductions in lean body mass are seen later in DR [125–127]. In the NIA animals that were still growing, the DR animals showed a slowed rate of weight gain resulting in smaller animals [127,128].

Metabolic rate as estimated by indirect calorimetry shows a significant reduction with the onset of DR [123] that is no longer different when adjusted for body size, following adjustment to the diet regimen [124,125]. These findings are consistent with rodent studies showing a transient adaptation period with lower energy expenditure [129].

Rodent studies have repeatedly shown the positive effects of DR on glucoregulation [130,131]. Both of the nonhuman primate DR studies have also shown this positive effect of lower serum insulin and glucose levels and higher insulin sensitivity [122,132]. The improvements in body composition and glucoregulation that occur with DR in nonhuman primates make these animals less susceptible to age-related diseases like diabetes and cardiovascular disease.

Rhesus macaques are the subject of another study similar to the DR studies just discussed. In this study [133], animals have been weight-clamped, or maintained at a certain given body weight, for approximately nine years, and were examined for glucoregulatory parameters. As in the other two primate dietary restriction studies, this study showed a decreased ability to glucoregulate in the control group when compared to the weight-clamped group. Another recent study of DR uses cynomolgus monkeys [134]. These monkeys are being fed an atherogenic diet either *ad libitum* or restricted by 30%.

11.6.2   Biomarkers

Conventionally, in short-lived species, the effectiveness of aging interventions is determined through the examination of survival curves. When aging interventions are employed in a nonhuman primate model, the life span time compression factor is not great enough to reasonably allow for the analysis of survival curves. For example, to use this technique in a study of Old World monkeys, the study would have to last for a minimum of 15 years. To examine survival curves from macaques, the most commonly used primate in aging research, a study would have to last approximately 35 years [5,135]. As an alternative, the variable of interest could be the rate of aging.

The rate of aging presumably indicates an individual's susceptibility to death from normal aging processes. It should be possible to determine the rate of aging within a much shorter time frame than is necessary to establish a survival curve. However, since aging is a complex process involving an organism's interaction with its environment, the determination of a rate of aging should be based upon the least environmentally plastic phenotypic markers available. In addition, different markers could be changing at different rates within the same individual, making it important to measure a panel of markers rather than an isolated characteristic, to place confidence in the resulting estimated aging rate [136, 137].

The National Institute on Aging mandated a drive to develop research approaches to extending the productive years of life in humans. In response to this mandate, ways to explore the effectiveness of aging interventions had to be developed [138]. This led to the development of a colony of three mouse genotypes and three rat genotypes by the National Institute on Aging for the purpose of biomarker exploration. In addition, contract support was given for the establishment of this type of study in a nonhuman primate model, the pigtailed macaque.

Much information has been gathered on the rate of aging in pigtailed macaques (*Macaca nemestrina*) at the Washington Regional Primate Research Center, via the longitudinal assessment of a large panel of biomarkers of aging [135,139,140]. A large-scale, cross-sectional study of a broad panel of potential biomarkers was an initial step. Of this group of approximately 300 characteristics, 12 showed significant longitudinal change. Three of the remaining twelve variables were dropped due to dependencies with other variables, leaving a panel of nine biomarkers of aging [135]. Work with this species continues, most recently with the analysis of the role of antioxidants as biomarkers of aging [141]. This study added understanding to the concept that a rate of biological aging can be measured in a long-lived species, with the aid of appropriate biomarkers of aging. It is important to keep in mind that the panel of biomarkers that is established for a given study is not necessarily universal. The species, sex, and housing conditions, among other variables, are likely to influence the appropriateness and usefulness of biomarkers of aging.

11.7 Future Directions

In the past three decades much information on primate aging has been accumulated. The information we now have and the techniques that have been developed can be used to discover more about primate aging and to develop even better nonhuman primate models for human aging and age-associated diseases.

We can still do much to improve the nonhuman primate model of aging. Specific pathogen-free (SPF) colonies have recently been established. These colonies can be used for research without fear of disease transmission, and without the concern of pathogens internal to the colony interfering with the system under study. Ever-advancing techniques for genetic mapping, leading to the mapping of the entire nonhuman primate genome, will greatly enhance our knowledge of primate genetics, disease inheritance pattern, and genetic component of disease progression.

The development of a broad base of knowledge regarding nonhuman primate nutritional, housing, enrichment, and health care requirements is an important area that needs more detailed exploration. These factors are all essential for the maintenance of healthy captive populations of nonhuman primates and for exploring their maximum life spans.

Because aging nonhuman primates are a limited resource, much can be learned from sharing blood samples, tissues, knowledge, and experience. The set-aside colonies for studies of aging at the Regional Primate Research Centers present one such opportunity.

Acknowledgments

The authors gratefully acknowledge the support of NIH grants RR00167 and AG11915, as well as supplements to RR00167 from the NIA to characterize normative aging in rhesus monkeys. This is publication number 37-007 of the Wisconsin Regional Primate Research Center.

References

1. Schneider, E. L. and Guralnik, J. M., The aging of America: impact on health care costs, *J. Am. Med. Assoc.*, 263, 2335, 1990.
2. Golini, A. and Lori, A., Aging of the population: demographic and social changes, *Aging*, 2, 319, 1990.
3. King, F. A., Yarbrough, C. J., Anderson, D. C., Gordon, T. P., and Gould, K. G., Primates, *Science*, 240, 1475, 1988.
4. Sibley, C. G. and Ahlquist, J. E., DNA hybridization evidence of hominoid phylogeny: results from an expanded data set, *J. Molec. Evol.*, 26, 99, 1987.
5. Bowden, D. M. and Williams, D. D., Aging, in *Research on Nonhuman Primates*, Cornelius, E. E., Simpson, C. F., and Hendricks, A. G., Eds., Academic Press, New York, 1984, 305.
6. Ordy, J. M., Neurobiology and aging in nonhuman primates, *Behav. Biol.*, 16, 575, 1975.

7. Tigges, J., Gordon, T. P., McClure, H. M., Hall, E. C., and Peters, A., Survival rate and life span of rhesus monkeys at the Yerkes regional primate research center, *Am. J. Primatol.*, 15, 263, 1988.
8. Terasawa, E., Nass, T. E,. Yeoman, R. R., Loose, M. D., and Schultz, N. J., Hypothalmic control of puberty in the female rhesus macaque, in *Neuroendocrine Aspects of Reproduction*, Norman, R. L., Ed., Academic Press, New York, 1983, 149.
9. Terasawa, E., Claypool, L. E., Gore, A. C., and Watanabe, G., The timing of the onset of puberty in the female rhesus monkey, in *Control of the Onset of Puberty III. Proceedings of the 3rd International Conference*, Delemarre-van de Waal, H. A., Plant, T. M., Van Rees, G. P., and Schoemaker, J., Eds., Elsevier, New York, 1989, 123.
10. Trotter, M., Hixon, B. B., and Deaton, S. S., Sequential changes in weight of the skeleton and in length of long limb bones of *Macaca mulatta*, *Am. J. Phys. Anthropol.*, 43, 79, 1975.
11. Walker, M. L., Menopause in female rhesus monkeys, *Am. J. Primatol.*, 35, 59, 1995.
12. Kemnitz, J. W., Uno, H., Eisele, S. G., Browne, M. A., and Kudia, S. C., Menopause in rhesus monkeys, *The North American Menopause Society Meeting*, 1996.
13. Gilardi, K. V. K., Shideler, S. E., Valverde, C. R., Robers, J. A., and Lasley, B. L., Characterization of the onset of menopause in the rhesus macaque, *Biology of Reproduction*, 1997.
14. Napier, J. R. and Napier, P. H., *The Natural History of the Primates*, MIT Press, Cambridge, 1985.
15. King, F. A., Yarbrough, C. J., Anderson, D. C., Gordon, T. P., and Gould, K. G., Primates, *Science*, 240, 1475, 1988.
16. Graham, C. E., Reproductive function in aged female chimpanzees, *Am. J. Phys. Anthropol.*, 50, 291, 1979.
17. Huffman, M. A., Some socio-behavioral manifestations of old age, in *The Chimpanzees of the Mahale Mountains; Sexual and Life History Strategies*, Nishida, T., Ed., University of Tokyo Press, Tokyo, 1990, 237.
18. Teleki, G., Hunt, E. E., and Pfifferling, J. H., Demographic observations (1963–1973) on the chimpanzees of Gombe National Park, Tanzania, J. Hum. Evol., 5, 559, 1976.
19. Goodall, J., *The Chimpanzees of Gombe, Patterns of Behaviour*, Belknap Press of Harvard University Press, Cambridge, 1986.
20. Hiraiwa-Hasegawa, M., Hasegawa, T., and Nishida, T., Demographic study of a large-sized unit-group of chimpanzees in the Mahale Mountains, Tanzania: a preliminary report, *Primates*, 25, 401, 1984.
21. Dyke, B., Gage, T. B., Mamelka, P. M., Goy, R. W., and Stone, W. H., Ademographic analysis of the Wisconsin Regional Primate Center Rhesus Colony, 1962–1982, *Am. J. Primatol.*, 10, 257, 1986.
22. Smith, D. G., A comparison of the demographic structure and growth of free-ranging and captive groups of rhesus monkeys (*Macaca mulatta*), *Primates*, 23, 24, 1982.
23. Dittus, W. P. J., Population dynamics of the toque monkey, *Macaca sinica.*, in *Socioecology and Psychology of Primates*, Tuttle, R.H., Ed., Aldine Publishing Company, Chicago, 1975, 125.
24. Rawlins, R. G., Kessler, M. J., and Turnquist, J. E. D., Reproductive performance, population dynamics and anthropometrics of the free-ranging Cayo Santiago rhesus macaques, *J. Med. Primatol.*, 13, 147, 1984.
25. Carey, K. D. and Rice, K., The aged female baboon as a model of menopause, paper presented at the 16th Congress of the International Primatological Society, 19th Conference of the American Society of Primatologists, 1996.
26. Kammerer, C. M., Sparks, M. L., and Rogers, J., Effects of age, sex, and heredity on measures of bone mass in baboons (*Papio hamadryas*), *J. Med. Primatol.*, 24, 236, 1995.
27. Jokinen, M. P., Clarkson, T. B., and Prichard, R. W., Animal models in atherosclerosis research, *Exp. Molec. Pathol.*, 42, 1, 1985.
28. Jones, M. L., Longevity of primates in captivity, *Int. Zoo Yearbook*, 8, 183, 1968.
29. Bowden, D. M. and Jones, M. L., Aging research in nonhuman primates, in *Aging in Nonhuman Primates*, D. M., Ed., Van Nostrand Reinhold Company, New York, 1979, 1.
30. Dukelow, M. R., *The Alpha Males: An Early History of the Regional Primate Research Centers*, University Press of America, Lanham, 1995.

31. DeRousseau, C. J., Rawlins R. G., and Denlinger, J. L., Aging in the musculoskeletal system of rhesus monkeys: I. Passive joint excursion, *Am. J. Phys. Anthropol.*, 61, 483, 1983b.

32. Peters, A., Rosene, D. L., Moss, M. B., Kemper, T. L., Abraham, C. R., Tigges, J., and Albert, M. S., Neurobiological bases of age-related cognitive decline in the rhesus monkey, *J. Neuropathol. Exp. Neurol.*, 55, 861, 1996.

33. Panel on Nonhuman Primate Nutrition, National Research Council, Nutrient Requirements of Nonhuman Primates, National Academy of Sciences, Washington, D.C., 1978.

34. Albert, M. S., Cognitive function, in *Geriatric Neuropsychology*, Albert, M. S., and Moss, M. B., Eds., Guilford Press, New York, 1988, 33.

35. Albert, M. S., Neuropsychological and neurophysiological changes in healthy adult humans across the age-range, *Neurobiol. Aging*, 14, 623, 1993.

36. King, F. A. and Yarbrough, C., Nonhuman primates in research: a review of their critical role, *Lab. Anim.*, 24, 28, 1995.

37. Mishkin, M., Neural circuitry underlying behavioral deficits in aging, *Neurobiol. Aging*, 14, 615, 1993.

38. Zola-Morgan, S. and Squire, L. R., Medial temporal lesions in monkeys impair memory on a variety of tasks sensitive to human amnesia, *Behav. Neurosci.*, 99, 22, 1985.

39. Voytko, M. L., Functional and neurobiological similarities of aging in monkeys and humans, *Age*, 20 (1), 29, 1997.

40. Bartus, R. T., General overview: past contributions and future opportunities using aged nonhuman primates, *Neurobiol. Aging*, 14, 711, 1993.

41. Dean, R. L. and Bartus, R. T., The nonhuman primate. Behavioral and pharmacological models of geriatric cognitive dysfunction, in *Handbook of Psychopharmacology*, Iverson, L., Iverson, S. D., and Snyderm S. H., Eds., Plenum Press, New York, 1988, 325.

42. Heilbroner, P. L. and Kemper, T. L., The cytoarchitectonic distribution of senile plaques in three aged monkeys, *Acta Neuropathol.*, 81, 60, 1990.

43. Uno, H., The incidence of senile plaques and multiple infarction in aged macaque brain, *Neurobiol. Aging*, 14, 673, 1993.

44. Rapp, P. R., Visual discrimination and reversal learning in the aged monkey (*Macaca mulatta*), *Behav. Neurosci.*, 104, 876, 1990.

45. Lai, Z. C., Moss, M. B., Killiany, R. J., Rosene, D. L., and Hemdon, J. G., Executive system dysfunction in the aged monkey: Spatial and object reversal learning, *Neurobiol. Aging*, 16, 947, 1995.

46. Moss, M. B., The longitudinal assessment of recognition memory in aged rhesus monkeys, *Neurobiol. Aging*, 14, 635, 1993.

47. Gearing, M., Rebeck, G. W., Hyman, B. T., Tigges, J., and Mirra, S. S., Neuropathology and apolipoprotein E profile of aged chimpanzees: Implications for Alzheimer disease, *Proc. Natl. Acad. Sci. U.S.A.*, 91, 9382, 1994.

48. Selkoe, D. J., Bell, D. S., Podlisny, M. B., and Price, D. L., Conservation of brain amyloid proteins in aged mammals and humans with Alzheimer's disease, *Science*, 235, 873, 1987.

49. Walker, L. C., Kitt, C. A., Schwam, E., Buckwald, B., Garcia, F., Sepinwall, J., and Price, D. L., Senile plaques in aged squirrel monkeys, *Neurobiol. Aging*, 8, 291, 1987.

50. Walker, L. C., Masters, C., Beyreuther, K., and Price, D. L., Amyloid in the brains of aged squirrel monkeys, *Acta Neuropathol.*, 80, 381, 1990.

51. Bons, N., Mestre, N., and Petter, A., Senile placques and neurofibrillary changes in the brain of an aged lemurian primate, *Microcebus murinus, Neurobiol. Aging*, 13, 99, 1991.

52. Bons, N., Mestre, N., Ritchie, K., Petter, A., Podlisny, M., and Selkoe, D., Identification of amyloid beta protein in the brain of the small, short-lived lemurian primate *Microcebus murinus, Neurobiol. Aging*, 15, 215, 1994.

53. Uno, H., Alsum, P. B,. Dong, S., Richardson, R., Zimbric, M. L., Thieme, C. S., and Houser, W. D., Cerebral amyloid angiopathy and placques, and visceral amyloidosis in aged macaques, *Neurobiol. Aging*, 17, 275, 1996.

54. Struble, R. G., Price, D. L., Cork, L. C., and Price, D. L., Senile plaques in cortex of aged normal monkeys, *Brain Res.*, 361, 267, 1985.

55. Cork, L. C., Placques in prefrontal cortex of aged, behaviorally-tested rhesus monkeys: incidence, distribution, and relationship to task performance, *Neurobiol. Aging*, 14, 675, 1993.

56. Cork, L. C., Masters, C., Beyreuther, K., and Price, D. L., Development of senile plaques. Relationships of neuronal abnormalities and amyloid deposits, *Am. J. Pathol.*, 137, 1383, 1990.

57. Uno, H., The incidence of senile plaques and multiple infarction in aged macaque brain, *Neurobiol. Aging*, 14, 673, 1993.

58. DeRousseau, C. J., Comparative aspects of aging in nonhuman primate, *Assoc. Anthropol. Gerontol.*, 4, 3, 1983a.

59. Bito, L. Z., DeRousseau, C. J., Kaufman, P. L., and Bito, J. W., Age-dependent loss of accomodative amplitude in rhesus monkeys: an animal model for presbyopia, *Invest. Ophthalmol. Visual Sci.*, 23, 23, 1982.

60. Kaufman, P. L. and Bito, L. Z., The occurrence of senile cataracts, ocular hypertension and glaucoma in thesus monkeys, *Exp. Eye Res.*, 34, 287, 1982.

61. DeRousseau, C. J., Bito, L. Z., and Kaufman, P. L., Age-Dependent impairments of the rhesus monkey visual and musculoskeletal systems and apparent behavioral consequences, in *The Cayo Santiago Macaques: History, Behavior & Biology*, Rawlins, R. G. and Kessler, M. J., Eds., State University of New York Press, Albany, 1986, 233.

62. Uno, H., Age-related pathology and biosenescent markers in captive rhesus macaques, *Age*, 20, 1, 1997.

63. Nakamichi, M., Aging and behavioral changes of female Japanese macaques, *Res. Rep. Arashiyama West East Groups Japanese Monkeys*, 87, 1988.

64. Hauser, M. D. and Tyrrell, G., Old age and its behavioral manifestations: a study on two species of macaque, *Folia Primatol.*, 43, 24, 1984.

65. Heydecke, H., Schwibbe, M., and Kaumanns, W., Studies on social behaviour of aging rhesus monkeys (*Macaca mulatta*), *Primate Rep.*, 15, 41, 1986.

66. Davis, R. T. and Leathers, C. W., Eds., *Behavior and Pathology of Aging in Rhesus Monkeys*, *Monographs in Primatology*, Vol. 8, Alan R. Liss, New York, 1985.

67. Lukas, J. J., Baum, S. T., and Kemnitz, J. W., Older monkeys have lower energy expenditure and less lean body mass, *Gerontologist*, 33, 235, 1993.

68. Van Wagenen, G., Menopause in a subhuman primate, *Anat. Rec.*, 166, 392, 1970.

69. Van Wagenen, G., Vital statistics from a breeding colony, *J. Med. Primatol.*, 1, 3, 1972.

70. Hodgen, G. D., Goodman, A. L., O'Connor, A., and Johnson, D. K., Menopause in rhesus monkeys: model for study of disorders in the human climacteric, *Am. J. Obstet. Gynecol.*, 127, 581, 1977.

71. Graham, C. E., Reproductive function in aged female chimpanzees, *Am. J. Phys. Anthropol.*, 50, 291, 1979.

72. Graham, C. E., Kling, O. R., and Steiner, R. A., Reproductive senescence in female nonhuman primates, in *Aging in Nonhuman Primates*, Bowden, D. M., Ed., Van Nostrand Reinhold Company, New York, 1979, 183.

73. Gould, K. G., Flint, M., and Graham, C. E., Chimpanzee reproductive senescence: a possible model for evolution of the menopause, *Maturitas*, 3, 157, 1981.

74. Dierschke, D., Temperature changes suggestive of hot flushes in rhesus monkeys: Preliminary observations, *J. Med. Primatol.*, 14, 271, 1985.

75. Short, R., England, N., Bridson, W. E., and Bowden, D. M., Ovarian cyclicity, hormones, and behavior as markers of aging in female pigtailed macaques (*Macaca nemestrina*), *J. Gerontol.*, 44, B131, 1989.

76. Aufdemorte, T. B., Fox, W. C., Miller, D., Buffum, K., Holt, G. R., and Carey, K. D., A nonhuman primate model for the study of osteoporosis and oral bone loss, *Bone*, 14, 581, 1993.

77. Walker, M. L., Menopause in female rhesus monkeys, *Am. J. Primatol.*, 35, 59, 1995.

78. Phoenix, C. H. and Chambers, K. C., Sexual behavior in aging male rhesus monkeys, in *Advanced Views in Primate Biology*, Chiarelli, A. B. and Corruccini, R. S., Eds., Springer-Verlag, Berlin, 1982, 95.

79. Johnson, R. L. and Kapsalis, E., Ageing, infecundity and reproductive senescence in free-ranging female rhesus monkeys, *J. Reprod. Fertil.*, 105, 271, 1995.

80. Tardif, S. D. and Clapp, N. K., Reproductive longevity in tamarin females (*Saguinus* sp.), *Lab. Anim. Sci.*, 34, 504, 1984.

81. Tardif, S. D., Histologic evidence for age-related differences in ovarian function in tamarins (*Saguinus* sp. Primates), *Biol. Reprod.*, 33, 993, 1985.

82. Tardif, S. D. and Ziegler, T. E., Features of female reproductive senescence in tamarins (*Saguinus* spp.), a New World primate, *J. Reprod. Fertil.*, 94, 411, 1992.

83. Hamilton, C. L. and Ciaccia, P. J., The course of development of glucose intolerance in the monkey (*Macaca mulatta*), *J. Med. Primatol.*, 7, 165, 1978.

84. Hansen, B. C. and Bodkin, N. L., Heterogeneity of insulin responses: phases leading to type 2 (non-insulin-dependent) diabetes mellitus in the rhesus monkey, *Diabetologia*, 29, 713, 1986.

85. Kemnitz, J. W. and Francken, G. A., Characteristics of spontaneous obesity in male rhesus monkeys, *Physiol. Behav.*, 38, 477, 1986.

86. Kemnitz, J. W., Roecker, E. B., Weindruch, R., Elson, D. F., Baum, S. T., and Bergman, R. N., Dietary restriction increases insulin sensitivity and lowers blood glucose in rhesus monkeys, *Am. J. Physiol. Endocrinol. Metabol.*, 266, E540, 1994.

87. Bodkin, N. L., Ortmeyer, H. K., and Hansen, B. C., Long-term dietary restriction in older-aged rhesus monkeys: effects on insulin resistance, *J. Gerontol.*, A50, B142–147, 1995.

88. Tonna, E. A., Arthritis, in *The Encyclopedia of Aging*, Maddox, G. L., Ed., Springer-Verlag, New York, 1987.

89. DeRousseau, C. J., Aging in the musculoskeletal system in rhesus monkeys: degenerative joint disease, *Am. J. Phys. Anthropol.*, 67, 177, 1985a.

90. DeRousseau, C. J., Osteoarthritis in rhesus monkeys and gibbons, A locomotor model of joint degeneration, in *Contributions to Primatology*, Vol. 25, Szalay, F. S., Eds., Karger, New York, 1988.

91. Carlson, C. S., Loeser, R. F., Jayo, M. J., Weaver, D. S., Adams, M. R., and Jerome, C. P., Osteoarthritis in cynomolgus macaques: a primate model of a naturally occurring disease, *J. Orthopaed. Res.*, 12, 331, 1994.

92. DeRousseau, C. J., Aging in the musculoskeletal system in rhesus monkeys: Bone loss, *Am. J. Phys. Anthropol.*, 68, 157, 1985b.

93. Smith, D. M., Johnston, C. L., Jr., and Yu, P., In vivo measurement of bone mass, *J. Am. Med. Assoc.*, 219, 325, 1972.

94. Morgan, D. B., The metacarpal bone: A comparison of the various indices for the assessment of the amount of bone and for the detection of loss of bone, *Clin. Radiol.*, 24, 77, 1973.

95. Reff, M. E. and Schneider, E. L., biological markers of aging, *Proceedings of Conference on Nonlethal Biological Markers of Physiological Aging*, NIH Publication No. 82-2221, Washington, D.C., 1982.

96. Williams, D. D. and Bowden, D. M., A nonhuman primate model for the osteopenia of aging, in *Comparative Pathobiology of Major Age-Related Diseases: Current Status and Research Frontiers*, Scarpelli, D. G. and Migaki, G., Eds., Alan R. Liss, New York, 1984, 207.

97. Bowden, D. M., Teets, C., Witkin, J., and Young, D. M., Long bone calcification and morphology, in *Aging in Nonhuman Primates*, Bowden, D. M., Ed., Van Nostrand Reinhold Co., New York, 1979, 335.

98. Przybeck, T. R., Histomorphology of the rib: Bone mass and cortical remodeling, in *Behavior and Pathology of Aging Rhesus Monkeys*, Davis, R. T. and Leatheus, C. W., Eds., Alan R. Liss, New York, 1985, 303.

99. Kessler, M. J., Rawlins, R. G., and London, W. T., The hemogram, serum biochemistry, and electrolyte profile of aged rhesus monkeys (*Macaca mulatta*), *J. Med. Primatol.*, 12, 184, 1986.

100. Pope, N. S., Gould, K. G., Anderson, D. C., and Mann, D. R., Effects of age and sex on bone density in the rhesus monkey, *Bone*, 10, 109, 1989.

101. Grynpas, M. D., Huckell, C. B., Reichs, K. J., DeRousseau, C. J., Greenwood, C., and Kessler, M.J., Effect of age and osteoarthritis on bone mineral in rhesus monkey vertebrae, *J. Bone Miner. Res.*, 8, 909, 1993.

102. Jerome, C. P., Carlson, C. S., Register, T. C., Bain, F. T., Jayo, M. J., Weaver, D. S., and Adams, M. R., Bone functional changes in intact, ovariectomized, and ovariectomized, hormone-supplemented adult cynomolgus monkeys (*Macaca fascicularis*) evaluated by serum markers and dynamic histomorphometry, *J. Bone Miner. Res.*, 8, 527, 1994.

103. Champ, J. E., Binkley, N., Havighurst, T., Colman, R. J., Kemnitz, J. W., and Roecker, E. B., The effect of advancing age on bone mineral content of the female rhesus monkey, *Bone*, 19, 485, 1996.

104. Sumner, D. R., Morbeck, M. E., and Lobick, J. J., Apparent age-related bone loss among adult female gombe chimpanzees, *Am. J. Phys. Anthropol.*, 79, 225, 1989.

105. Power, R. A., Age-related effects of life-history variables on bone mineral status in saddleback tamarins, *Am. J. Phys. Anthropol.*, Suppl. 12, 145, 1991.

106. Kohrt, W. M., and Holloszy, J. O., Loss of skeletal muscle mass with aging: Effect on glucose tolerance, *J. Gerontol.*, 50A, 68, 1995.

107. Evans, W. J. and Campbell, W. W., Sarcopenia and age-related changes in body composition and functional capacity, J. Nutr., 123, 465, 1993.

108. Hudson, J. C., Baum, S. T., Frye, D. M. D., Roecker, E. B., and Kemnitz, J. W., Age and sex differences in body size and composition during rhesus monkey adulthood, *Aging Clin. Exp. Res.*, 8, 197, 1996.

109. Hegreberg, G. A. and Hamilton, M. J., An age series study of skeletal muscle morphology of rhesus monkeys, in *Behavior and Pathology of Aging in Rhesus Monkeys, Monographs in Primatology*, Vol. 8, Davis, R. T., and Leathers, C. W., Eds., Alan R. Liss, New York, 1985, 327.

110. Makinodan, T. and Kay, M. M. B., Age influence on the immune system, *Adv. Immunol.*, 29, 287, 1980.

111. Walford, R. L., Immunologic theory of aging. Current status, *Fed. Proc.*, 33, 2020, 1974.

112. Ershler, W. B., Coe, C. L., Gravenstein, S., Schultz, K. T., Klopp, R. G., Meyer, M., and Houser, W. D., Aging and immunity in nonhuman primates: I. Effects of age and gender on cellular immune function in rhesus monkeys (*Macaca mulatta*), *Am. J. Primatol.*, 15, 181, 1988.

113. Eylar, E. H., Molina, F., Quinones, C., Zapata, M., and Kessler, M., Comparison of mitogenic responses of young and old rhesus monkey T cells to lectins and interleukins 2 and 4, *Cell. Immunol.*, 121, 328, 1989.

114. Weindruch, R. and Walford, R. L., *The Retardation of Aging and Disease by Dietary Restriction*, Charles C. Thomas, Springfield, IL, 1988.

115. Weindruch, R., Caloric restriction and aging, *Sci. Am.*, 274, 46, 1996.

116. McCay, C. M., Crowell, M. F., and Maynard, L., The effect of retarded growth upon the length of life and upon ultimate size, *J. Nutr.*, 10, 63, 1935.

117. Tannenbaum, A., The initiation and growth of tumors I. Effects of underfeeding, *Am. J. Cancer*, 38, 335, 1940.

118. Visscher, M. B., Ball, Z. B., Barnes, R. H., and Sivertsen, I., The influence of caloric restriction upon the incidence of spontaneous mammary carcinoma in mice, *Surgery*, 11, 48, 1942.

119. Yu, B. P., Food restriction research: Past and present status, *Rev. Biol. Res. Aging*, 4, 349, 1990.

120. Ingram, D. K., Cutler, R. G., Weindruch, R., Renquist, D. M., Knapka, J. J., April, M., Belcher, C. T., Clark, M. A., Hatcherson, C. D., Marriott, B., and Roth, G. S., Dietary restriction and aging: the initiation of a primate study, *J. Gerontol.*, 45, B148, 1990.

121. Kemnitz, J. W., Weindruch, R., Roecker, E. B., Crawford, K., Kaufman, P. L., and Ershler, W. B., Dietary restriction of adult male rhesus monkeys: Design, methodology, and preliminary findings from the first year of study, *J. Gerontol.*, 48, B17, 1993.

122. Kemnitz, J. W., Roecker, E. B., Weindruch, R., Elson, D. F., Baum, S. T., and Bergman, R. N., Dietary restriction increases insulin sensitivity and lowers blood glucose in rhesus monkeys, *Am. J. Physiol.*, 266, E540, 1994.

123. Ramsey, J. J., Roecker, E. B., Weindruch, R., and Kemnitz, J. W., Energy expenditure of adult male rhesus monkeys during the first 30 months of dietary restriction, *Am. J. Physiol.*, 272, E901, 1997.

124. Lane, M. A., Ingram, D. K., and Roth, G. S., Beyond the rodent model: calorie restriction in rhesus monkeys, *Age*, 20, 45, 1997.

125. Ramsey, J. J., Roecker, E. B., Weindruch, R., Baum, S. T., and Kemnitz, J. W., Thermogenesis of adult male rhesus monkeys: Results through 66 months of dietary restriction, *FASEB J.*, 10, A726, 1996.

126. Lane, M. A., Baer, D. J., Tilmont, E. M., Rumpler, W. V., Ingram, D. K., Roth, G. S., and Cutler, R. G., Energy balance in rhesus monkeys subjected to long-term dietary restriction, *J. Gerontol.*, 50A, B295, 1995a.

127. Lane, M. A., Ingram, D. K., Cutler, R. G., Knapka, J. J., Barnard, D. E., and Roth, G. S., Dietary restriction in nonhuman primates: progress report on the NIA study, *Ann. N.Y. Acad. Sci.*, 673, 36, 1992.

128. Lane, M. A., Resnick, A. Z., Tilmont, E. M., Lanir, A., Ball, S. S., Read, V., Ingram, D. K., Cutler, R. G., and Roth, G. S., Aging and food restriction alters some indices of bone metabolism in male rhesus monkeys (*Macaca mulatta*), *J. Nutr.*, 125, 1600, 1995c.

129. McCarter, R. J. M., Energy utilization, in *Handbook of Physiology, Section 11: Aging*, Masoro, E. J., Ed., Oxford University Press, New York, 1995, 95.

130. Koizumi, A., Weindruch, R., and Walford, R., Influences of dietary restriction and age on liver enzyme activities and lipid peroxidation in mice, *J. Nutr.*, 117, 361, 1987.

131. Masoro, E. J., Katz, M. S., and McMahon, C. A., Evidence for the glycation hypothesis of aging from the food restricted rodent model, *J. Gerontol.*, 44, B20, 1989.

132. Lane, M. A., Ball, S. B., Ingram, D. K., Cutler, R. G., Engel, J., Read, V., and Roth, G. S., Diet restriction in rhesus monkeys lowers fasting and glucose-stimulated glucoregulatory end points, *Am. J. Physiol.*, 268, E941, 1995b.

133. Hansen, B. C., Ortmeyer, H. K., and Bodkin, N. L., Prevention of obesity in middle-aged monkeys: food intake during body weight clamp, *Obesity Res.*, 3, 199s, 1995.

134. Cefalu, W. T., Wagner, J. D., Wang, Z. Q., Bell-Farrow, A. D., Collins, J., Haskell, D., Bechtold, R., and Morgan, T., A study of caloric restriction and cardiovascular aging in cynomolgus monkeys (*Macaca fascicularis*): A potential model for aging research, *J. Gerontol.*, 52A(1), B10, 1997.

135. Short, R. A., Williams, D. D., and Bowden, D. M., Modeling biological aging in a nonhuman primate, in *Practical Handbook of Human Biologic Age Determination*, Balin, A. K., Ed., CRC Press, Boca Raton, FL, 1994, chap. 24.

136. Comfort, A., Test-battery to measure ageing-rate in man, *Lancet*, 2, 1411, 1969.

137. Shock, N. W., Longitudinal studies of aging in humans, in *Handbook of the Biology of Aging*, Finch, C. E. and Hayflick, L., Eds., Van Nostrand Reinhold, New York, 1977, 639.

138. Sprott, R. L. and Schneider, E. L., Biomarkers of Aging, *Drug-Nutrient Interactions*, 4, 43, 1985.

139. Short, R. A., Williams, D. D., and Bowden, D. M., Cross-Sectional evaluation of potential biological markers of aging in pigtailed macaques: effects of age, sex, and diet, *J. Gerontol.*, 42, 644, 1987.

140. Bowden, D. M., Short, R., and Williams, D. D., Constructing an instrument to measure the rate of aging in female pigtailed macaques (*Macaca nemestrina*), *J. Gerontol. Biol. Sci.*, 45, B59, 1990.

141. Short, R. A., Williams, D. D., and Bowden, D. M., Circulating antioxidants as determinants of the rate of biological aging in pigtailed macaques (*Macaca nemestrina*), *J. Gerontol. Biol. Sci.*, 52A, B26, 1997.

Section D

Methods of Assessing Aging Processes

12

Dietary Restriction

Helen A. Bertrand, Jeremiah T. Herlihy, Yuji Ikeno, and Byung P. Yu

CONTENTS

12.1 Introduction

More than 60 years ago, McCay et al. [1] first reported that rats whose growth was experimentally retarded by limiting their food intake lived longer than control animals fed *ad libitum*. Since that seminal report, the antiaging actions of dietary restriction have been confirmed and reconfirmed repeatedly. Today, dietary restriction (DR) is widely recognized

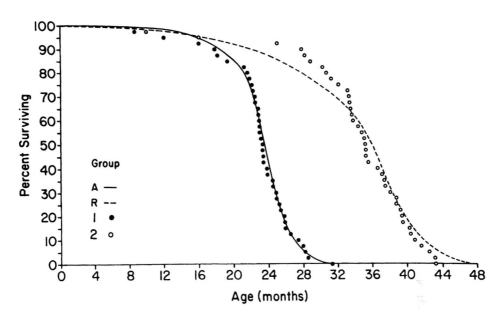

FIGURE 12.1

Survival curves for male Fischer 344 rats eating *ad libitum* throughout their lifetime (Groups A and 1) or eating 60% the amount consumed by the rats eating *ad libitum* (Groups R and 2). The data are from 2 experiments conducted 4 years apart. Rats in Groups A and R are from the first experiment, and Groups 1 and 2 are in the later experiment. There were 115 rats each in Groups A and R [2] and 60 rats each in Groups 1 and 2 [3]. The rats eating *ad libitum* ate a diet composed of 21% casein, 0.15% DL-methionine, 15% sucrose, 10% corn oil, 3% Solka Floc, 5% vitamin mix and 2% vitamin mix, 0.2% choline chloride with the remainder being dextrin (Purina Test Diets, Richmond, IN). The diet eaten by the DR rats was fortified in vitamins and choline chloride (3.33% and 0.33%, respectively) so that vitamin intake was the same for all rats. (From Yu, B. P., Masoro, E. J., and McMahan, C. A., *J. Gerontol.*, 40, 657, 1985. With permission.)

as the most effective means of altering and, therefore, studying the aging processes and is considered by many as the gold standard against which the efficacy of putative antiaging paradigms should be compared.

DR possesses several unique and desirable characteristics that make it suitable for the exploration of the mechanisms of aging. First, it increases life span, both median as well as maximum. The survival curves from 2 independent studies are depicted in Figure 12.1, which demonstrate that a 40% decrease in calorie ingestion leads to a 50% increase in both median and maximum life spans [2,3]. Second, the effect of DR on life span can be graded. As shown in Figure 12.2, the extent to which life span is prolonged is inversely proportional to the level of dietary intake [4,5]. Third, the frequency of the occurrence and the degree of severity of many age-related pathologies are decreased by DR [6–9]. Moreover, their onset of appearance is delayed [10,11]. Thus, it represents an effective model by which to explore the pathogenesis and course of progression of these diseases. Fourth, it is generally held that DR reduces the rate of age-related functional decline, thereby slowing the aging process, independent of its attenuation of age-related disease [12–15]. Finally, two characteristics that render DR an exceptionally useful paradigm for studying the aging processes are its reproducibility and general efficacy. For example, the studies shown in Figure 12.1 were performed nearly four years apart; yet, the curves are nearly superimposable. In fact, nearly every study that employed DR as a probe of the aging processes has reported antiaging actions, and these effects have been observed in a wide variety of species from rotifers to primates (Table 12.1).

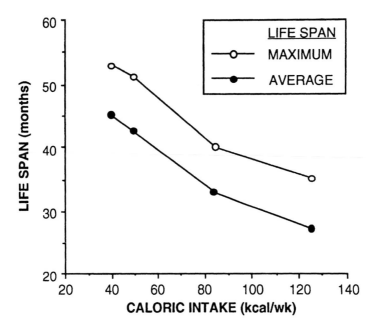

FIGURE 12.2
These data were derived from a study in which female B6C310F1 mice were given graduated levels of diet. The data are derived from survival studies and show how the level of food intake influences both maximum and average life spans. The range of calorie intake was achieved by restricting calorie intake prior to weaning as well as after weaning. Restricting calorie intake prior to weaning was accomplished by increasing the number of suckling pups per dam from 5 to 9 and by separating the suckling pup from its nursing dam every other day beginning at 1 week of age [4]. (From Weindruch, R., *Handbook of Physiology, Section II, Aging*, Masoro, E. J., Ed., Oxford Press, New York, 1995, chap. 3. With permission.)

The purpose of this chapter is to familiarize the reader with the various experimental paradigms used to restrict food intake and to present sufficient methodological detail to allow the reader to initiate such studies and avoid certain pitfalls.

12.2   Paradigms

Although DR has been applied to a number of animal species (Table 12.1), it has been most widely used in rodent model studies of aging. This section will first concentrate on three paradigms used for the study of aging in rodents and will then discuss special considerations that apply to other animal models.

12.2.1   DR by Controlled Growth or Body Weight

This method of DR, originally called the "stair-step growth" approach, was first used by McCay et al. [1]. Rats were allowed to consume only as much food as was necessary to maintain body weight constant. The restriction was initiated at weaning, and as rats became moribund, they were allowed to consume enough food to gain 10 g of body weight. After gaining 10 g, the animal's weight was then maintained constant by altering the food intake. The result was a stair-step pattern of growth. Since the time of McCay [1], this feeding paradigm has been used successfully in rodents [21] and more recently in adult monkeys [54].

TABLE 12.1
DR Paradigms

Animal	Species/Strain[a]	Gender	Animals/Cage	DR Paradigm[b]	% Restriction	Type of Diet[c]	Dietary Supplement	Age at Start of DR[d]	Feeding Time	Light Cycle	Comments	Ref.
Rat	F	Male	1	% AL	60	SS	Vitamins, calcium and phosphorus	6 w	1 h before dark	12:12	Figure 12.1 shows survival characteristics	2,3
	(–)	Both	(–)	BW	(–)	SS	Beef liver	Weaning	NA	(–)	DR rats had stair-step growth with beef liver added during growth	1
	(–)	Both	(–)	% AL	54 & 67	O	None	Weaning	(–)	12:12		16,17
	(–)	Male	(–)	% AL	50	SS	None	Weaning, 1 y, or 1st year only	(–)	(–)	Life span: DR in 1st y > DR in 2nd y > DR in all years > AL	18
	SD	Male	(–)	% AL	41 & 63	SS	None	Weaning	(–)	Natural	AL rats fed open formula diet	19
	SD	Male	(–)	PW	50 & 70	SS	None	Weaning	(–)	Natural	Control fed 10% less than ad lib	19
	SD	Male	2-4	% AL & EOD	60(–)	O	None	2 m	(–)	12:12	Both DR paradigms prevented the age-related increase in atrial ANP content	20
	SD	Female	1	BW	50	O	(–)	Weaning	(–)	12:12		21

TABLE 12.1 (continued)
DR Paradigms

Animal	Species/ Strain[a]	Gender	Animals/ Cage	DR Paradigm[b]	% Restriction	Type of Diet[c]	Dietary Supplement	Age at Start of DR[d]	Feeding Time	Light Cycle	Comments	Ref.
Rat (cont.)	W	Male	(–)	EOD	(–)	O	None	(–)	(–)	(–)	(–)	22
	W	Both	(–)	EOD	(–)	SS	Lettuce	42 d	(–)	(–)	Three DR groups by fasting: 1 d in 4; 1 d in 3; 1 d in 2	23
	W	Male	(–)	% AL	66	SS	None	32 d	(–)	(–)	Low protein shortens life of DR rats (AL protein = 12, 20, & 28%; DR protein = 18, 30, & 42%)	24
	W	Male	5	EOD	55 at 2.5 m; 67 at 12 m	O	None	30 d, 1 y, 1st year only	(–)	12:12	DR group = AL between 5 p.m. and 8 a.m. on alternate days	25
	L-W	Male	(–)	PW	70	SS	None	6 w	(–)	(–)	SS diet contained natural ingredients and was steam sterilized; DR rats fed no more than 8 g/d	26

Species	Strain	Sex		Method	%	O & SS	Supplement	Age	Feeding	Light	Comments	Ref
	F	Female	(-)	% AL	60	O & SS	Vitamins	14 w	During dark	12:12		27
	F	Male	1	% AL	60	O	(-)	14 w	During dark	12:12	Diet: NIH-31; NCTR (e); dark: 0600–1800	28
	F	Male	1	% AL	60	SS	None	6 w	(-)	12:12	AL permitted free access to diet between 1430 & 0630 only; lights on between 1630 & 0830	29
	BN & BN × F	Female	(-)	% AL	60	O	Vitamins, minerals	18 & 24 m	(-)	(-)	Diet: NIH-31; NCTR	30
	BN × F	Male	1	% AL	70	O	(-)	14 w	During dark	12:12	Diet: NIH-31; NCTR	31
Mouse	(-)	Male	(-)	% AL	50	SS	None	Weaning, 1 y, 1st year only	(-)	(-)	Life span: DR in 1st y > DR in 2nd y > DR in all yrs > AL	18
	C57BL/6J	Female	(-)	PW & EOD	50	SS	Vitamins, minerals	Weaning	(-)	(-)	Controls fed 14.5 cal/d; DR fed 14.5 cal/EOD	32
	C57BL/6	Male	1	PW	75	SS	None	12 m	(-)	(-)	Control fed 12 kcal/d (~15% < AL); DR fed 9 kcal/d (~55% AL); AL mice ate 16 kcal/d, weighed 38 g & were obese	33
	C57BL/6J & CBA/HT6J	Both	(-)	% AL & EOD	67	O	Vitamins, minerals	(-)	(-)	(-)	Diet = AIN 76	34

TABLE 12.1 (continued)

DR Paradigms

Animal	Species/Strain[a]	Gender	Animals/Cage	DR Paradigm[b]	% Restriction	Type of Diet[c]	Dietary Supplement	Age at Start of DR[d]	Feeding Time	Light Cycle	Comments	Ref.
Mouse (cont.)	C57BL/6	Female	(−)	PW	50	O	Vitamins, minerals	21 d	(−)	(−)	Controls fed 14.5 cal/d; DR fed 14.5 cal/EOD	35
	C57BL/6J	Female	3	% AL	71 ± 4	O	None	2 m	(−)	12:12	Diet = Purina Rodent Chow #5001	36
	C57BL/6J & B10C3F1	Male	PW: 4–6 DR: 1	PW	56	O & SS	Casein, vitamins, salts, Brewer's yeast & zinc oxide	12 m	(−)	(−)	PW = 160 kcal/w; 1st m DR = 115 kcal/w; 90 kcal/w thereafter	37
	MRL/lpr & C57BL/6	Male	1	PW	50	O & SS	Vitamins, minerals	6 & 12 w	(−)	12:12	Control mice fed open formula diet; DR mice fed supplemented SS diet	38
	B6C3F1	Female	1	% AL	60	O	None	(−)	(−)	(−)	Diet = NIH-31	34
	B6C3F1	Female	1	% AL	60	O	(−)	14 w	(−)	(−)	Diet = NIH-31 supplements = 1.67 × NIH-31	39
	B6C3F1	Male	1	% AL	60	O	Vitamins	14 w	mid-dark	12:12	DR ramped: week 1 = 90% AL; week 2 = 70% AL; week 3 = 60% AL	40

Strain	Sex		Type	No.	SS	Supplement	Timing			Description	Ref
B10C3F1	Female	1	PW	57	SS	Vitamins, minerals	Before weaning; weaning; 61 d	(–)	12:12	Control fed a single portion M–Th and a triple portion on F; DR fed a single portion on M&W & a double portion on F; DR before weaning = 9 pups/dam and pups separated from dam EOD after 1st w; Control = 5 pups/dam	4
C3B10RF1	Female	(–)	PW	59	SS	Yeast, protein, vitamins, minerals	Weaning; before & after weaning	a.m.	(–)	Control: 85 kcal/w; DR: 50 kcal/w; mice fed one portion M–Th & 3 portions on F	41
C3B10RF1	Female	1	% AL	60	SS	Antioxidants, vitamins, minerals	21–28 d	(–)	(–)	Dietary antioxidants did not increase life span	42
CH3/SHNF1	(–)	(–)	PW	50	SS	None	Weaning	a.m.	(–)	Control mice fed 95 kcal/w; DR fed 48 kcal/w; fed a single allotment M–Th & 3 allotments on F	43

TABLE 12.1 (continued)
DR Paradigms

Animal	Species/ Strain[a]	Gender	Animals/ Cage	DR Paradigm[b]	% Restriction	Type of Diet[c]	Dietary Supplement	Age at Start of DR[d]	Feeding Time	Light Cycle	Comments	Ref.
Mouse (cont.)	Emory	Male	(−)	PW	70	SS	Vitamins, minerals, ascorbate	(−)	(−)	(−)	Supplements = 40% excess	22
	Emory	(−)	(−)	PW	60	(−)	(−)	(−)	(−)	(−)	Emory mice show grade 3 cataracts by 10 m	44
	Balb/c	Female	1	% AL	60	O	Vitamins, minerals	4 w	(−)	(−)	Diet = Joint Stock Diet, Milling Industries, Adelaide	45
	MRL/lpr & MRL/n	Male	5	% AL	50–56	SS	Vitamins, minerals	4 w	(−)	12:12	Supplements = 2 × AL	46
	(NZB × NZW)	Both	1	PW	50	SS	Vitamins, minerals	Weaning	a.m.	12:12	Controls fed 20 kcal/d; DR fed 10 kcal/d	47
	(NZB × NZW) & DBA/2f	(−)	4	PW	50	SS	Vitamins, minerals	Weaning	(−)	12:12	Supplements = 2 × Cont; Control fed 5 g/mouse/d; DR fed 2.5 g/mouse/day	48
	DBA/2f & (NZB × NZW)	Both	4	PW	50	SS	Vitamins, minerals	3 w	a.m.	12:12		48

Species	Strain									Comments	Ref
	B6+/+ & B6ob/ob	Both	4	% AL	see com.	O	None	Weaning	Noon & 2 p.m.	12:12	ob/ob weight controlled to +/+ AL [approx 2/3 AL (+/+) & 1/2 AL (ob/ob) food intake]; dietary effect on life span not altered by body composition; DR of ob/ob mice does not eliminate obesity	49
Hamster	Golden	(-)	% AL	50	SS	None	Weaning, 1 y, 1st year only	(-)	(-)		Life span: DR in 1st y > DR in 2nd y > DR in all years > AL	18
Primate	Human	NA	PW	80	(-)	None	35–50 y	4/day	(-)		DR diet is low fat and low sucrose; duration of DR = 10 w	50
	Rhesus	1 (feeding only)	PW	70	O	Vitamins, minerals	Juvenile, adult & old	a.m. & p.m.	(-)		DR monkeys ramped by 10%/mo; old = 18–25 y; adult = 3–5 y; juvenile = 0.6–1.0 y; Controls fed 130 g/1.5 kg to 319 g/18 kg monkey	13

TABLE 12.1 (continued)
DR Paradigms

Animal	Species/ Strain[a]	Gender	Animals/ Cage	DR Paradigm[b]	% Restriction	Type of Diet[c]	Dietary Supplement	Age at Start of DR[d]	Feeding Time	Light Cycle	Comments	Ref.
Primate (cont.)	Rhesus	Male	(–)	% AL	71 to 82	SS	None	8–14 y	(–)	(–)		51
	Rhesus	Both	(–)	% AL	70	SS	Vitamins, minerals	Juvenile, adult & old	p.m.	0700–1400	Juvenile = 1–2 y; adult = 3–5 y; old = 18–25 y	52
	Rhesus	Female	(–)	% AL	70	O	Protein, vitamins, minerals	Juvenile, adult & old	a.m. & p.m.	(–)	Diet = NIH primate diet; DR = ramped 10%/mo; juvenile = 1–2 y; adult = 3–5 y; old = 18–25 y	14
	Squirrel monkey	Male	(–)	% AL	70	O	Protein, vitamins, minerals	Juvenile, adult & old	a.m. & p.m.	(–)	See above	14
	Rhesus	Male	1	% AL	70	SS	(–)	(–)	(–)	(–)	AL = free access to food for 6–8 h/d	15
	Rhesus	Male	1	PW	70	SS	Fruit	8–14 y	8 h/d	(–)	Control = (baseline + 100 g)/d; DR rampec to 70% baseline; animals caged singly in a room with other monkeys	53

Rhesus	Male	1	BW	42 to 83	O	(–)	Adult	3/d	(–)	DR: adult males weighed weekly; food intake adjusted to maintain constant body weight	54
Cynomolgus	Male	2 (see com.)	PW	70	55	Vitamins, minerals, β-sitrosterol, cholesterol	Adult	1000–1500	(–)	Number of monkeys reduced to 1 during feeding; diet was high fat (30%) with 0.25 mg cholesterol/cal; AL fed 100 cal/kg; DR fed 70 cal/kg; uneaten food weighed	55
Others											
Drosophila	(–)	(–)	Media dilution	see com.	(–)	None	Newly emerged adults	(–)	(–)	Media dilutions: 1/2, 1/4, 1/8, 1/16; 1/8 & 1/16 dilutions shortened life span	56
Drosophila	Both	5 mated or 10 virgin	Yeast dilution	(–)	S 101 medium	Live yeast	Newly emerged adults	p.m. daily	0800h–2000h	No increase in longevity was observed	57
Brine shrimp	(–)	(–)	EOD	(–)	(–)	(–)	(–)	(–)	(–)	DR: fed every third day; Control: fed every day	21

TABLE 12.1 (continued)
DR Paradigms

Animal	Species/Strain[a]	Gender	Animals/Cage	DR Paradigm[b]	% Restriction	Type of Diet[c]	Dietary Supplement	Age at Start of DR[d]	Feeding Time	Light Cycle	Comments	Ref.
Others (cont.)	Guppies	Female	(–)	EOW	(–)	(–)	(–)	(–)	(–)	(–)	DR: fed once a fortnight; Control: fed once a week	58
	Rotifers	(–)	(–)	EOD	(–)	(–)	(–)	Hatch-lings	(–)	(–)	Control: new media daily & fed algae daily; DR1: new media daily; DR2: new media M, W, F; death at 34 days (C), at 45 days (DR1), at 55 days (DR2)	59
	Water flea	(–)	(–)	EOD	(–)	(–)	(–)	(–)	(–)	(–)	Control = well fed; DR = starved lifelong or until 6th, 9th, 12th, or 15th instar	21

[a] F, Fischer 344; SD, Sprague-Dawley; W, Wistar; L-W, Lobund-Wistar; BN, Brown-Norway; BN × F, F₁ cross of BN & F.

[b] % AL, DR animals fed a percentage of the amount consumed by animals eating *ad libitum*; BW, constant body weight; PW, control and DR animals given a preweighed amount of diet; EOD, every other day; EOW, every other week.

[c] SS, semisynthetic; O, open formula.

[d] w, week; y, year; m, month; d, day(s)

[e] NCTR, National Center for Toxicological Research, Jefferson, AR.

12.2.1.1 Advantages

In this paradigm, the animals can be treated as individuals, i.e., the body weight of each animal can be monitored and maintained constant [54], or the animals can be treated as members of a group, i.e., the average body weight of a group of animals can be maintained constant [1] or within a predetermined range that allows for growth [24]. When animals are treated as individuals, the level of DR can be selected to accommodate the needs of the individual. For example, the amount of diet needed to maintain constant weight may differ from animal to animal even though the targeted body weight is the same for all members of the DR group. The information gained, i.e., calories ingested to maintain constant weight, will not be identical from animal to animal and provides information on the variability of the DR population.

12.2.1.2 Disadvantages

If each animal is monitored and maintained, the time commitment necessary to closely monitor the individual animals is very high. Moreover, if the preset level of growth is too low, as appeared to occur in McCay's work [1], the possibility arises that malnutrition with its associated morbidity may ensue. McCay et al. [1] monitored groups of animals rather than individuals, and it is possible, but not likely, that morbidity resulted from treating groups of rats rather than individual animals. Because of these issues, the controlled growth paradigm has been generally supplanted by other methods.

12.2.2 DR by Predetermined Restriction of Food Allotment

The most commonly used DR paradigms today involve feeding rodents subjected to food restriction a predetermined amount of calories. Based on the manner of determining the amount of food to be offered to the restricted rats, studies can be divided into two groups: (1) food determination based on a percentage of the intake of the *ad libitum* fed control group, and (2) predetermined amounts fed to both the control and restricted groups.

12.2.2.1 Intake Based on Consumption of ad libitum-Fed Controls

In this paradigm, control animals are allowed free access to food. The amount of food consumed by the *ad libitum* control group is measured for defined intervals (e.g., daily or weekly). This measurement is made by weighing the chow prior to offering to the animal and again at the end of the timing period. The intake of all members of the *ad libitum*-fed group can be measured or, if this group is large (e.g., greater than 30 animals), a representative sample of animals within the group can be used to monitor food intake.

Our group in San Antonio has traditionally used 27 rats in the control group to measure food consumption and the same rats are used throughout the study (Section 12.5.4.3.1). Alternatively, the control animals used to monitor food consumption could be rotated so that all animals eating *ad libitum* would be included in the group whose food intake is measured. Animals in the restricted group are fed a defined fraction of the intake of the control group. That fraction is determined from the mean daily intake of the *ad libitum*-fed animals. We feed DR animals by offering food approximately one hour prior to the dark period but we have divided the food allotment and offered half early in the light phase and half an hour before dark with no effect on longevity [60]. Other researchers have fed DR animals at different times of the day (Table 12.1).

12.2.2.1.1   Advantages

This paradigm offers several advantages. First, if the same *ad libitum*-fed animals are used throughout the study, the data obtained will constitute an impressive longitudinal data base for future analyses. Second, when animals in the study are housed singly in cages outfitted with wire bottoms, food spillage can be easily detected and often corrected for (e.g., censoring the data by deleting the food consumption data from the animals who spilled food) during calculation of the restricted amount. Moreover, such arrangements can minimize ingestion of animal waste and bedding. Such studies then benefit from a design in which the calorie intake of each animal is carefully assessed.

12.2.2.1.2   Disadvantages

The major disadvantage of this paradigm is that it is labor intensive, particularly with a large population of animals, and, therefore, expensive to conduct. The investigator must weigh the cost effectiveness of a design that requires such labor. Another concern is that the daily single feeding mode tends to transform the restricted animals from nibblers (as seen in animals eating *ad libitum*) to meal eaters.

12.2.2.2   Intake Predetermined for Control and Restricted Groups

This paradigm is similar to the previous paradigm in that DR animals are offered a percentage of what the control animals ingest. The paradigm differs from the previous one in that the control group is also given a predetermined amount of diet. The amount of food preweighed for control animals is sometimes based on the amount consumed by a small group of animals run in parallel eating *ad libitum* in which case control animals have been fed 10–15% less than these *ad libitum* animals [19] or, based on the experience of the investigator, is sometimes less than the amount which would yield obese animals [33]. The restricted animals are fed a percentage of the control amount, depending on the extent of restriction to be achieved. The frequency of providing the preweighed food allotment varies, usually occurring three or more times per week (Table 12.1).

12.2.2.2.1   Advantages

This paradigm has three major advantages. First, the feeding regimen can be scheduled at the convenience of the investigator. Second, the development of obesity, which is often observed in *ad libitum*-fed animals of certain strains, can be minimized [33]. Third, this paradigm lends itself to multiple caging of animals, especially mice. Thus, by avoiding the single caging of animals, the investigator can substantially reduce the per diem fees incurred for animal maintenance.

12.2.2.2.2   Disadvantages

The major disadvantage of this paradigm is that the control animals are not truly *ad libitum* fed. They, therefore, can also be considered restricted, albeit to a lower extent than the experimental DR group. If multiple caging is employed, longitudinal data on the food intake of individuals will be lost. Finally, with this feeding paradigm, the normal metabolic circadian pattern could be disturbed. For example, circadian patterns which are entrained, in part, by the act of eating itself or by the absorption of nutrients (e.g., glucocorticoid, insulin, glucagon, and glucose concentrations in plasma) will differ between control and DR animals when these two groups are not given their food allotment at identical times of the day.

12.2.3 DR by Every Other Day (EOD) Feeding

Another relatively popular way of imposing DR is to allow restricted animals to eat *ad libitum* every other day. In this paradigm, the food cup is removed from the cage (or the food hopper is emptied) on alternate days. Whereas predetermination of food intake (see 12.2.2.2. above) is the more common method for applying dietary restriction to multiply caged mice, the EOD feeding paradigm is more often used when rats are multiply caged.

12.2.3.1 Advantages

This method has several advantages. First, and most important, it is far less labor intensive than the previously described methods and therefore is much less expensive to perform. Second, it lends itself to studies in which a large number of animals are utilized and allows for the multiple caging of animals. Finally, the restricted population exhibits a variance in body weight similar to that found in the *ad libitum*-fed population.

12.2.3.2 Disadvantages

The principal shortcoming of EOD feeding is that the degree of restriction is not totally under the control of the investigator. For example, in young, rapidly growing rodents, the degree of restriction is on the order of 50 to 60% of the control group. In adults, however, this percentage rises to 70 to 80% [37]. Another disadvantage is that the food intake history of individual rats is unavailable if multiple housing is employed. Finally, the temporal pattern of food intake in the EOD paradigm differs substantially between the restricted and control animals. This last point is of some concern for those investigators interested in the effect of food intake on metabolic regulation. For example, the absorption of nutrients stimulates the secretion of the hormone, insulin. The metabolic consequences of this change include increased glucose utilization, protein synthesis, and fat deposition. With the EOD paradigm, the changes will occur at 48-h intervals rather than at 24-h intervals as seen in AL animals and as seen in DR paradigms in which DR animals are fed daily.

12.2.4 Degree and Initiation of Restriction

Other factors must be considered when selecting or designing a DR protocol. They are the degree or percentage to which food is restricted, the time in life at which restriction of food intake will occur, and the duration of food restriction, i.e., the fraction of the lifetime during which DR is maintained, e.g., early life, adult life, lifelong, etc. The percentage of DR used in a study determines, in part, the extent to which life span is increased (Figure 12.2). It has ranged from approximately 40%–80% of *ad libitum* or control animal food intake (Table 12.1).

DR is usually initiated sometime after weaning, but no optimal age has been determined. Studies have initiated restriction at weaning, at 6 weeks of age (2 weeks postweaning), at 3 months of age, at 6 months of age, and even later (Table 12.1). When DR is initiated in adult animals, the reduction in food intake is imposed gradually in steps over a period of weeks (rodents) or months (primates) as the full restriction of intake is accomplished (Table 12.1). In all cases, DR has increased life span. The magnitude of the diet-induced life extension appears directly proportional to the fraction of the lifetime over which the restriction is applied [3,18,28,40].

12.2.5   Special Considerations that Apply to Nonrodent Animal Models

12.2.5.1   Primates

Because of the reproducible and pronounced beneficial effects of dietary restriction on aging in rodents, the logical question arises as to whether its effects can be seen in humans. To this end, several laboratories have initiated studies designed to test the effects of DR in nonhuman primates (Table 12.1). The use of this animal model presents several major challenges to the investigator. First, because multiple births among primates are rare and matings are not easily synchronized, the animals in a birth cohort are few. In addition, the gene pool of the primate is heterogeneous relative to that of rat and mice strains. Thus, the combination of few animals in a cohort and the heterogeneity of the gene pool substantially increase the variability of the data. Second, primate husbandry itself is challenging. Totally free access to food should be avoided as primates often play with excess food, even to the point where it becomes unfit for consumption. Measurement of food intake then becomes impossible. Primates should be subjected to an eating pattern similar to humans, i.e., they should be meal-fed with meal size portions. Therefore, the investigator must either allow control animals to eat *ad libitum* during a set time period [15] or else feed them a predetermined amount of diet based on age and body weight [52, 53,55]. Usually, the level of restriction imposed on primates is less severe than that used for rodents because the same level of DR in primates as that employed in rodents can lead to malnutrition (unpublished observation).

At present there is only one published report [50] examining the effects of DR in humans. The duration of DR was 10 weeks. DR was, in fact, calorie restriction which was accomplished by reducing dietary fat and carbohydrate.

12.2.5.2   Nonmammals

DR has been tested in a number of nonmammalian species (Table 12.1) with results indicating a beneficial effect on aging. Unfortunately, it is almost impossible to quantitate the degree of restriction. Moreover, when different levels of restriction have been applied, severe restriction resulted in shortened life spans indicative of malnutrition [56,57].

12.3   Diet

The choice of diet is an extremely important aspect of DR regimens and one which deserves substantial consideration prior to the initiation of the experiment. DR must be achieved without producing malnutrition. Several facets of the optimal choice of diet for the rodent are discussed below. Similar principles would be of concern to other animal models. Quantitation of the amount of macronutrients and supplementations as well as the source of those macronutrients are of special concern. In addition, certain practical aspects of diet storage and preparation are worthy of mention.

12.3.1   Macronutrients

Protein, fat, and carbohydrate make up the bulk of the food consumed by the experimental animals. The decision as to the proportions of these macronutrients used in the experiment rests with the investigator and must be determined from the exigencies of the experimental goals. Literature values of the percentage of the diet comprised of protein range from 20–30%. Both too little protein, as well as too much protein, may lead to undesirable effects. In animals subjected to dietary restriction, a decrease in the percentage of protein in the diet was shown to shorten life span [24]. Conversely, high levels of protein have been shown to accelerate renal problems under conditions favoring renal pathologies [7]. With regard to the lipid composition of the diet, the amount varies from 4–25%. Diets high in lipid have been frequently used to induce pathological states, including obesity, hypertension, and atherosclerosis. After determination of the protein and lipid composition, the remainder of the diet consists mainly of carbohydrate, usually the main component of the diet. There are no absolute rules for deciding the percentages of the macronutrients to be employed; the decision should be based on the goals of the study and the animal model to be used in the study.

12.3.2   Supplementation

The topic of supplementation of the diet represents an especially bothersome aspect of diet composition. Earlier studies tended to "fortify" diets fed to restricted animals with liver or yeast extracts. The reasons for fortification and the choices of which substances should be fortified have never been convincingly defended. In a more recent study [61], it has been shown that the food consumed per gram of body weight is the same whether rats are eating *ad libitum* or are restricted, suggesting that the concentration of nutrients available to tissues is the same in DR animals and animals eating *ad libitum*. Therefore, it has become unclear whether supplementation is at all necessary.

The question of supplementation becomes especially important, however, when considering the minimum daily requirements for vitamins and minerals. In the past, the diets of animals subjected to dietary restriction have often been fortified with vitamins and/or minerals to somehow ensure that this group does not suffer a deficiency in any of these nutrients (Table 12.1). Our experience [61] suggests that, since the amount of food consumed per kilogram of rat is the same whether the animal eats *ad libitum* or is subjected to dietary restriction, supplementation is not necessary when based on body weight, provided chow is well balanced. No differences were observed in the effect of DR on calcium regulatory hormone levels between restricted rats that consumed a calcium-supplemented diet [3,62] and those offered a nonsupplemented diet [63].

12.3.3   Dietary Sources

The methods of preparing the chow can be divided into two main groups: commercial (or open-formula diet) and semisynthetic. In both, the amount of nutrient (protein, fat, and carbohydrate) is defined. However, only in the semisynthetic diet is the source of the macronutrient defined. Thus, although both open-formula and semisynthetic diets may contain, say, 21% protein, only the semisynthetic diet will identify the protein source (e.g., casein, soy, etc.). Open-formula diets utilize several natural sources for their macronutrient components, often making the identification of sources and their contributions to the diet difficult or impossible to assess.

A major advantage of using the semisynthetic diet comes when long-term studies are undertaken. In aging studies where rodents may live for two or more years, the semisynthetic diet allows investigators to know precisely the sources for the ingredients for the duration of the study. This information is not available in studies utilizing the open-formula. An example of the advantage of the semisynthetic diet is found in previous studies from San Antonio [7,9]. Switching from the traditional casein-based protein source to that of a soy-based protein source elicited a number of interesting changes in aging animals. Renal disease, a hallmark of the Fischer 344 rat, was drastically reduced [7], longevity was increased [7], and the age-related changes in the plasma levels of calcium regulatory hormones were attenuated [63]. The importance of the protein source on aging would have been missed had the experiment utilized an open-formula or commercial diet.

12.3.4   Storage of Diet

Several important details should be attended to with regard to storage of the chow. First, with regard to the open-formula, the USDA specifies that it be used within 6 months of the milling date, which is provided on the package. Although these diets are formulated to withstand storage at room temperature, storage in the cold room is preferred. This storage not only preserves the ingredients of the diet, but also minimizes the likelihood of insect infestation and consumption by nonexperimental rodents. The chow should be placed on a pallet to keep the contents dry.

Second, with regard to semipurified diets, these are generally not designed to be stored at room temperature and must be refrigerated or even frozen to protect the components from chemical changes that can occur at higher temperatures. This is especially important for the lipid component, which is very vulnerable to oxidation. Rats will not eat a diet in which the lipid component has become rancid (unpublished observation).

12.3.5   Physical Form of Diet

Finally, the physical form of the chow, pelleted vs. powder, also deserves consideration. Open-formula diets purchased from commercial vendors are usually prepared in pelleted form; semisynthetic diets are usually, but not always, supplied in powdered form. The pelleted form of the semisynthetic diet can be prepared by modifying the fat and carbohydrate constituents. For the *ad libitum*-fed animals, the handling of the pelleted and powdered forms are equally convenient. However, the powdered chow is preferred when preparing preweighed portions for restricted rats, as small quantities can be weighed with greater accuracy.

Another consideration is the dentition of aging rodents. It is widely thought that a pelleted diet is required to control the length of the rat incisors because it provides a hard substance on which to gnaw and wear down the teeth. Our experience with powdered chow suggests that, unless an individual Fischer 344 rat is born with some malocclusion, incisors need not be trimmed and the teeth are maintained at a healthy length.

12.4   Animals

Chapter 10 of this book describes in detail the selection of animal models for aging research, but certain factors need consideration when embarking on aging studies which include dietary restriction. These factors are the species, strain, and sex of the animal model, the number of birth cohorts in the study, and selected features of animal husbandry related to DR.

12.4.1   Species, Strain, and Gender

Studies on DR and aging have been performed on a wide variety of species and strains and on both sexes (Table 12.1). The choice of animal model will be dictated mainly by the interests and needs of the investigator because, as mentioned above, the effects of DR have been observed universally, regardless of the animal model chosen for study. The investigator should be aware, however, that the action of DR on longevity may be unrelated to the specialized characteristics of the animal model. For example, studies [49] designed to examine the relationship between obesity and longevity utilized the ob/ob mouse. Although DR did indeed extend the life span in this animal, body composition was not altered by this dietary manipulation. Thus, DR acted to extend life span by means other than changes in body composition, which is dictated by genotype.

12.4.2 Animal Cohorts

Both single and multiple birth cohorts have been used in previous studies and each design has advantages and disadvantages. From the point of view of feeding regimens, the single birth cohort possesses a certain advantage over multiple birth cohorts when a predetermined amount of chow is offered to the animals (see Section 12.2.2). The former design minimizes the diversity of chow measurements, whereas, with the multiple birth design, the degree of food offered varies among the number of cohorts in the study.

12.4.3 Husbandry

Whether the study should employ single or multiple caging depends upon which DR paradigm is utilized (Table 12.1). If EOD feeding is used, then animals may be housed in groups and fewer resources in time and space would be required for the study. On the other hand, if the amount of chow offered to the DR group is determined by the amount consumed by the *ad libitum*-fed group, then individual housing is preferred. Another complication in group vs. single housing is illustrated by a phenomenon observed with C57BL/6J male mice where the amount eaten by each mouse housed individually is greater than the amount consumed by each mouse when housed in groups (personal observation).

Another practical consideration is whether raised wire floors or soft bedding should be used in animal cages. The raised wire flooring facilitates the detection of chow spillage. Thus, in studies that utilize the food consumption of the *ad libitum* control group to determine the amount to be fed to the DR group, a more accurate assessment of the actual amount eaten by the control animals that spill food can be obtained with raised wire flooring. Soft bedding can be used in these types of studies; however, appropriate methods must be adopted to estimate spillage.

12.5. Standard Operating Procedures Used for Dietary Restriction Studies in San Antonio

12.5.1 Animals

12.5.1.1 Rat Strain and Gender

The Fischer 344 rat strain has always been used and was chosen because it is a genetically inbred strain. Also, the Fischer 344 rat does not become excessively obese at older ages, as do some strains. Only male rats have been utilized to avoid the confounding influence of the reproductive senescence seen in females.

12.5.1.2 Rat Supplier

Charles River (Wilmington, MA) was chosen because this vendor could supply 600 Cesarean-derived, male Fischer 344 rats whose birth dates were within 4 days of each other. The following were constantly monitored.

12.5.1.2.1 Animal Variability

Rats are purchased from only one Charles River facility to minimize variation between cohorts. Rat weight is measured within 2 weeks of arrival and every 2 weeks thereafter. Examination of initial body weight and growth characteristics is important for two reasons. First, it helps in the detection of rats who fail to thrive, allowing time for corrective procedures (e.g., inability to drink). Second, it allows comparison with other cohorts in the present and previous studies.

12.5.1.2.2 Pathogen Status

Two sentinel rats are sacrificed immediately after arriving in San Antonio and tested for the presence of rat pathogens which would mean that the supplier's colony is suspect. Three weeks later another four rats are tested to determine whether pathogens were encountered in shipment.

12.5.1.3 Rat Age and Weight

Rat age upon arrival at San Antonio is 28 ± 2 (range) days of age. Rat weight upon arrival at San Antonio is approximately 50 g, but this varies from cohort to cohort.

12.5.1.4 Shipping

Usually rats are shipped to San Antonio in temperature-controlled trucks or via air freight. Difficulties have arisen in the past when rats have spent excessive periods waiting on loading docks. Extremes of temperature, either cold or hot, should be avoided. Rats are shipped in filtered cartons (10–13 rats per carton) which contain moist food.

12.5.1.5 Transfer to Barrier Facilities

12.5.1.5.1 Quarantine

Newly arrived rats are quarantined for 3 weeks to insure that no infectious diseases have entered the facility. Thus, a room is set aside in the barrier to hold the rats for the quarantine period.

12.5.1.5.2 Transfer to Barrier and Caging

Upon arrival the shipping cartons are placed next to the barrier entrance. They are decontaminated individually by spraying with a bactericidal/fungicidal/viricidal agent (@Amphyl) before being passed into a vestibule (dressing room) of the barrier. There the carton is opened. The rats are transferred from the outer chamber to an inner, more aseptic chamber and placed randomly in individual cages premarked with identification numbers.

12.5.1.5.3 Pathology Testing

At the end of unloading, two sentinel rats from the same cohort are sacrificed and the plasma collected. Pathological analysis is performed on the carcass to check for birth defects/abnormalities and evidence of infectious disease. The plasma is sent to a commercial laboratory for testing of specific pathogens to which the rats may have been exposed in the supplier's colony. The procedure is repeated 3 weeks later on additional sentinel rats to determine if they were exposed to pathogens in shipment. Serum is obtained every 3 months thereafter and tested for the presence of rat pathogens in the barrier.

12.5.2   Animal Maintenance

12.5.2.1   *Housing*

Rats are housed throughout their life in barrier facilities under constant positive pressure. Temperature is monitored by a central control room in the Health Science Center and maintained at 75 ± 2°F. Lights are on from 0400 to 1600 hours.

12.5.2.2   *Cages*

Rats are housed singly in cages (10.5"w × 9.5"d × 8.0"h) that are half-sized, individual cages. Rats that exceed 500 g are transferred to larger cages (10.5"w × 19"d × 8.0"h). All cages are equipped with wire bottoms. Rather than soft bedding, deazotized cageboards are used to cover the bottom of the cage. The cages are hung on environmental racks whose tops are covered with filter paper.

12.5.2.3   *Water Bottles*

Standard 500 ml bottles equipped with sipper tubes are sterilized and the pH of the filtered water set at 3.1.

12.5.2.4   *Food Cups*

12.5.2.4.1   *Description*

Wide-mouth, half-pint mason jars are used. Some rats turn food cups over causing massive spillage. Overturning is usually avoided by placing food cup in metal ring attached to the wire bottom.

12.5.2.4.2   *Spillage*

When spillage is minimal, it is ignored. When it becomes substantial any data on food consumption for that animal for that feeding period are discarded. Determination of what is substantial spillage is a subjective judgment, but the technicians have arrived at a "feeling" of what is substantial and this "feeling" is consistently applied throughout the study.

12.5.3   Diet

12.5.3.1   *Supplier*

Purina Test Diets (Richmond, IN) is responsible for formulating and mixing the rat chow.

12.5.3.2   *Composition of the SR Vitamin-Fortified RP101*

RP101 soy protein isolate	21%
Sucrose	15%
Solka floc	3%

RP vitamin mix	3.3%
Mineral mix with reduced sodium	5%
DL-methionine	0.35%
Choline chloride	0.33%
Corn oil	6%
Dextrin	45.99%

12.5.3.3   Type

The diet is semisynthetic. The advantage of the semisynthetic diet was discussed in Section 12.3.3.1 of this chapter. The quality control of the diet is insured by analysis performed by an independent laboratory. Soy, rather than casein, is the source for protein.

12.5.3.4   Form

The powdered form of chow is utilized for three reasons: (1) the semisynthetic diets are usually prepared as powders; (2) powders can be weighed more accurately; and (3) less rat spillage occurs with the powder.

12.5.3.5   Supplementation

In earlier studies, food-restricted rats were fed a chow enriched in vitamins, calcium, and phosphorus. This has proved to be unneccesary as discussed in Section 12.3.2 of this chapter and has been discontinued.

12.5.3.6   Shipping and Storage

The chow was shipped from Purina via 2-day air freight and upon arrival stored at –20°C. Transit times longer than 2 days are not permitted.

12.5.4   Feeding Regimen

12.5.4.1   Accommodation

After their arrival in San Antonio, all rats are fed *ad libitum* for two weeks to allow them to recover from the stress of shipment and to accommodate to their new surroundings.

12.5.4.2   Ad Libitum Feeding

Feeding cups are filled with chow, placed in the rat cage, and replaced with a replenished cup one week later. The addition of replenished cups is synchronized with cage changing.

12.5.4.3   Ad Libitum-Fed "Monitored" Rats

12.5.4.3.1   Description

A week after arrival, 27 rats are chosen from each cohort. Their chow intake is closely monitored and the amount eaten is considered the food intake for the *ad libitum*-fed group. The amount fed to restricted rats is based on the intake of these "monitored" *ad libitum*-fed controls. The same rats serve this function throughout their lifetime.

12.5.4.3.2   *Food Intake Determination*

These "monitored" rats are treated identically to the *ad libitum*-fed rats, except that their food intake is closely monitored. Food cups are filled with chow and weighed. After either 3 or 4 days of feeding, the cup is removed from the cage and weighed a second time. The difference between weighings is used to calculate the mean daily food intake by each "monitor" rat. Although the consumption varies with age, the *ad libitum* rats consume approximately 12–15 g of chow per day. Spillage is handled as described above (Section 12.5.2.4.2).

12.5.4.3.3   *Limitations*

Two major limitations of the use of "monitor" rats occur. First, although the food intake of the *ad libitum*-fed group is relatively constant over most of their life span, in advanced age this group decreases its food intake. When this trend is first observed, the food offered to the restricted group is no longer calculated from the "monitor" group, but is the "usual" amount of food which has been given over the preceding months. Second, determination of *ad libitum* food intake becomes somewhat complex when more than one cohort is present in the barrier since the intake of all cohorts is measured biweekly. Usually, four to six cohorts are present in our barriers simultaneously.

12.5.4.4   **Restricted Rats**

12.5.4.4.1   *Description*

Restriction of food intake is routinely initiated at 6 weeks of age, i.e., 2 weeks after arrival in the barrier facility. The composition of the diet given to food restricted rats is the same as described in Section 12.6.3.2. However, the restricted rats are allowed to eat only 60% of that consumed by the *ad libitum* "monitor" rats. The rats appear to withstand very well the 40% decrease in food intake. The body and organ weights are generally 40% lower than the *ad libitum* group and remain at that level through most of the life span, which is about 30% longer than the *ad libitum* group.

12.5.4.4.2   *Practical Considerations*

The amount of chow is weighed daily and placed in the food cup of the restricted rats at 1500 hours. This feeding paradigm requires the presence of barrier technicians 7 days a week. The chow that is not consumed by a restricted rat is removed from the food cup, weighed, and then discarded.

The situation becomes complex when several cohorts are present in the barrier since each DR cohort is offered a particular weight of chow. Therefore, several different amounts of rat chow must be weighed out each day.

12.5.5   Rat Population and Usage

Both longevity (survival), longitudinal (repeated measures), and cross-sectional studies have been performed here and each paradigm has special uses and unique demands (Chapter 2).

12.5.5.1   **Longevity Studies**

Since dietary restriction is considered the "gold standard" against which other antiaging strategies are gauged, we routinely run *ad libitum*-fed control and restricted groups simultaneously with cross-sectional studies and studies designed to examine some putative anti-aging strategy. In this way, any effects of other treatments on the aging process can be compared to the basic dietary restriction paradigm.

12.5.5.2 Longitudinal Studies

In these studies, multiple measurements are made on the same animal over a period of time and, therefore, the animal is not sacrificed for the measurements.

12.5.5.3 Cross-Sectional Studies

These studies can be performed on individuals from the same cohort or from different cohorts. In either case, the problem arises that at older ages only a select group of "survivors" are available and, therefore, the data gathered from these "survivors" are skewed. The additional problem of possible "cohort" effects arises when individuals from the same cohort are used. This problem can be avoided by utilizing individuals from different cohorts.

12.5.6   Pathology

Gross and microscopic analyses are performed on all rats used in longevity studies; selected rats from longitudinal and cross-sectional studies are examined. Scrupulous records are kept on every rat examined and the data are available to researchers who have utilized organs or tissues from these rats.

12.5.7   Record Keeping

Rat weight and food intake measurements are entered into the barrier duplicate data sheets as they are gathered. When the data sheets are completed, the carbon copies of the data sheets are removed from the barrier and the data entered into a computer program. The original copy remains in the barrier.

Any change in the condition of the rat upon inspection is noted (along with the date) on the back of its identification card. This notation aids the pathologist when performing post-mortem analysis.

When a rat is removed from the barrier for any reason, its weight, the date, and the reason for removal is noted on the identification card which accompanies the animal out of the barrier. This information is also entered into the "Removal Log" which remains within the barrier for the duration of the study.

12.6 Summary

DR is the most effective known means of altering the aging processes. Its antiaging actions have been consistently confirmed by many investigators garnering the reputation as the "gold standard" against which other putative antiaging perturbations are compared [64]. A number of different restriction paradigms have been successfully applied to the study of aging, with each paradigm possessing advantages and disadvantages over other methods.

Similarly, various dietary compositions and restrictions of diverse animal models have been successfully employed in aging studies. These selective characteristics give investigators substantial versatility in designing dietary restriction/aging studies capable of answering their unique questions.

Acknowledgments

The authors wish to thank Ms. Kimberley Kennedy for her help in preparing this manuscript and acknowledge the NIA (AG01188) for years of support. Finally, the authors wish to thank those who have worked diligently on the animal maintenance, which has given our work the reputation it has earned.

References

1. McCay, C. M., Crowell, M. F., and Maynard, L. A., The effect of retarded growth upon the length of life span and upon the ultimate body mass, *J. Nutr.*, 10, 63, 1935.
2. Yu, B. P., Masoro, E. J., Murata, I., Bertrand, H. A., and Lynd, F. T., Life span study of SPF Fischer 344 male rats fed *ad libitum* or restricted diets: longevity, growth, lean body mass and disease, *J. Gerontol.*, 37, 130, 1982.
3. Yu, B. P., Masoro, E. J., and McMahan, C. A., Nutritional influences on aging of Fischer 344 rats: I. Physical, metabolic, and longevity characteristics, *J. Gerontol.*, 40, 657, 1985.
4. Cheney, K. E., Liu, R. K., Smith, G. S., Meredith, P. J., Mickey, M. R., and Walford, R. L., The effect of dietary restriction of varying duration on survival, tumor patterns, immune function and body temperature in B10C3F$_1$ female mice, *J. Gerontol.*, 38, 420, 1983.
5. Weindruch, R., Animal models, in *Handbook of Physiology, Section 11, Aging*, Masoro, E. J., Ed., Oxford Press, New York, 1995, chap. 3.
6. Maeda, H., Gleiser, C. A., Masoro, E. J., Murata, I., McMahan, C. A., and Yu, B. P., Nutritional influence on aging of Fischer 344 rats. II. Pathology, *J. Gerontol.*, 40, 671, 1985.
7. Iwasaki, K., Gleiser, C. A., Masoro, E. J., McMahan, C. A., Seo, E. J., and Yu, B. P., The influence of dietary protein source on longevity and age-related disease processes of Fischer rats, *J. Gerontol.*, 43, B5, 1988.
8. Masoro, E. J., Iwasaki, K., Gleiser, C. A., McMahan, C. A., Seo, E. J., and Yu, B. P., Dietary modulation of the progression of nephropathy in aging rats: An evaluation of the importance of protein, *Am. J. Clin. Nutr.*, 49, 1217, 1989.
9. Ikeno, Y., Bertrand, H. A., and Herlihy, J. T., Effects of dietary restriction and exercise on the age-related pathology of the rat, *Age*, 20, 107, 1997.
10. Shimokawa, I., Yu, B. P., Higami, Y., Ikeda, T., and Masoro, E. J., Dietary restriction retards onset but not progression of leukemia in male F344 rats, *J. Gerontol.*, 48, B68, 1993.
11. Higami, Y., Yu, B. P., Shimokawa, I., Bertrand, H., Hubbard, G. B., and Masoro, E. J., Anti-tumor action of dietary restriction is lesion-dependent in male Fischer 344 rats, *J. Gerontol.*, 50A, B72, 1995.
12. Weindruch, R. and Walford, R. L., *The Retardation of Aging and Disease by Dietary Restriction*, Charles C Thomas, Springfield, IL, 1988.
13. Nakamura, E., Lane, M. A., Roth, G. S., Cutler, R. G., and Ingram, D. K., Evaluating measures of hematology and blood chemistry in male rhesus monkeys as biomarkers of aging, *Exp. Gerontol.*, 29, 151, 1994.

14. Lane, M. A., Ingram, D. K., Cutler, R. G., Knapka, J. J., Barnard, D. E., and Roth, G. S., Dietary restriction in nonhuman primates: Progress report on the NIA study, *Ann. N. Y. Acad. Sci.*, 673, 36, 1992.

15. Kemnitz, J. W., Roecker, E. B., Weindruch, R., Elson, D. F., Baum, S. T., and Bergman, R. N., Dietary restriction increases insulin sensitivity and lowers blood glucose in rhesus monkeys, *Am. J. Physiol.*, 266, E540, 1994.

16. Berg, B. N., Nutrition and longevity in the rat. I. Food intake in relation to size, health and fertility, *J. Nutr.*, 71, 242, 1960.

17. Berg, B. N. and Simms, H. S., Nutrition and longevity in the rat. III, Food restriction beyond 800 days, *J. Nutr.*, 74, 23, 1961.

18. Stuchlikova, E., Juricova-Horakova, A. M., and Deyl, Z., New aspects of the dietary effect of life prolongation in rodents. What is the role of obesity in aging?, *Exp. Gerontol.*, 10, 141, 1975.

19. Ross, M. H., Length of life and nutrition in the rat, *J. Nutr.*, 75, 197, 1961.

20. Cavallini, G., Clerico, A., Del Chicca, M., Gori, Z., and Bergamini, E., Effects of different types of anti-aging dietary restrictions on age-related atrial natriuretic factor changes: An immuno-histochemical and ultrastructural study, *Aging Clin. Exp. Res.*, 7, 117, 1996.

21. Merry, B. J. and Holehan A. M., Onset of puberty and duration of fertility in rats fed a restricted diet, *J. Reprod. Fert.*, 57, 253, 1979.

22. Barrows, C. H., Jr. and Kokkonen, G. C., Diet and life extension in animal model systems, *Age*, 1, 130, 1978.

23. Carlson, A. J. and Hoelzel, F., Apparent prolongation of the life span of rats by intermediate fasting, *J. Nutr.*, 31, 363, 1946.

24. Davis, T. A., Bales, C. W., and Beauchene, R. E., Differential effects of dietary caloric and protein restriction in the aging rat, *Exp. Gerontol.*, 18, 427, 1983.

25. Beauchene, R. E., Bales, C. W., Bragg, C. S., Hawkins, S. T., and Mason, R. L., Effect of age of initiation of feed restriction on growth, body composition, and longevity of rats, *J. Gerontol.*, 41, 13, 1986.

26. Snyder, D. L. and Wostmann, B. S., The design of the Lobund aging study and the growth and survival of the Lobund-Wistar rat, in *Dietary Restriction and Aging*, Snyder, D. L., Ed., Alan R. Liss, New York, 1989, chap. 3.

27. Lehman, P. A. and Franz, T. J., Effect of age and diet on stratum corneum barrier function in the Fischer 344 female rat, *J. Invest. Dermatol.*, 100, 200, 1993.

28. Duffy, P. H., Feuers, R. J., Leakey, J. A., Nakamura, K. D., Turturro, A., and Hart, R. W., Effect of chronic caloric restriction on the physiological variables related to energy metabolism in the male Fischer 344 rat, *Mech. Ageing Dev.*, 48, 117, 1989.

29. Birchenall-Sparks, M. C., Roberts, M. S., Staecker, J., Hardwick, J. P., and Richardson, A., Effect of dietary restriction on liver protein synthesis in rats, *J. Nutr.*, 115, 944, 1985.

30. Gunness, M. and Hock, J. M., Anabolic effect of parathyroid hormone is not modified by supplementation with insulin-like growth factor I (IGFI) or growth hormone in aged female rats fed an energy-restricted or *ad libitum* diet, *Bone*, 609, 124, 1993.

31. Luhtala, T. A., Roecker, E. B., Pugh, T., Feuers, R. J., and Weindruch, R., Dietary restriction attenuates age-related increases in rat skeletal muscle antioxidant enzyme activities, *J. Gerontol.*, 49, B231, 1994.

32. Gerbase-DeLima, M., Liu, R. K., Cheney, K. E., Mickey, R., and Walford, R. L., Immune function and survival in a long-lived mouse strain subject to undernutrition, *Gerontologia*, 21, 184, 1975.

33. Ershler, W. B., Sun, W. H., Binkley, N., Gravenstein, S., Volk, M. J., Kamoske, G., Kloop, R. G., Roecker, E. B., Daynes, R. A., and Weindruch, R., Interleukin-6 and aging: blood levels and mononuclear cell production increase with advancing age and in vitro production is modified by dietary restriction, *Lymphokine Cytokine Res.*, 12, 225, 1993.

34. Bronson, R., Lipman, R. D., and Harrison, D. E., Age-related gliosis in the white matter of mice, *Brain Res.*, 609, 124, 1993.

35. Walford, R. L., Liu, R. K., Gerbase-DeLima, M., Mathies, M., and Smith, G. S., Long-term dietary restriction and immune function in mice: response to sheep red blood cells and to mitogenic agents, *Mech. Ageing Dev.*, 2, 447, 1973.

36. Goodrick, G. J. and Nelson, J. F., The decidual cell response in aging C57BL/6J mice is potentiated by long-term ovariectomy and chronic food restriction, *J. Gerontol.*, 44, B67, 1989.

37. Weindruch, R. and Walford, R. L., Dietary restriction in mice beginning at 1 year of age: Effect on life-span and spontaneous cancer incidence, *Science*, 215, 1415, 1982.

38. Kubo, C., Day, N. K., and Good, R. A., Influence of early or late dietary restriction on life span and immunologic parameters in MPL/mp-lpr/lpr mice, *Proc. Natl. Acad. Sci. U.S.A.*, 81, 5831, 1984.

39. Desai, V. G., Weindruch, R., Hart, R. W., and Feuers, R. J., Influences of age and dietary restriction on gastrocnemius electron transport system activities in mice, *Arch. Biochem. Biophys.*, 333, 145, 1996.

40. Lu, M. H., Hinson, W. G., Turturro, A., Sheldon, W. G. and Hart, R. W., Cell proliferation by cell cycle analysis in young and old dietary restricted mice, *Mech. Ageing Dev.*, 68, 151, 1993.

41. Weindruch, R., Walford, R. L., Fligiel, S., and Guthrie, D., The retardation of aging in mice by dietary restriction: Longevity, cancer, immunity and lifetime energy intake, *J. Nutr.*, 116, 641, 1986.

42. Harris, S. B., Weindruch, R., Smith, G. S., Mickey, M. R., and Walford, R. L., Dietary restriction alone and in combination with oral ethoxyquin/2-mercaptoethanolamine in mice, *J. Gerontol.*, 45, B141, 1990.

43. Koizumi, A., Wada, Y., Tsukada, M., Kamiyama, S., and Weindruch, R., Effects of energy restriction on mouse mammary tumor virus mRNA levels in mammary glands and uterus and on uterine endometrial hyperplasia and pituitary histology in C3H/SHNF$_1$ mice, *J. Nutr.*, 120, 140, 1990.

44. Mura, C. V., Roh, S., Smith, D., Palmer, V., Padhye, N., and Taylor, A., Cataract incidence and analysis of lens crystallins in the water-, urea-, and SDS-soluble fractions of Emory mice fed a diet restricted by 40% in calories, *Curr. Eye Res.*, 12, 1081, 1993.

45. Dempsey, J. L., Pfeiffer, M., and Morley, A. A., Effect of dietary restriction on in vivo somatic mutation in mice, *Mutat. Res.*, 291, 141, 1993.

46. Mark, D. A., Alonso, D. R., Quimby, F., Thaler, T., Kim, Y. T., Fernandes, G., Good, R. A., and Weksler, M. E., Effects of nutrition on disease and life span. I. Immune responses, cardiovascular pathology, and life span in MRL mice, *Am. J. Pathol.*, 117, 110, 1984.

47. Fernandes, G., Friend, P. Yunis, E. J., and Good, R. A., Influence of dietary restriction on immunologic function and renal disease in (NZB X NZW)F$_1$ mice, *Proc. Natl. Acad. Sci. U.S.A.*, 75, 1500, 1978.

48. Fernandes, G., Yunis, E. J., and Good, R. A., Influence of diet on survival of mice, *Proc. Natl. Acad. Sci. U.S.A.*, 73, 1279, 1976.

49. Harrison, D. E., Archer, J. R., and Astle, C. M., Effects of food restriction on aging: separation of food intake and adiposity, *Proc. Natl. Acad. Sci. U.S.A.*, 81, 1835, 1984.

50. Velthuis-te Wierik, E. J. M., van den Berg, H., Schaafsma, G., Hendriks, H. F. J., and Brouwer, A., Energy restriction, a useful intervention to retard human aging? Results of a feasibility study, *Eur. J. Clin. Nutr.*, 48, 138, 1994.

51. Roecker, E. B., Kemnitz, J. W., Ershler, W. B., and Weindruch, R., Reduced immune response in rhesus monkeys subjected to dietary restriction, *J. Gerontol.*, 51A, B276, 1996.

52. Grossmann, A., Rabinovitch, P. S., Lane, M. A., Jinneman, J. C., Ingram, D. K., Wolf, N. S., Cutler, R. G., and Roth, G. S., Influence of age, sex and dietary restriction on intracellular free calcium responses of CD4$^+$ lymphocytes in rhesus monkeys (*Macaca mulatta*), *J. Cell. Physiol.*, 162, 298, 1995.

53. Kemnitz, J. W., Weindruch, R., Roecker, E. B., Crawford, K., Kaufman, P. L., and Ershler, W. B., Dietary restriction of adult male rhesus monkeys: Design, methodology, and preliminary findings from the first year of study, *J. Gerontol.*, 48, B17, 1993.

54. Hansen, B. C. and Bodkin, N. L., Primary prevention of diabetes mellitus by prevention of obesity in monkeys, *Diabetes*, 42, 1809, 1993.

55. Cefalu, W. T., Wagner, J. D., Wang, Z. Q., Bell-Farrow, A. D., Collins, J., Haskell, D., Bechtold, R., and Morgan, T., A study of caloric restriction and cardiovascular aging in cynomolgus monkeys (*Macaca fascicularis*): a potential model for aging research, *J. Gerontol.*, 52A, B10, 1997.

56. David, J., Van Herrewege, J., and Fouillet, P., Quantitative under-feeding of *Drosophila*: effects on adult longevity and fecundity, *Exp. Gerontol.*, 6, 249, 1971.

57. Bourg, E. L. and Minois, N., Failure to confirm increased longevity in *Drosophila melanogaster* submitted to a food restriction procedure, *J. Gerontol.*, 51A, B280, 1996.

58. Comfort, A., Effect of delayed and resumed growth on the longevity of a fish (*Lefistes reticularis* Peters) in captivity, *Gerontologia*, 8, 150, 1963.

59. Fanestil, D. D. and Barrows, C. H., Jr., Aging in the rotifer, *J. Gerontol.*, 20, 462, 1965.

60. Masoro, E. J., Shimokawa, I., Higami, Y., McMahan, C. A., and Yu, B. P., Temporal pattern of food intake not a factor in the retardation of aging processes by dietary restriction, *J. Gerontol.*, 50A, B48, 1995.

61. Masoro, E. J., Yu, B. P., and Bertrand, H. A., Action of food restriction in delaying aging processes, *Proc. Natl. Acad. Sci. U.S.A.*, 79, 4239, 1982.

62. Kalu, D. N., Hardin, R. H., Cockerham, R., and Yu, B. P., Aging and dietary modulation of rat skeleton and parathyroid hormone, *Endocrinology*, 115, 1239, 1984.

63. Kalu, D. N., Masoro, E. J., Yu, B. P., Hardin, R. R., and Hollis, B. W., Modulation of age-related hyperparathyroidism and senile bone loss in Fischer 344 rats by soy protein and food restriction, *Endocrinology*, 122, 1847, 1988.

64. Kristal, B. S. and Yu, B. P., Aging and its modulation by dietary restriction, in *Modulation of Aging Processes by Dietary Restriction*, Yu, B. P., Ed., CRC Press, Boca Raton, FL, 1994, chap. 1.

13

Study Design for Nutritional Assessments in the Elderly

Hélène Payette, Yves Guigoz, and Bruno J. Vellas

CONTENTS

13.1 Introduction

13.1.1 Research Issues in Geriatric Nutrition

The assessment of nutritional status in older people is important because of its prognostic, therapeutic, and health-maintenance implications. The objectives of nutrition research in the elderly are to (1) describe the distribution and magnitude of nutritional problems, (2) understand the etiology of malnutrition, (3) clarify energy and nutritional requirements in old age, and (4) provide the information needed to manage and plan services for the prevention and treatment of malnutrition, that is, evaluate the effectiveness of various strategies for preventing or correcting poor nutritional status.

In the last 20 years, increased interest in geriatric nutrition has induced researchers to document the distribution and magnitude of nutritional problems in the elderly

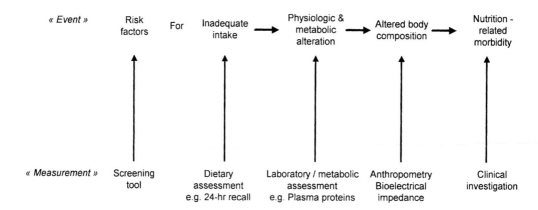

FIGURE 13.1
Sequence of events leading to clinically significant undernutrition in the elderly.

population. It has been observed that the prevalence of nutrition-related problems is greatly affected by the living arrangements and characteristics of older populations because these aspects are indicative of health and functional status. Among the autonomous and ambulatory elderly, the prevalence of protein-calorie undernutrition is very low [1–4]. However, as health and functional capacities deteriorate, the prevalence increases dramatically. In home support programs [5], low body weight was observed in 30% of the elderly clients and involuntary weight loss in 38%, while 45% were consuming less than the recommended daily minimum amount of protein [6].

In nursing homes and long-term care facilities, up to 30% of the elderly have a low dietary intake of protein, half are underweight and 15 to 60% have substandard midarm muscle circumference, serum albumin level or both, indicating clinical protein-calorie undernutrition [7]. The prevalence of protein-energy malnutrition in nursing home residents ranges from 23 to 85% [8]. This large range is related to population heterogeneity as regards reason for need for support or chronic institutionalization, mode of feeding, functional capacities, and degree of cognitive impairment. Therefore, the study design for the nutritional assessment of elderly subjects will have to take into account the living arrangements and characteristics of the various sub-groups.

Primary malnutrition, caused by inadequate dietary intake, appears to be a frequent phenomenon among high-risk elderly sub-groups. Adequate dietary intake is a basic requirement for nutritional health and, consequently, general health and function (see Figure 13.1). In the elderly, deterioration in nutritional status appears to be a rapid and hard-to-reverse process. Previous experimental studies have shown that advanced malnutrition is much more difficult to correct in the elderly than in younger adults [9–11]. It is therefore preferable to intervene before clear malnutrition becomes firmly established. Decreased food intake and weight loss are early warning signs that are prevalent in high-risk sub-groups and should be considered as key issues in research in geriatric nutrition. Furthermore, the benefits of early nutrition intervention as regards recovery from illness, rehabilitation, immune status, body strength and well-being have been demonstrated in the free-living and hospitalized elderly [12–16].

This chapter will focus on methodological issues regarding the design of studies to evaluate the impact of nutritional interventions in the elderly, particularly field trials among targeted sub-groups at risk of undernutrition. Clear answers to this last objective of

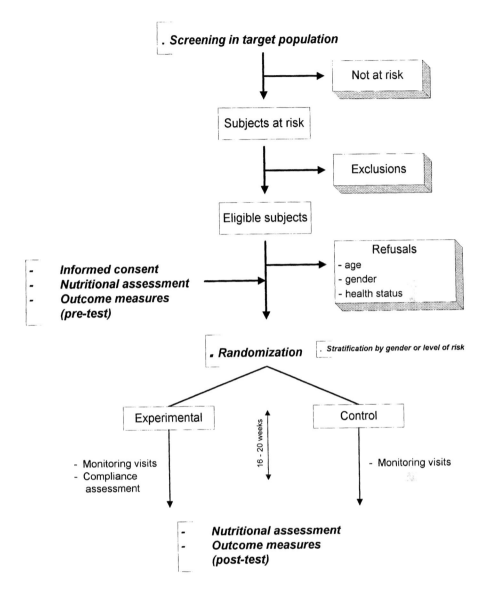

FIGURE 13.2
Study design for the evaluation of nutritional intervention in the elderly.

nutrition research in the elderly are best provided by experimental studies (see Figure 13.2). Research designs to answer the study question, namely, the effectiveness of nutrition intervention, require that the nutritional assessments (exposure and outcome measures of interest) are clearly defined and measured with the required accuracy. The evaluation of nutritional intervention in the community-living elderly requires that valid and reliable indicators as well as practical and non-invasive methods are available. Issues regarding selection of study subjects, determination of nutritional status, planning of the nutritional intervention, and measurement of the impact of the intervention will be discussed in terms of validity, reliability, avoidance of potential bias and confounding effects.

13.2 Selection of Study Subjects

13.2.1 Selection Criteria

In order to enable the researcher to detect nutrition, health, or functional improvements following nutrition intervention, subjects at risk of malnutrition should be carefully selected in order to avoid the "ceiling effect", a plateau where no further linear relationship can be observed. This effect is likely to occur in well-nourished or marginally malnourished elderly people. On the other hand, if nutritional status has deteriorated too far, it could be very difficult to reverse the malnutrition process [9,10]. Furthermore, very malnourished subjects are likely to be lost to follow-up because of worsening physical and mental health. Keeping lost to follow-up to a minimum is very important in experimental trials. If the drop out rate is high, this will affect the statistical power of the study, i.e. the capacity of the study to detect a significant impact of the nutritional intervention (see Section 13.6.1). In addition, if it is not proportional in both groups (for instance, if death occurs significantly more frequently in the experimental as compared to the control group), the drop-out rate may also affect the internal validity of the study. Internal validity refers to the capacity to ascribe the observed improvement, if any, to the treatment under study. Therefore, selection criteria for study subjects will be a compromise solution between these very different aspects regarding nutrition status, carefully taking into consideration the characteristics of the target population.

The selection criteria should also be valid. In other words, the measure used to select the study subjects should be both sensitive (truly at-risk subjects classified as such) and specific (truly not at-risk subjects classified as such). Two valid instruments are available in community and clinical settings and will permit the selection of subjects suitable for evaluating the effectiveness of nutritional intervention.

13.2.2 In a Community Setting

In this context, a screening tool based on risk factors for inadequate food intake is appropriate. The assumptions underlying this methodology are as follows: insufficient food intake is a primary cause of the malnutrition that is highly prevalent in the community-living frail elderly because of functional limitations in food-related activities. It is also the first step in the deterioration of the elder's nutritional status.

The *Questionnaire to Assess the Need for Help with Food in the Elderly* (Figure 13.3) is a multi-item questionnaire developed from factors associated with dietary intake identified through multivariate analyses of all the potential determinants of nutritional intake described in previous studies of community-living elderly people [17].The format is one page, both sides, where the professional ticks or circles the statements that apply to the elderly individual as reported by the latter. A total score is calculated and clear recommendations are provided for each risk level.

The preliminary version of the screening tool was developed after a consensus of experts, comprising field clinicians and dietitians, regarding the weight to be given to each item, the classification of the risk level and the appropriate recommendation. Validation studies showed that the screening tool could identify 78% of the elderly subjects at moderate risk of malnutrition. Its specificity for high risk levels was 77%.

This multi-item questionnaire is short, easy to use, and presents many advantages. It can be administered by any health or social service professional working in the community

Name : ______________________

Record N° : ______________________

Weight : ________ lbs

Adult height : ______ ft ______ in

Centre de recherche en gérontologie et gériatrie

QUESTIONNAIRE TO ASSESS THE NEED FOR HELP WITH FOOD IN THE ELDERLY

CIRCLE THE NUMBER CORRESPONDING TO THE STATEMENT THAT APPLIES TO THE CLIENT.

THE CLIENT :

is very thin	Yes	2
	No	0
has lost weight in the last year	Yes	1
	No	0
suffers from arthritis to the point of interfering with daily activities	Yes	1
	No	0
has a quality of vision with glasses (if necessary) that is	Good	0
	Medium	1
	Poor	2
has a good appetite	Often	0
	Sometimes	1
	Never	2
recently suffered a stressful life event (e.g. personal illness/death of loved one)	Yes	1
	No	0

The client USUALLY consumes the foods below for breakfast :

fruit or fruit juice	Yes	0
	No	1
eggs, cheese or peanut butter	Yes	0
	No	1
bread or cereals	Yes	0
	No	1
milk (1 cup or more than ¼ cup in coffee)	Yes	0
	No	1

TOTAL : ________

TOTAL SCORE		RECOMMANDATIONS
	Nutritional Risk	
6-13	High	Help with meal and snack preparation and referral to a dietitian
3-5	Moderate	Regular monitoring of diet (checking on food intake, providing advice and encouragement)
0-2	Low	Regular monitoring for changes in risk factors (i.e., changes in health or weight loss)

FIGURE 13.3
Questionnaire to assess the need for help with food in the elderly.

with the target population, and the screening procedure does not require biological specimens that are difficult to obtain in a community setting.

13.2.3   In a Clinical Setting

The objective of the *Mini-Nutritional Assessment* (MNA) (Figure 13.4) is to provide a rapid and simple assessment of the nutritional status of elderly people at risk of malnutrition in order to facilitate nutrition intervention [18,18a]. It had been designed for easy use in general medical practice, at admission to nursing homes or in geriatric assessment programs. The test comprises simple measurements and a brief questionnaire involving general, anthropometric and dietary assessments, as well as self-perceived health and nutrition.

The MNA has been validated in elderly populations ranging in age from 56 to 97 and above, from the very frail to the very active. Cross-validation indicated that the assessment was sensitive enough to allow 70 to 75% of the subjects to be classified as well-nourished or malnourished without further biochemical tests or clinical evaluation. When using the MNA, the malnourished category corresponds to a determination of low serum albumin that has been shown to be associated with increased morbidity [19], mortality [20], disability [21], and *sarcopenia* [22] (loss of skeletal muscle mass) in the elderly.

13.2.4   External Validity

The investigator wants the results obtained from his or her study to be generalized to the target population. This is referred to as external validity. For this purpose, the study sample must be representative of the population from which it is drawn because the impact of nutrition intervention for those who participate could be different from that for those who are eligible to participate from the target population but do not. Any factors that will affect the inclusion or the participation of eligible subjects at the beginning of the study might introduce a bias. For example, refusal to participate is likely to be more frequent among the less healthy or more frail elderly individuals. To avoid a selection bias of those who are more healthy and autonomous, a special effort must be made to reach ill or frail individuals through formal and informal care networks. Solutions include obtaining the cooperation of organizations such as nursing homes or home care programs, providing special training to interviewers so that they can deal with problems presented by the elderly, using other modes of interviewing or proxy respondents, when needed, as well as having relatives present to assist [23]. Furthermore, a way to obtain key information (age, gender, health status) on those people who refuse to participate or who subsequently drop out should be planned in the design of the study (see Figure 13.2).

13.3   Determination of Nutritional Status

The selection of the appropriate method and its correct use must be considered at the design stage of any study since it is determined by the objective of the study and the characteristics of the subjects being assessed. One might ask which measures or combination of measures might be used to make a satisfactory assessment of the nutritional status of elderly individuals in relation to the hypotheses being tested? Traditionally, nutritional assessment techniques are classified into four types: dietary, anthropometric, laboratory and

MINI NUTRITIONAL ASSESSMENT
MNA®

ID# _____

Last Name: _____ First Name: _____ M.I. _____ Sex: _____ Date: _____

Age: _____ Weight, kg: _____ Height, cm: _____ Knee Height, cm: _____

Complete the form by writing the numbers in the boxes. Add the numbers in the boxes and compare the total assessment to the Malnutrition Indicator Score.

ANTHROPOMETRIC ASSESSMENT

	Points
1. Body Mass Index (BMI) (weight in kg) / (height in m)^2 a. BMI < 19 = 0 points b. BMI 19 to < 21 = 1 point c. BMI 21 to < 23 = 2 points d. BMI ≥ 23 = 3 points	☐
2. Mid-arm circumference (MAC) in cm a. MAC < 21 = 0.0 points b. MAC 21 ≤ 22 = 0.5 points c. MAC > 22 = 1.0 points	☐.☐
3. Calf circumference (CC) in cm a. CC < 31 = 0 points b. CC ≥ 31 = 1 point	☐
4. Weight loss during last 3 months a. weight loss greater than 3kg (6.6 lbs) = 0 points b. does not know = 1 point c. weight loss between 1 and 3 kg (2.2 and 6.6 lbs) = 2 points d. no weight loss = 3 points	☐

GENERAL ASSESSMENT

	Points
5. Lives independently (not in a nursing home or hospital) a. no = 0 points b. yes = 1 point	☐
6. Takes more than 3 prescription drugs per day a. yes = 0 points b. no = 1 point	☐
7. Has suffered psychological stress or acute disease in the past 3 months a. yes = 0 points b. no = 2 points	☐
8. Mobility a. bed or chair bound = 0 points b. able to get out of bed/chair but does not go out = 1 point c. goes out = 2 points	☐
9. Neuropsychological problems a. severe dementia or depression = 0 points b. mild dementia = 1 point c. no psychological problems = 2 points	☐
10. Pressure sores or skin ulcers a. yes = 0 points b. no = 1 point	☐

DIETARY ASSESSMENT

	Points
11. How many full meals does the patient eat daily? a. 1 meal = 0 points b. 2 meals = 1 point c. 3 meals = 2 points	☐

Ref.: Guigoz Y, Vellas B and Garry PJ. 1994. Mini Nutritional Assessment: A practical assessment tool for grading the nutritional state of elderly patients. *Facts and Research in Gerontology.* Supplement #2: 15-59.

©1994 Nestec Ltd (Nestlé Research Center)/Nestlé Clinical Nutrition

	Points
12. Selected consumption markers for protein intake • At least one serving of dairy products (milk, cheese, yogurt) per day? yes ☐ no ☐ • Two or more servings of legumes or eggs per week? yes ☐ no ☐ • Meat, fish or poultry every day? yes ☐ no ☐ a. if 0 or 1 yes = 0.0 points b. if 2 yes = 0.5 points c. if 3 yes = 1.0 points	☐.☐
13. Consumes two or more servings of fruits or vegetables per day? a. no = 0 points b. yes = 1 point	☐
14. Has food intake declined over the past three months due to loss of appetite, digestive problems, chewing or swallowing difficulties? a. severe loss of appetite = 0 points b. moderate loss of appetite = 1 point c. no loss of appetite = 2 points	☐
15. How much fluid (water, juice, coffee, tea, milk,...) is consumed per day? (1 cup = 8 oz.) a. less than 3 cups = 0.0 points b. 3 to 5 cups = 0.5 points c. more than 5 cups = 1.0 points	☐.☐
16. Mode of feeding a. Unable to eat without assistance = 0 points b. self-fed with some difficulty = 1 point c. self-fed without any problem = 2 points	☐

SELF ASSESSMENT

	Points
17. Do they view themselves as having nutritional problems? a. major malnutrition = 0 points b. does not know or moderate malnutrition = 1 point c. no nutritional problem = 2 points	☐
18. In comparison with other people of the same age, how do they consider their health status? a. not as good = 0.0 points b. does not know = 0.5 points c. as good = 1.0 points d. better = 2.0 points	☐.☐

ASSESSMENT TOTAL (max. 30 points): ☐☐.☐

MALNUTRITION INDICATOR SCORE		
≥ 24 points	well-nourished	☐
17 to 23.5 points	at risk of malnutrition	☐
< 17 points	malnourished	☐

FIGURE 13.4
The Mini Nutritional Assessment form.

clinical (Figure 13.1). Measures that are inexpensive and easy to obtain in a wide variety of settings and that have been pre-tested and validated in elderly populations will be selected.

13.3.1   Dietary Assessment

In the context of evaluative studies, a detailed assessment of diet is required in order to provide an appropriate representation of the exposure and outcome variables of interest and to enable the researcher to detect any impact of the nutrition intervention on the subjects' usual diet.

In elderly persons, dietary intake information is preferably obtained through interviews because this causes less burden on the subjects and avoids problems such as writing difficulties related to physical (e.g., visual impairment, rambling) or mental (e.g., cognitive decline) capacities. A food diary or food record, although usually considered as a valid estimation of dietary intakes, is not always appropriate for the elderly since it requires highly motivated and educated subjects who are physically and mentally able to write and estimate portion sizes accurately.

The 24-hour recall is probably the most appropriate methodology to collect dietary data in elderly subjects for research purposes. Literacy or numeracy skills are not needed and subject involvement is kept to a minimum. It is designed to obtain data on the intake of all foods and beverages consumed within the preceding 24 hours. The interview is conducted by trained dietitians. Training is essential in order to circumvent some limitations in this methodology among the elderly. For example, since decreased short-term memory is often reported in this population, the interviewer will probe carefully for food consumption meal by meal, starting with the last meal going back 24 hours. The accuracy of assessment is improved by first measuring the cups and plates used by the participant on a regular basis, and then referring to these utensils to estimate portion sizes.

Because of day-to-day variations in food intake, one day is insufficient for determining an individual's usual dietary intake [24–27]. It was shown that 3 non-consecutive 24-hour dietary recalls were appropriate to characterize the usual food intake in free-living autonomous [4] and frail [28] elderly people. These homogeneous sub-groups were shown to have routine dietary patterns and a limited variety in food choices. For these reasons, 3-day repeated measures of food intake provide reliable estimates of the usual intake of energy and most nutrients. Taking into account that resources are often limited when carrying out research projects, the first interview is done face-to-face and the other two by telephone. Telephone dietary recall has been shown to provide a valid measure of the actual observed intake in elderly subjects [29].

Finally, a limitation often reported with the use of the 24-hour recall methodology, namely overestimation of low intakes and underestimation of high intakes, will have to be taken into account when interpreting the results.

Another method for estimating usual dietary intake includes the food frequency questionnaire. This method, either self-administered or administered during a personal interview, was designed to obtain information on the usual intake of a fixed list of foods and minimizes the effects of day-to-day variations in nutrient intake. Thus, it could represent an alternative to repeated 24-hour recalls. However, one must take into account that it is also dependent on memory recall, provides only semiquantitative data and is vulnerable to the tendency of subjects to exaggerate or minimize self-described behavior. Moreover, the level of abstraction required to estimate the frequency of intake of certain foods as well as portion size over long periods, such as four weeks, may be particularly difficult, if not impossible, for elderly subjects.

Food consumption data can also be obtained by a trained interviewer using the dietary history method, consisting of a combination of a 3-day record and a food checklist as used in the SENECA study [30,31]. The development of diet history methods for studies of older persons follows a series of sequential steps: (1) selecting the food items, (2) choosing the appropriate portion sizes for the visual aids, (3) estimating the usual frequencies, (4) collecting data on food preparation, (5) selecting the diet reference period, and (6) pretesting the questionnaire and interviewer training [32,33].

In practice, a full diet history is likely to be difficult and time-consuming to obtain, particularly in less healthy or frail elderly individuals. However, some specific questions can help the investigator to identify those at greater risk of malnutrition: (1) number of meals a day, (2) number of foods per meal, (3) patterns of eating, (4) capacity to purchase and prepare food, and (5) difficulties in eating alone [11].

Whatever the method used to collect dietary information, the conversion of food consumption into energy and nutrient intake is carried out using computer software and a nutrient database. Care should be taken to ensure that the software chosen meets the needs of the study adequately [30]. The accuracy of the final dietary intake data also requires that errors are kept to a minimum during collection, coding, aggregation, transcription and storage.

Interpretation of quantitative dietary intake data is carried out using standards such as the Recommended Dietary Allowances (RDA) [6] in the USA or *Apports Nutritionnels Conseillés* (ANC) in France (see Table 13.1). These standards are guidelines for adequate nutrition providing recommended intakes of specific levels of several essential nutrients by age group and sex. The levels are established on the basis of available scientific evidence to meet the known nutritional needs of practically all healthy persons in the target population. As regards the elderly, most recommendations are set up for ages 50 or 65 years and over, with the exception of the Canadian guidelines which are based on two age groups, namely 50 to 74 and 75 and over. Nutrient recommendations for older adults are based on extrapolations from younger adults. It is therefore preferable to interpret dietary data in the light of recent updates reviewing studies conducted in elderly populations [34–37]. Indeed, the needs and uses of nutrients are likely to be quite different at different levels of energy intake, and under different physiological states, in states of infection, and in the presence of the chronic diseases most prevalent in the elderly.

The recommended levels for protein, vitamins and minerals are values corresponding to two standard deviations above the mean of the distribution of the nutrient needs of healthy individuals. It is therefore generally understood that the probability of individual intakes being inadequate increases as the dietary intake moves away from the recommended level. The cut-off points most frequently used for interpreting observed individual intake are 67% or 75% of the recommended intake. However, although 100% of the nutrition recommendations overestimate the requirements of most healthy individuals, this is less likely to occur in a unhealthy, highly medicated elderly population recruited to participate in nutrition intervention studies. As a consequence, observed vitamins and minerals intake should probably also be compared to 100% of the recommended allowances.

13.3.2   Anthropometric Assessment

It is necessary to collect body composition information before, during and at the end of nutrition intervention trials in order to establish status and changes in energy stores and muscle mass. Therefore, the next component of nutritional assessment is anthropometry because the measures are easily performed in different settings, are non-invasive and

Recommended Dietary Allowances for Elderly in USA, UK, Germany, Canada, and France

	Units per day	1989 USA RDA	1991 UK DRV	1991 Germany ENZ	1990 Canada RNI / ANR		1992 France ANC
Age		51 +	50 +	65 +	50 - 74	75 +	65 +
Energy	kcal	1900 / 2300	1900 / 2330	1700 / 1900	1800 / 2300	1500 / 2000	1500 / 2100
Protein	g	50 / 63	46.5 / 53.3	47 / 55	47 / 60	47 / 57	60
	g / kg	0.75	0.75	0.8	0.86		1
	% of cal				15 - 20	15 - 20	12
Lipids	% of cal	30	33	25 - 30	30	30	30 - 35
Sat. fatty acids	% of cal	< 10	10		10	10	
Vitamin A	ug RE	800 / 1000	600 / 700	800 / 1000	800 / 1000	800 / 1000	800
Vitamin D	ug	5	10	5	5	5	12
Vitamin E	mg	8 / 10	> 3 / > 4*	12	6 / 7	5 / 6	12
Vitamin K	ug	65 / 80	62 / 71*	65 / 80			35
Thiamin	mg	1.0 / 1.2	0.8 / 0.9	1.1 / 1.3	0.8 / 0.9	0.8	1.3
Riboflavin	mg	1.4 / 1.2	1.1 / 1.3	1.5 / 1.7	1.0 / 1.3	1	1.5
Niacin	mg NE	13 / 15	12 / 16	15 / 18	14 / 16	14	15
Panthotenic acid	mg	4 - 7*	3 - 7*	6*	5 - 7*	5 - 7*	10
Vitamin B6	mg	1.6 / 2.0	1.2 / 1.4	1.6 / 1.8	1.1 / 1.8*	1.1 / 1.8*	2.0
Biotin	ug	30 - 100*	10 - 200*	30 - 100*			
Vitamin B12	ug	2.0	1.5	3	1	1	3
Folate	ug	180 / 200	200	300	195 / 230	200 / 215	300
Vitamin C	mg	60	40	75	30 / 40	30 / 40	80
Iron	mg	10	8.7	10	8 / 9	8 / 9	10
Calcium	mg	800	700	800	800	800	1200
Phosphorus	mg	800	550	1200	850 / 1000	850 / 1000	1000
Magnesium	mg	280 / 350	270 / 300	300 / 350	210 / 250	210 / 230	420
Zinc	mg	12 / 15	7.0 / 9.5	12 / 15	9 / 12	9 / 12	12
Iodine	ug	150	140	180	160	160	150

RDA :   Recommended Daily Allowances   *(National Research Council, RDA, 10th edition, 1989)*
DRV :   Dietary Reference Value   *(Report on Health and Social Subjects N° 41, HMSO, London, 1991)*
ENZ :   Empfehlungen für die Nährstoffzufuhr   *(DGE, ENZ, 5. Überarbeitung, 1991)*
RNI / ANR :   Recommended Nutrient Intakes / Apports   *(The Report of the Scientific Committee Review, 1990)*
    Nutritionnels Recommandés
ANC :   Apports Nutritionnels Conseillés   *(ANC pour la population française, 2ème édition, 1992)*

* Estimated safe adequate daily dietary intake (ESADDI) or tentative requirements.
When two values, a/b : a = recommended allowances for women, and b = recommended allowances for men

Adapted from Guigoz (1994); Guizoz, Y., Recommended dietary allowances (RDA) for the free-living elderly. Facts Res Gerontol, 15 (Suppl. : Nutrition), 113, 1994.

inexpensive and can be used to estimate body composition. However, some measurements and their interpretation present specific problems in elderly subjects.

The measure of stature is essential for the estimation of the different compartments of the body, such as % body fat or fat-free mass. As an example, the body mass index, a ratio of weight (kg) to height (m)^2, is generally considered to be a good indicator of adiposity. However, stature measurement poses real practical problems in older adults. First, height decreases with advancing age [38,39], almost entirely because of shrinkage of the vertebral column. Furthermore, many older persons have difficulty standing or maintaining an erect posture because of spinal deformity or the inability to stretch their knees with the result that stature measurements are spurious. An indirect estimate of stature from a measure of knee height has been described and shown to be accurate [40]. Appropriate equations and a nomogram were developed to facilitate the use of knee height to estimate stature [41]. Appropriate equations have also been developed for French elderly [42]. Some authors have suggested the use of a new index, namely the ratio of body weight (kg) to knee height (cm)^2, as an alternative to body mass index [43]. Further studies are required in order to establish the validity and predictive ability of this index among elderly populations. Recalled height, very easily obtained from each study subject, has been shown to be highly correlated with measured or estimated height in different elderly population groups and can be used as a proxy for stature measurement [44,45].

In a community setting with ambulatory elderly subjects, weight can easily be obtained with a portable electronic scale which gives precise measurements if it is used according to standardized techniques and calibrated before each use. Problems may arise, however, with older people who have difficulty standing erect without help. The validity of self-reported weight has been studied and a high correlation was observed between measured and recalled weight in elderly people [45,46].

Similar difficulties arise with the measurement of tricep skinfold that is used together with arm circumference to estimate % body fat and mid-arm muscle area which reflects muscle mass [47]. The measure is altered because of tissue flaccidity, increased compressibility of subcutaneous tissue, and decreased elasticity of the skin so that isolating adipose tissue from muscle is difficult. Training the interviewers as well as preliminary pilot studies are essential in order to ensure valid and reliable measures of skinfolds. Investigators should also be aware of some limitations of the use of arm circumference and tricep skinfold measurements to estimate subcutaneous fat deposits and muscle and bone area. The assumption underlying the estimation is that the size and proportion of fat and fat-free mass measured at the mid-arm are good estimates for the body as a whole [47]. Normal aging is associated with loss of skeletal muscle mass [48,49] and redistribution of adipose tissue from peripheral to central and intra-abdominal deposition [50]. These changes limit the ability of measurements such as arm circumference and triceps skinfold to correctly predict muscle, bone or fat area in the elderly since predictive equations and reference data are derived from measurements in young and middle-aged adults.

Trunk circumference has been suggested as a better index of stores of body fat than skin-folds in the healthy elderly [51,52] while calf circumference was reported to be significantly correlated with fat-free mass in the elderly [53]. Calf circumference is recommended as a sensitive measure of the loss of muscle mass in the elderly [54] and is recommended as a measurement of nutritional status in the elderly by the World Health Organization [55]. These measures, although not currently used, may represent good and sensitive indicators of changes in body composition.

Anthropometric indices of nutritional status in the elderly can lead to spurious interpretation because of the alteration in the composition and physical properties of certain tissues as well as the use of invalid assumptions in deriving body composition. However, in the absence of more accurate measures, anthropometry, when appropriately used, is very useful in showing changes in nutritional status following nutritional intervention.

A recent report from the World Health Organization recommends the use of data collected at local levels due to the limitations and absence of generally available reference data [55,56]. Reference data for anthropometric indices in elderly persons living in the USA are provided by the NHANES studies [57], by Chumlea and colleagues [58] and by the New Mexico Aging Process Study [59]. Corresponding European reference data have been reported in France [60,61] and the United Kingdom [62] and can be obtained from the SENECA study [63,64].

Other methods exist for precisely determining muscle mass, fat-free mass or fat mass that are suitable for the elderly. These include computed tomography, magnetic resonance imaging, and dual energy x-ray absorptiometry [65]. However, these methodologies require laboratory settings and support, and involve considerable cost and investment of time in the context of field or epidemiological studies.

New advances, such as bioelectric impendance, could allow an accurate quantification of body composition in elderly subjects. Bioelectrical impedance analysis has practical utility in elderly subjects in terms of safety, convenience to the subject and non-invasiveness. The development of a less expensive and portable instrument adds to its applicability in a variety of settings. The technology is simple to use and takes little time to perform. The technique can be particularly useful in differentiating whether weight gain following dietary modifications is due to an increase in muscle mass or body fat reserve.

The application of electrical impedance as a tool for evaluating body composition is fairly recent; it was developed by Lukaski and co-workers in 1985 [66]. The rationale and the detailed application of this methodology can be found in excellent recent reviews [67–69]. The estimation of fat-free mass is carried out using population-specific regression equations that are derived from statistical relationships of impedance to total body water or fat-free mass measured by independent techniques. In elderly populations, body composition estimates from impedance variables using age-specific prediction equations demonstrated high correlations with and relatively low prediction errors of estimates using densitometry [70], deuterium oxide (D_2O) dilution [71], tritiated water (^3H_2O) method [72] or potassium bromide (KBr) dilution [73]. The reliability of multi-frequency impedance measurements in estimating total body water, extra-cellular fluid volume, and fat-free mass has been demonstrated in young adults [74] and frail institutionalized elderly people [75].

Like other methods of assessing body composition, bioelectrical impedance depends upon several assumptions as regards the level of hydration and the density of the fat-free mass as well as the relationship among different body compartments. These assumptions may not hold true in the elderly. Possible edema and fluid accumulation accompanying degenerative disease, changes in skin texture, and changes in the amounts and structure of muscle, fat, and bone in the limbs with aging are likely to affect impedance measures and their interpretation. In addition, many conditions such as using diuretics or wearing a pacemaker or artificial protheses preclude or invalidate bioelectrical impedance measurements. For these reasons, despite its advantages, the method is not yet widely applicable in the elderly.

13.3.3   Laboratory Measurements

Some laboratory tests are useful indicators of recent nutrient intake and can be used in conjunction with dietary methods [76,77]. The most familiar and abundant of the serum proteins, as well as the most readily available clinically, is albumin. Decreased serum albumin level has been considered to be an indicator of depleted protein status and decreased protein intake. Low concentrations of serum albumin are associated with increased morbidity and mortality in institutionalized elderly [78] and elderly patients [79,80]. Several factors other than nutritional status affect serum albumin levels, such as inflammation or injury. Inflammatory response induces a decreased synthesis of albumin with a concomitant increased synthesis in acute phase proteins [77]. In the non-institutionalized elderly, serum albumin values < 40 g/L are associated with increased mortality [78] and values < 38 g/L are associated with increased morbidity, mortality and disability [22]. Albumin has also been reported to be associated with muscle mass, indicating that age-related decreases in albumin levels are associated with sarcopenia in the elderly [22]. Albumin may be considered as a non-specific indicator of health. Accepted normal ranges for albumin levels are 35–50 g/L; however, a newer cut-off in the normal range might be indicated for long-term nutritional studies (38 or 40 g/L) [22,78]. Reference values for the healthy elderly were also reported by Garry et al. (1989) [81].

Transthyretin (prealbumin), because of its short half-life (2 to 3 days) and small body pool (0.01 g/Kg body weight), is considered a more sensitive indicator of protein status and one that responds more rapidly to changes than albumin [82].

Alpha1-acid glycoprotein (AGP) and C-reactive protein (CRP) measurements will also be useful in assessing inflammatory processes that can affect albumin and transthyretin values. The measurement of acute phase proteins improve the interpretation of results in the evaluation of nutritional status [77].

The ratio of albumin and prealbumin to the acute phase proteins (AGP and CRP), refered to as Prognostic Inflammatory and Nutritional Index (PINI) [83], should be considered as an index for the prognosis of outcomes related to inflammation and nutrition in elderly patients [84].

Recommended methods to be used for micronutrient status have been published in the *Flair Concerted Action No. 10 Status Papers* (1993) [85]. Methods for vitamin A, carotenoids, vitamin E, vitamin D, riboflavin, vitamin B_6, vitamin B_{12}, folate, vitamin C, selenium, iron, zinc, and copper as well as clinical chemistry reference intervals for the healthy elderly are available [81,86].

Monitoring nutritional status requires the use of several laboratory tests used in conjunction with each other and with data derived from dietary and anthropometric methods, as well as the observation of clinical signs and symptoms of nutritional deficiencies [76]. Compared to other assessment methods, laboratory tests have the advantage of being relatively objective and quantitative.

13.3.4 Clinical Assessment

Clinical assessment should recognize the symptoms of malnutrition and include medical and social history in order to detect possible nutritional-related problems [76]. Specific attention should be given to difficulty in chewing (dysphagia), denture problems, and loss of appetite.

13.4 Nutritional Intervention

A two arm randomized controlled trial design is the best design to determine the impact of nutritional intervention on undernourished elderly individuals living in the community (Figure 13.2). The control group is used to determine whether any observed change is due to the intervention or some other factor that has also changed.

13.4.1 Allocation of Treatment Regimen

Randomly allocating subjects to different treatment groups ensures that any differences between the groups are only a chance occurrence and thus avoids as much as possible differences between the characteristics of the participants in the different groups. How-ever, if differences do arise, they must be taken into account in the analyses (see Section 13.6). If the investigator wants to make sure that the groups are exactly alike on certain key variables, it is possible to stratify on these variables (Figure 13.2). For example, elderly subjects can be allocated to groups that contain the same number of women and men at different levels of nutrition risk. This procedure enables the investigator to control for the fact that any dissimilarities between the experimental and control groups on these vari-ables introduce confusion in the analysis of the relationship between the intervention and the resulting nutritional status or other outcome variables. Tables of random numbers and computer programs are widely available for the random allocation of subjects in experi-mental studies.

Subjects should be selected for inclusion in the trial and pre-tested before it is known, by the investigator and by themselves, which group they will be assigned to (Figure 13.2).

13.4.2   Intervention

The goal of the nutritional intervention is to increase total energy and nutrient intake. However, renourishing undernourished elderly persons requires intensive and sustained supplementation in order to improve body weight, specific nutritional parameters and, ideally, functional and health status. Previous studies have shown that a considerable num-ber of excess calories are needed to restore optimal nutritional status in undernourished elderly individuals [9,10]. It is therefore necessary to use an oral commercial supplement, easily available on the market, that provides a high level of energy and nutrients, that tastes good and is appreciated by the elderly subjects. In order to improve compliance and pre-vent taste fatigue, a variety of different flavors should be offered throughout the duration of the study. Because of the wide variability in individual energy requirements, subjects should be encouraged to attain the maximal tolerable energy intake to gain 0.5 kg of weight per week or maintain their weight, if they were losing weight prior to entering the study. Prescribing a commercial nutrient-rich oral supplement will help overcome the anorexia of malnutrition frequently observed among the undernourished elderly and will lead to increased total food intake while providing sufficient amounts of proteins and micronutri-ents. The amounts of supplement recommended will be individually adjusted over the fol-low-up period according to total energy intake and weight changes.

The duration of supplementation should be long enough to enable the investigator to detect an improvement in the nutritional and health status of the experimental subjects as compared to the controls. Elderly persons may require more prolonged nutritional support than do younger patients to achieve nutrition replenishment. A recent 12-week trial period of oral supplementation among undernourished free-living frail elderly persons suggested that a longer follow-up may be necessary to observe very gradual improvements in func-tional status [87]. A serious limitation, however, is the increased number of lost to follow-up in this high risk population with a longer observation period. Nevertheless, it appears that a period of 16 to 20 weeks could be appropriate for showing detectable changes and keeping lost to follow-up to a minimum. Extending periods of data collection may also allow those who feel ill temporarily to have an opportunity to comply more closely with the study regimen when they feel better.

Deviation from the protocol needs to be documented in all subjects. Dietary intake assessment will be carried out regularly to ascertain that oral supplements do not replace usual dietary intake. Intake should be measured in the control group as well because they can alter their food behavior so as to make them more like the treatment group in their "exposure" status. An independent measure of compliance is essential in such a trial. Weight gain is a good indicator of increased energy intake. It is generally accepted that on average an extra 7,700 kcal (32,000 KJ) will provide a weight gain of 1 kg, although higher amounts could be necessary in undernourished elderly persons [9,10]. In addition transthyretin may be used as an indicator of short-term changes in protein status. Particular vitamins or minerals status can also be used to address specific research issues.

13.5 Outcome Measures

Valid, reliable, and clinically significant outcome measures for use in the evaluation of the effectiveness of nutrition interventions conducted in the elderly are greatly needed. These indicators must be uncomplicated, reproducible and sensitive to changes in body composition. Standardized methods or techniques are not fully applicable in elderly subjects especially those who are frail and undernourished and in the context of field trials. In community-living elderly subjects, better response rates are achieved if physiological data are collected and the physical examination is done at the respondent's residence rather than at a central location involving transportation. We have recently studied the validity of various functional status indicators. Results suggest that handgrip strength, as measured with the Martin Vigorimeter (Gebruder Martin, Ludwigstaler Strabe 132, Postfach 60, D-7200, Tuttlingen, Germany), and elbow flexion or knee extension strength, as measured with the Microfet2™ hand-held dynamometer (Hoggan Health Industries, Draper, Utah), are valid and reliable indicators of muscle mass in free-living frail elderly subjects [88].

When using questionnaires, it is essential that they be short, pre-tested and validated in elderly subjects. For example, the elderly are more likely than young adults to have difficulty placing themselves in predetermined response categories or on the type of visual analog scales traditionally used for health measurements [89]. Considering the fatigue factor with elderly study subjects, interviews should last no longer than 45 to 60 minutes. Alternating questionnaires and physiological data collection may decrease the burden of participation.

If interviewers assessing outcome variables know whether the subject is allocated to the experimental or the control group, they may solicit the information differently. It is therefore very important that the interviewers are blinded as to the group allocation of the study subjects. Furthermore, they should be carefully trained so that they collect information in a standardized manner throughout the study.

On the other hand, if the subject knows his or her treatment group — which is usually the case in dietary interventions — this may affect his or her response to treatment. In addition, improved nutrition or health status may result from the social aspect of the intervention rather than increased energy or nutrient intake. This potential bias can be partially circumvented by visiting and offering small, non-nutrition related, gifts to control subjects.

13.6 Statistical Analyses

The similarity of the baseline characteristics of the groups should be tested using unpaired t tests or chi-square tests. The study groups will be compared with respect to changes over time in nutrition status, functional or health status measures while controlling for baseline values. This can be achieved by ANOVA for repeated measures or paired t tests on the means of changes over the duration of the study. Analyses of covariance will be used to take into account baseline characteristics showing the different distributions between the groups.

The "intent to treat" principle will be applied to all analyses. This implies that all randomized subjects are included in the analyses whether they were compliant or not. In a subsequent step or secondary analyses, it may be acceptable to use a restricted subset of the data on the basis of compliance. However, the investigator should be aware that the direction of the relationship cannot be determined in these sub-group analyses because the potential confounders cannot be adequately controlled. For instance, it may be that those who comply sufficiently well to be included in this sub-group are different in other important characteristics (leading to greater compliance) from those not adequately complying.

13.6.1   Statistical Power

Calculating sample size requires the investigator to have an idea of the magnitude of the effect the treatment is likely to have on the outcome measures so that the number of subjects required in the treatment group can be estimated. The reader is referred to an extensive and comprehensive discussion of the principles and techniques underlying the concept of statistical power as related to sample size in the context of experimental studies [90,91]. Briefly, the principle of the tests of statistical significance is usually that there is a difference between groups that can be measured and that the variance of the estimates of the differences can be estimated. For a given variance, the bigger the difference between groups, the more likely that there will be a statistically significant result. On the other hand, if the sample size is very large, small differences will be statistically significant, but these differences may not have any biological importance. It is therefore reasonable to equate statistical significance with the clinical importance of the effect in order to maximize the ability of the study to detect a statistically *and* clinically significant result while keeping costs to a minimum.

References

1. Garry, P. J., Goodwin, J. S., Hunt, W. C., Hooper, E. M., and Leonard, A. G., Nutritional status in a healthy elderly population: dietary and supplemental intakes, *Am. J. Clin. Nutr.*, 36, 319, 1982.
2. McGandy, R. B., Russell, R. M., Hartz, S. C., Jacob, R. A., Tannenbaum, S., and Peters, H. et al., Nutritional status survey of healthy non-institutionalized elderly: energy and nutrient intakes from three-day diet records and nutrient supplements, *Nutr. Res.*, 6, 785, 1986.
3. Horwath, C. C., Dietary survey of a large random sample of elderly people: energy and nutrient intakes, *Nutr. Res.*, 9, 479, 1989.
4. Payette, H. and Gray-Donald, K., Dietary intake and biochemical indices of nutritional status in an elderly population, with estimates of the precision of the 7-d food record, *Am. J. Clin. Nutr.*, 54, 478, 1991.
5. Payette, H. and Gray-Donald, K., Risk of malnutrition in an elderly population receiving home care services, *Facts and Research in Gerontology*, Suppl. No 2, 71, 1994.

6. National Academy of Sciences, *Recommended Dietary Allowances*, 10th ed., National Academy Press, Washington DC, 1989.
7. Rudman, D. and Feller, A. G., Protein-calorie undernutrition in the nursing home, *J. Am. Geriatr. Soc.*, 37, 173, 1989.
8. Thomas, D. R., Outcome from protein-energy malnutrition in nursing homes residents, *Facts Res. Gerontol.*, Suppl. No 2, 87, 1994.
9. Shizgal, H. M., Martin, M. F., and Gimmon, A., The effects of age on the caloric requirement of malnourished individuals, *Am. J. Clin. Nutr.*, 55, 783, 1992.
10. Abbasi, A. A., Basu, S., and Rudman, D., Caloric requirements for weight gain in malnourihed nursing home patients, *J. Am. Ger. Soc.*, 40, SA24, 1992.
11. Vellas, B. and Guigoz, Y., Nutritional assessment as part of geriatric evaluation, in *Geriatric Assessment Technology: The State of Art*, Rubenstein, L. Z., Wieland, D., and Bernabei, R., Eds., Editrice Kurtis s.r.l., Milano (Italy), 1995.
12. Efthimiou, F., Fleming, J., Gomes, C., and Spiro, S. G., The effect of supplementary oral nutrition in poorly nourished patients with chronic obstructive pulmonary disease, *Am. Rev. Respir. Dis.*, 137, 1075, 1988.
13. Whittaker, J. S., Ryan, C. F., Buckley, P. A., and Road, J. D., The effects of refeeding on peripheral and respiratory muscle function in malnourished chronic obstructive pulmonary disease patients, *Am. Rev. Respir. Dis.*, 142, 283, 1990.
14. Chandra, R. K., Effect of vitamin and trace-element supplementation on immune responses and infection in elderly subjects, *Lancet*, 340, 1124, 1992.
15. Bogden, J. D., Bendich, A., and Kemp, F. W., et al., Daily micronutrient supplements enhance delayed-hypersensitivity skin test responses in older people, *Am. J. Clin. Nutr.*, 60, 437, 1994.
16. Delmi, M., Rapin, C. H., Bengoa, J. M., Delmas, P. D., Vasey, H., and Bonjour, J. P., Dietary supplementation in elderly patients with fractured neck of the femur, *Lancet*, 335, 1013, 1990.
17. Payette, H., Gray-Donald, K., Cyr, R., and Boutier, V., Predictors of dietary intake in a functionally dependent elderly population in the community, *Am. J. Public Health*, 85, 677, 1995.
18. Guigoz, Y., Vellas, B., and Garry, P. J., Mini Nutritional Assessment: a practical assessment tool for grading the nutritional state of elderly patients, *Facts Res. Gerontol.*, 15 (Suppl.: Nutrition), 59, 1994.
18a. Guigoz, Y., Vellas, B., and Gary, P.J., Assessing the nutritional status of the elderly: The Mini Nutritional Assessment as part of the geriatric evaluation, *Nutr. Rev.*, 54, S59, 1996.
19. Gillum, R. F., Ingram, D. D., and Makuc, D. M., Relationship between serum albumin concentration and stroke incidence and death: the NHANES I Epidemiologic Follow-up Study, *Am. J. Epidemiol.*, 140, 876, 1994.
20. Klonoff-Cohen, H., Berrett-Conor, L. E., and Edelstein, S. L., Albumin levels as a predictor of mortality in the healthy elderly, *J. Clin. Epidemiol.*, 45, 207, 1992.
21. Corti, M. C., Guralnik, J. M., Salive, M. E., and Sordin, J. D., Serum albumin level and physical disability as predictors of mortality in older persons, *JAMA*, 272, 1036, 1994.
22. Baumgartner, R. N., Koehler, K. M., Romero, L., and Garry, P. J., Serum albumin is associated with skeletal muscle in elderly men and women, *Am. J. Clin. Nutr.*, 64, 552, 1996.
23. Magaziner, J., The use of proxy respondents in health studies of the aged, in *The Epidemiologic Study of the Elderly*, Wallace, R. B. and Woolson, R. F., Eds., Oxford University Press, Inc., New York, 1992.
24. Beaton, G. H., Milner, J., and Corey, P. et al., Sources of variance in 24-hour dietary recall data: implications for nutrition study design and interpretation, *Am. J. Clin. Nutr.*, 32, 2546, 1979.
25. Hunt, W. C., Leonard, A. G., Garry, P. J., and Goodwin, J. S., Components of variance in dietary data for an elderly population, *Nutr. Res.*, 3, 433, 1983.
26. Marr, J. W. and Heady, J. A., Within- and between-person variation in dietary surveys: number of days needed to classify individuals, *Hum. Nutr. Appl. Nutr.*, 40A, 347, 1986.
27. Nelson, M., Black, A. E., Morris, J. A., and Cole, T. J., Between- and within-subject variation in nutrient intake from infancy to old age: estimating the number of days required to rank dietary intakes with desired precision, *Am. J. Clin. Nutr.*, 50, 155, 1989.
28. Payette, H., Unplublished data, 1993.

29. Dubois, S. and Boivin, J. F., Accuracy of telephone dietary recalls in elderly subjects, *J. Am. Diet. Assoc.*, 90, 1680, 1990.
30. Cameron, M. E. and van Staveren, W. A., *Manual on Methodology for Food Consumption Studies*, Oxford University Press, Inc., Oxford UK, 1988.
31. Van't Hof, M. A., Hautvast, J. G. A. J., Schroll, M., and Vlachonikolis, I. G., Design, methods and participation, *Eur. J. Clin. Nutr.*, 45, 5, 1991.
32. Garry, P. J., and Chumlea, W. C., Epidemiologic and methodologic problems in determining nutritional status of older persons, *Eur. J. Clin. Nutr.*, 50, 1121, 1989.
33. Hankin, J. H., Development of a diet history questionnaire for studies of older persons, *Am. J. Clin. Nutr.*, 50, 1121, 1989.
34. Russell, R. M. and Suter, P. M., Vitamin requirements of elderly people: an update, *Am. J. Clin. Nutr.*, 58, 4, 1993.
35. Campbell, W. W., Crin, M. C., Dallal, G. E., Young, V. R., and Evans, W. J., Increased protein requirements in elderly people: new data and retrospective reassessments, *Am. J. Clin. Nutr.*, 60, 501, 1994.
36. Wood, R. J., Suter, P. M., and Russell, R. M., Mineral requirements of elderly people, *Am. J. Clin. Nutr.*, 62, 493, 1995.
37. Roberts, S. B., Energy requirements of older individuals, *Eur. J. Clin. Nutr.*, 50, S112, 1996.
38. Chumlea, W. C., Garry, P. J., Hunt, W. C., and Rhyne, R. L., Serial changes in stature and weight in a healthy elderly population, *Hum. Biol.*, 60, 918, 1988.
39. Cline, M. G., Meredith, K. E., Boyer, J.T., and Burrows, B., Decline of height with age in adults in a general population sample: estimating maximum height and distinguishing birth cohort effects from actual loss of stature with aging, *Hum. Biol.*, 61, 415, 1989.
40. Chumlea, W. C., Roche, A. F., and Steinbaugh, M.L., Estimating stature from knee height for persons 60 to 90 years of age, *J. Am. Geriatr. Soc*, 33, 116, 1985.
41. Chumlea, W. C., Roche, A. F., and Mukherjee, D., Nutritional Assessment of the Elderly Through Anthropometry presented at The Ross Medical Nutritional System, Ross Laboratories, Columbus, Ohio, July 1987.
42. Guo, S., Wu, X., Vellas, B., Guigoz, Y., and Chumlea, W. C., Prediction of stature in the French elderly, *Age. Nutr.*, 5, 169, 1994.
43. Roubenoff, R. and Wilson P. W. F., Advantage of knee height over height as an index of stature in expression of body composition in adults, *Am. J. Clin. Nutr.*, 57, 609, 1993.
44. Heany, R. P. and Ryan, R., Relation between measured and recalled body height, *New Engl. J. Med.*, 319, 795, 1988.
45. Boutier, V. and Payette, H., Validité du poids et de la taille obtenus de mémoire chez des personnes âgées recevant des services d'aide à domicile, *Age Nutr.*, 6, 17, 1994.
46. Rowland, M. L., Self reported weight and height, *Am. J. Clin. Nutr.*, 52, 1125, 1990.
47. Gibson, R. S., *Principles of Nutritional Assessment*, Oxford University Press, Inc., New York, 1990, chap. 11.
48. Cohn, S. H., Vartsky, D., Yasumura, S., Savitsky, A., Zanzi, I., Vaswani, A., and Ellis K. J., Compartmental body composition based on total-body, potassium, and calcium, *Am. J. Physiol.*, 239, E524, 1980.
49. Kehayias, J. J., Zhuang, H., Fiatarone, M., Roubenoff, R., and Evans, W. J., Body composition in aging: new techniques for the study of lean tissue depletion, *FASEB J.*, 5, A1447, 1991.
50. Schwartz, R. S., Shuman, W. P., and Bradbury, V. L. et al., Body fat distribution in healthy young and older men, *J. Gerontol.*, 45, M181, 1990.
51. Chumlea, W. C., Roche, A. E., and Webb, P., Body size, subcutanious fatness and total body fat in older adults, *Int. J. Obes.*, 8, 311,1984.
52. Conceicao, J., Vellas, B., Ghisolfi-Marque, A., Suc, I., Lauque, S., Ficat-Faisant, C., Fagen, M. C., Bodin, T., Chumlea, W. C., and Albarede, J. L., Étude des marqueurs anthropométriques au sein d'une population de 224 sujets âgés vivant en maison de retraite, in *L'année Gérontologique*, Albarede, J. L., Vellas, B., and Garry, P., Eds., Serdi, Paris, 1994, 157.
53. Patrick, J. M., Bassey, E. J., and Fentem, P. H., Changes in body fat and muscle in manual workers at and after retirement, *Eur. J. Appl. Physiol.*, 49, 187, 1982.

54. Chumlea, W. C., Guo, S. S., Vellas, B., and Guigoz, Y., Techniques of assessing muscle mass and function (Sarcopenia) for epidemiological studies of the elderly, *J. Gerontol.*, 50A (Special Issue), 45, 1995.

55. World Health Organization, Physical status: the use and interpretation of anthropometry, Report of a WHO Expert Committee, WHO, *Tech. Rep. Ser.*, 854, 1995.

56. de Onis, M. and Habicht, J. P., Anthropometric reference data for international use: recommendations from a World Health Organization Expert Committee, *Am. J. Clin. Nutr.*, 64, 650, 1996.

57. Najjar, M. F., and Rowland, M., Anthropometric reference data and prevalence of overweight, United States, 1976–1980, Data from the Health Examination Survey. Vital and health Statistics Series 11, DHEW Publication (PHS), National Center for Health Studies, Rockville, MD, USA, 1987.

58. Chumlea, W. C., Roche, A. F., and Mukherjee, D., Some anthropometric indices of body composition for elderly adults, *J. Gerontol.*, 41, 36, 1986.

59. Garry, P. J., Goodwin, J. S., Hunt, W. C., Hopper, E. M., and Leonard, A. G., Nutritional status in a healthy elderly population: dietary and supplemental intakes, *Am. J. Clin. Nutr.*, 36, 319, 1982.

60. Delarue, J., Constans, T., Malvy, D., Pradignac, A., Couet, C., and Lamisse, F., Anthropometric values in an elderly French population, *Brit. J. Nutr.*, 71, 295, 1994.

61. Vellas, B. J., Chumlea, W. C., Béziat, F., Ghisolfi-Marque, A., Guo, S., Conceicao, J., Sédeuilh, M., Dufetelle, R., Albarède, J. L., Étude des normes anthropométriques dans une population française de 492 personnes âgées en bonne santé et vivant à domicile, *L'Année Gérontologique*, 7, 203, 1992.

62. Burr, M. L., Phillips, K. M., Anthropometric norms in the elderly, *Brit. J. Nutr.*, 51, 165, 1984.

63. de Groot, L. C. P. G. M., Sette, S., Zajkas, G., Carbajal, A., Amorin-Cruz, J. A., Nutritional status: anthropometry, *Eur. J. Clin. Nutr.*, 45 (suppl. 3), 31, 1991.

64. de Groot, L. C. P. G. M., Enzi, G., Perdigao, A. L., Deurenberg, P., Longitudinal changes in anthropemetric characteristics of elderly Europeans, *Eur. J. Clin. Nutr.*, 50 (suppl. 2), S9, 1996.

65. Lukaski, H. C., Methods for the assessment of human body composition: traditional and new, *Am. J. Clin. Nutr.*, 46, 537, 1987.

66. Lukaski, H. C., Johnson, P. E., Bolonchuk, W. W., and Lykken, G. I., Assessment of fat-free mass using bioelectrical impedance measurements of the human body, *Am. J. Clin. Nutr.*, 41, 810, 1985.

67. Baumgartner, R. N., Chumlea, W. C., and Roche, A. F., Bioelectric impedance for body composition, *Exerc. Sport. Sci. Rev.*, 18, 193, 1990.

68. Kushner, R. F., Bioelectrical impedance analysis: a review of principals and applications, *J. Am. Coll. Nutr.*, 11, 199, 1992.

69. Chumlea, W. C. and Guo, S. S., Bioelectrical impedance and body composition: Present status and future directions, *Nutr. Rev.*, 52, 123, 1994.

70. Deurenberg, P., Van der Kooij, K., Evers, P., and Hulshof, T., Assessment of body composition by bioelectrical impedance in a population aged > 60 y, *Am. J. Clin. Nutr.*, 51, 3, 1990.

71. Schols, A. M. W. J., Wouters, E. F. M., Soeters, P. B., and Westerterp, K. R., Body composition by bioelectrical-impedance analysis compared with deuterium dilution and skinfold anthropometry in patients with chronic obstructive pulmonary disease, *Am. J. Clin. Nutr.*, 53, 421, 1991.

72. Sergi, G., Baggio, B., and Perini, P. et al., Body composition in normal-weight elderly subjects: preliminary observations, *Age. Nutr.*, 3, 30, 1992.

73. Visser, M., Deurenberg, P., and van Staveren, W. A., Multi-frequency bioelectrical impedance for assessing total body water and extracellular water in elderly subjects, *Eur. J. Clin. Nutr.*, 49, 256, 1995.

74. Chumlea, W. C., Guo, S., Baumgartner, R. N., Bellasari, A., and Siervogel, R. M., Reliability for multifrequency bioelectric impedance. *Am. J. Hum. Biol.*, 6, 195, 1994.

75. Payette, H. Unpublished data, 1995.

76. Morrow, F. D., Sahyoun, N., Jacob, R. A., and Russell, R. M., Clinical assessment of the nutritional status of adulte, in *Nutritional Biochemistry and Metabolism with Clinical Applications*, 2nd ed., Linder, M. C., Ed., Elsevier, New York, 1991, 391.

77. Shenkin, A., Cederblad, G., Elia, M., and Isaksson, B., International federation of Clinical Chemistry — Laboratory assessment of protein-energy status, *Clin. Chim. Acta*, 253, S5, 1996.

78. Sahyoun, N. R., Jacques, P. F., Dallal, G., and Russell, R. M., Use of albumin as a predictor of mortality in community-dwelling and institutionalized elderly populations, *J. Clin. Epidemiol.*, 49, 981, 1996.

79. Agarwal, N., Acevedo, F., Leighton, L. S., Cayten, C. G., and Pitchumoni, C. S., Predictive ability of various nutritional variables for mortality in elderly people, *Am. J. Clin. Nutr.*, 48, 1173, 1988.

80. Ferguson, R. P., O'Connor, P., Crabtree, B., Batchelor, A., Mitchell, J., and Cappola, D., Serum albumin and prealbumin as predictors of clinical outcomes of hospitalized elderly nursing home residents, *J. Am. Geriatr. Soc*, 41, 545, 1993.

81. Garry, P. J., Hunt, W. C., Vanderjagt, D. J., Rhyne, R. L., Clinical chemistry reference intervals for healthy subjects, *Am. J. Clin. Nutr.*, 50, 1219, 1989.

82. Bernstein, L., Bachman, T. E., Meguid, M., Ament, M., Baumgartner, T., Kinosian, B., Martindale, R., and Siekman, M., Measurement of visceral protein status ins assessing protein and energy malnutrition: standard of care, *Nutrition*, 11, 169, 1995.

83. Ingenbleek, Y. and Carpentier, Y. A., A prognostic inflammatory and nutritional index scoring critically ill patients, *Intern. J. Vit. Nutr. Res.*, 55, 91, 1995.

84. Alix, E., Papin, A., Fromont, P., Queinec, S., Vieron, M. C., Coutant, P., Vetel, J. M., and Mazaud, P., Lesourd, B., Index pronostique nutritionnel et inflammatoire (PINI): évaluation en court séjour gériatrique chez 260 personnes âgées de plus de 75 ans, *Age. Nutr.*, 4, 63, 1993.

85. Flair Concerted Action No 10 Status Papers, Internat, *J. Vit. Nutr. Res.*, 63, 247, 1993.

86. Faulkner, W. R., and Meites, S., *Geriatric Clinical Chemistry Reference Values*, American Association for Clinical Chemistry Press, Washington, 1994.

87. Gray-Donald, K., Payette, H., and Boutier, V., Randomized clinical trial of nutritional supplementation shows little effect on functional status among free-living frail elderly, *J. Nutr.*, 125, 2965, 1995.

88. Payette, H., Hanusaik, N., Boutier, V., Morais, J. A., and Gray-Donald, K., Muscle strength and functional mobility in relation to lean body mass in free-living frail elderly women, *Eur. J. Clin. Nutr.*, 52, 45, 1998.

89. McDowell, I. and Newell, C., Measuring Health: A Guide to Rating Scales and Questionnaires, Oxford University Press, Inc., (2nd ed.), New York, 1996.

90. Friedman, L. M., Furberg, C. D., and DeMets, D. L., *Fundamentals of Clinical Trial*, PSG Publishing Co., Inc., (2nd ed.), Massachussets, 1985, chap. 7.

91. Overall, J. E. and Doyle, S. R., Estimating sample sizes for repeated measurement designs, *Controlled Clinical Trials*, 15, 100, 1994.

14

Exercise

Li Li Ji and Mitch Gore

CONTENTS

14.1 Introduction

The relationship between exercise and aging may be viewed in two perspectives: first, aging significantly affects mobility, metabolic responses, and efficiency of exercise [1]; and second, acute and chronic exercise may alter physiological, biomechanical, and biochemical changes normally seen in the aging process, or longevity itself [2]. As organisms age they often become more sedentary. Sometimes it is difficult to separate the effects of aging *per se* from the effects of lack of exercise. In keeping with the focus of this book, we will concentrate primarily on how exercise can be used to study the aging process.

During evolution, higher organisms have developed specialized organs and tissues to ensure the speed, endurance, and strength that allow for the capture of prey, the gathering of food, and escape from predators [3]. Because of this, the locomotive apparatus and accessory organs constitute the bulk of the body mass. While it is not surprising that most studies of exercise focus on musculo-skeletal systems, exercise is highly integrated in that it requires participation and coordination of virtually all physiological systems and mobilizes numerous biochemical, neuroendocrinal, and hemodynamic processes. Therefore, in order to understand the impact of exercise on one of the most basic biological phenomena, aging, we not only need to investigate muscles and bones, but also other tissues such as heart, liver, kidney, and endocrine glands. Furthermore, observations should not only be limited to organs and tissues, but also extended to the cellular, subcellular, and molecular levels.

The free radical theory of aging has allowed the establishment of a powerful link between exercise and aging research [4]. A fundamental premise for this theory is that reactive oxygen species (ROS) generated in normal metabolic processes are the underlying reason for cell and tissue oxidative damage seen throughout the aging process [5]. Since exercise increases metabolic rate, which is reflected by a greater amount of oxygen uptake, ROS production is also expected to increase during physical exertion [6]. While this may seem to be detrimental to the elderly who are physically active, exercise is also known to cause adaptive responses such as improvement in antioxidant defense capacity [7]. Therefore, an important consideration in aging and exercise studies is to be able to separate the influences of exercise from those that occur solely due to aging [8,9]. Despite decades of effort, there is no uniform theory which can accurately describe the influence of exercise on aging.

In this chapter, we intend to highlight the major factors to be considered in designing and conducting exercise and aging research. Because of the focus of the book and the limitation of our experience, discussions are primarily focused on the physiological and biochemical information gained from animal studies.

14.2   Importance of Exercise in Aging Research

14.2.1   Morbidity, Mortality, and Exercise

In the human population, morbidity is concentrated in the last two decades of life, beginning on the average at age 55 and increasing in frequency until the average age of death at 75 [10], with an increase of approximately two years in longevity in physically active people as compared to less active people [11,12]. At first glance this may seem to add to the health care problem in that more people would be living longer with chronic illness, but that is not the case. A recent longitudinal study noted that disability levels in a vigorously exercising population remained below that of non-exercisers and that significant increases in disability were delayed by approximately 15 years [10]. These data indicate that engaging in regular physical activity would increase the age of onset of chronic illness and shorten the time between the onset of morbidity and death. This compression of the period of morbidity as a result of physical exercise would represent a significant improvement in the quality of life of the elderly and result in major reductions in the cost of treating the medical conditions of the elderly [10]. Despite these clear benefits, little is known about the adaptive mechanisms involved and the time period where major protection offered by exercise occurs. Furthermore, it is not clear whether aged individuals are more susceptible to some of the harmful effects of rigorous exercise reported in recent literature as a result of increased exposure to ROS [6,7].

14.2.2 Aging and Sarcopenia

The loss in aerobic capacity in the elderly has an significant impact on the quality of life. As an individual's aerobic capacity declines, the oxygen requirement for the daily activities come closer and closer to the maximal oxygen consumption (VO_2 max), making it increasingly more difficult to perform. Evans [13] illustrated this point vividly by showing that VO_2 max of 75 to 80-year old men and women could be exceeded by walking upstairs. Activities such as taking a shower or walking 2 mph could require over 50% VO_2 max. The decline in aerobic capacity in the elderly may be caused by multiple factors such as decreased cardiac output and contractility, lower maximal heart rates, and muscle mass [14]. The decline in muscle mass with aging has been proposed to play an especially important role in the age-associated decline in aerobic capacity, partially due to the losses of muscle mitochondrial content. [15,16,17]. Aerobic training in older humans can increase skeletal muscle oxidative capacity to the same level as seen in younger muscle, however, the adaptive mechanisms are more likely to be of a peripheral nature such as increased mitochondrial volume and protein rather than central adaptations such as changes in maximal cardiac output [15,18]. Studies with animals have provided additional insights. Mitochondrial enzyme activities increase similarly in young and old endurance trained rats, indicating that aged animals are capable of maintaining the capacity to adapt to aerobic training over most of their life span [16]. It has been shown that exercise training can induce the proliferation of skeletal muscle satellite cells in old rats, which may be an important mechanism for the maintenance of muscle mass and repair of damaged muscle tissue [19].

Between the ages of 50 and 80, the loss of muscle strength is associated with a 30 % loss of muscle mass [16]. Thus, considerable attention has been paid to resistance training as a way to preserve or improve muscle mass in the elderly. Klitgard et al. [20] has clearly shown that muscle mass and strength in the elderly can be maintained with resistance training. Evans [13] has indicated that even the oldest old maintain their capacity to adapt to resistance training. In addition to the loss of muscle mass, loss of α-motor neurons and motor units could also contribute to the loss of strength [21]. Studies with rats have shown that atrophy and muscle weight loss occur mostly in the weight-bearing muscles that contain a high proportion of type IIb fibers and exercise training prevents the age-associated atrophy of these muscles [20,22–24].

14.2.3 Degenerative Disease and Aging

Adequate levels of exercise are beneficial in delaying or reducing the deleterious effects of aging-associated diseases such as cancer, cardiovascular disease, declines in the immune system, and diabetes [25]. For example moderate exercise training has been shown to improve immunity, but extensive training has been shown to be detrimental to the immune system [26,27]. Because of its potential role in modulating the immune system, the effect of exercise on the incidence and progression of cancer has been investigated extensively. Animal studies have indicated that moderate exercise can reduce tumor growth [28] and reduce susceptibility to carcinogens [29,30]. It is unclear what an optimal dose of exercise should be to elicit the protective effect, but several studies indicate that it may be relatively low [29,31–33]. The effects of exercise on the spread of metastasizing cells is unclear, with one study reporting slightly reduced [34] and another showing increased [35] numbers of tumors in animals that were exercised after the injection of neoplastic cells. It should be noted that one confounding factor in these animal studies relates to the fact that obesity promotes the development of several types of neoplasms and that exercise may act to protect against these types of cancers by controlling obesity [36].

The risk of cardiovascular disease increases with age [37,38] and a large component of this risk in humans is caused by increasing levels of plasma triglyceride, total cholesterol, and LDL cholesterol after the age of 20 years [39]. Numerous studies have shown that exercise can reduce cardiovascular disease risk by raising the plasma HDL concentrations [40], lowering plasma triglyceride levels [41], lowering heart rate and blood pressure [42], and reducing body fat [43]. Results from studies with rats are consistent with those of humans showing that sedentary older rats have higher plasma triglyceride levels and that exercise training can lower triglyceride concentrations [44]. These lower circulating triglyceride levels are likely brought about by increasing fatty acid oxidation during exercise, resulting in less fatty acid available for triglyceride synthesis by the liver [45] and by increased adipose lipoprotein lipase activities [46].

Non-insulin dependent diabetes mellitus (NIDDM) is common among the elderly and is associated with other health problems such as blindness, renal failure, and cardiovascular disease (National Diabetes Data Group). Aging-associated glucose intolerance is a complicated issue because factors such as physical inactivity, reduction in lean body mass, and low dietary intake of carbohydrate can influence carbohydrate metabolism [47]. However, when corrected for these influences, the results have shown that the glucose intolerance resulting from aging is due primarily to an impairment of the insulin-mediated glucose transport process such as a decreased amount of glucose transporter proteins or a reduction in the translocation of these proteins from the cytosol to the plasma membrane [48]. Physical training has been observed to improve insulin sensitivity in older humans and elevated levels of the GLUT-4 glucose transporter proteins in skeletal muscle [49,50]. Interestingly, exercise training in rats has been shown to increase GLUT-4 concentrations in young adult and mid-aged rats but not in older rats [51]. In contrast to the studies with rats, at least one study has shown that increases in skeletal muscle GLUT-4 concentrations in older (64 years) humans does not appear to be dependent on training intensity [52]. Therefore, while the rat provides a good model for the study of training-induced alterations in carbohydrate metabolism, the use of rodent models requires particular attention to the animal's age and how that may affect the capacity to alter training responses.

14.2.4   Exercise and the Free Radical Theory of Aging

The free radical theory of aging first advanced by Harmon [5] states that aging is the result of cellular damage resulting from the production of ROS generated during normal metabolism. Free radicals are derived from oxygen as it is reduced to water during oxidative metabolism in the mitochondria. The free radicals produced can damage DNA, lipids, and proteins [53]. Several lines of evidence indicate that oxidative stress due to free radical formation can accumulate with age and contribute to muscle wasting in the elderly [8]. Aged skeletal muscle has been shown to contain higher levels of lipid peroxidation products

[54,55,56], decreased enzyme activities in the electron transport chain [57], and increased numbers of mitochondrial DNA deletions [58]. Free radicals and oxidative stress may play a role in the neurodegenerative diseases of aging such as Alzheimer's [59] and Parkinson's [60]. Age-associated oxidative stress has been proposed to cause antioxidant adaptation in some tissues. Data from our laboratory has revealed increased antioxidant enzyme activities and glutathione in the skeletal and cardiac muscles of aged rats [55,56,61]. ROS and oxidative stress play an especially important role during exercise not only because total body oxygen consumption is increased, but also because unaccustomed and strenuous exercise are more likely to induce muscle injury in aged individuals [62]. This may provide a secondary source of free radical generation due to neutrophil activation as part of the acute phase response and exasperate the already increased level of oxidative stress due to aging [4]. Despite exercise-induced adaptations of the antioxidant defense system, higher levels of oxidative stress are observed in aged skeletal muscle, suggesting that antioxidant defenses are not adequate [4,55,56]. Supplementation of antioxidants, such as vitamin E, has been shown to be especially effective in the aged muscle [4,63]. Thus, it is of importance to elucidate the role of exercise in the protection against age-related tissue oxidative damage.

14.3 Designs and Procedures of Exercise Studies

14.3.1 General Considerations

A great challenge to exercise scientists and gerontologists is to distinguish whether the decreased functional capacity observed during aging is caused by aging *per se* or by lack of exercise [1]. For example, if we observe that senescent skeletal muscle displays a reduced mitochondrial oxidative potential measured by maximal enzyme activities in the Krebs Cycle, is it because there is a programmed reduction of muscle mitochondrial volume and protein concentration, or because aged muscles are used less, resulting in a deterioration of mitochondrial function? In human studies, habitual physical activity may be recorded or surveyed and taken into consideration. Although recording voluntary exercise provides some insight into this controversy in animal research [64], a clear conclusion is still lacking.

A second challenge to exercise and aging research is to select the appropriate organ, tissue, or cellular system which may reveal a potential interaction and allow for meaningful interpretation. Exercise can exert different effects on the various tissues and cellular components due to differential neuromuscular recruitment, changes in blood flow, and variations in hormonal and/or receptor distributions. Therefore, the exercise response is quite complex, and the study of exercise and aging together can be sometimes perplexing. For example, most organs and tissues in the mammals show a decrease in antioxidant enzyme activities with age [65]. However, there appears to be an up-regulation of these enzyme systems during aging in skeletal and cardiac muscles [55,61,66]. Furthermore, the age-related changes appear to be muscle fiber specific and demonstrate adaptations to physical training that are dependent on intensity [56,67]. Therefore, the tissue-specific responses to aging as well as the degree of involvement of those tissues should be considered when studying the effect of exercise on aging.

Perhaps the biggest challenge to exercise gerontologists is the difficulty to obtain a true picture of the parameters of interest *during* exercise, particularly when the measurements are at cellular and subcellular levels. Tissues, cells, and organelles are prepared *after* exercise ceases and therefore may not be representative of what occurs *in vivo*. Sometimes the parameters measured "immediately after exercise" can provide a reasonably accurate estimate of what happens during exercise. However, when the measured variables are short-lived, such as free radicals and ROS, intracellular redox status and redox-sensitive substances, transmembrane flux of ions, and certain metabolites, great caution should be paid to minimize measurement error due to time delay, or alternative *in situ* or *in vitro* approaches may have to be considered.

14.3.2 Levels of Investigation

Although exercise studies in a true sense have to be conducted *in vivo*, i.e., with intact animals, the majority of the current knowledge about the biochemical and metabolic consequences of exercise and aging have been obtained with *in vitro* studies [68]. In the past three decades, increasing numbers of exercise physiologists and gerontologists are using biochemistry and molecular biology to determine the cellular basis for the clinical challenges of their fields. Thus, before we discuss the various methods of conducting exercise studies, it is useful to familiarize ourselves with the various levels of investigation.

14.3.2.1 *Whole-Body Approach*

This *in vivo* method is suitable primarily for physiological measurements during exercise, such as whole-body oxygen uptake, body composition, energy expenditure, muscle contractile function, dietary intake, and certain hemodynamic parameters (heart rate, blood pressure, cardiac output, etc.). The advantage of the whole-body approach is that the physiological relevance of the measurements is readily seen. The major short-coming of the whole-body approach is that the cellular mechanisms that caused the observed changes are difficult to examine [68]. However, if designed properly, whole-body *in vivo* experiments can be a valuable way to obtain critical information. For example, the roles of antioxidants in aging and exercise have been delineated mostly using dietary depletion and supplementation in animal or human models [7,69]. Modern magnetic resonance instruments (NMR, MRI) have provided valuable non-invasive methods to study biochemical and molecular events at the subcellular levels during exercise [70]. With the emergence of transgenic animal models and gene knockout models, biochemical, molecular and genetic mechanisms may be investigated using intact animals and the precise physiological implication may be clearly revealed [71].

14.3.2.2 *Isolated Organs and Tissues*

Isolated perfused heart, liver, kidney, and skeletal muscle provide a valuable method to understand the function of each organ under controlled conditions. Heart (e.g., Langandoff method) may be paced to mimic the heart rate and work output occurring during exercise [72]. Isolated muscle(s) (e.g., soleus) may be stimulated electrically to contract at a given workload [73]. Isolated tissues such as diaphragm and epitrochlearis muscles have been studied when bathed in controlled media and stimulated to contract against a given load [74]. Functional capacities (e.g., contractility in myocardia; force production and fatigue in striated muscles) can be precisely measured under various experimental conditions. The biggest limitation of these *in vitro* approaches is the lack of neural and/or hormonal input which may limit some of the physiological relevance. To compensate for this shortcoming, some organs and tissues may be prepared *in situ*, such that blood flow and neurohumoral/hormonal input are similar to those *in vivo*, while independent variables are carefully controlled for or intervened with using surgical procedures. The open-chest heart model and the perfused hindlimb model are good examples of this approach [75].

14.3.2.3 Isolated Cells

Isolated erythrocytes, hepatocytes, and adipocytes are good examples of this method. Since the cell membrane is intact, this approach is particularly useful in the study of intracellular and transmembrane biochemical processes. However, application of isolated cells to exercise studies has limitations, mainly because the time required to isolate cells is relatively long (except erythrocytes) and the cells may not remain in the "exercised state", i.e., once isolated they may no longer preserve the exercise-induced effects. While exercise effects can be recapitulated to some degree in isolated cell preparations, the lack of *in vivo* hormonal, neural and metabolic influences make this approach somewhat indirect in the context of exercise.

14.3.2.4 Cell-free Homogenate and Isolated Organelles

These systems are suitable for studying metabolic and biochemical changes affected by aging and chronic exercise. Maximal enzyme activities measured in tissue homogenate may be used to estimate the maximal flux through a pathway, and the capacity to generate or remove chemical intermediates (including ROS) [68]. Concentration of certain metabolites may be indicative of the steady-state production of a pathway. A common approach in preparing homogenates involves removing tissues as quickly as possible after animals are killed and freezing tissues rapidly in liquid nitrogen. Needle biopsy is another useful method in exercise studies which significantly shortens the time span for tissue collection [76]. Once frozen, tissues can be either pulverized, homogenized, or freeze-dried to obtain cell-free homogenates.

Isolated organelles (such as mitochondrial, microsomes, etc.) are particularly useful because they play a significant and specific role during aging [77,78,79]. Unfortunately, their value in examining acute exercise effects is rather limited mainly because it usually takes a long time to isolate organelles (>30 min) as compared to tissue homogenates (a few minutes). As mentioned for isolated cells, the influence of acute exercise (see Section 14.3.3.3) may not be preserved after tissues are placed in controlled media.

14.3.2.5 Purified Enzyme, Protein, and Nuclear Systems

The major advantage of these systems is to provide direct insight into the precise mechanisms by which they control metabolic processes of aging and exercise response and adaptation. For example, studies of a key enzyme (e.g., δ-aminolevulinate synthase), including its kinetics, transcription, translation, and synthesis in a biochemical pathway (e.g., heme synthesis) or process (e.g., mitogenesis) may be an indispensable way to understand how exercise training induces mitochondrial cytochrome concentration and related enzyme activities [80]. However, caution is required when knowledge obtained from these purified systems is extrapolated to *in vivo* situations because of the different environments of these systems *in vivo* vs. *in vitro*.

14.3.3 Major Factors Determining the Exercise Response

14.3.3.1 Voluntary vs. Forced Exercise

Most animal models that study the effects of exercise utilize some form of forced exercise such as treadmill running or swimming. Voluntary exercise has the advantage of lower stress levels for the animal and particular utility for aging studies because exercise activity can merely be recorded without constant monitoring [81]. If the investigator is interested in the impact of exercise on the biological systems during the entire life-span, voluntary exercise is probably the only choice. Wheel running in rats is the most common form of voluntary exercise studied. However, using voluntary wheel running has an obvious shortcoming, i.e., the intensity and duration can not be selected to reflect different metabolic demands. Furthermore, rodents are active primarily in the dark cycle, therefore the effects of habitual exercise and other factors such as feeding and diurnal cycle could potentially confound each other.

The use of voluntary exercise in the study of aging rodents also has a special appeal because regularly performed exercise is known to cause a number of adaptations that may slow down the aging process [64] and improve average longevity [81,82]. Since wheel runners consume less food and it is well established that food-restriction results in increased maximal life span in rodents, it has been suggested that the improved survival of wheel runners may be due to decreased food intake rather than a specific adaptation that can be attributed to exercise, [83,84]. Specifically, male rats do not increase their food intake to compensate for the increased energy expenditure due to exercise, therefore exercised male rats can exhibit characteristics similar to rats subjected to food restriction [81,82,85,86]. In contrast, female rats will increase their food intake to compensate for the increased energy expenditure [87]. Therefore, the use of females rats has certain advantage in aging and exercise studies (see below).

14.3.3.2 Exercise Mode

Treadmill running and swimming are two commonly used exercise regimens in rodent studies. While for canine, porcine, and equine studies, treadmill running is the predominate form of exercise, a variety of exercise modes have been used for non-mammal (e.g., fish) and non-vertebrate (e.g., flies) studies. Human subjects can use a unique form of exercise, bicycling, during which body weight is supported by the seat and workload is provided primarily by pedaling resistance. Specially trained animals, including avian, may be used to perform resistance exercise [20,88].

In general the exercise mode selected should involve the large muscle groups and be of sufficient intensity and duration such that cardiovascular system is fully mobilized so that fuel supply and waste removal in the locomotive muscles does not become a limiting factor. For treadmill running, a certain workload is created when the body mass is moved over a level surface. Thus, speed is the main determinant of workload [3]. Workload increases considerably when the body weight has to be lifted up an incline, therefore higher workloads can be achieved with combinations of speed and slope. An advantage of treadmill running is that workload can be accurately determined based on speed and slope (see Section 14.3.3.4). One must keep in mind, however, that animals need to be acclimatized to the locomotive pattern of the exercise and that differences in ability exist between different species, strains and ages. This consideration is of particular importance in aging and exercise studies when one intends to compare the effects of exercise on old and young animals.

During swimming, body weight is supported largely by water buoyancy, so the metabolic demand on locomotive muscles is lower as compared to treadmill exercise. However, static water pressure puts an extra load on cardiopulmonary systems and therefore swimming training is an especially effective way to promote adaptations in heart function [89]. It should be noted that vigorous swimming exercise is known to stimulate release of stress hormones such as cortisol and adrenaline, therefore results regarding training effects should be interpreted with caution. A major shortcoming of swimming is the uncertainty of exercise intensity (see below).

14.3.3.3 Acute vs. Chronic Exercise

In conducting acute exercise studies, animals are subjected to a single bout of exercise and biological variables of interest are measured either during exercise (mostly physiological variables) or after animals are killed immediately after exercise (biochemical variables). Tissue samples should be collected and treated as quickly as possible. Sedentary control animals should be deprived of access to food and water for the same amount of time as their counterparts have exercised, and killed at a time as close to the exercised animals as possible.

Treadmill running involves a locomotive pattern and environment (noise, handling, etc.) unfamiliar to a laboratory animal. To ensure that an untrained animal can run at a predetermined workload for a given period of time or until exhaustion, animals need to be acclimatized for at least one week at a moderate intensity. Workload should be gradually increased to a level approaching the final assigned workload. Prolonged (>4 weeks) acclimatization periods may elicit a training effect which will confound acute exercise outcomes and should be avoided.

Exercise may be terminated at a given time, or prolonged until animals are exhausted. The former protocol is often used when the investigator is interested in the differential effects of exercise intensities on biological systems [90,91,92]. The latter protocol allows examination of responses of a system to extreme exercise stress for a prolonged period [93,94]. Exhaustion can be defined as the loss of the righting reflex, i.e., an animal is unable to right itself when laid on its back. It has been a common practice to equip a treadmill with an electrical shocking apparatus in rodent exercise studies. However, in addition to ethical considerations, repeated electrical shocking is stressful and can elicit catecholamine and stress hormone release, confounding the exercise effect. When one is interested in the maximal metabolic stress of exercise, but wants to avoid a sustained exercise duration, a maximal graded exercise protocol like those used for human VO_2max test may be considered [95]. Animals are typically run on a treadmill at a low intensity initially, and then speed and slope are increased gradually at given intervals (e.g., every 4 to 5 min), with a steady state reached for each workload, until exhaustion.

Although not precisely defined, exercise training usually refers to a long-term period of daily exercise bouts (or multiple exercise bouts per day) at progressively increasing workloads. To elicit a prominent training effect, training is usually performed for at least 5 days/week, for 8 weeks at a relative workload greater than 50% VO_2max (see below). Using mitochondrial oxidative enzymes (such as citrate synthase and cytochrome c) as markers, several early investigators have shown that training effects were a direct function of exercise intensity, frequency, and duration in rat skeletal muscle [96,97]. Interestingly, different muscle fibers demonstrated differential adaptability to training, with type 2a (red vastus lateralis) showing more prominent increases in oxidative capacity in response to an endurance training protocol [97]. Cytochrome c concentration in the type 1 (soleus) and type 2b (while vastus lateralis) was decreased and increased, respectively, as treadmill speed increased, possibly reflecting a recruitment pattern during exercise. In addition to muscle oxidative enzymes, cardiac hypertrophy and body weight reduction can also be used as training markers although they have considerable limitations (see below). To obtain true training effects not confounded by influences resulting from acute exercise bouts, animals should be sacrificed at least 24 hours after their last training bout in the resting state.

14.3.3.4 Exercise Intensity

Exercise intensity is measured in terms of total energy expenditure or work output per unit of time (power). At steady state of aerobic exercise 1 liter of oxygen is required to generate 5 Kcal of energy, therefore the amount of oxygen consumed per min is often used to measure workload or exercise intensity (liter of O_2 min^{-1} or VO_2). Maximal oxygen uptake (VO_2max) is often used to indicate an individual's maximal ability to perform exercise, however, when anaerobic metabolism becomes a significant energy source (such as sprinting exercise and maximal voluntary contraction), use of VO_2 to measure workload may be misleading and hence power output is a more appropriate measurement of exercise intensity. During treadmill exercise, body weight is an important determining factor of workload, and hence, total body VO_2 should be adjusted for body weight as representation of workload (O_2 min^{-1} kg body wt^{-1}).

Exercise intensity may also be defined in two terms: absolute workload (VO_2) and relative workload (% VO_2max). The latter takes into consideration an individual's maximal work capacity and can be used to adjust for difference due to age, gender, and training level. Thus, when a young and an old, a man and a woman, a trained and untrained work at the same relative work load (same percentage of their VO_2max), the metabolic demands imposed on the body are the same even though the absolute work loads (VO_2) could be quite different. In most circumstances, relative workload is a more useful index to measure an individual's exercise intensity.

For treadmill running, exercise intensity measured by VO_2 is directly dependent upon treadmill speed and slope [3]. Heart rate may be used as an indication of workload during submaximal exercise in humans. Using heart rate as an indication of exercise intensity in rodents is difficult because of the high heart rate (>400 bpm) in these animals. However, Brooks and White [98] established a linear correlation between VO_2 and heart rate vs. treadmill speed and slope, respectively, in rats which can be used as guidelines for determining workload for rats running on treadmill (Figure 14.1). These relationships hold true regardless of the body weight of the animals. Thus, rats working at 25 m/min with 5% grade represent 60 ml O_2 min^{-1} kg body wt^{-1} (~80% VO_2max). It is worthy of pointing out that these relationships were established using young adult rats. Since old rats have lower VO_2max, the same speed and slope may represent a much higher work load. At the present time, there is no published data recording treadmill speed and slope with respect to oxygen uptake in aged animals.

Exercise intensity for swimming is more difficult to determine. Several factors have to be taken into consideration in controlling exercise intensity when swimming rodents. The depth of water needs to be at least three times the length of the animal's body so that they do not develop a "diving" skill, i.e., animals descend to the bottom of the container while holding their breath and then ascend to the surface of water to breath. Swimming 3 to 4 animals in the same container will result in a wavy water surface which will increase the work effort. Rodent fur contains a layer of grease which traps air and increases buoyancy. Degreasing fur with detergent before swimming and periodically de-bubbling during swimming will help keep workload constant. Water temperature must be regulated at 33 to 35°C to avoid eliciting effects due to hypo- or hyperthermia. To increase animal's workload, a small weight equivalent to 5 to 10% of body weight may be attached to the tail of the rat or mouse. However, The quantitative relationship of tail weight and workload is unknown.

Because an individual's maximal work capacity (VO_2max) is continuously improved during the training period, a given level of absolute work intended to represent a particular relative work load (%VO_2max) at the beginning of training may no longer impose the same

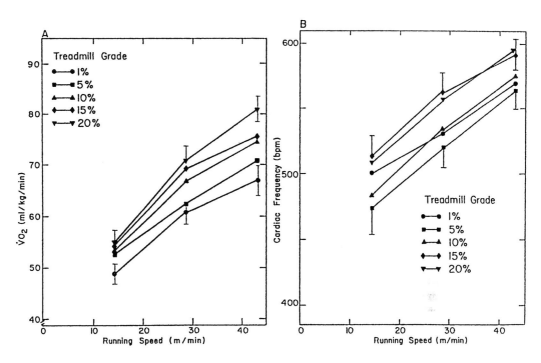

FIGURE 1

(A) represents the O_2 consumption of rats at different treadmill running speeds and grades. Graph (B) depicts the cardiac frequency in beats per minute at different running speeds and grades. Both graphs represent the mean of five trials. (Brooks and White 1978.)

relative workload after several weeks of training. If not corrected, this will result in a reduced stimulus for adaptation as training progresses. Therefore, in a chronic training program workload needs to be continuously increased over time. Table 14.1 provides a typical training protocol for rats running on a motor driven treadmill. In our hands this protocol has resulted in increases in muscle citrate synthase activity by 70% [99].

14.3.4   Interactions of Exercise and Other Biological Factors

14.3.4.1   Age

The decline of physical work capacity with aging at both maximal and submaximal workloads have been known since the earliest times. The reason for this decline may be multifaceted. A compromised cardiovascular function plays an important role as older human and animals demonstrate significantly lower maximal heart rates, stroke volumes, cardiac outputs, and VO_2 at different work loads and during exercise bouts of a given intensity [1]. Furthermore, muscle contractile functions indicated by maximal isometric and dynamic strength, maximal contractile velocity, titanic contraction force, and fatigue resistance, show a considerable deterioration during aging [16]. Aged individuals also have a greater risk of encountering acute heart problems during exercise than young individuals. All of these parameters are important considerations in designing studies involving aged subjects [1]. Although no specific speed and incline for aged rodents are established for which workloads can be compared with those of young animals, it is advisable that speed and slope be reduced to accommodate the reduced work capacity for the old. For example,

TABLE 14.1

Training Protocol on Treadmill in Rats

Week	Speed (m/min)	Slope (%)	Duration (min)	Frequency (day/wk)
1–2	20	0	20	5
3–4	25	5	40	5
5–6	25	10	60	5
7–8	30	15	75	5
9–10	30	15	90	5

Note: Exercise is performed on a motor-driven rodent treadmill by male or female rats at the age of 10 to 12 weeks when training starts.

young rats running on treadmill at 25 m/min and 5% grade represents approximately 75 to 80% VO_2max, assuming a VO_2max of 75 to 80 ml O_2 min^{-1} kg^{-1} [98,100]. For old rats, 75 to 80% VO_2max is equivalent to ~15 m/min, 0% grade if one would like to exercise the old and young at the same relative workload. For these conditions rats of both ages have been shown to be able to run for ~1 h [55].

Conducting physical training initiated at old age is more difficult, as aged animals are reluctant to run and are more prone to injury. Although it is generally accepted that the greater the exercise intensity, the greater the training effect, and there does not seem to be a minimal training intensity below which a training response will be absent, an exercising heart rate greater than 60% of maximal heart rate is recommended in elderly human subjects involved in chronic training [1]. In aged rodents, a variety of combinations of treadmill speeds and slopes have been used in training studies. Table 14.2 summaries some of the training protocols used for old rodents reported in the literature which successfully elicited a training response.

14.3.4.2 Gender

Although direct influence of animal gender on exercise-elicited physiological and biochemical effects have only been scarcely reported (see below), gender may confound interpretation of data on certain variables because of indirect influences. For example, male and female rats grow at different rates. Decreases in growth rate during training have been reported in male rats, but it occurs in female rats only when work load is excessively

TABLE 14.2

Training Intensity on Treadmill in Rodents

Species	Strain	Age	Speed & Slope	Frequency	Effects	Ref.
Rat	Fischer 344	24.0 mo	15 m/min, 15%	1 h/d, 10 wk	MDH* 48%	19
Rat	Fischer 344	24.0 mo	13 m/min, 15%	1 h/d, 10 wk	Ht wt* 21%	56
Rat	Fischer 344	26.5 mo	15 m/min, 5%	1 h/d, 10 wk	GPX* 50%	56
Rat	Fischer 344	27.5 mo	15 m/min, 15%	1 h/d, 10 wk	CS* 71%	99
Rat	Fischer 344	26.0 mo	20 m/min, 15%	1 h/d, 8 wk	Ht/bw* 25%	100

* All the studies exercised animals 5 days/wk.
MDH, malate dehydrogenase; GPX, glutathione peroxidase; CS, citrate synthase; Ht/bw, heart/body weight ratio.

high [89]. This is because male rats decrease food intake and reduce body weight during training, whereas trained female rats increase food intake and maintain body weight [85,87]. Thus, in designing studies wherein control of body weight is an important consideration, female rats may be more desirable than male rats. For example, if one uses male rats to study the effect of training on cardiac hypertrophy, a false conclusion may be reached based upon the observation that heart/body weight ratio is significantly increased after training.

Animal gender may also interact with exercise mode and intensity and give confounding results. Male rats generally do not show a training-induced cardiac hypertrophy with either treadmill or swim training. However, in female rats running does not usually cause an increase in heart mass, whereas swimming almost always leads to increased heart size and heart/body weight ratio [89].

Direct gender differences during exercise may also be observed when the variables measured are related to sex-specific hormonal status. An acute bout of strenuous exercise has been shown to elicit significant oxidative tissue damage in male rats due to increased ROS production, however, female rats do not exhibit this detrimental effect [101]. Likewise, male rats have demonstrated adaptation of antioxidant defense capacity after aerobic exercise training, whereas female rats show little response [102]. It was proposed that the increased release of female reproductive hormone estrodiol during exercise may serve a protective function in scavenging ROS [103]. These results are based on relatively few studies and recent findings of training adaptation of antioxidant enzymes in female rats seem to argue against this hypothesis [67]. However, these investigations serve to caution the researcher to seriously consider gender differences when designing experiments and interpreting data.

14.3.4.3 Diet and Feeding Patterns

Exercise training may significantly alter feeding patterns of experimental animals, especially in studies utilizing rodents [89,104]. Male rats show a suppressed appetite during training, and this will not only result in decreased body weights, but also reduced intake of micronutrients essential for biological functions. Therefore, food intake of the sedentary control animals often needs to be restricted in order to match food intakes of the trained animals. However, since trained rats also consume more energy, matching food intake does not usually result in identical body weight. Thus, in practice, daily food supplied to the sedentary animals may be reduced by 5% compared to the trained animals initially and thereafter adjusted every 2 to 3 days to match the body weight gain (or loss) of the two exercised groups [105]. The disadvantage of this design is that the arbitrarily reduced food intake in the sedentary animals will cause a mild retardation of growth which can seriously affect research outcomes, especially those involving protein and nucleic acid synthesis. For these reasons, female rats are preferred to male rats in training studies.

Fasting exerts a profound effects on metabolic, nutritional and hormonal responses to exercise. Fasting changes fuel utilization patterns especially during prolonged exercise largely because of changes in plasma insulin and glucagon concentrations [106]. Exercise normally increases glucose uptake by skeletal muscle, but in the fasted state muscle glucose utilization is attenuated due to diminished hepatic glucose output [107]. Food deprivation is also known to result in oxidative stress to the liver and extrahepatic tissues, along with alterations of antioxidant enzyme status [108,109]. Hepatic glutathione content is reduced in fasted animals because of decreased hepatic synthesis. This is thought to be due to diminished plasma insulin and cysteine supplies, elevated plasma glucagon concentration, and increased hepatic output [110]. Thus, an acute bout of intense exercise in the fasted animals can decrease liver GSH to less than 50% of the levels seen in the fed state [111]. Catalase activity in the various tissues has been shown to undergo dramatic changes with starvation [112] and liver and skeletal muscle lipid peroxidation increases significantly in fasted animals [109]. Since exercise and fasting have many similarities, most exercise studies are conducted in the fed state such that metabolic effects elicited by exercise are not confounded by fasting.

Dietary (or caloric) restriction is a well known intervention of aging in rodents [84,113]. Dietary restriction increases both average and maximal life span in rodents [83]. In aging rodent studies, the level of physical activity can influence survival patterns and other age-sensitive outcomes and may be confounded with the effect of dietary restriction [114]. For example, if sedentary male rats are food restricted in order to keep their body weights the same as wheel runners they showed increased maximal life spans compared to wheel runners, but offering male rats the opportunity to exercise leads to increases in average life span but not maximal life span [2]. This suggests that exercise and dietary restriction affect aging by different mechanisms. Dietary restriction has also been shown to attenuate age-related increases in antioxidant enzyme activities in rat skeletal muscle [78]. If both food restriction and exercise affect aging by different mechanisms, one might surmise that the combination of both would result in greater improvements in the health of aged rats. Several studies have shown that this is not the case. In one study, food restricted voluntary runners had increased mortality rates in the first 50% of their mortality curve as compared to food-restricted sedentary controls [104]. For the rest of the mortality curve (after age 900 days), survival of food-restricted runners did not differ from food-restricted sedentary controls and the runners exhibited the expected increase in maximal life span attributed to food restriction alone. A recent reevaluation of this issue reported similar results although there was not an increased mortality rate prior to 900 days of age [115].

14.3.4.4 *Diurnal Variation*

Biological variables exhibit rhythmic changes according to the length of exposure to day light and are reflected in the diurnal variation in activity levels of an animal. During the active period, animals eat, drink water, and are physically more active, whereas during the inactive period they sleep or rest. Many physiological and metabolic variables, such as hormones, are influenced profoundly by the diurnal cycle and must be taken into consideration when designing exercise studies. Most rodents are nocturnal animals. Therefore, exercise studies are preferably conducted during the dark cycle when animals are fed and active. However, in order that researchers do not have to do experiments at night, animals can be adjusted to a reversed light-dark cycle (e.g., 7:00 to 19:00 dark; 19:00 to 7:00 light) at least two weeks before the experiment. When exercising animals in the dark cycle, the lights of the room should be dimmed. If treadmill exercise is performed, the inner sides of the individual animal chambers may be painted black to reduce light exposure.

Diurnal effects should also be considered when animals are sacrificed. Whenever possible, animals from all experimental groups should be sacrificed at the same time of the day. In conducting experiments which includes multiple groups and large number of animals in each group, it may not be practical to sacrifice each animal at the same time of the day. In this case a randomized design should be adopted to distribute the diurnal variation among all groups. In other words, animals in a given experimental group should be killed at different time of the day (may be on different days) in random fashion, so that the sum of possible diurnal effects is minimal.

14.4   Concluding Remarks

The study of the role of exercise in the aging process is highly integrated in that it involves numerous physiological systems. Because of this, the study of the influence of exercise on the aging process is complex and constitutes a considerable challenge. Nevertheless, it is clear that exercise can positively influence both the process and outcome of aging, improve the quality of life of the elderly, and lower our national health care burden. It is also evident that major gaps in the knowledge exist regarding the influence of exercise on the aging process. The followings are several questions which might be worthy of considering in designing future studies in this field: (1) What is the most desirable model(s) to use in order to reveal the specific cellular mechanism(s) by which exercise influences normal process of aging? (2). What are the appropriate intensities and duration's of exercise that will allow beneficial adaptations but minimize exercise-induced damage and injury in aging studies? (3) What effect does exercise have on the nutritional requirements of the elderly and on the effectiveness of dietary supplements of essential nutrients (e.g., antioxidants) aimed at countering the effects of aging?

References

1. Stamford, B.A., Exercise and the elderly, in *Exercise and Sport Sciences Reviews*, Vol. 16,Pandolf, K.B., Ed., Macmillan Publishing Company, New York, 341, 1988.
2. Holloszy, J.O. and Smith, E.K., Effects of exercise on longevity of rats. *Fed. Proc.* 46, 1850, 1987.
3. Astrand, P. O. and Rodahl, K., *Textbook of Work Physiology*, McGraw-Hill Book Co., New York, 1977.
4. Meydani, M. and W.J. Evans., Free radicals, exercise, and aging. Yu, B.P. Ed. *Free Radical in Aging*. CRC Press, Boca Raton, FL, 183, 1993.
5. Harman, D., Aging: a theory based on free radical and radiation chemistry. *J. Gerontol.* 11, 298, 1956.
6. Jenkins, R. R., Exercise, oxidative stress and antioxidants: A review. *Intl. J. Sports Nutr.* 3, 356, 1993.
7. Ji, L.L., Exercise and oxidative stress: role of the cellular antioxidant systems, in *Exercise Sports Science Reviews*, Holloszy, J.O., Ed., Baltimore, Williams & Wilkins, 135, 1995.
8. Yu, B.P., Ed., *Free Radicals and Aging*. CRC Press, Boca Raton, FL, 1993.
9. Nohl, H., Involvement of free radicals in aging: a consequence or cause of senescence. *Brit. Med. Bull.* 49, 653, 1993.
10. Fries, J.F., Physical activity, the compression of morbidity, and the health of the elderly. *J. Royal Soc. Med.*, 89, 64, 1996.
11. Paffenbarger, R. S., Hyde, R. T., Wing, A. L., Lee, I. M., Jung, D. L., and Kampert, J. B. G., The association of changes in physical activity level and other lifestyle characteristics with mortality among men. *N. Eng. J. Med.* 328, 538, 1993.
12. Sandvick, L., Erikssen, J., Thaulow, E., Erikssen, G., Mundal, R., and Rodahl, D., Physical fitness as a predictor of mortality among healthy, middle-aged Norwegian men. *N. Eng. J. Med.* 328, 533, 1993.
13. Evans, W., Effects of exercise on body composition and functional capacity of the elderly, *J. of Geront.*, 50A, 147, 1995.
14. Flegg, J.L. and Lakatta, E.G., Role of muscle loss in the age-associated reduction V0₂max. *J. Appl. Physiol.* 65, 1147, 1988.

15. Coggan, A.R., Spina, R.J., King, D.Sl., Rogers, M.A., Brown, M., Nemath, P.M., and Hollosqy, J.O., Skeletal muscle adaptations to endurance training in 60- to 70 year-old men and women. *J. Appl. Physiol.* 72, 1780, 1992.

16. Rogers, M.A. and Evans, W.J., Changes in skeletal muscle with aging: Effects of exercise training, in *Exercise and Sport Sciences Reviews*, Vol. 21, Holloszy, J.O., Ed., Williams & Wilkins, Baltimore, 65, 1993.

17. Holloszy, J.O. and Coyle, E.F., Adaptations of skeletal muscle to endurance exercise and their metabolic consequences. J. Appl. Physiol. 56, 831, 1984.

18. Seals, D.R., Hagberg, J.M., Hurley, B.F., Ehsani, A.A., and Holloszy, J.O., Endurance training in older men and women: cardiovascular responses to exercise. *J. Appl. Physiol.* 57, 1024, 1984.

19. McCormick, K.M. and Thomas, D.P., Exercise induced satellite cell activation in senescent soleus muscle. *J. Appl. Physiol.* 72, 888, 1992.

20. Klitgaard, H., Brunet, A., Maton, B., Samaziere, Lesty, C., and Monond, H., Morphological and biochemical changes in old rat muscles: Effects of increased age. *J. Appl. Physiol.* 67, 1409, 1989.

21. Anssved, T. and Larsson, L., Quantitiative and qualitattive morphological properties of the soleus motor nerve and the L5 ventral root in young and old rats. Relation to the number of soleus muscle fibres. *J. Neurol. Sci.* 96, 269, 1990.

22. Holloszy, J.O., Chen, M. Cartee, G.D., and Young, J.C., Skeletal muscle atrophy in old rats: differential changes in the three fiber types. *Mech. Aging Dev.* 60, 199, 1991.

23. Daw, C.K., Starnes J.W., and White, T.P., Muscle atrophy and hypoplasia with aging: Impact of training and food restriction, *J. Appl. Physiol.* 64, 2428, 1988.

24. Brown, M., Ross, T.P., and Holloszy, J.O., Effects of aging and exercise on soleus and extensor digitorum longus muscles of female rats. *Mech. Age. Dev.*, 63, 69, 1992.

25. Lowenthal, D.T., Krischner, D.A., Scarpace, N.T. Pollock, M., and Graves, J., Effects of exercise on age and disease. *South. Med. J.* 87, S5, 1994.

26. Mackinnon, L.T., *Exercise and Immunology*, Human Kinetics Publishers, Champaign, IL, 1992.

27. Shephard, R.J., Rhind, S., and Shek, P.N., Exercise and the immune system: Natural killer cells, interleukins, and related responses. *Sports Med.* 18, 340,1994.

28. Shephard, R.J., Exercise and malignancy. *Sports Med.* 3, 235, 1986.

29. Andrianopoulos, G., Nelson, R.L., Bombeck, C.T., and Souza, G., The influence of physical activity in 1-2 dimethylhydrazine induce colon carcinogenesis in the rat. *Anticancer Res.*, 7, 849, 1987.

30. Kritchevsky, D., Influence of caloric restriction and exercise on tumorigenesis in rats, *Proc. Soc. Exp. Biol.* 193, 35, 1990.

31. Baracos, V.E., Exercise inhibits progressive growth of the Morris heptoma 7777 in male and female rats, *Can. J. Physiol. Pharmacol.*, 67, 864, 1989.

32. Thompson, H.J., Ronan, A.M., Ritacco, K.A., Tagliaferro, A.R., and Meeker, L.D., Effects of exercise on the induction of mammary carcinogenesis. *Cancer Res.* 48, 2720, 1988.

33. Thompson, H.J., Ronan, A.M., Ritacco, K.A., and Tagliaferro, A.R., Effect of type and amount of detary fat on the enhancement of rat mammary tumorigenesis by exercise. *Cancer Res.* 49, 1904, 1989.

34. MacNeil, B. and Hoffman-Goetz, L., Exercise training and tumour metastasis in mice: Influence of time of exercise onset. *Anticanc. Res.* 13, 2085, 1993.

35. MacNeil, B. and Hoffman-Goetz, L., Effect of exercise on natural cytotoxicity and pulmonary tumor metastases in mice. *Med. Sci. Sports Exerc.* 25, 922, 1993.

36. Shephard, R.J. and Shek, P.N., Cancer, immune function, and physical activity. *Can. J. Appl. Physiol.* 20, 1, 1995.

37. Gordon, T., Kannel, W.B., Castelli, W.P., and Dawber, T.R., Lipoproteins, cardiovascular disease and death. The Framingham Study. *Arch. Int. Med.* 141, 1128, 1981.

38. Kannel, W.B., Castelli, W.P., and Gordon, T., Cholesterol in the prediction of atherosclerotic disease. New perspectives based on the Framingham Study. *Ann. Intern. Med.* 90, 85, 1979.

39. Heiss, G., Tamir, I., Davis, C.E., Tyroler, H.A., Rifkind, B.M., Schonfeld, G., Jacobs, D., and Frantz, I.D., Lipoprotein-cholesterol distributions in selected North American Populations: the Lipid Research Clinics Program Prevalance Study, *Circulation*, 61, 302, 1980.

40. Wood, P.D. and Haskell, W.L., The effect of exercise on plasma high density lipoproteins. *Lipids*, 14, 417, 1979.
41. Oscai, L.B., Patterson, J.A. Bogard, D.L., Beck, R. and Roothermel, B., Normalization of serum triglycerides and electrophoretic patterns by exercise. *Am J. Cardiol.* 30, 775, 1972.
42. Hagberg, J.M., Graves, E., Limacher, M.C., Cardiovascular responses of 70-79 year old men and women to endurance and strength training. *J. Appl. Physiol.* 656, 2589, 1989.
43. Larson, E.B. and Bruce, R.A., Health benefits of exercising in an aging society. *Arch. Intern Med.* 147, 353, 1987.
44. Barakat, H.A., Dohm, G.L., Shukla, N., Marks. R.H.L., Kern, M., Carpenter, J.W., and Mazzeo, R.S., Influence of age and exercise training on lipid metabolism in Fischer-344 rats. *J. Appl. Physiol.*, 67, 1638, 1989.
45. Barakat, H.A., Kasperek, G.J., and Dohm, G.L., Progressive changes in fatty acid metabolism in rat liver and muscle during exercise. *Biochem. Med.*, 29, 298, 1983.
46. Borensztajn, J.M., Rone, M.S., Babirak, S.P., McGarr, J.A., and Oscai, L.B., Effect of exercise on lipoprotein lipase activity in rat heart and skelteal muscle. *Am. J. Physiol.*, 229, 394, 1975.
47. Coon, P. J., Rogus, E. M., Drinkwater, D., Muller, D. C., and Goldberg, A. P., Role of body fat distribution in the decline in insulin sensitivity and glucose tolerance with age. *J. Clin. Endo. Metab.* 75, 1125, 1992.
48. Jackson, R. A., Mechanisms of age related glucose intolerance. *Diabetes Care* 13 (suppl.2), 9, 1990.
49. Seals, D.R., Hagberg, J.M., Allen, W.K., Hurley, B.F., Dalski, G.P., Ehsani, A.A., Holloszy, J.O., Glucose tolerance in young and older athletes and sedentary men. *Am. J. Physiol.* 56, 1521, 1984.
50. Houmard, J.A., Egan P.C., and Neufer, P.D., Elevated skeletal muscle glucose transporter levels in exercise-trained middle-aged men. *Am. J. Physiol.*, 261, E437, 1991.
51. Kern, M., Dolan, P.L., Mazzeo, R.S. Wells, J.A. and Dohm, G.L., Effect of aging and exercise on GLUT-4 glucose transporters in muscle. *Am. J. Physiol.* 263, E362, 1992.
52. Hughes, V.A., Fiatarone, M.A., Fielding, R.A., Kahn, B.B., Ferrara, C.M., Shepherd, P., Fisher, E.C., Wolfe, R.R., Elahi, D., and Evans, W.J., Exercise increases muscle GLUT-4 levels and insulin action in subjects with impaired glucose tolerance. *Am. J. Physiol.* 27:E855-E862; 1993.
53. Ames, B. N., Shigenaga, M.K., and Hagen, T. M., Oxidant, antioxidants, and degenerative diseases of aging. *Proc. Natl. Acad. Sci.* 90, 7915, 1993.
54. Starnes, J. W., Cantu, G., Farrar, R. P., and Kehrer, J. P., Skeletal muscle lipid peroxidation in exercise and food-restricted rats during aging. *J. Appl. Physiol.* 67, 69, 1989.
55. Ji, L.L., Dillon, D., and Wu, E., Alteration of antioxidant enzymes with aging in rate skeletal muscle and liver. *Am. J. Physiol.* 258, R918, 1990.
56. Leeuwenburgh, C., Fiebig, R., Chandwaney, R., and Ji, L.L., Aging and exercise training in skeletal muscle: response of glutathione and antioxidant enzyme systems, *Am. J. Physiol.*, 267, R439, 1994.
57. Feuers, R.J., Weindruch, R.L., and Hart, R.W., Caloric restriction, aging, and antioxidant enzymes, *Mutat. Res.* 295, 191, 1993.
58. Chung, S.S., Weindruch, R., Schwarze, S.R., McKenzie, D.I., and Aiken, J.M., Multiple age-associated mitochondrial DNA deletions in skeletal muscle of mice. *Aging*, 6, 193, 1994.
59. Parker, W.D. Jr., Parks, J., Filley, C.M., and Kleinschmidt-DeMasters, B.K., Electron transport chain defects in Alzheimer's disease brain. *Neurology*, 44, 1090, 1994.
60. Shoffner, J.M., Watts, R.L., Juncos, J.L., Torroni, A., and Wallace, D.C., Mitochondrial oxidative phosphorylation defects in Parkinson's disease. *Ann. Neurol.* 30, 332, 1991.
61. Fiebig, R., Leeuwenburgh, C., Gore, M., and Ji, L.L., The interactive effects of aging and training on myocardial antioxidant enzymes and oxidative stress, *Age*, 19, 83, 1996.
62. Evans, W. and Cannon, J.G., The metabolic effect of exercise-induced muscle damage, in *Exercise and Sports Sciences Review*, Holloszy, J.O., Ed., Williams & Willkins, London, 1991, 99.
63. Zerba, E., Komorowski, T.E., and Faulkner, J.A.. Free radical injury to skeletal muscle of young, adult and old mice. *Am. J. Physiol.* 258, C429, 1990.
64. Beyer, R.E., Starnes, J.W., Edington, D.W., Lipton, R.J., Compton, III, R.T., and Kwasman, M.A., Exercise-induced reversal of age-related declines of oxidative reactions, mitochondrial yield, and flavins in skeletal muscle of the rat. *Mech. Ageing Dev.*, 24, 309, 1984.

65. Matsuo, M. Age-related alterations in antioxidant defense, in *Free Radicals in Aging*, Yu, B. P., Ed., CRC Press, Boca Raton, FL, 143, 1993.

66. Ji, L. L., Dillon, D., and Wu, E., Myocardial aging: antioxidant enzyme systems and related biochemical properties. *Am. J. Physiol.* 261, R386, 1991.

67. Leeuwenburgh, C., Hollander, J., Leichtweis, S., Griffiths, M., Gore, M., and Ji, L.L., Adaptations of glutathione antioxidant system to endurance training are tissue and muscle fiber specific. *Am. J. Physiol.* 272, R363, 1997.

68. Newsholme, E. and Leech. A. R., *Biochemistry for the Medical Sciences*, John Wiley & Sons, Chichester, 1983.

69. Kanter, M.M., Free radicals and exercise: effects of nutritional antioxidant supplementation, in *Exercise and Sport Science Reviews*, Holloszy, J.O., Ed., Baltimore, Williams & Wilkins, 375, 1995.

70. Sapega, A. A., Sokolow, D. P., Graham, T. J., and Chance, B., Phosphorus nuclear magnetic resonance: a non-invasive technique for the study of muscle bioenergetics during exercise. *Med. Sci. Sports. Exercise.* 25, 656, 1993.

71. Tsika, R.W., Transgenic animal models, in *Exercise and Sport Sciences Reviews*, Vol. 22, Holloszy, J.O., Ed., Williams & Wilkins, Baltimore, 361, 1994.

72. Bowles, D.K., Farrar, R.P., and Starnes, J.W., Exercise training improves cardiac function after ischemia in the isolated, working rat heart, *Am. J. Physiol.*, 263, H804, 1992.

73. Fitts, R.H. and Widrick, J.J., Muscle mechanics: Adaptations with exercise-training, in *Exercise and Sport Sciences Reviews*, Vol. 24, Holloszy, J.O., Ed., Williams & Wilkins, Baltimore, 427, 1996.

74. Reid, M. B., Haack, K. E., Franchek, K. M., Valberg, P. A., Kobzik, L., and West, M. S., Reactive oxygen in skeletal muscle I. Intracellular oxidant kinetics and fatigue *in vitro*. *J. Appl. Physiol.* 73, 1797, 1992.

75. Ji, L. L., Fu, R.G., Mitchell, E.W., Griffiths, M., Waldrop, T.G., and Swartz, H.M., Cardiac hypertrophy alters myocardial response to ischemia and reperfusion *in vivo*. *Acta Physiol. Scand.* 151, 279, 1994.

76. Coggan, A.R., Muscle biopsy as a tool in the study of aging, *J. Gerontol. Series A*, 50A, 30, 1995.

77. Hansford, R,G., Bioenergetics in aging, *Biochem Biophys Acta*, 726, 41, 1983.

78. Luhtala, T., Roecher, E.B., Pugh, T., Feuers, R.J., and Weindruch, R., Dietary restriction opposes age-related increases in rat skeletal muscle antioxidant enzyme activities. *J. Gerontol.* 49, B231, 1995.

79. Kim, J.D., McCarter, R.J.M., and Yu, B.P., Influence of age, exercise, and dietary restriction on oxidative stress in rats. *Aging Clin. Exp. Res.*, 8, 123, 1996.

80. Essig, D.A., Contractile activity-induced mitochondrial biogenesis in skeletal muscle, in *Exercise and Sport Sciences Reviews*, Vol. 24, Holloszy, J.O., Ed., Williams & Wilkins, Baltimore, 289, 1996.

81. Holloszy, J.O., Smith, E.K., Vining, M., and Adams, S.A., Effect of voluntary exercise on longevity of rats. *J. Appl. Physiol.* 59, 826, 1985.

82. Holloszy, J.O. and Schechtman, K.B. Interaction between exercise and food restriction: effects on longevity of male rates, *J. Appl. Physiol.* 70, 1529, 1991.

83. Masoro, E.J., Food restriction in rodents: an evaluation of its role in the study of aging. *J. Gerontol. Biol. Sci.* 43, B59, 1988.

84. Weindruch, R. and Walford, R.L. *The Retardation of Aging and Disease by Dietary Restriction*, Charles C. Thomas, Springfield, IL, 1988.

85. Holloszy, J.O., Exercise increases average longevity of female rats despite increased food intake and no growth retardation, *J. of Geront.* 48, B97, 1993.

86. Craig, B. W., Garthwaite, S. M., and Holloszy, J. O., Adipocyte insulin resistance: effects of aging, obesity, exercise, and food restriction. *J. Appl. Physiol.* 62, 95, 1987..

87. Oscai, L.B., Mole, P.A., Krusack, L.M., and Holloszy, J.O., Detailed body composition analysis in female rats subjected to a program of swimming. *J. Nutr.* 103, 412, 1973.

88. Laurent, G.J., Sparrow, M.P., Pates, P.C., and Millward, D.J., Turnover of muscle protein in the fowl. Collagen content and turnover in cardiac and skeletal muscles of the adult fowl and the changes during stretch-induced growth. *Biochem J.* 176, 419, 1978.

89. Schaible, T. F. and Scheuer, J., Cardiac adaptations to chronic exercise. *Prog. Cardiovasc. Dis.* 27, 297, 1985.

90. Ji, L. L., Fu, R.G., and Mitchell, E.W., Glutathione and antioxidant enzymes in skeletal muscle: effects of fiber type and exercise intensity, *J. Appl. Physiol.*, 73, 1854, 1992.

91. Powers, S.K., Criswell, D., Lawler, J., Ji, L.L., Martin, D., Herb, R., and Dudley, G., Influence of exercise intensity and duration on antioxidant enzyme activity in skeletal muscle differing in fiber type. *Am. J. Physiol.*, 266, R375, 1994.

92. Powers, S.K., Criswell, D., Lawler, J., Martin, D., Ji, L.L., and Dudley, G., Training-induced oxidative and antioxidant enzyme activity in the diaphragm: influence of exercise intensity and duration. *Resp. Physiol.* 95, 226, 1994.

93. Ji, L. L. and Fu, R. G., Responses of glutathione system and antioxidant enzymes to exhaustive exercise and hydroperoxide. *J. App. Physiol.* 72, 549, 1992.

94. Davies, K. J. A., Quintanilha, T.A., Brooks, G. A., and Packer, L., Free radical and tissue damage produced by exercise. *Biochem. Biophys. Res. Comm.* 107, 1198, 1982.

95. Ji, L.L. and Mitchell, E.W., Effects of adriamycin on heart mitochondrial function in rested and exercised rats. *Biochem. Pharmacol.*, 47, 877, 1994.

96. Hickson, R., Skeletal muscle cytochrome c and myglobin, endurance, and frequency of training, *J. Appl. Physiol.* 51, 746, 1981.

97. Dudley, G. A., Abraham, W. M., and Terjung, R. L., Influence of exercise intensity and duration on biochemical adaptation in skeletal muscle. *J. Appl. Physiol.* 53, 844, 1982.

98. Brooks, G.A. and White, T.P., Determination of metabolic and heart rate responses of rats to treadmill exercise. *J. Appl. Physiol.* 45, 1009, 1978.

99. Ji, L. L., Wu, E., and Thomas, D.P., Effect of exercise training on antioxidant and metabolic functions in senescent rat skeletal muscle. *Gerontol.*, 37, 317, 1991.

100. Mazzeo, R.B., Brooks, G.A., and Horvath, S.M., Effects of age on metabolic responses to endurance training in rats, *J. Appl. Physiol.* 57, 1369, 1984.

101. Amelink, G.J., Wal, W.A., Wokke, J.H., van Asbeck, B.S., and Bar, P.R., Exercise-induced muscle damage in the rat: the effect of vitamin E deficiency, *Pflugers Arch.*, 419, 304, 1991.

102. Tiidus, P. M. and Houston, M.E., Vitamin E status does not affect the responses to exercise training and acute exercise in female rats, *J. Nutr.* 123, 834, 1993.

103. Bar, P.R., Amelink, G.L., Oldenburg, B., and Blankenstein, M.M., Prevention of exercise-induced muscle membrane damage by oestradiol. *Life Science.* 42, 2677, 1988.

104. Holloszy, J.O., Exercise and food restriction in rats, *J. Nutr.* 122, 774, 1992.

105. Ji, L. L., Stratman, F.W., and Lardy, H.A., Antioxidant enzyme systems in rat liver and skeletal muscle. *Arch Biochem Biophys.*, 263, 150, 1988.

106. Galbo, H., *Hormonal and Metabolic Adaptation to Exercise*, Thieme-Stratton Inc., New York, 1983, 30-39, 64-86.

107. Hers, H.G. and Hue, L., Gluconeogenesis and related aspects of glycolysis. *Annu. Rev. Biochem.* 52, 617, 1983.

108. Bray, T.M. and Taylor, C.G., Enhancement of tissue glutathione for antioxidant and immune functions in malnutrition. *Biochem. Pharmacol.*, 47, 2113, 1994.

109. Godin, D.V. and Wohaieb, S.A., Nutritional deficiency, starvation, and tissue antioxidant status, *Free Radic. Biol. Med.*, 5, 165, 1988.

110. Teteishi, N., Higashi, T., Naruse, A., Nakashima, K., and Sakamoto, Y., Rat liver glutathione: possible role as a reservoir of cysteine. *J. Nutr.* 107, 5, 1977.

111. Leeuwenburgh, C. and Ji, L.L., Alteration of glutathione and antioxidant status with exercise in unfed and refed rats. *J. Nutr.*, 126, 1833, 1996.

112. Lammi-Keefe, C. L., Swan, P. B., and Hegarty, P.V. J., Effect of level of dietary protein and total or partial starvation on catalase and superoxide dismutase activity in cardiac and skeletal muscle in young rats. *J. Nutr.*, 114, 2235, 1984.

113. Yu, B.P., Modulation of oxidative stress as a means of life-prolonging action of dietary restriction, in *Oxidative Stress and Aging, Molecular and Cell Biology Updates*, Cutler, R.G., Packer, L., Bertram, J., and Mori, A., Eds., Birkhauser Verlag Basel, Switzerland, 1995, 331.

114. Weindruch, R., Animal models for aging research, in *Handbook of Physiology*, Volume on Aging, Masoro, E.J., Ed., Oxford Univ. Press, New York, 1995, 37-52.

115. Holloszy, J.O., Mortality rate and longevity of food- restricted exercising male rats: a reevaluation. *J. Appl. Physiol.* 82, 399, 1997.

15

Changes in Neurotransmitter Exocytosis: A Target for Regulation During Aging

Hoau-Yan Wang, David L. Snyder, Jay Roberts, and Eitan Friedman

CONTENTS

15.1 Introduction

The exocytosis of neurotransmitter from presynaptic neurons, one of the most important processes in neuronal function, is the principal mechanism through which nerves communicate and modulate the activity of target cells. In the central and peripheral nervous systems, control of synaptic concentrations of neurotransmitters is essential for initiating and coordinating physiological functions that are essential for survival. Aging in mammals is associated with changes in neurotransmission that may underlie some age-related deficits, such as those in cognitive functions, psychomotor performance, and many other essential functions subserved by the peripheral autonomic nervous systems. The mechanism involved in the regulation of neurotransmitter release and its regulation in response to dynamic intracellular and extracellular changes are, however, not well understood. During aging, neuronal functions such as transmitter synthesis and release and receptor status in the brain undergo marked changes which may be related to alterations in the synaptic concentrations of neurotransmitter [1-6]. Several studies have demonstrated that neurotransmitter exocytosis, both in central and peripheral nervous systems, is altered during aging [3,4,7–15]. Other studies suggest that age-related changes in neurotransmitter release occur concurrently with alterations in one or more modulatory mechanisms [3,4,11,16–18]. Thus, there is substantial information to suggest that senescence is associated with changes in neurotransmitter release and its regulation. Investigations concerning the effect of age on neurotransmitter exocytosis and its regulation are extremely important not only for understanding the physiological changes that occur during aging of the organism but such studies may also help in designing specific strategies aimed at counteracting or correcting age-related dysfunctions.

One of the most important methods for examining neurotransmitter release involves the measurement of radiolabelled neurotransmitter efflux after its uptake or synthesis in neuronal tissue. In these studies, tissue slices or synaptosomes, which are resealed nerve terminals derived from specific brain regions or peripheral organs are used. These tissue preparations can be loaded with radiolabelled neurotransmitter or radiolabelled precursor of the neurotransmitter through the unique transport mechanisms intrinsic to the specific nerve cell of interest. The tissue preparation (slice or synaptosome) is then washed and equilibrated in physiological solution. Determination of neurotransmitter exocytosis is generally carried out in superfusion systems with filters suitable to retain tissue slices or synaptosomes. After establishing a stable basal efflux of radiolabelled neurotransmitter, Ca^{2+}-dependent exocytosis can be elicited by a depolarizing stimulus such as high K^+ or by electrical field stimulation. The perfusate containing radiolabelled neurotransmitter is collected in sequential fractions and quantified by liquid scintillation spectrophotometry. Thus, this technique can be applied to investigate the effect of age on both basal and depolarization-induced neurotransmitter release. In addition to measuring radiolabelled neurotransmitter in the perfusate, release of endogenous neurotransmitter and its metabolites of interest can also be assessed using high performance liquid chromatography.

These *in vitro* procedures using tissue slices and synaptosomes have also been applied to assess the influence of presynaptic receptors and intracellular regulatory mechanisms on exocytoic processes [19,20]. In this regard, receptor agonists, ions, potential neuromodulators and other chemical agents may be introduced into the perfusion medium prior to or during depolarization so that their effects on basal and depolarization-stimulated neurotransmitter exocytosis can be determined. The specificity of these agents can be verified by appropriate receptor antagonists or inhibitors. In addition to these pharmacological agents, the dependence of the exocytotic process on neuronal proteins which play important roles in regulating neurotransmitter release can also be examined using this technique. In these investigations, specific target proteins may be modified or eliminated via introduction of transgenes into mice and using these as the tissue source for the investigation of exocytosis.

15.2 Background

15.2.1   Regulation by Presynaptic Autoreceptors and the Effect of Age

The existence of presynaptic (prejunctional) autoreceptors that regulate the release of neurotransmitter has been demonstrated in both central and peripheral nervous systems for norepinephrine, serotonin, dopamine, acetylcholine, γ-aminobutyric acid and other neurotransmitters [19,20,21]. Activation of these presynaptic autoreceptors by exogenous receptor agonists generally inhibits exocytosis of the neurotransmitter, whereas receptor antagonists counteract the effects of agonists, and when given alone frequently increase its release. The magnitude of the effects of receptor stimulation and antagonism on neurotransmitter release is related to the frequency of nerve stimulation and to the duration of depolarization and may depend on the amount of transmitter released into the synaptic cleft [22–25.] Presynaptic autoreceptors appear to operate as a feedback mechanism through which released neurotransmitter modulates its own release when a critical concentration of neurotransmitter is attained in the synaptic cleft. Thus, presynaptic autoreceptors may be a site that modulates age-related changes in transmitter exocytosis.

The effect of age on the sensitivity of presynaptic autoreceptor has been tested in both central and peripheral neurotransmitter systems. In hippocampal and parietal cortical brain slices, comparable concentration-response curves for lysergic acid diethylamide (LSD)-mediated inhibition of [^{3}H]5-HT (serotonin) release were obtained in 6-, 12- and 24-month-old rats [4]. This was accompanied by an age-dependent decline in K^+-stimulated exocytosis of serotonin in parietal cortex but not in the hippocampus [4]. Similarly, electrical stimulation-evoked release of [^{3}H]5-HT but not [^{3}H]norepinephrine was found to be reduced age-dependently. This is accompanied by comparable autoreceptor-mediated modulation of the release of either monoamine [2]. These data indicate that prejunctional serotonin autoreceptor sites are not altered during aging. Similar to presynaptic serotonin autoreceptors, dopaminergic autoreceptor functions in striatum and mesolimbic areas were not affected by age [26]. On the other hand, in the rat hippocampus, an age-related reduction in presynaptic α_2-adrenergic receptor function, the autoreceptor that regulates release from noradrenergic neurons has been observed [27]. An impaired muscarinic autoreceptor function has also been observed in hippocampus and cerebrocortex but not in striatum of old rats [28]. The age-related reduction in cholinergic autoreceptor function occurred concurrently with a marked decrease in K^+-evoked acetylcholine release in all three brain regions examined [28].

Analogous to these findings in the central nervous system, age-dependent alterations in autoregulation were also observed in peripheral nervous system. In hearts from old male but not female Fischer-344 rats, the presynaptic α_2-adrenoceptor-mediated autoregulation of norepinephrine release was diminished with age [29]. In contrast, an age-related reduction in depolarization-induced [^{3}H]norepinephrine release was found in the rat vas deferens without apparent changes in the sensitivity of presynaptic α_2-adrenergic autoreceptor [30]. Together, these reports suggest that presynaptic autoreceptors are potential sites for changes during senescence. The age-related alterations in presynaptic autoregulation may occur concomitantly with changes in exocytosis of neurotransmitter and are neurotransmitter-, tissue- or gender-specific.

15.2.2 Presynaptic Regulation of Exocytosis by Other Neurotransmitter Systems and the Effect of Age.

In addition to regulation mediated by presynaptic autoreceptors, exocytosis of neurotransmitters may also be modulated by heteroreceptors located presynaptically. The activation of presynaptic heteroreceptors may be mediated via the co-release of neuropeptides or other neuromodulators, by transmitter released from adjacent terminals, or by locally produced or blood-born substances which either inhibit or facilitate Ca^{2+}-dependent neurotransmitter exocytosis. The relevance of presynaptic heteroreceptors in the control of nervous system function has been demonstrated under both *in vitro* and *in vivo* experimental conditions in central as well as peripheral systems [17,31–36]. Since presynaptic heteroreceptors play a major role in modulating the release of neurotransmitters, age-dependent changes in these receptors may significantly impact on the quality and strength of neurotransmission and may result in age-associated alterations in cross-talk between neurotransmitter systems.

The effect of age on presynaptic heteroreceptor-mediated regulation of neurotransmitter release has been examined in various neuronal systems. In rat brain, the release of dopamine is known to be enhanced by activating muscarinic receptors. During aging, the efficacy of the muscarinic response is reduced in an age-related manner [17]. In contrast, the modulation of striatal acetylcholine neurotransmission mediated by D_2 dopamine receptors is not altered during senescence, but the release of acetylcholine from the cholinergic interneurons is significantly lower in the aged brain [27]. Dopamine D_2 receptor stimulation was also shown to inhibit K^+-stimulated glutamate release in rat striatum, a response which is lost with age [37]. In the heart, inhibition of norepinephrine release from cardiac adrenergic nerves occurs when presynaptic receptors for adenosine are activated. Adenosine is produced in myocardial cells as a byproduct of ATP degradation and interstitial concentrations can be particularly high when oxygen supply does not meet oxygen demand. Presynaptic adenosine receptor capacity to inhibit norepinephrine release seems to decline with age [38]. These reports demonstrate that in addition to presynaptic autoregulation, the interaction between neurotransmitter systems may also be changed in an age-related fashion.

15.2.3 Regulation of Intracellular Signaling Systems, Exocytosis and the Effect of Age

Intracellular signaling mechanisms are known to regulate various presynaptic functions including the exocytotic process. It is well recognized that depolarization of the nerve terminal membrane initiates opening of voltage-sensitive calcium channels, resulting in the influx of Ca^{2+}. The increase in the intracellular Ca^{2+} concentration is critical for stimulus-induced exocytotic neurotransmitter release- a process which involves the fusion of neurotransmitter containing synaptic vesicles with nerve cell membrane [39]. Intracellular signal transduction mechanisms have important roles in modulating neurotransmission. It has been noted that depolarization of brain slices or synaptosomes result in activation of guanine nucleotide regulatory binding proteins (G proteins) [40–42] and increases in the phosphorylation of particular neuronal proteins by protein kinases such as protein kinase C. These signal transducers are also known to regulate Ca^{2+} channels and presynaptic receptors [40,42,43–45]. Age-related alterations in one or more of these intracellular signaling systems may affect neurotransmitter exocytosis and ultimately influence neuronal functions.

15.2.3.1   Calcium

Elevation of cytosolic Ca^{2+} is required for neurotransmitter exocytosis [39]. Calcium entering the nerve terminal via voltage-gated channels is required to trigger fusion of synaptic vesicles to presynaptic membranes and for the release of neurotransmitter into the synaptic cleft. The existence of multiple voltage-sensitive calcium channels on nerve terminals was also demonstrated [46]. Synaptosomal release of neurotransmitter may therefore tightly correlate with the rate of Ca^{2+} entry [47–49]. Thus, factors that affect both calcium entry through voltage-gated calcium channels and intracellular calcium homeostasis will potentially influence the exocytotic process.

Calcium-dependent release of many neurotransmitters has been shown to decline during senescence. Both K^+- and electrically evoked acetylcholine release in brain and peripheral tissue decrease with age, whereas non-calcium dependent efflux is not altered [50–52]. Age-related reductions in exocytosis of acetylcholine parallel decreases in neuronal calcium uptake [53–55]. The addition of 3,4-diaminopyridine which promotes calcium entry, reversed the age-related deficit in calcium uptake and calcium-dependent acetylcholine release [53,56]. Similarly, K^+-evoked release of dopamine in the brain was lower in old rats than in young rats and the reduction was shown to be dependent on calcium concentration. The age-dependent decrease in stimulated dopamine release can be attenuated by increasing extracellular concentration of calcium [16]. In cardiac synaptosomes, the decrease in norepinephrine release which is observed in the senescent male rat heart seems to be due to a reduction in Ca^{2+} entry through neuronal Ca^{2+} channels that leads to norepinephrine release [12]. In other neuronal tissue, calcium sensitivity may be limiting. For instance, higher concentrations of calcium ionophore were required to elicit neurotransmitter release in brains of old than of young animals [3,52]. Together, these reports suggest that changes in either calcium entry or intracellular calcium homeostasis may mediate some of the age-related alterations in neurotransmitter release. Alterations in calcium signaling are therefore important processes through which age may influence neurotransmission.

15.2.3.2   Protein Kinase C

Protein kinase C, a Ca^{2+}-activated, phospholipid-dependent phosphorylating enzyme that has been implicated in the regulation of neuronal excitability by catalyzing the phosphorylation of ion channels, membrane- and microtubular-associated proteins which are closely related to the depolarization-exocytosis process [57]. Protein kinase C is highly localized in presynaptic nerve endings [58,59] and modulates depolarization-induced neurotransmitter release for a number of neurotransmitter substances in the central and the peripheral nervous systems [60–64]. This enzyme is distributed in cells in both cytosolic and membranous fractions. Upon activation by diacylglycerol, which is generated by hydrolysis of phosphatidylinositides [65], the enzyme translocates from the cytosol to the cell membrane [3,4,66]. Since a close relationship exists between protein kinase C, Ca^{2+}, lipid metabolism and the cellular membrane, any age-related alterations in membrane components, calcium levels [50] or lipid metabolism [67] may potentially influence the activation of protein kinase C, and thereby its modulation of the exocytotic process and thus synaptic transmission.

In brain cortical and hippocampal slices, stimulation of protein kinase C with the dia-cylglycerol analogues, phorbol esters, results in a facilitated release of serotonin in 6-month-old Fischer-344 rats but this facilitation of release was attenuated in cortical and hippocampal slices of 12-month-old rats [3,4]. Furthermore, in 24-month-old rats, inhibitory effects on cortical and hippocampal serotonin release were observed when tissues were stimulated with phorbol esters [3,4]. Concomitant with the age-dependent changes in serotonin release, phorbol ester-induced protein kinase C translocation was also reduced in an age-related manner in these tissues [3,4,68]. However, the lack of apparent correlation between changes in phorbol ester-mediated release and enzymatic activity of protein kinase C in these tissues may suggest that age-related changes in substrate proteins may be involved in neurotransmitter release or its regulation. In line with this notion, Barnes and his co-workers [69] reported that protein kinase C-mediated phosphorylation of protein F1 (GAP-43) was age-dependently reduced in cortex and hippocampus from Fischer-344 rats. These data provide compelling evidence that the activation of protein kinase C and protein phosphorylation mediated by this enzyme are altered in an age-related fashion. The results furthermore, suggest that changes in protein kinase C substrate may also accompany senescence. Although a more complete identification of specific phosphoproteins that regulate exocytosis is needed, changes in protein kinase C-mediated phosphorylation and neurotransmitter release may be used as specific biomarkers for aging in neuronal tissue.

15.2.3.3 G Proteins

The involvement of heterotrimeric G proteins in regulating neurotransmitter release has been suggested by a number of previous reports [40–42]. Injection of guanine nucleotide analogues into neuronal tissues has demonstrated that activation of G proteins is necessary for neurotransmitter release [70,71]. Heterotrimeric G proteins have been shown to regulate the amount of transmitter released at the synapse by modulating potassium and calcium channels and by regulating the exocytotic response to calcium [41,71,72]. A specific regulatory role for Go protein in modulating neurotransmitter release has been demonstrated in experiments which used pertussis toxin or antisera directed against $G\alpha o$ [40–42]. In addition to their role in modulating exocytosis, G proteins may also mediate presynaptic receptor functions. This is supported by the fact that intracellular application of non-hydrolyzable analogues of GDP and GTP, GDPβS and GTPγS, interfere with agonist effects at presynaptic sites. While GDPβS blocks the action of agonists, GTPγS converts the normally reversible responses of agonists into irreversible ones. In line with these data, a pertussis toxin-sensitive G protein, in particular Gi, was demonstrated to regulate the inhibitory responses of presynaptic autoreceptors and heteroreceptors [40,42,44,45]. There is also evidence indicating a role for Gs in regulating β_2-adrenoceptor mediated potentiation of norepinephrine release. This Gs-mediated effect seems to involve an increase in intracellular cAMP level that in turn activates protein kinase A and subsequently facilitates Ca^{2+} channel opening [73]. An age-related reduction in G protein function has been demonstrated [74–77], and therefore G protein regulated neurotransmitter release and presynaptic receptor function may be altered in an age-related manner. Investigation of age-dependent changes in G protein function and exocytotic release of neurotransmitter during senescence is therefore needed to fully understand age-related alterations in synaptic neurotransmission.

15.3 General Design

Superfusion methods have become an important tool in demonstrating changes in calcium-independent, spontaneous efflux and calcium-dependent, depolarization-evoked release of various neurotransmitters during aging. Using this approach, the age-related changes in presynaptic regulatory mechanisms can also be assessed. This is usually accomplished by using *in vitro* preparations containing pinched-off nerve terminals so that changes in nerve activity are avoided. The accurate determination of neurotransmitter release in tissues is hampered by the very low quantities of transmitter released in a given time period and by efficient inactivation processes. Biogenic amines (dopamine, norepinephrine and serotonin), amino acids (γ-aminobutyric acid, glutamate and aspartate) and precursor (choline) are selectively taken up in low concentrations by the nerve terminals where they normally are used as neurotransmitters. By taking advantage of such selective and efficient transport systems, tissue slices and synaptosomal preparations containing nerve terminals can be loaded with radiolabelled exogenous neurotransmitters, and the exocytosis of radiolabelled neurotransmitters can then be measured with precision. However, the specificity of the transport mechanism should be verified by using specific pharmacological agents that block the uptake and consequently the release of the neurotransmitter. To preserve the integrity of the transmitter under study, drugs that inhibit the catabolism of the transmitters are often included during the prelabelling period. The spontaneous (basal) efflux of labelled exogenous transmitters is not calcium-dependent. Thus, the effect of various conditions or drugs on depolarization-induced exocytosis have been examined on transmitter release evoked by high concentrations of K^+ or by electrical field stimulation. In these assays, the effects of the agents or pathophysiological conditions may depend on the duration of the stimulation and the concentration of K^+ or on the characteristics of the electrical stimulation used. In addition, the duration of exposure to each of the drugs being tested may also critically affect the responses. Moreover, the proportion of the radiolabelled transmitter vs. metabolites in the total radioactivity measured in the effluent may vary depending on the experimental conditions. Accordingly, monoamine oxidase inhibitors and reuptake blockers are often included in the superfusion medium when the release of monoamines is under study.

Neurotransmitter release is determined in a superfusion system in which the volume of perfusate around the tissues is small enough to allow a maximal efficiency for the exchanges of O_2 and glucose with CO_2 and metabolic byproducts that are important for achieving relative metabolic stability in tissues. In some studies, tissue slices or synaptosomes are resting stationary on a nylon or a glass-fiber filter, while in other systems the tissues are freely floating in oxygenated medium with continuous mixing. Korpi and Oja [78] compared these two types of superfusion systems in brain slices and found that while the floating system may result in some mechanical release thereby increasing both basal and depolarization-induced release, an impaired energy metabolism, in particular a reduction in ATP, was noted in the stationary system. Thus, tissue viability is different in these two superfusion systems and may critically affect neurotransmitter release. Such differences may have a greater impact on studies using synaptosomes since long-term maintenance of synaptosomes *in vitro* has proven more difficult than that of tissue slices [79,80].

Superfusion systems have also been used extensively to study changes in presynaptic receptors and various intracellular regulatory mechanisms that may modulate both basal efflux and depolarization-evoked neurotransmitter release. The former condition simply requires introduction of the drug to the tissue in the resting situation, while the latter condition requires exposure to drugs prior to and during each of the depolarizing stimuli. To allow the construction of a concentration-response curve, the concentrations of drugs are generally given in an increasing order. In systems with multiple chambers, a different drug concentration can be perfused through each chamber. The functional properties of agonists/activators and antagonists/inhibitors may be analyzed further using conventional pharmacological measurements such as efficacy, PA_2 or PD_2. This technique therefore determines not only the effect of drugs on the exocytosis process per se, but also their potency and efficacy.

To study the role of membrane G proteins in regulating exocytosis and mediating pre-synaptic receptor function, brain slices can be incubated in oxygenated medium for pro-longed periods of time with pertussis toxin (ADP-ribosylates Gi/Go) or cholera toxin (ADP-ribosylates Gs) [40]. Brain synaptosomes can be prepared in the presence of pertus-sis toxin, cholera toxin or various antisera directed against the specific G protein subunits [42]. Employing either approach, the heterotrimeric G proteins were found to modulate both neurotransmitter release and to regulate the function of presynaptic receptors. Thus, following proper treatment with toxins or antisera, this *in vitro* procedure can be useful in investigating the identity and the function of G protein(s) in modulating presynaptic activities.

Since calcium plays an essential role in regulating depolarization-evoked exocytotic release, the role of voltage-gated calcium channels may be tested with drugs or toxins which modify Ca^{2+} transport via membrane channels. Moreover, changes in the sensitivity of exocytosis to calcium can be examined using calcium ionophores, such as ionomycin, A23187 in the presence of physiological concentrations of calcium. Thus, this method can also be used to investigate whether changes in calcium sensitivity and/or calcium move-ment through calcium channels mediate the altered neurotransmitter release during aging.

15.4 Procedures

15.4.1 Preparation of Brain Slices and Use in a Superfusion System

To characterize the functions of presynaptic receptors and to study the effects of age on the exocytosis of neurotransmitter, specific brain areas are dissected and chopped into 300 × 300 µm (thickness around 3mm) slices using a McIlwain tissue chopper. The tissue is then incubated at 37°C for 30 min in Krebs-Ringer's solution which has the following composi-tion: 118 mM NaC1, 4.8 mM KC1, 1.3 mM $CaC1_2$, 1.2 mM KH_2PO_4, 1.2 mM $MgSO_4$, 25 mM $NaHCO_3$, 10 mM glucose, 100 µm ascorbic acid, and gassed to PH 7.4 with 5% CO_2/95% O_2. The solution also contains 0.1 µm radiolabelled neurotransmitter or precursor in the presence of degrading enzyme inhibitors such as the monoamine oxidase inhibitor, par-gyline when using monoamines.

Following incubation with radiolabelled neurotransmitter or precursor, the slices are washed three times with ice-cold Krebs-Ringer's solution, and suspended in 6 ml physio-logical solution. Equal aliquots of tissue suspension are placed in each of 16 parallel super-fusion chambers of 250 µl volume (Swinnex, 13 mm, Millipore Co.). A circular piece of nylon mesh (pore size 250 µm) is placed just below the outlet in order to prevent the loss of tissue slices. The chambers are perfused against gravity with oxygenated Krebs-Ringer's solution at a rate of 1 ml/min. 10 min fractions are collected starting 30 min after the begin-ing of the perfusion (T_0). Release of labelled transmitters is evoked by superfusing tissues for 30 sec with Krebs-Ringer's solution containing 65 mM K^+ (made by isomolar replace-ment of NaCl with KCl) at 40(S_1), 70(S_2), 100(S_3) and 130(S_4) min after T_0. Assuming uniform dilution in the chambers, a maximal K^+ concentration of 56 mM was calculated. No signif-icant deterioration in response is observed during the 4 consecutive stimuli applied, thus enabling the testing of multiple concentrations of agents in the same tissue preparation. 20 min before S_2, S_3 or S_4 and for the following 10 min (collection time after each stimulus), tissues are superfused with media that contained pharmacological agents. For constructing the dose response curves, various drug concentrations are used. PMA and other phorbol esters are solubilized with 0.01% dimethylsulfoxide (DMSO). Control stimulus is in the presence of 0.01% DMSO. This concentration of DMSO does not alter the release of radiolabelled neurotransmitter tested. Using this procedure, we found that the radiolabelled neurotransmitter-overflow induced by high K^+ and the tissue radoactivity from unmetabolized radiolabelled neurotransmitter can be determined chromatographically by HPLC. In the case of monoamines, a high percentage of released unmetabolized amine is obtained by using a long acting MAO inhibitor such as pargyline [81]. At the end of the experiments, the tissues are sonicated in 1 ml ethanol with a Kontes Micro-ultrasonic cell disrupter. Aliquots of superfusates (0.5 ml) and tissue homogenates (0.1 ml) are added to 8 ml of Liquescent (National Diagnostics) and analyzed for radioactivity by liquid scintillation spectrometry.

Radioactivity efflux into the superfusate is calculated as the fraction of tritium content in the slices at the onset of the respective collection periods. For calculation of the stimulated tritium overflow, the estimated basal efflux (the average of 10-min fractions before and after the stimulation) is subtracted from the total radioactivity efflux during stimulation. This difference is employed in calculating the percent of released neurotransmitter. In order to quantify the *in vitro* effect of drugs on stimulated tritium overflow, the ratios of the fractions released by S_2 (or S_3, S_4) to that evoked by S_1 are determined.

15.4.2 Preparation and Superfusion Procedure for Brain Synaptosomes

Synaptosomes (P_2 fraction) are prepared from specific rat brain region according to the method reported by Gray and Whittaker [82]. Briefly, brain tissues are homogenized in oxygenated 10 mM Tris HCl (pH 7.4), 0.1 mM EDTA, 0.32 mM sucrose and 0.1% dimethyl sulfoxide (DMSO) in the presence of cholera toxin ($25\mu g$/mg tissues), pertussis toxin ($5\mu g$/mg tissues), anti-Gαs antiserum (1:250 dilution), anti-Gαi [1,2] antiserum (1:250 dilution), anti-Gαo antiserum (1:250 dilution) or normal rabbit serum (1:250 dilution). The specificity for these antisera is defined by Spiegel and his co-workers [83]. Homogenates are centrifuge at $1000 \times g$ for 10 min at 4°C. The supernatants are incubated at 37°C for 1 hour under constant shaking and 95% O_2/5% CO_2 flush. Synaptosomes are pelleted by centrifugation at $15,000 \times g$ for 30 min. The synaptosomal fraction is washed and incubated at 37°C for 30 min in Krebs-Ringer's solution and 0.1 µM radiolabelled neurotransmitter or precursor in the presence of the proper inhibitors of enzymes capable of degrading the neurotransmitter. After incubation, the synaptosomes are diluted with three volume of ice-cold physiological solution and centrifuged at $15,000 \times g$ for 15 min. Equal aliquots of the tissue suspension are placed in each of 16 parallel superfusion chambers (Swinnex, 13mm × 250 µl, Millipore Co.) between two circular pieces of GF/B glass fiber (Whatman) in order to prevent the loss of tissue. The chambers are perfused against gravity with oxygenated Krebs-Ringer's containing the proper enzyme inhibitors at a flow rate of 0.5ml/min. The onset of superfusion is defined as time zero (t_0). Starting at 30 min after t_0, 10-min fractions of the effluent are collected.

Release of radiolabelled neurotransmitter is evoked by superfusing tissues for 30 sec with 65 mM K^+-Krebs-Ringer's (made by isomolar replacement of NaCl with KCl). Three potassium pulses are given respectively at 40 min (S_1), 70 min (S_2), 100 min (S_3) after t_0. Using this protocol, there is generally no significant deterioration in fractional release observed during the 3 consecutive stimuli. Twenty minutes preceding S_2 or S_3, tissues are superfused with Krebs-Ringer's solution that contain various concentrations of the drugs to be studied or vehicle (control).

Following perfusion, the tissues (on filter) and aliquots of superfusates (0.2ml) are added to 5 ml of Liquiscint (National Diagnostics). The radioactivity is counted by liquid scintillation spectrometry.

15.4.3 Use of Cardiac Synaptosomes to Examine Age-Related Changes in Norepinephrine Release in the Heart

In order to examine the basis for the age-related reduction in norepinephrine release in the heart, a method of harvesting cardiac synaptosomes from rat hearts was developed in our laboratory [84]. Synaptosomes are used primarily to investigate the biochemical basis for neurotransmitter release at the synapse and have the advantage of being free of interference from surrounding cells. Synaptosomes (resealed presynaptic nerve endings) retain most of the structural and functional characteristics of the nerve terminal from which they are derived [85]. Neurotransmitter is released in a Ca^{2+}-dependent manner when synaptosomes are depolarized by raising extra-cellular K^+ or by electrical stimulation [86]. While the isolation of subcellular fractions containing synaptosomes is routinely done with brain tissue, similar procedures were not found to yield functional nerve terminals from peripheral tissues such as the heart [85]. We discovered that collagenase pretreatment of the heart before fractionation allows the preparation of a crude synaptosomal fraction [84]. This fraction shows norepinephrine accumulation which is attenuated by the neuronal uptake inhibitor desmethylimipramine (DMI) but is insensitive to the non-neuronal uptake blocker metanephrine. Norepinephrine uptake into cardiac synaptosomes is also sensitive to incubation temperature, sodium ion concentration and buffer osmolality. These findings are characteristic of a preparation containing intact synaptosomes [85] and indicates that only adrenergic neurons are accumulating norepinephrine.

Whittaker [85] has emphasized the use of synaptosomes over tissue slices for studying transmitter release and the effect of drugs on release mechanisms, since reuptake of transmitters is minimized by prompt removal of neurotransmitter during superfusion. Due to the irregular thicknesses and shapes of slices, rates of perfusion are not consistent between chambers and there is a greater variation in fractional release of K^+-induced norepinephrine release as indicated by the larger coefficients of variation in sliced atria preparations [10]. Dawson [87] found large variations in release of dopamine from superfused quinea pig brain slices that were due to variations in the thickness of slices within chambers. Synaptosome preparations avoid problems with tissue thickness and allow for uniform perfusion in each chamber thereby reducing the variation between chambers.

15.4.3.1 *Preparation of Cardiac Synaptosomes*

The method developed in our laboratory for obtaining cardiac synaptosomes has been published [83]. In order to prevent interference of neurotransmitter exocytosis from anesthesia, rats are decapitated. The hearts are then rapidly removed and dissected free of surrounding fat and connective tissue, rinsed free of blood using 0.32 M sucrose, and then weighed. Hearts are individually minced in 0.32 M sucrose containing 1 mM EGTA. The minced heart tissue is digested in HEPES-buffered saline solution (HBS) containing 12 units of collagenase per mg of heart tissue (class II, Worthington Biochemicals). The collagenase is dissolved in 10 ml of HBS per gram wet heart weight and the digestion proceeds for 40 min at 37°C. The actual quantity of collagenase (mg/10 ml HBS) varies depending on the activity of each batch of collagenase. HBS (pH 7.4 at 22°C) contains 20 mM HEPES, 144 mM NaCl, 5 mM KCl, 1.2 mM $CaCl_2$, 1.2 mM $MgCl_2$ and 10 mM glucose. The buffer also contains 1mM pargyline to inhibit enzymatic destruction of norepinephrine and 1 mM ascorbic acid to prevent oxidation of norepinephrine. The buffer is oxygenated prior to and during digestion. The collagenase is decanted after low speed centrifugation. The resulting pellet is suspended in 10 volumes (by heart weight) of 0.32 M sucrose and homogenized with 20 strokes of a Teflon/glass homogenizer. The homogenate is centrifuged for 10 min ($650 \times g$ at $4°C$). The pellet (P1) is resuspended in 5 volumes (by heart weight) of 0.32 M sucrose, homogenized with 10 strokes of the Teflon/glass homogenizer and then centrifuged for 10 min ($650 \times g$ at $4°C$). The supernatants from each 10 min centrifugation are combined and recentrifuged for 20 min ($21,000 \times g$ at $4°C$). The final pellet (P2) is resuspended in 2 ml ice-cold HBS. The P2 pellet contains the cardiac synaptosomes [84].

15.4.3.2 [³H]norepinephrine Uptake, Superfusion and [³H]norepinephrine Release from Cardiac Synaptosomes

To load the synaptosomes with radiolabelled norepinephrine, the P_2 is diluted to a concentration of 1 mg protein/ml of HBS and incubated with 300 nM [³H]norepinephrine (42 Ci/mmole, NEN) for 1 hour at $37°C$. Two hundred μl aliquots of the [³H]norepinephrine-P_2 preparation are pipetted into the chambers of a superfusion system. Our laboratory uses the Brandel 12-chamber superfusion system with automated fraction collector (Model SF-12, Biomedical Research and Development Laboratories, Gaithersburg, MD). Whatman GF/D filters retain the synaptosomes in the chambers. Oxygenated HBS with 0.2% bovine serum albumin is perfused through the system at a rate of 250 μl/min. In our system after 60 minutes of buffer perfusion a steady state of [³H]norepinephrine release is reached. 5 min fractions are then collected for the remainder of an experiment. In a typical experiment the synaptosomes are stimulated by changing the superfusion buffer to one containing a high [K⁺] for a period of 7 minutes followed by a return to the regular oxygenated HBS. Perfusion with buffer containing [³H]norepinephrine to act as a tracer has shown that changing the buffer for 7 minutes causes the [K⁺] within the superfusion chambers to rise and fall over a 15-minute period and that the peak [K⁺] reached in the chambers equals the K⁺ concentration in the high [K⁺] buffers. We typically expose the synaptosomes only once to K⁺ concentrations of 30 to 70 mM when determining K⁺ concentration-responses. In our system a second K⁺ exposure usually results in a 10% to 30% reduction in norepinephrine release and therefore we prefer the single exposure approach. The perfusion buffer is changed to deionized water for the last 30 minutes of the fraction collection. This causes the synaptosomes to rupture and release the remaining [³H]norepinephrine held within the synaptosomes. In a typical experiment 20 fractions are collected from each chamber. After the addition of liquid scintillation fluid to each fraction, the fractions are counted in a scintillation counter.

In the cardiac synaptosomes, experiments were conducted using tyramine to induce release of the [³H]norepinephrine for up to 3 hours. This result indicate that the entire [³H]norepinephrine pool from cardiac synaptosomes is releasable [10]. Tyramine produces a non-calcium dependent norepinephrine release.

15.4.3.3 Examining Basal [³H]norepinephrine Release from Cardiac Synaptosomes

The [³H]norepinephrine present in extra synaptosomal sites contributes to basal release but presumably not to K⁺-induced release. This assumption was tested in synaptosome preparations of hearts from a 6-month and a 24-month old in which desipramine (DMI) was added to the [³H]norepinephrine/P2 incubation mixture (1 μM final concentration) in order to prevent [³H]norepinephrine uptake by the synaptosomes [10]. Aloyo et al. [84]

have shown that this concentration of DMI completely blocks [³H]norepinephrine uptake by cardiac synaptosomes. These [³H]norepinephrine/P2 preparations were then placed in the superfusion system and norepinephrine release was induced by K^+ and then by perfusion with 160 µM tyramine. The presence of DMI during P2 loading with [³H]norepinephrine significantly reduced the amount of [³H]norepinephrine present in the fractions collected from the superfusion chambers during HBS perfusion (see Figure 15.1, A and B). Subsequent K^+-induced stimulation and tyramine perfusion failed to induce [³H]norepinephrine release from DMI treated P2 of either 6-month or 24-month old rats. Therefore the net fractional release of [³H]norepinephrine due to K^+ stimulation and the release of [³H]norepinephrine during tyramine perfusion appears to be solely due to the release of [³H]norepinephrine from cardiac synaptosomes.

To verify that [³H]norepinephrine release from the cardiac synaptosome preparation originated only from adrenergic nerve terminals, 6-month-old male Fischer-344 rats were injected with 6-hydroxydopamine (6-OHDA) or vehicle on day 1 and day 8 and killed on day 9 [10]. 6-OHDA selectively destroys peripheral adrenergic nerve terminals of the heart and depletes the norepinephrine content of the heart [85]. 6-OHDA injections reduced the endogenous norepinephrine content of the P2 by 85% when expressed as ng norepinephrine/mg protein. K^+ depolarization failed to induce [³H]norepinephrine release from P2 of 6-OHDA treated rats, whereas [³H]norepinephrine release was present in vehicle treated rats (see Figure 15.2). Water perfusion released approximately 48,000 DPM of [³H]norepinephrine from chambers containing P2 from vehicle treated rats and only 7,000 DPM from chambers containing P2 from 6-OHDA treated rats. Since 6-OHDA is not 100% effective in eliminating adrenergic nerve terminals the water-induced release of [³H]norepinephrine in 6-OHDA treated rats is from the small number of adrenergic nerve terminals still remaining in the preparation.

We have used these procedures for measuring norepinephrine release from cardiac synaptosomes in a superfusion system to show that exocytosis of norepinephrine in the heart is calcium dependent, can be attenuated by calcium channel blockers such as Mg^{2+} and omega-conotoxin (a N-type calcium channel blocker), and can be inhibited by activation of presynaptic adenosine A_1 receptors [13,36]. These experiments have also shown that neuronal calcium channels and presynaptic adenosine A_1 receptors have been altered during aging. Future studies using cardiac synaptosomes and superfusion systems will hopefully allow us to determine the underlying cellular basis for these age-related changes.

15.4.4   Calculation

The radioactivity released into each fraction is converted to fractional release to correct for variation in total DPM between experiments due to differences in tissue slice or synaptosome content in the P2. Fractional release is the percent of the radiolabelled neurotransmitter contained within the synaptosomes that is released into a single fraction. Exposure to high [K^+] stimulates neurotransmitter release from the tissue slices or synaptosomes and increases the fractional release. Fractional release is calculated by dividing the DPM collected in a fraction by the DPM in the chamber at the time the fraction was collected. The following formula is used:

$$\% \text{ Fractional Release from Fraction `A'} = \frac{\text{DPM in Fraction `A'}}{\text{(Total DPM Collected)} \pm \text{(DPM Collected Prior to Fraction `A')}} \times 100$$

Net fractional release of radiolabelled neurotransmitter due to high [K^+] exposure is calculated as the fractional release above the basal fractional release. Basal fractional release is

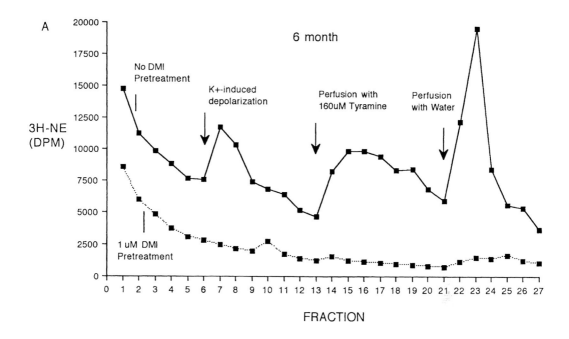

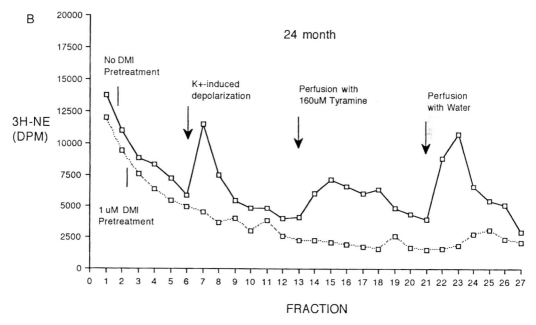

FIGURE 15.1

(A) Effect of desmethylimipramine (DMI) pretreatment on K⁺- and tyramine-induced [³H]norepinephrine release from cardiac synaptosomes from a 6-month-old male F344 rat. The cardiac synaptosome preparation was incubated for 1 h with [³H]norepinephrine and buffer with and without 1 μM DMI. These preparations were then placed in a superfusion system. The system was perfused with HEPES buffered saline for 30 min at 250 μl/min before fraction 1 was collected. Each fraction was 5 min in duration. During fraction 7 and 8 the K⁺ concentration in the chamber rose to 60 mM and induced release of [³H]norepinephrine from the cardiac synaptosomes. Buffer containing 160 μM tyramine was perfused continuously through the system starting at fraction 13. The perfusion was switched to deionized water at fraction 21. Each point on the graph is the mean DPM released from three chambers during a single experiment. **(B)** Effect of DMI pretreatment on K⁺- and tyramine-induced 3H-norepinephrine release from cardiac synaptosomes from a 24-month-old male F344 rat. Same procedure as A. (From Snyder et al., 1992 with permission.)

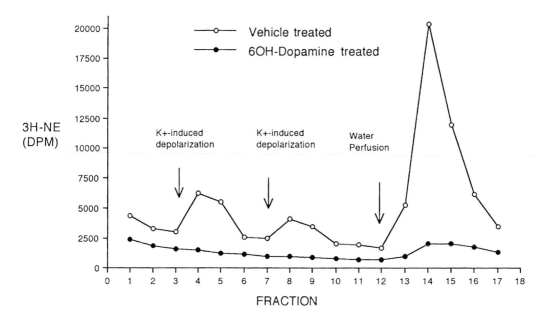

FIGURE 15.2
Representative experiment showing the effect of 6 hydroxy-dopamine (6-OHDA) injections on K⁺-induced
[³H]norepinephrine release from cardiac synaptosomes from 6-month-old male F344 rats. Rats were injected with
6-OHDA on day 1 and day 8 and killed on day 9. Cardiac synaptosomes were prepared, placed in a superfusion
system and depolarized with 60 mM K⁺. Each point on the graph is the mean DPM released from 3 chambers.
The experiment was performed three times with cardiac synaptosomes from a vehicle treated and a 6-OHDA
treated rat perfused simultaneously in the same experiment. The almost complete destruction of adrenergic
nerve terminals in the 6-OHDA treated rats resulted in no K⁺-induced [³H]norepinephrine release and greatly
reduced water perfusion release of [³H]norepinephrine. (From Snyder et al., 1992 with permission.)

obtained immediately before and after the fractions collected during the high [K⁺] exposure. Experiments using varying amounts of tissue or P2 protein per superfusion chamber,
which changes the number of synaptosomes and total radioactivity per chamber, show that
the fractional releases are unchanged [3,4,10,39]. Thus expressing radiolabelled neurotransmitter release as fractional release corrects for any differences in the amount of tissue slices and purity of the synaptosome preparation (synaptosomes/mg protein) or
radiolabelled neurotransmitter loading of the synaptosome preparations from different
age groups.

15.5 Conclusion

One of the most widely used *in vitro* methods for the investigation of neurotransmitter exocytosis is the superfusion of tissue slices or synaptosomes preloaded with radiolabelled
neurotransmitter or precursor. The techniques allow convenient methods for investigation
of the role presynaptic receptors and intracellular regulatory mechanisms responsible for
modulating neurotransmitter release. Superfusion systems are particularly useful for
determining drug efficacy and affinity for a receptor or a specific intracellular regulatory
machinery when multiple depolarization pulses can be introduced. This is a rapid, highly

sensitive method which may be applied to study neurotransmitter release and its regulation in tissues which are sparely innervated by the neurotransmitter system of interest. The technique is extremely efficient since it is feasible to process hundreds of samples per day.

However, the most important issue in using the *in vitro* release method is whether it measures the same exocytotic process that operates in regulating endogenous neurotransmitter release. Release of radiolabelled and endogenous monoamines do not always occur in parallel [89]. This may be caused, in part, by asymmetrical uptake and distribution of radiolabelled neurotransmitters or precursors within the releasable transmitter pools [90,91]. These differences, nevertheless, are generally quantitative rather than qualitative. This is especially true regarding the functions of presynaptic receptors or intracellular regulatory mechanisms since they appear to be independent of the methodology.

As changes in neurotransmission play an important role in mediating functional alterations of nervous systems, the *in vitro* superfusion method provides comprehensive and reliable information regarding exocytosis of neurotransmitters, a vital presynaptic function that may be altered by the aging process. Furthermore, results obtained using this procedure can also be used to suggest plausible mechanisms that contribute to the underlying changes in neurotransmitter release. With this technique, an extensive and thorough study can be conducted to determine underlying molecular mechanisms which accompany the aging process. In addition, the *in vitro* superfusion method may furnish us with information to help understand the effectiveness of newly designed drugs for the treatment of age-related alterations in presynaptic functions, as well as the impact such agents have on neurotransmission following drug exposure. Ultimately, this technique may be used in fresh and postmortem human tissues to study age-dependent changes in presynaptic functions and to assess the pharmacological profile of drugs that may be relevant for treatment of nervous system disorders [2–4,9,56]. Information derived from such studies will greatly enhance our understanding of changes and underlying mechanisms responsible for age-related changes in neurotransmission, in particular, presynaptic function and regulation. This will allow subsequent design of more specific and effective pharmacotherapeutic agents to counter exocytotic dysfunctions associated with aging of the nervous system.

Acknowledgments

This work is supported in part by grants AG07700 and AG11060 from the National Institute on Aging.

References

1. Navarro, H.A., Aloyo, V.J., Rush, M.E. and Walker, R.F. Serotonin pharmacodynamics in hypothalamic tissues from young and old female rats. *Brain Res.* 421:291–296, 1987.
2. Schlicker, E., Betz, R. and Gothert, M. Investigation into the age-dependence of release of serotonin and noradrenaline in the rat brain cortex and of autoreceptor-mediated modulation of release. *Neuropharmcol.* 31: 1–18, 1988.
3. Friedman, E. and Wang, H.Y. The effect of age on brain cortical protein kinase C and its mediation of serotonin release. *J. Neurochem.* 52: 187–192, 1989.

4. Wang, H.Y. and Friedman, E. Age-related changes in brain protein kinase C and serotonin release. In: *Biomedical Advances in Aging* (Allan L. Goldstein ed.), Plenum Publishing Co., New York, pp. 469–476, 1990.

5. Laping, N.J., Dluzen, D.E., and Ramirez, V.D. Aging alters opiate inhibition of potassium (K+)-stimulated dopamine release from the corpus striatum of male rats. *Neurobiol. Aging* 11: 395–399, 1990.

6. Casamenti, F., Scali, C. and Pepeu, G. Phosphatidylserine reverses the age-dependent decrease in cortical acetylcholine release:a microdialysis study. *Eur. J. Pharmacol.* 194: 11–16, 1991.Leslie, S.W., Chandler, L.J., Barr, E. and Farrar, R.P. Reduced calcium uptake by rat brain mitochondria and synaptosomes in response to aging. *Brain Res.* 329: 177–183, 1985.

7. McIntosh, H.H. and Westfall, T.C. Influence of aging on catecholamine levels, accumulation, and release in F-344 rats. *Neurobiol. Aging* 8: 233–239, 1987.

8. Freeman, G.B., and Gibson, G.E. Selective alteration of mouse brain neurotransmitter release with age. *Neurobiol. Aging* 8: 147–152, 1987.

9. Vannucchi, M.G., Casamenti, F. and Pepeu, G. Decrease of acetylcholine release from cortical slices in aged rats: Investigation into its reversal by phospatidylserine. *J. Neurochem.* 55: 819–825, 1990.

10. Snyder D.L., Aloyo V.J., McIlvain B. Johnson M.D. and Roberts J. Effect of age on potassium- and tyramine-induced release of norepinephrine from cardiac synaptosomes in male F344 rats. *J. Gerontol. Biol. Sci.* 47: B190-B197, 1992.

11. Snyder, D.L., Johnson, M.D., Eskin, B.A., Wang, W., Roberts, J. Effect of age on cardiac norepinephrine release in the female rat. *Aging* 7: 210–217, 1995.

12. Snyder, D.L., Johnson, M.D., Aloyo, V., Eskin, B.A., Roberts, J. Age-related changes in cardiac norepinephrine release: role of calcium movement. *J. Gerontol.* 50: B358–367, 1995.

13. Daly, R.N., Goldberg, P.B. and Roberts, J. The effect of age on presynaptic alpha 2 adrenoceptor autoregulation of norepinephrine release. *J. Gerontol.* 44: B59–66, 1989.

14. Roberts, J., Mortimer, M.L., Ryan, P.J., Johnson, M.D., Tumer, N. Role of calcium in adrenergic neurochemical transmission in the aging heart. *J. Pharmacol. Exp. Ther.* 253: 957-964, 1990.

15. Tumer, N., Ryan, P.J., and Roberts, J. Action of potassium on neurochemical transmission at the cardiac adrenergic neuroeffector junction with aging. *Mech. Ageing Dev.* 52: 87-91, 1990.

16. Raiteri, M., Bonanno, G., Calderini, A., Caviglia, M., Marchi, M. and Maura, G. Calcium dependence of transmitter release in nerve endings isolated from the brain of aged rats. *Soc. Neurosci. Abstr.* 12: 271, 1986.

17. Joseph, J.A., Dalton, T.K. and Hunt, W.A. Age-related decrements in the muscarinic enhancement of K+-evoked release of endogenous striatal dopamine: an indicator of altered cholinergic-dopaminergic reciprocal inhibitory control in senescence. *Brain Res.* 454: 140–148, 1988a.

18. Joseph, J.A., Dalton, T.K. Roth, G.S. and Hunt, W.A. Alterations in muscarinic control of striatal dopamine autoreceptors in senescence: a deficit at the ligand-muscarinic receptor interface? *Brain Res.* 454: 149–155, 1988.

19. Chesselet, M.-F. Presynaptic regulation of neurotransmitter release in the brain:facts and hypothesis. *Neuroscience* 12: 347–375, 1984.

20. Middlemiss, D.N. and Hutson, P.H. Measurement of the *in vitro* release of endogenous monoamine neurotransmitters as a means of identification of prejunctional receptors. *J. Neurisci. Methods.* 34: 23–28, 1990.

21. Starke, K., Gothert, M. and Kilbinger, H. Modulation of neurotransmitter release by presynaptic autoreceptors. *Physiol. Rev.* 69: 864–989, 1989.

22. Baumann, P.A. and Koella, W.P. Feedback control of noradrenaline release as a function of noradrenaline concentration in the synaptic cleft in the cortical slices of the rat. *Brain Res.* 189: 437–448, 1980.

23. Wemer, J. and Mulder, A.H. Postnatal development of presynaptic alpha-adrenoceptors in rat cerebral cortex studies with brain slices and synaptosomes. *Brain Res.* 208: 299–310, 1981.

24. Story, D.F., McCulloch, M.W., Rand, M.J. and Standford-Starr, C.A. Conditions required for the inhibitory feedback loop in noradrenergic transmission. *Nature* 293: 62–65, 1981.

25. Reichenbacher, D., Reiman, W. and Starke, K. α-adrenoceptor-mediated inhibition of norad-renaline release in rabbit brain cortex slices. Receptor properties and role of biphase concentration of noradrenaline. *Naunyn-Schmiedebergs Arch. Pharmacol.* 319: 71–77, 1982.

26. Carfagna, N., Trunzo, F., Moretti, A. Brain dopamine autoreceptors in aging rats. *Exp. Gerotol.* 21: 169–175, 1986.

27. Zsilla, G., Zelles, T., Mike, A., Kekes-Szabo, A., Milusheva, E., Vizi, E.S. Differential changes in presynaptic modulation of transmitter release during aging. *Internat. J. Develop. Neurosci.* 12: 107–115, 1994.

28. Araujo, D.M., Lapchak, P.A., Meaney, M.J., Collier, B., Quirion, R. Effects of aging on nicotinic and muscarinic autoreceptor function in the rat brain: relationship to presynaptic cholinergic markers and binding sites. *J. Neurosci.* 10: 3069–3078, 1990.

29. Tumer, N., Mortimer, M.L., Roberts, J. Gender differences in the effect of age on adrenergic neurotransmission in the heart. *Exp. Gerontol.* 27: 301–307, 1992.

30. De Avellar, M.C. and Markus, R.P. Age-related changes in norepinephrine release and its modulation by presynaptic alpha-2 adrenoceptors in the rat vas deferens. *J. Pharmacol. Exp. Ther.* 267: 38–44, 1993.

31. Giorguieff-Chesselet, M.-F., Le Floc'h, M.-L., Glowinski, J. and Besson, M.-J. Involvement of cholinergic presynaptic receptors of nicotinic and muscarinic types in the control of the spontaneous release of dopamine from striatal dopaminergic terminals in the rat. *J. Pharmacol. Exp. Ther.* 200: 535–544, 1977.

32. Muscholl, E. Presynaptic muscarinic receptors and inhibition of release. In *The Release of Catecholamines from Adrenergic Neurons* (Paton, D.M. ed.) Pergamon Press, Oxford Press, Oxford, 1979.

33. Lavallee, M., De Champlain, J. and Nadeau, R.A. Reflexly induced inhibition of catecholamine release through a peripheral muscarinic mechanism. *Can. J. Physiol. Pharmacol.* 58: 1334–1341, 1980.

34. Taube, H.D., Starke, K. and Borowski, E. Presynaptic receptor system on the adrenergic neurons of rat brain. *Naunyn-Schmiedebergs Arch. Pharmacol.* 299: 123–141, 1977.

35. Langer, S.Z. Presence and physiological role of presynaptic inhibitory alpha$_2$-adrenoceptors in guinea-pig atria. *Nature* 294: 671–672, 1981.

36. Raiteri, M., Marchi, M. and Maura, M. Presynaptic muscarinic receptors increase striatal dopamine release evoked by quasi-physiological depolarization. *Eur. J. Pharmacol.* 83: 127–129, 1982.

37. Donzanti, B.A., Hite, J.F. and Yamamoto, B.K. Extracellular glutamate levels increase with age in the lateral striatum: potential involvement of presynaptic D-2 receptors. *Synapse* 13: 376–382, 1993.

38. Snyder, D.L., Roberts, J. The effect of age on the capacity of adenosine to modulate norepinephrine release in the heart. *FASEB J.* 10: A335, 1996.

39. Llinas, R.R. Calcium and transmitter release in squid synapse. *Neuroscience Symposium, Vol II, Approaches to Cell Biology of Neurons.* (Cowan, W.M. and Ferrendelli, J.A. eds.) Bethesda: Society for Neuroscience, pp.139–160, 1977.

40. Hertting, G., Wurster, S. and Allgaier, C. Regulatory proteins in presynaptic function. *N. Y. Acad. Sci.* 604: 289–304, 1990.

41. Ohara-Imaizumi, M., Kameyama, K., Kawae, N., Takeda, K., Muramatsu, S., Kumakura, K. Regulatory role of the GTP-binding protein, G(o), in the mechanism of exocytosis in adrenal chromaffin cells. *J. Neurochem.* 58: 2275–2284, 1992.

42. Wang, H.Y., Yue, T.-L., Feuerstein, G. and Friedman, E. Platelet-activating factor: diminished acetylcholine release from rat brain is mediated by a G$_i$ protein. *J. Neurochem.* 63: 1720–1725, 1994.

43. Dolphin, A.C., Huston, E., Pearson, H., Menon-Johanssen, A., Sweeney, M., Adams, M.E. and Scott, R.H. G protein modulation of calcium entry and transmitter release. *N.Y. Acad. Sci.* 635: 139–152, 1991.

44. Passarelli F., Costa T., Almeida, O.F.X. Pertussis toxin inactivates the presynaptic serotonin autoreceptor in the hippocampus. *Eur. J. Pharmacol.* 155: 297–299, 1988.

45. Fredholm, B.B., Duner-Engstorm, M., Fastbom, J., Hu, P.-S. and Van Der Ploeg, I. Role of G proteins, cyclic AMP, and ion channels in the inhibition of transmitter release by adenosine. *Ann. N. Y. Acad. Sci.* 604: 276–288, 1990.

46. Tareilus, E., Breer, H. Presynaptic calcium channels: pharmacology and regulation. *Neurochem. Internat.* 26: 539–558, 1995.

47. Drapeau, P. and Blaustein, M.P. Initial release of ^{3}H dopamine from rat striatal synaptosomes: correlation with calcium entry. *J. Neurosci.* 3: 703–713, 1983.

48. Floor, E. Substance P release from K$^+$-depolarized rat brain synaptosomes at one second resolution. *Brain Res.* 279: 321–324, 1983.

49. Suszkiw, J.B. and O'Leary, M.E. Temporal characteristics of potassium-stimulated acetylcholine release and inactivation of calcium influx in rat brain synaptosomes. *J. Neurochem.* 41: 868–873, 1983.

50. Gibson, G.E. and Peterson, C. Calcium and the aging nervous system. *Neurobiol.* Aging 8: 329–344, 1987.

51. Sastry, B.V., Janson, V.E., Jarswal, N. and Tayels, D.S. Changes in enzymes of the cholinergic system and acetylcholine release in the cerebra of aging Fischer rats. *Pharmacol.* 26:61–72, 1983.

52. Meyer, E.M., Baker, S.P., Crews, F.T. and Larsen, K. Aging and acetylcholine-release from cortical synaptosomal and atrial minces. *Soc. Neurosci. Abstr.* 11: 980, 1985.

53. Peterson, C. and Gibson, G.E. Aging and 3,4-diaminopyridine alter synaptosomal calcium uptake. *J. Biol. Chem.* 258: 11482–11485, 1983.

54. Leslie, S.W., Chandler, L.J., Barr, E. and Farrar, R.P. Reduced calcium uptake by rat brain mitochondria and synaptosomes in response to aging. *Brain Res.* 329: 177–183, 1985.

55. Peterson, C., Nicholls, D.G. and Gibson, G.E. Subsynaptosomal calcium distribution during hyposia and 3,4-diaminopyridine treatment. *Neurobio.* Aging 6: 297–304, 1985.

56. Peterson, C. and Gibson, G.E. Amelioration of age-related neurochemical and behavioral deficits by 3,4-diaminopyridine. *Neurobiol.* Aging 4: 25–30, 1983.

57. Nestler E.J., Greengard P. Neuron-specific phosphoproteins in mammalian brain. *Advance in Cyclic Nucleotide & Protein Phosphorylation Research* 17:483–488, 1984.

58. Wood, J.G., Girard, P.R., Mazzei, G.J., and Kuo, J.F. Immunocytochemical localization of protein kinase C in identified neuronal compartments of rat brain. *J. Neurosci.* 6: 2571–2580, 1986.

59. Worley, P.F., Baraban, J.M., Snyder, S.H. Heterogeneous localization of protein kinase C in rat brain: Autoradiographic analysis of phorbol ester receptor binding. *J. Neurosci.* 6: 199–207, 1986.

60. Wakade, A.R., Malhotra, R.K. and Wakade, T.D. Phorbol ester, an activator of protein kinase C, enhances calcium-dependent release of sympathetic neurotransmitter. *Naunyn-Schmiedebergs Arch. Pharmacol.* 133: 122–127, 1985.

61. Wang, H.Y. and Friedman, E. Protein kinase C: Regulation of serotonin release from rat brain cortical slices. *Eur. J. Pharmacol.* 141: 15–21, 1987.

62. Nichols, R.A., Haycock, J.W., Wang, J.K.T. and Greengard, P. Phorbol ester enhancement of neurotransmitter release from rat brain synaptosomes. *J. Neurochem.* 48: 615–621, 1987.

63. Zurgil, N. and Zisapel, N. Phorbol ester and calcium act synergistically to enhance neurotransmitters release by brain neurons in culture. *FEBS Lett.* 185: 257–262, 1985.

64. Coffey, E.T., Shira, T.S., Nicholls, D.G., Pocock, J.M. Phosphorylation of synapsin I and MARCKS in nerve terminals in mediated by Ca^{2+} entry via an Aga-GI sensitive Ca^{2+} channel which is coupled to glutamate exocytosis. *FEBS Lett.* 353: 264–268, 1994.

65. Berridge, M.J. Inositol triphosphate and diacylglycerol as second messengers. *J. Biochem.* 220: 345–360, 1984.

66. Wolf, M., Cuatrecasas, P., and Sahyoun, N. Interaction of protein kinase C with membranes is regulated by Ca^{2+}, phorbol esters and ATP. *J. Biol. Chem.* 260: 15718–15722, 1985.

67. Gatti, C., Noremberg, K., Brunetti, M., Teolato, S., Calderini, G. and Gaiti, A. Turnover of palmitic and arachidonic acids in the phospholipids from different brain areas of adult and aged rats. *Neurochem. Res.* 11: 241–252, 1986.

68. Battaini, F., Elkabes, S., Bergamaschi, S., Ladisa, V., Lucchi, L., De Graan, P.N., Schuurman, T., Wetsel, W.C., Trabucchi, M., and Govoni, S. Protein kinase C activity, translocation, and conventional isoforms in aging rat brain. *Neurobiol.* Aging 16: 137–148, 1995.

69. Barnes, C.A., Mizumori, J.Y., Lovinger, D.M., Sheu, F.-S., Murakami, K., Chan, S.Y., Linden, D.J., Nelson, R.B. and Routtenberg, A. Selective decline in protein F1 phosphorylation in hippocampus of senescent rats. *Neurobiol. Aging* 9: 393–398, 1988.

70. Vitale, N., Aunis, D., and Bader, M.F. Distinct heterotrimeric GTP-binding-proteins act in series to control to exocytotic machinery in chromaffin cells. *Cell. Mol. Biol.* 40: 707–715, 1994.

71. Fang, Y., Durgerian, S., Basarsky, T.A., and Haydon, P.G. GTP-binding proteins: necessary components of the presynaptic terminal for synaptic transmission and its modulation. *Adv. Second Messenger & Phosphoprotein Res.* 29: 121–132, 1994.

72. Walsh, K.B., Wilson, S.P., Long, K.J., and Lemon, S.C. Stimulatory regulation of the large-conductance, calcium-activated potassium channel by G proteins in bovine adrenal chromaffin cells. *Mol. Pharmacol.* 49: 379–386, 1996.

73. Majewski, H., Costa, M., Foucart, S., Murphy, T.V. and Musgrave, I.F. Second messengers are involved in facilitatory but not inhibitory receptor actions at sympathetic nerve endings. *Ann. N.Y. Acad. Sci.* 604: 266–275, 1990.

74. Wang, H.Y. and Friedman, E. Receptor-mediated activation of G proteins is reduced in post-mortem Alzheimer's disease brains. *Neurosci. Lett.* 173: 37–39, 1994.

75. Greenwood, A.F., Powers, R.E. Jope, R.S. Phosphoinositide hydrolysis, G alpha q, phospholipase C, and protein kinase C in postmortem human brain: effects of postmortem interval, subject age, and Alzheimer's disease. *Neurosci.* 69: 125–138, 1995.

76. Gurdal, H., Friedman, E. Johnson, M.D. Beta-adrenoceptor-G alpha S coupling decreases with age in rat aorta. *Mol. Pharmacol.* 47: 772–778, 1995.

77. Joseph, P.A., Cutler, R., Roth, G.S. Changes in G protein-mediated signal transduction in aging and Alzheimer's disease. *Ann. N. Y. Acad. Sci.* 695: 42–45, 1993.

78. Korpi, E.R. and Oja, S.S. Comparison of two superfusion systems for study of neurotransmitter release from rat cerebral cortex slices. *J. Neurochem.* 43: 236–242, 1984.

79. Bradford, H.F. Cerebral cortex slices and synaptosomes: *in vitro* approaches to brain metabolism. In *Methods of Neurochemistry, vol. 3* (Fried R. ed.) Marcel Dekker, New York, pp155–202, 1972.

80. Campbell, C.W.B. The Na$^+$, K$^+$, Cl$^-$ contents and derived membrane potentials of presynaptic nerve endings *in vitro*. *Brain Res.* 101: 594–599, 1976.

81. Hellerman, L. and Erwin, V.G. Mitocondrial monoamine oxidase II. Actions of various inhibitors for the bovine kidney enzyme. Catalytic mechanism. *J. Biol. Chem.* 243: 5234–5243, 1968.

82. Gray E.G. and Whittaker U.P. The isolation of nerve endings from brain: an electron microscopic study cell fragments derived by homogenization and centrifugation. *J. Anat.* 96: 79–87, 1962.

83. Spiegel A.M. Immunologic probes for heterotrimeric GTP-binding proteins. In: *G Proteins* (Iyengar R. and Birnbaumer L., eds), San Diego: Academic Press pp. 115–143, 1990.

84. Aloyo V.J., McIlvain H.B., Bhavsar V.H.m and Roberts J. Characterization of norepinephrine accumulation by a crude synaptosomal-mitochondrial fraction isolated from rat heart. *Life Sci.* 48: 1317-1324, 1991.

85. Whittaker, V.P. The synaptosome. In: *Handbook of Neurochemistry Vol. 7.* (Lajtha, A., ed.) New York: Plenum Press, pp 1-39, 1984.

86. de Belleroche, J.S. and Bradford, H.F. The stimulus-induced release of acetylcholine from synaptosome beds and its calcium dependence. *J. Neurochem.* 19: 1817-1819, 1972.

87. Dawson, R.M. Factors influencing the calculation of results from studies of the release of titrated neurotransmitters from superfused slices of quinea pig striata. *J. Pharmacol. Meth.* 22: 65-75, 1989.

88. Tumer, N., Houck, W.T., and Roberts, J. Upregulation of adrenergic beta receptor subtypes in the senescent rat heart. *Mech. Ageing Dev.* 49: 235-243, 1989.

89. Herdon, H., Strupish, J. and Nahorski, S.R. Differences between the release of radiolabelled and endogenous dopamine from superfused rat brain slices:effects of depolarising stimuli, amphetamine and synthesis inhibition. *Brain Res.* 348: 309–320, 1987.

90. Leviel, V. and Guibert, B. Involvement of intraterminal dopamine compartments in the amine release in the cat striatum. *Neurosci. Lett.* 76: 197–202, 1987.

91. Zumstein, A., Karduck, W. and Starke, K. Pathways of dopamine metabolism in the rabbit caudate nucleus *in vitro*. *Naunyn-Schmiederbergs. Arch. Pharmacol.* 316: 205–217, 1981.

16

Pathological Analysis in Aging Research

Gene B. Hubbard and Yuji Ikeno

CONTENTS

16.1 Introduction

Pathology is the science dealing with disease in living organisms with special attention to causes, responses, and morphological and functional changes [1–4]. Essentially all disease starts at the cellular level and can lead to tissue, organ, systemic and whole body disease if the organism cannot resist the cause and heal. General causes of cell injury are hypoxia, physical, chemical and infectious agents, immunological reactions, genetic defects, and nutritional imbalances. Cells can respond to reversible or irreversible injury in a variety of ways including atrophy, hypertrophy, hyperplasia, metaplasia, formation of subcellular alterations and inclusions, mineralization and death [3]. Morphological changes that are reversible include cellular swelling, hydropic change or vacuolar degeneration and fatty change. Morphological changes that are not reversible include fibrosis, necrosis, and apoptosis. Ultrastructural changes beyond light microscopic evaluation can best be evaluated with electron microscopy and include membrane, nuclear, and mitochondrial changes. Functional changes are best evaluated with clinical observation and clinical pathology [2].

Aging, the process of growing old, begins at conception and ends with death. Aging is associated with years of sublethal cell injury, causing cell loss, gradual deterioration and irreversible physiological and structural changes, which lead to a decreased ability to cope with injury and eventually death. Cellular aging is multifactorial and involves endogenous and exogenous influences. The morphological changes in aged cells include abnormal nuclei, pleomorphic vacuolated mitochondria, decreased endoplasmic reticulum, distorted Golgi apparatus, and concomitant accumulation of pigmented lipofuscin. These morphological changes can be seen and recorded by pathological evaluation of animals and tissues and serve as useful biomarkers of aging. Major lesions of disease are of no more importance to the evaluation of aging than are benign lesions [2,5].

It is the intent of this chapter to convince the reader of the importance of pathology to the study of aging and provide the information necessary to make maximum use of research materials using pathological methods. A rather extensive reference list is part of the chapter so the interested individual can seek out more information. It is not the intent, nor is it possible, to provide a complete review of all pathological data accumulated from animals used in aging research. Nor is it designed to teach, in other than a "common sense" manner, pathology, anatomy, histology, or experimental design. The persons doing aging research and the related pathology should be well trained in these disciplines. This chapter is designed as a "how-to-do" pathology, specifically in aging research, but really for all research or work requiring pathology.

16.2 Background

It is widely acknowledged that pathological information is an essential component of studies involving aging animals [5–7]. Not only must one know the survival data on the colony of animals used in aging studies, but it is also critical that the pathological lesions associated with senescence be characterized. Much of the early research in gerontology suffered greatly from the lack of well-documented pathological data. It is now clear that survival data without pathological data can provide only limited information [6–8]. This deficit becomes more serious when interventions of any sort are implemented to study aging as it is impossible to conclude from survival data alone if changes in life span arise because aging has been altered by the experimental manipulation.

16.2.1 Historical Perspectives

Animal models are used to study aging utilizing pathology. They are generally used when human issues require research evaluation but the research cannot be done in humans because of legal, ethical, or technical restraints. However, it is not obvious how age-related disease can be related to problems in old people. Therefore, animal models are used to study primary aging processes that occur in all mammals and possibly all animals [9]. While a variety of species have been used in aging research, this methods text will deal primarily with mammals. General criteria that must be considered when choosing an animal model should include life table data, short life spans, defined environmental conditions, defined health status, knowledge of pathologic lesions, genetic characteristics, and availability [9]. Excellent reference material is available on animal models in aging research including useful data on the pathology of various species [10–18]. The researcher will soon discover that all animal models have certain idiosyncrasies that have to be considered when choosing a model, e.g., nephropathy in Fischer 344 rats [7,18]. Complicating factors can be used to the investigator's advantage when evaluating experimental interventions, e.g., the use of soy protein and diet restriction alleviates the nephropathy of Fischer 344 rats [6]. This example illustrates and reinforces the need to define the test animal to the greatest extent possible; and, pathologic data on normal and test animals are especially valuable.

16.2.1.1 Rats

Laboratory rats have been used extensively in aging research. They are readily available, well-characterized, defined pathologically, large, easily maintained in the laboratory environment and short-lived [9,10,13,14,17,19–23]. Because rats are relatively large, they can easily be sampled for clinical, gross, and histologic evaluation. The pathologic changes most commonly associated with aging in rats, which may vary by genotype, are nephrosis, periarteritis nodosa, myocardial and skeletal muscle degeneration, radiculoneuropathy, and cancer [19]. Most pathology and probably most aging research in the rat has been done in the Fischer 344 strain. In fact, since 1979, more than half of all the literature about the pathological study of the aging process in the Fischer 344 rat was published by the scientists of the aging research group in San Antonio, Texas. Three outbred rat stocks, the Sprague-Dawley, Wistar, and the Long-Evans, have been used extensively in aging research. New genotypes are becoming available for research use and may solve some of the concerns about inherent research-complicating characteristics in the other established genotypes [9].

16.2.1.2 *Mice*

The laboratory mouse is probably the most commonly used animal in aging research and the animal model best characterized pathologically. Mice are well adapted to laboratory use, economical to use, well-characterized, readily available, and have short life spans. Because the mouse is small, it is generally difficult to sample for clinical and gross pathologic evaluation. It is very easy to evaluate histologically. The mouse is used extensively by many research disciplines in addition to aging research; therefore, massive information is available on the pathology of the mouse [9,10,12,14,16]. More than 500 inbred mouse strains are available for research use. The National Institute of Aging supports the use of 10 mouse genotypes: five inbred strains, three hybrid strains, one congenic strain, and an outbred stock. Abundant information on age-associated pathology is available on mice [9]. Any baseline or age-related pathology data must be evaluated on the basis of the specific mouse genotype so generalizations about the pathology of mice cannot be made [19,24-28].

16.2.1.3. *Other Animals*

Other animals used in aging research are numerous and range from protozoa and nematodes to carnivores. An excellent review has been written for the interested reader [9]. There are advantages and disadvantages to most models but the other animal models besides rats and mice that have been used most include: Peromyscus, hamsters, gerbils, cats, dogs, and nonhuman primates [9,15,28,29]. The hamster is best defined pathologically followed by the dog, cat, and nonhuman primate. None of these species is well-defined pathologically for aging, except possibly the hamster. Generally the larger animals are not readily available to the researcher, are long-lived, and more expensive to maintain in the laboratory [9,12]. However, their pathologic assessment would be essentially the same as for any other mammalian species used in aging research.

16.2.1.4 *Nonhuman Primates*

Nonhuman primates have gained some attention as animals with promise for studying aging because they are very similar to humans. They are moderately well-defined pathologically but only minimally for aging research. Any data derived from nonhuman primates that is to be related to aging pathology must be used carefully [9].

16.2.2 Importance to Aging Research

16.2.2.1 *Advantages*

Pathology is important to the study of aging to establish normal baseline pathology of any animal used experimentally. Most lesions become important biological markers of aging. This is especially important in longevity studies. If baseline information is not available, the investigator cannot determine if differences observed in research animals are caused, for example, by infectious disease, sex, species, environment, genetics, or nutrition or if lesions are actually due to aging [6,10,19,29,30]. Pathology has the added advantage that one can see, grade, and characterize lesions as to severity, prevalence, location, organ system, and probable mechanisms to name a few. Experimental interventions can also be evaluated pathologically when they occur. While new techniques have been developed to evaluate and confirm that experimental animals are actually what they are represented or thought to be, pathologic characterization helps to promote standardization of the animal genotypes. With good pathological data, it is possible to predict pathology for living animals and use this information for experimental evaluation and planning [12]. Good baseline information can actually reduce the numbers of animals needed to perform certain studies. This generally reduces the cost of doing research. An added advantage of pathologic evaluation is that tissues can be saved and stored indefinitely by chemical preservation, freezing, paraffin embedding, and on glass slides. Any interested scientist can have access to these materials at anytime, even after many years. This allows for good peer review, retrieval, and reevaluation of information. Special techniques such as immunohistochemistry, electron microscopy, chemical stains, *in situ* hybridization, and polymerase chain reaction to name a few can also be done on these tissues. Generally the tissues can be cut very thinly and used extensively. Pathological baseline and experimental data can be compared between species, including man, which is the main thrust of aging research. All in all, pathology is essentially indispensable to aging research and at the very least a valuable adjunct.

16.2.2.2 Disadvantages

The primary disadvantage of pathologic evaluations is the cost. It is expensive to do pathology because of the high quality and expense of personnel, materials, and equipment that are required to produce a quality product. The evaluation can be a major drawback if the adequate evaluation of pathology specimens cannot be accomplished because of monetary restraints. A second possible disadvantage is the difficulty in determining the status of lesions found pathologically. There is considerable concern and controversy as to exactly how an investigator determines if a pathologic lesion or disease is age related or age dependent. A helpful working definition of disorders of aging is the distinction between age-dependent and age-related disease. Age-dependent diseases are those where the pathogenesis is related to the aging process. Good examples are coronary heart disease and Alzheimer's disease, which increase exponentially with increasing age. Age-related diseases are temporal and not necessarily related to the aging process. These diseases occur at specific ages and then decline in frequency or do not increase exponentially. An example of this would be multiple sclerosis, which in humans, occurs mainly in the third and fourth decades of life [32]. Sometimes the disease status distinction is quite obvious but at other times is extremely difficult and controversial. Of course, this is the crux of aging research; why controls are used must ultimately be decided by individual investigators and the scientific process. Abundant literature is available for the interested individual to study [9-12,19,31,32].

16.3   General Description and Principles and Techniques

The basic principles and implementation of pathologic evaluation of animals are essentially the same for any animal species or test procedure. It is extremely important that any research involving pathology be designed and managed to produce reliable and useful information. The essence of aging studies is to observe animals during their life spans, document the pathologic lesions that occur, and relate the impact of the these lesions on the aging process. This knowledge may provide insight into the mechanisms of aging and determine if lesions are associated with, or independent of, underlying conditions. The requirements for long-term research and testing are well known, expensive, and require adequate materials, equipment, physical facilities, and qualified personnel. These requirements have remained essentially unchanged for decades. However, it is still important that good basic guidelines be established and adhered to throughout any long-term study. These standards will be reaffirmed here.

16.3.1   Materials and Methods

Personnel are the most important part of any endeavor, and the study of aging pathology is no exception. It is imperative that one person is in charge of the study. The pathologist should work directly for this individual. The pathologist should be well-trained and experienced in general pathology as well as in the basic pathology of the laboratory animal used in the study. The ideal is a board certified veterinary pathologist with experience in long-term animal studies and the study of animal species. Animal support staff should also be competent and, ideally, certified in laboratory animal care by a recognized accreditation agency such as the American Association for Laboratory Animal Science and The American College of Laboratory Animal Medicine [33]. Certified clinical pathology personnel should perform the clinical pathology support [34]. Certified histotechnologists and technicians should be available to handle and process tissues for histologic evaluation [35]. The specific duties of all these individuals should be clear and be monitored by the project leader. Of primary importance is the monitoring of the health of the study animals and the animal surveillance to preclude the loss of tissue for histologic evaluation [15]. Good commercial pathology companies are available to support pathology studies.

Statisticians should be involved initially and throughout the aging study to ensure the study is designed properly so that useful data will be collected for analysis. It is especially important that enough animals are used, and that they be individually marked or identified, randomized, and properly assigned to study or control groups [15].

The laboratory animal selected for the study should be well-defined pathologically and, ideally, should have been proven useful in prior aging studies. If adequate background data are not available for a study animal, it must be established either prior to the study or by the use of adequate control groups. It is best to obtain animals from reputable animal vendors who will provide both background and current health and genetic information on the animals. After receipt and quarantine, the health status of the experimental animals must be monitored throughout the study to ensure study results are not influenced by natural disease. Euthanasia may be accomplished by carbon dioxide asphyxiation, anesthesia with pentobarbital (1-mg/kg IP), followed by cervical dislocation, pentobarbital (100 mg/kg IP) or just cervical dislocation, depending on the research. Ideally, these test animals will be maintained in a facility that is accredited by the American Association for Accreditation of Laboratory Animal Care (AAALAC) or at least be managed in compliance with the Guide for the Care and Use of Laboratory Animals and other laws governing the care and use of laboratory animals [36].

Abundant information is available concerning the methods for obtaining optimal results when performing pathologic evaluations of study animals [15]. It is extremely important that all requirements of the study are followed consistently throughout the study. This means that check lists and standard operating procedures are established, followed, and monitored. It must be possible to identify the animals submitted to pathology and that all pertinent information, tissues, and lesions be noted or collected and documented. Personnel must be available during weekends, holidays, and at night if necessary to perform necropsies, harvest tissues, or formalin fix or freeze the animal for future necropsy evaluation.

Information recorded on the pathology submission form for each animal or specimen submitted for pathologic evaluation can be tailored to individual needs but should include at least the following: identification number, accession number, genus and species, common name, sex, body weight, birth date, date of death, time of death, date of necropsy, project identification, funding source if known, name and phone number of person submitting specimen, clinical history to include any pertinent information, special instructions, tissue harvesting instructions, and name of person to receive the pathology report [15].

Tissues to be harvested should be stipulated in a research protocol and be consistently harvested if at all possible. The following is a general list of tissues that can be harvested or selected: adrenals, accessory sex organs, aorta, bone, bone marrow, brain, esophagus, eyes, gall bladder, gonads, gastrointestinal tract, heart, injection sites, kidneys, liver, lung, lymph nodes, mammary gland, middle ears, nasal turbinates, nerve, pancreas, parathyroid, pituitary, salivary glands, skeletal muscle, skin, spinal cord, spleen, tattoos or identification tags, thyroids, tongue, tonsils, trachea, urinary bladder, uterus/oviducts, and veins. Specific anatomical sites, regions and paired organs can be stipulated if more thorough sampling is required. Other tissues and fluids can be harvested to include blood, cerebrospinal fluid, joint fluid, urine, lymph, and semen to mention the major ones. A specific tissue list can be developed and special handling described. Examples of these would include tissue for electron microscopy, both transmission and scanning; tissue for microbiologic sampling for agents such as bacteria, viruses, fungi, protozoa, and arthropods; frozen sections for immunologic procedures; decalcification; radiology, or perfusion/infusion.

Specific written diagnostic and grading criteria should be developed and used when evaluating histologic samples. Good examples and guidance are available in the literature [37]. Basically, a grading system appropriate for statistical evaluation of lesions should be used. This is usually 0 for normal or no significant lesion, 1 for minimal, 2 for moderate, and 3 for severe or marked. This is just a guide and essentially any numerical system can be instituted. It is best if one person does the grading and, at all costs, it should be done consistently. If it is a large study and several people are doing the evaluation, it is imperative that they communicate and make sure they are using the same grading system. It is a good idea to statistically evaluate the different graders to see if they are indeed grading the same. If obvious problems occur, the grading system should be altered, and all lesions that have been graded should be regraded for final evaluation. Some tissues or lesions may not be worth evaluating and can be dropped for grading. Again, the main point is to establish a method of numerical evaluation of lesions that is consistent, supportive of clinical evaluations, reproducible, and useable for statistical evaluation [15].

Preservation of tissue for microscopic evaluation can be accomplished by chemical fixation or freezing. Formalin is the best fixative for general usage. Fixation volumes to tissue should be approximately 15 to 20 parts fixative to one part tissue. Formalin is generally prepared as a 10%, neutral buffered solution. Other chemical fixatives can be used for special purposes but generally are more difficult to handle, expensive, and labor intensive. Among the common fixatives are ethyl alcohol and Zenker's and Bouin's solutions. Glutaraldehyde is generally used to fix tissues for electron microscopic evaluation although other fixatives can also be used if glutaraldehyde is not available. These fixatives can be prepared in the laboratory or purchased from commercial sources. Freezing of tissues can be accomplished with liquid nitrogen, dry ice, or by simply placing tissues in either a –20 or –70°C freezer. Freezing of tissues and fluids is generally not recommended for those tissues to be evaluated histologically but is used for those samples to be analyzed using immunohistochemistry, *in situ* hybridization, and chemical evaluation [38–41].

Basic necropsy instruments, equipment, and supplies can be tailored to each project's needs and should include the following general categories: clothing to protect the prosectors and other persons in the necropsy room; equipment to facilitate data collection; instruments and equipment to do the physical necropsy; materials to preserve and store the specimens taken; materials to clean and disinfect the necropsy facility to include instruments and supplies; materials to handle emergencies and safety hazards; and materials and methods to dispose of necropsy materials. All these items are available through vendors and generally are off-the-shelf [42–44].

Protective clothing needs will vary depending on the specimen, but generally in aging studies, one works with clean, biologically nonhazardous animals. An obvious exception would be the nonhuman primates. The basic list of protective garb should include gloves, gowns or aprons, masks, scrubs, head covers, shoe covers, and eye protection. Probably most rodent necropsies can be performed with only an apron and gloves.

Data retrieval supplies and equipment include cameras, scales and balances, a computer, forms, film, photo stand, and moisture-resistant writing instruments for marking items.

Necropsy equipment should include a stainless steel down draft necropsy table that is large enough to accommodate any reasonably sized animal, good light fixtures, plenty of counter space, good cabinets for storage, carts, access to an autoclave, directed flow hoods, and perfusion/infusion equipment.

Instruments can be quite specialized, but for the purposes of this chapter, basic stainless steel instruments for routine necropsies include scalpel handles and blades, scissors, forceps, hemostats, bone saws, rulers, probes, needle forceps, needles with suture, instrument trays, containers for chemical disinfection of instruments, dissection boards, and possibly protective stainless steel gloves. Of course, there are many sizes and shapes to these instruments, and most of the choices can be personnel preferences and are not generally critical to the performance of the necropsies [45].

Preservation and storage of specimens, as discussed earlier, can be chemical or frozen. Adequate preservative and good durable containers are necessary to handle tissues. These containers should be big enough to contain 1 part tissue to 20 parts chemical fixative and durable enough to last indefinitely. Various types of freeze preservation containers are available, but all must be freeze durable, and ideally, air tight. These materials include metal foil and freeze-durable test tubes, which are the most desirable.

Disinfection of the necropsy facility should also be a routine procedure using good off-the-shelf, all purpose agents that kill both bacteria and viruses. Copious hot water, soap, and manual labor are necessary to remove most materials prior to application of disinfectants. A good inexpensive general disinfectant is Clorox® and water [45].

Disposable supplies and instruments are useful to control disease and contamination of facilities. Disposal of sharps and hazardous or controlled substances requires specially marked containers and established methods of disposal. These items are generally managed by a facility safety officer. At the minimum, safety requires sharps containers for disposal of needles, syringes and blades; chemical containers for bulk disposal; and specially marked bags and containers for hazardous tissues. It is desirable to have first aid kits, eye flush equipment, and forms for documentation of exposure available in the laboratory. Animal tissues and carcasses, if not handled by contract to accredited facilities that dispose of biohazardous materials, can be incinerated. Although this can be expensive, it prevents bodies, body parts, and medical items from being found or seen by the public [46–47].

16.3.2   Laboratory Requirements

The necropsy facility should be physically separated from the animal facility to prevent spread of disease to the live animals. This separation should not be so great as to hamper use of the facility. Ideally a change facility with shower and lockers should be designated for the necropsy personnel. The facility should be large enough to accommodate at least two people and room for eight, approximately 15 × 15 feet, would be desirable. The walls, floors, and ceiling should be water resistant so the facility can be cleaned with water. Good drains, safe electrical fixtures, and good lighting are very important. The facility should have gas available to run Bunsen burners. Equipment should be built with a good grade of rust-resistant, stainless steel [15,45].

Clinical pathology is often required for the performance of aging studies and health monitoring of the laboratory animals and should be of the highest quality. This means the laboratory should be accredited and staffed by certified medical technologists and technicians. Personnel should be supervised by a board certified pathologist. Good references and laboratories are available [33,34,48–50].

Anatomic pathology should also be under the supervision of a board certified pathologist and staffed by certified histotechnologists and technicians. The laboratory should be able to perform all routine procedures generally required for histopathology. These procedures would include production routine paraffin embedded tissue sections, special stains, frozen sections, decalcification, immunocytochemistry, and electron microscopy. Good references and laboratories are available [33,35,38–41].

A microbiology laboratory is essential to the proper evaluation of disease outbreaks and health monitoring of the laboratory animals. It is also necessary for proper etiologic diagnosis of microbiologically induced inflammatory lesions. These laboratories should be staffed by competent personnel, ideally, supervised by a Ph.D. microbiologist. Generally, commercial microbiological support can be obtained, therefore an in-house laboratory is not necessary [50,51].

Photographs or video of procedures or lesions may be required. Good cameras and lighting systems are very useful and help to document any pathologic lesions found. These items can be kept in the necropsy facility or maintained by the parent facility.

An electron-microscopy facility with both transmission and scanning capabilities is a valuable adjunct to any anatomic pathology work. This facility should be staffed by qualified personnel. Ideally a Ph.D. or equivalent electron microscopist is desirable. There are generally good electron microscopy laboratories available [41,43].

16.3.3 Support Requirements

Ancillary support requirements are centered around the administration of the parent institution. The top administrative personnel must support the aging research effort both financially and actively. The handling of all support personnel including security, environment, laboratory animal resources, communication, procurement, engineering, clerical, and custodial is essential to the long-term success of any aging study [35,46,47,52].

16.3.4 Experimental Design

Aging research utilizing pathology is long-term, time-consuming, expensive, and requires adequate facilities, healthy defined animals, and qualified and experienced personnel. A detailed written plan should be developed by all personnel involved in the study, including laboratory animal specialists, statisticians, physiologists, molecular biologists, chemists, and geneticists. The plan, once established, should not be altered unless absolutely necessary and changes must be agreed to by all involved parties. Any changes should be explained and documented. The aim of any pathology study is to find and document the existing pathology. This means that all animals and tissues must be harvested and evaluated. This is not always possible because of cannibalism, post mortem autolysis, or other unforeseen occurrences. It is imperative that all necropsies and tissue harvests be done in a thorough and consistent manner. All loss of tissue must be noted and justified on the pathology forms and steps taken to prevent any loss of pathology material. Prime elements of any experimental design are assigning specific personnel responsibilities and determining which specific animal to use, the study dates, number of animals needed and randomization method, information to collect, data handling and analysis, and specific aims [15,53].

16.3.4.1 *Longevity Study*

Longevity and longitudinal study methods are essentially identical and may differ only by a shorter designated time period in the case of longitudinal studies. The animals are followed throughout their lives in these studies to see what pathology develops. A pathological profile that includes but is not limited to prevalence and severity of neoplastic and nonneoplastic disease, effect on longevity, probable cause of death, and effect of experimental manipulations is developed by pathological analysis of all animals that die spontaneously. The end-point in longevity studies is natural death. There is some controversy that animals may be euthanized when moribund and still be used in longevity studies with good results [5]. Until this issue is resolved scientifically, the only rational approach is to use death as the end-point. These studies have the disadvantage that it is not always possible to tell when lesions develop or to follow their progression. The spontaneous death data may answer some questions, but generally these questions are answered in cross-sectional studies [12,53,54].

16.3.4.2 *Cross-Sectional Study*

These studies are conducted on all animals to be evaluated at one place in time. They have the advantage of being planned, relatively less time consuming, and because the animals are euthanized there is no loss of data due to cannibalism and post mortem autolysis. These studies are used to evaluate associations between disease and risk or experimental manipulations [54]. All criteria for necropsy, evaluation of lesions and data collection are the same as for the longevity/longitudinal studies [4,8,9,10,44].

16.4 Procedures

This is the section where the actual application and data acquisition occurs by doing pathology. The object or substrate of this work is a dead animal or specimen removed from that animal. The work effort is the necropsy, and all procedures follow the death of the animal and the physical evaluation or necropsy. It is important that all the procedures be done properly, diligently, consistently, and, generally as soon as possible after death or sampling. It is also important that the individuals that do the work are well-trained, experienced, and conscientious. These procedures can discover, diagnose, document, enlighten, teach, and preserve information. In many experimental studies the pathologic evaluation is the only tangible data that can be gathered by the research effort.

16.4.1 Necropsy

The examination of a dead body can be called a necropsy, autopsy, or postmortem examination. Necropsy is the term generally reserved for the examination of animals other than humans. Autopsy is generally reserved for the examination of humans. Postmortem examination or "post" is used comfortably for all animals including humans. The necropsy is a scientific investigation to determine what pathology is present and its relationship to the clinical history, background, or experimental status of the animal. The necropsy should be regarded as a pleasurable experience as it represents a challenge to determine the cause of death in spontaneous cases, to find and identify lesions, and perhaps to identify a pathogenic entity in infectious diseases. It is possible to use good necropsy techniques to determine the causes of the pathologic changes and learn about the processes of disease in animals [55]. There are ample publications concerning design, procedures, and recommendations for proper construction and use of necropsy facilities [15]. The necropsy facility and equipment need not be expensive but should be large enough to work in comfortably and well enough equipped not to compromise safety or the necropsy procedure. The facility should be well lighted, properly ventilated, and easy to clean and disinfect. It should be properly supplied with protective clothing, adequate instruments, sampling and storage supplies, and data recording materials [42,43]. Necropsies should be conducted as soon after death as possible. If it is not possible to necropsy an animal immediately after death, the carcass should be refrigerated. Carcasses can be kept for periods up to 72 hours if not too large and if the refrigerator capacity is adequate. Contrary to general opinion, freezing the carcasses works quiet well with limited tissue damage. These tissues can usually, except in special circumstances, be adequately evaluated by an experienced pathologist. Freezing is a last resort to be used when other methods are not available or the necropsy cannot be done within a reasonable time. For small animals such as rodents that cannot be necropsied immediately, the body cavities can be opened and the skin over the skull excised and the whole animal immersed in formalin. Generally, carcasses are considered fresh for 1 hour at room temperature except for special studies. In most cases, animals may remain at environmental temperatures for approximately 12 hours and still be acceptable for evaluation. Post mortem autolysis can be recognized and evaluated for duration or degree of odor, discoloration, consistency of the tissue, and gas formation. It should be noted that different organs decompose at different rates. For example, the eye, pancreas, stomach, and intestines autolyse faster than connective tissues. Of course, extremely high temperatures accelerate decomposition; small animals such as mice dehydrate rapidly; and large or fat animals do not cool well. If carcasses are insulated in bags, wraps, or other materials or devices they will not cool properly. Conversely, small animals may dehydrate excessively; and if this is a concern, they should be placed in an air tight container. The refrigeration unit should be large enough to easily accommodate the carcass. The carcass should be placed in the unit so there is adequate ventilation around the body. The prosector should not despair if there seems to be advanced post mortem autolysis or decomposition of the tissues. It is often possible to get very good and useful information from tissues that are not fresh.

Necropsies conducted for aging studies for all practical purposes should meet standards equivalent to those expected of toxicologic necropsies. The animals are extremely valuable by the time they are submitted for necropsy in lifetime studies, and any lesion or observation that is missed or tissue that is lost is serious and could severely compromise the final evaluation of the study. This means the studies should be conducted, as much as possible, with Good Laboratory Practices (GLP) standards [15,52].

16.4.1.1 *Anatomy*

Anatomy is one of the basic foundations of biomedical science and familiarity with it is imperative for anyone performing necropsies. Anatomic and necropsy terminology are based primarily on the human body, but for quadrupeds, the Nomina Anatomica Veterinaria is also used [56–58]. Anatomy is really a guide map to the structure of a body and provides a basis of communication between people in and within different disciplines. The intent of this chapter is not to teach pathology, anatomy, or histology but simply to rein-force the idea that anyone performing necropsies must know the anatomy of the animal to do an effective necropsy and tissue sampling. It should be noted that there is considerable anatomic variation among animal species, including man. There are good anatomy texts for most commonly used laboratory animals, but it is possible that no information is avail-able on lesser used species. In these cases people must teach themselves by using anatomy texts on similar species or by seeking assistance from people who have experience with the species [42,43,56,57,59,60].

16.4.1.2   Gross Pathology

Gross pathology is a very important part of the research process. If not done properly or if samples are not taken, very valuable information can be lost that will severely impact the final product. The prosector must know what is normal to be able to identify what is abnormal. This is not always easy as there are many variables that can occur if the speci-men is not fresh and the prosector is inexperienced or poorly trained. Agents that cause variation in tissues at necropsy include temperature, handling, trauma, age or sex of spec-imen, time since death, method of euthanasia, husbandry, refrigeration, and diet to name a few. The prosector must be able to describe structures or lesions and locations of the lesions or structures so another person using the same terminology and standards can rec-ognize the lesion or structure and be able to locate it. This chapter cannot cover all gross necropsy techniques for all species and it is not necessary as good texts and publications are available [56–60]. All necropsies are basically the same and can be done properly many different ways if standard accepted procedures are followed. First, all the necropsies should be done in the same manner to ensure that all necessary tissues are harvested and evaluated. This is best done by having and following checklists. To help control sampling variation, it is best if the same people do the necropsies, as will performing them in the same location each time with the same equipment, instruments, and supplies. The more time spent, dissections made, and samples taken, the more thorough the necropsy. How-ever, the amount of time and techniques used should be essentially the same for each ani-mal. The prosector may wish to inject fixative into the lungs, urinary bladder or other organs to improve fixation. Tissue samples do not generally have to be large. If they are approximately 0.25 to 0.5 cm in thickness and 1 cm in diameter, they will fix in approxi-mately 12 hours and be excellent for evaluation. Of course, individual requirements of an experimental protocol will dictate the size and number of samples and how they should be taken. If it is possible to lose a sample during processing because of its small size or con-sistency, then these samples should be placed in tea bags or placed between sponges so they cannot be washed away during handling and processing. In the case of small animals, whole organs can be fixed; but if the organ has a capsule, it should be cut to allow the fix-ative to permeate the tissue.

It is possible and very convenient to place samples for conventional processing and his-tologic evaluation directly into cassettes for processing. This saves time and ensures that all samples are taken. Backup reserve tissues should be taken when the primary sample is taken.

The following specific steps are only guidelines for the gross pathologic examination and can be modified to fit individual protocols and personal preferences:

Step 1   Confirm the animal's identity and the accuracy and completeness of the submitted paperwork. When the animal is placed on the necropsy table and the paperwork is pro-vided, the necropsy can begin.

A. Confirm that all information on the submittal request is correct.

B. Confirm the identification of the animal according to sex, body weight, and other identifying features. It is often useful to place an animal's identifying tattoo, tag, chip, clipped ear, or foot in the reserve formalin container.

C. If the animal is to be perfused and euthanized, cannulate and perfuse at this time. When sterile techniques are required, disinfect incision sites first; then use sterilized instruments and supplies from this point forward for the harvesting of sterile tissue.

D. If tissues are to be shared with other investigators, have a list of which researcher gets which tissue. This allows a controlled sampling to avoid one individual from taking all tissues, thereby ensuring that some tissue will be available for histologic evaluation. Tissue sharing can be very disruptive to a necropsy procedure, as it prevents the normal routine which can lead to missed lesions and omissions.

Step 2 Do an external examination. This is essentially the same as a clinical examination of a live animal — the findings should be essentially the same as those noted in the clinical history.

A. Evaluate the condition, degree of hydration, posture, exudates from body organs, swellings, color changes, oral cavity appearance, joint mobility and range, tissue consistency, etc., for deviations at this time.

B. This is a good time to collect ectoparasites, urine, feces, and cerebrospinal fluid and to harvest the eyes, injection sites, tattoos, skin, or other external lesions or tissues not requiring a major invasive skin incision.

C. If it becomes apparent that special techniques such as aspiration of fluid, culture, scrapings, or impression smears must be made, this is the time to prepare to take the samples or to take the samples.

D. Record all lesions on tape, paper, or film or on all three.

E. Starting now, preserve all tissues; the saving process should continue throughout the post mortem procedure.

Step 3 Do an internal examination. There are many methods and approaches to the internal examination of animals; it is up the prosector to chose the best approach for each study or species. Again, it is important to use the same technique or approach for all necropsies within that study. A method recommended by the authors that can be used for all species is as follows:

A. Simply lay the animal on its back and make a midline incision through the skin from the mandibular symphysis or chin to the perineum or pubic symphysis. Extend the incision down each hind limb to the femorotibial/femoropatellar or knee joint.

B. At this point, cut the skin from the head and neck and reflect the pectoral muscles and skin down to the axillary spaces. If necessary to examine the whole subcutaneous area and subjacent tissues, you may need to completely skin the body.

C. At this time, you may either open the abdominal cavity or leave it intact. Open the joints, examining them and the body for abnormalities. This is the best time to sample regional lymph nodes generally to include the inguinal, axillary, cervical, and submandibular nodes. Dissect the tongue from the head, then pull it and the attached structures, including the esophagus and trachea, down to the thoracic inlet.

D. Now, or later if it is more beneficial to wait, take the bone marrow from the long bones by fracturing them. You may also sample and preserve skeletal muscle, testicles, mammary gland, salivary glands, and other visible tissues or lesions at this time.

Step 4   Do an internal examination of the body organs, structures, and cavities. Remember, if sterile harvesting is required, disinfect the sites to be incised and use sterile instruments and supplies from this point on.

A. Sever the cervical spinal column and remove the head, setting aside for examination later.

B. Open the abdomen, if not already opened, by cutting on the midline and reflecting the abdominal muscles laterally.

C. Open the thoracic cavity by cutting the ribs bilaterally on either side of the sternum approximately at the costochondral junctions. In young or small animals this can often be done with a scalpel blade or scissors. Whatever instrument is used, care must be taken not to cut the underlying structures including the lung.

D. Cut and remove the ventral pelvis to view the pelvic cavity. Harvest the sternum and ribs. Now, make a prior-to-handling examination of the thoracic, abdominal and pelvic organs and structures to check for any abnormalities such as swellings, color change, fluid accumulation, gas formation, tumors, hemorrhages, etc. This also is the time to take samples for culture of exudates and organs that may be colonized by microorganisms.

E. The prosector has a choice to make at this point in the necropsy. Either grasp the trachea and esophagus and pull the heart and lungs from the thoracic cavity and sever these tissues from the body at the diaphragm or continue to pull all tissues and organs from the body including the anus. Dissect out and examine all the viscera.

F. Another technique is to remove each organ or organ system separately. The most used method is to first remove and sample the easy-to-see and -to-get tissues such as the thymus and spleen.

G. After these are taken out, remove the liver and intestines, followed by the urogenital tract.

H. At all stages, thoroughly examine all tissues and structures to detect possible variations from normal.

I. Extract the brain by removing the skull cap and harvest the pituitary and spinal nerves. Evaluate each, individual organ and tissue by opening lumens, examining contents, surfaces, etc.

J Now examine the intestinal tract. To prevent contamination of the rest of the tissues, this is generally the last of the organs opened.

    K. Harvest the vertebral column and spinal cord at this time. Special samples, such as frozen samples for tissue culture media, fresh tissue, electron microscopy, etc., can also be harvested at this time.

Step 5   Review the necropsy to check for any lesions that may have been missed and the checklists to make sure all samples were taken and all procedures accomplished.

Step 6   Write up the necropsy, check all entries to make sure they were done accurately, and sign the report. Make sure that the designated laboratories and researchers receive the proper specimens.

Step 7   Clean and disinfect the necropsy facility, including all equipment and instruments. Restock supplies and prepare for the next necropsy.

Step 8   If the tissues are not cut (grossed) directly into cassettes at necropsy, the prosector must cut all formalin-fixed tissues into cassettes. This is done as soon as tissues are fixed, which generally requires a minimum of 12 hours (see above). Ideally, all cassettes should be labeled prior to the placement of tissues into them. A list should be available that defines which tissues go into each cassette. Of course, each cassette should have the correct accession number on it.

16.4.2   Histology

Histology is really the study of microanatomy or the structure of cells, tissues, or organs that cannot be seen with the naked eye. Because histology is not only the study of individual cells, but also of organs and tissues, it is valuable to study not only structure but function and correlations between the two. Gross pathology is complemented by and dependent on microscopic anatomy or histology. Generally, any lesion seen grossly that the prosector feels needs more evaluation should be examined microscopically to confirm or further correlate the changes observed with cause of death, function, or pathogenesis. Histology is also concerned with the handling and preparation of tissues for microscopic evaluation and the type of microscopic unit technique that will be used to examine the specimens. In a relatively short time, major advances have been made in the handling, processing, techniques, stains, and histology equipment. Good quality histopathology requires good quality histotechnology. Although beyond the scope of this chapter, this information is well presented in published material. The pathologic evaluation of tissue for aging studies or any other pathology study can readily use any of this technology when it makes sense to use it. Therefore, this section covers only the basic methods of histology, highlighting some things of which the nonpathologist researcher should be aware. If the parent facility has a good histology laboratory, there should be limited problems with tissue preparation for microscopic evaluation [61–64].

16.4.2.1 *Tissue Handling*

The proper, conscientious handling of tissues from harvesting at necropsy to final microscopic evaluation is critical to producing a good product. Tissue handling for aging studies is not different from that for any other pathology study or evaluation. The physical handling is especially important, and several things that must be kept in mind are as follows: Use the freshest tissue as possible prior to preservation; do not introduce artifact by crushing, freezing, staining, or using improper fixative or fixative volumes; assure the identity of the specimen; have backup tissue available; cut specimens for proper placement into processing cassettes — a thickness less than the height of the cassette and of a size that will allow proper processing and cutting, which is about half of the cassette surface area; if the lesions can be seen grossly, include the lesion to be examined with the specimen and some adjacent normal tissue; use correct and standardized processing; and ensure that processing chemicals, temperatures, times, pH, concentrations, and pressures are correct and consistent [38–41].

16.4.2.2   *Paraffin Sections*

With few exceptions, the bulk of any tissue for histopathologic evaluation in aging studies or any other pathology study will be conventionally processed and embedded in paraffin. It is the most commonly used embedding medium because it is inexpensive, durable, available, and easy to use. Paraffin embedded tissues can be used for many procedures and techniques including special stains and immunocytochemistry. The tissue blocks can be stored essentially indefinitely in climate controlled facilities in a state that takes little space, are easily accessible, and are neither hazardous to personnel nor the environment. With newly improved technologies, these paraffin embedded tissues are useful for a variety of special procedures for which they could not be used in the not so distant past. The researcher or pathologist should ensure the embedding is done with attention given to placing the cut surfaces containing lesions (or areas of interest) facing the microtome blade, and properly oriented and convenient for cutting with the microtome. The correct placement of tissues can often be facilitated by marking the surface of interest with a cut or notch on the opposite side of the cutting surface or marking it with India ink. The paraffin blocks can be cut with the microtome at any thickness that is desirable for further use. This is generally at 5 microns for routine tissue staining with hematoxylin and eosin and routine evaluation. Again, a good histology laboratory will produce good slides for histopathologic evaluation [38–41].

16.4.3   Special Techniques

There are no special pathologic techniques available, to the best of our knowledge, that are specifically used only for aging research. This is a significant fact because it means that all special techniques available for pathologic evaluation of specimens are available for aging research. These techniques are generally used to aid in evaluating lesions to determine if there is a difference between aging groups and other groups or controls. The special techniques are expensive, time consuming, difficult to use and interpret, and require experienced personnel to produce and evaluate. The use of these special techniques are often no more valuable than the evaluation of routine tissue sections by an experienced pathologist or research scientist. The person doing the aging research should keep this in mind prior to investing limited resources into special histologic techniques.

16.4.3.1   *Frozen Sections*

This technique requires that tissue be frozen when fresh, which generally means within 15 minutes of harvesting. The sooner the tissue is frozen after sampling the better. Although the technique is valuable for speed of diagnosis on surgical specimens, it is not a good technique for the evaluation of tissue structure because of the freeze damage to the tissue. Its value to research is that it has the advantage of not harming tissues chemically as most fixatives do. The technique is useful for identifying fats and lipids that are removed by conventional tissue processing. Today the frozen section is invaluable for the immunocytochemical evaluation of tissues and the identification of specific tissue antigens. The variable expression of these antigens could be useful to separate experimental changes from control data. The production of good frozen sections requires special techniques, equipment, and a good histology laboratory. The size of the tissue to be sampled should be small enough to cut easily and fit on a glass slide and fix easily and rapidly. Tissue pieces approximately 0.5 by 0.5 cm are generally satisfactory. The tissues must be frozen with liquid nitrogen or dry ice and kept at -70°C. Handling these cold products can be hazardous, so safety precautions should be taken [38–41].

16.4.3.2   Special Stains

The use of special stains probably is the easiest, most useful, and cost effective special technique available to the person doing aging research. It should be kept in mind that these stains can be expensive and difficult to interpret. Besides being useful for diagnostic work, the stains can be useful in identifying tissue, which is often useful when dealing with neoplasia. Stains can also be used to evaluate differences between tissue responses in various experimental groups. The great variety of special stains available is overwhelming, so the individual researcher will need to determine which stain (or stains) is the most useful [38–41].

16.4.3.3   Immunohistochemistry

The term immunohistochemistry is used interchangeably with the term immunocytochemistry. This special technique has become a major field, producing diagnostic advances and definitions of pathologic lesions, becoming especially useful in the diagnosis of neoplasia. Immunohistochemistry also has the potential to be used to indicate differences between experimental groups. Again, this is an expensive, time consuming, specialized technique, requiring good laboratory procedures, support, and trained, experienced personnel. If the parent organization has the capability for immunohistochemistry, it can be a valuable research tool. The individual investigator can become proficient in limited aspects of immunohistochemistry and can do the work in his or her own laboratory [38–41,65,66].

16.4.3.4   Other Special Techniques

There are numerous other special techniques that could be tailored to individual aging research, but to discuss them here would be beyond the scope of this chapter. There are adequate references available for the interested reader. These special techniques include transmission and scanning electron microscopy, enzyme histochemistry, polymerase chain reaction, and *in situ* hybridization [3841].

16.4.4   Microscopic Evaluation

The microscopic evaluation of the tissues harvested from the aging research animals is the most important and rewarding part of the experimental process, if all the supporting steps have been properly done prior to the evaluation. The data gathered from the tissue sections can be evaluated for lesions, graded for statistical evaluation, manipulated with special techniques, and kept for future use. It is imperative that this work be done by consciousness, capable, experienced, and dedicated people. It is not possible to teach pathology in this methods text as it requires approximately 15 years to learn. Therefore, this section will reaffirm the basic needs for the successful use of pathology in aging research. This will perhaps aid the nonpathologist, who may be a project leader, to understand pathology and what is required to produce good data through its use. The basic steps in any tissue evaluation are not firm and can be modified; they minimally include the following:

1. Visually scan each slide in the slide box to ensure they are all there and are properly labeled and placed for correct microscopic evaluation. It is not unusual for slides to be misplaced, mislabeled, and placed in containers incorrectly. This can lead to evaluation errors if one is not careful.

2. Inspect each slide individually prior to placing it on the microscope. This is a good time to evaluate the quality of the histology laboratory. All tissues should be in the proper cassette and the sections satisfactory for evaluation. If tissues are missing, it is possible that they were not embedded in the same plane as the other tissues and did not get cut by the microtome. It is also possible the tissue was not cut into the cassette and must therefore be found and cut in from the formalin tissue. It is also possible the tissue was not harvested or saved.

3. Place the slide on the microscope and evaluate with magnification to determine again, that the correct tissues are present and cut properly. This is a good time to look for major lesions that need to be evaluated at higher magnifications.

4. Record your observations on a standard form. It is imperative that standardized pathology terminology be used.

5. Review your work and sign the form.

16.4.4.1 *Morphologic Diagnosis*

The histopathology is recorded in the form of morphologic diagnoses. These morphologic diagnoses are ideally based on standardized, accepted terminology so that anyone can understand and interpret the information. There are, however, variations in the interpretation of lesions and between and within species. Therefore, a decision should be made at the beginning of the study as to which terminology to use and the diagnostic criteria to follow. It is best to write down standards for lesions that may be significant or controversial. It is not unusual to diagnose a lesion incorrectly either because of a difference in interpretation or because the lesion is new and unique and has not been properly defined or understood. In these cases, if the descriptions are good, the lesions can be reviewed and the correct and acceptable morphologies applied to the lesions. It also very desirable that the pathologists review the lesions together and decide on how they are going to handle the situations. At the end of studies, it is a good idea to have the material peer reviewed [15,73]. Abundant literature is available for review and for well-characterized species such as rats and mice, there are few if any lesions that have not been well defined [2,10,13,14,16,17,28,37,68–73].

16.4.4.2 *Collection of Data*

It is imperative that all data be collected completely, correctly, and consistently. The best way to do this is to use standardized forms by filling in the blanks. This can be done with a pencil or using a computer. Of course, the computer is the most desirable, especially if large volumes of data are collected. The computer also facilitates handling the data for retrieval and analysis. Software programs are readily available for the collection and handling of data generated by pathological analysis of animal tissues used in research.

16.4.4.3 Grading Criteria

The establishment of grading criteria for any organ or lesion can be very valuable for statistical analysis. The criteria should be developed realistically and put in writing. Any system that makes sense can be used, but each grade should be assigned a numerical value. Generally, a normal tissue with no lesions can be given a numerical grade of 0; minimal or mild lesions, a grade of 1; moderate lesions, a grade of 2; and severe or marked lesions, a grade of 3. Of course, more grades can be used if deemed necessary. This is a good issue to discuss with your statistician. A peer review of the grading is always a good idea. Good examples of grading are available in the literature [15,37,67].

16.4.4.4 Neoplastic Disease

A neoplasm is virtually the same as a tumor and the terms may be used interchangeably. A neoplasm may be benign or malignant. The term cancer is generally reserved for malignant tumors. Neoplastic or "new growth" is the presence of abnormal tissue that, while dependent on the host for nourishment, tends to be independent of normal growth constraints and has no function. Although impossible to give a complete review of neoplasia and cancer in this chapter, abundant reference material is available on the subject and much of this material is species-specific, which is ideal for aging studies.

It is extremely important that neoplastic disease be diagnosed when it occurs and not be diagnosed when it is not present. This would seem simple, and generally is, but neoplasia is often difficult to diagnose because of poor sampling, post mortem autolysis, or differentiation. Also, neoplasia does not often occur suddenly and has many transitional stages or preneoplastic stages. This comes into play when small cells go through stages of selection or development on the way to neoplasia. For example, a small nodular mass in the liver, if followed in time, might first be a focus of hyperplasia, then an adenoma, and finally a carcinoma. It is unlikely all these stages would be seen in one animal, unless all were clearly present at one time, but could be seen in several animals. This would be a situation in which cross-sectional harvesting would perhaps show the progression of a particular neoplasm. Also, considerable inflammation can be associated with neoplasia and mask the cancer. It is also possible for neoplasia to appear quite normal but still be a neoplasm, and for the neoplasm to be so poorly differentiated that the tissue cannot be identified with routine light microscopy. This is when special diagnostic procedures are especially important to determining the identity of a neoplasm. It may be necessary to send the tumor tissue to specialized laboratories or consultants for additional evaluation. In any pathological evaluation, the effort should be made to do the best possible job and get a diagnosis in every case. Although possible to have the inability to make a specific diagnosis, at least one can take solace in the fact that no one else can determine the identity of the neoplasm either. In many studies, several undifferentiated neoplasms may occur in different animals; and at some point, adequate differentiation may lead to a definitive diagnosis. The diagnosis, "undifferentiated neoplasm" is completely acceptable. In many aging and toxicological studies, the difference between benign and malignant is not nearly as important as the change itself and the diagnosis of nonneoplastic or neoplastic change. It is very useful to have written criteria and grades, if possible, for each neoplastic condition. Sometimes the grading will simply designate the neoplasm as benign or malignant. The grading may help to differentiate between experimental groups, but generally the difference between groups will be in the number of tumors. The study should have standardized, morphologic diagnostic terminology with peer review used to maintain the standard. This allows for retrospective evaluation if there are mistakes in interpretation or a specific diagnosis cannot be made. Again, it should be stressed that the determination of neoplasia is generally more important than the determination of whether the neoplasm is benign or malignant [2,13,14,27,28,69,70,72,74].

16.4.4.5 Nonneoplastic Disease

This category includes everything that is not neoplastic disease, so essentially the same criteria to determine if the lesion is neoplastic are used in reverse to place a lesion in the nonneoplastic category. This is a very broad and fruitful category to evaluate for age-dependent disease. The challenge here is to separate age-dependent disease from age-related/spontaneous disease. Spontaneous disease, for example, would include trauma and infectious and congenital disease. The best approach to this problem is to use an adequate number of control animals and be very conscientious about proper data collection. A thorough evaluation of data on the animal used in the study is also very important, so that this information can be compared with the study results. Grading of lesions is especially valuable in the evaluation of non-neoplastic lesions. Adequate references are available to researchers when dealing with this situation [10,11,13,14,16,17,27,67-70,72–76].

16.4.4.6 Cause of Death

The determination of the cause of death in animals used in aging studies and other pathologic studies can be very useful information when comparing experimental groups. Determination of the cause of death can also be used to evaluate and monitor the health status of the study animals. For example, if a subclinical infectious disease was present in the colony animals, it would quickly become obvious by the determination of the cause of death. The determination of the specific cause of death can be easy and clear or it can be difficult and unclear. Generally, must decide which is the primary or major lesion attributing to the cause of death. Problems occur if there are multiple lesions, each of which could be responsible for the death of the animal. A set of criteria should be established to aid in the determination of a consistent cause of death. There should also be an adequate number of animals to evaluate in all experimental and control groups; it will then become obvious which causes of death are the most common or significant. The person or persons who determine the cause of death should be consistent, making sure all decisions are based on the same criteria. In some studies the animals may be euthanized if moribund, but the cause of death in these animals would still be based on the pathology found in the animals and be valid in the determination of the cause of the morbidity and ultimately the cause of death [15,77].

16.4.4.7 Evaluation of the Data

At the end of the aging study or any other pathologic study, the information collected should be in an acceptable form for proper statistical evaluation. This means using consistent morphologic terminology and numerical grades assigned to lesions where possible. Here, a good statistician is invaluable. Some statistical methods used in prior aging studies are readily available. The chi-square test, Fisher's exact tests, and chi-square test for trends in proportions, analysis of variance, multicomparison procedures [78], and a Huber

M-estimator [79] are useful for the analysis of pathological data obtained in both longitudinal and longevity studies [80,81]. For the analysis of discrete variables, such as pathology data, linear models modified for categorical variables are used [82–87]. Survival distributions in experimental groups can be compared using both parametric and nonparametric statistical methods. Survival distributions and predictions can be evaluated using a Gompertz model and a proportional hazards model [88–90].

Conclusions

Several basic and encompassing conclusions can be made about the use of pathology and its relationship to aging research. First, the pathological analysis of animals used in aging research, although expensive and demanding, is essentially a necessity as the advantages far outweigh the disadvantages. Second, all the pathological methods for collection and analysis of specimens are the same and are available for aging research as they are for other studies using pathologic techniques. Third, rats and mice are the best and most used laboratory animals for the pathological analysis of aging.

Acknowledgments

We would like to thank Dr. Yu, Jo Fletcher, and Corinne Price for their individual contributions to the writing of this text and chapter.

References

1. *Stedman's Medical Dictionary*, 24th Edition, Williams and Wilkins, Baltimore, 1982, 1041.
2. Cotran, R. S., Kumar, V., and Robbins, S. L., Eds., *Robbins Pathologic Basis of Disease*, 5th Edition, W. B. Saunders Company, Philadelphia, 1994.
3. Robbins, S. L, and Angell, M., *Basic Pathology*, W. B. Saunders Company, Philadelphia, 1971.
4. Griner, L. A., *Pathology of Zoo Animals. A Review of Necropsies Conducted over a Fourteen-year Period at the San Diego Zoo and San Diego Wild Animal Park*, Zoological Society of San Diego, [San Diego], 1983, xxxiii.
5. Bronson, R. T., and Lipman, R. D., The role of pathology in rodent experimental gerontology, *Aging Clin. Exp. Res.*, 5, 253, 1993.
6. Weindruch, R., and Masoro, E. J., Concerns about rodent models for aging research, *J. Gerontol.*, 46, B87, 1991.
7. Hazzard, D. G., Bronson, R. T., McClearn, G. E., and Strong, R., Selection of an appropriate animal model to study aging processes with special emphasis on the use of rat strains, *J. Gerontol.*, 47, B63 (editorial), 1992.
8. Cohen, B. J., Anver, M. R., Ringler, D. H., and Adelman, R. C., Age-associated pathological changes in male rats. *Fed. Proc.*, 37, 2848, 1978.
9. Masoro, E. J., Animals models in aging research, in *Handbook of the Biology of Aging*, 3rd Edition, Schneider, E. L. and Rowe, J. W., Eds., Academic Press, Inc., San Diego, 1990, chap. 5.

10. Bronson, R. T., Rate of occurrence of lesions in 20 inbred and hybrid genotypes of rats and mice sacrificed at 6 month intervals during the first years of life, in *Genetic Effects on Aging II*, Harrison, D. E., Ed., Telford Press, Caldwell, N.J., 1990, chap. 17.

11. Bronson, R. T., and Lipman, R. D., Reduction in rate of occurrence of age related lesions in dietary restricted laboratory mice, *Growth Dev. Aging*, 55, 169, 1991.

12. Zurcher, C., van Zwieten, M. J., Solleveld, H. A., and Hollander, C. F., Aging research, in *The Mouse in Biomedical Research*, Vol. IV, *Experimental Biology and Oncology*, Foster, H. L., Small, J. D., and Fox, J. G., Eds. (Am. Coll. Lab. Anim. Med. Ser.), Academic Press, New York, 1982, chap. 2.

13. Boorman, G. A., Eustis, S. L., Elwell, M. R., Montgomery, C. A., Jr., and MacKenzie, W. F., Eds., *Pathology of the Fischer Rat. Reference and Atlas*, Academic Press, San Diego, 1990.

14. Altman, P. L., Ed., *Pathology of Laboratory Mice and Rats* (Biology Databook Ser.), Pergamon Infoline, McLean, Va., 1985.

15. International Agency for Research in Cancer. *Long-Term and Short-Term Screening Assays for Carcinogens: A Critical Appraisal*, IARC Monographs, Supplement 2, International Agency for Research on Cancer, Lyon, 1980.

16. Foster, H. L., Small, J. D., and Fox, J. G., Eds., *The Mouse in Biomedical Research*, Vol. IV: *Experimental Biology and Oncology* (Am. Coll. Lab. Anim. Med. Ser.), Academic Press, New York, 1982.

17. Baker, H. J., Lindsey, J. R., and Weisbroth, S. H., Eds., *The Laboratory Rat*, Vol. I: *Biology and Diseases*, Academic Press, New York, 1979.

18. Shimokawa, I., Higami, Y., Hubbard, G. B., McMahan, C. A., Masoro, E. J., and Yu, B. P., Diet and the suitability of the male Fischer 344 rat as a model for aging research, *J. Gerontol.*, 48, B27, 1993.

19. Cohen, B. J., and Anver, M. R., Pathological changes during aging in the rat, in *Special Review of Experimental Aging Research*, Elias, M. F., Eleftheriou, B. E. and Elias, P. K., Eds., EAR (Experimental Aging Research), Inc., Bar Harbor, Maine, 1977.

20. Thurman, J. D., Bucci, T. J., Hart, R. W., and Turturro, A., Survival, body weight, and spontaneous neoplasms in *ad libitum*-fed and food-restricted Fischer-344 rats, *Toxicol. Pathol.*, 22, 1, 1994.

21. Iwasaki, K., Gleiser, C. A., Masoro, E. J., McMahan, C. A., Seo, E. -J., and Yu, B. Y., Influence of the restriction of individual dietary components on longevity and age-related disease of Fischer rats: the fat component and the mineral component, *J. Gerontol.*, 43, B13, 1988.

22. Bertrand, H., Higami, Y., Shimokawa, I., and Hubbard, G., Nutrition, longevity and pathology, *Age Nutr.*, 3, 165, 1992.

23. Charles River Breeding Laboratories. The Fischer 344 laboratory rat. *Charles River Digest*, XVI(3), 1977.

24. Sugimura, Y., Sakurai, M., Hayashi, N., Yamashita, A., and Kawamura, J., Age-related changes of the prostate gland in the senescence-accelerated mouse, *Prostate* 24, 24, 1994.

25. Blackwell, B. -N., Bucci, T. J., Hart, R. W., and Turturro, A., Longevity, body weight, and neoplasia in *ad libitum*-fed and diet restricted C57BL6 mice fed NIH-31 open formula diet, *Toxicol. Pathol.*, 23, 570, 1995.

26. Sheldon, W. G., Bucci, T. J., Hart, R. W., and Turturro, A., Age-related neoplasia in a lifetime study of *ad libitum*-fed and food-restricted B6C3F1 mice, *Toxicol. Pathol.*, 23, 458, 1995.

27. Schmidt, R. E., Eason, R. L., Hubbard, G. B., Young, J. T., and Eisenbrandt, D. L., Eds. *Pathology of Aging Syrian Hamsters*, CRC Press, Boca Raton, FL, 1983.

28. Jones, T. C., and Hunt, R. D., *Veterinary Pathology*, 5th Edition, Lea & Febiger, Philadelphia, 1983.

29. Lipman, R. D., Chrisp, C. E., Hazzard, D. G, and Bronson, R. T., Pathologic characterization of brown Norway, brown Norway X Fischer 344, and Fischer 344 X brown Norway rats with relation to age, *J. Gerontol.*, 51A, B54, 1996.

30. Coleman, G. L., Barthold, S. W., Osbaldiston, G. W., Foster, S. J., and Jonas, A. M., Pathological changes during aging in barrier-reared Fischer 344 male rats, *J. Gerontol.*, 32, 258, 1977.

31. Brody, J. A., and Schneider, E. L., Diseases and disorders of aging: an hypothesis, *J. Chronic Dis.*, 39, 871, 1986.

32. Thompson, M. K., The need for a new biological model in geratology, *Age Ageing*, 25, 168, 1996.

33. *1996 AVMA Directory and Resource Manual*, 46th ed., American Veterinary Medical Association, Schaumburg, Illinois, 1996.

34. American Society of Clinical Pathologists, 2100 W. Harrison St., Chicago, Illinois 60612.

35. National Society for Histotechnology, Bowie, MD.

36. Institute of Laboratory Animal Resources (U.S.) *Guide for the Care and Use of Laboratory Animals*, National Academy Press, Washington, D. C.,1996.

37. Maeda, H., Gleiser, C. A., Masoro, E. J., Murata, I., McMahan, C. A., and Yu, B. P., Nutritional influences on aging of Fischer 344 rats: II. Pathology, *J. Gerontol.*, 40, 671, 1985.

38. Carson, F. L., *Histotechnology: A Self-Instructional Text*, ASCP, Chicago, 1990.

39. Luna, L. G., Ed., *Manual of Histologic Staining Methods of the Armed Forces Institute of Pathology*, 3rd Edition, McGraw-Hill, New York, 1968.

40. Sheehan, D. C., and Hrapchak, B. B., *Theory and Practice of Histotechnology*, 2nd Edition, C. V. Mosby Co., St. Louis, 1980.

41. Prophet, E. B., Mills, B., Arrington, J. B., and Sobin, L. H., Eds., *Laboratory Methods in Histotechnology*, American Registry of Pathology, Washington, D.C., 1992.

42. Feldman, D. B., and Seely, J. C., *Necropsy Guide: Rodents and the Rabbit*, CRC Press, Boca Raton, 1988.

43. Jones, T. C., and Gleiser, C. A., Eds., *Veterinary Necropsy Procedures*, J.B. Lippincott Co., Philadelphia, 1954.

44. *U.S. Departments of The Army, The Navy, and The Air Force Autopsy Manual* (TM 8-300, NAVMED P-5065, AFM 160-19), U.S. Govt. Printing Office, Washington, D.C., 1960.

45. *Buyers Guide, Lab. Anim.*, 25(11), 1996.

46. *Biosafety in Microbiological and Biomedical Laboratories*, 3rd Edition, (HHS Publ. No. (CDC) 93-8395), U.S. Department of Health and Human Services, U.S. Govt. Printing Office, Washington, D.C., 1993.

47. National Research Council (U.S.), Committee on Hazardous Biological Substances in the Laboratory, *Biosafety in the Laboratory*, National Academy Press, Washington, D.C., 1989.

48. Sanderson, J. H., and Phillips, C. E., *An Atlas of Laboratory Animal Haematology*, Clarendon Press, Oxford, 1981.

49. Loeb, W. F., and Quimby, F. W., Eds., *The Clinical Chemistry of Laboratory Animals*, Pergamon Press, New York, 1989.

50. Baron, E. J., and Finegold, S. M., *Bailey and Scott's Diagnostic Microbiology*, 8th Edition, C. V. Mosby Co., St. Louis, 1990.

51. Lennette, E. H., Balows, A., Hausler, W. J., Jr., and Shadomy, H. J., Eds., *Manual of Clinical Microbiology*, 4th Edition, American Society for Microbiology, Washington, D.C., 1985.

52. Fox, J. G., Cohen, B. J., and Loew, F. M., Eds., *Laboratory Animal Medicine*, (Am. Coll. Lab. Anim. Med. Ser.) Academic Press, Orlando, 1984.

53. Altman, D. G., *Practical Statistics for Medical Research*, Chapman & Hall, London,1991.

54. Salsburg, D., The effects of lifetime feeding studies on patterns of senile lesions in mice and rats, *Drug Chem. Toxicol.*, 3, 1, 1980.

55. College of American Pathologists, College of American Pathologists Conference XXIX: Restructuring Autopsy Practice for Health Care Reform, *Arch. Pathol. Lab. Med.* 120, 1996.

56. Rohen, J. W., and Yokochi, C., *Color Atlas of Anatomy, A Photographic Study of the Human Body*, 3rd Edition, Igaka-Shoin, Tokyo, 1983.

57. Hebel, R., and Stromberg, M. W., *Anatomy of the Laboratory Rat*, Williams & Wilkins Co., Baltimore, 1976.

58. Prichard, R. W., Descriptions in pathology, avoiding pathological descriptions, *Pathol. Vet.* 3, 169, 1966.

59. Sisson, S., *The Anatomy of the Domestic Animals*, 4th Edition, W. B. Saunders Co., Philadelphia, 1953.

60. Swindler, D. R., and Wood, C. D., *An Atlas of Primate Gross Anatomy. Baboon, Chimpanzee, and Man*, University of Washington Press, Seattle, 1973.

61. Di Fiore, M. S. H., *An Atlas of Human Histology*, 2nd Edition, Lea & Febiger, Philadelphia, 1963.

62. Fawcett, D. W., *A Textbook of Histology*, 11th Edition, W. B. Saunders Co., Philadelphia, 1986.

63. Leeson, T. S., and Leeson, C. R., *Histology*, 2nd Edition, W. B. Saunders Co., Philadelphia, 1970.

64. Ham, A. W., and Lesson T. S., *Histology,* 4th Edition, J. B. Lippincott Co., Philadelphia, 1961.
65. Larsson, L. I., *Immunocytochemistry: Theory and Practice,* CRC Press, Boca Raton, 1988.
66. Taylor, C. R., M. B., Chir, B., and Phil. D., Immunoperoxidase techniques. Practical and theoretical aspects, *Arch. Pathol. Lab. Med.,* 102, 113, 1978.
67. Hardisty, J. F., Pathology peer review in assessing carcinogenicity studies, *American College of Veterinary Pathologists, 47th Annual Meeting,* Seattle, Washington, 1996, [program and abstracts], p. 112.
68. Frith, C. H., and Ward, J. M., *Color Atlas of Neoplastic and Non-neoplastic Lesions in Aging Mice,* Elsevier Science, Amsterdam, 1988.
69. Tseng, C. H., *Color Atlas of Diagnostic Histopathology,* CRC Press, Boca Raton, 1986.
70. Turusov, V., and Mohr, U., Eds., *Pathology of Tumours in Laboratory Animals,* Volume 3, *Tumours of the Hamster,* 2nd Edition, (IARC Sci. Publ. No. 126), International Agency for Research on Cancer, Lyon, France, 1996.
71. Rhodin, J. A. G., *An Atlas of Ultrastructure,* W. B. Saunders Co., Philadelphia, 1963.
72. Mohr, U., Dungworth, D. L., and Capen, C. C., Eds., *Pathobiology of the Aging Rat,* Volume I, *Blood and lymphoid, respiratory, urinary, cardiovascular, and reproductive systems,* ILSI Press, Washington, D. C., 1992.
73. Stinson, S. F., Schuller, H. M., and Reznik, G. K., Eds., *Atlas of Tumor Pathology of the Fischer Rat,* CRC Press, Boca Raton, 1990.
74. Flynn, R. J., *Parasites of Laboratory Animals,* Iowa State University Press, Ames, 1973.
75. Hottendorf, G. H., *The Pathology of Spontaneous Diseases of Small Laboratory Animals,* Pharmaceutical Research and Development Division, Bristol-Myers Company, Syracuse, N.Y., 1985.
76. Hamm, T. E., Jr., Ed., *Complications of Viral and Mycoplasmal Infections in Rodents to Toxicology Research and Testing,* Hemisphere Publishing Corp., Washington, 1986.
77. Chrisp, C. E., Turke, P., Luciano, A., Swalwell, S., Peterson, J, and Miller, R. A., Lifespan and lesions in genetically heterogeneous (four-way cross) mice: a new model for aging research, *Vet. Pathol.,* 33, 735, 1996.
78. Chew, V., Comparing treatment means: A compendium, *Hortscience* 11(4): 348-357, 1976.
79. Huber, P. J. *Robust Statistical Procedures.* Society for Industrial and Applied Mathematics, Philadelphia, 1977.
80. Fleiss, J. L. *Statistical Methods for Rates and Proportions.* Wiley, New York, 1973.
81. Siegel, S. *Nonparametric Statistics for the Behavioral Sciences.* McGraw-Hill, New York, 1956.
82. Schrader, R. M., and Hettmansperger, T. P., Robust analysis of variance based upon a likelihood ratio criterion, *Biometrika* 67, 93, 1980.
83. Grizzle, J. E., Starmer, C. F., and Koch, G. G., Analysis of categorical data by linear models, *Biometrics* 25, 489, 1969.
84. Bishop, Y. M. M., Feinberg, S. E., and Holland, P. W. *Discrete Multivariate Analysis: Theory and Practice.* MIT Press, Cambridge, MA, 1975.
85. Walker, S. H., and Duncan, D. B., Estimation of the probability of an event as a function of several independent variables, *Biometrika* 54, 167, 1967.
86. Morrison, D. F. *Multivariate Statistical Methods.* McGraw-Hill, New York, 1967.
87. Grizzle, J. E., and Allen, D. M., Analysis of growth and dose response curves, *Biometrics* 25:357, 1969.
88. Gross, A. J., and Clark, V. A. *Survival Distributions: Reliability Applications in the Biomedical Sciences.* Wiley, New York, 1975.
89. Kalbfleish, J., and Prentice, R. L. *The Statistical Analysis of Failure Time Data.* Wiley, New York, 1980.
90. Marinez, Y. N., McMahan, C. A., Barnwell, G. M., and Wigodsky, H. S., Ensuring data quality in medical research through an integrated data management system, *Stat. Med.,* 3, 101, 1984.

Section E

Molecular and Evolutionary Probes of Senescence Alterations

17

Transgenic Manipulation of the Mouse Genome

Christi A. Walter

CONTENTS

Dedicated to Barbara H. Bowman, Ph.D.

0-8493-3112-9/99/$0.00+$.50
© 1999 by CRC Press LLC

17.1 Introduction

Transgenic mice can be powerful genetic tools for studying and modulating specific genes in attempts to elucidate the physiological, biochemical and molecular mechanisms involved in aging. Descriptive research on aging has identified many parameters that change with age and has lead to several hypotheses regarding the fundamental mechanisms of aging. Transgenic technology provides the opportunity to more directly test some of these hypotheses. It is clear from the application of transgenic technology to disease processes, such as cancer, and life cycle processes, such as embryology, that transgenic mice can be utilized to address basic biological questions. The time is now appropriate to test hypotheses concerning aging by using transgenic mouse technology. To date, only a few examples of the direct application of transgenic technology to aging research are found in the literature [1–4].

    This chapter is intended to provide sufficient detail for a novice to produce and maintain transgenic mouse lines. Extensive descriptions of alternative techniques used to produce transgenic mice can be found in procedure manuals [5]. The techniques described herein are routinely used to successfully produce transgenic mice. Special consideration has been given to issues relevant to aging research.

17.2 Historical Perspectives

A cloned gene was first introduced into the mouse genome by pronuclear microinjection in 1980 [6]. Since then numerous laboratories have successfully utilized the technology to produce transgenic mice. The basic protocol remains essentially unaltered from its original format. To date, aging studies utilizing rodent models have been predominantly descriptive and have demonstrated that many physiological and molecular changes correlate with aging. Until the advent of transgenic mouse technology the principle means of experimentally manipulating biological processes known to correlate with aging, has been through pharmacological methods. Transgenic technology provides a genetic approach to address a complex process by systematically testing the role of individual genes in aging. With transgenic methodologies, the effects of genes thought to be involved in extending and/or determining lifespan can be tested individually and in combinations.

    There are of course, limitations and potential pitfalls associated with any model system, including transgenic mouse models. For example, there is the possibility that integration of the transgene into the genome will affect a gene at the site of integration that will in turn produce a phenotype that is due to changes in the endogenous gene rather than the transgene. This problem is largely overcome by utilizing multiple independent lines carrying the same transgene. Furthermore, it may be difficult to alter the activity of complex pathways unless rate-limiting steps can be identified. However, transgenic mice represent a largely under-utilized method to address mechanisms involved in aging.

17.3 Basic Transgenic Methodology

17.3.1 Embryo Collection, Pronuclear Microinjection, and Embryo Transfer

17.3.1.1 Media and Reagents

Whittens Medium (for 100ml)

0.400 gm	NaCl
0.036 gm	KCl
0.016 gm	KH_2PO_4
0.190 gm	$NaHCO_3$
0.100 gm	Glucose
0.029 gm	$MgSO_4$ $7H_2O$
0.053 gm	Ca lactate $5H_2O$
20 µl	0.5% Phenol red
370 µl	Na lactate 60% syrup
0.001 gm	Gentamicin sulfate powder

pH to 7.35 to 7.4. Sparge with 5% CO_2/95% air gas mixture until pH is stable.

Osmolality = 245-250.

Filter sterilize, gas the airspace above the medium, and store tightly capped for no more than 1 week at 4°C.

Supplement with 0.004 gm of bovine serum albumin (BSA) and 10 ml of Na pyruvate stock per ml of medium on the day of use.

Na Pyruvate Stock

0.035 gm pyruvic acid

10 ml H_2O

Filter sterilize and freeze in 1 ml aliquots at –20°C.

Modified Whitten's Medium (for 100 ml)

0.400 gm	NaCl
0.036 gm	KCl
0.016 gm	KH_2PO_4
0.034 gm	$NaHCO_3$

0.100 gm	Glucose
0.029 gm	$MgSO_4 \cdot 7H_2O$
0.053 gm	Ca lactate·$5H_2O$
0.500 gm	HEPES, Na salt (f.w. = 260.3)
20 ml	0.5% Phenol Red
370 ml	Na lactate 60% syrup
0.002 gm	Gentamicin sulfate powder pH to 7.3.

Osmolality = 234–245.

Filter sterilize and store for no more than one week at 4°C

Supplement with 0.003 gm of BSA and 10 ml of Na pyruvate st°Ck per ml of medium on the day of use

Hylauronidase

0.0004 gm	Hyaluronidase
1 ml	Modified Whitten's medium

Pregnant Mare's Serum Gonadotropin (PMSG)

Add 10 ml of sterile 0.9% NaCl to 1000 IU of PMSG (Calbi°Chem # 367222).

Aliquot into microfuge tubes and store at –20°C.

After thawing, the PMSG must be used or thrown away.

Note that PMSG has a limited shelf life when stored frozen at –20°C, although it is good for at least one or two months.

Human Chorionic Gonadotropin (hCG)

Add 50 ml of sterile 0.9% NaCl to 5000 IU of hCG (Sigma #CG 5).

Aliquot into microfuge tubes and store at –20°C protected from light.

Upon thawing, the hCG must be used or thrown away.

Anesthesia Cocktail

Rompum: Combine 1 ml of concentrated Rompum (20mg/ml) with 4 ml of sterile 0.9% NaCl or lactated Ringer's solution.

Ketamine: Combine 1 ml of concentrated Ketamine (100mg/ml) with 4 ml of sterile 0.9% NaCl or lactated Ringer's solution.

Mix diluted rompum and diluted ketamine 1:1. The mixture can be stored at room temperature.

Paraffin oil

Lacri-Lube

70% ethanol

17.3.1.2 *Supplies*

Inverted microscope equipped with right- and left-hand micromanipulators and a 40× objective. Surgical quality instruments: Mayo scissors, Iris scissors, thumb forceps, a fine forceps, Pierce forceps, watchmakers forceps, micro scissors, needle holders, a Serrefine clamp. Hamilton syringe and tubing filled with fluorinert attached to a micromanipulator

Dissection microscope	Surgical microscope
Needle puller	Microforge
Sterile 35-mm culture dishes	Sterile 100-mm culture dishes
Pasteur pipette and rubber bulb	Disposable transfer pipettes
Gauze 4 × 4 pads	1-cc syringes
3-cc syringes	25-gauge needles
Pulled glass pipette	Thumb pump
Plastic-backed absorbent paper	Microscope slide with an injection chamber
Bunsen burner with pilot light	Micro sponges
Holding pipette	Microinjection pipettes
Nitrogen gas, tubing, Y or T valve	Glass transfer pipettes
Animal clippers	Dental roll or cotton roll
Suture, 7-0 or 8-0	Wound clips
Ear punch	Tape

TABLE 17.1

Glass Stock Information for Glass Microtools

Pipette	Glass Stock	Final Dimensions	
		Inner Diameter	Outer Diameter
Pulled	Coagulation capillary tube	110–180 μm	130–200 μm
Transfer	Coagulation capillary tube	100–120 μm	120–140 μm
Holding	Pyrex, I.D. 0.6 mm, O.D. 1.0 mm	10–15 μm	100–140 μm
Microinjection	Kwik-Fil capillary with filament	1–3 μm	3–6 μm

17.3.1.3 Animals

Superovulated and plugged donor female mice

Vasectomized male mice

Pseudopregnant female recipient mice

Intact stud male mice

17.3.1.4 Overview of Procedures

Donor female mice are hormonally induced to superovulate because large numbers of one-cell embryos are required to make transgenic founder mice. PMSG is administered i.p. to mimic follicle stimulating hormone while hCG is subsequently administered i.p. to mimic lutenizing hormone. Many factors are known to affect the success of superovulation: age of the donor mice, strain of donor mice, dose of gonadotropins, time of gonadotropin administration and weight of the donor females [5]. The number of one-cell embryos obtained is further dependent on the performance of the intact stud males to which the donor females are mated. In general, prepubescent females (3- to 5-weeks of age) are superovulated and mated with intact males that have a demonstrated reproductive performance. However, young (5- to 6-week old) donor females from hybrid strains, often superovulate as well as prepubescent females. The mice need to be maintained under a 14/10 light/dark cycle. Commercially supplied animals will need a few days to adjust to the light/dark cycle of the mouse room before they are treated with PMSG and hCG to achieve efficient superovulation and developmental uniformity. After administration of hCG, one hormone-treated female is placed in a cage with one proven fertile stud male.

Early on the following morning, mated donor females are humanely euthanized, the abdomen opened, and oviducts excised. Embryos with surrounding cumulus cells are liberated and hyaluronidase is used to dissociate the cumulus cells and render the embryos visible by microscopic observation. Each embryo is assessed for quality and for fertilization. Afterward, appropriate one-cell embryos are microinjected with the transgene construct then transferred to pseudopregnant females. A detailed protocol for each manipulation can be found in Section 17.4.

17.3.2   Mice Needed for Transgenic Founder Mouse Production

Four categories of mice are needed for the production of transgenic founder mice. One category is donor female mice. Most transgenic mice are made in a hybrid background simply because the efficiency is much greater than in an inbred background. However, for aging studies an inbred line of mice may be most appropriate. Lifespan and spontaneous pathology studies have been largely restricted to inbred lines and F1 hybrid progeny. Because transgenic studies usually require the propagation of progeny over many generations, the F1 hybrid data cannot be used as a standard. Furthermore, maintenance of hybrid transgenic lines leads to much greater genetic variability and to a greater possibility for genetic drift [7]. Finally, due to the genetic variability in hybrid lines, each animal is unique. Consequently, experimental control over genetic variability is lost. While inbred lines may be more appropriate for aging studies, they are unfortunately more difficult to work with. Indeed the frequency of transgenic mice among potential founder mice is only 8 to 10% on average while the frequency of transgenic mice among potential founder mice in hybrid lines is approximately 30% [8]. Thus, the first and one of the most important decisions to make when embarking on transgenic mouse experiments is selection of mouse strain. In general, prepubescent female mice (3- to 5-weeks old) are used for superovulation and donation of embryos because they superovulate better than sexually mature females. Unless plans include a core facility, it will probably be easiest to purchase weanling female mice from a supplier.

Donor female mice are mated with intact fertile males after they have been hormonally treated to induce superovulation. The females are mated 1:1 with intact males. To maximize efficiency, it is best to test the presumptive male mice for reproductive performance prior to including them among the stud males. This can be done by allowing them to mate with female mice and checking for the presence of copulation plugs. If plugs are present and the females become pregnant, the males have sufficient reproductive performance to use as stud males. If multiple microinjection sessions are planned for consecutive days, it may be wise to have a set of males that can be used for each day because the sperm count will decrease after they have mated. If the availability of stud males is limited, allow a minimum of one or two days in between matings to ensure maximum fertilization. When the plugging drops below 60 to 80%, a male should be replaced. Replacement could be required in as little as six months.

After one-cell embryos are microinjected they will be transferred to the oviduct of a pseudopregnant female that has been mated in such a manner that she is synchronous with the donor female and with the embryos. Pseudopregnant females are obtained by mating with vasectomized stud males at the same time donor females are mated with intact stud males. If desired, recipient females can be of a strain that has a different coat color than the donor mice. Any mice that are born as the result of a failed vasectomy can then be readily identified by coat color. It will be necessary to mate a sufficient number of recipient females to transfer all microinjected embryos. Recipient mice are generally introduced as potential recipients when they are 6- to 8-weeks of age. If a plugged female is not used for a transfer on one day, she can be recycled and be used in a subsequent transfer.

As noted above, vasectomized stud males are required to generate pseudopregnant recipient females. Vasectomies are performed when the mice are about 8-weeks old. Any strain with good reproductive performance is suitable. Success of the vasectomy can be tested by mating with three or four females one month after surgery. One month allows the reproductive tract to be cleared of spermatozoa that had been generated prior to the vasectomy. If any of the females become pregnant, the vasectomy was not successful and the male should not be used to produce pseudopregnant females. When the plugging efficiency begins to drop below 60-80% a male should be replaced. As with pseudopregnant females, a mouse strain with a coat color different from the donor strain of mice can be used to identify pups born as a result of failed vasectomies.

Recipient female mice can be housed together as long as the embryos they received were injected with the same transgene construct and the embryos were transferred on the same day. Similarly, recipient females that give birth and their progeny can be multiply housed. Multiple housing often reduces cannibalism of newborn pups.

17.3.3 Transgene Design and Preparation for Microinjection

17.3.3.1 Supplies

Microinjection Buffer
10 mM Tris, HCl

0.1 mM $Na_2EDTA \cdot 2H_2O$

Adjust to pH 7.4 (pH is critical). Filter sterilize using a 0.2 µm filter. Store at room temperature.

Hoechst Dye Stock
Hoechst 33258 1mg/ml H_2O.

Store at 4°C for six months in the dark (light sensitive).

Ultrapure Calf Thymus DNA (Sigma # D4764)
25 µg/ml TNE (see below)

100 µg/ml TNE

250 µg/ml TNE

Standard DNAs can be stored at 4°C.

10X TNE buffer

12.1 gm Tris, HCl

3.7 gm Na$_2$EDTA.2H$_2$O

58.4 gm NaCl

pH to 7.4. Filter through a 0.2 mm filter to remove debris. Can be stored at room temperature.

Hoeschst Working Dye

10 µl Hoechst stock

100 ml 1X TNE buffer

The working dye solution should be prepared fresh each day and protected from light.

10X TE Buffer

6.06 gm   Tris, HCl (f.w. 121.1)

1.87 gm   Na$_2$EDTA

pH to 8.0. Bring to 500 ml. Filter through a 0.2 mM filter to remove debris. Can be stored at room temperature in a plastic container.

1X TES Buffer

3.63 gm   Tris, HCl (f.w. 121.1)

1.86 gm   Na$_2$EDTA·2H$_2$O

2.92 gm   NaCl

pH to 8.0 Bring to 1000 ml.

Electrophoresis power supply	Transilluminator equipped with a camera
Sea Kem GTG agarose	Horizontal gel electrophoresis unit
Fluorometer	Agarose
Millipore type VS, 0.025 mM dialysis membrane	

17.3.3.2 Transgene Design

The anatomy of the transgene is dependent on the type of study that is being performed. However, certain components are found in most transgene constructs: cis-acting transcriptional regulatory elements, DNA coding sequences and polyadenylation signals (Figure 17.1 A). The cis-acting transcriptional regulatory elements consist of promoters and enhancers. In a few instances, locus control regions have been included and shown to impart copy number dependent and position independent expression [9,10]. It is important to include a transcription start site for appropriate transcription of the coding sequences that are located downstream of the transcriptional regulatory elements. A variety of promoters and enhancers are available and can be used to direct tissue-specific, developmental specific or "ubiquitous" expression. Furthermore, some cis-acting transcriptional elements direct robust expression of transgenes while others impart modest expression. Thus, the selection of cis-acting transcriptional regulatory elements is based on the type of experiment to be performed. Finally, many transgenic experiments are designed to characterize

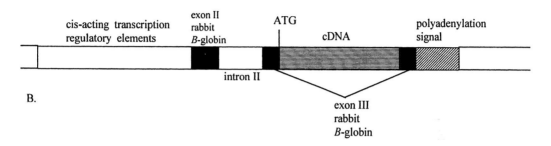

FIGURE 17.1

Schematic representations of typical transgene constructs. **(A)** The simplest transgene consists of transcriptional regulatory elements fused with a cDNA and a polyadenylation signal. **(B)** In a second typical transgene design, the cDNA has been inserted downstream of an intron in an attempt to enhance expression. Translation is initiated at the ATG in the inserted cDNA.

the cis-acting transcriptional regulatory elements of a given gene. A traditional approach employs several constructs containing serial deletions of the 5′-flanking region that are used to identify the amount of 5′-flanking DNA required for appropriate expression. Afterward, more extensive characterization of the 5′-flanking region can be performed using site-directed mutagenesis of putative cis-acting transcriptional regulatory elements.

The DNA coding sequence is usually a cDNA of the gene of interest. However, in some instances, the genomic sequence or a portion of the genomic sequence is employed. Expression of coding sequences is directed by the cis-acting transcriptional regulatory elements located upstream of the coding sequences. Heterologous combinations of cis-acting transcriptional regulatory elements and coding sequences are possible and actually quite common (Figure 17.1 A,B). Reporter genes are commonly employed when the promoter of a gene is being characterized. The rationale is that reporter gene products are normally not found in mouse cells and therefore, are easy to identify and quantify. Secondly, they are not involved in mammalian metabolic pathways so that expression of a reporter should not alter mouse cell physiology. Chloramphenicol acetyltransferase (CAT), luciferase (luc), β-galactosidease (β-gal) and green fluorescent protein GFP are frequently used as reporter genes.

Polyadenylation is important for normal cellular processing of most eukaryotic mRNAs. Furthermore, cleavage signals appear to be required [11]. Therefore, polyadenylation and cleavage signals will need to be added to the transgenes if they are not already present. A fragment of SV40 DNA containing polyadenylation and cleavage signals are commonly utilized. Commercially available vectors can be used for this purpose.

Because introns have been shown to enhance expression of some transgene constructs, many transgenes contain introns [12]. If genomic DNA fragments are used, introns already present in the fragment are usually suitable and no additional manipulation is required. However, if a cDNA is used, a heterologous intron may prove useful. There are several sources of introns that are amenable to inclusion in transgene constructs. Among the most frequently used is the SV40 intron, which is readily liberated from several commercial vectors using restriction endonucleases that will also liberate the cleavage signal in the same restriction fragment. Thus, this fragment can be fused at the 3′-end of the cDNA and will provide an intron, a polyadenylation signal and a cleavage signal for appropriate RNA processing. In some instances, it has been thought that the SV40 intron is too small to facilitate efficient splicing and does not enhance expression in this situation. However, the SV40 intron has been used with great success by many laboratories. A plasmid containing exon II, intron II and exon III of rabbit β-globin is another frequently used vector for transgene construction [13–15]. The vector was modified by Howe et al., to make it more amenable to transgene construction [16]. The modified vector, pBSpKCR3⁺, contains a multiple cloning sequence upstream of exon II and BamHI and EcoRI sites, in exon II and exon III respectively, that can be used to insert cDNAs. The multiple cloning sequence upstream of exon II can be used to insert cis-acting transcriptional regulatory elements. While many transgene constructs are engineered to contain introns, very good expression can be obtained in the absence of an intron [17]. Overall, it is recommended that an intron be included in the transgene construct if feasible; however, the absence of an intron does not necessarily mean the transgene will not be expressed robustly.

17.3.3.3 *Preparation of the Transgene for Microinjection*

The quality of the transgene preparation will have a strong impact on the success of transgenic founder mouse production. Therefore, it is worth while to prepare high-quality DNA for microinjection. Cesium chloride density centrifugation is used to purify the vector. The transgene is then liberated from the vector such that the fragment is relatively free of vector sequences. Agarose gel electrophoresis is employed to separate the transgene fragment from vector fragments. Afterward, the transgene fragment is removed from the agarose, quantified, the quality of the preparation confirmed and diluted to concentrations appropriate for microinjection. The quality of agarose that is used to separate the transgene fragment from vector fragments is critical. It must be free of impurities that could be toxic to the embryos. Furthermore, the electrophoresis is performed in the absence of ethidium bromide and the gel box must be free of ethidium bromide as well. Careful preparation of the transgene fragment for microinjection will be rewarded in later stages of transgenic mouse production.

17.3.4 Identification of Transgenic Mice

17.3.4.1 *Supplies*

10% Sodium dodecyl sulfate (SDS)	1 M Tris, pH 7.5
100 mg Proteinase K/ml H$_2$O stored aliquots at –20°C in nonfrost-free freezer	Chloroform/isoamyl alcohol 24:1 volume to volume)
Ice-cold 70% ethanol	Ice-cold 100% ethanol
5 M NaCl	0.5 M EDTA, pH 8.0
10× TNE buffer (Section 17.3.3.1)	Micro pipettes
Hoechst dye stock (Section 17.3.3.1)	Fluorometer
10 mg Ethidium bromide/ml H$_2$O.	

Note: **Ethidium bromide is a potent mutagen.** Wear gloves whenever it is handled and a mask if you must weigh it!!

Tail Lysis Buffer Stock

50 ml	10% SDS
5 ml	1 M Tris, pH 7.5
5 ml	0.5 M $Na_2EDTA \cdot 2H_2O$
440ml	H_2O

May be stored at room temperature after filtering through a 0.2 μM filter to remove debris.

Working Tail Lysis Buffer

43 ml	Tail lysis buffer stock
3.8 ml	5 M NaCl
188 μl	1 mg Proteinase K/ml

Proteinase K should be added just prior to use. Until then, working tail lysis buffer can be stored at room temperature. After proteinase K is added the buffer should be used or discarded.

NaCl-Saturated Phenol

| 45 ml | Tris-saturated phenol |
| 4 ml | 5M NaCl |

Can be prepared in advance and stored at 4°C. Addition of hydroxyquinoline to the Tris solution will serve as a radical scavenger and indicate changes in quality of the phenol via a color change, and aid in the identification of the phenol phase during DNA isolation procedures.

17.3.4.2 Methodology

The method used to identify transgenic mice, especially founder mice, must be extremely accurate. Typically, Southern blot analysis, slot blot analysis and/or PCR amplification protocols are used with DNA obtained from potential transgenic animals to identify transgenic animals. However, on occasion, an enzyme assay or some other technique can be used just as reliably. Only commonly used protocols will be covered here.

Small tail biopsies (approximately 1 cm) are obtained from potential transgenic pups when they are weaned from their mothers, usually at 3- to 4-weeks of age. DNA is prepared from the tail samples and used to detect the presence of the transgene. DNA samples are immediately subjected to PCR amplification using primers specific for the transgene simultaneously with primers for an endogenous murine gene for rapid identification of founder mice. Afterward, if the efficiency of the transgene primers is unknown or is not optimal, the negative DNA samples are subjected to Southern blot analysis using a transgene-specific probe followed by a probe for a murine gene. All founder mouse DNA samples positive by PCR are subjected to quantitative Southern blot analysis after transgenic pups have been born and identified. A quantitative Southern blot allows characterization of the number of copies of the transgene that are integrated as well as, whether multiple integration sites are present. Because a significant proportion of transgenic founder mice are mosaic, it is recommended that DNA obtained from transgenic progeny be used to determine how many copies of the transgene are integrated. Furthermore, analysis of transgenic progeny DNA can assist in evaluating the number of integration sites. If multiple integration sites are present in the founder, this should be considered when establishing breeding schemes for that transgenic line. Southern blot or slot blot analyses can be used to identify transgenic progeny, but PCR is more efficient when well-designed primers are available.

17.4. Production and Identification of Transgenic Mice

17.4.1 Preparation of Transgene DNA for Microinjection

Cesium chloride ethidium bromide density gradient centrifugation is the method of choice for preparing high quality DNA for subsequent microinjection. A detailed description of the procedure can be found in standard molecular biology procedure manuals and will not be given here [18]. Note, it is extremely important that the DNA not be nicked or otherwise damaged while extracting the ethidium bromide. It is equally important to remove all of the ethidium bromide which is toxic to the embryos and can damage the DNA if exposed to the appropriate wavelengths of light. After extracting the ethidium bromide, the DNA should be dialyzed against four liters of 4°C TES buffer (Section 17.3.3.1) with two changes of TES buffer. This method will remove cesium chloride, which can be toxic to the embryos, better than precipitation methods. An aliquot of the dialyzed DNA should be subjected to agarose gel electrophoresis to ascertain the quality of the DNA. At least 95% of the DNA should be in a supercoiled form. Open circular or linear DNA above 5% indicates that too much DNA damage has occurred for the sample to be used with a high degree of confidence.

After determining that the DNA is of sufficient quality to proceed, the vector carrying the transgene fragment is subjected to restriction endonuclease digestion to liberate the transgene as a single fragment relatively free of vector sequence. An aliquot of the digested DNA should be subjected to agarose gel electrophoresis to ascertain whether restriction endonuclease digestions went to completion. The digested DNA is precipitated with ammonium acetate and ethanol. Completely dry the DNA in a speed vac to remove all of the ammonium. The dried DNA is resuspended in 0.1X TE (Section 17.3.3.1) and subjected to agarose gel electrophoresis in the absence of ethidium bromide. The desired band can be excised if a small aliquot of the digest is run adjacent to the preparative lane and subsequently cut away from the remainder of the gel. The gel slice, also containing a lane of size standards, is then stained with ethidium bromide. The distance the desired fragment migrated is measured and used to excise the fragment from the preparative lane. Remember, ethidium bromide is a powerful mutagen and should be handled with gloves at all times. Furthermore, the UV light used to fluoresce ethidium bromide is also mutagenic. Appropriate face wear or shields should be used to prevent UV exposure.

The transgene fragment can be removed from the gel slice using any of several methods: Schleicher and Schuell Elu-Quick kit, Gene Clean, electroelution (again without ethidium bromide), etc. If the fragment is electroeluted from agarose, the DNA should then be applied to a column (e.g., Elutip D or NACS pack) to remove all traces of agarose. If Elu-Quick or Gene Clean was used, be sure to spin the sample several times at the end of the procedure to remove all glass beads which will clog the microinjection pipettes. At this point the transgene is dialyzed against microinjection buffer (Section 17.3.3.1) for four hours at room temperature or overnight at 4°C. To perform the dialysis, float a piece of Millipore type VS, 0.025 μM dialysis membrane on top of microinjection buffer in a sterile plastic petri dish. The DNA solution can be gently dropped onto the dialysis membrane and allowed to float. Cover the petri dish to maintain a clean and relatively dust-free environment. After dialysis is complete, the transgene solution is collected and transferred to a sterile microfuge tube. Microfuge at full speed for 10 min to sediment debris. Gently transfer the top 4/5 of the solution to another sterile microfuge tube.

A fluorometer in conjunction with Hoescht 33258 is a reliable method for quantifying the amount of DNA prepared for microinjection. Hoecsht 33258 has an excitation spectrum that peaks at 365 nm and emission that peaks at 458 nm. Minifluorometers are commercially available that are significantly less expensive than a fluorometer and can be dedicated to DNA quantification. Two milliliters of Hoechst working dye is delivered into a cuvette and can be used to zero the fluorometer. Then 2.0 μl of DNA sample, standard or transgene construct, is added to 2.0 ml of working Hoecsht dye, mixed well and read. After establishing a standard curve, transgene sample concentrations can be determined. Once the concentration of transgene DNA has been determined, dilutions can be prepared using microinjection buffer (Section 17.3.3.1). Prepare multiple concentrations for microinjection because the size of the construct and other unknown factors can influence the optimal concentration. DNA concentrations of 3ng/μl, 5ng/μl and 7.5ng/μl are prepared and utilized. Aliquots of 25- to 30μl can be dispensed into 0.2ml microfuge tubes and stored at –20°C in a nonfrost-free freezer until ready for use.

17.4.2 Vasectomized Stud Male Mice and Pseudopregnant Female Recipient Mice

17.4.2.1 Vasectomized Male Mice

Young adult (at least 8-weeks old) male mice are prepared for surgery by anesthetizing the mice with a 0.2 cc i.p. injection of anesthesia cocktail (Section 17.3.1.1) and the lower abdomen shaved using small animal clippers. Wipe the site with alcohol-wetted gauze to remove excess hair. Position the mouse on its back on a clean and firm surface that has been covered with a clean absorbent, plastic-backed paper. Using scissors open the skin with a 1.5 cm transverse incision so that the opening will be approximately level with the top of the rear legs. A longitudinal incision is then made along the lina alba starting at the abdominal fat pad and continuing caudally approximately 1.5 cm. Use forceps to pull out the fat pad on one side (Figure 17.2 A). The testis, vas deferens and epididymis will come out with the fat pad. The vas deferens will be found underneath the testis and will appear as a tube (Figure 17.2 B) while the epididymis will be smaller and coiled. Forceps with red hot tips can be used to cauterize two points of the vas deferens and the intervening section subsequently cut out (Figure 17.2 C). A cautery unit can be used similarly. Use blunt forceps to gently place the tissues back inside the body wall. Place two or three stitches to sew up the body wall then use wound clips (Figure 17.2 D) or additional sutures to close the skin. Clips can be removed one week after surgery. To be sure the vasectomy was successful, a test mating can be performed one month after the surgery.

17.4.2.2 Pseudopregnant Female Mice

The female reproductive system becomes able to support embryos after mating with a sterile male although her own oocytes will not be fertilized and will degenerate. Mating with a vasectomized male should be timed so that the pseudopregnant females reproductive status is synchronous with donor females. The females should be reproductively mature, at least six- to eight-weeks old, and not fat. Increased fat content makes embryos transfers much more difficult. Two females are placed in a cage with one vasectomized male the day before a planned transfer and checked for a copulation plug the next morning. Only females with copulation plugs should be used. A record of the vasectomized males' reproductive activities can be maintained and when the plugging frequency begins to decline

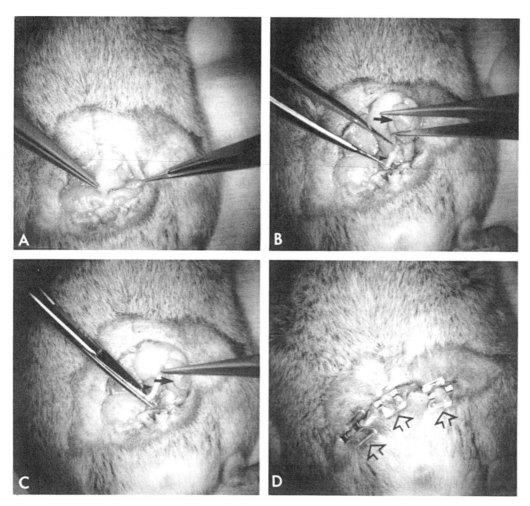

FIGURE 17.2
Photomicrographs of techniques involved in performing a vasectomy. **(A)** The lower abdomen is opened and fat pads located. **(B)** Pulling on the fat pad will reveal the testis (arrow), vas deferens (arrowhead) and epididymis. **(C)** After the vas deferens (arrowhead) is cauterized in two places, the intervening section is excised. Testis (open arrow). **(D)** The abdominal wall is closed with stitches and the skin with wound clips (open arrows).

they should be replaced. Vasectomized males can be used every day, but if possible give at least one day of rest between uses. Two randomly selected females placed with each of ten vasectomized males should result in three or four pseudopregnant females.

17.4.3   Fertile Stud Male Mice and Superovulated Female Donor Mice

17.4.3.1   *Fertile Stud Male Mice*

Fertile stud males should be sexually mature, at least 8-weeks old, to optimize fertilization. Stud mice are housed individually to prevent fighting and injury. A single superovulated female is placed in the cage with one stud male to obtain one-cell embryos the next morning. The males and females can be housed together immediately after the female receives the hCG injection or later in the afternoon of the day the female receives the hCG injection. The following morning the female is checked for the presence of a copulation plug. In some strains of mice the copulation plug is more difficult to see. For example, in C57BL/6J mice the copulation plug is further up the vagina than for most strains of mice. The reproductive performance of each male can be monitored by scoring how often a copulation plug is found when he is placed with a superovulated female. When the reproductive activity declines, the male should be replaced. The number of stud males required is dependent on the strain of mice used and the number of embryos that will be needed. For example 10 to 15 stud mice may be needed for each day of injection with inbred mice while only 5 to 10 may be needed for hybrid mice.

17.4.3.2 Superovulated Female Donor Mice

One female mouse is treated hormonally to superovulate for each stud mouse that will be used per injection session. Thus, for inbred lines 10 to 15 females will be superovulated while five to ten females will be superovulated when hybrid donors are used. A typical schedule of events is present in Table 17.2. There is some room for adjustment as to when PMSG and hCG are given as long as hCG continues to be administered 46- to 48 hours post PMSG. Such adjustments can make scheduling procedures more convenient for laboratory personnel. Five IU of PMSG delivered i.p. is standard for most strains of mice while hCG doses are typically 5 IU given i.p. A few injections at various doses will quickly identify optimal hormone doses. All superovulated females should be humanely euthanized.

TABLE 17.2

Timetable for Manipulations Involved in the Production of Transgenic Mice

Day	Event
–2	PMSG injection
0	hCG injection. Mate donor females and recipient females with appropriate males
1	Early a.m. collect embryos
	Early p.m. perform microinjection
	Mid- to late-p.m. transfer embryos
19–21	Birth of potential founder pups

17.4.4 Embryo Retrieval and Preparation for Microinjection

Prior to retrieving one-cell embryos, all media and reagents should be prepared for use. Thus, BSA and pyruvate should be added to Whitten's, modified Whitten's and hyaluronidase. Whitten's medium should be loosely capped and kept in a CO_2 incubator while modified Whitten's and hyaluronidase can be kept capped at room temperature. Label one 35-mM sterile culture dish on the side as "non-injected" and one "injected." Place two large and one small drop of Whitten's medium in the dish labeled "non-injected," and cover with paraffin oil (Figure 17.3 A). Next place two large drops of Whitten's medium in the dish labeled "injected" and cover with paraffin oil (Figure 17.3 A). Wipe down surgical instruments with ethanol and flame. Finally, place a small drop of modified Whitten's medium along the edge of a sterile 100-mm culture dish that will be used to hold harvested oviducts, one large drop of hyaluronidase and three small drops of modified Whitten's medium that will be used to rinse away hyaluronidase (Figure 17.3 B).

Pulled pipettes, used for transferring embryos from one medium to another, should be made well in advance. Following gas sterilization the pipettes should be allowed to air at

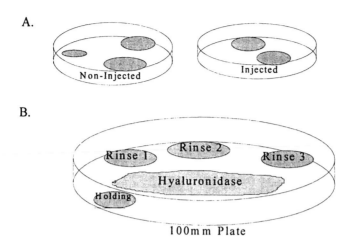

FIGURE 17.3
Schematic representation of plates used to collect oviducts and one-cell embryos. **(A)** Drops of Whitten's medium are placed in a 35-mm dish labeled "non-injected" and covered with paraffin oil. A second dish is prepared for "injected" embryos, as shown. **(B)** A 100-mm dish is used to collect oviducts, treat cumulus masses with hyaluronidase and rinse embryos free of hyaluronidase.

least one week so that the toxic gas is lost from the pipettes before they are used to handle embryos. Pulled pipettes can be washed, sterilized and reused many times. Coagulation capillary tubes are pulled over the pilot light of a Bunsen burner to make two pieces of glass of equal size. Each piece of pulled glass can make one pulled pipette. The large end of the pulled tube is placed into the holder on a microforge and the position adjusted until the glass is vertical and to the right of the filament. Adjust the pipette up and down until the portion of the pipette with inner diameter measurements between 120- to 150 µm is centered in the field of view. Move the filament close to, but not touching, the glass with heat at a moderate setting. Apply heat near the glass until the filament is about 1/3 to 1/2 through the pipette. Turn off the heat. It will be helpful to place a small hook, made from Pyrex glass, on the right side of the glass while it is being heated to prevent it from bending to the right. If the pipette fails to break on its own after removing the heat, place the hook around the pipette just above the filament and pull gently to the right. The break should be smooth. If it is not smooth the process can be repeated on the jagged edge or the glass can be broken again further up. Fire polish the broken end of the pipette by heating the filament at a high setting and lower the glass until it is near the filament. When the pipette begins to melt, turn off the heat. The pipette should be removed from the holder and the large end cut with a diamond pen so that the total length of the pipette does not exceed 1.25 inches. The large end can now be fire polished using the pilot light of a burner. Residual glass should be removed from the filament of the microforge each time glass is broken. This can be accomplished by heating the filament to a high setting and touching the large end of the hook to the glass when it becomes molten. The hook is pulled slowly from the filament. While the pipette is slightly warm it can be inserted into a Unopette holder, capped and placed in a rack with a cover. When completed, the pulled end of the pipette should have an outer diameter between 150 and 300 µm and an inner diameter of 120 to 200 µm. Patience, perseverance and practice will allow the production of high-quality glass microtools that are essential to the production of transgenic mice.

Sacrifice the donor females one at a time by cervical dislocation, place on their backs on the absorbent paper and wipe the abdominal area with 70% ethanol and gauze 4 × 4's.

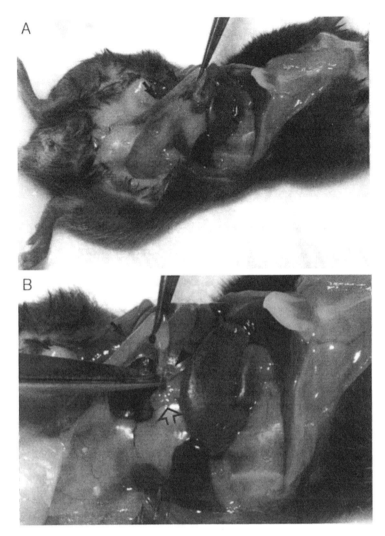

FIGURE 17.4
Oviduct collection. **(A)** The abdominal cavity is opened and the reproductive tract located: uterine horn (arrows). **(B)** Fine scissors are used to excise the oviduct which appears as a small coil (arrowhead) between the uterine horn (arrow) and the ovary (open arrow).

Open the abdomen with Mayo scissors and pull upward on the skin to reveal the entire abdominal cavity (Figure 17.4 A). Locate the reproductive tract and use one forceps to gently, but firmly, hold up one uterine horn. Use another forceps to create an incision between the uterine horn and the mesovarium (Figure 17.4 A). Retract the mesovarium towards the ovary. This will create a separation between the ovary and the oviduct which will facilitate excising the oviduct with a fine scissors (Figure 17.4 B). The oviduct will be smaller than the uterine horn and will be coiled. Place the oviduct in the holding drop of modified Whitten's medium on the edge of the 100-mm culture dish. Follow the same protocol to remove the other oviduct.

Use a dissection microscope to collect one-cell embryos by removing one oviduct at a time from the holding drop and placing it in the hyaluronidase. Pinch off the segment of oviduct containing the cumulus mass using the fine forceps. Use the 25-gauge needle on the 1cc syringe to tear a hole in the segment of oviduct containing the cumulus mass. The mass should ooze out, but it can be teased out with the 25-gauge needle if help is needed. At this point the hyaluronidase should have caused most of the cumulus cells to fall away from the one-cell embryos. Quickly remove the embryos and rinse them off in small drops of modified Whitten's medium in the 10-mm culture dish. All of the embryos can be picked up with a pulled pipette and a thumb pump as soon as the cumulus cells are off. Transfer as little medium and cellular debris as possible to the rinsing drops. Once all the embryos are in the first rinse drop they can be graded. All good embryos should be transferred to the second rinse drop, magnification increased and the embryos graded again. Transfer all of the good embryos to the third rinse drop then to the drop of Whitten's medium in the 35-mm dish labeled "non-injected" and store in the incubator until microinjection.

17.4.5   Pronuclear Microinjection of One-Cell Embryos

Holding pipettes and microinjection needles will need to be prepared in advance. Holding pipettes will be used to hold the embryos in position for microinjection. Glass (Section 17.3.1.1) is initially tooled as for a pulled pipette except that the heat settings will be slightly higher to break and fire polish the Pyrex glass that is used to make holding pipettes. In addition, when applying heat the filament should be heated until it is 1/2 to 3/4 of the way through the glass. The filament should be heated until the inner diameter of the pipette melts down to 10- to 15 μm when fire polishing the broken end of the holding pipette. The opening should not be large enough to encompass the diameter of a one-cell embryo. It is also important that the end of the holding pipette is very flat. Otherwise, the embryos will roll when microinjection is attempted. Finally, the large end of the pipette is fire polished. As with pulled pipettes, careful washing after each experiment will render the pipette reusable for several injection sessions. Microinjection needles are made with Kwik-Fil capillary with filament glass (Section 17.3.2). This glass should not be handled with bare hands. The glass is pulled following instructions supplied with the pipette puller. Individual settings may need to be optimized and can be determined with practice. Microinjection pipettes are preferentially made the day of use or the day prior to use. Old pipettes will become clogged more readily than fresh pipettes. Microinjection pipettes can be stored in covered petri dishes and held in place with ridges of Plasticine or rolled-up tape. Microinjection pipettes cannot be washed and reused. Have at least six microinjection pipettes prepared for each injection session.

In addition to the glass microtools described above, a Hamilton syringe and tubing filled with flourinert should be placed in the holder of the left micromanipulator (Figure 17.5 A). The microinjection chamber should be prepared just prior to starting microinjections. Several options are available for the microinjection chamber. These include petri dishes, depressions slides or flat slides. Place a microdrop (approximately 150 μl) of modified Whitten's medium in the center and cover with paraffin oil (Figure 17.5 C). Lastly, micro-injection pipettes should be filled with the transgene construct by capillary action. The tube containing the DNA should be pulse-spun in a microfuge after it is thawed on ice to bring down condensation from the lid of the tube. Gently mix the microfuged DNA suspension before filling the microinjection pipettes. Transgene DNA can be stored on ice until micro-injections are completed for the day and should then be returned to a –20°C nonfrost free freezer.

Thirty to 40 one-cell embryos from the 35-mm culture dish labeled "non-injected" are collected and placed slightly above the center of the drop of modified Whitten's medium using a pulled pipette (Figure 17.5 D). Individual embryos are picked up and moved to the

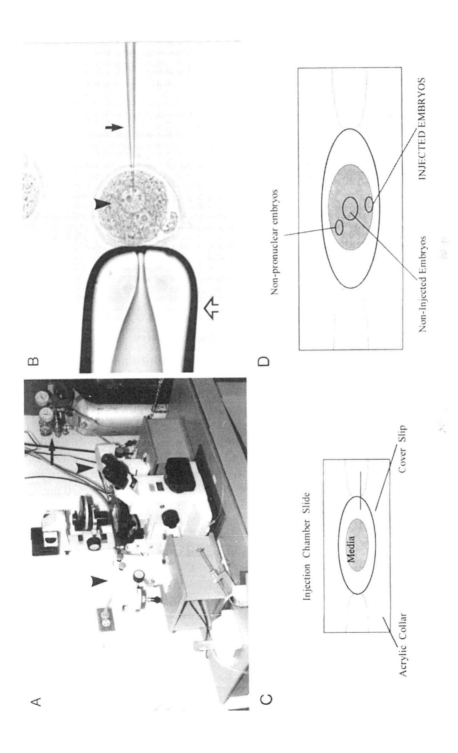

FIGURE 17.5

Microinjection of one-cell embryos. **(A)** A typical microinjection station includes an inverted microscope, right- and left-hand micromanipulators (arrowheads), and nitrogen gas (arrow) used to expel the transgene solution into a pronucleus of the one-cell embryo. **(B)** Microinjection of the male pronucleus (arrowhead) in a one-cell embryo. The holding pipette (open arrow) is shown on the left while the microinjection pipette (arrow) is on the right side. **(C)** Schematic representation of a microinjection chamber. **(D)** Schematic representation showing the placement of embryos in the injection chamber with regard to microinjection status.

center of the medium drop using the holding pipette and the left micromanipulator. The microinjection pipette is used to gently pierce the embryo and deliver the DNA solution to a pronucleus of the one-cell embryo (Figure 17.5 B). The male pronucleus will appear slightly larger and is a slightly better target for microinjection. DNA is delivered by placing your right thumb over the open arm of the Y valve that is used to connect the microinjection pipette to nitrogen gas. The slight pressure exerted after covering the open arm of the valve will push DNA solution into the pronucleus. The pronucleus will visibly swell if it is successfully microinjected. An approximate doubling of size is a good indication that a sufficient volume of DNA suspension was delivered. If the pronucleus does not swell either the pipette is not in the pronucleus or the pipette is clogged. If the pipette is clogged, it will be necessary to fill another one and replace the clogged pipette in the right manipulator. Once DNA has been delivered in the pronucleus, quickly but smoothly remove the microinjection pipette. Embryos that are successfully injected are moved to the bottom of the medium drop. Embryos that are not pronuclear at the time of microinjection are moved to the far top of the medium drop (Figure 17.5 D). When the embryos have been injected, they are transferred to a drop of Whitten's medium in the 35-mm dish labeled "injected," and the dish is returned to the incubator. Non-pronuclear embryos are placed in a small drop in the dish labeled "non-injected," and returned to the incubator. These embryos can be checked for pronuclei later. Another group of embryos is transferred to the microinjection chamber and the process repeated until all of the suitable embryos have been injected. Injected embryos are stored in the incubator until they are transferred to a pseudopregnant female.

17.4.6   Transfer of Injected Embryos to Pseudopregnant Females

Coagulation capillary tubes are heated and pulled over a pilot flame as previously described (Section 17.4.4). The pulled glass is placed in the microforge as described for pulled pipettes and similarly broken and fire polished, but the inner diameter should be 100 to 120 μm. Lower the pipette through the filament until the tip of the pipette is approximately 3.6- to 4mm below the filament. Adjust the heat setting to high and heat the pipette while holding the lower end with a hook, until the inner diameter narrows to 10 to 20 μm. Immediately turn off the heat. The pipette is removed and the large end fire polished with a pilot light from a gas burner. These can be stored in covered petri dishes and held in place with Plasticine or rolled tape. Sterilization of transfer pipettes is not required. Used pipettes should be disposed and not reused. All surgical instruments used for embryo transfers should be sterilized by dipping them in ethanol and then flaming them with a small gas or alcohol burner.

The recipient pseudopregnant mouse is anesthetized and the sides of the mouse are shaved with animal clippers. 70% alcohol-wetted gauze 4 × 4's are used to wipe the surgical site and remove excess hair. The animal is placed under a surgical microscope on its abdomen on top of a dental roll (Figure 17.6 A) and the eyes are lubricated with lacri-lube. Tape may be used to secure the mouse's feet. Remove 25 to 30 embryos from the "injected" dish using a pulled pipette and place in a "transient" dish. Modified Whitten's is used to fill the transfer pipette to just past the shoulder then 25 to 30 injected embryos are loaded. The transfer pipette is attached to tubing and a mouth pipetting apparatus. Store the transfer pipette with the mouth pipetting device attached while the recipient female is prepared for embryo transfer.

A <1cm transverse incision is made on one side approximately 1cm down from the spinal cord at the level of the last rib (Figure 17.6 A). Use watchmaker's forceps to pick up the body wall over the ovary and fat pad. Fine dissection scissors can be used to make a small

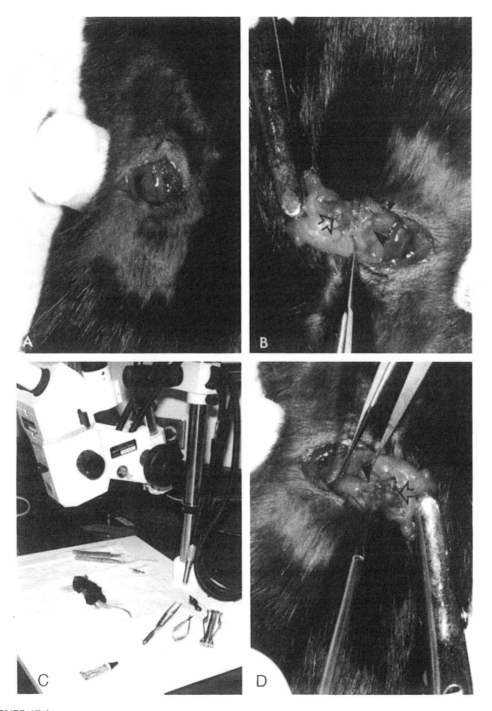

FIGURE 17.6
Surgical procedures involved in embryo transfer. **(A)** Placement of the animal on a dental roll to elevate the embryo and the incision to obtain access to the oviduct is shown. **(B)** Pulling on the fat pad will extricate the ovary (open arrow), oviduct (arrowhead) and uterus (arrow). **(C)** A surgical microscope is used to visualize the opening of the oviduct for embryos transfer. Also shown are surgical instruments utilized in the procedure. **(D)** A transfer pipette is used to place the microinjected embryos into an oviduct (arrowhead). Ovary (open arrow); uterus (arrow).

incision in the body wall. A cautery unit can be used to stop bleeding and clear the surgical field. Pick up the fat pad with blunt forceps and pull until the ovary, oviduct, and uterus are outside the body wall (Figure 17.6 B). A Serrefine clamp can be used to hold the oviduct and ovary outside the body wall by clipping it onto the fat pad and draping the clamp down over the middle of the back. Locate the opening of the oviduct, the infundibulum, using the surgical microscope (Figure 17.6 C). Tear the bursa over the infundibulum with fine forceps being careful to miss large blood vessels. The edge of the infundibulum is located and the transfer pipette inserted (Figure 17.6 D). Gently blow into the mouth pipetting device to expel the embryos. Remove the Serrefine clamp and use blunt forceps to place the tissues back inside the body wall. The body wall can be closed with one or two stitches prior to closing the skin with wound clips. If only one side is used per transfer, recipient females that go on to produce litters can be used for a second transfer on the opposite side at a later time. However, embryo transfers can also be made to both sides in one injection session. After surgery, the animal can be placed under a heat lamp and monitored until it recovers from the anesthetic. Recipient females are monitored periodically for signs of pregnancy and the date of birth of litters recorded.

17.4.7   Isolation of DNA From Mouse Tail Biopsies

It is convenient to obtain a tail biopsy and to mark the animals when they are separated from the mothers at three- to four-weeks old. Isofluorane is used to quickly anesthetize the pups. However, care must be taken because pups are easily overdosed with any anesthetic. Immediately excise a 1cm piece of tail from the tip using a sharp scalpel. Silver sticks can be used to help stop bleeding. The tail biopsy is placed in a 15 ml conical polypropylene tube containing 3 ml of tail lysis buffer without proteinase K. After all tail biopsies have been collected proteinase K is added to each tube and the tops tightly closed. The tubes are placed in a shaker pre-warmed to 55°C and incubated with shaking for 1 hour. Samples are checked to determine if the tissue has been largely dissociated and lysed as evidenced by the presence of cartilage and hairs remaining visible in the tubes. If samples are not completely dissolved, more proteinase K can be added and the samples incubated an additional amount of time (up to 1 to 2 more hours).

Sodium chloride-saturated phenol (1.5 ml) and chloroform/isoamyl alcohol (1.5 ml) are added to each tube using glass pipettes. The organic solutions are mixed with the samples by rocking back and forth approximately 5 min. Samples are centrifuged to separate the phases and the aqueous phase (top layer) is removed taking care not to disturb the interface and delivered into a new polypropylene snap top tube. Organic extractions are repeated at least one more time. If the interface is still large, a third organic extraction can be performed. As a final organic extraction, an equal volume of chloroform/isoamyl alcohol is added to the aqueous phase in a clean snap-top tube, mixed, and centrifuged. The aqueous layer (top layer) is carefully removed and transferred to another snap top tube. Ice-cold 100% ethanol (2.5 volumes) is added and gently mixed to precipitate the DNA. Precipitated DNA is removed from the ethanol using a Pasteur pipette that has been hooked on the end by heating in the flame of a small gas burner. Hooked pipettes can be prepared in advance and a new hook should be used for each DNA sample. The hooked DNA is transferred to a 1.5 ml microfuge tube containing 300ml of ice-cold 70% ethanol. The samples are microfuged for 10 to 15 min to pellet the DNA. The ethanol is removed without disturbing the pellet using a pipette. The pellet is washed with 200µl of 100% ice-cold ethanol and microfuged for 1 min. DNA samples should be dried in a speed vac to ensure the complete removal of ethanol. Two hundred µl of H_2O is used to dissolve the DNA. Several micrograms of DNA are routinely obtained for each sample with this one day isolation protocol.

The DNA is of sufficient quality for Southern blot analysis, slot blot analysis and PCR analysis. DNA samples should be stored at 4°C until needed. If long term storage is anticipated, the samples should be stored at –20°C in a nonfrost-free freezer after ethanol precipitation. DNA is quantified using the Hoechst 33258 protocol (Section 17.4.1).

17.4.8   Identification of Transgenic Mice

Because a significant amount of work and expense is involved in producing transgenic founder mice, an accurate test for the presence of the transgene is warranted. Southern blot analysis is an accurate method to identify transgenic mice. Detailed protocols for Southern blot analysis can be found in techniques books for molecular biology [18]. Slot blot analysis can also be used to identify transgenic animals. This technique is also found in molecular biology technique manuals [18]. Another approach is to use PCR amplification to identify transgenic founder animals. The advantage of the PCR approach is that a rapid turn around time can be realized. However, no matter what the initial identification protocol, transgenic founder mice and some of their progeny should be analyzed by quantitative Southern blot analysis to determine (1) the approximate number of copies of the transgene that were integrated, (2) whether there are multiple integration sites, and (3) whether gross alterations have occurred in the transgene during the integration process. An example of a quantitative Southern blot is shown in Figure 17.7 A.

If a large colony of transgenic mice and/or many transgenic lines are going to be maintained, Southern blot analysis or slot blot analysis of DNA from each potential transgenic pup becomes labor intensive. The use of PCR amplification to identify transgenic pups can result in time and money savings. The actual PCR protocol will be dependent on the primers being utilized. A detailed protocol can be found in the literature and an example is shown in Figure 17.7 B [19].

17.5   Establishing, Analyzing, and Preserving Transgenic Mouse Lines

17.5.1   Establishing Transgenic Mouse Lines

Once a founder mouse is obtained, it should be immediately housed with a nontransgenic mouse of the opposite sex to begin generating transgenic progeny. Although nontransgenic littermates can be used for this purpose, there is always a small chance that a transgenic animal was incorrectly diagnosed as a nontransgenic animal. Mating the two animals would then cross the two lines and pure lines from each founder would be lost. To avoid this potential complication, unquestionable nontransgenic animals should be used. For aging studies, it is extremely important to record accurate dates of birth for each mouse born in the colony. Nontransgenic littermates can be used as control animals if you are willing to accept the low risk of misdiagnosing transgenic progeny as nontransgenic. As hemizygous transgenic progeny are obtained, they can be entered into breeding schemes until sufficient numbers of animals have been born to secure the propagation of the line and to fulfill experimental demands. Each founder and its resulting offspring should be maintained as a separate and unique line. In a few instances, transgenic founder mice will have multiple sites of transgene integration. These will be detected by Southern blot analysis of the founder and some of its progeny. It will be necessary to separate the animals into sublines depending on which transgene integration pattern is inherited. Progeny in each of the

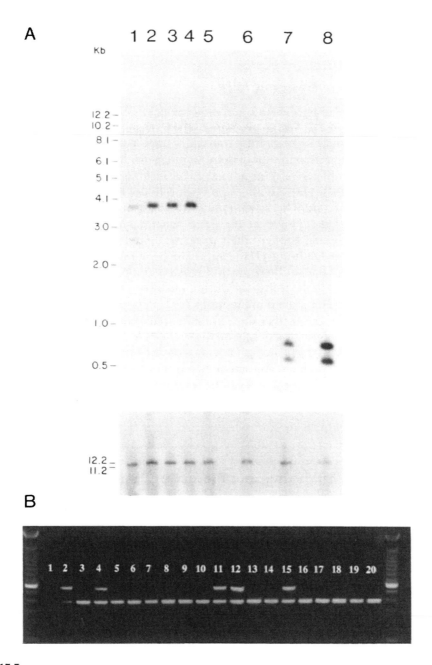

FIGURE 17.7
Identification of transgenic mice by analysis of DNA samples. **(A)** An example of a quantitative Southern blot used to characterize integration of the transgene is shown. The top panel shows hybridization with a transgene-specific probe. A founder (lane 1), progeny (lanes 2–4), a negative control (lane 5) and copy controls (lanes 6–8) are included. After transgene-specific bands are quantified the membrane is probed with an endogenous murine gene, lower panel, to correct for differences in loading or transfer. **(B)** An example of PCR amplification to identify transgenic mice. The lower band represents amplification of a portion of an endogenous murine gene. The upper band represents amplification of a portion of the transgene. Lane 1, water blank; lane 2 positive transgenic control; lane 3, negative control; lanes 4–20, sample DNAs. Lanes 4, 11, 12 and 15 are positive for the presence of the transgene.

sublines can then be used to propagate that particular subline. Failure to separate the integration sites will invalidate experimental results because some progeny will have one integration, some another integration and some both integrations of the transgene. The site and number of transgene integrations can impact on transgene expression. Thus, a greater variability in expression will be observed in experimental results if the individual integrations are not segregated. Furthermore, the experiments are not genetically controlled if the individual integrations are not segregated into sublines. Finally, it should be noted that it is not unusual for a founder to be mosaic for the transgene. This can sometimes be detected by monitoring the ratio of transgenic to nontransgenic progeny that are born from matings involving the founder. Inheritance of transgenes will follow Mendelian inheritance; thus, 50% of offspring produced by a hemizgyous transgenic animal would be expected to be transgenic. Significantly less than 50% transgenic offspring is an indication of mosaicism in the founder. Significantly greater than 50% transgenic offspring is an indication of multiple integration sites. All hemizygous progeny should transmit the transgene to 50% of their offspring, with equal distribution among male and female progeny, when mated with a nontransgenic animal. Exceptions are encountered if the transgene integrated into a sex chromosome.

Mice for aging studies should be specific-pathogen free and maintained in microisolater topped cages or in laminar flow cage racks. The presence of sentinel animals in each mouse room is essential. The sentinel animals should be regularly monitored for the presence of specific-pathogens and parasites. All experimental manipulations in which the animals will be returned to their cages should be performed in laminar flow workbenches. These precautions help prevent pathogen-mediated confounds such as an inflammatory response, that may invalidate experimental analyses.

17.5.2 Analysis of Transgene Expression

Methodologies for analyzing transgene expression are dependent on the system being developed. However, transgene expression can usually be analyzed with multiple approaches. Expression of transcripts from the transgene can be performed using northern blot analysis, RNase protection assays or reverse-transcription coupled PCR (RT-PCR) amplification. Northern blot analysis is particularly useful to ascertain the size of the transgene transcript. RNase protection assays can be quantitative and are useful when the transgene is expressed as a low abundance message. RT-PCR amplification is fast and most useful when the transgene is expressed as an extremely low abundance message or when transgene expression is limited to a subset of cells within an organ/tissue.

Reporter gene expression can be assayed at the protein level by several methods. First, the activity of the protein can be measured. Second, ELISAs can often be used to quantify the level of expressed protein. Third, antisera/antibodies are available for most reporter proteins and can be used in immunohistochemical protocols to indirectly visualize protein expression in specific tissues/cells. Methods for analyzing non-reporter gene products will have to be determined individually. In summary, it is usually possible to analyze transgene expression at the RNA and protein levels. Protein level analyses can include measurements of the amount of protein product and activity assays. Finally, a complete pathological assessment of transgenic animals used in aging studies is recommended. There is a paucity of information on the effects of transgene expression and pathological findings in aging studies employing transgenic mice.

17.5.3 Establishing Lines Homozygous for the Transgene

Before animals are bred to homozygosity for the transgene, they are correctly called hemizygous transgenic animals because for each integration event, only one of the two homologous chromosomes carries the transgene. It is not necessary to breed animals to homozygosity to study transgene expression; however, making a line homozygous for the transgene has several benefits. First, all animals born from homozygous parents will be homozygous and the DNA from each animal does not need to be checked for the presence of the transgene. It is recommended that one animal from each litter be tested to validate the presumed transgenic genotype of the litter. Second, expression of the transgene is usually doubled in the homozygous state. Third, embryonic studies are more readily performed because DNA testing for the transgene does not have to be performed to identify transgenic mice. Fourth, fewer embryos will have to be frozen to cryopreserve the line.

Homozygous animals are generated by crossing hemizygous animals. Inheritance of the transgene will follow Mendelian inheritance so that 1/4 of the resulting progeny will be homozygous for the transgene, 1/2 will be hemizgyous for the transgene and 1/4 will be nontransgenic. PCR amplification can be used to quickly identify the transgenic animals in a litter. Afterward, DNA from each of the transgenic progeny is subjected to quantitative Southern blot analysis. DNA from a known hemizygous animal, a nontransgenic animal and a homozygous animal (if available) should be included as controls. After the blot is probed with a transgene-specific probe the blot should be stripped and reprobed with an endogenous murine gene to correct for differences between lanes in loading and/or transfer. Hybridizing bands from the transgene probe and from the endogenous murine gene should be quantified. Potential homozygous progeny are identified by comparing the ratio of the signals from a transgene band to the murine gene band. The ratio should be approximately two times greater in lanes containing DNA from animals that are potentially homozygous. Animals identified as potentially homozygous should be mated immediately to a nontransgenic mouse. This process needs to happen quickly because female mice can have a very limited reproductive life. DNA obtained from the resulting progeny should be tested for the presence of the transgene. If the potential homozygous parent is truly homozygous, all of the progeny will be transgenic. If a single pup is nontransgenic, the putative homozygous animal is not homozygous. A minimum of 10 to 12 pups should be analyzed before labeling an animal homozygous. Homozygous male and female mice can then be mated to propagate the line in a homozygous fashion.

17.5.4   Crossing Transgenic Lines

On occasion it may be desirable to cross transgenic lines carrying different transgenes to study the effect of expressing the transgenes together. This can be accomplished by crossing animals from the two lines using hemizygous, homozygous or a combination of genetic statuses. Inheritance will follow Mendelian patterns, so it will be possible to predict the number of offspring that will carry two transgenes (i.e. doubly transgenic). It could be important to monitor the obtained genetic ratios because some combinations may be lethal and this can be ascertained by comparing the ratios of the offspring.

17.5.5   Cryopreservation of Transgenic Lines

17.5.5.1   *Supplies*

1/4-cc freezing straws	Controlled-rate alcohol freezer
Alcohol-resistant labeling pen	Timer
Sterile petri dishes	Ruler
1cc syringes	Liquid nitrogen dewar

Thumb pump/mouth pipetting device   Metal forceps and spatula

Liquid nitrogen   Heat sealer

1M sucrose filter, sterilized

Cryoprotectant agent (CPA)

1.5 M Glycerol for 8-cell embryos

17.5.5.2 *Cryopreservation Procedure*

Female mice should be treated hormonally to superovulate and mated with fertile stud males (Section 17.4.3.2). Females that have copulation plugs the morning after being placed with a male mouse are sacrificed at 3-days post coitus to collect 8- to 32-cell embryos. The abdominal cavity is opened (Section 17.4.4). The uterus is removed by grasping it with fine forceps just above the cervix and cutting across the cervix with fine scissors. Pull up on the uterus and use fine scissors to cut away the membrane that lies close to the uterine horns. Cut each uterine horn just below the junction with the oviduct. Do not include the uterine-oviduct junction because flushing the uterus will be impaired if this structure is present. Transfer excised uteri to a small drop of modified Whitten's medium contained in a 35-mm sterile culture dish. Insert a 25-gauge needle attached to a 1- or 2-cc syringe into the cervical opening of the uterus and slide the needle up to the juncture of the uterine horns. Flush each horn by sliding the needle to the base and expel about 0.2 ml of modified Whitten's medium. The uterus can be held below the juncture of the uterine horns with a pair of fine forceps to prevent slipping off of the needle during flushing. Hold the flushed embryos in modified Whitten's medium until they are used for cryopreservation.

Prior to flushing embryos, 1/4 cc freezing straws should be marked with an alcohol resistant labeling pen 1 and 7 cm from the open end. The plug should be pushed down from the opposite end approximately 1 cm to allow room for heat sealing the straw. Label all straws. Select a straw, insert the plug end into the tip of a 1cc syringe, draw up 1M sucrose to the 7 cm mark, then draw 1 cm of air and 1 cm of CPA. When embryos are transferred to a sterile 35mm culture dish containing CPA start the timer. Once the embryos have settled in the bottom of the dish, they should be transferred to the CPA column of the prepared straw using a pulled pipette. Afterward draw up 1cm of air, 1cm of CPA then pull up more air until the sucrose wets the plug. Heat seal both ends of the straw. When the 8-cell embryos have been in CPA for 12 minutes, the straw should be placed in the alcohol freezer that is pre-chilled to –7.5°C. After all of the straws are in the freezer the sucrose and the CPA (without the embryos) columns are seeded by touching with forceps or spatulas that have been prechilled to liquid nitrogen temperature. *Caution:* Do not seed the embryo column itself. When ice crystals are present throughout the straw an automated freezing program can be initiated. Leave the straws in the freezer for 15 min after the final temperature is achieved then plunge them immediately into liquid nitrogen.

Embryos are thawed by removing the straw from liquid nitrogen and letting it stand at room temperature until surface condensation is gone. Do not allow the column containing the embryos to touch any surface as this could alter the rate of embryo thawing and decrease viability. Grasp the 1/4cc straw at the end opposite of the plug. Shake the straw downward so that the CPA and sucrose solutions mix. A single column of solution and one air bubble should be visible in the straw if correctly performed. Submerge the straw in a 37°C water bath with the plug end inserted first. Incubate for 3 minutes. Remove the straw from the water bath and wipe dry. Use a dissection microscope to identify the location of the embryos in the straw, then cut the heat sealed ends with scissors. The contents of the straw are expelled into a sterile 35mm culture dish by using the style/plunger to push the plug to the end of the straw. Do not push the plug out of the straw. Alternatively, the straw can be cut below the plug and a 1cc syringe used to gently expel the medium and embryos. Transfer the embryos to modified Whitten's medium to rehydrate.

Pseudopregnant females can be obtained using the procedure in Section 17.4.2.2. Progression of development is delayed after embryos are frozen and thawed. Consequently, recipient females are mated the night prior to a transfer and they are not in synchrony with the embryos. Transfers are made to the oviducts (Section 17.4.6).

17.6 Summary of Transgenic Technology in Aging Research

The application of transgenic mouse technology to aging studies is clearly feasible. Such an approach will provide a new avenue for exploring the mechanisms involved in aging and will facilitate a more direct testing of theories of aging. The greater ability to temporally regulate transgene expression with the tetracycline binary system of gene regulation will enhance efforts to study the effects of altered gene expression on aging [20]. Transgenic technology can be applied to build on the wealth of information that descriptive studies have provided and consequently move aging research forward in an additional dimension of scientific endeavor.

Acknowledgments

It is a pleasure to thank and recognize Ms. Kim Hildreth who has been instrumental in having a successful transgenic animal facility at The University of Texas Health Science Center at San Antonio (UTHSCSA) and for her invaluable assistance in the preparation of this chapter. Drs. Damon Herbert and Frank Weaker should also be recognized for their significant contributions to the production and maintenance of transgenic mice for aging research at The UTHSCSA over the past decade. Finally, NIH should be recognized for contributions that largely supported the use of transgenic mice in aging research through grants PO1AG06872, PO3AG13319, and RO1AG13560.

References

1. Adrian, G. S., Bowman, B. H., Herbert, D. C., Weaker, F. J., Adrian, E. K., Robinson, L. K., Walter, C. A., Eddy, C. A., Riehl, R., and Pauerstein, C. J. Human transferrin. Expression and iron modulation of chimeric genes in transgenic mice, *J. Biol. Chem.*, 265, 1344, 1990.
2. Adrian, G. S., Herbert, D. C., Robinson, L. K., Walter, C. A., Buchanan, J. M., Adrian, E. K., Weaker, F. J., Eddy, C. A., Yang, F., and Bowman, B. H., Expression of a human chimeric transferrin gene in senescent transgenic mice reflects the decrease of transferrin levels in aging humans, *Biochim. Biophys. Acta*, 1132, 168, 1992.

3. Boland, E. J., Liu, Y. C., Walter, C. A., Herbert, D. C., Weaker, F. J., Odom, M. W., and Jagadeeswaran, P., Age-specific regulation of clotting factor IX gene expression in normal and transgenic mice, *Blood*, 86, 2198, 1995.

4. Adrian, G. S., Seto, E., Fischbach, K. S., Rivera, E. V., Adrian, E. K., Herbert, D. C., Walter, C. A., Weaker, F. J., and Bowman, B. H., YY1 and Sp1 transcription factors bind the human transferrin gene in an age-related manner. *J. Gerontol.* 51,B66, 1996.

5. Hogan, B., Beddington, R., Costantini, F., and Lacy, E., *Manipulating the Mouse Embryo; A Laboratory Manual*, 2nd ed., Cold Spring Harbor Labortory Press, Cold Spring Harbor, 1994.

6. Gordon, J. W., Scangos, G. A., Plotkin, D. J., Barbosa, J. A., and Ruddle, F. H., Genetic transformation of mouse embryos by microinjection of purified DNA, *Proc., Natl., Acad., Sci., U.S.A.*, 77, 7380, 1980.

7. Richardson, A., Heydari, A. R., Morgan, W. W., Nelson, J. F., Sharp, D. Z., and Walter, C. A., The use of transgenic mice in aging research, *ILAR J.*, in press, 1997.

8. Brinster, R. L., Chen, H. Y., Trumbauer, M. E., Yagle, M. K., and Palmiter, R. D., Factors affecting the efficiency of introducing foreign DNA into mice by microinjecting eggs, *Proc. Natl. Acad. Sci., U.S.A.*, 82, 4438, 1985.

9. Palmiter, R. D., Sandgren, E. P., Koeller, D. M., and Brinster, R. L., Distal regulatory elements from the mouse metallothionein locus stimulate gene expression in transgenic mice, *Molec. Cell. Biol.*, 13, 5266, 1993.

10. Bonifer, C., Yannoutsos, N., Kruger, G., Grosveld, F., and Sippel, A. E., Dissection of the locus control function located on the chicken lysozyme gene domain in transgenic mice, *Nucl. Acids Res.*, 22, 4202, 1994.

11. Wahle, E., The end of the message: 3'-end processing leading to polyadenylated messenger RNA, *BioEssays*, 14, 113, 1992.

12. Brinster, R. L., Allen, J. M., Behringer, R. R., Gelinas, R. E., and Palmiter, R. D., Introns increase transcriptional efficiency in transgenic mice, *Proc. Natl. Acad. Sci. U.S.A.*, 85, 836, 1988.

13. O'Hare, L., Benoist, C., and Breathnach, R., Transformation of mouse fibroblasts to methotrexate resistance by a recombinant plasmid expressing a prokaryotic dehydrofolate reductase, *Proc. Natl. Acad. Sci. U.S.A.*, 76, 1527, 1981.

14. Nishi, M., Ishida, Y., and Honjo, T., Expression of functional interleukin-2 receptors in human light chain/Tac transgenic mice, *Nature*, 331, 267, 1988.

15. Matsui, Y., Halter, S. A., Holt, J. T., Hogan, B. L. M., and Coffey, R. J., Development of mammary hyperplasia and neoplasia in MMTV-TGFa transgenic mice, *Cell*, 61, 1147, 1990.

16. Howes, K. A., Ransom, N., Papermaster, D. S., Lasudry, J. G. H., Albert, D. M., and Windle, J. J., Apoptosis or retinoblastoma: alternative fates of photoreceptors expressing the HPV-16 E7 gene in the presence or absence of p53. *Genes & Develop.*, 8, 1300, 1994.

17. Walter, C. A., Lu, J., Bhakta, M., Mitra, S., Dunn, W., Herbert, D. C., Weaker, F. J., Hoog, T., Garza, P., Adrian, G. S., and Kamolvarin, N., Brain and liver targeted overexpression of O^6-methylguanine-DNA methyltransferase in transgenic mice, *Carcinogen.*, 14, 1537, 1993.

18. Maniatis, T., Fritsch, E. F., and Sambrook, J., *Molecular Cloning: A Laboratory Manual*, Cold Spring Harbor Press, Cold Spring Harbor, 1982, 363.

19. Walter, C. A., Nasr-Schirf, D., and Luna, V. J., Identification of transgenic mice carrying the CAT reporter gene with PCR amplification, *BioTechniques*, 7, 1065, 1989.

20. Gossen, M. and Bujard, H., Tight control of gene expression in mammalian cells by tetracycline-responsive promoters. *Proc. Natl. Acad. Sci. U.S.A.*, 89, 5547, 1992.

18

Telomeres and Replicative Senescence

R.C. Allsopp, S.-S. Wang, N.W. Kim, and C.B. Harley

CONTENTS

18.1 Introduction

Normal eukaryotic somatic cells can only undergo a finite number of divisions *in vitro*, also known as the Hayflick limit. This phenomenon was originally described as replicative senescence at the cellular level over 30 years ago [1], and has now been established as a senescence process in higher eukaryotes [2–6]. Numerous models have been proposed to explain the cause(s) of replicative senescence (reviewed in [4,5]). One of these models is based upon the loss of telomeric DNA that occurs during aging of somatic cells, providing an intrinsic biological clock to explain replicative senescence [7].

Telomeres are the physical ends of eukaryotic chromosomes. In most eukaryotes, including protozoa, fungi, algae, slime molds, nematodes, mice, humans, and flowering plants, the ends of their chromosomes are capped by telomeric structures which are composed of tandem reiterations of simple repeats. These simple repetitive sequences have a G- and C-rich strand with a 5' to 3' orientation of the G-rich strand toward the telomere. The common structural features likely confer upon the telomeres the capacity for replication, nuclear architecture, meiotic pairing, and genome stability [8,9](reviewed in [10]).

The telomeric DNA sequence of vertebrates is (TTAGGG)n [11]. Although the terminal restriction fragment (TRF; Figure 18.1) length of human somatic cell DNA is typically in the 8–10 kbp range in young adults [12–14], about 4–5 kbp of the TRF in normal cells is estimated to be non-telomeric DNA (Figure 18.1; [15]). Thus the length of the terminal TTAGGG tract (ie. telomeric DNA length) in these cells is approximately 4-5 kbp. In addition to non-telomeric DNA sequences, the portion of the TRF proximal to the terminal TTAGGG tract is composed of degenerate TTAGGG sequences as well as short blocks of TTAGGG sequences (Figure 18.1) [12,16–18].

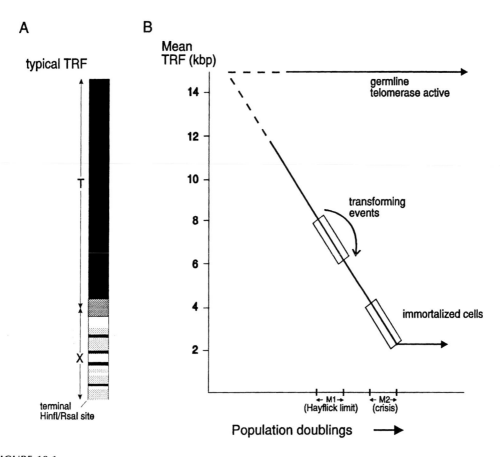

FIGURE 18.1

Structure of a typical TRF. **(A)** The TRF, generated by digesting high molecular weight DNA with restriction enzymes, is composed of a terminal (TTAGGG)n tract (T) and a proximal sequence that is devoid of cleavable restriction sites (X). The X portion of the TRF is likely composed of nontelomeric sequences (open boxes) as well as degenerate telomeric sequences (shaded boxes) and interrupted stretches of pure telomeric sequences (black boxes). **(B)** The telomere hypothesis of cell aging and immortalization. See text for details. (Reproduced from the *The Journal of Cell Biology,* 1996, Vol. 134, p.4, by copyright permission of The Rockefeller University Press.)

The continuous stretch of terminal repeats is heterogeneous on different chromosomes and in different cells. The length is thought to be determined by a dynamic equilibrium between the loss and addition of terminal repeats [19,20]. Terminal repeats can be polymerized onto the end of the chromosomes by the specialized ribonucleoprotein telomerase (reviewed in [21]). Telomerase activity has been detected in several species including ciliated protozoa, yeast, mice and humans. In humans, telomerase can be detected in germ line cells but not in most of the normal somatic cells. It is the lack of telomerase is most somatic cells that causes gradual telomere loss in these cells.

18.2 Background and Historical Perspective

The need for a special mechanism to complete telomere replication was independently recognized by Olovnikov [22,23] and Watson [24]. This prediction was based upon the inability of the DNA replication machinery to completely replicate the ends of linear DNA molecules (the "end-replication problem" [22,23]) (reviewed in [10]). Recently, this prediction has been shown to be correct in studies by us and others which have revealed that the amount of telomeric DNA decreases during replicative aging of various human somatic cell types both *in vitro* and *in vivo* [12–14,25–30]. These observations led to the telomere hypothesis of cell aging which proposes that the shortening of 1 or more telomeres below a certain length which is critical to proper telomere structure and function (the critical telomere length or Tc) will induce irreversible cell cycle arrest (cell senescence or the M1 checkpoint; Figure 18.1 B) [7].

However, there must exist some mechanism to maintain telomere length during species propagation. This mechanism is provided by telomerase, a protein/RNA complex capable of synthesizing telomeric DNA *de novo* ([31]; reviewed in [10]). The RNA component of telomerase contains a short motif which corresponds to the telomeric DNA sequence and which provides a template for the addition of telomeric DNA onto the 3′ terminus of chromosomal ends [32]. Telomerase has been detected in various human cell lines and tumor tissue [27,33–35] as well as human germ line tissue [35] and appears to be either absent or present at very low levels in most adult somatic tissues [33,35–39]. After transformed cells have bypassed cell senescence, for example by tumor suppressor gene inactivation, telomere length continues to shorten and the frequency of dicentric chromosomes increases until cells reach crisis (M2) during which time there is massive genomic instability including telomere associations (Figure 18.1 B) [27,34]. In transformed human cells which have acquired the ability to bypass the M1 checkpoint, the expression of telomerase has been shown to coincide with the maintenance of telomere length, suggesting that telomerase is essential for immortalization (i.e., the survival through crisis where most cells die) [27, 34]. However, recently a small number of apparently immortal human cell lines which have no detectable telomerase activity have been identified [35,40,41]. These observations have led to the suggestion that there may be alternative mechanisms for telomere maintenance. Overall, the data point to a critical role of telomeres in both senescence and cell crisis [7,42].

Two predictions of the telomere hypothesis are (1) the shortening of 1 or more telomeres below a critical length, Tc, will trigger cell senescence and (2), the accumulation of critically shortened telomeres to a threshold level will initiate crisis in transformed cells which have bypassed cell senescence [7]. We have previously calculated the mean telomere length at senescence, which provides an upper estimate of Tc, to be approximately 2–4 kbp [15,43]. However, if interchromosomal heterogeneity in telomere length exists as suggested in a recent study [44], then Tc will be significantly less than the mean telomere length. Since there is a large distribution of lengths for each individual telomere in a cell population at any given population doubling level (PDL) [25,40], it is even possible that cell senescence is induced by the complete loss of telomeric DNA from the ends of 1 or a few chromosomes.

In this chapter we describe different methods to measure telomere length. These methods include, (1) in-gel hybridization analysis on total genomic DNA using a telomeric probe; (2) fluorescent *in situ* hybridization (FISH) to evaluate the relative amount of telomeric DNA for individual chromosomal ends on at the cellular level; (3) DNA dot or slot blotting, a high throughput measurement of total telomere signal to monitor the total amount of telomeric DNA in a large number of genomic DNA samples.

18.3 Methods for Telomere Analysis

18.3.1 DNA Blotting

The DNA blotting method is a high throughput method to measure the total amount of TTAGGG repeat in the genome. Partial, non-specific degradation of DNA during sample handling or incomplete DNA digestion is not a concern to this method. Probed blots can be stripped and rehybridized with other probes. The amount of genomic DNA loaded in each well is normalized using human alpha satellite DNA as a probe. The background is extremely low and the hybridization signal is consistent and clear.

Protocol

1. Digest 0.5 to 2 ug of total genomic DNA from the cells of interest with restriction enzymes. Alternatively, high molecular weight DNA can be mechanically sheared. Although DNA can be blotted directly, we find that results are more reproducible with cut and re-purified DNA.

2. Purify the digested DNA by phenol/chloroform extraction followed by ethanol precipitation using standard procedures.

3. Dissolve the digested genomic DNA in 100 ul of 1× TE. Bring the volume to 300 ul with 200 ul of 5× SSC. Mix the solution thoroughly.

4. Assemble a slot blot apparatus (e.g., Schleicher & Schuell Minifold II) as recommended by the manufacturer. Wet the blotting and nitrocellulose papers with 5× SSC before placing onto the apparatus. Rinse each slot by loading 300 ul 5× SSC in wells and applying house vacuum to the apparatus to filter the liquid through slowly. Turn the vacuum off before loading the DNA samples into the slots.

5. Load all the digested genomic DNA samples (each in 300 ul solution) onto the designated slots.

6. Take the apparatus apart and using forceps place the blot on a piece of 3 MM paper and air dry for 30 min.

7. Denature the DNA by placing the blot on a stack of 3 MM paper soaked with 0.5 N NaOH/1.0 M NaCl and let rest for 30 min.

8. Neutralize the DNA by transferring the blot to another stack of 3MM paper soaked with 0.5 M Tris at pH8.0/1.0 M NaCl for 30 min.

9. Cross-link the denatured DNA to the blot with UV light (eg. the Stratalinker 1800, Stratagene).

10. Prehybridize the blot at 37°C with 15 ml of 5× SSC/5× Denhardt's solution/0.02 M phosphate pH6.5/0.1 mg/ml salmon sperm DNA/0.5% SDS/50% formamide for 2 hours.

11. Hybridize the blot with 15 ml of the prehybridization buffer plus ^{32}P-end labeled TTAGGG 18mer as a probe (10^6 cpm/ml). Incubate the blot at 37°C overnight.

12. Wash the blot with 500 ml 1× SSC/0.1% SDS once for 10 min at room temp followed by two washes with 500 ml of 0.1×SSC/0.1%SDS at 37°C for 20 min each time.

13. Expose the blot to a PhosphoImager screen overnight (e.g., PhosphoImager, Molecular Dynamics).

14. Analyze the intensity of the hybridization signal.

15. Strip the blot by immersing in 0.5L of boiling water and allow to cool to ~37°C. Rehybridize the same blot with labeled human alpha satellite DNA to normalize the amount of genomic DNA [43] (repeat steps 10 through 14).

Typical Results

High molecular weight DNA was isolated from BJ cells, a human diploid fibroblast (HDF) strain established from neonatal foreskin, at various PDLs and 1.5 ug was digested with *Eco*RI. The digested DNA was blotted, hybridized, washed and exposed as described above. The signal intensity for the TTAGGG hybridization was normalized to the signal from human alpha satellite hybridization and plotted as a function of PDL (Figure 18.2). The slot blotting method shows a decrease in average telomere length with increasing PDL, which is consistent with the results from TRF length analysis for this same HDF strain using Southern hybridization (Figure 18.3 A). The difference in telomere length from cells at various replicative ages is distinguished clearly. Additionally, a linear dilution series of a plasmid containing a (TTAGGG)n insert can also be blotted to allow construction of a standard curve. The standard curve, together with the knowledge of the total DNA applied, can be used to convert the normalized TTAGGG signal intensity of the genomic DNA samples to an estimate of the mean telomere length for all chromosome ends.

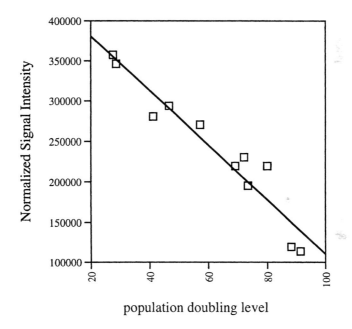

FIGURE 18.2
Analysis of telomere signal intensity for a human diploid fibroblast (HDF) strain at various population doublings using DNA blotting. High molecular weight DNA was isolated from strain BJ at various PDLs ranging from 20 to 90. DNA blotting was carried out as described in the text using a slot blot apparatus. The normalized telomeric signal intensity in arbitrary units is plotted as a function of PDLs. A tight, linear decrease in telomere length is observed. Note that the axes do not begin at zero.

18.3.2. TRF Length Analysis

The terminal restriction fragment or TRF is generated by digesting high molecular weight genomic DNA with restriction enzymes. To reduce the amount of nontelomeric DNA in the TRF (Figure 18.1), it is best to digest genomic DNA with two or more 4-base cutting enzymes. TRF length analysis involves resolving the digested DNA by electrophoresis in 0.5% agarose gels. The gels are then dried down and probed with 32P-labeled TTAGGG

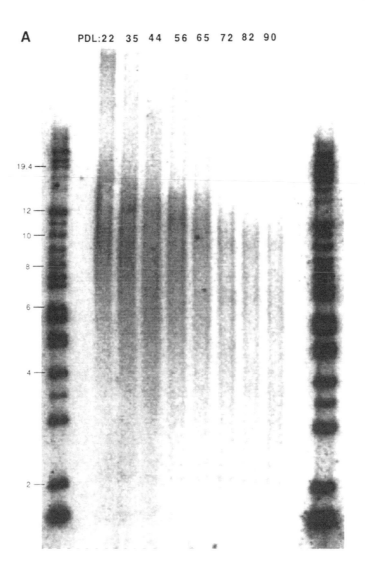

FIGURE 18.3
Calculation of mean TRF length and approximation of mean telomere length. **(A)** Sample image of TRF lengths from DNA isolated from the fetal HDF strain BJ at various PDLs. The DNA was digested with the restriction enzymes HinfI and RsaI and 1 ug of digested DNA was loaded per lane. Position of size markers are indicated on the side. **(B)** Positioning of a typical grid used in mean TRF length analysis. Li is the length in kbp corresponding to the center of row i, ODi is the total signal intensity for a given grid cell i, and n is the number of rows. See text for more details. **(C)** Plot of mean TRF length as a function of population doubling level for the image shown in A and B. **(D)** Plot of the total telomeric signal intensity for each sample as a function of population doubling level.

oligonucleotide to allow detection of the TRFs. Internal restriction fragments containing TTAGGG blocks are also detected, but when 4-base cutters are used to digest the genomic DNA, these fragments are typically small and do not interfere with TRF length analysis. After hybridization, the gel is washed and exposed to autoradiographic film or a PhosphorImager screen (Molecular Dynamics). The detection limit for this method of analysis is approximately 0.1 ug for human cells having a mean TRF length in the range of 5–7 kbp (see below). An example of the image obtained is presented in Figure 18.3 which shows the TRF length distribution at increasing population doublings for the HDF strain BJ.

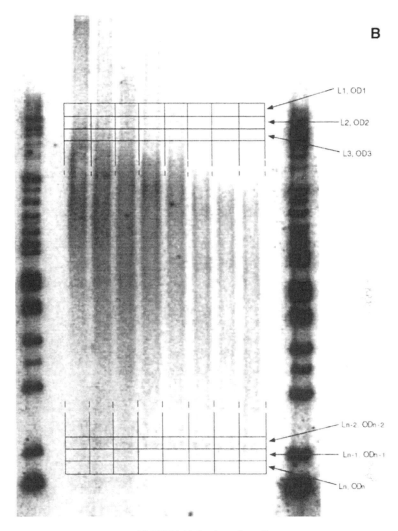

B

L1, OD1
L2, OD2
L3, OD3

Ln-2, ODn-2
Ln-1, ODn-1
Ln, ODn

FIGURE 18.3 (continued)

For all human cell populations, there is a broad distribution of TRF lengths, generally ranging from 20 kbp to 2 kbp (see Figure 18.3). To allow quantitative analysis of different DNA samples, the mean of each TRF distribution is calculated as will be described (see Figure 18.3 B). The mean TRF length and telomeric signal intensity as a function of population doublings are shown in Figure 18.3 C and 18.3 D, respectively. For a given cell type, the limit of sensitivity for detection of a change in telomere length by this method is on the order of ≈0.5 kbp, but can be less for cells with short telomeres.

Protocol

1. Digest total genomic DNA with restriction enzymes to generate terminal restriction fragments.
2. Load 0.5–2.0 ug of each digested DNA sample on 0.5% agarose gel.
3. Perform electrophoresis for 600-800 V-hr, typically at 25 V for 30 hrs.

4. Slide gel onto Whatman paper and dry gel on a gel-dryer. It is important to thoroughly dry the gel to prevent background problems.

5. Carefully remove gel from Whatman paper. Soaking the Whatman and gel with ddH2O can facilitate this. Pouring the agarose gel on Gel-Bond can also help stabilize the gel. Denature the gel in 0.5 M NaOH/ 1.0M NaCl for 10 min. followed by neutralization in 0.5 M Tris pH 8/ 1.0M NaCl for 10 min.

6. Transfer the gel to a plastic bag, add hybridization solution (see below), seal bag and hybridize over night at 37 C.

7. Remove gel from bag and wash twice at room temperature in 0.3X SSC and once at 37 C in 0.5X SSC (10 min. each).

8. Expose gel to autoradiographic film or PhosphorImager screen. For quantitative analysis when doing an autoradiographic exposure, it is best to use pre-flashed X-ray film.

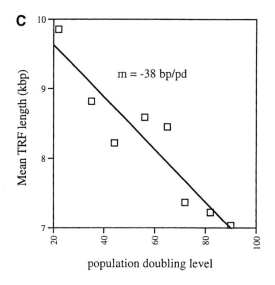

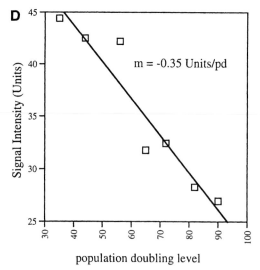

FIGURE 18.3 **(continued)**

Hybridization Solution

5X SSC buffer

5X Denhardt's solution

0.01M sodium pyrophosphate

≈1 pmol/ml 32P end-labeled (TTAGGG) oligonucleotide (specific activity ≈ 1600 Ci/mmol)

For a 25 × 25 cm² gel, use 20 to 30 ml of hybridization buffer.

The telomere length for each sample is estimated by computing the mean TRF length. Mean TRF length analysis is most easily done on the scanned image, but can also be done a densitometric scan of the autoradiogram. In addition, quantitation is more accurate when a PhosphorImager is used because of improved sensitivity and dynamic range. The mean TRF length for each sample is calculated by integrating the signal intensity over the entire TRF distribution as a function of TRF length and using the formula,

$$L = \sum(ODi{\cdot}Li)/\sum(Li)$$

where ODi and Li are the signal intensity and TRF length respectively at position i on the gel image. We use this formula, which gives equal weight to all TRFs for a given sample, since previous work by us and others has shown that the large distribution of TRF lengths for human cells within one population is mostly due to variability in size of the sub-telomeric portion of the TRF [15].

Calculation of Mean TRF Length

1. A grid consisting of X columns, where X is the number of samples, and many rows is overlayed on the gel image (see Figure 18.3 B). Because the mean TRF length calculation is an integrative process, the number of rows (typically 30) used in the grid needs to be large enough to ensure accuracy.

2. The top and bottom rows are used to calculate the background above and below the TRF distribution respectively and thus are positioned such that they fall in the regions above and below the TRF distribution. If there are large differences in telomere length between samples on the same gel then it is best to use more than one grid.

3. For each sample, OD and L are computed for each cell, where OD is the total signal intensity within a cell and L is the size, in kbp, at the mid-point of the cell. Mean TRF length is then calculated using the above equation. The determination of L must be done accurately with a standard curve and known size markers.

TRF length should not be confused with telomere length, the length of the terminal TTAGGG tract. However, TRF length analysis can be used to calculate mean length of the TTAGGG portion of the TRF if quantitative measurements of both the rate of decrease of mean TRF length and telomeric signal intensity as a function of PDL are available, for example digestion with Bal31 or other exonuclease (Figure 18.3 C and Figure 18.3 D; also see references [15], [28] and [44]). The ratio of these rates can then be used to convert the signal intensity at any particular PDL to the mean telomere length by the following formula:

$$\text{mean telomere length} = (\Delta bp/\Delta U)^*Ui$$

where Δbp is the rate of mean TRF length decrease, ΔU is the rate of signal intensity decrease and Ui is the telomeric signal intensity of the DNA sample at a given PDL. For the example shown in Figure 18.3, Δbp is –38 bp/pd and ΔU is –0.35 Units/pd, yielding a value for the ratio Δbp/ΔU of 108 bp/Unit. This ratio can be directly used to calculate mean telomere length. For example, at senescence (PDL 90; see Figure 18.3), the estimate for the mean telomere length for the HDF strain BJ is ≈108 bp/Unit × 27 Units or ≈ 2.9 kbp. This corresponds to an 'X' region (Figure 18.1) of ≈4.5 kbp for this donor.

Similar methods for performing TRF length analysis was incorporated into the TeloQuant™ Telomere Length Assay Kit (PharMingen) which is now commercially available.

18.3.3. Fluorescence *In Situ* Hybridization (FISH)

FISH provides a means for interchromosomal and intercellular comparison of telomere length. The fluorescent signal intensity of telomeres can be quantitatively analyzed using a suitable CCD camera and image analysis software system. Telomeres can be analyzed either in interphase nuclei or in metaphase chromosome spreads (Figure 18.4 A). The latter method allows analysis of telomere length for a specific chromosome or set of chromosomes. For example, DAPI/distamycin A staining of the metaphase spread shown in Figure 18.4B allows the identification of chromosomes 1, 9, 15 and 16. We and others have used FISH to detect telomere shortening during division of normal human cells and for analysis of interchromosomal variability in telomere length [44,45]. We have found FISH to be a sensitive method for detection of telomere shortening.

Protocol—Standard Probes

1. To prepare metaphase spreads, cells are harvested, incubated in hypotonic KCl (0.075M) for 20–30 min., fixed in Carnoy's solution (3:1 methanol/glacial acetic acid) and dropped onto clean microscope slides. The slides are allowed to air dry over night. A quick check of the quality of the metaphase spreads by Giemsa staining should be done 1–2 hours after dropping the cells onto slides. FISH should be carried out within a few days after preparing the slides. For long term storage, the slides can be kept at –20°C under nitrogen gas.

2. Slides are immersed for 2–5 min. in 50% formamide/2X SSC (pH 7.2) pre-heated to 72°C to denature chromosomal DNA.

3. Slides are immediately tranferred from the denaturing buffer to ice cold 70% ethanol.

4. Slides are incubated for 3 min. in the ice-cold 70% ethanol followed by two more washes in ice cold 95% ethanol and then 100% ethanol (3 min. each).

5. Slides are allowed to air dry at room temp. While the slides are drying, the labeled telomeric DNA probe is dried down in a rotary evaporator (e.g., Speed-Vac). The probe label can either be a biotin- or digoxygenin (DIG)-tagged nucleotide (from Sigma or Boerhinger Mannheim respectively) for indirect probe detection, or a fluorescein (FITC)- or rhodamine-tagged nucleotide (Sigma) for direct detection (other fluorophors can also be used). The labeled-nucleotide may be incorporated into the telomeric DNA probe either by nick translation or random primer extension. In either case, the average size of the ssDNA fragments subsequent to labeling should be between 200 and 500 bp.

We will describe here the protocol we use when performing FISH with a DIG labeled probe.

6. Formamide is added to the dried probe to yield a final concentration of 20–30 ng/ul followed by incubation at 75°C for 10 min to denature the probe.

7. The probe is immediately transferred to ice. After 1 min. on ice, an equal volume of hybridization mix is added to the denatured probe. The hybridization mix is composed of 4× SSC buffer, 8× Denhardt's solution and 100mM Tris-HCl (pH 7.2).

8. Slides are tranferred to a heating block set at 37°C. After a few minutes of heating, a drop (20 ul) of the probe/hybridization mix is added to each slide over the region containing the chromosome spreads and a 18 × 18 mm² cover slip is laid on top of the drop. The cover slip is sealed onto the slide with rubber cement and the slides are incubated in a humid chamber set at 37°C for 16-24 hours.

9. The coverslips are removed from the slides and the slides are washed twice in 50% formamide/2X SSC buffer at 42°C (10 min. each).

10. Slides are subsequently washed twice in 2X SSC at room temp. (30 min. each).

11. Fifty ul of blocking buffer consisting of 0.1 M Na_2HPO_4 pH 8.0/0.5% NP40 is added to the slides. The buffer is overlayed with a 30 × 20 mm² coverslip followed by a 5 min. incubation at room temp. The coverslip is removed and 50 ul of blocking buffer containing FITC labeled anti-DIG Fab fragments (final concentration of 25 ug/ml) (Boerhinger Mannheim) is added. The buffer is overlayed with a coverslip followed by a 20 min. incubation at room temp. Alternatively, a monoclonal anti-DIG antibody can be used at his step followed by successive staining with DIG labeled anti-mouse antibody and FITC labeled anti-DIG Fab fragments (Boerhinger Mannheim) [44]. This protocol provides further amplification of the signal.

12. The coverslip is removed and the slide is washed consecutively in 4X SSC, 4X SSC/0.1% Triton X100, 4X SSC, 0.1 M Na_2NPO_4 pH 8.0/0.5% NP40 at room temp (5 min. each).

13. At this point, 10 ul of antifade (e.g., 9:1 glycerol:PBS containing 0.1% p-phemylenediamine buffered to pH 8.0 with 0.5 M carbonate/bicarbonate buffer [44]) is added to the slide. Also, the chromosomes are generally counterstained at this time. Common counterstains are propidium iodide and DAPI which can be added directly to the antifade at final concentrations of 0.2–1 ug/ml. The antifade is overlayed with a 18 × 18 mm² coverslip which should be gently squeezed against the slide to force excess antifade out from under the coverslip. The slide is now ready to be viewed under the fluorescent microscope.

Telomeres can also be detected by FISH using recently developed peptide nucleic acid (PNA) oligomers as a probe. PNA oligomers consist of a peptide backbone with bases as the side chains. These oligomers bind to DNA with higher affinity and specificity than a corresponding DNA oligomer of the same size and sequence because of the neutral backbone of the PNA molecule (Figure 18.5 A and B). Since telomeres are composed of simple repetitive DNA, PNA oligomers are well suited as a probe for *in situ* detection of telomeres. A sample image of an interphase nucleus and metaphase chromosomes of the human cell strain MA is shown in Figure 18.5 C and D respectively. A protocol for detection of telomeres by FISH using a PNA probe [45] is described below.

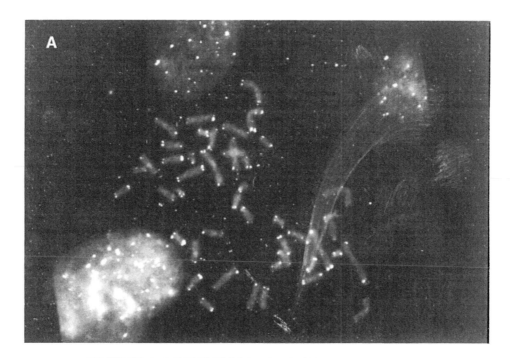

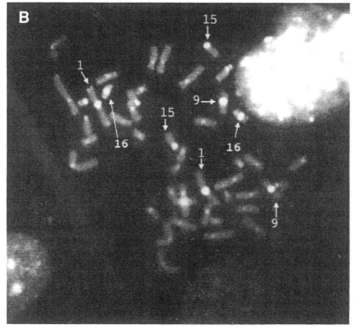

FIGURE 18.4

Analysis of telomeres on individual chromosomes using FISH. **(A)** Detection of telomeres on metaphase chromosomes from the HDF strain IMR90 (PDL 24). Cloned telomeric DNA labeled by nick translation with DIG-tagged dUTP was used as a probe. Telomeric signals are also detected in an adjacent interphase nuclei (upper right). DNA is counter-stained with propidium iodide. **(B)** Simultaneous staining with DAPI/distamycin A allows the identification of chromosomes 1, 9, 15 and 16 [51]. The chromosome is absent because this fibroblast strain was established from a female donor.

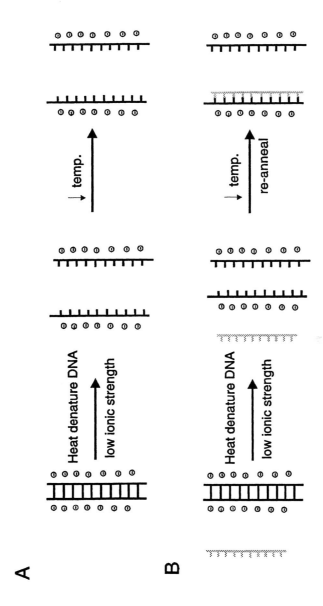

FIGURE 18.5

Hybridization of a PNA oligomer to DNA. **(A)** When double-stranded DNA (black) is denatured in a low-ionic strength solution, subsequent reannealing of the complementary single strands is not favored because of the electromagnetic repulsion of the negatively charged phosphate backbones. **(B)** If double-stranded DNA is denatured in the same solution in the presence of PNA oligomers complementary to one DNA strand, the PNA oligomer will anneal to the complementary DNA strand due to the neutral peptide backbone of the PNA. **(C)** Detection of telomeres in an interphase nucleus and metaphase chromosomes **(D)** from the human cell line MA using a FITC-tagged PNA oligomer (18mer) that is complementary to the G-rich telomeric strand.

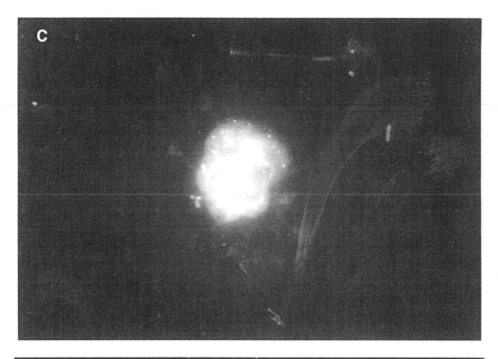

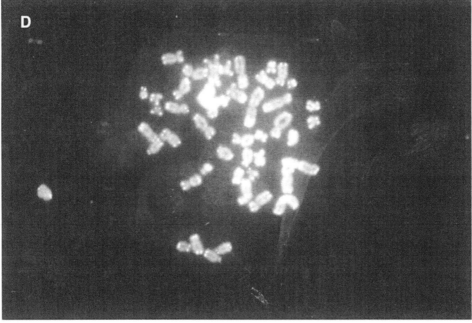

FIGURE 18.5   (continued)

Protocol—PNA Probes

1. Metaphase spreads are prepared as described above. After drying overnight, the slides are washed briefly in phosphate buffered saline (PBS) followed by further fixation in 4% formaldehyde in PBS for 2 min.

2. Slides are washed 3 times in PBS, for 5 min. and then immersed in 1 mg/ml pepsin (Sigma), pH 2.0, for 10 min at 37°C.

3. Slides are washed briefly in PBS followed by a repeat of the fixation in 4% formaldehyde and the PBS washes.

4. Slides are allowed to drain briefly and 10 ul of hybridization mix is added to each slide. The hybridization mix is composed of 70% formamide, 0.3 ug/ml FITC-$(C_3TA_2)_3$ PNA probe (Perseptive Biosystems, Bedford, MA), 1% blocking reagent (Boerhinger- Mannheim) and 10 mM Tris pH 7.2. The hybridization mix is cover with a 18×18 mm^2 coverslip and sealed with rubber cement.

5. Slides are heated to 80°C for 3 min. followed by incubation at room temp. for 2 hours.

6. Slides are washed twice in 70% formamide/10 mM Tris pH 7.2 (15 min. each) and then washed 3 times with 0.05M Tris, 0.15M NaCl pH 7.5 containing 0.05% Tween 20 (5 min. each).

7. Slides are drained briefly and 10 ul of antifade is added to the slide as described above. The slides are now ready for viewing.

18.4 Applications of Telomere Length Analysis

For many normal somatic cells, the length of telomeres present a unique record of their replicative history. Like the odometer in an automobile, telomere length reflects how much a cell has replicated in the past, and predicts how long the cell will replicate in to the future. This characteristic presents a novel opportunity for utilizing telomere length in studying the replicative history of certain cell populations in the body, and for diagnosis and prognosis of a wide range of diseases.

In normal cells, the main cell population that retains its telomere length is the germ line cells [26]. This observation is logical since the maintenance of telomere length, and thereby passing the complete set of genetic material to the next generation, is a critical factor in the propagation of eukaryotic organisms. It is observed that during development, most somatic cells experience repression of telomerase which triggers a gradual loss in their telomere length as they continue to grow and differentiate. The relationship between age of human subjects and telomere length has been shown in number of studies [12–14,25–30,44]. These results indicate that length of telomeres is a marker for biological age of an individual or tissue, which is an application in determining whether a person has a higher risk of age-related disease, such as cardiovascular disease, and in forensic applications Telomere lengths are shorter in vascular tissues with high cellular turnover due to hemodynamic stress, thus are more susceptible to atherosclerosis in than vascular tissues with less cellular turnover [30]. Telomere lengths of subpopulations of CD8+ T cells from patients with AIDS were also recently shown to be shorter than telomere lengths of T cells from a control group [46,48]. These data demonstrate that a diagnostic and prognostic test based on telomere length measurement for predicting the onset of cardio-vascular diseases and AIDS could have significant impact on the treatments of these diseases. Furthermore, since there is a gradual decrease in telomere lengths of cells from fetal tissue, to cord blood, to adult bone marrow [29], the difference in telomere lengths between fetal and maternal cells can be used to identify fetal cells circulating in maternal blood. This telomere-based fetal also has application in non-invasive blood tests to detect genetic abnormalities in the fetus.

Most stem cell populations in the body possess longer telomeres than other somatic cells, probably relating to a large proportion of stem cells that are in quiescent state, which experience gradual shortening of telomeres as they enter the proliferative pool, divide and differentiate [29,38]. This characteristic shows that telomere length can also be a differentiation marker which could be used to identify and purify rare population of pleuripotent stem cells. Identification and purification of pleuripotent stem cells or early progenitor cells by telomere length has utility in tissue regeneration and bone marrow transplantation.

The average telomere length of cancer cells is generally shorter than that of their normal counterparts and short telomere lengths in tumors can reflect more advanced cancers [49,50]. Thus, a cancer diagnostic assay based on the measurement of telomere length, especially in an *in situ* hybridization format, may be of significant interest. Furthermore, telomere length measurement will provide important information in determining the maximum replicative capacity of cancer cells that are being subjected to cancer therapies, including a telomerase inhibitor therapy. This parameter will determine how long telomerase inhibitor therapy or other therapeutics have to be given to a patient, and will aid in determining the best course of treatment for the disease.

18.5 Summary and Conclusion

Telomeres are essential genetic elements that cap chromosome ends, distinguishing them from chromosome breaks and preventing abnormal recombination. Thus, anything that alters telomere length or integrity can impact human health through effects on cell function. It is clear that telomeres gradually shorten during aging of normal somatic cells *in vitro* and *in vivo*, while germ line and cancer cells maintain telomeres through expression of telomerase. Thus, it is important that both basic and applied researchers utilize sensitive, reproducible, and robust methods of measuring telomere length. We have previously promoted the use of standardized telomerase assays [35]. This chapter has reviewed three methods currently in use for assessing telomere length: (1) dot,- or slot-blot methods for total telomeric signal in a DNA preparation; (2) gel-based methods for analysis of the mean terminal restriction fragment length; and (3) fluorescent *in situ* hybridization for detection of individual telomere signals on interphase or metaphase chromosomes. The detailed protocols provided should allow greater reproducibility in results, and more reliable comparisons of data between different laboratories. Accomplishment of these goals will accelerate the discovery of novel therapeutic and diagnostic application of telomere length measurements in the diagnosis and treatment of age-related diseases, including cancer.

Acknowledgments

The authors would like thank Steve Sherwood for providing the prints for Figure 18.5, and Robin Scearce and Jill Regoezi for technical assistance. This work was funded by Geron Corporation and a scholarship from the MRC of Canada (R.A.).

References

1. Hayflick, L. and P. Moorhead, The serial cultivation of human diplid cell strains. *Exp. Cell Res.*, 25, p. 585, 1961.
2. Martin, G.M., C.A. Sprague, and C.J. Epstein, Replicative life-span of cultivated human cells. *Laboratory Investigation*, 23, p. 86, 1970.
3. Dell'orco, R.T., J.G. Mertens, and J. Kruse, P.F., Doubling potential, calendar time, and senescence of human diploid cells in culture. *Exp. Cell Res.*, 77 p. 356, 1973.
4. Stanulis-Praeger, B., Cellular senescence revisited: a review. *Mech. Ageing Devel.*, 38, p. 1, 1987.
5. Goldstein, S., Replicative senescence: the human fibroblast comes of age. *Science*, 249, p. 1129, 1990.
6. Dimri, G.P., X. Lee, G. Basile, M. Acosta, C. Scott, C. Roskelley, E. Medrano, M. Linskens, I. Rubelj, O. Pereira-Smith and J. Campisi, A biomarker identifies senescent human cells in culture and in aging skin in vivo. *Proc. Natl. Acad. Sci. U.S.A.*, 92, p. 9363, 1995.
7. Harley, C., Telomere loss: mitotic clock or genetic time bomb? *Mut. Res.*, 256, p. 271, 1991.
8. Muller, H.J., The remaking of chromosomes. collecting *Net.*, 13, p. 182, 1938.
9. McClintock, B., *The stability of broken chromosomes in Zea mays. Genetics*, 41, p. 234, 1941.
10. Blackburn, E.H., Structure and function of telomeres, *Nature*, 350, p. 569, 1991.
11. Meyne, J., R. Ratliff, and R. Moyzis, Conservation of the human telomere sequence (TTAGGG)n among vertebrates. *Proc. Natl. Acad. Sci, U.S.A.*, 86, p. 7049, 1989.
12. de Lange, T., L. Shiue, R.M. Myers, D.R. Cox, S.L. Naylor, A.M. Killery, and H.E. Varmus, *Structure and variability of human chromosome ends. Molecular and Cellular Biology,* 10, p. 518, 1990.
13. Harley, C.B., A.B. Futcher, and C.W. Greider, Telomeres shorten during ageing of human fibroblasts. *Nature*, 345, p. 458, 1990.
14. Hastie, N.D., M. Dempster, M.G. Dunlop, A.M. Thompson, D.K. Green, and R.C. Allshire, Telomere reduction in human colorectal carcinoma and with aging. *Nature*, 346, p. 866, 1990.
15. Levy, M.Z., R.C. Allsopp, A.B. Futcher, C.W. Greider and C.B. Harley, Telomere end-replication problem and cell aging. *J. Mol. Biol.*, 255, p. 951, 1992.
16. Allshire, R.C., M. Dempster, and N.D. Hastie, Human telomeres contain at least three types of G-rich repeats distributed non-randomly. *Nucl. Acids Res.*, 17, p. 4611, 1989.
17. Brown, W.R.A., P.J. MacKinnon, A. Villasante, N. Spurr, V.J. Buckle and M.J. Dobson, Structure and polymorphism of human telomere-associated DNA. *Cell*, 63, p. 119, 1990.
18. Weber, B., C. Collins, C. Robbins, R.E. Magenis, A.D. Delaney, J.W. Gray and M.R. Hayden, Characterization and organization of DNA sequences adjacent to the human telomere associated repeat (TTAGGG)$_n$. *Nucl. Acids Res.*, 18, p. 3353, 1990.
19. Blackburn, E.H. and J.W. Szostak, The molecular structure of centromeres and telomeres. *Ann. Rev. Biochem*, 53, p. 163, 1984.
20. Shampay, J. and E. Blackburn, Generation of telomere-length heterogeneity in Saccharomyces cerevisiae. *Proc. Nat. Acad. Sci. U.S.A.*, 85, p. 534, 1988.
21. Blackburn, E.H., Telomerases. *Ann. Rev. Biochem.*, 61, p. 113, 1992.
22. Olovnikov, A.M., Principle of marginotomy in template synthesis of polynucleotides. *Dokl. Acad. Nauk. S.S.S.R.*, 201, p. 1496, 1971.

23. Olovnikov, A.M., A theory of marginotomy: the incomplete copying of template margin in enzymatic synthesis of polynucleotides and biological significance of the phenomenon. *Journal Theoretical Biology,* 41, p. 181, 1973.

24. Watson, J., Origin of Concatemeric T7 DNA. *Nature New Biology,* 239, p. 197, 1972.

25. Lindsey, J., N.I. McGill, L.A. Lindsey, D.K. Greene and H.J. Cooke, *In vivo* loss of telomeric repeats with age in humans. *Mutat. Res.,* 256, p. 45, 1991.

26. Allsopp, R.C., H. Vaziri, C. Patterson, S. Goldstein, E.V. Younglai, A.B. Futcher, C.W. Greider and C.B. Harley, Telomere length predeicts replicative capacity of human fibroblasts, 89, p. 10114, 1992.

27. Counter, C.M., A.A. Avilion, C.E. LeFeuvre, N.G. Stewart, C.W. Greider, C.B. Harley and S.B. Bacchetti, Telomere shortening associated with chromosome instability is arrested in immortal cells which express telomerase activity, *EMBO J.,* 11, p. 1921, 1992.

28. Vaziri, H., F. Schachter, I. Uchida, L. Wei, X. Zhu, R. Effros, D. Cohen, and C.B. Harley, Loss of telomeric DNA during aging of normal and trisomy 21 human lymphocytes, *Am. J. Human Genetics,* 52, p. 661, 1993.

29. Vaziri, H., W. Dragowska, R.C. Allsopp, T.E. Thomas, C.B. Harley and P. Lansdorp, Evidence for a mitotic clock in human hematopoietic stem cells: Loss of telomeric DNA with age, *Proc. Natl. Acad. Sci. U.S.A.,* 91, p. 9857, 1994.

30. Chang, E. and C.B. Harley, Telomere length and replicative aging in human vascular tissues. *Proc. Natl. Acad. Sci. U.S.A.,* 92, p. 11190, 1995.

31. Greider, C. and E. Blackburn, Identification of a specific telomere terminal transferase activity in tetrahymena extracts. *Cell,* 43, p. 405, 1985.

32. Greider, C. and E. Blackburn, A telomeric sequence in the RNA of tetrahymena telomerase required for telomere repeat synthesis. *Nature,* 337, p. 331, 1989.

33. Counter, C.M., H.W. Hirte, S. Bacchetti and C.B. Harley, Telomerase activity in human ovarian carcinoma. *Proc. Natl. Acad. Sci.,* 91, p. 2900, 1994.

34. Counter, C.M., F.M. Botelho, P. Wang, C.B. Harley and S.Bacchetti, Stabilization of short telomeres and telomerase activity accompany immortalization of Epstein-Barr virus-transformed human B lymphocytes. *J. Virol.,* 68, p. 3410, 1994.

35. Kim, N.W., M.A. Piatyszek, K.R. Prowse, C.B. Harley, M.D. West, P.L. Ho, G.M. Coviello, W.E. Wright, S.L. Weinrich and J.W. Shay, Specific association of human telomerase activity with immortal cells and cancer [see comments]. *Science,* 266, p. 2011, 1994.

36. Counter, C.M., J. Gupta, C.B. Harley, B. Leber and S. Bacchetti, Telomerase activity in normal leukocytes and in hematologic malignancies. *Blood,* 85, p. 2315, 1995.

37. Broccoli, D., J.W. Young, and T. de Lange, Telomerase activity in normal and malignant hematopoietic cells. *Proc. Natl. Acad. Sci. U.S.A.,* 92, p. 9082, 1995.

38. Chiu, C.-P., W. Dragowska, N.W. Kim, H. Vaziri, J. Yui, T.E. Thomas, C.B. Harley and P.M. Lansdorp, Telomerase expression in human cells and tissues. *Aging Clinical and Exper. Res.,* 7, p. 460, 1995.

39. Hiyama, K., Y. Hirai, S. Kyiozumi, M. Akiyama, E. Hiyama, M.A. Piatyszek, J.W. Shay, S. Ishioka and M. Yamakido, Telomerase activity in human peripheral blood and bone marrow cells. *J. Immunol.,* 155, p. 3711, 1995.

40. Murnane, J.P., L. Sabatier, B.A. Marder and W.F. Morgan, Telomere dynamics in an immortal human cell line. *EMBO J.,* 13, p. 4953, 1994.

41. Bryan, T.M., A. Englezou, J. Gupta, S. Bacchetti and R.R. Reddel, Telomere elongation in immortal human cells without detectable telomerase activity. *EMBO J.,* 14, p. 4240, 1995.

42. Wright, W.E. and J.W. Shay, Time, telomeres and tumours: is cellular senescence more than an anticancer mechanism? *Cell. Biol.,* 5, p. 293, 1995.

43. Allsopp, R.C. and C.B. Harley, Evidence for a critical telomere length in senescent human fibroblasts. *Exper. Cell Res.,* 219, p. 130, 1995.

44. Henderson, S., R.C. Allsopp, D. Spector, S.-S. Wang and C.B. Harley, *In situ* analysis of changes in telomere size during replicative aging and cell transformation. *J. Cell Biol.,* 134, p. 1, 1996.

45. Lansdorp, P.M., N.P. Verwoerd, F.M. van de Rijke, V. Dragowska, M.-T. Little, R.W. Dirks, A.K. Raap and H.J. Tanke, Heterogeneity in telomere length of human chromosomes. *Human Molecular Genetics,* 5, p. 685, 1996.

46. Effros, R.B., R.C. Allsopp, C.-P. Chiu, M.A. Hausner, K. Hirji, L. Wang, C.B. Harley, B. Villeponteau, M.D. West and J.V. Georgi, Shortened telomere in the expanded CD28-CD8+ cell subset in HIV disease implicate replicative senescence in HIV pathogenesis, *AIDS*, 10, p. F17, 1996.

47. Saltman, D., R. Morgan, M.L. Cleary and T. de Lange, Telomeric structure in cells with telomeric associations, *Chromosoma*, 102, p.121, 1993.

48. Palmer, L.D., N. Weng, B.L. Levine, C.H. June, H.C. Lane and R.J. Hodes, Telomere length, telomerase activity, and replicative potential in HIV infection: analysis of CD4+ and CD8+ T cells from HIV-discordant monozygotic twins, *J. Exp. Med.*, 185, p. 1381, 1997.

49. Hiyama, K., S. Ishioka, Y. Shirotani, K. Inai, E. Hiyama, I. Murakami, T. Isobe, T. Inamizu and M. Yamakido, Alterations in telomeric repeat length in lung cancer are associated with loss of heterozygosity in p53 and Rb. *Oncogene*, 10, p. 937, 1995.

50. Hiyama, E., T. Yokoyama, N. Tatsumoto, K. Hiyama, Y. Imamura, I. Murakami, T. Kodama, M.A. Piatyszek, J.W. Shay and Y. Matsuura, Telomerase activity in gastric cancer. *Cancer Res.*, 55, p. 3258, 1995.

51. Schweizer, D., P. Ambros, and M. Andrle, Modification of DAPI banding on human chromosomes by prestaining with a DNA-binding oligopepetide antibiotic, distamycin A, 111, p. 327, 1978.

19

Evolutionary Approaches to Probing Aging Mechanisms

Steven N. Austad and Donna J. Holmes

CONTENTS

19.1 Introduction

In purely conceptual terms, the evolution of senescence is well understood as a genetic consequence of natural selection's declining power as a function of advancing postpubertal age [1,2]. Specifically, these genetic consequences are: (1) mutation accumulation, the genomic accumulation, by drift, of alleles with neutral effects in early life and deleterious effects in late life, and (2) antagonistic pleiotropy, adaptive selection for alleles with salubrious effects early in life but injurious effects later on. It is also understood that aging mechanisms due to mutation accumulation are expected to be idiosyncratic, differing in type and degree among species and even among genetic lineages within species. By contrast, assuming that there are some manageable number of processes whereby early life benefits lead to late life damage, aging mechanisms associated with antagonistic pleiotropy are expected to be more general in nature [3], spanning a wide array of species.

Conceptual understanding, however, does not necessarily signify a similar understanding of the physiological or molecular mechanisms involved in aging. Such aging mechanisms are still largely unknown, although a multitude of hypothetical processes exist. However, evolutionary biology can provide more than a conceptual understanding of natural processes. Particularly in its embodiment as comparative biology, an evolutionary approach is indeed capable of providing useful tools for probing physiological and molecular mechanisms of aging.

We define comparative biology as the assumption that differences or similarities among species or populations can provide insight into both patterns and processes of nature. As such, comparative biology has a long and distinguished history, most notably Darwin's discovery of the evolutionary mechanism of natural selection. Even within the field of aging research, much of the early thinking about aging processes sprang from explicit comparisons among species [4–7]. Yet the utility, procedural details, and potential pitfalls of the comparative approach, especially for investigating aging mechanisms, remain little appreciated within the gerontological community [8]. It is the purpose of this paper to provide general guidelines for the implementation of comparative biology in the investigation of physiological or molecular mechanisms of aging.

The general uses of the comparative approach in aging research have been explored elsewhere [9]. However, for the specific purpose of probing mechanisms of aging experimentally, comparative biology can play at least three major roles. First, by surveying patterns of mechanisms among a range of species with a range of aging rates, it can help evaluate the potential importance of putative aging processes [10,11]. It is important to note, however, that because such studies can only be correlative in nature, these processes should be validated by experiments before being generally accepted. Second, comparative biology can be used to assess the generality of aging mechanisms identified in one or a few strains or species. As previously mentioned, some aging mechanisms are expected to be idiosyncratic, that is, confined to a narrow range of species, and others more general in nature. In using animal models to investigate aging processes, researchers make the implicit assumption that findings with their particular model will have relevance beyond the species studied. Yet an explicit methodology for assessing the general relevance of findings with a particular model has seldom been specified. Again, a comparative perspective can provide such a methodology [9].

A third use of a comparative evolutionary approach is to assist in identifying key mechanisms of aging from an array of possible mechanisms. For instance, a common research tactic for investigating putative aging mechanisms is to choose, or create by some means, experimental populations that age at different rates. Then by searching for parallel differences in physiological, biochemical, or molecular processes among these populations, we try to identify key aging mechanisms. However, a common result of such a research design is that the populations differ in so many aspects of their physiology, biochemistry, or gene expression that determining which of the differences is causally involved in aging proves difficult. It is this phenomenon of multiple differences, for instance, which has contributed to the difficulty in determining the mechanism(s) by which caloric restriction retards aging in laboratory rodents, even though that effect has been investigated for more than 60 years. Comparative biology can assist in this endeavor, as will be shown below.

19.2 Background

19.2.1 Historical Use of Comparative Method in Aging Research

Comparative biology has a long history in aging research, although its use has been chiefly confined to formulating, or empirically evaluating, general hypotheses about aging processes from survey data rather than from specifically designed experiments. Typically, such research has consisted of determining how some measure of aging rate, such as maximum captive longevity, correlates across species with some parameter thought to be important in determining aging rate, such mass-specific metabolic rate [5] or relative brain size [12]. These types of analyses have been useful even though they have not led directly to identification of aging mechanisms. For instance, comparative work on the putative effect of metabolic rate on longevity by a number of authors has likely led to our current, more experimental, focus on aging as mediated by by-products of metabolism such as reactive oxygen species and glycated proteins.

19.2.2 Potential Problems in the Comparative Method

Comparative approaches must always be implemented with certain caveats in mind. Simply enumerating variables for an array of available species is generally not sufficient. To illustrate the potential difficulties of utilizing a comparative approach, we will select examples from the aging literature of what we interpret as flawed analyses, including some of our own. Our examples are not chosen because they represent particularly weak studies, but because they depict particularly insidious complications of comparative analysis.

First, as every elementary statistics book cautions, causation can not be inferred from correlation alone, regardless of the strength of that correlation. Misunderstandings on this point at one time led aging researchers to the conclusion that aging processes were influenced largely by brain size, simply because relative brain size in mammals was apparently more strongly correlated with maximum captive longevity than was total body mass itself [12–14]. Besides representing a lapse in statistical logic, the empirical credibility of such a claim was seriously undercut when it was noted that spleen, heart, liver, and kidney mass all correlate as closely, or more closely, with maximum longevity than does brain mass [15].

Second, one must ensure than any two variables to be correlated in studies such as those described above will not be confounded by *necessary lack of independence* between the variables. For instance, Harvey and Zammuto [16] noted a striking correlation between age at first reproduction and life expectancy in a sample of mammal species. This correlation remained even after statistically removing the effects of body size. However, since one component of life expectancy consists of the time to reproductive maturity, animals with a late age at maturity cannot be very short-lived, therefore the correlation is not really surprising or informative. A second example of the same insidious problem can be found in Tolmasoff et al. [17]. They investigated superoxide dismutase activity in several tissues and attempted to correlate their results with maximum reported longevity in a sample of 14 mammal species. No correlation was found, unless they divided superoxide dismutase (= SOD) activity by mass-specific metabolic rate. While this makes intuitive sense (it is metabolism after all which accounts for superoxide ion production), it also leads to an automatic confound of the results: well-known negative correlation already exists between mammalian longevity and mass-specific metabolism. Therefore, the positive correlation observed was virtually assured a priori; unless SOD activity exhibited a compensatory increase as activity as metabolic rate declined.

Third, one must beware of artifacts of phylogenetic sampling. Phylogeny, or the evolutionary relationship among species or populations, should never be ignored in comparative studies, whether those studies are observational or experimental. Phylogenetic relationships can be most easily visualized as branching diagrams representing speciation events (Figure 19.1). Ignoring phylogeny has probably been the greatest failing of past comparative research in evolutionary biology. However during the past decade, several advances have made the consideration of phylogeny in comparative studies possible as well as more feasible. For one thing, the advent of molecular techniques for estimating

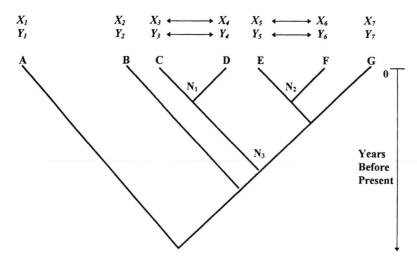

FIGURE 19.1
Phylogeny of seven hypothetical species (A–G). N_1 and N_2 are nodes, which represent the ancestral species from which species C, D and E, F, respectively, diverged, as well as the time of divergence. X and Y notation above the phylogeny refers to hypothetical values for two traits observed in each of the seven species. The significance of these trait values is discussed in Section 19.3.1.

phylogeny [18] have greatly increased our scope of knowledge about phylogenetic relationships throughout the animal kingdom. Also, a range of new analytical techniques have been developed specifically for statistical tests of hypotheses from comparative survey data [19,20].

Phylogenetic sampling artifacts can be of two types. First, a biased sample of species (with respect to, say, aging rate) can be selected for analysis. For instance, imagine an investigator hypothesizes that a particular gene product is causally involved in aging. One preliminary test of this might be to see how mRNA levels for this product vary as a function of some measure of aging (such as longevity or Gompertzian mortality acceleration) in a sample of mammalian species. If the investigator chose only a few rodent and a few primate species, then any gene product that differed consistently between rodents and primates would appear to be correlated with aging rate, (even if it was not even weakly related to aging rate within either group), for no other reason than that primates are generally longer-lived than rodents (Figure 19.2). Note that this problem does not depend on the number of species used in the analysis, but only on the breadth of their phylogenetic distribution.

One alternative way to interpret the above problem is to note that because of evolutionary affinity, each species can not necessarily be treated as an independent data point for statistical analysis and therefore some sort of phylogenetic correction may be needed. The same problem may take other guises, but the problem is ultimately whether there are inflated statistical degrees of freedom in the analysis because of the evolutionary relationships among species used. For instance, consider the hypothesis that arboreal rodents are longer-lived than their terrestrial relatives (Figure 19.3). A quick graphical analysis seems to support the hypothesis, in that all eleven arboreal species are longer-lived that expected for rodents. Any statistical test which treated species as independent data points would likely confirm the validity of this hypothesis. However, six of the eleven arboreal species come from a single rodent family, and four of these six from North American representatives of that family. Therefore, arboreality and long life likely evolved in concert far fewer than eleven times, and some phylogenetic correction is called for.

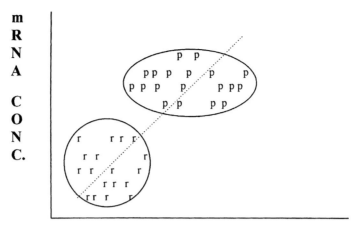

LONGEVITY

FIGURE 19.2

Example of phylogenetically biased sampling. Note that any variable (in this example mRNA concentration for a specific protein) that differs consistently between rodents and primates will be correlated with longevity even though there is no correlation within primates or rodents, if only because primates generally are longer-lived. Very little can be inferred from data with this sort of sampling bias. (r = rodent, p = primate)

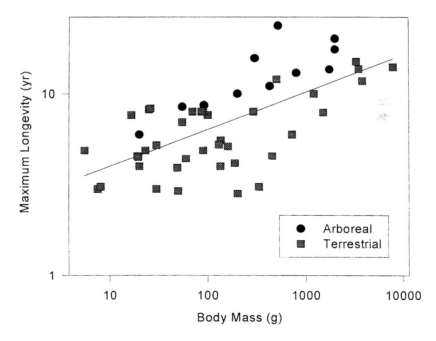

FIGURE 19.3

Another type of phylogenetically biased sampling. Longevity of terrestrial versus arboreal rodent species (Austad, unpublished data). This figure illustrates how lack of phylogenetic independence creates the appearance of a trend where none may exist. Statistically the probability that all eleven arboreal species are longer-lived than expected for rodents as a whole is unlikely to occur due to chance alone. However, nine of the eleven arboreal species are New World squirrels. Because they are all closely related, they probably represent only a single evolutionary transition to long life. Therefore, they should not be represented as independent data points. If the statistical degrees of freedom were appropriately reduced, this trend would lose its statistical significance.

A more subtle problem, which does indeed depend primarily upon the number of species used, is that a single outlier species can have a disproportionate effect on one's inference. For example, one of us (SNA) recently reanalyzed published data [21] on cellular replicative senescence in eight species of mammals, attempting a reasonable correction for the potential effects of age at sampling (adult vs. fetal) and species body size (on the assumption that a certain number of cellular proliferative ability may be associated with developing a larger or smaller body from an original fertilized egg). After the corrections, the relation between number of cell population doublings and maximum captive longevity was still statistically significant (r =.84, p < 0.01). However, this significance resulted from a single outlying species (humans), which lived more than twice as long and had more than twice as many population doublings as any of the other species. Hence removal of humans from the analysis altered the statistical outcome dramatically (now r =.38, p =.40). No other species had this statistical effect.

19.3 Procedures for Implementing Comparative Methods

Bearing these caveats in mind, let us examine how comparative biology can best be implemented for the three uses in aging research that we have outlined.

19.3.1 Formulating and Evaluating Hypotheses about Aging Mechanisms

Formulating and evaluating mechanistic hypotheses in aging research generally entails the analysis of surveys from a number of species and searching for correlations between some variable thought to be mechanistically involved in the aging process with some measure of aging rate, such as longevity or Gompertzian mortality rate. These studies are subject to all the caveats mentioned above. That is, one must be attentive to the following potential problems:

1. Only correlation, not causation, should be inferred from correlational analyses. Although the case for a causative link may be considerably strengthened by the analysis, validation of a causative relationship must await experimental manipulation of the variable thought to be causing differences in aging rate.

2. The variable of interest should not be expressed in units that ensure a correlation with aging rate for other reasons.

3. Care must be taken to avoid phylogenetic sampling bias.

The first two problems are straightforward to avoid and require no special techniques. We will now concentrate on problems with species sampling.

Problems with false correlations due to single outliers are only relevant to studies in which a small number of species are used. In practical terms, studies of a dozen or more species should be relatively immune to such problems, unless the outlier is spectacularly distant from the other data points, or unless the hypothesis under test relies on a more refined outcome than observing a statistical association between the variables. If the hypothesis depends, for instance, on finding a certain slope in the regression of the putative independent variable on aging rate, then outliers can still unduly affect the outcome unless samples become quite large indeed. If the form of the relationship, (e.g., linear versus exponential), is critical then an exploratory technique such as LOWESS (Locally Weighed Sum of the Squares) [22] may be used to determine whether the underlying relationship indeed has the required form.

With respect to more subtle aspects of phylogenetic sampling bias, such as data heavily weighted by too narrow a range of species, there are now a number of analytic techniques which have been developed precisely to deal with this problem [19]. The most widely used of these techniques invokes phylogenetically independent contrasts (= PICs) to replace data from individual species [19,20].

PICs are calculated as differences in trait values between individual species sharing an immediate common ancestor and thus represent independent evolutionary events. These events can then be used as legitimate independent data points in statistical treatments of comparative analyses. For instance, in Figure 19.1, the combination of $X_4 - X_3$ and $Y_4 - Y_3$ represents a single evolutionary change from the past represented by the ancestral species N_1. Similarly, $X_6 - X_5$ and $Y_6 - Y_5$ represent another independent evolutionary event since divergence from N_2. The details of implementing this type of analysis are lucidly explained in several sources [19,20].

However, a legitimate issue is when PIC, as well as other phylogenetically based analyses not described here, should be used in the place of more traditional analyses utilizing individual species as data points [23]. PIC analysis is in principle clearly more compatible with the strict assumptions of conventional statistics and the problem of phylogenetic bias deserves to be taken seriously. However, there are some drawbacks to PIC analysis. First and foremost, it requires a phylogeny, and to the extent that the phylogeny used is in error, so will be the analysis. Also, some forms of PIC analysis make assumptions about the form of evolutionary change which are themselves problematic. Furthermore, because the error variance associated with PIC analysis is necessarily larger than that based on simple species analysis, statistical power is reduced. Finally, a recent review [23] comparing PIC versus simple species analysis of 30 published correlations between body size and various other parameters in birds and mammals, found that analytic method made little difference in the conclusions reached.

At present then, we recommend that comparative studies utilize phylogenetically-based analytical methods only when a reasonably reliable phylogeny exists. Otherwise more traditional non-phylogenetic comparative studies can still be useful, and probably not misleading, if the species used are phylogenetically diverse and no large fraction of the data are clustered within one or a few groups.

19.3.2 Assessing Generality of Specific Aging Mechanisms

A common problem in the use of animal models for investigating mechanisms of aging is assessing the relevance of one's findings to other animal species, particularly with respect to humans. That is, to what extent is an aging mechanism identified in, say, a mouse or *Drosophila* likely to function similarly in humans? Of course, the most direct way to determine whether an aging mechanism is operative in humans would be to do the appropriate experiments on humans or another closely related primate. However, it is often logistically unfeasible, prohibitively expensive, or ethically objectionable to do such work on humans or other primates. Another way to approach this question is to assess how general (meaning conserved and therefore applicable to a wide range of species) or idiosyncratic (meaning recently evolved and therefore confined to one or a few closely related species) is a particular aging mechanism? This issue leads to a consideration of the best sorts of animal models one might employ for making inferences about human aging mechanisms.

Ironically, a similar question has for years bedeviled evolutionary biologists attempting to construct proper phylogenetic relationships among species. In taxonomy, determining phylogenetic relationships depends upon being able to distinguish conserved general traits from idiosyncratic, recently evolved, traits. In fact in modern taxonomy, groups are defined by the idiosyncratic traits they share [24].

Modern taxonomists generally determine ancestral, or general, traits by utilizing one or more "outgroups." Outgroups are species or groups of species that are as closely related as possible to the group of interest, but clearly "outside" (meaning more distantly related) the group of interest [25]. The logic of using outgroups is that traits shared by the species of interest and an outgroup(s) are more likely to be general, whereas those not shared with outgroup(s) are likely to be idiosyncratic to the group studied. Thus in Figure 19.1, if the group of interest were species E, F, and G, then C and D will compose the best outgroup for determining the generality of traits exhibited by E, F, and G.

Translating this conceptual argument into concrete terms, any aging mechanisms observed to be shared among the nematode, *C. elegans*, *Drosophila* spp., and mice are possibly highly conserved, very general aging mechanisms and therefore likely to be operative in humans as well. On the other hand, imagine a mechanism of aging discovered in mice, but not operative in either *C. elegans* or *Drosophila*. How could we determine the likelihood this mechanism were operative in other mammals besides mice, such as humans?

An answer can be determined by inspecting Figure 19.1. If G represents humans, E and F can logically represent the murine rodent species, house mice and Norway rats, the laboratory mammal species of choice. What would be the mammalian equivalent of C and D, that is, the appropriate outgroup? Although the recent evolutionary history of many mammalian groups is increasingly well understood, mammalian taxonomy at higher levels, such as among orders, is generally not resolved [26]. It is this higher level taxonomy that is required to determine the appropriate outgroup(s) for analyzing the phylogenetic branch which includes rodents and humans. Fortunately, it is uncontroversial that one mammalian order, the Marsupials, do indeed represent a legitimate out-group to the other living mammals (excluding the extant 3 monotreme species). For this reason, the possibility of rapid assessment of the generality of aging (and other biomedically important) mechanisms discovered in laboratory rodents we have advocated the addition of one or more marsupials to the bestiary of commonly used laboratory animals [27,28].

Two marsupials immediately present themselves as possibilities to fill the role of outgroups to laboratory rodents. The first is *Monodelphis domestica* (gray, short-tailed opossum), which is a small (90 g) New World marsupial that has already been developed for laboratory use [29,30]. These animals reproduce easily in captivity [29], live only 2 to 2.5 years on average, and die from a range of pathologies, including a variety of neoplasias and congestive heart failure [31]. The second group composes several species from the Australian genus *Sminthopsis* (dunnarts) — small marsupials that are easily maintained in the laboratory [32,33]. The most commonly used species is the mouse-sized *Sminthopsis crassicaudatus*, which breeds continuously in the laboratory [34], and lives up to 1.5 years in the wild and as long as four years in captivity. The slightly larger *S. macroura* has also adapted well to the laboratory, and has lived as long as 4.8 years in captivity [35].

19.3.3   Identifying Key Aging Mechanisms from an Array of Candidates

A standard of the design of scientific experiments is controlling as many variables as possible besides the variable(s) purposely manipulated. In terms of research into aging mechanisms, such logic has often translated into the potentially useful experimental approach of comparing the operation of hypothetical aging mechanisms in populations that are as similar as possible to one another except for aging rate. Several current paradigms of aging research make use of this logic.

One such paradigm is manipulation of aging rate by altering the environmental — caloric restriction. It is well known that restricting the caloric intake of laboratory rodents and various poikilotherms below their ad lib feeding rate retards aging by virtually any known measure [36]. The popularity of the caloric restriction paradigm in searching for primary mechanisms of aging is that genetics and environment are maximally controlled between experimental and control populations, and the difference in aging rate is entirely attributable to the restricted feeding itself. However, this simple dietary manipulation has not to date led to any dramatic breakthroughs in our understanding of aging mechanisms, primarily because caloric restriction leads to a bewildering range of changes in the physiological systems of the experimental animals. Thus more than 60 years after the discovery of caloric restriction's anti-aging effect, and after many, many studies of potential mechanisms underlying this effect, there has emerged no generally accepted explanation of restriction's mechanism of action.

Similarly, several laboratories have used artificial directional selection of the timing of reproduction in *Drosophila melanogaster* stocks to produce strains that are long-lived relative to unselected controls [37,38]. Reared under identical conditions, these populations only differ in whatever alleles were differentially favored during the process of selection. The aims of such research is to identify specific alleles and processes causally involved in modulating aging. This approach appears so promising that there are currently at least two long-term selection studies currently being carried out in mice with the express aim of creating long-lived mouse stocks for similar analyses. Yet, in the *Drosophila* experiments, as with the caloric restriction experiments, so many processes and alleles are altered by selection that it has proven difficult to isolate the key changes involved in modulating aging rate [39,40].

Finally, in the past few years, the alteration of aging rate in genetically engineered animals has become a more common style of research into aging mechanisms [41], and this mode of research will no doubt become even more common in the future [42,43]. But even this highly targeted approach is likely to lead to alterations in multiple biochemical pathways and therefore have multiple phenotypic effects, as virtually all known gene products function in multiple pathways. Thus the problem of identifying key mechanistic changes from accidental byproducts will likely continue to plague aging research.

There are at least two explanations for the manifold changes in physiological systems in these types of experiment. One possible explanation is that aging itself is multifarious, occurring due to the combined action of so many mechanisms that the number of identifiable underlying processes is experimentally intractable. An alternative explanation is that aging is primarily the result of a few key processes, but that modulation of these processes can lead to adventitious changes in a host of other processes.

We believe that a comparative approach, performing identical experiments on a judiciously chosen range of existing strains or species, can assist in sorting out these alternative explanations, and that if there are indeed a manageable number of primary aging processes, a comparative approach can help identify them.

The rationale for performing identical experimental manipulations on a variety of strains, stocks or species is that each experiment can be interpreted as a partially to completely independent replication of the experimental protocol, and an analysis of the variation in the results can therefore be instructive. For instance, assuming that a consistent senescence-retarding effect is observed across the strains or species studied, one can then look for consistency in the response of variables associated with hypothesized mechanisms of aging. Variables which do not respond similarly across these replications are unlikely to reflect underlying general mechanisms of aging. On the other hand, if a similar treatment leads to a highly variable degree of retarded senescence across study strains or species, parameters reflecting hypothesized mechanisms of aging will likely change in a parallel fashion.

The array of laboratory mammals currently in common use for aging studies provides a powerful comparative framework for investigating aging mechanisms operating in murine rodents. An outline of such an analysis for investigating the mechanism(s) by which caloric restriction retards aging is given below.

19.4 Example: Comparative Assessment of Metabolic Rate and Body Temperature as Contributors to the Anti-Aging Effect of Caloric Restriction

As previously mentioned, more than sixty years of research on how caloric restriction retards aging in laboratory rodents has provided no consensus on the mechanism(s) involved. A key problem has been sorting the large range of variables that are altered by caloric restriction. A comparative approach to evaluating mechanisms of the retardation of aging has occasionally been used implicitly to dismiss candidate mechanisms, although no focused comparative studies of mechanisms have been performed. For example, an early hypothesis was that caloric restriction retarded aging by reducing mass-specific metabolic rate and therefore the production of reactive oxygen species [44]. However, Masoro et al. [45] demonstrated that caloric restriction in weanling F344 rats did not, in fact, reduce mass-specific metabolic rate except during a brief (4-week) initial period immediately after restriction was initiated. This lack of specific metabolic effect was later confirmed and extended to lifetime energy consumption in F344 rats [46,47]. Note that no similar finding in mice not necessary to demonstrate that metabolic rate reduction was superfluous to the senescence-retarding effect of caloric restriction. This sort of reasoning highlights what sort of results would be informative in a thorough comparative analysis using available strains and species of rodents.

It is worthwhile to reemphasize that comparative studies are best implemented when a reliable phylogeny of the groups studied is available. Fortunately, excellent molecular phylogenies exist for house mice and Norway rats, the rodent species commonly used in biomedical research [48], and for inbred strains and substrains most commonly used of both species [49,50]. The array of mouse strains represents one end of the utility spectrum of resources for comparative biology — a number of very closely-related genotypes with a known relationship to one another. Rats provide the other end of the utility spectrum by representing a reasonable outgroup for the mice. The tools for a comparative analysis of caloric restriction are therefore already in place.

The reason that a proper comparative analysis cannot be performed on data already available in the literature is inconsistency in the design of caloric restriction experiments among laboratories. Some laboratories compare *ad lib* fed animals with those fed 60% of *ad lib*. Other laboratories do not use *ad lib* feeding for a control but compare various degrees of restricted feeding. Sometimes restriction is initiated at weaning, sometimes at sexual maturity, sometimes even later in life. Some rodent colonies live in specific pathogen-free conditions, some in clean conventional colonies. An ideal comparative design would be centralized rearing and maintenance of animals of a variety of genotypes under identical conditions with identical caloric restriction regimes. Researchers with expertise in specific mechanistic interests and hypotheses could then receive animals from the centralized colonies and investigations among the various genotypes could then be realistically compared.

Ironically, The National Institute on Aging (NIA) has provided the precise logistical background for just such a comparative study of aging mechanisms, yet advantage has yet to be taken of this unique opportunity. Since 1985, NIA has supported four genetically defined mouse genotypes and three genetically defined rat genotypes under identical caloric restriction regimes in specific pathogen-free conditions at the National Center for Toxicological Research. For each genotype, restricted animals are fed 60% of the *ad lib* diet and are calorically restricted from the age of 16 weeks throughout life. The mouse genotypes include the most phylogenetically distant common laboratory strains [49], and the mice are demographically diverse as well. Median longevities differ by as much as one year (88 to 140 weeks) among the ad lib fed genotypes [28]. The rat genotypes are also diverse genetically [50] although they are somewhat less divergent demographically than the mice (range of median longevity: 103 to 145 weeks).

Even a crude comparison of the comparative effects of caloric restriction across species and genotypes reveals some intriguing trends (Figure 19.4). Specifically, irrespective of absolute longevity, the *proportional* increase in median longevity with this caloric restriction regime is remarkably consistent (16 to 21%) in males across all rat genotypes and both mouse strains. By contrast, the effect was approximately doubled (36%) in both F_1 mouse genotypes. If median longevity can be considered a provisional metric for the senescence-retarding effect of caloric restriction, then one way to assess the effects of putative mechanisms of aging is to compare how key parameters associated with such mechanisms change across strains and species.

No research has been specifically performed along these lines, although suggestive evidence comes from a comparison that can be made from the published information on one mouse and one rat genotype. Duffy et al. [51,52] measured a number of the same parameters in *ad lib* fed and restricted males in F344 rats and $B_6C_3F_1$ mice from the NCTR colonies. One

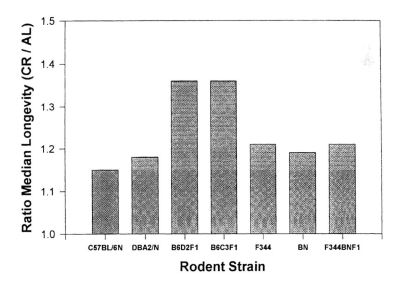

FIGURE 19.4
Life-extending responses to identical caloric restriction regimes (restricted diet = 60% ad lib calories initiated at 14 weeks of age) among male rodents from two species (*Mus musculus* and *Rattus norvegicus*) and a variety of strains. Animals reared for the National Institute on Aging at the National Center for Toxicological Research (data from Table II in 28). Median longevity for the *ad lib* mouse populations: C57BL/6N = 120 weeks; DBA2/N = 88 weeks; B6D2F1 = 138 weeks; B6C3F1 = 140 weeks. Median longevity for *ad lib* rat populations: F344 = 103 weeks; BN = 129 weeks; F344BNF1 = 145 weeks.

might expect that variables causally involved in caloric restriction's effect might vary substantially between the mice and rats and that variation should be larger in the mice than the rat genotype, since the enhancement of longevity in mice was almost twice as large (Table 19.1).

Confirming and extending the previous work on F344 rats, these studies find also a trivially, and statistically nonsignificant, *higher* mass-specific metabolic rate in the calorically restricted rats. A similar metabolic response is found in the mice [53]. Since the change in metabolism is nonsignificant in both species, and even the raw numbers are in the opposite direction from that predicted by the hypothesis that metabolic rate reduction is responsible for restriction's enhanced longevity effect, metabolism itself apparently plays no role in the effect.

TABLE 19.1

Comparison of Potentially Significant Parameter Changes in Male F344 Rats and $B_6C_3F_1$ Mice Subjected to Identical Caloric Restriction Regimens*

Parameter	Change in:	
	F344 Rats	**$B_6C_3F_1$ Mice**
Percent median longevity (CR/AL)	21.3	36.4
Mean oxygen consumption per gram lean body mass (% CR/AL)	1.4	3.0
Mean body temperature, °C (AL/CR)	0.82	1.24

* Data from Ref. 28, 51, 52.

Note: Percent changes given as (larger value)/(smaller value). AL = ad lib, CR = calorically restricted.

From the data in Table 19.1, the change in body temperature observed could remain a candidate for the cause of caloric restriction's effects given the data presented here. In fact, mean temperature variation does not capture the different patterns between these rats and mice. Rats have relatively little change in the circadian oscillation of body temperature which varies by about 1.5°C in both control and restricted animals, with the restricted animals' temperatures ranging about 1°C below that of the ad lib animals [51]. By contrast, the restricted mice experience a dramatic alteration in their body temperature regulation. Control mice have a circadian rhythm of body temperature variation similar to rats, only a bit larger (range about 2°C). Restricted mice have a dramatically larger temperature range (about 6°C), with their high temperature remaining about the same as controls, but their low temperature considerably lower than controls [52]. However, this pattern shows how simple two-strain or species comparisons can be misleading. Other studies using inbred mouse strains including C57BL/6 show that extreme body temperature variation (as much as 13+°C) is the rule in calorically restricted mice [36,53]. Therefore, because mice generally do not respond to caloric restriction with greater longevity enhancement, body temperature is not likely to be causally involved in that enhancement.

The above analysis was not meant to be definitive, only illustrative of the potential of the comparative approach even to familiar problems. A proper comparative analysis would require cognate data for all strains and F_1 hybrids of both species. An important additional point is that a proper comparative analysis would employ a more highly resolved measurement of caloric restriction's effect on mortality patterns. Measures of longevity generally are poor tools for describing alterations in mortality patterns. Longevity can be extended by a variety of effects on age-specific mortality patterns and therefore analysis of longevity alone may mask different effects of different aging mechanisms. Much more useful is the study of age-specific mortality patterns [54,55], including changes in the rate at which mortality increases with time (aging in a strict sense) versus a generally reduced mortality rate throughout life with no altered rate of increase, or some combination of these two. Several distinctive changes in mortality patterns have already been reported in the caloric restriction literature, using different rodent species and strains, and different restriction regimes [56,57]. Teasing apart these different effects, combined with a comparative approach, may help overcome the intractability of understanding general mechanisms of aging.

19.5 Conclusions

Evolutionary biology in the guise of comparative biology has a continuing role to play in probing mechanisms of aging, regardless of whether research is conducted at the role of the population, the individual, the cell, or the gene. Comparative biology can play three roles in the search for aging mechanisms: (1) hypothesis formulation and evaluation, (2) assessment of the generalizability to other species of particular research findings, and (3) isolation of key mechanisms of aging from a host of potential mechanisms. Certain caveats must be borne in mind when implementing a comparative approach, however. First, causation should not be inferred from correlative evidence alone, but correlative evidence is very useful for formulating hypotheses to be experimentally evaluated. A particularly danger in correlative analyses is that the variables under study may be correlated for reasons having nothing to do with the question being addressed. Moreover, comparative analyses are subject to artifacts associated with biased phylogenetic sampling. Phylogeny can never be ignored when employing comparative analyses and a number of recently developed techniques have become widely used to manage the effects of phylogenetic bias in the formulation and evaluation of specific hypotheses.

In order to assess the generality of aging mechanisms, one searches for evidence of common mechanisms in evolutionary outgroups from the group in question. For the study of eutherian mammals, which include rodents and humans, an underutilized outgroup is the marsupials. For this reason, the development of a small, short-lived marsupial as a laboratory model of the aging process would be a tactical research innovation. Finally, comparative biology is useful for sorting among a range of variables that have been associated with modulated aging rate. Potentially causal, or at least general, variables will be those that correlate with aging rate across species and genotypes. Variables which are noncausally associated with modulation of aging rate within particular species or genotypes correlate less well across species and genotypes. Because a range of genotypes in two rodent species separated by 10 million years of evolutionary change are already commonly used as animal models of aging, a comparative analysis of aging mechanisms within rodents could be easily undertaken with the resources already in place. A preliminary analysis suggests that alteration of body temperature is unlikely to be involved in the senescence-retarding effect of caloric restriction on laboratory rodents.

References

1. Rose, M. R., *The Evolutionary Biology of Aging*. Oxford University Press, Oxford, U. K., 1991
2. Charlesworth, B. *Evolution in Age-Structured Populations*, Cambridge University Press, Cambridge, U.K., 1994.
3. Martin, G. M., Austad, S. N., and Johnson, T. E., Genetic analysis of ageing: role of oxidative damage and environmental stresses. *Nature Genetics*, 13:25, 1996.
4. Weismann, A. The duration of life, in *Essays Upon Heredity and Kindred Biological Problems*, Poulton, E. B., Schonland, S., and Shipley, A. E., Eds., Clarendon Press, Oxford, U. K., 1889, 1.
5. Rubner, M. *Das Problem der Lebensdauer und seine Beziehungen zum Wachstum und Ernahrung*, Oldenbourg, Munich, 1908.
6. Metchnikoff, E., *The Prolongation of Life*, G. P. Putnam's & Sons, New York, 1910.
7. Pearl, R. *The Rate of Living*, Knopf, New York, 1928.
8. LeBourg, E. Correlational analysis in comparative gerontology: an examination of some problems. *Exp. Gerontol.* 31, 645, 1996
9. Austad, S. N., Comparative aging and life histories in mammals, *Experimental Gerontology*, 32, 23, 1997.
10. Ku, H.-H., Brunk, U. T., and Sohal, R. S. Relationship between mitochondrial superoxide and hydrogen peroxide production and longevity of mammalian species. *Free Radical Biology & Medicine*, 15, 621, 1993.
11. Sell, D. R., Lane, M. A., Johnson, W. A., Masoro, E. J., Mock, O. B., Reiser, K. M., Fogarty, J. F., Cutler, R. G., Ingram, D. K., Roth, G. S., and Monnier, V. M., Longevity and the genetic determination of collagen glycoxidation kinetics in mammalian senescence, *Proc. Natl. Acad. Sci. U.S.A.*, 93, 485, 1996.
12. Sacher, G. A., Relation of lifespan to brain weight and body weight in mammals, in *Ciba Foundation Colloquia of Ageing*, vol. 5, *The Lifespan of Animals*, Wolstenholme, G. E. W. and O'Connor, M. O., Eds. Churchill, London, 1959, 115.
13. Mallouk, R. S. Longevity in vertebrates is proportional to relative brain weight. *Federal Proceedings*, 34, 2102, 1975
14. Hofman, M. A. Energy metabolism, brain size and longevity in mammals. *Quarterly Review of Biology*, 58, 495, 1983
15. Austad, S. N. and Fischer, K. E. Primate longevity: its place in the mammalian scheme. *American Journal of Primatology*, 28, 251, 1992.
16. Harvey, P. H. and Zammuto, R. M. Patterns of mortality and age at first reproduction in natural populations of mammals. *Nature*, 315, 319, 1985.
17. Tolmasoff, J. M., Ono, T., and Cutler, R. G. Superoxide dismutase: correlation with life span and specific metabolic rate in primate species. *Proc. Natl. Acad. Sci. U.S.A.*, 77, 2777, 1980.
18. Hillis, D. M., Moritz, C., and Mable, B. K. (Eds), *Molecular Systematics*, 2nd Ed., Sinauer Associates, Sunderland, MA, 1996.
19. Harvey, P.H. and Pagel, M. D., *The Comparative Method in Evolutionary Biology*, Oxford University Press, Oxford, UK, 1991.
20. Garland, T. Jr., Harvey, P. H., and Ives, A. R., Procedures for the analysis of comparative data using phylogenetically independent contrasts. *Systematic Biology*, 41, 18, 1992.
21. Röhme, D., Evidence for a relationship between longevity of mammalian species and lifespans of normal fibroblasts *in vitro* and erythrocytes *in vivo*. *Proc. Natl. Acad. Sci. U.S.A.* 78, 5009, 1981.
22. Cleveland, W. S., Robust locally weighted regression and smoothing scatterplots. *J. Am. Stat. Assoc.*, 74, 829, 1979.
23. Ricklefs, R.E. and Starck, J. M., Applications of phylogenetically independent contrasts: a mixed progress report. *Oikos*, 77, 1, 1996.
24. Hennig, W., *Phylogenetic Systematics*. University of Illinois Press, Urbana, IL, 1979.

25. Stevens, P. F., Evolutionary polarity of character states, *Annual Review of Ecology and Systematics*, 11, 333, 1980.
26. Honeycutt, R. L. and Adkins, R. M., Higher level systematics of eutherian mammals: an assessment of molecular characters and phylogenetic hypotheses. *Annual Review of Ecology and Systematics*, 24, 279, 1993.
27. Austad, S. N., The comparative perspective and choice of animal models in aging research. *AGING, Clinical and Experimental Research* 5, 259, 1993.
28. Sprott, R. L. and Austad, S. N., Animal models for aging research. in *Handbook of the Biology of Aging*, 4th Ed. Schneider, E. L. and Rowe, J. W., Eds. Academic Press, San Diego, CA, 1996, 3.
29. Fadem, B. H., Trupin, G. L., Maliniak, E., Vandeberg, J. L., and Hayssen, V., Care and breeding of the gray, short-tailed opossum (*Monodelphis domestica*). *Lab. Anim. Sci.*, 32, 405, 1982.
30. Vandeberg, J. A., The grey short-tailed opossum: a new laboratory animal. *Inst. Lab. Anim. Res. News*, 26, 9, 1983.
31. Hubbard, G. B., Mahaney, M. C.., Gleiser, C. A., Taylor, D. E., and Vandeberg, J. L., Spontaneous pathology of the gray short-tailed opossum (*Monodelphis domestica*), *Lab. Anim. Sci.*, 47, 19, 1997.
32. Hope, P. J., Wittert, G. A., Horowitz, M., and Morley, J. E., Feeding patterns of *S. crassicaudata* (Marsupialia: Dasyuridae): role of gender, photoperiod, and fat stores. *Am. J. Physiol.* 272, R78, 1997.
33. Kennedy, G. A., Coleman, G. J., and Armstrong, S. M., Daily restricted feeding effects on the circadian activity rhythms of the stripe-faced dunnart, *Sminthopsis macroura*. *J. Biol. Rhythms*, 11, 188, 1996.
34. Godfrey, G. K. and Crowcroft, P., Breeding the fat-tailed marsupial mouse *Sminthopsis crassicaudata* in captivity. *Internatl. Zoo Yearbook*, 11, 33, 1971.
35. Jones, M. L. Longevity of mammals in captivity, *Zool. Garten*, 52, 113, 1982.
36. Weindruch, R. and Walford, R. L., *The Retardation of Aging and Disease by Dietary Restriction*, Charles C Thomas, Springfield, IL, 1988.
37. Rose, M. R., Laboratory evolution of postponed senescence in *Drosophila melanogaster*. *Evolution*, 49, 649, 1984.
38. Luckinbill, L. S., Arking, R., Clare, M. J., Cirocco, W. C., and Buck, S. A., Selection for delayed senesence in *Drosophila melanogaster*. *Evolution*, 38, 996, 1984.
39. Service, P. M., Hutchinson, E. W., and Rose, M. R., Multiple genetic mechanisms for the evolution of senescence in *Drosophila melanogaster*. *Evolution*, 42, 708, 1988.
40. Arking, R., Force, A. G., Dudas, S. P., Buck, S., and Baker, G. T. III, Factors contributing to the plasticity of the extended longevity phenotypes of *Drosophila*. *Exp. Gerontol.* 31, 623, 1996.
41. Orr, W. C. and Sohal, R. C., Extenstion of life span by overexpression of superoxide dismutase and catalase in *Drosophila melanogaster*. *Science*, 263, 1128, 1994.
42. Tower, J., Aging mechanisms in fruit flies. *Bioessays*, 18, 799, 1996.
43. Andersen, J. K. and Jurma, O. P., Use of genetically engineered mice as models for understanding human neurodegenerative disease. *J. Am. Geriatr. Soc.*, 44, 717, 1996.
44. Sacher, G. A., Life table modfication and life prolongation. in *Handbook of the Biology of Aging*, 1st Ed., Finch, C. E. and Hayflick, L. (Eds). Van Nostrand, New York, 582, 1977.
45. Masoro, E. J., Yu, B. P., and Bertrand, H., Action of food restriction in delaying the aging processes. *Proc. Natl. Acad. Sci. U.S.A.*, 79, 4239, 1982.
46. McCarter, R., Masoro, E. J., and Yu, B. P., Does food restriction retard aging by reducing the metabolic rate? *Am. J. Physiol.* 248, E488, 1995.
47. McCarter, R. J. and Palmer, J. Energy metabolism and aging: a lifelong study in Fischer 344 rats. *Am. J. Physiol.* 263, E448, 1992.
48. Boursot, P., Auffray, J.-C., Britton-Davidian, J., and Bonhomme, F., The evolution of house mice, *Annual Review of Ecology & Systematics*, 24, 119, 1993.
49. Fitch, W. M. and Atchley, W. R., Evolution in inbred strains of mice appears rapid. *Science* 228: 1169, 1985.
50. Canzian, F., Phylogenetics of the laboratory rat, *Rattus norvegicus*. *Genome Res.*, 7, 262, 1997.

51. Duffy, P. H., Feuers, R., Leakey, J. A.,, Nakamura, K. D., Turturro, A., and Hart, R. W., Effect of chronic caloric restriction on physiological variables related to energy metabolism in the male Fischer 344 rat. *Mech. Age. Develop.*, 48, 117, 1989.

52. Duffy, P. H., Feuers, R. J., and Hart, R. W., Effect of chronic caloric restriction on the circadian regulation of physiological and behavioral variables in old male $B_6C_3F_1$ mice. *Chronobiology International* 7: 291, 1990.

53. Koizumi, A., Tsukada, M., Wada, Y., Masuda, H., and Weindruch, R., Mitotic activity in mice is suppressed by energy restriction-induced torpor. *J. Nutrition*, 122, 1446, 1992.

54. Finch, C. E., *Longevity, Senescence, and the Genome*, University of Chicago Press, Chicago.

55. Gavrilov, L. A. and Gavrilova, N. S., *Biology of Lifespan: a Quantitative Approach*, Harwood Academic Publishers, London, 1991.

56. Harris, S. B., Weindruch, R., Smith, G. S., Mickey, M. R., and Walford, R. L., Dietary restriction alone and in combination with oral ethoxyquin/2-mercaptoethylamine in mice. *J. Gerontol.*, 45, B141, 1990.

57. Masoro, E. J., Dietary restriction, *Exp. Gerontol.*, 30, 291, 1995.

20

Apoptosis and Programmed Cell Death

Ivor D. Bowen

CONTENTS

20.1 Introduction

The cells of an organism may die either by accident or design. Accidental cell death is usually the result of traumatic insult or disease and is termed necrosis by pathologists. Death by design is usually part of a normal physiological program and indeed, has been called programmed cell death [1,2,3]. Such a programmed cell death generally forms part of the normal development of multicellular organisms and indeed, is essential for their continued life [4,5]. Programmed cell death, an altruistic suicide in this respect, is employed to shape and structure tissues, organs and organisms and forms an essential part of normal embryological development, morphogenesis and metamorphosis. These processes are clearly under genetic control since the relevant mutants affecting cell death result in specific disruptions of body plan.

It is also becoming clear that programmed cell death plays an important role in maintaining the population balance of cells in fully grown or mature organisms. In terms of tissue kinetics, such organisms are usually more or less in dynamic equilibrium, cell division being broadly balanced by cell death. This programmed homeostatic cell death, led to the introduction of a special term " apoptosis", functionally defined as an equal and opposite force to mitosis [6]. It was concluded, for example, that disturbances in the subtle balance between mitosis and apoptosis could lead to the formation of cancer, where cell division exceeded cell death.

How do the concepts of programmed cell death and apoptosis have a bearing on aging? In a general sense aging leads to death and in plants the senescence of cells clearly leads to cell death, fruit ripening, leaf fall, etc. Such a connection may be less obvious in animals, although Hayflick [7] has shown that *in vitro* animal cells do appear to age to death and have a finite life span.

In terms of tissue dynamics, a good example is the loss of dead cornified cells from the surface of the skin. Here, cells are born in the basal granular layer, then specialize in the production of the fibrous protein keratin as they move up through the intermediate layers, finally differentiating to death as flattened dead keratinocytes at the cornified surface. A similar example is seen in the production of xylem in plants. During the process of xylogenesis the cells enlarge and elongate forming long tubular elements that will aquire vascular function. During this process the cells are strengthened by rings of lignin, the nuclei are lost and the cells die. Although individual cells differentiate to death, the dead xylem cells are nonthelesss essential for the life of the plant.

It could thus, be argued that specialization, and cytodifferentiation leads to a largely preordained cellular aging and death. Specialized cells become post-mitotic and lose reproductive viability and finally differentiate to death. Such changes are paradoxically essential for the normal development and maintenance of the organism as a whole. Differentiation to death falls into the category of programmed cell death in that it is dependant on a program of selective gene expression, but does not always fit comfortably into the category of "apoptosis". Apoptosis as originally defined [8] presented a well-described range of morphological symptoms including cell shrinkage, chromatin margination followed by DNA fragmentation and a florid break-up of the cell into spherical apoptotic bodies. Not all cells differentiating to death display these exact symptoms.

20.1.1   Necrosis and Apoptosis

Since the end-point of apoptosis and necrosis is the same, it may not always be easy to distinguish between them, although more diagnostic tools are being developed and will be presented later in this chapter. Wyllie et al.[9] have argued that both forms of cell death may represent the extremes of one continuous spectrum.

From a comparative point of view apoptosis leads to cell condensation and shrinkage, whilst necrosis leads to cellular swelling. Apoptotic cells loose water, whilst necrotic cells gain water. The membrane pumps continue functioning during apoptosis and such cells exclude vital dyes like Nile Blue, Trypan Blue and Nigrosine. Membrane pumps fail early on in necrosis and there is an immediate influx of water, sodium and calcium. One of the earliest detectable changes in necrosis is calcium overload[10].

Energy production is maintained during apoptosis and there is usually, although not always, a synthetic surge accompanying the appearance of new mRNA and *de novo* synthesis of protein. In necrosis, calcium influx activates phospholipase A, which compromises the internal mitochondrial membranes, leading to a catastrophic drop in ATP production. Since ATP is no longer available to energize the membrane pumps there is an autocatalytic deterioration in necrotic cells following mitochondrial disfunction. Necrotic cells experience an increasing acidosis which leads to the precipitation or "salting-out" of the chromatin giving rise to dense or "pycnotic nuclei". In apoptosis there is a dilation of the nuclear envelope leading to nuclear blebbing and the blebs tend to fill up with marginating chromatin. In this case, the chromatin fragments under the influence of endonuclease an enzyme which cleaves the DNA at the internucleosomal linker regions. When run on an agar gel the oligonucleotide fragments produce what has been calied a "DNA ladder" (see pages 462).

During the later stages, the lysosmes of necrotic cells fragment and hydrolytic enzymes are released which further lyse the cell leading to a complete breakdown of subcellular compartmentalization. The necrotic cell eventually bursts often eliciting an inflammatory response. Apoptotic cells on the other hand usually undertake a discreet but florid fragmentation into membrane bound spheres or apoptotic bodies which are rapidly phagocytosed by neighboring cells or phagocytes, without eliciting an immune response. There is some controversy still as to the role of lysosomes and indeed extralysosomal hydrolases in apoptosis and more generally in programmed cell death. Some authors claim that lysosomal enzymes and hydrolases are employed only in the secondary digestion of apoptotic bodies after phagocytosis, others believe that extralysosomal hydrolases and proteases are involved in the early stages of apoptosis. Some types of cell death also appear to involve lysosomal hydrolases in an early autophagic build up. These percieved differences between apoptotic and non-apoptotic programmed cell death are analysed in greater detail below.

20.1.2 Definitions of Apoptosis and Programmed Cell Death

Apoptosis and programmed cell death are relevant to studies in aging research in that not only is cell death often the ultimate end-point of senescence, but the processes of aging and senescence may well employ selective gene expression parallel if not identical to those of differentiation and programmed cell death. Examples of such selective gene expression leading to cell death will be dealt with later. Meanwhile, there continues to be some confusion in the literature between the terms programmed cell death and apoptosis [11]. The term programmed cell death was originally coined by Lockshin [2] in the context of insect development and as such had an implied genetic basis. The term apoptosis was introduced by Kerr et al. [6] broadly as an equal and opposite force to mitosis and as such initially aquired a morphological and kinetic definition.

It has subsequently been shown that apoptosis can be induced by genetic means [12,13], but can also be engendered by non-genetic means [14,15].It is also emerging that not all kinds of apparently genetically programmed cell death are strictly speaking apoptotic in either morphological or biochemical terms. Clarke [15] has described three morphological types of programmed cell death in neurones and Bowen et al.[17] and Lockshin and Zakeri [18], have detailed non-apoptotic types of programmed cell death with diverging molecular symptoms such as lack of endonuclease activation and changes in subcellular hydrolase pattern (see pages 458).

Clarke [16] envisaged three types of cell death:

Type 1 Cell Death, which accords with classical apoptosis and is also described in morphological terms as Type 1 cell death by Schweichel and Merker [19].

Type 2 Cell Death, previously described by Lockshin and Williams [1,20] and Lockshin [2] as programmed cell death in insect metamorphosis is characterized by and increasing level of autophagic activity. The dying cell fills up with active secondary lysosomes. There is no clear chromatin margination and the nucleus survives until it is finally extruded from the cell or becomes autolysed. This vacuolar type of cell death has been extensively described in a range of meta-morphosing inverebrates including planarian worms, insects, and molluscs [17,21–27]. This type of programmed cell death usually involves changes in the subcellular distribution of hydrolases. There is an initial increase in the autoph-agic lysosomal activity of acid phosphatase, followed by a *de novo* synthesis of extracisternal hydrolase at ribosomal sites [17].

Type 3 Cell Death is described by Clarke [16] in neuronal cells. It can be induced by withdrawal of nerve growth factor [28,29,30] and as such, could be thought of as atrophic cell death. Morphologically it is characterized by dilation of the endoplasmic reticulum, Golgi and nuclear envelope leading to extensive vacu-olation. There is no chromatin margination as seen in apoptosis. Interestingly, this type of atrophic cell death has been induced in tumour cells following inhibition of tyrosine kinase [31].

In a recent review Lockshin and Zakeri [18] refer to apoptosis as Type 1 cell death and non-apoptotic programmed cell death as Type 2. If we consider senescence and differentiation as also leading to cell death then a more general classification [32] can be constructed as shown in Figure 20.1.

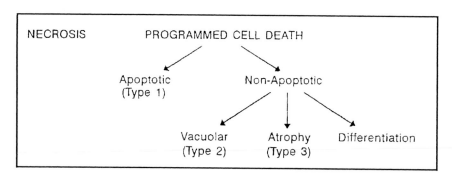

FIGURE 20.1
A comprehensive classification of cell death.

20.1.3 Senescence and Differentiation to Death

Postmitotic cells specialize and differentiate to death. In the first instance there is a loss of reproductive viability, post-mitotic cells do not usually regain the ability to reproduce and divide. All differentiating cells move along specialized pathways and their ultimate death is predictable, even necessary for the life of the organism as a whole. As with programmed cell death there is a point of commitment in time usually followed by differential gene activity, leading to the irrivisible appearance of new products. The differentiation of cells into somatic cell lines, as observed in metazoa, makes cell death inevitable, in contrast with the potentially immortal undifferentiated stem or germ cell line.

The search for genes that control entry into aging, senescence and even death must surely start with genes controlling the cell cycle itself. What controls exit from the cell cycle and entry into a resting phase or G_0, for example?

20.1.4   Aging Genes and the Cell Cycle

Many cell cycle control genes (*cdc*) were first elucidated in yeasts and Jazwinskin [33] has identified genes in yeast that appear to control aging and senesence. Mutation in the *LAG1*, longevity assurance gene has been shown to reduce the life-span of yeast by 40%. It has also been shown that overexpression of *RAS* extends the life span of yeast by 30%. It also appears to postpone aging by delaying generation time. In yeast, *RAS* integrates growth and cell aging and the response may be dependant on nutritional status. RAS can activate adenyl cyclase or inositol phospholipid turnover stimulating growth and division, or it can under different conditions cause an arrest at the G1/S boundary. Yeast is not a multicellular organism and does not appear to enter into cell suicide as such, however, primitive genes influencing senesence and longevity may ultimately prove to have some homology with genes that influence cell death in metazoa.

20.1.5   Differentiation and Senescence in Plants

In the context or plant development, the term senescence is used to signal maturation and differentiation which usually lead to death. Leaf fall is one potent example of a seasonally triggered cell death in trees. Similarly, the formation of bark, thorns, phloem, xylem, and heartwood all involve cytodifferentiation and a progressive senescence of cells to death. Although the fine structural changes do not always follow that shown by classical apoptosis, changes similar to those seen in apoptosis have been reported in the nuclei of protophloem [34] and during bioregulator induced xylogenesis [35]. Apoptotic-like behavior was also reported by Mittler et al.[36] in nuclei from the tobacco plant and DNA degradation has also been reported in developing tracheids, in induced lesions[37] and in Soybean *root necrosis* mutants [37].

Thomas [39], reviewing aging in plants and animals draws attention to the parallels that exist between programmed cell death in animals and cytodifferentiation and cellular senescence in plants. He stresses that whilst senescing plant cells are not technically dead, they are undergoing a phase in development which ends in death and that this differentiation to death requires the orchestration of a range of genes. He has also identified one such gene [40,41], the *Sid*, or"stay-green" gene in grass. The genetic basis of leaf senescence has also been recently reviewed [42].

20.1.6   The Genetic Basis of Apoptosis and Programmed Cell Death

20.1.6.1   Apoptosis

The genetic basis of apoptosis has been proven in only a few definitive cases which include the genes controlling neuronal development of the nematode *C. elegans* and the reaper gene controlling embryonic cell death in the fruit fly *Drosophila*. A total of 14 genes appear to control the process of apoptosis in the nematode and mutations in these genes exert specific effects on the cell death pathway [43]. Specifically, two genes, *ced-3* and *ced-4* induce apoptosis, whilst another, *ced-9* inhibits apoptosis in cells destined to live. Most of the other genes govern the reacognition of the apoptotic cells by phagocytes and regulate their ultimate digestion and disposal. Interestingly, two master genes *ces-1* and *ces-2* stipulate whether two specific motor neurone cells at the top of the cascade in the nematode pharynx will live or die.

A gene called *reaper* controls apopotosis in embryonic *Drosophila* [44]. The gene and its product clearly play a role in apoptosis since the cloned gene restores apoptotic capability to cell death defective embryos and *reaper* mRNA is specifically expressed in cells that are committed to die, indeed the early expression of reaper mRNA makes it an excellent specific marker for impending apoptosis in *Drosophila*.

Many other genes have, however, been identified as having a bearing on apoptosis and these include the mammalian oncogenes, some of which like *c-myc* also influence mitosis. These genes include *p53*, and a range of *ICE*-like proteases or caspases as well as inhibitory genes like those of the *Bcl-2* family [45].

Much future potential, thus, lies in the evolution of probes to detect the products and expression of such genes into cell death. Meanwhile, despite progress on the molecular and genetic basis of apoptosis the fact remains that the concept was originally developed using broadly morphological criteria [8,46], which can readily be established by histological and microscopical means. To this was added the incidence of endonuclease activation leading to charecteristic DNA laddering. Such a feature although of wide occurrence is, however, now known not to be ubiquitous and the method should not in the final analysis be regarded as diagnostic [47].

20.1.6.2   Programmed Cell Death

Ubiquitin has been claimed as a potential marker for impending programmed cell death in insects [48,49,50]. This type of cell death is more akin to Clarke type 2 than apoptosis in that usually there is no DNA fragmentation early on and no nuclear or cellular blebbing [48]. Ubiquitin is a highly conserved protein found in all eukaryotic cells and plays a role in the removal and degradation of cellular proteins including the cyclins involved in mitosis. During the process, ubiquitin is added to lysine residues on the target proteins which are lysed within the subcellular proteasomes by ATP-dependant non-lysosomal proteinases.

Selective gene expression has also been reported in a non-apoptotic type of programmed cell death seen in blow-fly salivary gland metamorphosis [17,24]. Here, *de novo* synthesis of mRNA gives rise to ten new proteins in a prelude to cell death. The first new proteins having a MWt between 30 and 100kDa appear into day 8 of the life cycle and a number persist until cell death later on in day 9. *In vitro* translation of the mRNA produced at this time shows a significant 53kDa protein appearing just before cell destruction. Could this be *p53* product?

20.2   Microscopic and Histological Analysis of Apoptosis

20.2.1   Histological Staining of Apoptosis

Light microscopy provides the easiest and most basic way of identifying apoptotic cells. A range of approaches are available. Vital dyes may be used to test the viability of cells in living tissues or cell suspensions. Alternatively, solid tissues may be fixed, embedded and sectioned for routine histological examination, apoptotic cells being differentiated using appropriate stains.

20.2.1.1 *Vital Dyes*

Saunders et al.[51], introduced an useful vital staining procedure for demonstrating regions of cell death in developing embryos. The technique consisted of exposing the embryo or living tissue for 15 to 30 min at room temperature to a 1:10,000 solution of Nile Blue Sulfate in Ringer's saline. The tissues are then rinsed with the saline, removing the Nile Blue Sulfate from normal cells. The dye appears to concentrate in dead and dying cells. Embryos may also be stained *in ovo* using Nile Blue Sulfate at concentrations of 1:40,000 in Ringer's saline. Microscopical examination of vitally stained tissues reveal numerous degenerating figures ranging from small cells possessing one or several blue "degeneration granules", thought to be materials lost from the nucleus to large bodies containing many granules.

Trypan Blue exclusion techniques may also be useful in estimating cytotoxicity in living tissues or cell suspensions [52,53]. Here again, live cells exclude the dye, which is differentially taken up by dying cells. The technique has been adapted for determination of cell death *in vivo* [54] in the thymus, where a density dependant pattern of mitotic and dying cells was demonstrated. The method has also been adapted for use with formalin fixed smears in an automated assay [55].

20.2.1.2 *Histological Staining*

Tissues for histological staining may be treated routinely with Hematoxylin and Eosin (see Table 20.1).

TABLE 20.1

Hematoxylin and Eosin Staining

Suitable for alcohol or formalin fixed material.

1. Stain in hematoxylin for 5 min.
2. Wash in tap water. Tap water is slightly alkaline and "blues" the tissue.
3. Examine microscopically and if overstained differentiate with 1% HCl in 70% ethanol.
4. Wash and "blue" in tap water.
5. Counterstain with a water soluble eosin, using a 1% solution of the dye in either water or 20% ethanol for about 1 min.
6. Differentiate in water or 70% ethanol.
7. Dehydrate, clear, and mount in DPX.

The nuclei will appear blue, apoptotic nuclei will appear black and cytoplasm will appear pink.

Material may be traditionally fixed in neutral buffered formalin and embeded in paraffin wax. Much better resolution can be achieved, however, through fixing in a 9:1 formalin : acetone mix followed by embeding in hydroxyethylmethacrylate [56], as detailed in Table 20.2, since such material allows thinner sections to be cut.

TABLE 20.2

A Hydroxyethyl Methacrylate (HEMA) Embedding Technique

1. Tissues or cells are fixed for approximately 2.5 hr in 1:9 v/v mixture of acetone and 10% neutral buffered formalin.
2. Place into hydroxyethyl methacrylate for 3hr with agitation and several changes.
3. Transfer into the following polymerization mix:

Monomer	80 ml 2-Hydroxyethyl methacrylate
	16 ml 2-Butoxyethanol
	0. 27g Benzoyl peroxide
Activator	15 ml Polyethylene glycol 200
	1 ml N,N-Dimethylaniline

    Polymerization is achieved by adding two drops of the activator per 5ml of the monomer. Polymerize overnight at room temperature, under carbon dioxide gas to eliminate air.
4. Allow the blocks to harden over a few days.
5. Cut thin sections 1–2um on a microtome with a glass knife.

Methacrylate sections are amenable to a very wide range of stains including H&E, enzyme cytochemistry and immunocytochemistry. Penetration of methacrylate is slow however, and routine incubation times need to be increased. Sections can be cut much thinner than with paraffin wax and this improves resolution for apoptotic studies.

Histological preparations need careful interpretation. Generally apoptotic cells occur singly (see Figure 20.2) or in small groups and characteristically show considerable cell shrinkage.

The most obvious aspect usually consists of dark dense chromatin material which initially accumulates at the margins of the nucleus. The nucleus eventually breaks up into several chromatin rich spheres or apoptotic bodies along with variabable amounts of cytoplasm. Such bodies may be seen inside neighbouring cells or macrophages which rapidly phagocytose them. Such profiles, must not be confused with necrotic cells, which if present, usually occur in extensive necrotic areas. Necrotic cells usually show symptoms of swelling rather than shrinkage, although eventually the cells do disintegrate scattering irregular nuclear and cytoplasmic fragments. Necrotic nuclei may also appear dense or "pycnotic" although they do not show distinctive chromatin margination or the break up into spherical apoptotic bodies. Since necrosis usually induces some immune response necrotic areas may show accumulation of neutrophils and cellular debris.

One very useful variant which helps resolve differences between necrosis and apoptosis has been described by Mofitt [57], where a methyl green-pyronin technique was used to demonstrate cell death in a murine S180 tumor. Here, methyl green highlights the enhanced chromatin by selectively staining DNA, whilst the pyronin Y acts as a selective stain for RNA which appears to be elevated into apoptosis and absent giving a negative response in necrotic areas (Table 20.3).

20.2.2 Fine Structural Changes

Confusion between histological profiles of apoptosis and necrosis is best resolved by means of electron microscopy [20,21] (see Table 20.4).

Fine structural differences between necrosis and apoptosis can usually readily be resolved. Apoptotic cells exhibit shrinkage (see Figure 20.3), surface blebbing and an

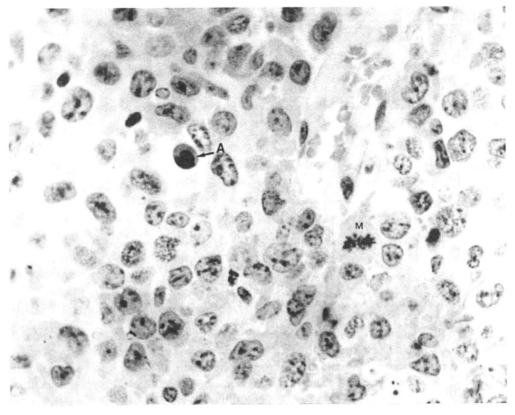

FIGURE 20.2
The localization of acid phosphatase in HEMA embedded S 180 murine tumor. Note spherical apoptotic cell (A) with very diffuse acid phosphatase activity (pale pink color) and dense chromatin bodies. Note also mitotic cell (M). Chromatin was counterstained with haematoxylin. Mag. × 950.

TABLE 20.3
Methyl Green Pyronin Method for DNA and RNA

Suitable for use with cells, paraffin wax/HEMA embeded tissues.

1. If using paraffin embeded tissues, de-wax and bring sections to water.
2. Make up 2% aqueous solutions of purified methyl green and pyronin Y.
3. Stain de-waxed sections for up to 5 min. Stain HEMA sections for 3hr.
4. Wash sections in distilled water for a few seconds.
5. Blot dry.
6. Mount in a suitable mountant such as DPX or Gurr's Xam.

Necrotic cells are negative, showing little or no pink cytoplasmic staining with pyronin, apoptotic cells have dark green to black nuclei. Early apoptotic cells tend to be more positive for RNA than normal cells and appear red rather than pink.

TABLE 20.4

Transmission Electron Microscopy

A range of different techniques are available, the procedure described here is suitable for combination with cytochemical analysis as described in Box 14.

1. Cells or small tissue blocks (50μm) are fixed in 3%, 0.1M cacodylate buffered glutaraldehyde at pH 7.2, at 0–4°C for 1–3 hr.
2. Washed overnight in the cacodylate buffer at 0–4°C.
3. Post-fixed in Millonigs 1% osmium tetroxide for 1hr at 0–4°C.
4. Dehydrated in a series of alcohols.
5. Embeded in Araldite or Epon.
6. Thin sections are then cut on an ultratome and counterstained routinely with uranyl acetate and lead citrate.

In terms of apoptosis the nuclei should show some margination of chromatin, the nuclear envelope may be expanded and the endoplasmic reticulum shows vacuolation. Plasma membrane and nuclear membrane should show some blebbing.

expansion of the nuclear envelope. Dense chromatin accumulates at the margins of the nucleus which also starts to blebb. The mitochondria appear normal. The endoplasmic cisternae expand and loose their ribosomes, forming a number of vesicles or vacuoles which eventually fuse with the plasma membrane, giving the cell a spongy appearance in the scanning electron microsope (Table 20.5). Scanning electron microscopy can be usefully extended to include the backscattered electron imaging of enzymatically released products, especially acid phosphatase active into programmed cell death [58].

Necrotic cells on the contrary swell, loose their subcellular compartmentalization and usually burst with the nucleus still intact. There may be some precipitation of chromatin in the nucleus but there is no extensive margination or rounding off of material as seen in apoptosis. The mitochondria swell and burst in necrosis and characteristic dense matrices containing calcium phosphate are laid down among the cristae. Finally, the lysosomes also break open releasing hydrolytic enzymes which result in further autolysis.

20.3 Electrophoretic and Cytochemical Methods

20.3.1 Electrophoresis

Conventional wisdom recommends methods that demonstrate the fragmentation products of DNA induced by endonuclease activity. It should, however, be realised that endonuclease activity may not always accompany apoptosis and certainly there are instances of programmed cell death established where DNA synthesis occurs [25]. Conventionally, endonuclease activity cuts double stranded DNA at the nucleosomal linker regions giving rise to regular oligonucleotide runs based on multiples of 180–200bp, which may be biochemically detected when run on an agarose gel, producing characteristic ladders (Table 20.6).

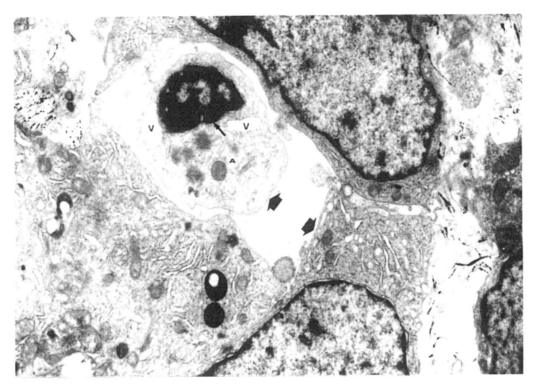

FIGURE 20.3

Electron micrograph of an early apoptotic cell (A) from an S180 murine tumor. Note vacuolating cytoplasm (v), marginating chromatin (small arrow) and extensive "shrinkage" gap (large arrows). The apoptotic cell is surrounded and almost engulfed by "normal" tumor cells. Mag. × 20,000.

TABLE 20.5

Scanning Electron Microscopy

Suitable for cells and tissues. Again a range of options are available for pretreatment. The option presented allows for the possibility of enzymatic localization by means of backscattered electron imaging.

1. Fix in 0.5% glutaraldehyde in 0.1M cacodylate buffer at pH 7.4 for 1hr at 0–4°C.
2. Wash in cacodylate buffer for 1–2hrs.
3. Incubate for acid phosphatatase or ATPase activity if required [58].
4. Wash in buffer for 1–2hrs, dehydrate and critical point dry from CO_2.
5. Mount specimens for backscattered electron imaging on carbon stubs.
6. Morphology specimens should not be incubated and should be sputtered or coated with gold.

Secondary electron imaging will provide topographical details of cells whilst backscattered imaging can detect subsurface distribution of enzyme released reaction product in dead and dying cells.

TABLE 20.6

Analysis of DNA Fragments on Agarose Gels

1. Lyse cells with a proteinase based digestion buffer.
2. Incubate at 50°C for 12 hrs.
3. Extract lysates once with phenol/chloroform/isoamyl alcohol and twice with chloroform/isoamyl alcohol.
4. Precipitate DNA by adding 2.5 M ammonium acetate, stir and add 2 vol absolute alcohol.
5. Incubate at 4 C for 2 hr and pellet the DNA by centrifugation at 4000 rpm.
6. Wash pellets with 70% ethanol, aspirate, and air dry.
7. Dissolve pellet in TE buffer.
8. Incubate 10ug of the DNA sample with 0.1 U of DNase-free RNase for 2 hr at 37°C.
9. Load DNA into the wells of a 2% agarose gel.
10. Electrophorese in 0.5x TBE buffer at 4V/cm for approx. 7 hr.
11. Stain gel for 30 min with ethidium bromide.
12. Destain the gel for a few hours in water.
13. Examine the gel on a UV illuminator and photograph.

A DNA ladder with 180 bp periodicity is produced if endonuclease activity is present, a common symptom of classical apoptosis.

It has also been shown that early on relatively large 50kbp and 300kpb fragments are produced probably representing chromatin loops detached from the nuclear matrix. The enzymes that catalyse DNA fragmentation are usually non-lysosomal nuclear endonucleases. They may be constitutive and can be spuriously activated, thus positive results need to be interpreted with caution. In certain cells they are activated by Ca^{++} and Mg^{++} and inhibited by Zn^{++}. Other cells appear to have Ca^{++} and Mg^{++}-independant endonucleases and some invertabrate cells such as those of *C. elegans* have a lysosomal endonuclease which is only activated in phagocytosing host cells, subsequent to apoptosis.

20.3.1.1 *Single Cell Gel DNA Comet Assay*

The single cell gel assay was originally developed for the detection of DNA double strand breaks[59] and was later modified to detect single strand breaks [60]. The assay is based on the migration of DNA in an electric field. Cells are embedded in an agarose gel, lysed and electrophoresed; the mobile DNA is then visualized with a fluorescent DNA binding dye. Broken DNA migrates towards the anode in the electric field creating the charateristic comet tail appearance. The extent of DNA damage appears to be related to comet tail length. The method has been developed to provide sensitive and accurate detection and measurment of DNA strand breaks in individual cells [61]. Although the technique reflects levels of DNA damage and can be useful in genotoxic studies it was not designed primarily for demonstrating apoptotic change. Nevertheless, it has been used to demonstrate enhanced double strand breaks in the context of apoptosis and future improvements in imaging and analysis may extend its application in this field.

20.3.2 Cytochemistry

20.3.2.1 *Molecular Basis*

Several biochemical kits are now available to assay DNA fragmentation e.g., Boehringer Mannheim and Oncor (Apotag kit), based on the TUNEL reaction (see Table 20.7) and these

TABLE 20.7

TUNEL Method for Fixed Tissue Samples

1. Fix tissues in 10% neutral buffered formalin.
2. Embed in paraffin wax.
3. Cut 4um sections of the embeded material.
4. Pick up sections on albuminized slides.
5. Dewax the sections with xylene and bring them to water via a graded series of ethanol.
6. Remove proteins from section with PK for 30 min at room temperature.
7. Wash slides several times in DDW.
8. Inactivate endogenous peroxidase with 2% H_2O_2 for 5min at room temperature.
9. Add TdT and biotinylated dUTP in TdT buffer and incubate at 37°C for 1hr.
10. Stop reaction by immersing slides in terminating buffer for 15 min.
11. Rinse slides in DDW and immerse in PBS for 5 min.
12. Place slides in DAB/H202 solution for 30 min at 37°C.
13. Counterstain sections with haematoxylin or methyl green.

As a result of endonuclease activity, sites of DNA fragmentation appear brown following
  DAB reaction.

may use fluoresein, alkaline phosphatase or peroxidase for inmmunocytochemical reporting of reaction.

Kits are available as biochemical ELISA techniques or for *in situ* histochemical reporting
(Table 20.8).

Not all methods are entirely dependant on tagging fragmented DNA with dUTP via terminal transferase (TdT), for example Boehringer Mannheim Cell Death Detection ELISA
use an antihistone antibody as well as an anti-DNA component. When the DNA fragments
it not only exposes cut-ends of DNA but also releases histone content from the enclosed
nucleosomes. This fact has been exploited recently in developing an anti-histone histochemical technique for detecting impending apoptosis [62] (see Table 20.9).

A method has been recently developed to detect "early" apoptosis. This is based on the
fact that cells appear to undergo a change in membrane phospholipid asymmetry. Negatively charged phospholipids such as phosphatidylserine (PS), usually found on the inner
surface of the plasma membrane bilayer, are exposed on the exterior of the membrane early
on in apoptosis, whilst membrane integrity is still maintained. A kit which employs
annexin V conjugated to FITC that specifically binds PS is marketed by R&D Systems
Europe Ltd., and Clonotech Inc., USA.

Lastly evidence of changes in the subcellular traffic of acid hydrolases during cell death
exists [63,64], although close scrutiny shows that most of these changes involving *de novo*
synthesis of acid phosphatase, occur during non-apoptotic forms of cell death and will be
dealt with separately at the end of this chapter. Some authorities [65,18 have noted enhanc-
ment in the levels of autophagic lysosomal hydrolases as a prelude to cell death and it is
generally accepted that extensive increases in hydrolase actvity occurs in the secondary
lysosomes of host cells and macrophages following phagocytosis of apoptotic bodies.

The histochemical and cytochemical localization of acid phospahatase can thus, be an
useful tool to track non-apoptotic cell death and the secondary stages of apoptotic cell
death. In this context, it is also worth noting that activation of proteases (caspases) e.g., ICE-
protease which is homologus to *ced-3*, plays a crucial and early role in the cascade of
changes that lead to apoptosis. Further attention should, therefore, be paid to the localization of proteases into apoptotic cell death.

TABLE 20.8

Apop-Tag Technique *in situ* Localization

The method is used to detect apoptotic cells by direct immunoperoxidase localization of digoxigenin-labeled genomic DNA in paraffin embeded tissue.

1. The tissue sections are deparaffinized by 3 changes of xylene and 3 changes of alcohol (5 min each change).
2. Washed in PBS for 5 min.
3. Proteinase K (20ug/ml) is applied to sections for 15 min at room temperature.
4. Washed in 4 changes of distilled water (2 min each wash).
5. Endogenous peroxidase is quenched in 2% H_2O_2 in PBS for 5 min.
6. After blotting, 2drops of 1X Equilibrium buffer is applied to sections. Plastic coverslips provided are used and sections incubated for 15 min at room temperature.
7. Coverslips are removed, and slides blotted and immersed in 54 ul of working strength TdT enzyme and incubated in a humid chamber at 37°C for 1 hr.
8. Specimens are transfered to working strength stop/wash buffer and incubated for 10 min at room temperature.
9. Sections were washed in 3 changes of PBS (5 mins each).
10. Two drops of anti-digoxigenin-peroxidase was applied to the slides, covered with plastic coverslips and incubated in a humid chamber for 30 min at room temperature. They were then washed in 3 changes of PBS.
11. Slides were transfered into DAB/H_2O_2 for 5 min at room temperature and washed extensively with distilled water.
12. Sections were counterstained in hematoxylin or methyl green.

Dark brown product appears at sites of endonuclease induced DNA fragmentation. The product may be intensified with silver enhancement.

TABLE 20.9

Anti-Histone Antibody Technique [62]

For use with paraffin wax, L R White embedded or cryosections.

1. Paraffin sections are dewaxed in 3 changes of xylene and 3 changes of alcohol.
2. Endogenous peoxidase blocked with a mixture of 75.5 ml methanol and 1.3 ml H_2O_2 for 30 min at room temerature.
3. Washed in tap water for 5 min and held in PBS pH 7.1.
4. Sections are incubated overnight in 200 μl of anti-histone antibody (conc. 1.5 ug/ml tris buffer) at 4°C and then rinsed in PBS.
5. 200 ul of mouse immunoglobulin conjugated with peroxidase is applied to each section and left for 1 hr at room temperature.
6. Sections are washed with PBS and transferred into DAB/H_2O_2 for 5 min.
7. Specimens are then washed in distilled water and counterstained with either haematoxylin or methyl green.

Peroxidase positive areas indicative of exposed nucleosomal histone appear brown. Contrast may be improved with silver enhancement.

20.3.3 Cell Viability Assays

Some classical vital staining methods commonly used in developmental biology have already been alluded to, however, a number of cell viability assays are in common biochemical usage and although not directly indicative of apoptosis they can provide a good estmate of consequential cell death. The most commonly used metabolic imairment assays include the neutral red viability assay, the MTT viability assay and the Fluorescein diacetate (FDA) membrane integrity assay:

20.3.3.1 Neutral Red Viability Assay

Neutral red normally accumulates in the lysosomes of living cells. Dead cells loose this ability to accumulate and hold neutral red. It thus, provides a simple colorimetric viability test which is widely used in cellular toxicology. Success with this assay depends on employing suitable dye incubation periods and this will vary with cell type and cell line; an avarage protocol is presented in Table 20.10, but preliminary experiments may be required to establish optimum incubation and washing times.

TABLE 20.10
Neutral Red Cell Viability Test

Suitable for cells in suspension or living tissues.

1. Make up a 0.4% stock solution of Neutral Red in distilled water and protect from direct light with silver foil.
2. Dilute stock solution 1:100 with saline or growth medium.
3. Warm solution to 37°C and centrifuge to remove undissolved Neutral Red.
4. Incubate cells or tissues for approx. 4 hr at 37°C (adjust as necessary).
5. Remove Neutral Red solution and wash cells quickly in 4% neutral buffered formalin containing 1% calcium chloride.
6. Examine the cells microscopically.
7. Solubilize the residual dye with 1ml glacial acetic acid in 50% ethanol.
8. Read absorbance at 540 nm.

20.3.3.2 MTT Viability Assay

Tetrazolium reactions are used widely to indirectly demonstrate dehydrogenase activity. The test relies on the ability of living cells to reduce tetrazolium salts to coloured insoluble formazan. The tetrazoliums do not react directly with the dehydrogenases, but rather through their products NADH and NADPH, thus the yellow-colored MTT is converted by reduction reaction into a purple formazan. The method which is widely used as a histochemical technique hs been adapted for use with microtitre plates as a cellular viability assay (see Table 20.11).

The method can give variable results and the reaction must be measured over a linear range in the presence of exess MTT under optimal conditions.

20.3.3.3 Fluorescein Diacetate Assay

Fluorescein diacetate is used extensively for testing the viability of cell lines and also isolated plant protoplasts (see Table 20.12). It is an electrically neutral non-fluorescent

TABLE 20.11

MTT Cell Viability Assay

1. Prepare a stock solution of MTT (2μg/ml in PBS).
2. Add 50μl of the MTT stock to 200μl of tissue culture medium.
3. Incubate for approx. 3 hrs at 37°C (adjust as necessary).
4. Remove medium and add 24μl of Sorenson's glycine buffer.
5. Add 200ul of DMSO and incubate for 15 min whilst shaking to solubilize the formazan.
6. Read absorbance at 570 nm.

TABLE 20.12

Fluorescein Diacetate Cell Viability Assay

1. Make up 10μg/ml stock solution of FDA in DMSO and store at –20°C, protect from light with silver foil.
2. Remove tissue culture medium from test plates and wash gently with PBS.
3. Add 200μl of warm FDA and incubate at 37°C for 1 hr (adjust as necessary).
4. Centrifuge at 200g for 5 min annd remove solutions from plates.
5. Add warm PBS to each well.
6. Read fluorescence quickly at 485/538 nm excitation/emission wavelengths.

molecule which accumulates in viable cells and is hydrolysed by intracellular esterase to release highly fluorescent fluorescein.

This is retained over a short period within viable cells. Dead cells do not accumulate fluorescein diacetate and lack the capability to hydrolyse it and therefore, appear non-fluorescent. The compounds are non-toxic and thus, provide an assay in which the cells can subsequently be reused for other purposes.

20.4   Flow Cytometric Assays

Because flow cytometry is able to measure several attributes of the cell and can indeed be adapted to use most of the biochemical and molecular markers outlined above it has become a major tool for demonstrating, quantifying and indeed, isolating apoptotic bodies and apoptotic cells [66]. This versatility stems from the fact that the method is able to accurately measure cell or particle size in terms of its light scattering properties whilst simultaneously measuring fluoresence, revealing molecular or chemical properties within the cell or fragment. In addition, modern instruments also permit particular populations of cells or their fragments to be sorted and harvested electronically for further study.

Since apoptosis usually involves cell shrinkage and subsequent fragmentation to apoptotic bodies of intermediate size and since necrosis initially involves swelling followed by a catastrophic dissembling to particles of minute size, it is not surprising that flow cytometry can readily distinguish between these two modes of cell death on light scattering properties alone. Thus, a reduced ability to scatter light in the forward direction and either an increase or no change on the 90' light scatter, characterize cells in the early phase of apoptosis. Such an assay could be combined with an analysis of surface immunofluorescence or uptake of a vital dye such as Hoechst 33342. Indeed, it is often essential to combine complementary techniques to obtain definitve answers [67,68]. A very wide range of immunofluorescent paramaters may be measured, only a sample of which can be detailed here.

One useful differentiating technique to apply consists of removal of long-dead cells by preincubation with DNase 1 and trypsin. Limited exposure to these reagents do not compromise live cells, whilst cells with damaged plasma membranes are digested, thus weeding out isolated nuclei, necrotic and very old apoptotic cells, leaving early-apoptotic and live cells unharmed.

As indicated previously, a wide range of viability and immunofluorescent markers may now be selected for including for example Trypan Blue, Fluorescein diacetate, Propidium iodide, Hoechst 33342, and also importantly, DNA strand breaks can be labelled in apoptotic cells using biotinylated dUTP(see Table 20.13) or alternatively digoxygenin conjugated dUTP (as in Table 20.8).

TABLE 20.13

Labelling DNA Strand Breaks for Flow Cytometry

Suitable for cells in suspension.

1. Fix cells in 1% neutral buffered formalin for 10 min.
2. Sediment cells and resuspend in 5 ml HBSS.
3. Resuspend in 70% ethanol at 0–4°C and store at –20°C (cells may be kept for a few weeks).
4. Rinse in HBSS and resuspend in 50μl of:
    10 μl reaction buffer
    1 μl biotin-16-dUTP
    0.2μl of TdT in storage buffer
    38.8μl distilled water.
5. Incubate in the mixture for 30 min at 37°C.
6. Rinse in rinsing buffer and centrifuge for 5 min at 100g.
7. Resuspend pellet in 100 ul of saline-citrate buffer containing fluoresceinated avidin.
8. Incubate at room temperature for 30 min.
9. Add 1.3 ml of the rinsing buffer and centrifuge as previous.
10. Rinse again and resuspend the pellet in 1ml of PI/RNase A solution.
11. Incubate for 30 min at room temperature and analyse the cells by flow cytometry.
12. Illuminate with blue (488 nm) light and measure fluoresceinated avidin at 330 nm. Measure PI fluorescence at 620 nm.

Using flow cytometry Sun et al. [68], describe an useful combination of markers using the vital dye Hoechst 33342 and the DNA intercalating agent, Propidium iodide, which enabled the identification of three distinct populations of cells to be identified and sorted. Dead cells flouresced red due to propidim iodide whereas normal and apoptotic cells fluoresced blue due to Hoechst 33343. Apoptotic cell were distinguished from normal cells both by their higher intensity of blue fluoresence and by their smaller size as determined by a reduction in forward light scatter. The population characteristics matched parallel microscopical studies.

20.5 Non-Apoptotic Programmed Cell Death

Not all forms of programmed cell death are apoptotic [17,18,31] as was emphasised in the introduction. Such conclusions are not only based on histological and histochemical findings. Careful flow cytometric studies [66] report profiles, "atypical apoptosis" and even raise the possibility of delayed reproductive or mitotic death. To some extent the norm for apoptosis has been based on thymocyte cell death [46]. This is a cell with very sparse cytoplasm, consisting almost entirely of nucleus. It is not surprising therefore, that endonuclease activation and DNA fragmentation, rather than protease or hydrolase activities, have until recently, received more attention.

As indicated in the introduction, at least three morphologically dictist types of programmed cell death have been described in the literature. These include Type 2 Cell Death where early changes involve an elevation of autophagic activity [1,18] and also in many instances the apperance of a characteristically extracisternal free acid phosphatase leading to a final cell autolysis. In this type of cell death the nucleus show no margination of chromatin and persists until the cytoplasm disintegrates. Examples of this type of cell death come mainly from invertebrates and have been described in starving planarian worms [20–22,69] in regenerating planarians [23], in developing mollusca [70,25] and in metamorphosing insects [25,71,72], including *Drosophila* [73] and indeed, in insect pathology [62].

In essence, most of the studies listed above draw as their base the cytochemical observation that *de novo* sythesis of extracisternal ribosomal acid phosphatase occurs very early on in Type 2 cell death, variously called autophagic cell death and single cell deletion. The utility of the cytochemical technique is based on the selection of a non-specific substrate, p-nitrophenyl phosphate as substrate for the ultrastructural localization of acid phosphatase and impending cell autolysis leading to a programmed cell death (Table 20.14).

TABLE 20.14

Localization of Acid Phosphatase in Programmed Cell Death [63,21]

Suitable for fixed tissues and cells.

1. Fix cells and tissues as described in Table 20.4.
2. After prefixation non-frozen sections (50 µm) arecut on a tissue chopper
3. Incubate slices in 0.1M acetate buffer, pH 5.0 containing 3.8 mM p-nitrophenyl phosphate (Sigma) and 2.64 mM lead acetate (BDH) for 45 min at 37°C
4. Tissues are then rinsed in cacodylate buffer and post-fixed in osmium tetroxide as indicated in Table 20.4.

Sites of lysosomal and free acid phosphatase are localized in the electron microscope as dark areas of lead phosphate reaction product. Specific extracisternal acid phosphatase appears on ribosomal surfaces preceding cell death.

Not only, therefore, are there multiple pathways to apoptosis [74] but, there are also multiple forms of programmed cell death, one of which may be described as apoptotic.

References

1. Lockshin, R. A., and Williams C. M., Programmed cell death ll. Endocrine potentiation of the breakdown of the intersegmental muscles of silk4oths, *J. Insect Physiol.*, 10, 643, 1964.
2. Lockshin, R. A., Programmed cell death: Activation of lysis by a mechanism involving the synthesis of protein, *J. Insect Physiol.*, 15, 1505, 1969.

3. Lockshin, R. A., Programmed cell death: Nature of the nervous signal controlling breakdown of intersegmental muscles, *J. Insect Physiol.*, 17, 149, 1971.
4. Glucksmann, A., Cell deaths in normal vertebrate ontogeny, *Biol. Rev.*, 26, 59, 1951.
5. Saunders, J. W., Death in embryonic systems, *Science*, 154, 604, 1966.
6. Kerr, J. F. R., Wyllie, A. H. and Currie, A.R., Apoptosis: a basic biological phenomenon with wide-ranging implications in tissue kinetics, *Br. J. Cancer.*, 26, 239, 1972.
7. Hayflick, L., Cell death *in vitro*, in *Cell Death in Biology and Pathology*, Bowen, I. D. and Lockshin, R. A., Eds., Chapman & Hall, London & New York, 1981, 243.
8. Wyllie, A. H., Cell death: a new classification separating apoptosis from necrosis,. in *Cell Death in Biology and Pathology*, Bowen, I. D. and Lockshin, R. A., Chapman & Hall, London & New York, 1981, 9.
9. Wyllie, A. H., Duvall, E. and Blow, J. J., Intracellular mechanisms in cell death in normal and pathological tissues, in *Cell Aging and Cell Death*, Davies, I. and Sigee, D. C., Cambridge University Press, Cambridge, London & New York, 1984, 269.
10. Trump, B. F., Berezesky, I. K. and Osoeino-Vargas, A. R. Cell death and the disease process: the role of calcium, in *Cell Death in Biology and Pathology*, Bowen, I. D. and Lockshin, R. A., Eds., Chapman & Hall, London and New York, 1981, 209.
11. Bowen, I. D., Apoptosis or Programmed Cell Death? *Cell Biol. Internat.*, 17, 365,1993.
12. Ellis, R. E., Yuan, J. and Horovitz, H. R., Mechanisms and functions of cell death, *Ann. Rev. Cell.Biol.*, 7, 663, 1991.
13. White, E., Life death and the pursuit of apoptosis, *Genes and Development*, 10, 1, 1996.
14. Duke, R. C., Sellins, K. S. and Cohen, J. J., Cytolytic lymphocyte-derived lytic granules do not induce DNA fragmentation in target cells, *J. Immunol.* 141, 2191, 1988.
15. Arends, M. J. and Wyllie A. H., Apoptosis mechanisms and roles in pathology, *Int. Rev. Exp. Pathol.*, 32, 223, 1991.
16. Clarke, P. G. H., Developmental cell death: morphological diversity and multiple mechanisms, *Anat. Embryol.*, 181, 195, 1990.
17. Bowen, I. D., Mullarkey, K. and Morgan, S. M., Programmed cell death in the salivary glands of the blow fly *Calliphora vomitoria*, *Microsc. Res. Tech.*, 34, 202, 1996.
18. Lockshin, R. A. and Zakeri, Z., The biology of cell death and its relationship to Aging, in *Cellular Aging and Cell Death*, Holbrook, N. J., Martin, G. R. and Lockshin, R. A., Eds., Wiley-Liss, New York, 1996, 167.
19. Schweichel, J. U. and Merker, H.J., The morphology of various types of cell death in pre-natal tissues, *Teratology*, 7, 253, 1973.
20. Lockshin, R. A. and Williams, C. M., Programmed cell death, V. Cytolytic enzymes in relation to the breakdown if intersegmental muscles of silkmoths, *J. Insect Physiol.*, 11, 831, 1965.
21. Bowen, I. D. and Ryder, T. A., Cell autolysis and deletion in the planarian *Polycelis tenuis* Iijima, *Tiss. Cell Res.*, 154, 265, 1974.
22. Bowen, I. D. and Ryder, T. A., Use of the p-nitrophenyl phosphate method for the demonstration of acid phosphatase during starvation and cell autolysis in the planarian *Polycelis tenuis* Ijima, *Histochem J.*, 8, 318, 1976.
23. Bowen, I.D., Ryder, T. A. and Dark, C., The effects of starvation on the planarian worm *Polycelis tenuis*, *Cell Tiss. Res.*, 169, 193, 1976.
24. Bowen, I. D., Den Hollander, J. E. and Lewis, G.H.J., Cell death and acid phosphatase activity in the regenerating planarian *Polycelis tenuis*. *Differentiation*, 21, 160, 1982.
25. Bowen, I. D., Morgan, S. M. and Mullarkey, K., Cell Death in the salivary glands of metamorphosing *Calliphora vomitoria*, *Cell Biol. Internat.*, 17, 13, 1993.
26. Jones, G. W. and Bowen, I.D., The fine structural localization of acid phosphatase in pore cells of embryonic and newly hatched *Deroceras reticulatum* (Pulmonata:Stylommatophora), *Cell Tissue Res.*, 204, 253, 1980.
27. Jones, H. E. and Bowen, I. D., Acid phosphatase activity in the larval salivary glands of developing *Drosophila melanogaster*, *Cell Biol. Internat.*, 17, 305, 1993.

28. Levi-Montacini, R. and Aloe, L., Mechanisms of action of nerve growth factor in intact and lethaly injured sympathetic nerve cells in neonatal rodents, in *Cell Death in Biology and Pathology,* Bowen, I. D. and Lockshin, R. A., Eds., Chapman and Hall, London & New York, 1981, 295.

29. Server, A.C., and Mobley, W. C., Neuronal cell death and the role of apoptosis, in *Apoptosis: the Molecular Basis of Cell Death,* Tomei, L. D. and Cope, F. O., Eds., Current communications in cell and molecular biology Vol. 3. Cold Spring Harbor Laboratory Press, 1991, 263.

30. Martin, D. P., Johnson, J., Programmed cell death in the peripheral nervous system, in *Apoptosis: the Molecular Basis of Cell Death,* Tomei, L. D. and Cope, F. O., Eds., Current communications in cell and molecular biology, Vol. 3.Cold Spring Harbor Laboratory Press, New York, 1991, 247.

31. Szende, B., Keri, Gy., Szegedi, Zs., Benedecsky, I., Csikos, A., Orfi, L. and Gazit, A., Tyrphostin induces non-apoptitic programmed cell death in colon tumour cells, *Cell Biol. Internat.,* 19, 903, 1995.

32. Bowen, I. D., Bowen S. M. and Jones A. H., *Mitosis and Apoptosis—Matters of Life and Death.* Chapman & Hall, London and New York, 1998.

33. Jazwinski, S. M., Genes of youth: Genetics of aging in Baker's yeast, *ASM News,* 172, (1992).

34. Eleftheriou, E.P., Ultrastructural studies on protophloem sieve elements in *Triticum aestivum* L. Nuclear degeneration, *J. Ultrastrucr. Mol. Struct. Res.,* 95, 47, 1986.

35. Gahan, P. B., Bowen, I. D. and Winters, C., Plant bioregulator-induced apoptotic-like behavior of nuclei from mesophyll cells of *Solanium avicular, Proc. Roy. Microsc. Soc.,* 30, 117, 1995.

36. Mittler, R. Shulaev, V. and Lam, E., Coordinated activation of programmed cell death and defence mechanisms in transgenic tobacco plants expressing a bacterial proton pump. *The Plant Cell,* 7, 29, 1995.

37. Garrieli, Y., Sherman, Y. and Ben-Sasson, S. A., Identification of programmed cell death *in situ* via specific labelling of nuclear DNA fragmentation, *J. Cell Biol.,* 119, 493, 1992.

38. Kosslak, R. M., Chamberlin, M. A., Palmer, R. G. and Bowen, B. A., Programmed cell death in the root cortex of soybean *root necrosis* mutants, *The Plant Journal,* 11, 729, 1997.

39. Thomas, H., Aging in the plant and animal kingdoms—the role of cell death. *Reviews in Clinical Gerontology,* 4, 5, 1994.

40. Thomas, H., *Sid:* a Mendelian locus controlling thylakoid membrane disassembly in senescing leaves of *Festuca pratensis, Theor. Appl. Genet.* 73, 551, 1987.

41. Thomas, H., Ougham, H.J. and Davies, T. G. E,. Leaf sensesence in a non-yellowing mutant of *Festuca pratensis* transcripts and translation products, *J. Plant Physiol.,* 139, 403, 1992.

42. Smart, C. M., Gene expression during leaf senescence. Tansley review 63, *New Phytol,* 126, 419, 1994.

43. Hengarter, M. O. and Horvitz H. R., Activation of *C. elegans* cell death protein CED-9 by an amino acid substitution in a domain conserved in Bcl-2, *Nature Lond.,* 369, 318, 1994.

44. White, K., Grether, M. E., Abrams, J. M. Young, L.,Farrell, K. & Steller, H., Genetic control of programmed cell death in *Drosophila. Science,* 264, 677, 1994.

45. Bellamy, C. O. C., Malcomson, R. D. G., Harrison, D. J. and Wyllie, A, H., Cell death in health and disease: the biology and regulation of apoptosis, *Cancer Biology,* 6, 3, 1995.

46. Wyllie, A. H., Glucocorticoid-induced thymocyte apoptosis is associated with endogenous endonuclease activation, *Nature, Lond.,* 284, 555, 1980.

47. Oberhammer, F., Frisch G, Schmied M., Pavelka, M., Printz, D., Purchio, T.,Lassmann, H. and Schulte-Hermann, R., Condensation of the chromatin at the membrane of an apoptotic nucleus is not associated with activation of an endonuclease, *J. Cell Science,* 104, 317, 1993.

48. Schwartz, L. M., Kosz, L. and Kay, B. K., Gene activation is required for developmentally programmed cell death, *Proc. Natl. Acad. Sc. U.S.A.,* 87, 6594, 1990.

49. Schwartz, L. M., The role of cell death genes during development, *BioEssays,* 13, 389, 1991.

50. Schwartz, L. M. and Osborne, B. A., Programmed cell death, apoptosis and killer genes, *Immunology Today,* 14, 582, 1993.

51. Saunders, J. W., Gasseling, M. T. and Saunders, L. C., Cellular death in morphogenesis of the avian wing, *Dev. Biol.,* 5, 147, 1962.

52. Brus, I. and Glass, G. B. J., Estimation of cytotoxic injury to gastric parietal cells by Trypan Blue exclusion test, followed by haematoxylin and eosin counterstaining of fixed smears, *Stain Technol.*, 48, 127, 1973.

53. Attalah, A. M. and Johnson, R. P., A simple, highly sensitive method for the determination of cell viability using electronic analyser, coulter counter, *J. Immunol. Methods*, 41, 155, 1981.

54. Hinsull, S. M., Bellamy, D. and Franklin, A., A quantitative histological assesment of cellular death in relation to mitosis in rat thymus during growth and age involution, *Age and Aging*, 6, 77, 1977.

55. Drake, W.P., Ungoro, P. C. and Mardinay, M. R., Formalin fixed cell preparations as standards for use in the automated trypan blue cytotoxic assay, *Transplantation*, 14, 127, 1972.

56. Lewis, G. H. J. and Bowen, I. D., A methacrylate embedding technique for combined auto-radiography and acid phosphatase histochemistry, *Histochem J.*, 17, 467, 1985.

57. Moffitt, P., A methyl green-pyronin technique for demonstrating cell death in the murine tumour S180, *Cell Biol. Internat.*, 18, 677, 1994.

58. Bowen, I. D. Worrill, N., Winters, C. A. and Mullarkey, K., The use of backscattered electron imaging, X-ray microanalysis and X-ray microscopy in demonstrating physiological cell death, *Scanning Microscopy*, 2, 1453, 1988.

59. Osterling, O. and Johanson, K. J., Microelectrophoretic study of radiation-induced DNA damages in individual mammalian cells, *Bioch. Biophys. Res. Comm.*, 123, 291, 1984.

60. Singh, N. P. McCoy, M. T., Tice, R. R. and Schneider, E. L., A simple technique for quantitation of low levels of DNA damage in individual cells, *Exp. Cell Res.*, 175, 184, 1988.

61. Olive, P. L., D., Durrand, R. E., Le Riche, J., Olivotto, L. A. and Jackson, S. M., Gel electrophoresis of individual cells to quantify hypoxic fraction in human breast cancers, *Cancer Res.*, 53, 733, 1993.

62. Gregorc, A. and Bowen, I. D., Programmed cell death in the honey bee (*Apis mellifera* L.) larvae midgut, *Cell Biol. Internat.*, 21, 151, 1997.

63. Bowen, I. D. and Bowen, S. M., Programmed Cell Death in Tumours and Tissues. Chapman & Hall, London and New York, 1990.

64. Bowen, I. D., Laboratory techniques for demonstrating cell death, in *Cell Aging and Cell Death*, Davies, I. & Sigee, D. C. Eds., Cambridge University Press, Cambridge, London & New York, 1984, 5.

65. Hinchliffe, J. R., Cell Death in Embryogenesis, in *Cell Death in Biology and Pathology*, Bowen, I. D. and Lockshin, R. A, Eds., Chapman & Hall, London and New York, 1981, 35.

66. Darzynkienicz, Z, Li, X., Gong, J., Hara, S. and Traganos, F., Analysis of cell death by flow cytometry, in *Cell Growth and Apoptosis—A Practical Approach*, Stuszinski, G. P., Ed., IRL Press at Oxford University Press, Oxford, New York & Tokyo, 1995, 143.

67. Ormerod, M. G., Collins, M., K., L., Rodriguez-Tarduchy, G., and Robertson, D., Apoptosis in interleukin-3-dependant haemopoietic cells — quantification by two flow cytometric methods, *J. Immunol. Methods*, 153, 57, 1992.

68. Sun, X-M., Snowden, R. T., Skilleter, D. N., Dinsdale, D., Ormerod, G. G. and Cohen, G. M., A flow-cytometric method for the separation and quantitation of normal and apoptotic thymocytes, *Analytical Biochemistry*, 204, 351, 1992.

69. Bowen, I. D., Phagocytosis in *Polycelis tenuis*, in *Nutrition in Lower Metazoa*, Smith, D. C. and Tiffon, Y., Eds.,Pergamon Press, Oxford, 1980, 1.

70. Bowen, I. D. and Davies, P., The fine structural distribution of acid phosphatase in the digestive gland of *Arion hortensis* (Fer.), *Protoplasma*, 98, 63, 1971.

71. Skelton, J. K. and Bowen, I. D., The cytochemical localisation and backscattered electron imaging of acid phosphatase and cell death in the midgut of developing *Calliphora vomitoria*, *Epithelia*, 1, 213, 1988.

72. Bowen, I. D., Mullarkey, K. and Worrill, N., Backscattered electron imaging, X-ray micro-analysis and X-ray microscopy in the study of cell death, *Micron & Microscopica Acta*, 23, 347, 1992.

73. Jones, H. E. and Bowen, I. D., Acid phosphatase activity in the larval salivary glands of developing *Drosophila* melanogaster, *Cell Biol. Internat.*, 17, 305, 1993.74.

74. Evans, V.G., Multiple pathways to apoptosis, *Cell Biol. Internat.*, 17, 461, 1993.

21

Mitochondrial DNA Deletions

Maria N. Gadaleta, Angela M. S. Lezza and Palmiro Cantatore

CONTENTS

0-8493-3112-9/99/$0.00+$.50
© 1999 by CRC Press LLC

21.1 Introduction

Mitochondria are complex organelles found in all eukaryotes. They are involved in many crucial metabolic processes and possess their own genetic information and protein synthesis machinery. The mitochondrial DNA (mtDNA) is usually a covalently closed circular molecule and is present in multiple copies in each mitochondrion. Its size varies from about 14 kb in the green alga *Chlamydomonas reinhardtii* to 16 kb in mammals up to 2400 kb in some higher plants [for review see Ref. 1–4]. The mtDNA codes in mammals for 13 subunits which, together with nuclear-encoded polypeptides, constitute four out of five mitochondrial oxidative phosphorylation complexes. Through oxidative phosphorylation mitochondria produce the major fraction of the cellular energy. This bioenergetic capacity, as other mitochondrial functions, shows an age-dependent deficit which might be the driving force behind the process of aging itself [5]. Such a functional decrease might be the result of macromolecular modifications that compromise the oxidative phosphorylation system as well as the mitochondrial genome [6].

Numerous reports have suggested that mtDNA is more prone to mutations than nuclear DNA, because of its physical proximity to the major cellular sources of oxygen radicals and of the lack of coverage by histones [7–9]. Following this hypothesis, many laboratories searched for age-dependent modifications of mtDNA. The initial studies, carried out with classical approaches such as restriction enzyme analysis, Southern blotting or sequencing of mtDNA regions from old individuals did not reveal any variation [10,11]. These studies were then continued by using more sensitive approaches. It was found that the level of one specific modified base, 8-hydroxydeoxyguanosine (8-OH-dG), was 16-fold higher in the mtDNA than in the nuclear DNA of rat liver [12,13] and that 8-OH-dG accumulated with age in rat liver as well as in human heart and brain [14,15]. The search for age-specific mutations in the mtDNA of old subjects did not provide any result, until Cortopassi and Arnheim [16] developed a novel strategy for detecting mutated mtDNA molecules present even at a very low concentration. The approach consisted in amplifying mtDNA by polymerase chain reaction (PCR) in conditions unfavouring the amplification of the longer, wild-type mtDNA with respect to the shorter, mutated molecule [17]. By using primers encompassing a 4977-bp mtDNA region, deleted in the mtDNA of patients with mitochondrial myopathies [18,19], Cortopassi and Arnheim [16] found that the same region was deleted, although at a very low level (around 0.1%), in various human tissues of healthy old individuals. Since this initial identification many studies were focused on age-dependent deletions of the mtDNA in different tissues from several species. Deletions of various length were reported: some were common for many tissues, as the previously mentioned 4977-bp deletion, then named "common deletion", some were tissue-specific [20–29]. In most of the cases the deletions were flanked by direct repeats, but deletions lacking such motifs were also reported [23,27,30]. The mechanism of formation of deleted mtDNA molecules has not yet been completely clarified; prevalent theories are based either on the slipped mispairing between direct repeats during replication [31] or on the direct repeats-linked homologous recombination [32]. Recently Chung et al. [33] proposed a new model based on the "replication jumping", which hypothesizes that oxidative damage to mtDNA results in replication errors, leading to deletion formation.

21.2 Background

The development of the polymerase chain reaction has represented a milestone in many fields of the biological research [34]. Just to remain in the area of mtDNA deletions in aging, this technique has revealed deletions present in a very small minority of molecules. MtDNA deletions in aging exhibit several peculiar characteristics, such as the high number of different deleted species, the very low percentage with respect to the total mtDNA and the mosaic distribution among tissues and in the same tissue as well [22,35–40]. The highest levels of mtDNA deletions are found late in life in postmitotic tissues such as brain, heart and muscle, with deletion levels increasing over 10,000-fold from young to old humans [37–39,41–44]. On the contrary, mtDNA deletions are rare in rapidly dividing tissues such as platelets [45] and peripheral and splenic limphocytes [36]. It was proposed that postmitotic tissues accumulate deleted mtDNA molecules because they are not diluted out by cell division and possibly also because the smaller deleted mtDNA is preferentially replicated in these tissues [6,16]. The predominance of the shorter deleted mtDNA over the longer wild-type molecule, is not due to a difference in the rate of completion of the the wild-type and deleted mtDNA synthesis [46]. Factors other than a differential rate of genome duplication are the likely causes of the accumulation of deleted mtDNA molecules during aging. On the other hand, the low level of deleted mtDNA in actively replicating cells has been attributed to the growth retardation of cells that harbor higher levels of deleted mtDNA, having as consequence an overgrowth of normal cells [36]. Another important feature of the mtDNA deletions is their mosaicism in the same tissue: a difference of two to three orders of magnitude was observed for the 4977 bp deletion level in different brain regions [37,39]. This result was suggested to be due to monoaminooxidase (MAO) activity which should induce H_2O_2 production in brain regions active in dopamine metabolism and higher level of deletions. Therefore, although the average proportion of 4977-bp deleted genomes is around 0.1%, a value well below the threshold level for the occurrence of a mitochondrial myopathy [19], this focal distribution of the deletion might lead to a tissue mosaic where some cells have an overall high level of the deletion with a grossly defective bioenergetic capacity. Moreover, considering that many other mtDNA deletions can occur (see above) the characterized mtDNA deletions could constitute just the tip of the iceberg of all possible kinds of DNA mutations accumulating with age [36,47]. This may result in a progressive decline in the efficiency of energy production and, according to the somatic mutation theory [48–50], under particular stress conditions, severe impairment or even death may be the consequence when a critical percentage of cells of a given organ (heart, brain) becomes defective [51].

An alternative theory considers mtDNA deletions an effect of aging rather than a cause. In particular, the correlation found by Lezza et al. [44] between the decline of the oxidative phosphorylation capacity and the increase in the level of the mtDNA[4977] during aging is supposed to be due to an increased generation of oxygen radicals, mainly produced by

Complex III [52]. The radicals might leak from the respiratory chain and hit the mtDNA molecules thus causing the somatic accumulation of deleted molecules. According to this view the deleted mtDNA should not represent the tip of the iceberg, but perhaps the iceberg itself [53]. In any case, independently of the possibility that mtDNA deletions may be among the direct causes of tissue aging, they must be considered a signal of mitochondrial damage and their accumulation may render the different tissues of an individual less prone to sustain stresses. Therefore, the analysis of the mtDNA deletions during aging represents an useful tool to evaluate the senescence of a tissue in an individual.

In this chapter we will focus our attention on the methods used to detect mtDNA deletions in aged healthy individuals. We will first consider the basic methods to identify deleted mtDNA. Then, we will describe more refined techniques for a more detailed study of mtDNA deletions. Finally, a section will be dedicated to the description of the methods set up to quantify mtDNA deletions in aged individuals. The organism in which these methods will be described is the human; however, whenever possible, procedures and results with other mammals will be described.

21.3 General Procedures

21.3.1 Materials and Information Needed

Obtain human bioptic or autoptic tissues with the informed consent of the patient or of the competent authority. Treat them as specified in Section 21.3.3. All reagents are of analytical grade. PCR primers are purchased from MedProbe (Oslo, Norway) or Tib-MolBiol (Genoa, Italy). Taq DNA polymerase is from Boehringer or Perkin Elmer. All radionuclides are from Amersham or NEN-Dupont. PCR is carried out in the Perkin Elmer DNA Thermal Cycler (mod. 480 or mod. 2400) or in the MJ Research DNA Engine Peltier Thermal Cycler (mod. PTC-200).

21.3.2 Buffers and Media Composition

RSB
  10 mM Tris-HCl pH 7.4
  10 mM NaCl
  25 mM EDTA

Lysis Buffer (LB)
  10 mM Tris-HCl pH 8.0
  25 mM EDTA
  0.5% SDS (w/v)

PBS
  2.7 mM KCl
  1.5 mM KH_2PO_4
  137 mM NaCl
  8 mM Na_2HPO_4

MSB-Ca²⁺

> 210 mM mannitol
> 70 mM sucrose
> 50 mM Tris-HCl pH 7.5
> 3 mM CaCl$_2$

1 × SDS buffer

> 120 mM NaCl
> 0.5 mM EDTA
> 5 mM Tris-HCl pH 7.4
> 1% SDS (w/v)

1 × SSC

> 150 mM NaCl
> 15 mM sodium citrate

1 × TBE

> 90 mM Tris-borate
> 2 mM EDTA pH 8.3

1 × Denhardt

> 0.02% Ficoll 400
> 0.02% polyvinylpyrrolidone
> 0.02% bovine serum albumin

TE

> 1 mM EDTA
> 10 mM Tris-HCl pH 7.4

1 × SSPE

> 0.17 M NaCl
> 10 mM NaP0$_4$ pH 7.7
> 1 mM EDTA

1 × TAE

> 40 mM Tris-acetate pH 7.8
> 1 mM EDTA

21.3.3 Isolation of Total DNA from Different Tissues

21.3.3.1 *Human Skeletal Muscle (whole tissue)*

Powderize frozen skeletal muscle samples (100–200 mg) under liquid nitrogen, let them thaw, resuspend in 0.5 ml/100 mg tissue of RSB and digest with 1 µg/µl proteinase K, 1%

SDS (w/v) at 50°C for 2 hrs. Add 50 µl of 5M NaCl/0.5 ml of RSB and extract the DNA once with phenol:chloroform:isoamyl alcohol (25:24:1) and once with two volumes of diethyl ether. Precipitate the DNA with ethanol (remove visible nuclear DNA from the sample immediately after ethanol addition), wash with 70% ethanol. Recover the precipitated DNA in 50-100 µl of double-distilled water.

21.3.3.2 *Human Skeletal Muscle (single fibers)*

Isolate single fibers segments under an inverted microscope from 30 µm-thick sections, by aerosol-protected mouth suction with siliconized microcapillaries. Transfer each single fiber segment to a 0.5 ml Eppendorf tube containing 10 µl of double-distilled water and store it at –20°C until being used. Just before amplification, heat the fiber to 96°C for 20 min, chill on ice for 20 min and centrifuge for 1 min at 12,000 × g. Take the supernatant, containing the total DNA from each single fiber, and immediately use it for PCR.

21.3.3.3 *Human Brain*

Take autoptic brain samples (100-200 mg) previously dissected from the contiguous areas and from the associated white matter and powderize under liquid nitrogen. Incubate the thawed samples in 1-2 ml of LB containing 100 µg/ml proteinase K for 2 hrs at 55°C. Extract with phenol:chloroform:isoamyl alcohol (25:24:1) and precipitate the nucleic acids with ethanol and 0.3 M sodium acetate pH 5.3. Wash the pellet with 70% ethanol and resuspend the DNA in 50-100 µl of double-distilled water.

21.3.3.4 *Human Heart*

Take 100-200 mg of frozen autoptic or bioptic cardiac muscle and extract the DNA as described for human brain (Section 21.3.3.3).

21.3.3.5 *Human Liver*

Take 0.5-1.0 g of frozen liver biopsies and extract the DNA as described for human brain (Section 21.3.3.3).

21.3.3.6 *Human Platelets*

Take 30 ml of blood samples and immediately add 1.5 ml of 11 mM EDTA pH 7.4 and centrifuge at 150 × g for 15 min. Transfer the supernatant into a new tube and centrifuge at 1000 × g for 15 min at 4°C. Suspend the pellet in 10 ml of PBS and centrifuge at 1000 × g for 15 min at 4°C. Resuspend this pellet in an adequate volume of TE, add 1.2% SDS (w/v) and 100 µg/ml of proteinase K and incubate at 37°C for 15 min. Extract the lysate once with phenol, then (1-3 times) with phenol:chloroform (1:1) and once with chloroform. Precipitate the nucleic acids with ethanol and 0.3 M sodium acetate pH 5.3. Rinse the pellet with 70% ethanol and resuspend in 20-40 µl of TE.

21.3.3.7 *Rat Skeletal Muscles*

Obtain tibial muscle samples (100–200 mg) from rats of different ages and immediately hand-homogenize in 5 volumes of lysis buffer containing: 4 M guanidine thyocyanate, 25 mM sodium citrate pH 7.0, 0.5% sodium lauroyl sarcosine, 0.1 M β-mercaptoethanol. Extract with 1.2 volumes of phenol:chloroform (1: 0.2) for 10 min at room temperature and incubate for 20 min at –20°C. Centrifuge the suspension at 4°C for 10 min and transfer the upper phase to a new tube. Perform an eventual second organic extraction, combine the upper phases and extract with an equal volume of chloroform. Ethanol-precipitate the nucleic acids in the presence of 0.3 M sodium acetate pH 5.3. Rinse the pellet with 70% ethanol, centrifuge and dry. Resuspend the nucleic acids in 50-100 µl of double-distilled water and incubate with 100 µg/ml RNase A at 37°C for 1 hr. Extract once with phenol and once with chloroform, precipitate with ethanol, rinse the pellet with 70% ethanol and redissolve the DNA in 50-100 µl of double-distilled water.

21.3.4 Isolation of DNA from Mitochondria of Different Tissues

21.3.4.1 *Human Brain*

Take a frozen dissected brain region, weigh and thaw. Homogenize in 2 ml/g tissue of MSBCa^{2+} buffer with a motor-driven glass homogenizer with a teflon pestle (A.H. Thomas Co., Philadelfia, PA, USA). Centrifuge the homogenate at 1,500 × g for 15 min in a swinging bucket rotor and centrifuge the supernatant at 20,000 × g for 20 min. Lysate the pelleted mitochondria by adding 1% SDS (w/v), 100 µg/ml proteinase K and digest at 37°C for 30 min. Extract with phenol-chloroform-isoamyl alcohol (25:24:1) and precipitate the DNA with ethanol and 0.3 M sodium acetate pH 5.3. After pelletting and washing in 70% ethanol, resuspend the DNA in 100-200 µl of double-distilled water.

21.3.4.2 *Rat Liver*

Kill rats of different ages by decapitation. Remove the livers immediately, weigh and chill in cold homogenizing medium (0.25 M sucrose, 1 mM EDTA and 10 mM Tris-HCl pH 7.4). Mince them, wash several times and homogenize in 10 volumes of the medium with a motor-driven glass homogenizer with a teflon pestle at 1,100 rpm by using four up-and-down-strokes. Centrifuge the homogenate at 1,000 × g for 5 min. Resuspend the nuclear pellet and recentrifuge at 1,000 × g for 5 min. Combine the supernatants and centrifuge at 8,000 × g for 10 min. Lysate the mitochondria at a protein concentration of 10 mg/ml in 120 mM NaCl, 10 mM Tris-HCl pH 7.4, 1 mM EDTA, 1.2% SDS (w/v) and extract the nucleic acids by two phenol:chloroform:isoamyl alcohol (25:24:1) treatments. Precipitate the nucleic acids phase with ethanol and 0.2 M NaCl. Wash the pellet with 70% ethanol and dissolve it in 0.5–1 ml/liver of double-distilled water.

21.3.4.3 *Rat Cerebral Hemispheres*

Kill rats of different ages by decapitation and remove their brains. Dissect the cerebella and the cerebral hemispheres. Weigh the cerebral hemispheres and immediately chill in cold homogenizing medium (0.32 M Sucrose, 1 mM EDTA, 10 mM Tris-HCl pH 7.4), then mince and wash them several times with the same medium. Hand-homogenize the suspension in about 20 volumes of medium with 10–12 up-and-down-strokes in a Dounce homogenizer. Centrifuge the homogenate at 3,000 × g for 5 min at 4°C. Discard the nuclear pellet and centrifuge the supernatant at 20,000 × g for 10 min at 4°C to yield a crude mitochondrial pellet. Remove the supernatant and the fluffy layer of mitochondria, resuspend the pellet by delicate hand-homogenization in a small volume of the homogenizing medium (3 ml/g fresh tissue). Lysate the mitochondria with an equal volume of 2 × SDS-buffer containing 100 µg/ml proteinase K for 30 min at room temperature. Extract the nucleic acids twice with phenol:chloroform:isoamyl alcohol (25:24:1) on ice for 15 min. Precipitate the nucleic acids with ethanol and 0.2 M NaCl. Wash with 70% ethanol and dissolve the pellet in 200-500 µl of double-distilled water.

21.3.4.4   Rat Heart

Kill rats of different ages by decapitation and remove their hearts, weigh and immediately chill in cold homogenizing medium (0.25 M Sucrose, 1 mM EDTA, 10 mM Tris-HCl pH 7.4). Mince the hearts, wash them several times and homogenize in 10 volumes of medium as described for the liver mitochondria. Centrifuge the homogenate at $1,000 \times g$ for 5 min at 4°C. Discard the nuclear pellet and centrifuge the supernatant at $8,000 \times g$ for 10 min at 4°C. Resuspend the pelleted mitochondria in a small volume of homogenizing buffer, hand-homogenize and lysate with an equal volume of $2 \times$ SDS-buffer containing 100 µg/ml proteinase K for 30 min at room temperature. Extract the nucleic acids twice with phenol:chloroform:isoamyl alcohol (25:24:1) on ice for 15 min. Precipitate the nucleic acids with 0.2 M NaCl and ethanol. Wash the pellet with 70% ethanol and dissolve it in 100-200 µl of double-distilled water.

21.3.5   Standard PCR

Unless differently specified, carry out the polymerase chain reaction in 50-100 µl of $1 \times$ PCR buffer (10 mM Tris-HCl pH 8.3, 1.5 mM $MgCl_2$, 50 mM KCl) containing 50 pmol of each primer, 200 µM of each dNTP and 1-10 µl of DNA. The amount of DNA depends on the source from which it is extracted. In general, to detect mtDNA deleted molecules, use from 100 ng to 1 µg of total DNA or 10 to 100 ng of DNA isolated from purified mitochondria. In any case, carry out preliminary experiments with different amounts of DNA to optimize the reaction conditions. Denature first the samples for 5 min at 94°C, then add the enzyme (2.5 U of Taq DNA polymerase) and perform the cycling reaction. The amplification conditions are indicated in each of the methods described below (see Section 21.4).

21.4   Experimental Design and Methods

21.4.1   *In Situ* Hybridization

This technique permits the identification of those cells carrying deleted mtDNA molecules even if they are present at a very low level in a cell population. The technique allows the study of the intracellular distribution of the deleted genomes versus the wild-type ones. It relies on isolating serial sections from a frozen tissue sample and on hybridizing them with two different mtDNA probes, one for the deleted molecules and the other for the wild-type species. The comparison of the differential results allows to localize the presence and the eventual accumulation of the two species of mtDNA inside the cells. The technique has been used by Müller Höcker in humans [54] for the study of the accumulation of the 4977 bp deletion in fibers from heart and skeletal muscles and in oxiphillic cells from the parathyroid glands of healthy aged individuals. A more recent development of the *in situ* hybridization technique, mainly used in pathological muscle specimens, utilizes a 45 bp-long "chimeric" probe including both ends of the deleted mtDNA region as well as the 13-bp direct repeat in its middle portion. Such a probe can detect specifically the deleted mtDNA species in the muscle fibers reducing the cost and the time-consumption required by two differential *in situ* hybridization procedures [55].

21.4.1.1   *Preparation of mtDNA Probes*

Prepare by PCR three distinct mtDNA probes [54] of 1087 bp (A), 1638 bp (B) and 843 bp (C). Purify the PCR products by Centricon 100 Microconcentrators (Amicon) and label them by random priming, with the digoxigenin labelling kit (Boehringer Mannheim) according to the suppliers instructions.

Probes A and B hybridize to total mtDNA (wild-type and deleted), whereas probe C can only hybridize to wild-type mtDNA, because its sequence is included in the deleted region, and this makes it specific for the "common deletion".

21.4.1.2 In Situ Hybridization

Obtain the specimens from human tissues such as extraocular muscles (m. rectus sup.), limb muscles, diaphragm and heart from subjects of various ages, presenting no sign of mitochondrial pathologies. The specimen must be frozen in liquid nitrogen vapors not later than 5 hours after the subjects death. From each specimen, cut at least two 8 mm-thick serial sections, lay on siliconized glass slides and air-dry for 30 min. Fix in 4% paraformaldehyde for 45 min, wash in water and dehydrate in graded alcohol. Then wash in PBS containing 5 mM $MgCl_2$, incubate for 10 min with 5 μg/ml proteinase K at room temperature, wash with PBS and treat with 0.25% acetic anhydride in 0.1 M triethanolamine for 5 min at room temperature. Before hybridization, add 50 μg/ml DNase-free RNase in 50 mM NaCl, 10 mM Tris-HCl pH 8.0 for 30 min at 37°C, wash with PBS and then cover with 2–5 ml of pre-hybridization solution (50% formamide, 0.6 mM NaCl, 20 mM Tris-HCl pH 7.5, 1 × Den-hardt, 1 mM EDTA, 0.5 mg/ml sonicated salmon sperm DNA, 10% dextran sulphate, 0.05 mg/ml total yeast RNA, 0.010 μg/ml yeast tRNA) for 2 hr at 42°C. Denature each probe at 92°C for 10 min, add to the prehybridization solution and cover the sample sections, previously denatured at 92°C for 10 min. Hybridize one section with the total mtDNA probe (A or B) and the other with probe C. Incubate at 42–44°C overnight and wash the slides in 2 × SSC for 1 hr at room temperature and in 0.2 × SSC for 3 hr at 50°C. Detect the signals according to the directions of the Digoxigenin-Detection kit by Boehringer Mannheim.

Carry out control hybridizations without denaturation of the tissue DNA or using a plasmid probe without any mtDNA insert.

21.4.1.3 Evaluation of the Results

Figures 21.1 A and B allow to compare the results of an *in situ* hybridization experiment performed on extraocular muscle serial sections using probe C (Figure 21.1 A) and probe B (Figure 21.1 B). The fiber marked with the X shows a relevant signal similar to that of the other fibers after the hybridization with the total mtDNA probe (Figure 21.1 B). The same fiber, tested with the probe for the deleted mtDNA, shows a marked absence of hybridization signal (Figure 21.1 A) suggesting a large fraction of deleted mtDNA molecules in its mtDNA population at the level of that specific section of the muscle fiber. However, it has been demonstrated that the distribution of the various species of mtDNA along each muscle fiber [56] is segmental and this makes very likely that the situation pictured by one section is not homogeneously reproduced along the same muscle fiber.

21.4.2 Identification of Mitochondrial DNA Deletions by Standard PCR

21.4.2.1 Detection of mtDNA Deletions

The detection of mtDNA deletions by PCR may be performed in different ways. To detect well-characterized deletions such as that of 4977 bp it is sufficient to use a couple of primers located outside the deletion break-points. On the other hand, when multiple deletions are searched, amplifications with multiple sets of primers can be carried out.

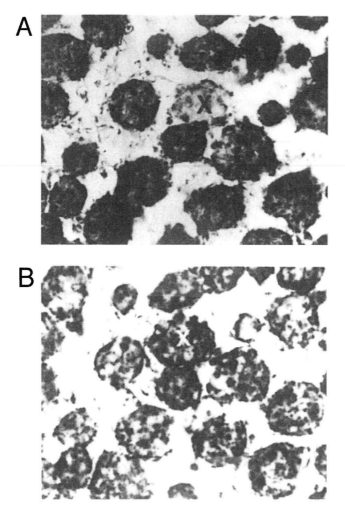

FIGURE 21.1
In situ hybridization of mtDNA in human extraocular muscle fibers. **(A)** Selective hybridization defect in a muscle fiber (X) with a "common deletion" probe (probe C) **(B)** Hybridization with a probe outside the "common deletion" region, showing an intense signal in the same fiber (X) [From Müller-Höcker, J., *Bull. Mol. Biol. Med.*, 18, 25, 1993. With permission.]

The 4977-bp deletion removes a section of human mtDNA between nucleotide positions 8470 and 13447 and is flanked by two perfect 13-bp direct repeats, which have probably a role in the generation of the deleted molecule [32]. To evidentiate deleted mtDNA molecules present at a very low level, a PCR assay, firstly set up by Cortopassi and Arnheim, is used [16]. It relies on designing two primers sufficiently close to the repeats to enable efficient amplification across the deleted region and on performing the PCR by using short cycle times (Figure 21.2). This prevents the amplification of the wild-type mtDNA, where the two primers are more than 5 kb far away, and permits the amplification of the shorter deleted mtDNA.

Carry out PCR as described in Section 21.3.5. Use primers ATP-For (L8282-8305)/13 Rev-Bis (H13928-13905) to detect human mtDNA molecules carrying the 4977 bp deletion (primer positions are according to Anderson et al.[57]).

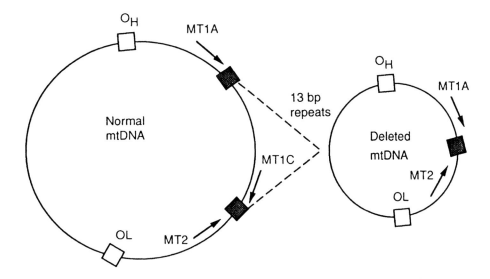

FIGURE 21.2
Schematic representation of the strategy adopted to visualize a 4977 bp deletion in human mtDNA from senescent individuals. By performing the PCR with short cycle times, primers MT1A and MT2 amplify preferentially deleted mtDNA molecules. Primers MT2 and MT1C amplify a section of undeleted mtDNA and are used for normalization of total mtDNA. Shaded boxes indicate the 13 bp direct repeats. O_H and O_L denote the mitochondrial origins of replication. [From Soong, N.-W., and Arnheim, N., *Methods in Enzymology*, 264, 421, 1996. With permission.]

Amplify for 30 to 40 cycles with 1 min denaturation at 94°C, 1 min annealing at 65°C and 1 min extension at 72°C. In these conditions, no amplification of undeleted mtDNA molecules occurs, even if such sequences are largely in excess. If needed, it is possible to increase the number of cycles, but in this case it is advisable to add, after 20–30 cycles, another aliquot of enzyme. At the end of cycling eventually perform a final extension step of 3–5 min at 72°C. Always include proper contamination controls since the amount of mtDNA deleted molecules in old healthy individuals is generally very low.

To visualize the reaction products (that in this case should consist in a unique band of 669 bp), load 10–15 μl of the reaction mixture on a 1–1.2% agarose gel, run at 150 Volts in 1 × TBE and visualize the DNA bands by ethidium bromide staining. If the amplification products are present at a very low level, ethanol-precipitate the DNA in the presence of 0.3 M sodium acetate pH 5.3, wash the pellet with 70% ethanol, resuspend in 10–15 μl of double-distilled water and load the DNA on the gel. Alternatively, include in the reaction mixture 10–50 μCi of $\alpha[^{32}P]dATP$ and detect the amplification products by autoradiography. A similar protocol, varying only the PCR conditions, can be used when other deleted molecules are searched. The amplification conditions used for detecting the most frequent human mtDNA deletions are reported in Table 21.1.

21.4.2.2 Characterization of PCR Products

The presence among the amplification products of one DNA band of the expected size for a given mtDNA deletion represents a first demonstration of the existence of such mutation. However, when the deletion is unknown and also to exclude that the amplified fragment is the result of primer misannealing, the PCR products have to be characterized. This can

TABLE 21.1

Human mtDNA Deletions More Frequently Observed in Different Tissues of Aged
Subjects

Deletion (bp)	Tissues[a]	Primers	PCR Conditions
4977	Liver, muscle, brain, heart, lung, testis, skin, oocyte[a]	ATP-For(L8282-8305)/ 13 Rev-Bis (H13928-13905)	1 min, 94°C 1 min, 65°C 1 min, 72°C for 30 cycles [see Ref. 44]
6063	Liver, muscle[b]	L7293-7316/H13928-13905	1 min, 94°C 1 min, 59°C 45 sec, 72°C for 30 cycles [see Ref. 90]
7436	Heart, muscle, brain[c]	L8531-8550/H400-381	15 sec, 94°C 15 sec, 50°C 80 sec, 72°C for 30 cycles [see Ref. 91]
10422	Heart[d]	L4308-4325/H15226-15207	30 sec, 94°C 1 min, 51°C 1 min, 72°C for 35 cycles [see Ref. 40]

[a] See Ref. [22,23; 36–44; 86–89].
[b] See Ref. [41,90].
[c] See Ref. [14,39,40,41,91,92].
[d] See Ref. [40].

be made with different methods: primer-shift PCR, Southern-blot hybridization and
sequencing of the PCR products.

21.4.2.2.1 *Primer-Shift PCR*

This method is based on the principle that, if there is not misannealing, amplification of the
same mtDNA with different couples of primers should produce mtDNA fragments of different sizes, corresponding to the distances between the primers. Sato et al. [58] used the
primer shift–PCR to detect multiple deleted mtDNA molecules in the DNA from the skeletal muscle of a patient with ocular myopathy. Also the first identification of the 7.4-kb
deletion in human heart affected by hypertrophic or dilated cardiomyopathy was made
through the primer–shift method [59].

The principle of this method is shown in Figure 21.3. To carry out this experiment perform at least two amplifications. After the first PCR, the second amplification uses as template either the product of the first or another aliquot of unamplified DNA. The second
couple of primers may still include one of the two primers of the first amplification (semi-nested PCR) or may be completely different from the first couple (nested-PCR). In the case
of the "common deletion" perform the first amplification with the primers couple ATP-For
(L8282-8305)/13 Rev-Bis (H13928-13905) in the conditions reported above (see Section
21.4.2.1), which generates a band of 669 bp. Then, amplify 1–10 μl of the first reaction or an
amount of unamplified DNA identical to that used for the first amplification with the primers 8.3-For (L8361-8380)/13 Rev-Bis (H13928-13905), in the same conditions of the first

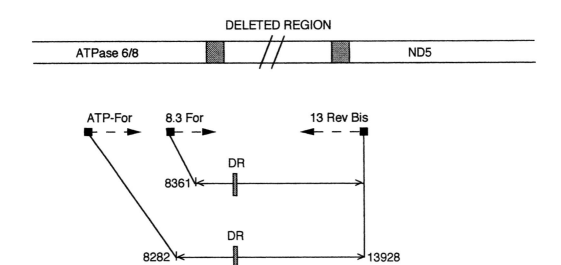

FIGURE 21.3

Schematic diagram illustrating the primer shift approach. The primers ATP-For (L8282-8305)/13 Rev-Bis (H13928-13905) amplify the 4977 bp-deleted mtDNA, generating a 669 bp product. The shift to a second primer pair 8.3-For (L8361-8380)/13 Rev-Bis (H13928-13905), in which the primer 8.3 For is 79 bp downstream ATP-For, shortens the amplification product (591 bp) of a length corresponding to the distance between the two primers.

PCR. Run the products on an agarose gel (see Section 21.4.2.1): the size of the product deriving from a correct amplification should be of 591 bp.

21.4.2.2.2 Southern-Blot Hybridization

Southern-blot hybridization experiments between amplification products and specific mitochondrial probes are performed to obtain preliminary information about the localization of mtDNA deletions. This method is particularly useful when there are already some indications on the location of the deletion break-points. In an application of this method, described by Ozawa et al. [59], the same PCR product is hybridized with three different probes, two of which (A and C) are located at either ends of the suspected deletion and the third (B) in the middle of the deletion (Figure 21.4). Probes A and C hybridize with the PCR band deriving from deleted mtDNA, whereas probe B detects only wild-type DNA.

After amplification, run the PCR products on a 1.2% agarose gel (see Section 21.4.2.1), transfer onto a nylon filter (Hybond-N, Amersham) in 2 × SSC. Bake the filter for 2 hr at 80°C and prehybridize for 2–3 hr at 65°C with 5–10 ml of a solution containing 5× Denhardt, 5× SSPE, 0.5% SDS, and 50 μg/ml calf thymus DNA. Remove the prehybridization solution and add 5 ml of the same solution containing 2–3 × 10⁷ cpm of denatured radioactive probe labelled by random priming [60]. Hybridize for 16–20 hr at 65°C, then wash the filter twice with 2× SSPE, 0.1% SDS for 10 min at room temperature, once with 1 × SSPE, 0.1% SDS for 15 min at 65°C and 1–2 times with 0.1× SSPE, 0.1% SDS at 65°C for 10 min. Dry the filter with Whatman paper and expose for autoradiography. In order to re-hybridize the same filter with different probes, after the autoradiography, soak the filter in 0.1% SDS for 30 min at 95°C.

21.4.2.2.3 Restriction Enzyme Mapping

The analysis of the fragment pattern obtained by the digestion with different enzymes allows to map the deletion, narrowing the region flanking the break-points. In order to do

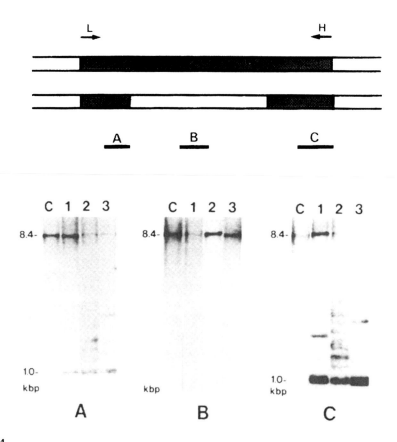

FIGURE 21.4
PCR-Southern method. Three probes A (320 bp), B (240 bp) and C (299 bp) were prepared by PCR on normal mtDNA as described by Ozawa et al. [59]. Probes A and C will hybridize with a full length mtDNA fragment, derived from the amplification with the primer pair L and H, and with a shorter fragment, derived from the deleted mtDNA. Probe B will hybridize only with the full length fragment (top). The bottom of the figure reports an application of the method to the mtDNA of a patient with cardiomyopathy. Probes A and C hybridize with a 8.4 kb fragment and with multiple shorter bands; probe B hybridizes only with a 8.4 kb fragment. [From Ozawa, T., Tanaka, M., Sugiyama, S., Hattori, K., Ito, T., Ohno, K., Takahashi, A., Sato, W., Takada, G., Mayumi, B., Yamamoto, K., Adachi, K., Koga, Y., and Toshima, H., *Biochem. Biophys. Res. Commun.*, 170, 830, 1990. With permission.]

this, elute the PCR-amplified fragment deriving from deleted mtDNA and digest with different restriction endonucleases. Run products on an agarose gel and stain with ethidium bromide. Alternatively label at the 5′ end the restriction products with γ[^{32}P]ATP and polynucleotide kinase [61], load on the gel and visualize the products by autoradiography.

21.4.2.2.4 Sequencing of PCR Products.

Although the methods previously described provide some information about the specificity of the PCR products and their location, the only way to fully characterize a mtDNA deletion is the determination of the sequence of the deletion break-points. The PCR products may be sequenced directly or, especially when there are multiple bands, they can be cloned and the sequence of the clones determined by the usual methods. The identification of the deletion break-points may be obtained by using as sequencing primers the same used for the amplification. If they are too far from these points, new primers for walking along the molecule must be designed.

Direct Sequencing of the PCR Products by Asymmetric Amplification

This method is based on the asymmetrical amplification of double-stranded PCR products. After amplification of the deleted molecule with two primers present at the same concentration, an aliquot of the double-stranded product is amplified with the same primers, which are now present at different molar concentrations. In these conditions, during the first 10–15 cycles of amplification the product is exclusively double-stranded DNA, which accumulates at an approximately exponential rate. After 12–15 cycles of amplification the concentration of one primer becomes limiting and the rate of production of the double-stranded product declines markedly. However, the second oligonucleotide continues to prime the synthesis of single-stranded DNA that is complementary to one of the two strands of the template DNA. Although the single-stranded product accumulates at a linear rate during successive rounds of extension and denaturation, it reaches concentrations more than adequate for conventional dideoxy-mediated sequencing. The following protocol refers to the sequencing of the "common deletion" of human mtDNA. It can be adapted to the sequencing of any amplification products, changing only the primer annealing temperature [62].

Carry out the first amplification in standard conditions (see Section 21.3.5) with the primers ATP-For (L8282-8305)/13 Rev-Bis (H13928-13905). Amplification profile is the following: 1 min denaturation at 94°C, 1 min annealing at 65°C, 1 min extension at 72°C for 30 cycles. Separate the double-stranded product from deleted mtDNA (669 bp for the "common deletion" of human mtDNA) by electrophoresis on a 2% NuSieve (FMC) agarose gel, containing 0.5 mg/ml ethidium bromide, in 1× TAE buffer. Cut out the product band and elute the DNA by incubating the gel slice at 65°C for 10 min in 1 ml of water. Dilute 1:100 an aliquot of the eluted DNA and use 1 μl of the diluted DNA for the second (asymmetric) amplification. For this PCR use conditions identical to the first one, except for the amounts of primers that are 50 pmol for one primer and 0.5 pmol for the other. At the end of the reaction check the synthesis of the single-stranded DNA, running an aliquot of the mixture on a 1.2% agarose gel. Precipitate the remaining amplified DNA with 90 μl of 4 M ammonium acetate and 540 μl of ethanol. Wash the resulting pellet three times with 1 ml of 70% ethanol and suspend in 7 μl of double-distilled water. Add to the DNA 1 μl (5 pmol) of sequencing primer, (which is the one present in the lower amount in the asymmetric amplification), 2 μl of Sequenase reaction buffer (200 mM Tris-HCl pH 7.5, 50 mM $MgCl_2$, 250 mM NaCl), incubate for 4 min at 65°C and slowly cool (for about 30 min) up to 35°C. Put the mixture on ice and add 1 μl of 0.1 M dithiothreitol (DTT), 2 μl of a 5-times diluted labelling mix (5 × labelling mix : 7.5 μM dGTP, 7.5 μM dCTP, 7.5 μM dTTP), 5 μCi of $\alpha[^{32}P]$dATP (3000 Ci/mmol), 1.5 U of Sequenase 2.0 (USB) and incubate at 25°C for 2–5 min. If it is necessary to read bases further than 200 nt from the annealing site, use an undiluted labelling mix and increase the extension time up to 15–20 min. At the end of the extension, transfer 3.5 μl of the mixture to four, 42°C-preheated, Eppendorf tubes each containing 2.5 μl of the appropriate termination mixture (each mixture contains 80 μM of dNTP, 50 mM NaCl and 8 μM of ddGTP or ddATP or ddTTP or ddCTP). Incubate the tubes for 10–20 min at 42°C and stop the reaction by adding 4 μl of stop solution (95% formamide, 20 mM EDTA, 0.05% bromophenol blue, 0.05% xylene cyanol). Denature for 5 min at 94°C and load 3-5 μl on a 40 cm × 20 cm × 0.2 mm 6% polyacrylamide sequencing gel. After the run, expose the gel for autoradiography.

Direct Sequencing of PCR Products by Linear Amplification of Double-Stranded DNA.

The second direct sequencing method of amplified DNA is based on the linear amplification of double-stranded DNA. According to this method (also known as cycle-sequencing

PCR), double-stranded amplification products are incorporated in a dideoxy-mediated sequencing reaction mixture containing a 5' end-labelled primer and incubated in a thermal cycler; the programmed temperature cycling repeatedly denatures the template and anneals the primer. The Taq polymerase synthesizes from the primer until incorporation of a dideoxynucleotide occurs. The successive rounds of denaturation increase the amount of DNA in a single-stranded form, providing a more efficient template than the standard protocol. Template renaturation, that may occur during the primer extension, is eliminated during the next cycle of denaturation. As this method determines also a modest, but significative linear amplification of the template, it provides the opportunity to use less DNA, at least 1/20th of the molar amount used in a conventional sequencing reaction [63].

In a typical reaction separate the double-stranded products deriving from the first amplification by electrophoresis on a 1% Sea-Plaque (FMC) agarose gel in 1 × TAE buffer. Visualize the bands by ethidium bromide staining and elute the DNA, with the Quiaex extraction kit (Quiagen), collecting in a volume of 12 μl. Dispense 3 μl aliquots of the DNA (0.01–0.1 pmol) in four 0.5 ml Eppendorf tubes containing also, in a final volume of 10 μl, 0.5 pmol of a 5' end-labelled primer (it can be one of the primers used for the amplification or an internal oligonucleotide), 30 mM Tris-HCl pH 9.0, 4 mM $MgCl_2$, 10 μM of each dNTP, 0.1 mM ddGTP or 0.3 mM ddATP or 0.4 mM ddTTP or 0.2 mM ddCTP and 0.5 U of Taq DNA polymerase. Incubate for 20 cycles (30 sec at 94°C, 30 sec at 55°C and 60 sec at 72°C) followed by 10 cycles (30 sec at 94°C and 60 sec at 72°C) and terminate the reaction by adding 5 μl of stop solution. Heat at 95°C for 5 min and load a 1 to 4 μl sample on a sequencing gel for electrophoresis.

Cloning and Sequencing of PCR products.

When the amplification products are present in amounts not sufficient for direct sequencing or when they consist of multiple bands not well separated on the gel it is preferable to clone them and then determine their sequence.

Early methods of cloning PCR products relied on the use of oligonucleotide primers containing at their ends the recognition sequence of a restriction enzyme. After the amplification the products were digested with that enzyme and then inserted in plasmid vectors linearized with the same enzyme. The enzyme to be used had to create sticky ends and not to cleave internally the fragment to be cloned. As the efficiency of digestion changes greatly among different enzymes [64], the yield of this method is rather variable. Recently, a new cloning method for PCR products, that does not require the creation of restriction enzyme sites, has been developed. The method takes advantage of the non-template-dependent activity of thermostable polymerases used in PCR, that adds a single deoxyadenosine at the 3' ends of all duplex molecules synthesized by PCR [65]. These A-overhangs are used to achieve direct ligation of PCR products into specifically designed vectors, such as the pMOS *Blue* T-vector (Amersham) which contains single 3' T-overhangs. After transformation of competent *E.coli* cells, the recombinant plasmid DNA is selected and the inserts are sequenced either by PCR-linear sequencing or with the thermostable T-7 DNA polymerase (ThermoSequenase).

In a typical experiment, carry out the PCR in standard conditions, recover the products from 1% Sea-Plaque (FMC) agarose gel and ligate 2–3 pmol of the amplification product with 0.25 pmol of pMOS *Blue* T-vector (Amersham) in a 10 μl reaction mixture containing also 5 μM DTT, 0.5 mM ATP, 6 mM Tris-HCl pH 7.5, 6 mM $MgCl_2$, 5 mM NaCl, 100 μg/ml bovine serum albumin, 7 mM β-mercaptoethanol, 1 mM spermidine and 3 U of T4 DNA ligase (Boehringer). Incubate for 16 hr at 14°C and use an aliquot of the ligation to transform MOS *Blue* competent cells in the presence of ampicillin. Select white colonies, prepare recombinant plasmid DNA, and sequence by the dideoxy-mediated Sanger method.

21.4.3 Long Extension PCR (LX PCR)

As reported above (see Sections 21.1 and 21.2) the study of a single mtDNA deletion does not provide a full qualitative or quantitative representation of the DNA damage. The visualization of multiple deletions would require either the simultaneous use of many primers pairs or the use of a couple of primers, sufficiently distant to cover more than one deletion. The first method, although it has been adopted in some applications [66], has the limitation that different mtDNA deletions may be amplified with variable efficiency according to the annealing temperature of the primers and that these might compete with each other for the DNA template. The second method provided some interesting data. Zhang et al. [22] used a set of primers distant up to 8.6 kb to detect at least 10 different mtDNA deletions in various tissues of a healthy aged individual. More recently, Cheng et al. [67] firstly described a PCR protocol that amplified virtually the entire mtDNA, and thus greatly enhanced the utility of PCR for finding deletions along the whole molecule. The method was based on the use of two adjacent primers oriented in opposite directions which are subjected to a two-step PCR with short denaturation and long extension times.

Different parameters must be optimized to obtain the amplification of long DNA molecules:

1. **DNA Integrity** — The first requirement for a successful amplification of the entire mitochondrial genome is the DNA integrity. We tested different methods for the preparation of intact mtDNA from tissues and we obtained the best results by using the procedure reported by Davis et al. [68] and reported in Section 21.3.3.1.

2. **Buffer** — The buffer used for the LX-PCR contains glycerol (50% v/v). Glycerol influences long amplifications by increasing the thermal stability of the polymerases and lowering melting and strand separation temperatures (2.5–3°C per 10% glycerol). This makes possible the use of short denaturation times, minimizing DNA deamination and depurination. The pH of the buffer plays also an important role in this technique: a higher pH is protective since depurination increases with a lower pH.

3. **Mg^{2+}** — The magnesium concentration strikingly influences the reaction. Higher Mg^{2+} concentrations often result in non specific bands, whereas lower Mg^{2+} decrease the yield of the reaction. We tested a range of magnesium concentrations from 0.5 to 2.0 mM and found better results with 1.15 mM magnesium acetate.

4. **DNA Polymerase** — The choice of the DNA polymerase is also important. Currently, many thermostable DNA polymerases are available. We obtained the best results by combining *rTth* polymerase (recombinant form of the *Thermus thermophilus* DNA polymerase from Perkin Elmer), which possesses also a 5'-3' exonuclease activity, with *Vent* DNA polymerase (New England Biolabs) which has the proof-reading 3'–5' exonuclease activity. The inclusion of this enzyme prevents the stable incorporation of mismatched nucleotides which could determine altered reaction products.

5. **Primers** — The primers must have compatible annealing temperatures and present the least possible number of false priming sites.

6. **PCR Conditions** — The length of each denaturation step is reduced to 10 sec to minimize DNA damages, whereas the annealing temperature is kept at the highest possible value to reduce aspecific products.

We defined the following protocol for LX-PCR of human mtDNA.

Procedure

Carry out the reaction in a GeneAmp PCR System 2400 (Perkin Elmer) or similar, in 0.2 ml thin-walled reaction tubes. Mix the following components in a final volume of 50 µl:

  50–250 ng of genomic DNA isolated from human skeletal muscle or from other organs (see Section 21.3.3)

  5 µl of 10 × XL buffer (Perkin Elmer) containing: 50% glycerol, 20 mM Tris-HCl pH

  8.0, 100 mM KCl, 0.1 mM EDTA, 1 mM DTT, 0.5% (v/v) Tween 20

  1.15 mM magnesium acetate (added after the first denaturation step)

  50–100 µg/ml bovine serum albumin (Serva)

  0.2 mM of each dNTP

  1–2 U of recombinant *Tth* (from *Thermus thermophilus*; Perkin Elmer)

  0.02–0.05 U Vent DNA polymerase (from *Thermus litoralis*; New England Biolabs)

  5–10 pmol of each primer

Primers, designed to point away from each other along the circular mtDNA genome, are L-14841-14866 and H15149-15124. Set up a PCR profile with an initial denaturation of 2 min at 94°C; then add the magnesium acetate and run the following protocol: 32 cycles of 10 sec at 94°C and 10 min at 72°C with a 20 sec/cycle autoextension in the cycles from 16 to 32. End the reaction with a final extension step of 10 min at 72°C. Separate the products on a 0.35% agarose SeaKem Gold gel (FMC). This matrix, which is the ideal gel for the resolution of DNAs between 1 and 50 kb, has a very high strength, forming easy-to-handle gels even at low agarose concentration. Run the gel at 4°C in 1 × TAE at 10-20 Volt for 16 hr, visualize products by ethidium bromide staining and characterize them by Southern blotting (see Section 21.4.2.2.2) or by sequencing (see Section 21.4.2.2.4).

The Figure 21.5 shows an example of LX-PCR on DNAs extracted from the skeletal muscle of two individuals, 20-yr and 70-yr old, respectively. The young subject shows a single 16.3-kb band, whereas the old one presents, together with this band, also other products with sizes ranging from 10 kb to 2 kb. In order to understand whether the low molecular weight products observed in the amplification of the DNA from the 70-yr old individual represent genuine amplification products of deleted mtDNA molecules, Southern-blot hybridization experiments were carried out. Figure 21.6 shows that a probe containing part of the 16S rRNA and ND1 genes (probe P1) detects a 16.3 kb band, corresponding to the entire mtDNA molecule, in both subjects and other bands, with a molecular weight from 10.8 to 5.4 kb, only in the old subject. The hybridization with probes covering other portions of the genome (probes P2-P4) allows a preliminary characterization of these bands. Table 21.2 shows that most of them correspond to deletions already characterized, whereas some others have not been reported previously. Recently, Melov et al. [69] have used this technique for studying age-dependent mtDNA deletions in postmitotic mice tissues and found also 2–4 kb minicircles containing the origin of replication of both strands.

21.4.4 Single-Fiber PCR

MtDNA deletions are distributed in the same tissue in a mosaic-like fashion [54,55]. Soong et al. [37] found a striking variation among different anatomical locations of the level of the 4977 bp deletion in 12 brain regions. The deletion was much more represented in the caudate, putamen, and substantia nigra with a difference between two and three orders of

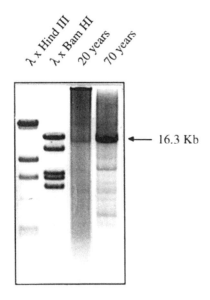

FIGURE 21.5

Long extension PCR (LX-PCR) of DNA extracted from the skeletal muscle of 20 yr and 70 yr old subjects. Amplification products were separated on a 0.35% agarose Seakem Gold (FMC) gel and visualized by ethidium bromide staining. Molecular weight of DNA markers are the following: λ DNA × Hind III, 23130, 9416, 6557, 4361, 2322, 2027 bp; λ DNA × Bam HI, 16841, 12275, 7233, 6527, 5626 bp. The arrow on the right indicates the molecular weight of the main amplification product.

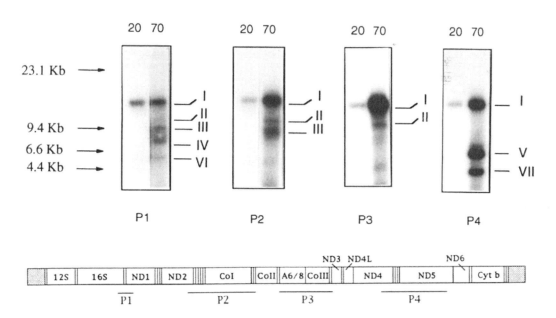

FIGURE 21.6

Southern-blot hybridization between LX-PCR products from skeletal muscle DNA of two subjects 20 and 70 years old (20 and 70) and different mtDNA probes. Amplification products were separated on an agarose gel, transferred onto a nylon membrane and sequentially hybridized with mtDNA probes P1–P4. On the right of each panel the position of hybridized bands (I-VII) is reported. Probe position is reported in the lower part of the figure.

TABLE 21.2

Results of Southern-Blot Hybridization Between LX-PCR Amplification Products From Skeletal Muscle DNA of Two 20- and 70-year Old Subjects (20 and 70) and Four mtDNA Probes (P1-P4).

Band	Size	P1 70	P1 20	P2 70	P2 20	P3 70	P3 20	P4 70	P4 20	Band Identity
I	16.3 kb	+	+	+	+	+	+	+	+	Tot. mtDNA
II	10.8 kb	+	–	+	–	+	–	–	–	Δ 5.0 kb[a] Common deletion (8468-13446)
III	9.6 kb	+	–	+	–	–	–	–	–	Δ 6.1 kb[b] (7841-13905)
IV	7.6 kb	+	–	–	–	–	–	–	–	Δ 8.2 kb[c] (5786-13923) (6023-14424)
V	6.2 kb	–	–	–	–	–	–	+	–	Δ 9.4 kb
VI	5.4 kb	+	–	–	–	–	–	–	–	Δ 10.9 kb[d] (3173-14161)
VII	4.5 kb	–	–	–	–	–	–	+	–	Δ 11.1 kb

Note: Presence or absence of a band in each hybridization is indicated by "+" or "–". Band identity (last column) derives from the comparative analysis of the four hybridizations; it reports the presumed position of the deletion responsible for a specific amplification product. In the case of band IV two possible alternatives, both consistent with band size and hybridization data, are presented. Bands V and VII derive from two not yet identified 9.4 kb and 11.1 kb deletions.

[a] See Ref. [22,23; 36-44; 86-89].
[b] See Ref. [41,90].
[c] See Ref. [93,94].
[d] See Ref. [95].

magnitude with respect to the cerebellum. This result raised the question about the real significance of the mtDNA deletions during aging. In particular, it was argued that mtDNA deletions might accumulate in specific regions of an organ or in specific cells of that organ where the deletion may reach a level as high to affect the cell function. Moreover, if we apply this concept to the large number of mtDNA deletions already characterized in human and in other mammals [33] and suppose that probably there are many other deletions to be discovered, the relevance of mtDNA deletions in the aetiology of the aging process increases greatly. Therefore, whereas the PCR analysis of DNA extracted from the whole tissue gives an average value of the content of the mtDNA deletions in the tissue, it does not provide any indication on the cellular distribution of the deleted molecules. To approach this problem it is necessary to carry out a specific analysis of the cellular distribution of the mtDNA deletions in a given tissue. This can be done either by *in situ* PCR or by performing the PCR on DNA extracted from a single cell. While the first method has not yet been set up for mtDNA deletions, we have recently adapted a technique, used for the search of mtDNA deletions on DNA extracted from single muscle fibers of patients with mitochondrial diseases [70], to healthy individuals of different ages.

In order to examine single muscle fibers for the presence of the "common deletion" perform the following protocol. Extract the total DNA from a single skeletal muscle fiber (see Section 21.3.3.2) and use it in PCR with the primers ATP-For (L8282-8305)/13 Rev-Bis

(H13928-13905), in the conditions reported above (see Section 21.4.2.1). At the end of the reaction precipitate the 90% of the reaction mixture and run the resuspended pellet on a 1% agarose gel in 1 × TAE buffer in the presence of ethidium bromide (see Section 21.4.2.1). If the amplification of DNA does not show the presence of a band deriving from the "common deletion" re-amplify 0.1% of the first PCR with the primers 8.3-For (L8361-8380)/13 Rev-Bis (H13928-13905) (see Section 21.4.2.2.1). In this case a band of 591 bp corresponding to the deleted mtDNA should be produced. Use a small aliquot of the DNA (about 1/100) for amplification with the primers ATP-For (L8282-8305)/ATP-1B (H 8628-8608) to detect undeleted mtDNA by using the following PCR profile: 1 min at 94°C, 1 min at 65°C, 1 min at 72°C is for 30 cycles. An example of this method, that allows to evidentiate the presence of deleted mtDNA molecules at the single cell level, is reported in Figure 21.7.

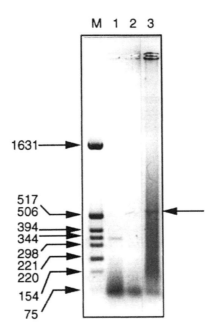

FIGURE 21.7
Single-fiber PCR. DNA isolated from a single fiber from human skeletal muscle was amplified with primers ATP-For (L8282-8305)/ATP-1B (H8628-8608) and ATP-For (L8282-8305)/13 Rev-Bis (H13928-13905). Aliquots of the amplification products were run on an agarose gel (lanes 1 and 2 respectively). 0.1% of the amplification product from lane 2 was reamplified with primers 8.3 For (L8361-8380)/13 Rev-Bis (H13928-13905) and run on the gel (lane 3). The arrow shows the position of the product deriving from the amplification of the 4977 bp-deleted mtDNA. M: molecular weight marker (pBR322 × Hinf I).

21.4.5 Quantitative PCR

Three major groups of methods are available for the quantitation of deleted mtDNA molecules: serial dilution PCR, kinetic PCR and competitive PCR. Every method has to be adopted only after a qualitative PCR, including proper contamination controls, has revealed the searched deleted species in the examined sample.

21.4.5.1 Serial Dilution Method

The serial dilution-PCR method was initially developed by Corral-Debrinski et al. [38], then it was used by several groups, included ours. With this method we quantitated the

4834-bp deletion of rat mtDNA in liver samples from aging animals [11]. The method is based on carrying out two separate PCR, one for the deleted mtDNA and the other for the total mtDNA, using for each of them a distinct range of dilutions and a different primers pair. The amplification products of both reactions are separated on agarose gel and quantitated by densitometry. The optical density of each band is plotted versus the log of weight of the DNA template. The two plots, which fit a sigmoidal curve, are examined to determine the values on the x axis at which the optical densities of the deleted and undeleted PCR products are equivalent. The amount of the deleted mtDNA is expressed as the ratio of deleted mtDNA to total mtDNA. To minimize differences in the amplification efficiency the primers have to be chosen so to produce similar size products.

The following protocol was set up for the quantitation of the 4977-bp deletion in different human brain regions [39]. It can be easily adapted to other deletions or other tissues.

Procedure

Isolate total DNA from tissue or from cells (see Section 21.3.3), digest with Pst I and Hind III, which cleave wild-type mtDNA molecules three times in the deleted regions. This, together with the peculiar PCR conditions, prevents the amplification of the long DNA fragment (5569 bp) from the wild-type mtDNA. Prepare serial dilutions of digested DNA to yield template amounts in a range such that the absorbance values of the products are within the linear range of the density curve. Set up the amplification reactions in the conditions reported above (see Section 21.3.5) using as primers: L8282-8305/H13851-13832 for the deleted mtDNA and L3108-3127/H3717-3701 for the total mtDNA. The last primer pair has been chosen because it amplifies a rarely deleted mtDNA region containing portions of the 16S rRNA and ND1 genes. Carry out the amplifications under the following conditions:

- For deleted mtDNA, 35 cycles made of 30 sec at 94°C, 1 min at 56°C and 1 min at 72°C. The initial denaturation step lasts 2 min rather than 30 sec. In these conditions a single band of 593 bp is obtained.

- For total mtDNA, 35 cycles made of 30 sec at 94°C, 1 min at 51°C and 1 min at 72°C. The initial denaturation step lasts 2 min rather than 30 sec. A band of 609 bp is the product of this reaction.

Separate the products on a 1.2% agarose gel in 1× TAE, stain with ethidium bromide.

Photograph and analize the picture with a densitometer as the LKB-Pharmacia Ultroscan Laser Densitometer, equipped with a XL evaluation software. Plot the values of OD versus the logarithm of the DNA concentration and determine the values on the x axis at which the optical densities of the deleted and undeleted PCR products are equivalent. The Figure 21.8 shows an example of this determination on the DNA extracted from the temporal lobe of 24-yr and 85-yr old healthy subjects. It can be seen that the relative level of the deletion differs of more than two orders of magnitude, reaching in the older individual a value of 2.6% of total mtDNA.

21.4.5.2 Kinetic Method

This method was previously described by Chelly et al. [71] and then designed by Wiesner et al. [72] for the measurement of the concentration of mtDNA and mtRNAs in different animal tissues. The method to measure the level of mtDNA deletions during aging was recently set up in our laboratory by Lezza et al. [44]. It is based on the measurement of the concentration of the amplification product accumulating in consecutive cycles during the

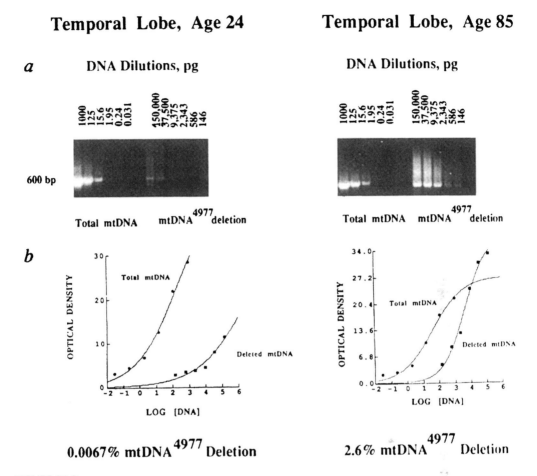

FIGURE 21.8
Estimate of the 4977 bp-deleted mtDNA in the temporal lobe of human brain with the serial dilution method.
(a) PCR products of the 4977 bp mtDNA deletion and total mtDNA reactions from the temporal lobe DNA of
a 24 (left panel) and a 85 years old (right panel) subject. **(b)** Graphs showing densitometric data points and fitted
curves for the total mtDNA (closed circles) and deleted mtDNA (closed squares) analyses of the 24 (left panel)
and 85 years old (right panel) subjects. [From Corral-Debrinski, M., Horton, T., Lott, M. T., Shoffner, J. M., Flint
Beal, M., and Wallace, D. C., *Nature Genet.*, 2, 324, 1992. With permission.]

exponential phase of the PCR. The initial concentration of the DNA template is determined
by plotting the concentration of the amplification product versus the cycle number and
extrapolating to 0 cycle the curve obtained by linear regression.

During the exponential phase of PCR, the amount of product, N_n, is related to the amount
of initial target, N_o, by the equation:

$$N_n = N_o \, (eff)^n \tag{21.1}$$

where *eff* is the efficiency factor, which has a theoretical maximum of 2 and declines during
the last cycles when the exponential phase is past, and *n* is the number of PCR cycles. Since:

$$\log N_n = \log N_o + n \log(eff) \tag{21.2}$$

a linear relationship between $\log N_n$ and *n* is expected in a plot of these parameters.

By performing the reaction in the presence of a labelled nucleotide precursor, such as $\alpha[^{32}P]$ dATP, the concentration of the PCR product (moles/μl) produced in consecutive cycles is calculated from the incorporated radioactivity (cpm/μl), determined in distinct aliquots of the reaction mixture, from the specific activity of the radioactive precursor (cpm/mol) and from the number of adenosine moieties in the product (nA) according to:

$$\text{moles}/\mu l = \frac{\text{cpm}/\mu l}{\text{cpm}/\text{mol} \times nA} \qquad (21.3)$$

By plotting the logarithm of this value, (which represents N_n) versus the cycle number, a straight line having *eff* as slope is obtained. By extrapolating to 0 cycle, the intercept of the line with the y-axis gives the initial amount of target N_o. This method calculates the absolute concentration of the template DNA and does not require the amplification of external or internal standards. However, since the level of the deleted mtDNA is always referred to total mtDNA, the quantitative determination of the deleted mtDNA is always accompanied by the measurement of the concentration of the total mtDNA. We have used this method to estimate the levels of the mtDNA deletions in some human and rat tissues [44, 73]. Hayakawa et al. have used a similar method to quantitate the 7.4 kb deletion in human heart [74].

Before beginning the assay, it is necessary to carry out preliminary tests to determine the conditions (cycle ranges and DNA template concentrations) where the reaction is in the exponential phase.

Set up a standard PCR assay (see Section 20.3.5) except for the use of 20 μM of each dNTP and 50 μCi of $\alpha[^{32}P]$dATP (3000 Ci/mmol). For the measurement of the mtDNA^{4977} concentration use the primers pair ATP-For (L8282-8305)/13 Rev-Bis (H13928-13905) and 100–500 ng of total DNA. Incubate for 5 min at 94°C, followed by addition of the enzyme (hot-start PCR), then perform 30 cycles each made of 1 min at 94°C, 1 min at 65°C and 1 min at 72°C. Determine the concentration of total mtDNA by using the primers pair L3007-3023/H3538-3520 and 2–5 ng of total DNA. Perform the PCR by incubating for 5 min at 94°C, then add the enzyme, and continue for 25 cycles made of 1 min at 94°C, 1 min at 55°C and 1 min at 72°C. Take 3.5 μl aliquots at different cycles, (starting from 19th up to 29th cycle for deleted mtDNA and from 13th to 24th cycle for total mtDNA) and load on a 5% polyacrylamide gel (0.75 mm × 10 cm × 8 cm) in 1 × TBE, run at 130 Volts for 1 hr. Expose the gel, cut out the amplification products (a band of 669 bp for the mtDNA^{4977} and of 531 bp for the total mtDNA), dry at 80°C for 4 hrs in scintillation vials and count in 2 ml of scintillation cocktail (Maxifluor-Baker). Plot the incorporated radioactivity, transformed in moles/μl [see Equation (21.3)] in a semilogarithmic scale versus the number of cycles, extrapolate to 0 cycle the experimental line and determinate the absolute initial concentration of the template DNA. The Figure 21.9 shows an example of such an assay for the quantitation of the mtDNA^{4977} in the skeletal muscle of a 58-year old individual [44].

21.4.5.3 *Competitive Method*

More recently, another development of the PCR quantitative technique, that is competitive PCR, has gained a large consensus and it has been used also for the determination of the mtDNA^{4977} percentages in various samples [43,75,76]. The general method relies on the coamplification of low amounts of target DNA with known quantities of added competitor DNA. The competitor is a DNA which shares with the target DNA the same primer sites and the near totality of the amplified sequence, so that they are both

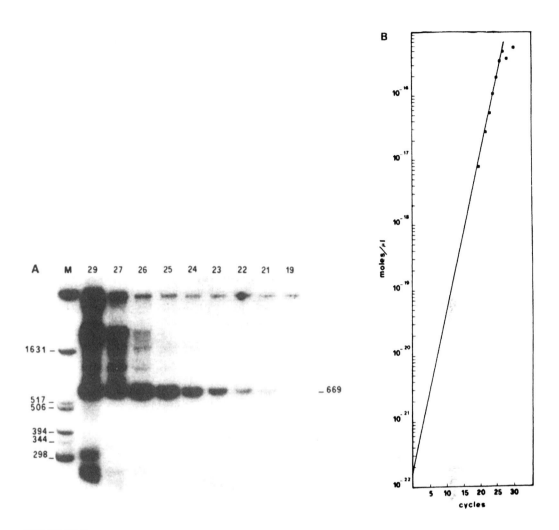

FIGURE 21.9

Quantitation of the mtDNA[4977] template concentration in the skeletal muscle of a 58 years old subject. **(A)** Autoradiogram showing the progressive increase of the 669 bp product. The numbers on the top of each lane indicate the cycle number. M: molecular weight marker (pBR322 × Hinf I). **(B)** Semilogarithmic plot of product concentration (moles/μl) vs. cycle number. Dots indicate the experimental values. [From Lezza, A. M. S., Boffoli, D., Scacco, S., Cantatore, P., and Gadaleta, M. N., *Biochem. Biophys. Res. Commun.*, 205, 772, 1994. With permission.]

amplified at the same rate. By adding increasing amounts of competitor DNA to a fixed amount of the target DNA, the ratio between the final amplification products for the two species can be evaluated for each point. By detecting the concentration of the competitor DNA at which the amplification signals of the target DNA and of the competitor DNA are the same, the concentration of the target DNA is obtained. To determine the amounts of both deleted and wild-type mtDNA two separate competitive PCR, each one involving a specific template and its related competitor, are carried out. Moreover, since it is necessary to differentiate between the PCR products from the target and competitor DNA templates, a unique restriction enzyme site can be added or removed from the competitor. Alternatively, a difference in the fragments size can be created by insertion or deletion of a short sequence.

21.4.5.3.1   Addition of a Restriction Site

An example of the addition of a restriction site has been presented by Wang et al. [75], who introduced a Dde I restriction site in two 280 bp and 394 bp fragments, resulting from the amplification of deleted and wild-type mtDNA, respectively. We here present the protocol developed for the measurement of the 4977-bp deletion. The protocol may be used also for other deletions, choosing proper primers (see Table 21.1) and restriction sites.

Extract the DNA from a tissue sample or from cells and amplify it in standard conditions (see Section 21.3.5) with the following primers:

1. for wild-type mtDNA amplification, MT-1 (L8339-8359) and MT-2 (H8732-8712).
2. for amplification of deleted mtDNA, MT-1 (L8339-8359) and MT3 (H13595-13577).

Perform the amplification for 30 cycles with this profile: 94°C for 30 sec, 55°C for 45 sec and 72°C for 90 sec.

Separate the amplification products, (394 bp and 280 bp from wild-type and deleted mtDNA, respectively) on an agarose gel, elute them and clone in the plasmid Bluescript KS(-) (Stratagene) originating the recombinant DNA (plasmid pMT400 and pMT280, Figure 21.10). Introduce a Dde I site by site-directed mutagenesis and use the plasmid obtained, (pMMT400 and pMMT280, Figure 21.10 A), as competitors. Set up this reaction in standard conditions in the presence of a fixed amount of target total DNA, a variable amount of each competitor and 1 μCi of α[^{32}P]dCTP. At the end of the reaction, digest the mixtures with Dde I and load 10–20 μl aliquots on a 9% polyacrylamide gel in 1 × TBE. Locate the reaction products by autoradiography, excise them and quantitate by scintillation counting. Plot the cpm of the target and the competitor DNA products vs. the logarithm of the competitor added (Figure 21.10 B): the point at which the curves intersect are where equal amounts of competitor and target DNA are present, enabling the calculation of the concentration of the target DNA.

Since it might be difficult to compare data between separate sets of experiments where different preparations of target DNA are evaluated, an internal nuclear standard can also be used. This can be any single copy gene whose sequence is available. Wang et al. [75] have used the Na/H exchanger gene, that is cloned in the plasmid pSK-4, as a 2.4 kb insert. They have removed a Nco I restriction site at bp 102 by digestion and filling-in with DNA polymerase I is generating the competitor pSK-40. Competitor and cellular DNA are then amplified with the same primer pair.

A similar quantitative procedure, involving the loss of a restriction enzyme site, was used for the determination of the 4834 bp deletion level in the liver mtDNA of aging rats [77].

21.4.5.3.2   Insertion of a Short Sequence

The second possibility of discrimination between DNA template and related competitor relies on the generation of a difference in the respective sizes [76]. This is performed as described in Figure 21.11. The competitors are two plasmids (pCZ24 and pCZ26) containing an amplified fragment from deleted and wild-type mtDNA, respectively. In addition, both plasmids contain a 156 bp fragment made by 148 bp of human mtDNA (nt 7293 to 7440) fused to an 8 bp fragment containing the Xba I recognition sequence. The coamplification of a fixed amount of total cellular DNA with serial dilutions of one competitor at a time allows the detection of the bands of target and competitor DNA having the same intensity, and thus the calculation of the target concentration.

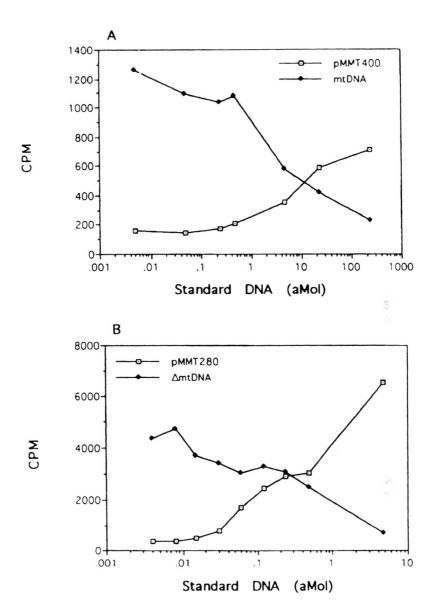

FIGURE 21.10
Competitive PCR, using as competitor a template containing an extra Dde I restriction site, to estimate the content of wild-type mtDNA and of 4977 bp-deleted mtDNA in a cell lysate from human fibroblasts. **(A)** Titration of wild-type mtDNA with the competitor pMMT400. **(B)** Titration of 4977 bp-deleted mtDNA with the competitor pMMT280. **(C)** Schematic diagram showing the construction of the competitors template for total (pMMT400) and deleted (pMMT280) mtDNA. Insertion of a Dde I site was carried out by site-directed mutagenesis using primers OM400A/OM400B and OM280A/OM280B, respectively. [Modified from Wang, H., Fliegel, L., Cass, C. E., Penn, A. M. W., Michalak, M., Weiner, J. H., and Lemire, B. D., *BioTechniques*, 17, 76, 1994. With permission.]

We describe a protocol, developed for the quantitation of the "common deletion" in human heart mtDNA [76]. Coamplify 1 µg of cellular DNA with different amounts of the competitors in two separate series of amplifications, one containing as competitor pCZ24 and the other pCZ26. The concentrations of the competitors should be in ranges to give product band intensities including those expected from the target DNA. Perform the

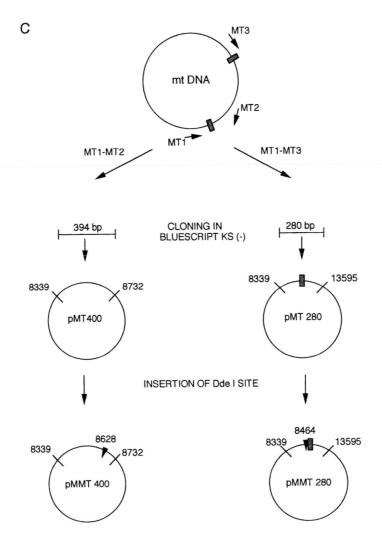

FIGURE 21.10   (continued)

amplification in standard conditions (see Section 21.3.5) for 30 cycles except that the primers amount is reduced to 40 pmoles. As primers pairs use L7901-7920 and H13650-13631 for the deleted mtDNA and L7901-7920 and H8540-8521 for the wild-type mtDNA. Incubate as follows: 94°C for 60 sec, 55°C for 90 sec, 72°C for 150 sec, preceded by an initial denaturation step of 300 sec at 94°C. To eliminate heteroduplexes, at the end of the reaction incubate the products at 85°C, in the presence of 1 M NaCl and 5 mM EDTA [78]. Chill the mixture in a dry-ice ethanol bath and resolve the treated fragments on a 1.2% agarose gel. Expected products are:

1. a band of 773 bp from the target and of 929 bp from the competitor, in the amplification of deleted mtDNA.
2. a band of 640 bp from the target and of 796 bp from the competitor in the amplification of wild-type mtDNA.

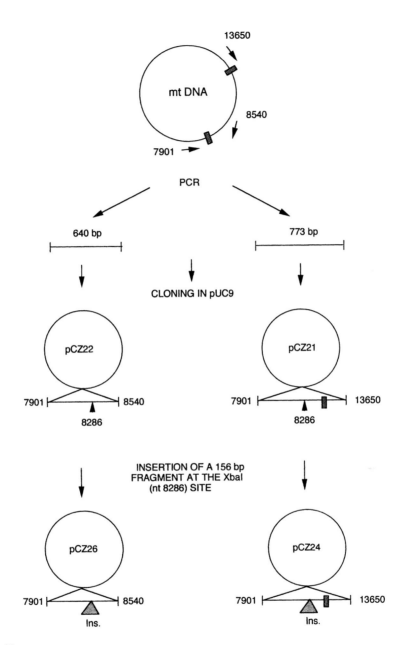

FIGURE 21.11

Schematic diagram illustrating the construction of a DNA competitor containing an insertion of a 156 bp fragment (Ins). Fragments of 640 bp and 773 bp, deriving from the amplification of wild-type and deleted mtDNA, respectively, were cloned in pUC9, digested with Xba I and ligated to a Xba I-treated 156 bp fragment. Template pCZ24 and pCZ26 were used as competitors for the quantitation of total and deleted human mtDNA, respectively. Dashed box represents the residual 13 bp direct repeat and dashed triangle indicates the 156 bp insert. [Modified from Zhang, C., Peters, L. E., Linnane, A. W., and Nagley, P., *Biochem. Biophys. Res. Commun.*, 223, 450, 1996. With permission.]

Photograph the gel and compare the intensities of all the bands within each lane, by scanning the negative film with a densitometer like the LKB-Pharmacia Laser Densitometer equipped with a XL-Evaluation Software. Plot the band absorbance of the target and of the competitor products vs. the logarithm of the competitor weight and determine the point at which the two curves intersect. This represents the mixture where equal amounts of competitor and target DNA are present, thus enabling to calculate the concentration of the target DNA. By choosing proper primers (see Table 21.1) this protocol can be adapted to other mtDNA deletions.

21.4.5.4   Evaluation of the Methods

We have presented the main quantitative methods available at present for estimating the proportion of age-dependent mtDNA deletions by means of their application to the 4977 bp deletion of the human mtDNA. All these methods allow the detection of very low levels of deleted mtDNA: for example, with the kinetic method we have been able to estimate a deletion level of 0.00005% in the skeletal muscle of a 34-yr old healthy subject [44], corresponding to one deleted mtDNA molecule in 2×10^6 molecules.

The serial dilution is probably the most technically convenient method, although the necessity to perform many simultaneous amplifications involves the use of large quantities of expensive Taq DNA polymerase. Moreover, given the propensity of PCR efficiency to vary according to several parameters such as primer composition, template sequence and concentration and fragment size, the major disadvantage of this method is its relying on a series of amplifications performed in different tubes and with two different sets of primers. To minimize these drawbacks, the efficiencies of the primers pairs used to amplify deleted and total mtDNA are tested and the two primers pairs are chosen so to generate PCR products of similar size. The kinetic PCR is an extremely fast and sensitive method, it does not require any standard and gives the absolute value of the mtDNA concentration. The calculation of the amplification efficiency with the kinetic method directly shows that total and deleted mtDNA are amplified with different efficiency, and this constitutes a serious problem for all the measures made with the serial dilution method. In order to directly compare the two methods we measured with both assays the level of the "common deletion" in adult and senescent rat liver. With the serial dilution method, we found a percentage of the "common deletion" of 0.0005% in the adult and 0.02% in the senescent individual [11]; with the kinetic method these values changed to 0.03% and 0.93%, respectively [73]. Moreover, the kinetic method is less expensive since it uses a small amount of Taq DNA polymerase and, relying on single-tube PCR, avoids all the reproducibility problems that can arise when the same reaction is carried out in different tubes. The method uses a small amount (about 50 μCi) of a radioactive precursor. This facilitates the calculation of the synthesized product. An alternative to the use of radioactive labelling for the detection of the PCR product may be pursued with densitometric or CCD camera analysis of the products stained with ethidium bromide or SYBR Green I stain [79]. In such cases, however, the quantitation is made more difficult by the standardization procedure that must be repeated for every experiment. The major inconvenients encountered with the kinetic PCR depend on the fact that it uses a regression line to calculate the ratio of deleted to total mtDNA. This line is extremely sensitive to outliers and, thus, especially the y-intercept point used to calculate the deleted product concentration can be significantly skewed. Therefore, it is crucial to perform multiple parallel experiments on the same sample to determine the most accurate estimate. Moreover, it is known that in the first cycles, especially at low template concentrations, the PCR is delayed by the search of the specific DNA template molecules. This might imply that at such cycles the equation (21.1) is not satisfied and the extrapolation to 0 cycle could give an underestimation of the template concentration. In order to check at which extent the extrapolation is allowed, it is recommended to carry out preliminary experiments with known amounts of templates.

The competitive method has the intrinsic advantage of the contemporaneous amplification of tissue and competitor DNAs with the same primer pair, which eliminates the susceptibility of the PCR to the different efficiencies of two primer pairs. The possibility to carry out two contemporaneous amplifications in the same tube eliminates also eventual variations in the PCR, caused by minor changes in PCR reagents and/or amplifications conditions, possible in different tubes. Moreover, the products have approximately the same size and this guarantees a similar amplification efficiency. Only with the method involving the insertion of a small fragment, the modest size difference (a 156 bp long insert added to the competitor) [76] might cause a slight difference in the amplification efficiencies. The possible formation of heteroduplexes between competitor and target products constitutes the main source of error inherent to this method. This is reduced by specific treatments of the samples after the amplification [78]. Other disadvantages are the fact that the method is time-consuming, especially for the construction of competitors, and expensive, since a relevant number of amplifications (corresponding to different mixtures of target DNA and competitor DNA), have to be performed for each group of analyzed samples. The method does not rely on any assumption and overall it seems the most accurate to quantify mtDNA deletions in normal aging tissues.

21.5 Conclusions

A variety of mtDNA mutations accumulate with age in different tissues. MtDNA deletions are the most frequently observed, although base modifications [5,13,74,80] and point mutations [53,81,82] have also been found. The deletion most commonly studied in human aging is the 4977 bp deletion, but many other deletions have been also reported (see Section 21.1) so to make the search for new deletions an exhausting and perhaps almost useless work. The PCR has revealed that mtDNA deletions are present in a very small minority of the total mtDNA molecules and they are distributed in a focal manner. Soong et al. [37] demonstrated that the proportion of the "common deletion" ranged from 0.0013% in the cerebellar grey matter to 0.46% in the substantia nigra of a healthy old individual.

Although the PCR has been of invaluable help for characterizing mtDNA deletions in aging, we still must await a unique approach to simultaneously answer to different questions concerning the high number of mtDNA deletions, their mosaic-like distribution and their low levels. The LX-PCR method promises to be an useful method to contemporaneously visualize many deletions. The method still needs to be optimized, since it depends on many variables (see Section 21.4.3); moreover, it is a qualitative method and its application to single fiber is still to be developed. The problem of the mosaic-like distribution of mtDNA deletions must be tackled with an approach which allows to discriminate among different cells. The *in situ* hybridization gives already some answers to this problem, but it has the limit of the low sensitivity. Alternative approaches might be the PCR on different regions of a tissue (an approach of this type led to the discovery of the regional distribution of mtDNA[4977] in the brain [37,39]), but this approach is time-consuming, especially if performed in different subjects and on different mtDNA deletions. Single-cell PCR seems a more convenient way to investigate the point: this method has been already set up in the skeletal muscle (see Section 21.4.4) where the analysis of a single fiber is relatively easy. In other tissues this procedure can be difficult. An alternative to this method is the *in situ* PCR. This technique, developed to evaluate the presence of small amount of viral DNA in biological samples [83], has not yet been applied to mtDNA deletions in aging.

The quantitative analysis of mtDNA deletions relies on three methods of which two, the kinetic method and the competitive method, seem the most reliable. The kinetic method is simpler [44,84,85] than the other, but has some limitations (see Section 21.4.5.4). Both methods allow the detection of very low levels of deleted mtDNA molecules; since the lowest measured value of the 4977 bp deletion is 0.00005% (in the skeletal muscle of a 34-yr old subject) [44] and assuming that each cell contains about 5000 mtDNA molecules, this corresponds to about 1 deleted molecule per 400 cells.

In conclusion, at the present, to study mtDNA deletions during aging, the following approaches can be undertaken. For a preliminary analysis it is sufficient to look for the presence of the "common deletion". This can be done by analyzing with PCR the DNA extracted from the whole tissue. A more detailed study will consider the distribution of the deletion in different parts of the tissue. This could be done by *in situ* hybridization and/or by PCR on a single cell. To answer to the question if there is a correlation between the level of a deletion and the age it is necessary to use a method that allows to quantify the deleted molecules. Finally, it must be considered that often the "common deletion" is not the most represented one and, therefore, to have a more complete picture of the distribution of the mtDNA damages it is necessary to look for the presence of other deletions. This can be done either by using one different set of primers for each deletion or with the LX-PCR. With these data at hand it will be finally possible to draw a complete picture of the phenomenon at the cell level and to draw some conclusions about the role of mtDNA deletions in aging.

Acknowledgments

This work has been accomplished with funds from Progetto Finalizzato "Invecchiamento" Code N. 971726 of National Research Council (CNR) and from Ministero dell'Università e della Ricerca Scientifica e Tecnologica (MURST). We thank A. Cormio, V. Pesce, F. Fracasso, and V. Cataldo for helping in preparing the manuscript and R. Longo for word processing.

References

1. Attardi, G. and Schatz, G., Biogenesis of mitochondria, *Annu. Rev. Cell. Biol.*, 4, 289, 1988.
2. Ryan, R., Grant, D., Chang, K. S., and Swift, H., Isolation and characterization of mitochondrial DNA from *Chlamydomonas reinhardtii*, *Proc. Natl. Acad. Sci. U.S.A.*, 75, 3268, 1978.
3. Cantatore, P. and Saccone, C., Organization, structure, and evolution of mammalian mitochondrial genes, *Int. Rev. Cytol.*, 108, 149, 1987.
4. Ward, B. L., Anderson, R. S., and Bendich, A.J., The mitochondrial genome is large and variable in a family of plants (Cucurbitaceae), *Cell*, 25, 793, 1981.
5. Shigenaga, M. K., Hagen, T. M., and Ames, B. N., Oxidative damage and mitochondrial decay in aging, *Proc. Natl. Acad. Sci. U.S.A.*, 91, 10771, 1994.
6. Wallace, D. C., Bohr, V. A., Cortopassi, G., Kadenbach, B., Linn, S., Linnane, A. W., Richter, C., and Shay, J. W., Group report: the role of bioenergetics and mitochondrial DNA mutations in aging and age-related diseases, in *Molecular Aspects of Aging*, 15, Esser, K., and Martin, G. M. (Eds), John Wiley & Sons Ltd., Chichester, UK, 1995, 199.

7. Harman, D., Free radical theory of aging: consequences of mitochondrial aging, *Age*, 6, 86, 1983.

8. Miquel, J., and Fleming, J. E., Theoretical and experimental support for an "oxygen radical-mitochondrial injury" hypothesis of cell aging, in *Free Radicals, Aging and Degenerative Diseases*, Johnson, J. E. Jr, Walford, R., Harman, D., and Miquel, J. (Eds), Liss, New York, 1986, 51.

9. Ames, B. N., Shigenaga, M. K., and Hagen, T. M., Mitochondrial decay in aging, *Biochim. Biophys. Acta*, 1271, 165, 1995.

10. Bunn, C. L. and White, F. A., Mitochondrial DNA in aging human cells, in *Molecular Biology of aging: gene stability and gene expression*, Sohal, R. S., Birnbaum, L. S., and Cutler, R. J. (Eds), Raven Press, New York, 1985, 77.

11. Gadaleta, M. N., Rainaldi, G., Lezza, A. M. S., Milella, F., Fracasso, F., and Cantatore, P., Mitochondrial DNA copy number and mitochondrial DNA deletion in adult and senescent rats, *Mutat. Res.*, 275, 181, 1992.

12. Richter C., Park J. W., and Ames B. N., Normal oxidative damage to mitochondrial and nuclear DNA is extensive, *Proc. Natl. Acad. Sci. U.S.A.*, 85, 6465, 1988.

13. Ames, B. N., Shigenaga, M. K., and Hagen, T. M., Oxidants, antioxidants, and the degenerative diseases of aging, *Proc. Natl. Acad. Sci. U.S.A.*, 90, 7915, 1993.

14. Hayakawa, M., Sugiyama, S., Hattori, K., Takasawa, M., and Ozawa, T., Age-associated damage in mitochondrial DNA in human hearts, *Mol. Cell. Biochem.*, 119, 95, 1993.

15. Mecocci, P., MacGarvey, U., Kaufman, A. E., Koontz, D., Shoffner, J. M., Wallace, D. C., and Flint Beal, M., Oxidative damage to mitochondrial DNA shows marked age-dependent increases in human brain, *Ann. Neurol.*, 34, 609, 1993.

16. Cortopassi, G. A., and Arnheim, N., Detection of a specific mitochondrial DNA deletion in tissues of older humans, *Nucleic Acids Res.*, 18, 6927, 1990.

17. Soong, N.-W., and Arnheim, N., Detection and quantification of mitochondrial DNA deletions, *Methods in Enzymology*, 264, 421, 1996.

18. Holt, I. J., Harding, A. E., and Morgan-Hughes, J. A., Deletions of muscle mitochondrial DNA in patients with mitochondrial myopathies, *Nature*, 331, 717, 1988.

19. Moraes, C. T., Di Mauro, S., Zeviani, M., Lombes, A., Shanske, S., Miranda, A. F., Nakase, H., Bonilla, E., Werneck, L. C., Servidei, S., Nonaka, I., Koga, Y., Spiro, A. J., Brownell, A. K. W., Schmidt, B., Schotland, D. L., Zupanc, M., De Vivo, D.C., Schon, E. A., and Rowland, L. P., Mitochondrial DNA deletions in progressive external ophthalmoplegia and Kearns-Sayre syndrome, *N. Engl. J. Med.*, 320, 1293, 1989.

20. Torii, K., Sugiyama, S., Tanaka, M., Takagi, K., Hanaki, Y., Iida, K.-I., Matsuyama, M., Hirabayashi, N., Uno, Y., and Ozawa, T., Aging-associated deletions of human diaphragmatic mitochondrial DNA, *Am. J. Respir. Cell Mol. Biol.*, 6, 543, 1992.

21. Yamamoto, H., Tanaka, M., Katayama, M., Obayashi, T., Nimura, I., and Ozawa, T., Significant existence of deleted mitochondrial DNA in cirrhotic liver surrounding hepatic tumor, *Biochem. Biophys. Res. Commun.*, 182, 913, 1992.

22. Zhang, C., Baumer, A., Maxwell, R. J., Linnane, A. W., and Nagley, P., Multiple mitochondrial DNA deletions in an elderly human individual, *FEBS Lett.*, 297, 34, 1992.

23. Baumer, A., Zhang, C., Linnane, A.,W., and Nagley, P., Age-related human mtDNA deletions: a heterogeneous set of deletions arising at a single pair of directly repeated sequences, *Am. J. Hum. Genet.*, 54, 618, 1994.

24. Pang, C. Y., Lee, H. C., Yang, J. H., and Wei, Y. H., Human skin mitochondrial DNA deletions associated with light exposure, *Arch. Biochem. Biophys.*, 312, 534, 1994.

25. Lee, C. M., Chung, S. S., Kaczkowski, J. M., Weindruch, R., and Aiken, J. M., Multiple mitochondrial DNA deletions associated with age in skeletal muscle of Rhesus Monkeys, *J. Gerontol.*, 48, B201, 1993.

26. Chung, S. S., Weindruch, R., Schwarze, S. R., McKenzie, D. I., and Aiken, J. M., Multiple age-associated mitochondrial DNA deletions in skeletal muscle of mice, *Aging Clin. Exp. Res.*, 6, 193, 1994.

27. Van Tuyle, G. C., Gudikote, J. P., Hurt, V. R., Miller, B. B., and Moore, C. A., Multiple, large deletions in rat mitochondrial DNA: evidence for a major hot spot, *Mutat. Res.*, 349, 95, 1996.

28. Melov, S., Hertz, G. Z., Stormo, G. D., and Johnson, T. E., Detection of deletions in the mitochondrial genome of *Caenorhabditis elegans, Nucleic Acids Res.,* 22, 1075, 1994.
29. Melov, S., Lithgow, G. J., Fischer, D. R., Tedesco, P. M., and Johnson, T. E., Increased frequency of deletions in the mitochondrial genome with age of Caenorhabditis elegans, *Nucleic Acids Res.,* 23, 1419, 1995.
30. Mita, S., Rizzuto, R., Moraes, C. T., Shanske, S., Arnaudo, E., Fabrizi, G. M., Koga, Y., Di Mauro, S., and Schon, E. A., Recombination via flanking direct repeats is a major cause of large-scale deletions of human mitochondrial DNA, *Nucleic Acids Res.,* 18, 561, 1990.
31. Shoffner, J. M., Lott, M. T., Voljavec, A. S., Soueidan, S. A., Costigan, D. A., and Wallace, D. C., Spontaneous Kearns-Sayre/chronic external ophthalmoplegia plus syndrome associated with a mitochondrial DNA deletion: a slip-replication model and metabolic therapy, *Proc. Natl. Acad. Sci. U.S.A.,* 86, 7952, 1989.
32. Schon, E. A., Rizzuto, R., Moraes, C. T., Nakase, H., Zeviani, M., and Di Mauro, S., A direct repeat is a hotspot for large-scale deletion of human mitochondrial DNA, *Science,* 244, 346, 1989.
33. Chung, S. S., Eimon, P. M., Weindruch, R., and Aiken, J. M., Analysis of age-associated mitochondrial DNA deletion breakpoint regions from mice suggests a novel model of deletion formation, *Age,* 19, 117, 1996.
34. Arnheim, N. and Erlich, H., Polymerase chain reaction strategy, *Annu. Rev. Biochem.,* 61, 131, 1992.
35. Linnane, A. W., Zhang, C., Baumer, A., and Nagley, P., Mitochondrial DNA mutation and the aging process: bioenergy and pharmacological intervention, *Mutat. Res.,* 275, 195, 1992.
36. Cortopassi, G. A., Shibata, D., Soong, N.-W., and Arnheim, N., A pattern of accumulation of a somatic deletion of mitochondrial DNA in aging human tissues, *Proc. Natl. Acad. Sci. U.S.A.,* 89, 7370, 1992.
37. Soong, N.-W., Hinton, D. R., Cortopassi, G., and Arnheim, N., Mosaicism for a specific somatic mitochondrial DNA mutation in adult human brain, *Nature Genet.,* 2, 318, 1992.
38. Corral-Debrinski, M., Stepien, G., Shoffner, J. M., Lott, M. T., Kanter, K., and Wallace, D. C., Hypoxemia is associated with mitochondrial DNA damage and gene induction, *JAMA,* 266, 1812, 1991.
39. Corral-Debrinski, M., Horton, T., Lott, M. T., Shoffner, J. M., Flint Beal, M., and Wallace, D. C., Mitochondrial DNA deletions in human brain: regional variability and increase with advanced age, *Nature Genet.,* 2, 324, 1992.
40. Corral-Debrinski, M., Shoffner, J. M., Lott, M. T., and Wallace, D. C., Association of mitochondrial DNA damage with aging and coronary atherosclerotic heart disease, *Mutat. Res.,* 275, 169, 1992.
41. Hsieh, R.-H., Hou, J.-H., Hsu, H.-S., and Wei, Y.-H., Age-dependent respiratory function decline and DNA deletions in human muscle mitochondria, *Biochem. Mol. Biol. Intl.,* 32, 1009, 1994.
42. Cooper, J. M., Mann, V. M., and Schapira, A. H. V., Analyses of mitochondrial respiratory chain function and mitochondrial DNA deletion in human skeletal muscle: effect of ageing, *J. Neurol. Sci.,* 113, 91, 1992.
43. Simonetti, S., Chen, X., DiMauro, S., and Schon, E. A., Accumulation of deletions in human mitochondrial DNA during normal aging: analysis by quantitative PCR, *Biochim. Biophys. Acta,* 1180, 113, 1992.
44. Lezza, A. M. S., Boffoli, D., Scacco, S., Cantatore, P., and Gadaleta, M. N., Correlation between mitochondrial DNA 4977-bp deletion and respiratory chain enzyme activities in aging human skeletal muscles, *Biochem. Biophys. Res. Commun.,* 205, 772, 1994.
45. Sandy M.S., Langston J. W., Smith M. T., and Di Monte D. A., PCR analysis of platelet mtDNA: lack of specific changes in Parkinson's disease, *Movement Disord.,* 8, 74, 1993.
46. Moraes, C. T., and Schon, E. A., Replication of a heteroplasmic population of normal and partially-deleted human mitochondrial genomes, in *Progress in Cell Research,* 5, Palmieri, F., Papa, S., Saccone, C., and Gadaleta, M. N. (Eds), Elsevier Science B.V., Amsterdam, 1995, 209.

47. Arnheim, N., and Cortopassi, G., Deleterious mitochondrial DNA mutations accumulate in aging human tissues, *Mutat. Res.*, 275, 157, 1992.

48. Harman, D., The biologic clock: the mitochondria?, *J. Am. Geriatr. Soc.*, 20, 145, 1972.

49. Hayflick, L., Current theories of biological aging, *Federation Proc.*, 34, 9, 1975.

50. Dice, J. F., Cellular and molecular mechanisms of aging, *Physiol. Rev.*, 73, 149, 1993.

51. Osiewacz, H. D., and Hermanns, J., The role of mitochondrial DNA rearrangements in aging and human diseases, *Aging Clin. Exp. Res.*, 4, 273, 1992.

52. Wallace, D. C., Diseases of the mitochondrial DNA, *Annu. Rev. Biochem.*, 61, 1175, 1992.

53. Pallotti, F., Chen, X., Bonilla, E., and Schon, E. A., Evidence that specific mtDNA point mutations may not accumulate in skeletal muscle during normal human aging, *Am. J. Hum. Genet.*, 59, 591, 1996.

54. Müller-Höcker, J., Mitochondria and ageing-enzyme-immunohistochemical and in situ hybridization studies, *Bull. Mol. Biol. Med.*, 18, 25, 1993.

55. Nakamura, N., Hattori, N., Tanaka, M., and Mizuno, Y., Specific detection of deleted mitochondrial DNA by in situ hybridization using a chimera probe, *Biochim. Biophys. Acta*, 1308, 215, 1996.

56. Shoubridge, E. A., Karpati, G., and Hastings, K. E. M., Deletion mutants are functionally dominant over wild-type mitochondrial genomes in skeletal muscle fiber segments in mitochondrial disease, *Cell*, 62, 43, 1990.

57. Anderson, S., Bankier, A. T., Barrell, B. G., de Bruijn, M. H. L., Coulson, A. R., Drouin, J., Eperon, I. C., Nierlich, D. P., Roe, B. A., Sanger, F., Schreier, P. H., Smith, A.J. H., Staden, R., and Young, I. G., Sequence and organization of the human mitochondrial genome, *Nature*, 290, 457, 1981.

58. Sato, W., Tanaka, M., Ohno, K., Yamamoto, T., Takada, G., and Ozawa, T., Multiple populations of deleted mitochondrial DNA detected by a novel gene amplification method, *Biochem. Biophys. Res. Commun.*, 162, 664, 1989.

59. Ozawa, T., Tanaka, M., Sugiyama, S., Hattori, K., Ito, T., Ohno, K., Takahashi, A., Sato, W., Takada, G., Mayumi, B., Yamamoto, K., Adachi, K., Koga, Y., and Toshima, H., Multiple mitochondrial DNA deletions exist in cardiomyocytes of patients with hypertrophic or dilated cardiomyopathy, *Biochem. Biophys. Res. Commun.*, 170, 830, 1990.

60. Feinberg, A. P., and Vogelstein, B., A technique for radiolabeling DNA restriction endonuclease fragments to high specific activity, *Anal. Biochem.*, 132, 6, 1983.

61. Sambrook, J., Fritsch, E. F., and Maniatis, T., *Molecular Cloning: A Laboratory Manual*, 2nd Edition, New York, NY, Cold Spring Harbor Laboratory Press, 1989.

62. Gyllensten, U., Direct sequencing of *in vitro* amplified DNA, in *PCR Technology: Principles and Applications for DNA Amplification*, Erlich H. A. (Ed), Stockton Press, New York, 1989, 45.

63. Adams, S. M., and Blakesley, R., Linear amplification DNA sequencing, *Focus*, 13, 56, 1991.

64. Moreira, R. F. and Noren, C. J., Minimum duplex requirements for restriction enzyme cleavage near the termini of linear DNA fragments, *Biotechniques*, 19, 56, 1995.

65. Clark, J. M., Novel non-templated nucleotide addition reactions catalyzed by procaryotic and eucaryotic DNA polymerases, *Nucleic Acids Res.*, 16, 9677, 1988.

66. Ernst, B. P., Wilichowski, E., Wagner, M., and Hanefeld, F., Deletion screening of mitochondrial DNA via multiprimer DNA amplification, *Mol. Cell. Prob.*, 8, 45, 1994.

67. Cheng, S., Higuchi, R., and Stoneking, M., Complete mitochondrial genome amplification, *Nature Genet.*, 7, 350, 1994.

68. Davis, L. G., Dibner, M. D., and Battey, J. F. (Eds), *Basic Methods in Molecular Biology*, New York, Elsevier, 1986, 47.

69. Melov, S., Hinerfeld, D., Esposito, L., and Wallace, D. C., Multi-organ characterization of mitochondrial genomic rearrangements in ad libitum and caloric restricted mice show striking somatic mitochondrial DNA rearrangements with age, *Nucleic Acids Res.* 25, 974, 1997.

70. Sciacco, M., Bonilla, E., Schon, E. A., Di Mauro, S., and Moraes, C. T., Distribution of wild-type and common deletion forms of mtDNA in normal and respiration-deficient muscle fibers from patients with mitochondrial myopathy, *Hum. Mol. Genet.*, 3, 13, 1994.

71. Chelly, J., Kaplan, J.-C., Maire, P., Gautron, S., and Kahn, A., Transcription of the dystrophin gene in human muscle and non-muscle tissues, *Nature*, 333, 858, 1988.

72. Wiesner, R. J., Rüegg, J. C., and Morano, I., Counting target molecules by exponential polymerase chain reaction: copy number of mitochondrial DNA in rat tissues, *Biochem. Biophys. Res. Commun.*, 183, 553, 1992.

73. Lezza, A. M. S., Rainaldi, G., Cantatore, P., and Gadaleta, M. N., Quantitative determination of a 4.8 Kb deletion in mtDNA of aging rat liver, *Bull. Mol. Biol. Med.*, 18, 67, 1993.

74. Hayakawa, M., Hattori, K., Sugiyama, S., and Ozawa, T., Age-associated oxygen damage and mutations in mitochondrial DNA in human hearts, *Biochem. Biophys. Res. Commun.*, 189, 979, 1992.

75. Wang, H., Fliegel, L., Cass, C. E., Penn, A. M. W., Michalak, M., Weiner, J. H., and Lemire, B. D., Quantification of mitochondrial DNA in heteroplasmic fibroblasts with competitive PCR, *BioTechniques*, 17, 76, 1994.

76. Zhang, C., Peters, L. E., Linnane, A. W., and Nagley, P., Comparison of different quantitative PCR procedures in the analysis of the 4977-bp deletion in human mitochondrial DNA, *Biochem. Biophys. Res. Commun.*, 223, 450, 1996.

77. Edris, W., Burgett, B., Colin Stine, O., and Filburn, C. R., Detection and quantitation by competitive PCR of an age-associated increase in a 4.8-kb deletion in rat mitochondrial DNA, *Mutat. Res.*, 316, 69, 1994.

78. Wenger, R. H., and Nielsen, P. J., Reannealing of artificial heteroduplexes generated during PCR-mediated genetic isotyping, *Trends Genet.*, 7, 178, 1991.

79. Becker, A., Reith, A., Napiwotzki, J., and Kadenbach, B., A quantitative method of determining initial amounts of DNA by polymerase chain reaction cycle titration using digital imaging and a novel DNA stain, *Anal. Biochem.*, 237, 204, 1996.

80. Hayakawa, M., Torii, K., Sugiyama, S., Tanaka, M., and Ozawa, T., Age-associated accumulation of 8-hydroxydeoxyguanosine in mitochondrial DNA of human diaphragm, *Biochem. Biophys. Res. Commun.*, 179, 1023, 1991.

81. Zhang, C., Linnane, A. W., and Nagley, P., Occurrence of a particular base substitution (3243 A to G) in mitochondrial DNA of tissues of aging humans, *Biochem. Biophys. Res. Commun.*, 195, 1104, 1993.

82. Münscher, C., Rieger, T., Müller-Höcker J., and Kadenbach, B., The point mutation of mitochondrial DNA characteristic for MERRF disease is found also in healthy people of different ages, *FEBS Lett.*, 317, 27, 1993.

83. Nuovo, G. J., In situ PCR: protocols and applications, *PCR Methods Appl.*, 4, S151, 1995.

84. Ikebe, S.-I., Tanaka, M., Ohno, K., Sato, W., Hattori, K., Kondo T., Mizuno, Y., and Ozawa, T., Increase of deleted mitochondrial DNA in the striatum in Parkinson's disease and senescence, *Biochem. Biophys. Res. Commun.*, 170, 1044, 1990.

85. Ozawa, T., Tanaka, M., Ikebe, S.-I., Ohno, K., Kondo, T., and Mizuno, Y., Quantitative determination of deleted mitochondrial DNA relative to normal DNA in parkinsonian striatum by a kinetic PCR analysis, *Biochem. Biophys. Res. Commun.*, 172, 483, 1990.

86. Yen, T.-C., Su, J.-H., King, K.-L., and Wei, Y.-H., Aging-associated 5 kb deletion in human liver mitochondrial DNA, *Biochem. Biophys. Res. Commun.*, 178, 124, 1991.

87. Lee, H.-C., Pang, C.-Y., Hsu, H.-S., and Wei, Y.-H., Differential accumulations of 4,977 bp deletion in mitochondrial DNA of various tissues in human ageing, *Biochim. Biophys. Acta*, 1226, 37, 1994.

88. Chen, X., Prosser, R., Simonetti, S., Sadlock, J., Jagiello, G., and Schon, E. A., Rearranged mitochondrial genomes are present in human oocytes, *Am. J. Hum. Genet.*, 57, 239, 1995.

89. Fahn, H.-J., Wang, L.-S., Hsieh, R.-H., Chang, S.-C., Kao, S.-H., Huang, M.-H., and Wei, Y.-H., Age-related 4,977 bp deletion in human lung mitochondrial DNA, *Am. J. Respir. Crit. Care Med.*, 154, 1141, 1996.

90. Yen, T.-C., Pang, C.-Y., Hsie, R.-H., Su, C.-H., King, K.-L., and Wei, Y.-H., Age-dependent 6Kb deletion in human liver mitochondrial DNA, *Biochem. Intl.*, 26, 457, 1992.

91. Hattori, K., Tanaka, M., Sugiyama, S., Obayashi T., Ito, T., Satake, T., Hanaki, T., Asai, J., Nagano, M., and Ozawa, T., Age-dependent increase in deleted mitochondrial DNA in the human heart: possible contributory factor to presbycardia, *Am. Heart J.*, 121, 1735, 1991.

92. Sugiyama, S., Hattori, K., Hayakawa, M., and Ozawa, T., Quantitative analysis of age-associated accumulation of mitochondrial DNA with deletion in human hearts, *Biochem. Biophys. Res. Commun.*, 180, 894, 1991.

93. Degoul, F., Nelson, I., Lestienne, P., Francois, D., Romero, N., Dubuc, D., Eymard, B., Fardeau, M., Ponsot, G., Paturneau-Jouas, M., Chaussain, M., Leroux, J. P., and Marsac, C., Deletions of mitochondrial DNA in Kearns-Sayre syndrome and ocular myopathies: genetic, biochemical and morphological studies, *J. Neurol. Sci.*, 101, 168, 1991.

94. Ota, Y., Tanaka, M., Sato, W., Ohno, K., Yamamoto, T., Maehara, M., Negoro, T., Watanabe, K., Awaya, S., and Ozawa, T., Detection of platelet mitochondrial DNA deletions in Kearns-Sayre syndrome, *Investigative Ophthalmology and Visual Science*, 32, 2667, 1991.

95. Miyabayashi, S., Hanamizu, H., Endo, H., Tada, K., and Horai, S., A new type of mitochondrial DNA deletion in patients with encephalomyopathy, *J. Inherit. Metab. Dis.*, 14, 805, 1991.

Section F

Techniques for Exploring Age-Related Intra- and Subcellular Changes

22

Intracellular Signal Transduction Pathways Involved in Hepatocyte DNA Synthesis Following Growth Factor Stimulation

Yusen Liu, Nikki J. Holbrook, Gertrude C. Kokkonen, Shoichi Kitano, and George S. Roth

CONTENTS

22.1 Introduction

22.1.1 Aging Phenotype

22.1.1.1 *In vitro Aging*

Prior to the observations of Hayflick [1] it was believed that cultured cells were essentially immortal and could divide indefinitely. His seminal studies established the ground work for the field now known as *"in vitro* aging." The underlying principle is that normal diploid cells placed in culture can undergo a finite number of cell divisions and loss of division potential occurring over this time is analogous to aging processes which occur *in vivo*. Support for this hypothesis is offered by the fact that cells (usually, but not limited to, skin fibroblasts) obtained from older donors do not divide as many times as those obtained from younger donors [2,3]. Prior to the cessation of division, the time it takes a cell to complete the cycle increases, primarily due to a lengthening of the G1 phase [4].

A great deal of effort has been devoted to elucidating those molecular mechanisms which are responsible for *"in vitro* aging" [4]. A comprehensive discussion is beyond the scope of the present article, but the most recent data suggest that the shortening of telomeres (chromosomal ends) with each cell division may be a primary cause [5].

22.1.1.2 *In vivo Aging*

Clearly, one of the principal values of *"in vitro* aging" models is to simplify studies of *" in vivo* aging" by better controlling experimental conditions. Most earlier studies of aging focused at the organismic level with particular emphasis on physiological and behavioral functions. The pioneering work of Shock and his colleagues documented the deterioration of a number of such functions [6], leading to a more refined focus at the organismic, and eventually the cellular levels. Cellular aging can then be examined using the *in vitro* models mentioned above or through use of a combined *in vivo/in vitro* model in which viable cells are obtained from donors of various ages, but assayed for altered functions directly or after *in vitro* manipulations. This approach offers the advantage of measuring changes that occurred during *in vivo* aging, but in a simplified model system outside the host.

22.1.2 Proliferative Response in Young and Aged Cells

22.1.2.1 *Cell Cycle Control*

Remarkable progress towards understanding the basic machinery driving the cell cycle has been made over the past several years [7–9], offering new hypotheses to explain the age-related loss in proliferative capacity, as well as reagents for their testing. The eukaryotic cell cycle progression is governed by sequential activation of cyclin-dependent kinases (CDK) which phosphorylate key regulatory proteins. In normal cycling cells, CDKs are regulated largely through their interaction with a specific cyclin. In most mammalian tissues, cells remain in a resting state known as G_0, which can be mimicked in cultured cells by growth factor starvation. In response to growth factor stimulation, cells increase cyclin D1 expression as an early response, resulting in the activation of CDK4 and/or CDK6 [9]. CDK4/CDK6 phosphorylates and inactivates the retinoblastoma protein Rb, resulting in the release of the transcription factor E2F from Rb. E2F binds to the promoters of a number of genes encoding proteins necessary for cell cycle progression including cyclin E and cyclin A, resulting in enhanced gene transcription. While activation of cyclin D/CDK4 complexes is critical for both the reentry of quiescent cells from the G_0 state into the cell cycle and progression of cells through early and mid G1, cyclin E/CDK2 activation is required for subsequent G1/S transition as well as DNA synthesis in S phase.

Most studies indicate that *in vitro* senescence is associated with a block in late G1 or at the G1 to S border [4]. In addition, both *in vivo* and *in vitro* aging are likewise associated with a reduction in the ability of quiescent cells to respond to proliferative stimuli and reenter the cell cycle. Thus, most aging studies including our own have concentrated on examining the activities of regulatory proteins associated with G1 and S phase.

22.1.2.2 Signaling Pathways Involved in the Proliferative Response

Changes in signal transduction elicited by growth factors, hormones, neurotransmitters, drugs and other related agents during aging have been reported for numerous systems [10,11]. Such alterations represent important mechanisms by which regulation of physiological and behavioral functions become impaired with increasing age. Initial studies focused on the receptors for these substances, and in many cases reported decreasing concentrations with minimal age changes in binding affinity [10]. However, in many other cases no receptor alterations of any kind were detected despite altered responsiveness.

As the areas of molecular endocrinology, neurobiology and pharmacology developed, it became obvious that interaction with receptor represents only the tip of the iceberg in the elicitation of biological responses. The term "signal transduction" was coined to define the series of components and events that transfer the initial hormone/neurotransmitter drug signal through the cell to produce the final biological response. This process includes "transducers" such as G proteins, "effectors" such as adenyl cyclase and phospholipase C, and "second messengers" such as cyclic AMP, Ca^{2+} and inositol triphosphate (the hormone or neurotransmitter is considered the "first messenger") which initiate subsequent molecular events in the signal transduction cascade. Age-associated changes in essentially all of these components and/or events have been reported in particular cell/tissue systems [10,11].

Probably the most important advance in our understanding of how diverse signals can initiate a proliferative response has been the identification of phosphorylation cascades which link environmental signals to the activation of one or more members of the group of so-called mitogen-activated protein (MAP) kinases, which include the extracellular signal regulated kinase (ERK), stress-activated protein kinase/c-Jun N-terminal kinase (SAPK/JNK) and p38 kinase families [12–15]. Although ERK appears to be the predominant MAP kinase involved in regulating the response to proliferative stimuli, SAPK/JNK and p38 can also contribute to the response. Activation of MAP kinases leads to the phosphorylation of downstream targets, including transcription factors that regulate the expression of genes necessary for cell cycle progression [16,17]. As indicated in Figure 22.1, diverse initiating events involving different transducers and second messengers can converge upstream of ERK [15,18]. Thus, the level of ERK activity seen can be influenced by changes in the activities of proteins involved at any step along the pathways.

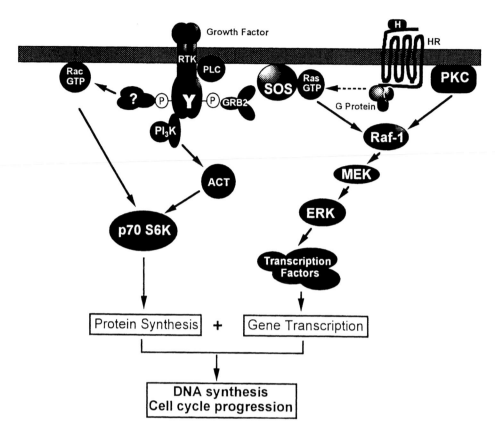

FIGURE 22.1

Model for signal transduction pathways involved in growth factor-stimulated DNA synthesis. Growth factor binding to its receptor triggers autophosphorylation on tyrosine residues, allowing the receptor to interact with SH2-domain-containing adaptor protein such as Grb2. In turn, Grb2 recruits guanosine nucleotide releasing factor SOS to the membrane to activate Ras. Activated Ras (Ras-GTP) then recruits Raf-1 to the cell membrane for activation, resulting in an activated ERK MAP kinase cascade. Activated ERK phosphorylates a group of transcription factors, resulting in enhanced gene transcription. Growth factor binding to its receptor can also activates PI3 kinase (PI3-K) and Rac, which in turn activate p70 S6 kinase. S6 kinase can inrease the rate of synthesis of a number of proteins required for entry into S phase. Activation of both ERK MAP kinase and S6 kinase contributes to entry into S phase and cell cycle progression.

Another signaling pathway which appears to play an important role in regulating cell proliferation in response to external signals involves p70 S6 kinase (p70[S6k]) [19]. This kinase is believed to influence translational processes through the phosphorylation of the S6 ribosomal protein, a component of the 40S subunit of eukaryotic ribosomes [20]. p70[S6k] is itself activated *via* phosphorylation, and this has been shown to occur in response to a variety of signals that influence cell cycle progression including growth factor or hormonal stimulation [19]. Treatment of cells with either anti-p70[S6k] antibodies or the immunosuppressive agent rapamycin, both of which block p70[S6k] activity, prevents the progression of cells through G1 into S phase and inhibits DNA synthesis [21,22]. Although the mechanisms controlling p70[S6k] activation are poorly understood, the molecular players which are known to participate in the p70[S6k] pathway are illustrated in Figure 22.1 [23,24]. That the ERK MAP kinase and p70[S6k] signaling pathways are distinct is evidenced by the fact that rapamycin inhibits p70[S6k] activation without affecting ERK activity [25]. Alterations in the activities of either or both the MAP kinase or p70^{S6k} signaling pathways could contribute to altered proliferative status with aging.

22.1.3 The Hepatocyte Primary Culture Model

The system studied jointly by our laboratories is the stimulation of DNA synthesis by catecholamine and growth factors in primary cultures of hepatocytes from rats of various ages.

The hepatocyte DNA synthesis model is particularly valuable for aging research for a number of reasons. First, impairments in the control of DNA synthesis and cell division are responsible for many disabilities and diseases of aging, including impaired wound healing [26] and tissue regeneration [27], decreased immune response [28], and cancer [29]. Second, it offers the opportunity to compare in young and aged cells signal transduction events that have been well defined but continues to offer new possibilities for studying "cross-talk", coordinate regulation and other mechanisms of information transfer. Finally, it offers a useful *in vivo* compliment to the fibroblast model of *in vitro* aging which has been under study for nearly three decades [1–5].

In this chapter, we discuss those methodological considerations which are particularly relevant to elucidating the mechanisms by which stimulation of hepatocyte DNA synthesis becomes impaired during aging. We conclude with a description of the current status of our observations and directions for future investigation.

22.2 Methodologies Employed to Examine Age-RelatedAlterations in Proliferative Response of Hepatocytes

22.2.1 Hepatocyte Isolation and Culture

22.2.1.1 General Description

The procedure described is a modification of the two-step method of Seglen [30]. The technique is based on liver perfusion with collagenase after removal of Ca^{2+} by perfusion with the chelator EGTA. Moderately high yields of viable, single hepatocytes (4 to 6 × 10^8 cells per adult rat), essentially free of nonparenchymal cells, can be obtained. Two hours after plating on collagen-coated dishes, primary cultures of metabolically active hepatocytes can be established which in our laboratory have been maintained for up to 3 to 5 days.

22.2.1.2 Materials

1. 1× Hank's Balanced Salt Solution + HEPES (pH 7.4): 140 mM NaCl, 5 mM KCl, 0.33 mM Na_2HPO_4, 0.44 mM KH_2PO_4, and 10 mM HEPES.

2. EGTA solution (pH 7.2): 0.5 mM EGTA, 1× Hank's Balanced Salt Solution + HEPES (see above), 5 mM glucose, and 4 mM $NaHCO_3$.

3. Collagenase solution (pH 7.5): 1× Hank's Balanced Salt Solution + HEPES (see above), 5 mM $CaCl_2$, 0.05% trypsin inhibitor (type 1-S, Sigma, Catalog #T6522), and 4 mM $NaHCO_3$. Just before using, add collagenase (type IV, Sigma, Catalog #C5138) to obtain a final concentration of 0.05%.

4. Phosphate-buffered saline (pH 7.4) without calcium and magnesium to be used for bathing the liver to maintain its temperature at 37°C.

5. Williams E culture media: Williams medium E 1× without glutamine (Quality Biological Inc., Catalog #112-033-100) supplemented with 2 mM L-glutamine, 1 nM dexamethasone, penicillin (100 U/ml), and streptomycin (98 μg/ml).

6. Peristaltic pump equipped with Teflon tubing and an 18-gauge cannula.

7. Water bath set at 39°C to equilibrate perfusion solutions. The higher temperature is required so that the perfusion solutions will be 37°C when entering liver.

8. Microscope and hemocytometer for counting cells.

22.2.1.3 Protocol

1. Place phosphate-buffered saline, 250 ml of EGTA solution and 500 ml of Collagenase solution into a 39°C water bath to equilibrate.

2. Set up perfusion apparatus and prime the peristaltic pump and lines with EGTA solution.

3. Sacrifice rat by CO_2 inhalation. Moisten abdomen with 70% ethyl alcohol and open peritoneal cavity by a mid-ventral incision. Deflect intestines to right side and gently push two main lobes of liver toward animal's head to uncover portal vein and inferior vena cava.

4. Keeping 18-gauge perfusion needle parallel to the portal vein, enter vein approximately 1.5 cm from point where it branches and maneuver until tip of needle is about 0.5 cm proximal to the branch point. Clamp needle to vein and immediately start perfusion with EGTA solution at a rate of 20 ml per min.

5. Immediately sever inferior vena cava below level of the kidney to enable perfusate to exit. A sign of adequate perfusion is when the liver clears immediately and completely.

6. Perfuse EGTA solution for at least 10 min (minimum of 200 ml), during which time continue to bath the liver with 37°C phosphate buffered saline.

7. Prepare 50 ml of the collagenase solution 5 min prior to ending the EGTA perfusion. Keep the solution at 37°C waterbath.

8. After perfusion with EGTA solution, begin perfusing the liver with collagenase solution.

9. Digest liver for approximately 15–20 min until liver begins to disintegrate.

10. Remove digested liver to iced beaker containing 100 ml of 1× Hank's Balanced Salt Solution + HEPES.

11. Break liver capsule apart with forceps and disperse the cells with a gentle stirring movement.

12. Filter cell suspension into iced beaker first through 40-micron nylon mesh and then through 100 micron nylon mesh.

13. Divide the cell suspension equally among four iced sterile 50 ml test tubes.

14. Centrifuge the cell suspension at $50 \times g$ for 2 min at 4°C.

15. Discard supernatant. To each test tube, add 25 ml of Williams E culture media containing 5% fetal bovine serum (FBS), mix gently and centrifuge as above.

16. Repeat previous washing step 2 more times for a total of 3 washes.

17. Combine all cell pellets and bring cell suspension to 40 ml with supplemented Williams E culture media containing 5% FBS.

18. Determine cell viability using trypan blue at a final concentration of 0.2%.

22.2.2   DNA Synthesis Measurements

22.2.2.1   *General Description*

Estimates of the rate of DNA synthesis in a cell population or in individual cells have often been used as a measure of the rate of cell proliferation. Uptake of ^{3}H-thymidine, an indicator of DNA synthesis, is rapid and equilibration of extracellular and intracellular thymidine usually occurs within a short time. Several laboratories have been successful in achieving hepatocyte DNA synthesis in primary culture. Maintenance of hepatocytes at low densities (0.1-0.4 × 10^5 cells/cm^2), which allows for adequate spreading on films of collagen, particularly Type I collagen, appears to be necessary to support adequate growth. To obtain substantial levels of growth in culture, media that meet basal requirements for supporting DNA synthesis as well as growth-stimulatory factors for hepatocytes (e.g., epidermal growth factor) need to be utilized. Even under these conditions, however, DNA synthesis in primary cultures of hepatocytes is limited.

22.2.2.2   *Materials*

1. Type I calf skin collagen (Sigma, Catalog #CA919) containing 76 μg/ml type I calf skin in sterile 0.1% acetic acid.

2. 12 well tissue culture plates.

3. Phosphate-buffered saline (pH 7.4) without calcium and magnesium.

4. Williams E culture media: Williams medium E 1× without glutamine (Quality Biological Inc., Catalog #112-033-100) supplemented with 2 mM L-glutamine, 1 nM dexamethasone, penicillin (100 U/ml), and streptomycin (98 μg/ml).

5. ^{3}H-thymidine (10 μCi/ml., 85 Ci/mmol, Amersham, Catalog #TRK758).

6. 10% trichloroacetic acid (TCA).

7. 100% trichloroacetic acid (TCA).

8. 1 M NaOH.

9. Coomassie Plus Protein Assay Reagent Kit (Pierce, Cat. #23236).

22.2.2.3   *Protocol*

1. The experiment is done in triplicate using 12 well tissue culture plates (area of well = 3.8 cm^2). Before plating cells, coat each well with 0.5 ml of type I calf skin collagen in 0.1% sterile acetic acid (Sigma, Catalog #CA919). Rinse each well with at least 1 ml of phosphate-buffered saline (pH 7.4) just before plating cells.

2. To each well add 1 ml of hepatocytes suspension containing 1.5 × 10^5 viable cells in Williams E culture media plus 5% FBS and incubate at 37°C in 5% CO_2.

3. Two h after plating, rinse each well once with Williams E culture media and replace the volume of each well with 1 ml of the same medium. Incubate cells at 37°C in 5% CO_2 for 18-20 h before stimulating the cells.

4. Stimulate cells with a reagent such as EGF by adding it directly into the culture medium.

5. Two hours before harvesting cultures, to each well add 0.5 ml of Williams E culture media containing 5 μCi of ³H-thymidine (10μCi/ml, 85 Ci/mmol), and continue to incubate cells until harvesting.

6. To harvest cell cultures, remove medium containing ³H-thymidine and wash each well twice with 1 ml of cold phosphate-buffered saline (pH 7.4).

7. To each well, add 1 ml of cold 10% TCA and let sit for 10 min at room temperature.

8. Remove the TCA from each well, then add 500 μl of 1M NaOH and incubate 30 min at 37°C to solubilize cells.

9. Transfer each sample of solubilized cells to its respective 1.5 ml microtube.

10. To each microtube, add 100 μl of 100% TCA, vortex and let sit overnight at 4°C to precipitate the acid-insoluble DNA and protein.

11. The following day, vortex and centrifuge tubes at 2,500 rpm for 20 min at 4°C. Discard the supernatant.

12. Wash each precipitate with 500 μl of 10% TCA by vortexing and centrifugation as above.

13. Add 500 μl of 10% TCA to each tube, vortex, and incubate at 90°C for 15 min to hydrolyze the precipitates.

14. Vortex each sample and let stand to cool at room temperature for 30 min.

15. Vortex each sample and centrifuge at 2,500 rpm for 20 min at 4°C (precipitate is protein; supernatant is acid-soluble DNA).

16. Aliquot 300 μl of each supernatant, then add into 4 ml of scintillant and count the radioactivity.

17. Take off any excess supernatant in the microtube. Solubilize each protein precipitate in 100 μl of 1 M NaOH and bring the final volume to 1 ml with H_2O.

18. Assay 900 μl of the solubilized precipitate for protein concentration using the Coomassie Plus Protein Reagent Kit (Pierce, Catalog #23236).

19. Express DNA synthesis as: dpm/mg protein

22.2.3   Cyclin-Dependent Kinase 2 (CDK2) Assay

22.2.3.1   General Description

CDKs play a crucial role in regulating cell growth. CDK2, in particular, is critical for the initiation of DNA synthesis and progression of cells into S phase. Since aged hepatocytes display reduced DNA synthesis following growth factor stimulation compared to young cells, it is reasonable to investigate whether CDK2 activation is impaired in aged cells. The described procedure to measure CDK2 activity is modified from that used by Gorospe et al.31 It relies on the specific interaction between an antibody and an antigen to partially purify CDK2 from soluble cell lysates. CDK2 can form complexes with a rabbit antisera against the carboxyl—terminal end of CDK2. Since protein A interacts with the Fc region of the rabbit antibody, the CDK2-antibody complexes can be pulled down by the protein A-immobilized Sepharose beads. Because the antibody utilized was generated to a peptide corresponding to the C-terminal region of CDK2 (far away from the kinase domain), the immune-complex retains its kinase activity which can be measured based on its ability to phosphorylate its substrates. In the presence of γ-^{32}P-ATP, phosphorylation of the substrate histone H1 can be visualized by autoradiography after SDS-polyacrylamide gel electrophoresis. This general technology is not limited to the analysis of CDK2, but in theory can be adapted to examine other CDKs important for cell growth and division by adjusting the particular antibodies and substrates utilized.

22.2.3.2 Materials

1. PBS (pH 7.4), without calcium and magnesium.
2. Lysis buffer: 50 mM Tris-HCl (pH 7.4), 250 mM NaCl, 0.1% Triton X-100, 5 mM NaF, 0.1 mM sodium orthovanadate. This basic buffer can be stored at 4°C. The following proteinase inhibitors are supplemented to the basic buffer immediately before use: 10 μg/ml leupeptin, 10 μg/ml aprotinin, 10 μg/ml pepstatin A, and 1 mM PMSF.
3. Histone H1 kinase buffer: 50 mM Tris-HCl (pH 7.4), 10 mM $MgCl_2$, 1 mM DTT, 2 mM EGTA.
4. ATP solution containing 0.6 mM ATP and 10 mM $MgCl_2$.
5. γ-^{32}P-ATP (~3000 Ci/mmol) from Amersham Life Science (Catalog #: AA0068).
6. Histone H1 solution containing 2 mg/ml histone H1 (Ambion, Austin, TX, Catalog #2630).
7. Rabbit polyclonal anti-CDK2 anti-sera available from Pharmingen, Inc. (San Diego, CA, Catalog #15536E), which is generated by immunizing with a synthetic peptide (DVTKPVPHLRL) corresponding to the carboxyl-terminal sequence of human CDK2.
8. Preimmune rabbit serum, or a rabbit polyclonal antisera against any proteins unrelated to CDK2.
9. Protein A-Sepharose 4B Fast Flow beads (Sigma, Catalog #P9424).
10. 4× Laemmli's SDS Loading buffer: 200 mM Tris-HCl, pH6.8, 8% SDS, 40% glycerol, 400 mM DTT, 0.4% bromophenol blue.
11. Bio Rad protein assay system (Bio Rad, Catalog #500-0006).

22.2.3.3 Protocol

1. Treatment:

    Prior to treatment, cells are placed in serum-free medium for at least 16 h. The majority of cells from liver tissue are in the G1 phase of the cell cycle. Serum starvation maintains them in a uniform state and/or leads to further enrichment of cells in G1 phase. Stimulation of cells with a proliferative stimulus can be carried out by direct addition of a stock solution to the medium. Serum stimulation can be achieved by adding FBS into the medium to a final concentration of 20%.

2. Harvesting of Cells:

    a. Wash plates with ice-cold PBS twice, suck all PBS out of the plates.

    b. Add 1 ml of ice-cold lysis buffer to each 100 mm plate. Scrap cells off the plate, and transfer the cell lysates into 1.5 ml Eppendorf tubes.

c. Pass through a 25-gauge needle several times to ensure complete lysis of the cells.

d. Remove insoluble cellular debris by centrifugation at 14,000 rpm for 10 min at 4°C.

e. Collect the soluble cell extracts. Determine the protein concentration using the Bio Rad protein assay system in duplicate, take the average value as the concentration.

3. Immunoprecipitation:

a. Take 200-500 μg of protein and add lysis buffer to make up to 1 ml.

b. Add 1-2 μg of rabbit preimmune serum or 1-2 μg of a rabbit antibody against a protein unrelated to CDK2, and protein A-Sepharose 4 fast flow beads (bed volume 20μl) into each tube.

c. Rotate at 4°C for 30 min to remove proteins unspecifically bound to antibody or protein A beads.

d. Collect the supernatant, and add 1-2 μg of the CDK2 antisera alone with the protein A-Sepharose 4 fast flow beads (bed volume 20 μl) into each tube.

e. Rotate at 4°C for 4 h to immunoprecipitate proteins specifically bound to CDK2 antisera.

f. Collect the pellet by pulse-centrifugation, dispose the supernatant.

g. Wash the pellets twice with 1 ml of lysis buffer.

h. Wash the immunoprecipitate 4 times with 1 ml of histone H1 kinase buffer. After the last wash, take all buffer out of the tube using the pipet tips. The immunocomplexes can be stored at –80°C if not used immediately for histone H1 kinase assay.

4. Determination of Kinase Activity:

a. Add 20 μl histone H1 kinase buffer, 5 μl histone H1 solution containing 10 μg histone H1, 1 ml γ-^{32}P-ATP into each tube containing the immunoprecipitates.

b. Incubate the mixture at 30°C for 30 min, vortex every 2-5 min.

c. Stop the reaction by adding 25 μl of 4X Laemmli's loading buffer (200 mM Tris-HCl, pH 6.8, 8% SDS, 40% glycerol, 400 mM DTT, 0.4% bromophenol blue). Vortex the mixture and boil for 5 min. Centrifuge the mixture briefly to pellet the beads, then load 10 μl of supernatant on to 15% SDS-polyacrylamide gels.

d. After electrophoresis, cut the front off the resolving gel. Rinse the gel in solution containing 10% acetic acid and 20% methanol for 5 min then dry the gel.

e. Expose the gel to X-ray film or phosphorImager. Histone H1 migrates as double bands of about 21 kilodalton. The incorporation of γ-^{32}P-ATP into histone H1 can be quantitated by the ImageQuant program.

22.2.4 ERK Mitogen-Activated Protein Kinase Assay

22.2.4.1 *General Description*

The immune complex procedure described here is modified from that developed by Kyriakis et al. [23]. The principal is essentially the same as that described above for immunoprecipitation and assay of CDK2 activity. Again, it relies on the specific interaction between an antibody and an antigen to enrich for ERK protein, thus high quality and specific antibodies must be used. Briefly, a rabbit antibody to the carboxyl-terminal end of ERK2 MAP kinase is used to immunoprecipitate ERK and protein A-immobilized Sepharose beads can be used to pull down the ERK2-antibody immune-complexes. The immune-complexes retain their kinase activity which can be measured based on their ability to phosphorylate a substrate, myelin basic protein. In the presence of γ-^{32}P-ATP, phosphorylation of the substrate can be visualized after SDS-polyacrylamide gel electrophoresis. The activation of ERK2 can be verified by Western blot analysis using a monoclonal ERK2 antibody. Activated ERK migrates slower than the inactive ERK due to its phosphorylation. Although the protocol described is for assaying the ERK MAP kinase activity, with modification it can also be used to assess the kinase activities of other members of the MAP kinase family [32–34].

22.2.4.2 Materials

1. PBS (pH 7.4) without calcium and magnesium.
2. Lysis buffer: 20 mM HEPES (pH 7.4), 50 mM β-glycerol phosphate, 2 mM EGTA, 1 mM DTT, 10 mM NaF, 1 mM sodium orthovanadate, 1% Triton-100, 10% glycerol. The buffer can be stored at 4°C for 3 months. It is supplemented with 10 μg/ml leupeptin, 10 μg/ml aprotinin, 1mM PMSF and 10 nM okadiac acid immediately before use.
3. Washing buffer: 500 mM LiCl, 100 mM Tris-HCl (pH7.6), 0.1% Triton X-100, and 1 mM DTT. This buffer can be stored at 4°C for several months.
4. Kinase reaction buffer: 20 mM MOPS (pH7.2), 2 mM EGTA, 10 mM NaF, 10 mM MgCl$_2$, 1 mM DTT, 0.1% Triton X-100. This buffer is stable at 4°C for a long time.
5. Rabbit antibody against the carboxyl-terminal amino acids 345-358 of ERK2 MAP kinase available from Santa Cruz Biotechnology, Inc. (Catalog #: SC-154).
6. Protein A-Sepharose 4B Fast Flow beads (Sigma, Catalog # P9424).
7. Myelin basic protein (Sigma, M1891). Myelin basic protein (MBP) is dissolved in the kinase reaction buffer to a final concentration of 0.3 mg/ml. Store this solution at –20°C.
8. γ-^{32}P-ATP (1 mCi, ~3000 Ci/mol.) (Amersham Life Science, Catalog # AA0068). Dilute original into 900 μl of cold ATP/MgCl$_2$ solution (100 μM ATP, 60 mM MgCl$_2$) to make up the γ-^{32}P-ATP working solution.
9. Bio Rad protein assay system (Catalog #500-0006).

22.2.4.3 Protocol

1. Treatment:
   Serum starved cells are treated with a given stimulating agent by direct addition of a stock solution to the culture medium. Serum stimulation is performed by adding fetal bovine serum to a final concentration of 20%.

2. Harvesting of Cells:
   a. Wash plates with ice-cold PBS twice, suck all PBS out of the plates.
   b. Lyse cells with 1 ml of lysis buffer (for 100 mm plate, 1.5 ml for 150 mm plate) and place the plates on ice for at least 15 min.

    c. Scrape cells, and transfer the lysates into Eppendorf tubes, then centrifuge at 14,000 rpm for 10 min at 4°C.

    d. Collect the supernatant. Determine the protein concentration using Bio Rad system in duplicate, taking average value as the concentration.

3. Immunoprecipitation of ERK2 MAP Kinase:

    a. Take 200-1000 μg of protein and add lysis buffer to make up to 1 ml.

    b. Add 1-2 μg of antibody against ERK2 and protein A-Sepharose 4 fast flow beads (bed volume 20 μl) into each tube.

    c. Rotate at 4°C for 4 h or overnight.

    d. Centrifuge up to maximum speed at 4°C briefly. Collect the pellets.

    e. Wash the pellets 3 times with 1 ml of lysis buffer.

    f. Wash the immunoprecipitate 3 times with 1 ml of washing buffer.

    g. Wash the precipitate 3 times with 1 ml of kinase assay buffer. After the last wash, take all buffer out of the tubes using the pipet tips. The immunocomplexes can be stored at –80°C for several months without loss of kinase activity.

4. Determination of Kinase Activity:

    a. Add 20 μl MBP solution and 15 μl γ-^{32}P-ATP working solution into the tube containing the immunoprecipitates.

    b. Incubate the mixture at 30°C for 20 min, vortex every 2–5 min.

    c. Stop the reaction by adding 25 μl of 4× Laemmli's loading buffer (200 mM Tris-HCl, pH 6.8, 8% SDS, 40% glycerol, 400 mM DTT, 0.4% bromophenol blue). Vortex the mixture and boil for 5 min. Spin down the pellet and then load 10 μl of supernatant onto 15% SDS-polyacrylamide gel.

    d. After electrophoresis, cut the front off the resolving gel. Rinse the gel in solution containing 10% acetic acid, 20% methanol for 5 min then dry the gel.

    e. Expose the gel to X-ray film or phosphorImager. Myelin basic protein migrates as a double band of about 16 kilodalton. The incorporation of γ-^{32}P-ATP into myelin basic protein can be quantitated by the ImageQuant program.

22.2.5 p70 S6 Kinase Assay

22.2.5.1 General Description

The procedure for immunoprecipitating p70^{S6k} from hepatocytes is similar to that utilized above for assaying CDK2 and ERK kinase activities. Kinase activity is determined by the phosphorylation of a peptide corresponding to the carboxyl-terminal region of S6 protein. This peptide can be separated from free ATP based on its binding to P81 phosphocellulose filters.

    The immunosuppressant rapamycin should be used to demonstrate the specificity of the immune-complex kinase assay for p70^{S6k}.

2.5.2 Materials

1. PBS (pH 7.4) without calcium and magnesium.

2. Lysis buffer: 10 mM KH$_2$PO$_4$, 1 mM EDTA, 5 mM EGTA, 10 mM MgCl$_2$, 2 mM DTT, 1 mM NaF, 50 mM β-glycerol phosphate, 1 mM sodium orthovanadate, 1% Triton-X 100. The buffer can be stored at 4°C for several months. Supplement

the buffer with: 10 µg/ml leupeptin, 10 µg/ml aprotinin, 1 mM PMSF and 10 nM okadaic acid, and 0.5 ng/ml microcystin-LR immediately before use.

3. RIPA buffer: 10 mM Tris-HCl (pH 7.2), 150 mM NaCl, 1% deoxycholic acid, 1% Triton X-100, 0.1% SDS. This buffer can be stored at 4°C for several months.

4. Kinase reaction buffer: 25 mM MOPS (pH 7.2), 5 mM EGTA, 10 mM NaF, 15 mM MgCl$_2$, 1 mM DTT, 60 mM β-glycerol phosphate, and 30 mM *p*-nitrophenol phosphate. This buffer is stable at 4°C for a long time.

5. Rabbit antibody against the carboxyl-terminal amino acids 485–502 of p70 S6 kinase available from Santa Cruz Biotechnology, Inc. (Catalog #: SC-230).

6. Protein A-Sepharose 4B Fast Flow beads (Sigma, Catalog #: P9424).

7. S6 pepetide (RRRLSSLRA) corresponding to amino acids 231–239 of human 40S ribosomal protein S6 (Santa Cruz Biotechnology, Catalog #: SC-3009).

8. Peptide inhibitor for cAMP-dependent protein kinase available from Santa Cruz Biotechnology, Inc (Catalog #: SC-3010), which corresponds to the carboxyl-terminal auto-inhibitory domain of PKA.

9. γ-^{32}P-ATP (1 mCi, ~3000 Ci/mmol) (Amersham Life Science, Catalog #: AA0068).

10. BioRad protein assay system (Catalog #500-0006).

22.2.5.3. Protocol

1. Treatment:

   Place cells in serum-free medium for at least 16 h prior to treatment with stimulating agent. Cells are treated with a given stimulating agent by direct addition of a stock solution to the culture medium. Serum stimulation is performed by adding FBS to a final concentration of 20%.

2. Harvesting of Cells:

   a. Wash plates with ice-cold PBS twice, suck all PBS out of the plates.

   b. Lyse cell with 1 ml of lysis buffer (for 100 mm plate, 1.5 ml for 150 mm plate) and place the plates on ice for at least 15 min.

   c. Scrape cells, and transfer the lysates to Eppendorf tubes, then centrifuge at 14,000 rpm for 10 min at 4°C.

   d. Collect the supernatants. Determine the protein concentration using the Bio-Rad system in duplicate, taking average value as the concentration.

3. Immunoprecipitation of p70^{S6k}:

   a. Take 200–1000 µg of protein and add lysis buffer to make up to 1 ml.

   b. Add 1–2 µg of antibody against p70 S6 kinase and protein A-Sepharose 4 fast flow beads (bed volume 20 µl) into each tube.

   c. Rotate at 4°C for 4 h or overnight.

   d. Collect the pellets by centrifugation and wash the pellets 3 times with 1 ml of lysis buffer.

   e. Wash the immunoprecipitate 3 times with 1 ml of RIPA buffer.

   f. Wash the precipitate 3 times with 1 ml of kinase assay buffer. After the last wash, take all buffer out of the tubes using the pipet tips. The immunocomplexes can be stored at –80°C if not used for kinase assay immediately.

4. Determination of Kinase Activity:

    a. Add 30 μl 2X kinase assay buffer supplemented with 130 nM PKA inhibitor and 25 μl of ATP solution containing 100 μM cold ATP and 20 μCi γ-³²P-ATP into the tubes containing the immunoprecipitates.

    b. Start reaction by adding 5 μl of S6 kinase substrate solution contain 5 μg of the substrate. Incubate the mixture at 30°C for 20 min. Vortex every 2–5 min.

    c. At two different times (usually 10 and 20 min after starting the reaction), take 10 μl of the reaction mixture supernatant out, and mix with 8 μl of 12% TCA to terminate the reaction.

    d. Load the above TCA mixture on to a P81 phosphocellulose SpinZyme unit (Pierce, Catalog #29520). Spin the filter unit for 10 sec at 14,000 rpm, then wash the filter 3 times with 400 μl of 150 mM phosphoric acid, once with ethanol, and count the radioactivity in a scintillation counter. The amount of 32P incorporated into the substrate in 20 min should be about twice of that in 10 min.

22.3 Findings in Hepatocytes and Future Directions

22.3.1 DNA Synthesis is Impaired in Aged Hepatocytes

Figure 22.2 shows results from a typical experiment in which cells of young (6 month), middle-aged (12 month) and old (24 month) Wistar rats were stimulated with epidermal growth factor (EGF) and DNA synthesis assessed. Two-hour 3H-thymidine pulses were administered just prior to harvesting the cells at the indicated times. The time course of DNA synthesis is similar in all 3 age groups, beginning after 12 h and reaching a maximum at 48 h. However, the magnitude of the response is progressively reduced with age. The concentration of EGF (100 ng/ml) was shown to be maximal for cells of all ages, indicating that the age differences are not merely a sensitivity problem [35]. Furthermore, more than 90% of the ³H-thymidine incorporation is sensitive to the DNA polymerase inhibitor, aphidicoline, and thus represents true DNA replication rather than repair [35].

22.3.2 Age-Associated Impairment in the Early Signaling Pathways

Cell proliferation in response to growth factor stimulation is regulated through a number of signaling pathways. The immediate event following stimulation with a growth factor such as EGF is tyrosine phosphorylation of growth factor receptors, which leads to the activation of a number of signaling pathways including the ERK MAP kinase and p70 S6 kinase pathways. ERK can influence transcription of proliferation-associated genes through phosphorylation of a group of transcription factors. p70 S6 kinase is believed to play an important role in regulating the rates of synthesis of numerous proteins required for entry into S phase. In aged hepatocytes, we have observed a significant decline in the activities of both protein kinases [33], providing an explanation for the age-associated

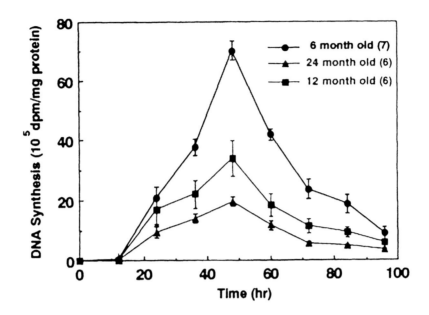

FIGURE 22.2

Time course of EGF stimulated DNA synthesis in hepatocytes from 6, 12, and 24 month-old rats. EGF (100 ng/ml) was added to the culture medium 20 h after cell inoculation. Cells were then incubated and labeled with 3H-thymidine for 2 h at the indicated time. DNA synthesis was calculated by subtracting unstimulated values at each time. Each point represents the mean ± SEM of 6 to 7 individual animals: ●, 6 month old; ▲, 12 month old; ○, 24 month old.

loss in proliferative capacity. These alterations could be either indicative of multiple deficiencies in the signaling cascades or reflect a change in an upstream event common to both signaling pathways (i.e., close to or at the receptor level). Although we have not detected any change in EGF receptor number or binding capacity, we can not rule out the possibility that the growth factor receptors may be deficient in activating downstream targets.

22.3.3 Future Directions

Future studies will be required to determine whether the EGF receptor in aged cells efficiently undergoes autophosphorylation on tyrosine residues and the subsequent conformational changes required for activating the downstream signaling cascades. Recent findings indicate that significant overlap exists between the signaling pathways triggered by growth factors and those activated by stress. Given that aged cells are often deficient in their ability to cope with environmental stress, future studies will investigate to what extent age-associated alterations in MAP kinase and S6 kinase signaling pathways contribute to this defect.

In addition, new signal transduction pathways continue to be discovered for growth factors, stress, and other agents which initiate cellular DNA synthesis. In light of the complexity of these schemes and their potential interrelationships, it is very likely that age related alterations in a number of such components and events will be established. Ultimately, we hope that such knowledge will allow design of rational interventions to help better control DNA synthesis in senescent cells.

References

1. Hayflick, L., The limited lifespan of human diploid cell strains, *Exp. Cell Res.*, 37, 614, 1965.
2. Martin, G.M., Sprague, C.A., and Epstein, C.J., Replicative lifespan of cultured human cells: effect of donor age, tissue, and genotype, *Lab. Invest.*, 23, 86, 1970.
3. Schneider, E.L., and Mitsui, Y., The relationship between *in vitro* cellular aging and *in vivo* human age, *Proc. Natl. Acad. Sci. U.S.A.*, 73, 3584, 1976.
4. Campisi, J., Dimri, G., and Hara, E., Control of replicative senescence, *Handbook of the Biology of Aging*, Scheider, E.L., and Rowe, J.W. Ed., Academic Press, San Diego, CA, 1996. (Part 3).
5. Harley, C.B. and Villeponteau, B., Telomeres and telomerase in aging and cancer, *Curr. Opin. Genet. Dev.*, 5, 249, 1995.
6. Shock, N.W., Greulich, R.C., Andres, R., Arenberg, D., Costa, P.T. Jr., Lakatta, E.G., and Tobin, J.T., *Normal Human Aging: The Baltimore Longitudinal Study*, U.S. Govt. Printing Office, Washington, D.C., 1984.
7. Morgan, D.O., Principles of CDK Regulation, *Nature*, 374, 131, 1995.
8. Pines, J., Cyclins, CDKs and Cancer, *Semin. Cancer Biol.*, 6, 63, 1995.
9. Sherr, C.J., D-type cyclins, *Trends in Biochem. Sci.*, 20, 187, 1995.
10. Roth, G.S. and Hess, G.D., Changes in the mechanisms of hormone and neurotransmitter action during aging: Current status of the role of receptor and post-receptor alterations, *Mech. Aging and Devel.*, 20, 175., 1995, Chap. 8.
11. Miyamoto, A. and Roth, G.S., Changes in transmembrane signaling mechanism during aging-cellular and molecular aspects, *Molecular Basis of Aging*, Macievia-Coelho, A., Eds., Crc Press, Inc., Boca Raton, FL., 1995, Chap. 8.
12. Seger, R. and Krebs, E.G., The MAPK signaling cascade, *FASEB J.*, 9, 726, 1995.
13. Kyriakis, J.M., Banerjee, P., Nikolakaki, E., Dai, T., Rubie, E.A., Ahmad, M.F., Avruch, J., and Woodgett, J.R., The stress-activated protein kinase subfamily of c-Jun kinases, *Nature*, 369, 156, 1994.
14. Han, J., Lee, J.D., Bibbs, L., and Ulevitch, R.A., A MAP kinase targeted by endotoxin and hyperosmolarity in mammalian cells, *Science*, 265, 808, 1994.
15. Davis, R.J., MAPKs, new JNK expands the group, *Trends Biochem. Sci.* 19, 470, 1994.
16. Davis R. J., Transcriptional regulation by MAP kinase, *Mol. Reprod. Dev.*, 42, 459, 1995.
17. Lavoie, J.N., L'Allemain, G., Brunet, A., Muller, R., and Ponyssegur, J., Cyclin D1 expression is regulated positively by the p42/p44 MARK and negatively by the p38/HOG MAPK pathway, *J. Biol., Chem.*, 271, 20608, 1996.
18. Avruch, J., Zhang, X.-F., and Kyriakis, J.M., Raf meets Ras: completing the framework of a signal transduction pathway, *Trends Biochem. Sci.*, 19, 279, 1994.
19. Prond, C.G., p70 S6 kinase: an enigma with variations, *Trends Biochem. Sci.*, 21, 181, 1996.
20. Brown, E.J. and Schreiber, S.L., A signaling pathway to translational control, *Cell*, 86, 517, 1996.
21. Lane, H.A., Fernandez, A., Lau, N.J.C., and Thomas, G., p70 S6 kinase function is essential for G1 progression, *Nature*, 363, 170, 1993.
22. Dumont, F.J. and Su, Q., Mechanisms of action of the immunosuppressant rapamycin, *Life Sci.*, 58, 373, 1996.
23. Chou, M.M. and Blenis, J., The 70 kDa S6 kinase complexes with and is activated by the Rho family G proteins Cdc42 and Rac1, *Cell*, 85, 573, 1996.
24. Grammer, T.C., Cheatham, L., Chou, M.M., and Blenis, J., The p70S6K signaling pathway: a novel signaling system involved in growth regulation, *Cancer Surv.*, 27, 27, 1996.
25. Price, D.J., Grove, J.R., Calvo, V., Avruch, J., and Bierer, B.E., Rapamycin-induced inhibition of the 70-kilodalton S6 protein kinase, *Science*, 257, 973, 1992.
26. Danon, D., Kowatch M.A., and Roth, G.S., Promotion of would repair in old mice by local injection of macrophages, *Proc. Natl. Acad. Sci. U.S.A.*, 86, 2018, 1989.
27. Bucher, N.L.R., Swaffield, M.N., and Ditroia, J.F., The influence of age upon the incorporation of thymidine-2-^{14}C into the DNA of regenerating rate liver, *Cancer Res.*, 24, 509, 1964.

28. Kay, M.M.B. and Makinoden, T., *Handbook of Immunology in Aging*, CRC Press, Inc., Boca Raton, FL., 1981.
29. Hartwell, L.H. and Kostern, M.B., Cell cycle control and cancer, *Science*, 266, 1821, 1994.
30. Seglen, P.O., Preparation of isolated liner cells, *Methods Cell Biol.*, 13, 29, 1976.
31. Gorospe, M., Liu, Y., Xu, Q., Chrest, F.J., and Holbrook, N.J., Inhibition of G1 cyclin-dependent kinase activity during growth arrest of human breast carcinoma cells by prostaglandin A_2, *Mol. Cell Biol.*, 16, 762, 1996.
32. Liu, Y., Guyton, K.Z., Gorospe, M., Xu, Q., Lee, J.C., and Holbrook N.J., Differential activation of ERK, JNK/SAPK and p38/CSBP/RK map kinase family members during the cellular response to arsenite, *Free Radic. Biol. Med.*, 21, 771, 1996.
33. Liu, Y., Guyton, K.Z., Gorospe, M., Xu, Q., Kokkonen, G.C., Mock, Y.D., Roth, G.S., and Holbrook, N.J., Age-associated decline in mitogen-activated protein kinase activity in epidermal growth factor-stimulated rat hepatocytes, *J. Biol. Chem.*, 271, 3604, 1996.
34. Liu, Y., Gorospe, M., Yang, C., and Holbrook, N.J., Role of mitogen-activated protein kinase phosphatase during the cellular response to genotoxic stress: Inhibits of c-Jun N-terminal kinase activity and AP-1-dependent gene activation, *J. Biol. Chem.*, 270, 8377, 1995
35. Ishigami, A., Reed, T.D., and Roth, G.S., Effect of aging on EGF stimulated DNA synthesis and EGF receptor levels in primary cultured rat hepatocytes, *Biochem. Biophys. Res. Commun.*, 196, 181, 1993.

23

Kinetic Measurement of Mitochondrial Oxygen Radical Production

G. Barja

CONTENTS

23.1 Introduction

There is growing evidence that reactive oxygen species (ROS) of biological origin are implicated in the development of many degenerative diseases and can also be causal agents of the normal aging process [1–4]. In healthy tissues, mitochondria are the main source of free radicals because these organelles are responsible for more than 90% of total oxygen consumption. During the electron flow in the mitochondrial respiratory chain, a small percent of electrons incompletely reduce oxygen to O_2^- and H_2O_2 before reaching Complex IV. The H_2O_2 produced by mitochondria probably comes from initial univalent reduction of oxygen to O_2^- at the free radical generator site, followed by dismutation of this O_2^- to H_2O_2. Those ROS (O_2^- plus H_2O_2) can subsequently generate the highly reactive hydroxyl radical (OH·), which can damage many kinds of biologically relevant macromolecules including lipids, proteins, and DNA.

Almost 30 years ago, it was proposed that mitochondria are implicated in the aging process [1]. Present information indicates that mitochondrial DNA has significantly higher levels of markers of oxidative damage like 8-oxo-7,8-dihydro-2'-deoxyguanosine, dele-tions, DNA-protein crosslinks, and mutation rates, than nuclear DNA, and that various of these deleterious modifications increase with age in post-mitotic tissues [5–8]. The high steady-state level of oxidative damage to mitochondrial DNA is probably mainly caused its proximity to the inner mitochondrial membrane, the place where the majority of ROS are produced in healthy cells, although other factors like the lack of protection by histones and polyamines or the presence of low repair activities can also be involved. The high rate of damage to mitochondrial DNA, together with its very high information density, proba-bly causes functional alteration of the organelles with increasing age. Although there is debate about the physiological relevance of those mitochondrial modifications, decreases in membrane potential, state 3 and 4 respiration, activities of respiratory Complexes, mem-brane fluidity, phosphate and pyruvate traslocators, and cardiolipin, have been described in mitochondria isolated from tissues of old animals (see Ref. 6 for review). Various of these changes are postponed by chronically subjecting rodents to caloric restriction [4,6], the only experimental manipulation which is known to decrease the rate of aging. Thus, it is impor-tant to develop methods allowing to appropriately measure the rate of mitochondrial ROS production in studies about degenerative diseases and aging.

23.2 Background

Spectrophotometric, fluorometric, chemiluminescent, and electron spin resonance (ESR) methods have been used to estimate ROS production in mitochondria or submitochondrial particles. Since the amount of ROS produced by intact mitochondria respiring with sub-strate alone is small, methods with high sensitivity and specificity are needed. Chemilumi-nescent methods are among those showing higher sensitivities [9], but they lack chemical specificity and are then scarcely used to assay mitochondrial ROS production. Further-more, they usually require long integration intervals, and are frequently limited to single end-point measurements [10]. Direct or spin-trap ESR techniques can have enough sensi-tivity and specificity, have the advantage of being direct methods of free radical detection, and have been occasionally used to assay ROS levels in brain mitochondria [11], frozen heart [12], heart submitochondrial particles [13], or cortical neurons [14]. The characteris-tics of the ESR techniques make them more suitable to single-point than to kinetic measure-ments of ROS production. Most importantly, ESR requires expensive equipment. Appropriate methods to measure ROS production using instruments commonly present in basic biochemistry laboratories are desirable.

The spectrophotometric measurement of O_2^- production in mitochondrial preparations has been most commonly performed by kinetic assays of superoxide dismutase-sensitive epinephrine reduction to adrenochrome or reduction of acetylated or succinylated cyto-chrome c [15–24]. These assays can be performed with basic equipment and have been applied to submitochondrial particles or to isolated Complex I [17] and Complex III [16].

ROS production by intact mitochondria needs methods which detect H_2O_2 rather than O_2^-. A classic method for the measurement of the rate of mitochondrial H_2O_2 production, originally developed by Britton Chance and co-workers [25,26], used double wavelength spectrophotometry to follow the enzyme-substrate complex between H_2O_2 and cyto-chrome c peroxidase. The method has been successfully applied by various authors to heart mitochondria [15,16,18]. However, cytochrome c peroxidase is not commercially available and its preparation from yeast [27] would complicate routine assays. Substitution of cytochrome c peroxidase by horseradish peroxidase has been performed [18,25,28], but it further decreases the intrinsically low sensitivity of the spectrophotometric ROS detection technique. A solution to this would be to turn to more sensitive fluorometric techniques, which can be performed also with basic laboratory equipment. One of the earliest ways in which mitochondrial H_2O_2 production was measured was the scopoletin method [29,30]. In this assay the fluorescent compound 6-methyl-7-hydroxy-1,2-benzopyrone (scopoletin) is oxidized by H_2O_2 to a nonfluorescent substance in the presence of horseradish peroxidase. While specific and sensitive, it is a "negative" method since it is based on the disappearance of fluorescence. In addition to the intrinsic disadvantages of methods that use differences as the basis of estimation, this method requires the use of graded quantities of scopoletin in order to obtain the best range for measurement [29,30].

Estimation of ROS levels in biological samples is nowadays frequently performed by methods in which the fluorescence of a substance appears as a function of assay time. Perhaps the most commonly used fluorometric method is based on the fluorescence derived from oxidation products of the non-fluorescent 2',7'-dichlorofluorescin (DCF). Nonpolar nonionic DCF-diacetate readily crosses cell membranes and is intracellularly hydrolysed by esterases to non-fluorescent DCF, which is rapidly oxidized by ROS to the fluorescent product 2',7'-dichlorofluorescein [31,32]. This reaction is unspecific for H_2O_2 since DCF can be oxidized also by other cellular oxidants, although not by O_2^{-} [32]. Intracellular fluorescence due to DCF oxidation has been considered either to be inversely related to antioxidant levels [33] or to represent intracellular free radical production [34]. But any antioxidant present between the free radical generator and DCF inside the cells will decrease the final fluorescence. Thus, the final values will be rather indicative of the overall degree of intracellular oxidative stress.

DCF is frequently used as an estimator of oxidative stress, most commonly inside isolated cells [33,34,35–40]. DCF-based assays have been occasionally applied also to the measurement of free radical production in isolated mitochondria [38]. Nevertheless, single-point measurements were performed in parallel after very long incubation times (60 to 125 min) [38]. In addition to the lack of specifity of DCF towards oxidants, the applicability of the DCF method to the measurement of ROS production in isolated mitochondria is seriously limited due to the following reasons: (1) in the case of isolated mitochondria, an appropriate detector of ROS production should stay outside the mitochondria instead of entering them; since DCF-diacetate readily crosses the plasma membrane, it is possible that it also crosses mitochondrial membranes; (2) lack of mitochondrial esterases would preclude hydrolysis of DCF-diacetate to non-fluorescent DCF; (3) most importantly, the non-catalyzed reaction between DCF-diacetate (e.g., 50 μM) and a pulse of externally added H_2O_2 (e.g., 3 nanomoles of H_2O_2 to a 2-ml cuvette) in a purely chemical system without mitochondria is a very slow process which does not reach completion after at least 30 min; this is not solved by substituting DCF-diacetate by DCF (not commercially available), which can be prepared by alkaline hydrolysis with NaOH. The very slow spontaneous reaction between DCF-diacetate and H_2O_2 obliged to use long incubation times (one hour or more) when assaying mitochondrial ROS production by a DCF technique [38]. This extremely slow response makes the DCF method inappropriate to study quick effects (in the order of fractions of seconds) of respiratory inhibitors, physiological modulators like ADP, or drugs, on the rates of mitochondrial ROS production. For the same reason, the DCF method will be also unappropriate for the kinetic study of basal rates of ROS production. Long incubation times are also needed when DCF-diacetate is used to assay ROS levels in intact cells by flow cytometry. This can possibly explain why the reported effects on the rate of ROS production of respiratory inhibitors like antimycin A or rotenone, or of uncouplers, are different and can even have an opposite sense when studied in intact cells (e.g., using dihydrorhodamine 123 as

ROS detector) [14] *versus* in isolated mitochondria (assayed by different methods) [13,15–22,25–30,41,42]. Long-term incubation of cells in the presence of those metabolic poisons can conceivably lead to many different secondary effects. The fluorescent horseradish peroxidase-homovanillic acid method described below, although appropriate for isolated mitochondria, will not be useful in intact cells due to the lack of permeability of the plasma membrane to the horseradish peroxidase enzyme. Thus, there is a need for the future development of more appropriate methods of ROS determination inside intact cells.

Kinetic assays are preferable to end-point measurements to estimate the rate of mitochondrial ROS production due to various reasons (see Section 23.4.3 below). Relatively short incubation times (in the order of 5 min) are also needed to perform various kinetic measurements in series while avoiding long-term loss of functionality of mitochondria. Fluorescent probes, specific for H_2O_2, are available for this purpose.

A specific indicator for enzymatic H_2O_2 determination uses 4-hydroxy-3-methoxy-phenylacetic acid (homovanillic acid) as substrate. Enzymatic oxidation of this kind of compound by H_2O_2 is currently used by various laboratories to detect H_2O_2 production in isolated mitochondria [2,3,13,24,41–45]. The method is specific for H_2O_2 due to the presence of horseradish peroxidase, and has been used to detect differences in mitochondrial ROS production during aging or between animal species with different longevities [2,3,41–44]. Originally designed for end-point measurements in polymorphonuclear leucocytes [46], the method has been adapted to the kinetic measurement of H_2O_2 production in intact mitochondria respiring with substrate alone [2,41,42]. When appropriate precautions are taken, as described below, the method can be used to quantify basal and quick changes in the rate of mitochondrial ROS production, as well as to localize the main sites of free radical generation in the respiratory chain, without altering the respiratory control ratio and without interference from endogenous mitochondrial antioxidants.

23.3 Isolation of Mitochondria

In order to obtain intact mitochondria for the assay of mitochondrial free radical production, they should be isolated as quickly as possible, and from fresh tissue. Tightly coupled intact mitochondria can only be isolated from fresh tissues. All the glassware and centrifuge tubes must be cleaned without detergents using ethanol, hot water, and distilled water and dried before use. All tools in contact with the mitochondria should be precooled to 0°C and be free of ice. Isolation of brain and heart mitochondria, two postmitotic tissues highly relevant for aging, is described here.

Isolation of heart mitochondria can be performed in a buffered isolation medium containing mannitol-saccharose to give osmotic support plus ethylenediaminetetraacetate (EDTA) or ethylene-bis(oxyethylenenitrilo)tetraacetate (EGTA), or fatty acid-free albumin as possible additions. Inclusion of a proteinase like nagarse is needed to liberate mitochondria from muscle myofibrils in good yield. Adult rats are sacrificed by decapitation, ventricles are separated from remaining vessels and rings of fat, chopped into small pieces, rinsed several times, and homogenized with a loose-fitting pestle in 10 ml of isolation buffer (220 mM mannitol, 70 mM sucrose, 1 mM EDTA, 10 mM Tris-ClH, pH 7.4) containing 5 mg of nagarse and 25 mg of fatty acid-free albumin. After standing for 1 min, 25 ml of additional isolation buffer containing 25 mg of albumin are added and homogenization is gently performed again with a tighter fitting pestle. After homogenization, the pH should be checked and readjusted to 7.4 if needed. The nuclei and cell debris are removed by centrifugation at 700 g during 10 min. Heart mitochondria are obtained after centrifugation of the first supernatant at 8,000 g during 10 min. The second supernatant is discarded together with the fluffy layer on top of the mitochondrial light-brown pellet found at the bottom. This second centrifugation can be repeated one or two times after resuspension of the mitochondrial pellet in isolation buffer to wash the mitochondria. The mitochondrial pellet is finally resuspended in 1 ml of isolation medium. All the above procedures are performed at 5°C. Mitochondrial protein is measured in the final suspensions by the Biuret method. The final mitochondrial suspensions are kept on ice and used immediately for the oxygen consumption and oxygen radical production measurements, since mitochondria are only stable for a few hours.

A classic procedure to obtain brain non-synaptic mitochondria free from contaminating membranes, using discontinuous Ficoll gradients [47], is recommended with minor modifications. Just after decapitation the brain is rinsed several times, chopped, and manually homogenized with a loose fitting pestle in 35 ml of isolation medium (250 mM sucrose, 0.5 mM K^+-EDTA, 10 mM Tris-HCl, pH 7.4). The homogenate is centrifuged at 2,000 g for 3 min and this centrifugation protocol is repeated in the first supernatant. The second supernatant is centrifuged at $12,500 \times g$ for 8 min to obtain the crude mitochondrial pellet. In order to obtain non-synaptic mitochondria, the crude mitochondrial pellet is resuspended in 6 ml of 3% Ficoll medium carefully underlayered with 25 ml of 6% Ficoll medium (6% Ficoll, 240 mM mannitol, 60 mM sucrose, 50 µM K^+-EDTA, 10 mM Tris-HCl, pH 7.4). Commercial Ficoll must be previously subjected to 24 hours of dialysis at 5°C against 3 changes of pure water and stored in the cold for posterior preparation of centrifugation media. The 3% Ficoll medium is obtained by 1:1 dilution of the 6% Ficoll medium with purified water and pH readjustment to 7.4. After centrifugation of the crude mitochondria in the discontinuous Ficoll gradient at 11,500 g for 30 min, the supernatant is sharply decanted. The brown pellet is resuspended in 5 ml of isolation medium and centrifuged at 11,500 g for 10 min. The pellet is resuspended in 35 ml of isolation medium followed by an additional centrifugation step at 11,500 g for 10 min. The final pellet is resuspended in 1 ml of isolation medium. Using various centrifugation protocols and different concentrations of Ficoll, synaptic mitochondria can also be isolated [47].

The oxygen consumption of the final mitochondrial suspensions is measured by polarography in a small volume of incubation medium in the presence of a mitochondrial substrate without (state 4 respiration) and with (state 3) saturant ADP (500 µM final concentration). This allows calculation of the respiratory control index RCI (state 3/state 4 oxygen consumption) as an indicator of the degree of coupling and metabolic activity of the mitochondrial preparations. Inclusion of 0.1% fatty acid-free albumin in the incubation medium is desirable in order to obtain high RCI and low state 4 values, typical of tightly coupled mitochondria. With the isolation methods described above, Complex I-linked pyruvate/malate (each at 2.5 mM final concentration) and Complex II-linked succinate (5 to 10 mM) are appropriate substrates for heart and brain mitochondria. Preparations showing a too high state 4 respiration or a too low RCI value should not be used to assay H_2O_2 production, since these are signs of uncoupled or damaged mitochondria.

23.4 Fluorometric Enzymatic H_2O_2 Detection of H_2O_2 Production

23.4.1 Procedure

The method measures mitochondrial H_2O_2 production kinetically following its reaction with homovanillic acid in the presence of horseradish peroxidase to form a dimer fluorescent at 312 nm excitation and 420 nm emission.

The basic incubation medium contains 145 mM KCl, 30 mM Hepes, 5 mM KH_2PO_4, 3 mM $MgCl_2$, 0.1 mM EGTA, 0.1% fatty-acid free albumin, pH 7.4. The following solutions are prepared in this incubation medium: 70 Units/ml of high purity horseradish peroxidase; 4 mM homovanillic acid; pyruvate/malate (125 mM each) and 250 mM succinate. To a standard fluorometric cuvette add first a large volume of incubation medium and then add small volumes of the reactants in the following order: mitochondria, horseradish peroxidase, homovanillic acid, superoxide dismutase (SOD, optional), and the substrate (pyruvate/malate or succinate) to start the reaction inside the cuvette compartment of the fluorometer (Figure 23.1; first part of the trace). The volume of incubation medium added should be around 85% of the total reaction volume. The volumes added of the rest of the reactants are those needed to reach the following final concentrations inside the cuvette: around 0.25 mg (rat heart) or 0.4 mg (rat brain) of mitochondrial protein per ml, 6U/ml of horseradish peroxidase, 0.1 mM homovanillic acid, 50 U/ml of SOD (optional), 5 mM (heart) or 10 mM (brain) succinate, or 2.5 mM pyruvate/2.5 mM malate (heart) or 5 mM pyruvate/2.5 mM malate (brain). In the absence of SOD, the rates represent H_2O_2 production. SOD added in excess converts O_2^- produced (if any) to H_2O_2. Thus, in the presence of SOD, the assay estimates the mitochondrial production of O_2^- plus H_2O_2 (ROS). Using three methods, including the one described here, to assay ROS production in macrophages and neutrophils, it was described that the addition of SOD could have the advantage of avoiding a possible backwards reductive interaction of the generated O_2^- with the phenoxyl (substrate) radicals formed during the peroxidase cycle; that interaction would lead to consumption of H_2O_2 by the peroxidase without net oxidation of the substrate and would thus decrease the rate of formation of the fluorescent dimer. SOD prevented this and allowed the true amount of H_2O_2 generated by the leucocytes to be measured [48]. Thus, if a particular mitochondrial preparation produces any O_2^- in addition to H_2O_2, SOD should be included in the incubation medium. Nevertheless, preliminary results from our laboratory indicate that at least rat heart mitochondria, prepared as described in section 3, show the same rate of H_2O_2 production with and without SOD. This means that they only produce H_2O_2, not O_2^-.

The reaction rate is kinetically followed for a few minutes and the slope is measured with a chart recorder or computer software to obtain the increment in arbitrary fluorescence units per minute, which is converted to nanomoles of H_2O_2 (see below) and referred to mg of mitochondrial protein.

23.4.2   Validity

Any useful assay of ROS production should not interfere with mitochondrial function. This can be checked by measuring mitochondrial respiration in the presence of the molecular probes constituting the chemical H_2O_2 detection system. Figure 23.2 shows that addition of horseradish peroxidase and homovanilic acid, at the same final concentration used in the fluorometric H_2O_2 assay, does not change the rate of state 4 respiration of heart mitochondria measured by polarography with an O_2 electrode. The same occurs when oxygen consumption is measured in the presence of SOD (50 U/ml). The figure also shows that addition of saturant ADP (500 μM) releases state 3 respiration (oxidative ADP phosphorylation) in the presence of homovanillic acid and horseradish peroxidase. Repetition of this kind of experiment in the presence and in the absence of horseradish peroxidase and homovanilic acid showed that these two chemicals do not alter either the state 4 oxygen consumption or the respiratory control index. High-purity horseradish peroxidase and superoxide dismutase should be used. Commercial solutions of these two enzymes

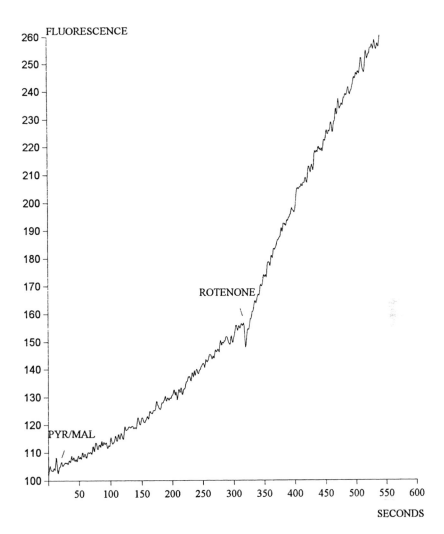

FIGURE 23.1

Increase in fluorescence (312 nm excitation 5 nm slit, and 420 nm emission 10 nm slit) due to H_2O_2 production after addition of pyruvate/malate to rat heart mitochondrial samples and effect of rotenone addition. Concentrations of reactants as described in the text. The slope of the tracings before or after the addition of rotenone is used to calculate the rate of H_2O_2 production in each case. Pyruvate/malate (PYR/MAL) are added at time 0.

stabilized with agents capable of damaging mitochondria [e.g., with 3.8 M $(NH_4)_2SO_4$] should be avoided. In any case, the appropriateness of the particular commercial reactants used can be easily checked by performing experiments like that described in Figure 23.2.]

23.4.3 State 4 and State 3 H_2O_2 Production

The described fluorometric assay of ROS production can be performed first in the presence of mitochondria and pyruvate/malate (state 4 ROS production). After some minutes, saturant ADP (500 µM) is added on line without stopping the kinetics. The new slope corresponds to the state 3 rate of ROS production. It is well known that ADP addition (state 3) essentially stops ROS production in heart mitochondria respiring with succinate

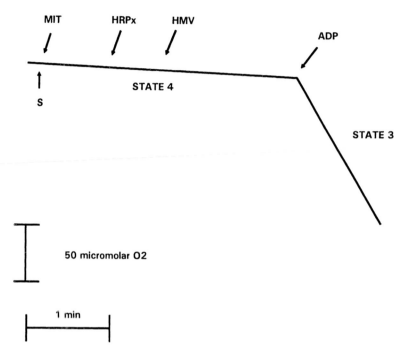

FIGURE 23.2
Oxygen consumption experiments show the absence of effect of horseradish peroxidase (6U/ml) and homovan-illic acid (0.1 mM) on the state 4 oxygen consumption and the respiratory control index of rat heart or non-synaptic brain mitochondria, measured by polarography with an oxygen electrode. ADP: 500 μM. The same is true for additions of 50 Units of SOD/ml. MIT = mitochondria; S = substrate; HRPx = horseradish peroxidase; HMV = homovanillic acid.

(a Complex II-linked substrate) [25,30,41]. However, ROS production continues unchanged in state 3, or can be even stimulated by 2 fold (but no more; Figure 23.3), when heart or non-synaptic brain mitochondria respire with pyruvate/malate (a Complex I-linked substrate) [41]. This indicates that Complex I of heart and non-synaptic brain mitochondria generates ROS both during state 4 and state 3 respiration, whereas Complex III would produce ROS only during state 4 [41]. Thus, if pyruvate/malate instead of succinate is used as substrate, addition of saturant ADP in the middle of a single kinetic fluorometric run allows the measurement of both the state 3 and state 4 rates of ROS production in the same mitochondrial sample.

23.4.4 H₂O₂ Pulse Experiments

Mitochondria contain endogenous antioxidants like SOD, glutathione peroxidase, or glutathione in the matrix and can have vitamin E in their membranes. The presence of catalase has also been described in heart mitochondria [49]. An accurate method measuring the rate of ROS production should be free from the interference of mitochondrial antioxidants. ROS are produced by respiratory Complexes situated at the inner mitochondrial membrane. Thus, matrix antioxidants would in principle not interfere with H_2O_2 diffusing out of the inner mitochondrial membrane to the external medium where the chemical H_2O_2 detection system is situated. In order to test this, H_2O_2 pulse experiments (internal standards) can be easily performed during a kinetic assay. Incubation medium, mitochondria, horseradish peroxidase, homovanillic acid, SOD (optional), and substrate are first added to the cuvette.

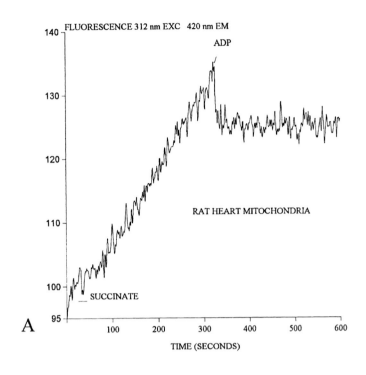

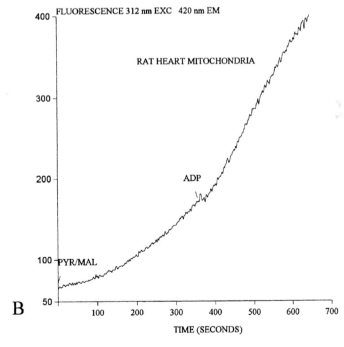

FIGURE 23.3

Effect of ADP on free radical production of rat heart mitochondria with Complex II- (A) or Complex I-linked (B) substrates during the energy transition from state 4 to state 3. Substrate was added to the reaction mixture and the kinetic was started (time 0). After some minutes, addition of 500 μM ADP stopped free radical production with succinate **(A)** but not with pyruvate/malate **(B)**. Transient perturbation of the tracings at the moment of ADP addition are due to slightly opening the sample compartment to add ADP. (From Herrero, A. and Barja, G., *J. Bioenerg. Biomembr.*, 29, 241, 1997. With permission.)

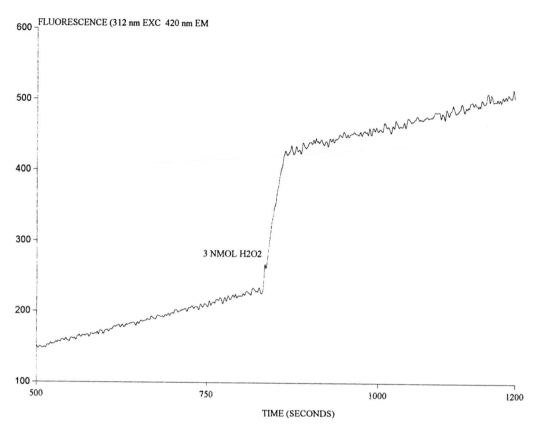

FIGURE 23.4

H_2O_2 pulse experiment. Free radical production of mouse heart mitochondria was followed by some minutes, after which a pulse of 3 nanomoles of H_2O_2 was added to the cuvette compartment of the fluorometer. The H_2O_2 pulse causes an instantaneous increase in fluorescence followed by a slope of free radical production identical to that shown before the H_2O_2 pulse. The amount of H_2O_2 found was equal to the amount of H_2O_2 added.

After an initial slope of ROS production is obtained, 3 nanomoles of H_2O_2 (pulse experiment) are added on line to the cuvette. The result is a quick upwards deflection of the fluorescence trace after which the original slope is resumed (Figure 23.4). The vertical increase in fluorescence due to H_2O_2 addition is used to calculate the amount of H_2O_2 found. The amount of H_2O_2 added is then compared to the amount of H_2O_2 found. In the experiment shown in Figure 23.4 H_2O_2 added was 3 nanomoles and H_2O_2 found was 3.1 nanomoles. This shows that endogenous mitochondrial antioxidants do not interfere with measurements of the rate of ROS production by the horseradish peroxidase-homovanillic acid method. The H_2O_2 pulse experiment also shows that the response of the detection system to an increase in H_2O_2 is almost instantaneous. This validates this method to study quick changes in the rate of mitochondrial ROS production induced by specific respiratory inhibitors, ADP, or other physiological or pharmacological modulators.

23.4.5 Standards

The arbitrary fluorescence units must be converted to amounts of H_2O_2. For this purpose, some authors use the fluorescence of standard solutions of H_2O_2 in the presence of the reactants. Since it is not always easy to obtain a good reproducibility when preparing dilute H_2O_2 solutions from stock commercial 30% hydrogen peroxide, an alternative is to construct an standard curve using a glucose-glucose oxidase reaction system. The system generates H_2O_2 kinetically (like the mitochondria), at a rate dependent on the amount of glucose oxidase added. For this purpose the following is added to a fluorometer cuvette: incubation medium, 6 Units/ml of horseradish peroxidase, 0.1 mM homovanillic acid, glucose oxidase, and 140 mM glucose. Glucose oxidase can be added in amounts generating 0, 0.5, 1 and 2 nanomoles of H_2O_2/min. Regressions of units of fluorescence/min against nanomoles of H_2O_2/min produced by glucose-glucose oxidase are performed. The linear equation obtained is used to calculate the final mitochondrial production of oxygen radicals, which is expressed in nanomoles of H_2O_2/min. mg of protein. However, when using this kind of standard, care should be taken that no limiting losses of activity have occurred during transport to or storage of glucose oxidase and horseradish peroxidase at the laboratory.

23.4.6   Precautions

The mitochondrial and standard reactions must be performed maintaining constant the temperature of the fluorometer cuvette compartment and with continuous agitation inside the cuvette in order to ensure appropriate mixing of the reactants and the mitochondria. Even though kinetic assays are more time consuming than the simultaneous run of various samples outside the fluorometer followed by end-point measurements, they offer various advantages: (1) initial lag (if present) can be eliminated; (2) flat readings at the end of the kinetics (rarely present at 0.1 mM homovanillic acid concentration) due to consumption of substrate in very active samples can also be eliminated; (3) instantaneous changes in absolute fluorescence due to the addition of fluorescent respiratory inhibitors or other fluorescent compounds during the kinetic assays is easily discarded in slope measurements (see the next section for examples in studies with respiratory inhibitors).

A chemical (in the absence of mitochondria) H_2O_2 pulse experiment can be easily performed each day before starting the measurements with mitochondria, or even before sacrificing the animal to isolate them. An alternative to this would be to run a standard with glucose-glucose oxidase-horseradish peroxidase and homovanillic acid, although in this case the chemical reaction will be also dependent on the activity of the glucose-glucose oxidase pair. Those standard reactions can be very useful to check the quality of the horseradish peroxidase and homovanillic acid reactants, in order to avoid erroneously attributing an absence or very low rate of ROS production to the mitochondrial preparations if a recently received or stored reactant is not in good condition or has lost enzymatic activity.

The measurement of ROS production and oxygen consumption using the same incubation medium (145 mM KCl, 30 mM Hepes, 5 mM KH_2PO_4, 3 mM $MgCl_2$, 0.1 mM EGTA, 0.1% fatty-acid free albumin, pH 7.4), temperature, and concentrations of substrates and respiratory inhibitors in each mitochondrial preparation allows to calculate the fraction of electrons out of sequence which reduce oxygen to oxygen radicals along the mitochondrial respiratory chain—the percent free radical leak (%FRL)—instead of reducing oxygen to water at the terminal cytochrome oxidase (Complex IV). Since two electrons are needed to reduce one molecule of oxygen to H_2O_2, whereas four electrons are needed to reduce one molecule of oxygen to water, the free radical leak is easily calculated by dividing the rate of ROS production by two times the rate of oxygen consumption, the result being multiplied by 100. The use of highly purified ion-free water is recommended for preparation of all solutions.

544 Methods in Aging Research

23.5 Localization of the Oxygen Radical Source

Using substrates and inhibitors specific for different segments of the respiratory chain, sites of oxygen radical generation have been localized mainly at respiratory Complexes I [17,19,41,42] and III [15,16,21,28,42]. The scheme of Figure 23.5 shows that the mitochondrial substrates described above, pyruvate/malate and succinate, feed electrons respectively at respiratory Complexes I and II. It also shows the approximate site of action of the specific respiratory chain inhibitors rotenone (ROT), thenoyltrifluoroacetone (TTFA), antimycin A (AA) and myxothiazol (MYX).

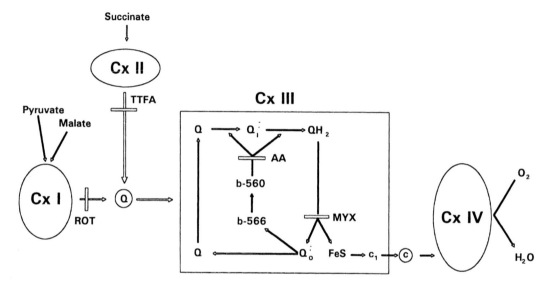

FIGURE 23.5
The scheme shows the sites of action of specific substrates and inhibitors of the respiratory chain in relation to the four mitochondrial electron transport Complexes (I to IV). Pyruvate and malate feed electrons at Complex I, whereas succinate introduces electrons at Complex II. Rotenone (ROT) inhibits electron flow from Complex I to the mobile ubiquinone pool (Q). Thenoyltrifluoroacetone (TTFA) inhibits electron flow from Complex II to Q. Antimycin A (AA) blocks electron transport from cytochrome b-560 to Q or Q_i^- at center "in" (matrix side). Myxothiazol (MYX) inhibits electron flow from ubiquinol (QH$_2$) to the FeS center of Complex III. The scheme inside Complex III corresponds to the Q cycle hypothesis [50]. (Q_i^- = center "in" semiquinone; Q_o^- = center "out" (intermembrane side) semiquinone; FeS = iron sulfur center of Complex III; c_1 = cytochrome c_1; c = mobile cytochrome c pool.

The rate of mitochondrial free radical production increases as a function of the degree of reduction of the autoxidizable electron carriers of the respiratory chain [25,26,30]. Blocking the respiratory chain with an specific inhibitor increases the reduction state of electron carriers on the substrate side of the inhibitor, whereas those on the oxygen (opposite) side change to a more oxidized state. An increase in free radical production following the addition of an inhibitor indicates that the free radical generation site/s are located on the substrate side. Conversely, if free radical production decreases after the addition of the inhibitor, the generator/s must be situated on the "oxygen" side. Thus, an increase in free radical production after rotenone addition to pyruvate/malate-supplemented mitochondria unequivocally demonstrates the capacity of Complex I for free radical generation (Figure 23.1). TTFA, which is considered to inhibit electron flow at a site situated at the end of the Complex II electron path from succinate to ubiquinone, can be used in succinate-supplemented mitochondria to check the capacity of Complex II for oxygen radical production. The same can be done for Complex III with appropriate combinations of substrates and antimycin A or myxothiazol [50]. The effect of myxothiazol addition to antimycin A-treated mitochondria can be also used to ascertain the capacities of Complex III cytochrome b and ubisemiquinone for oxygen radical generation. An assay with pyruvate/malate in the presence of antimycin A would be representative of Complex I + Complex III ROS production. Addition of myxothiazol to antimycin A-treated pyruvate/malate-supplemented mitochondria (double kill experiment) would stop the Complex III contribution to the ROS production of this preparation. The fluorometric assay described in this chapter is useful to study the effect of specific respiratory chain inhibitors on the rate of ROS production due to its almost instantaneous response to H_2O_2 (see Figure 23.4 and Section 23.4.4).

The localization experiments must be performed using inhibitor concentrations having specific effects on the mitochondria, but not perturbing the horseradish peroxidase-homovanillic H_2O_2 detection system. This kind of precaution must be taken, no matter which kind of H_2O_2 detection method is used. Table 23.1 shows appropriate final cuvette inhibitor concentrations for this purpose at 0.25 to 0.4 mg of mitochondrial protein per ml, as well as the concentrations used for preparation of the initial solutions in pure ethanol. Under these conditions, the small amount of ethanol added to the final reaction does not affect at all the kinetics of mitochondrial free radical production or the oxygen consumption or the RCI. Performing the assay in a kinetic way offers two important additional advantages. Rotenone is not fluorescent at the wavelengths of excitation and emission selected, and TTFA affects the fluorescence readings only slightly. But antimycin A and myxothiazol cause large instantaneous increases in fluorescence during addition, due to the intrinsic fluorescence of these molecules. Measuring the slope of the fluorescence trace after these instantaneous changes occur, avoids any interference in the calculation of the final values of H_2O_2 production, which would be present in end-point measurements. Second, the addition of the inhibitors during the kinetics allows to compare the rate of oxygen radical production with and without inhibitor in the same mitochondrial sample. Variability in oxygen radical production between mitochondrial samples, running in parallel with and without respiratory inhibitors during end-point measurements, would obscure the effects of the inhibitors especially when they are not very strong.

TABLE 23.1

Initial and Final Concentrations of Inhibitors of the Mitochondrial Respiratory Chain

Inhibitor	Initial Concentration	Final Concentration
Rotenone	100 μM	2 μM
TTFA	550 μM	11 μM
Antimycin A	500 μM	10 μM
Myxothiazol	500 μM	10 μM

Note: Solutions of inhibitors are first prepared at the described initial concentration in pure high grade ethanol. A small amount of these solutions is added to the reaction mixture reaching the final concentrations described in the table by 50-fold dilution. These final concentrations are appropriate for a concentration of 0.25 (heart) and 0.4 (brain) mg of mitochondrial protein per ml of reaction mixture. TTFA = thenoyltrifluoroacetone.

23.6 Conclusion

In summary, the inner mitochondrial membrane continuously generates reactive oxygen species during respiratory activity, which can the damage mitochondrial proteins and lipids implicated in electron transfer as well as mitochondrial DNA. These modifications can have deleterious consequences related to the development of degenerative diseases or to the normal aging process. A fluorometric kinetic method appropriate to measure the rate of generation of H_2O_2 or reactive oxygen species in intact mitochondria is described. Mitochondria are isolated from fresh tissues by direct or gradient-assisted differential centrifugation techniques, followed by the polarographic measurement of the state 4 and state 3 oxygen consumption and the respiratory control index as indicators of the quality of the mitochondrial preparations. Rates of generation of H_2O_2 are then kinetically measured by fluorometry in the presence of homovanilic acid and horseradish peroxidase. The method is at the same time specific for H_2O_2, sensitive enough to assay mitochondrial H_2O_2 generation in the presence of respiratory substrate alone (state 4), and free from the interference of mitochondrial antioxidants. The molecular H_2O_2 detector probes used do not alter the respiratory control index. The method can also be applied to assay state 3 H_2O_2 production if Complex I-linked substrates like pyruvate/malate are used. Its almost instantaneous response to H_2O_2 makes it appropriate to study the effects of specific respiratory inhibitors, ADP, or other physiological or pharmacological modulators, on the rate of mitochondrial oxygen radical generation.

Acknowledgment

The research described was supported by a grant (no. 96/1253) from the National Research Foundation of the Spanish Ministry of Health (FISss). I am grateful to Dr. Christoph Richter for previously suggesting the H_2O_2 pulse experiment and for encouraging me to write this paper.

References

1. Harman, D., The Biologic clock: the mitochondria? *J. Amer. Geriatr. Soc.*, 20, 145, 1972.
2. Barja, G., Cadenas,S., Rojas, C., Pérez-Campo, R., and López-Torres, M., Low mitochondrial free radical production per unit O_2 consumption can explain the simultaneous presence of high longevity and high aerobic metabolic rate in birds, *Free Rad. Res.*, 21, 317, 1994.
3. Sohal, R. S. and Weindruch, R., Oxidative stress, caloric restriction, and aging, *Science*, 273, 59, 1996.
4. Yu, B. P., Aging and oxidative stress: modulation by dietary restriction, *Free Rad. Biol. Med.*, 21, 651, 1996.
5. Richter, Ch., Role of mitochondrial DNA modifications in degenerative diseases and aging, *Curr. Topics Bioenerg.*, 17, 1, 1995.
6. Shigenaga, M. K. and Ames, B. N., Oxidants and mitochondrial decay in aging, in *Natural Antioxidants in Health and Disease*, Academic Press, New York, 1994, chap. 3.

7. Kadenbach, B., Münscher, Ch., Frank, V., Müller-Höcker, J., and Napiwotzki, J., Human aging is associated with stochastic somatic mutations of mitochondrial DNA, *Mut. Res.*, 338, 161, 1995.

8. Lee, C.M., Weindruch, R., and Aiken, J.M., Age-associated alterations of the mitochondrial genome. *Free Rad. Biol. Med.* 22, 269-290, 1997.

9. Chance, B. and Gao, G., *In vivo* detection of radicals in biological reactions, *Environm. Health Persp.*, 102, 29, 1994.

10. Bates, T. E., Heales, S. J. R., Davies, S. E. C., Boakye, P., and Clarck, J. B., Effects of 1-methyl-4-phenylpyridinium on isolated rat brain mitochondria: evidence for a primary involvement of energy depletion, *J. Neurochem.*, 63, 640, 1994.

11. Dykens J. A., Isolated cerebral and cerebellar mitochondria produce free radicals when exposed to elevated Ca^{2+} and Na^+: implications for neurodegeneration, *J. Neurochem.* 63, 584, 1994.

12. Ambrosio G., Zweier J. L., Duilio C., Kuppusamy P., Santoro G., Elia P. P., Tritto I., Cirillo, P., Condorelli M., Chiariello M. and Flaherty J. T., Evidence that mitochondrial respiration is a source of potentially toxic oxygen free radicals in intact rabbit hearts subjected to ischemia and reflow, *J. Biol. Chem.*, 268, 18532, 1993.

13. Giulivi C., Boveris, A., and Cadenas, E., Hydroxyl radical generation during mitochondrial electron transfer and the formation of hydroxydesoxyguanosine in mitochondrial DNA, *Arch. Biochem. Biophys.*, 316, 909, 1995.

14. Dugan, L. L., Sensi, S. L., Canzoniero, L. M. T., Handran, S. D., Rothman, S. M., Lin, T. S., and Goldberg, M. P., Mitochondrial production of reactive oxygen species in cortical neurons following exposure to *n*-methyl-D-aspartate, *J. Neurosci.*, 15, 6377, 1995.

15. Boveris, A., Cadenas E., and Stoppani O. M., Role of ubiquinone in the mitochondrial generation of hydrogen peroxide, *Biochem. J.*, 153, 435, 1976.

16. Cadenas, E., Boveris, A., Ragan, I., and Stoppani, A. O. M., Production of superoxide radical and hydrogen peroxide by NADH-ubiquinone reductase and ubiquinol-cytochrome c reductase from beef-heart mitochondria, *Arch. Biochem. Biophys.*, 180, 248, 1977.

17. Takeshige, K. and Minakami, S., NADH- and NADPH-dependent formation of superoxide anions by bovine heart submitochondrial particles and NAD-ubiquinone-reductase preparation, *Biochem. J.*, 180, 129, 1979.

18. Cadenas, E., and Boveris, A., Enhancement of hydrogen peroxide formation by protophores and ionophores in antimycin-supplemented mitochondria, *Biochem. J.*, 188, 31, 1980.

19. Turrens, J. F. and Boveris A., Generation of superoxide anion by the NADH dehydrogenase of bovine heart mitochondria, *Biochem. J.*, 191, 421, 1980.

20. Turrens, J. F., Freeman, B., Levi, H. J. G., and Crapo, J. D., The effect of hyperoxia on superoxide production by lung submitochondrial particles, *Arch. Biochem. Biophys.*, 217, 401, 1982.

21. Nohl, H. and Jordan, W., The mitochondrial site of superoxide formation, *Biochem. Biophys. Res. Comms.*, 138, 533, 1986.

22. Nohl, H., A novel superoxide radical generator in heart mitochondria, *FEBS Letts.*, 214, 269, 1987.

23. Veitch, K. and Hue, L., Flunarizine and cinnarizine inhibit mitochondrial Complexes I and II: possible implications for Parkinsonism, *Mol. Pharmacol.*, 45, 158, 1994.

24. Sohal, R. S., Sohal, B. H., and Orr, W. C., Mitochondrial superoxide radical and hydrogen peroxide generation, protein oxidative damage, and longevity in different species of flies, *Free Rad. Biol. Med.*, 19, 499, 1995.

25. Boveris, A., Oshino, N., and Chance, B., The cellular production of hydrogen peroxide, *Biochem. J.*, 128, 617, 1972.

26. Boveris, A. and Chance, B., The mitochondrial generation of hydrogen peroxide. General properties and effect of hyperbaric oxygen, *Biochem. J.*, 134, 707, 1973.

27. Prat, A. G., Bolter, C., Chavez, U., Taylor, C., Chefurka, W., and Turrens, J., Purification of cytochrome c peroxidase for monitoring H_2O_2 production, *Free Rad. Biol. Med.*, 11, 537, 1991.

28. Turrens, J. F., Alexandre, A., and Lehninger, A. L., Ubisemiquinone is the electron donor for superoxide formation by complex III of heart mitochondria, *Arch. Biochem. Biophys.*, 237, 408, 1985.

29. Loschen, G., Azzi, A., and Flohé, L., Mitochondrial H_2O_2 formation: relationship with energy conservation, *FEBS Lett.*, 33, 84, 1973.

30. Loschen, G., Flohé, L., and Chance, B., Respiratory chain linked H_2O_2 production in pigeon heart mitochondria, *FEBS Lett.*, 18, 261, 1971.
31. Keston, A. S., and Brandt, R., The fluorometric analysis of ultramicro quantities of hydrogen peroxide, *Anal. Biochem.*, 11, 1, 1965.
32. LeBel C. P., Ischiropoulos, H., and Bondy, S. C., Evaluation of the probe 2',7'-dichlorofluorescin as an indicator of reactive oxygen species formation and oxidative stress, *Chem. Res. Toxicol.*, 5, 227, 1992.
33. Hockenbery, D. M., Oltvai, Z. N., Yin, X. M., Milliman, C. L., and Korsmeyer, S. J., Bcl-2 functions in an antioxidant pathway to prevent apoptosis, *Cell*, 75, 241, 1993.
34. Kane, D. J., Sarafian, T. A., Anton, R., Hahn, H., Gralla, E. B., Valentine, J. S., Örd, T., and Bresden, D. E., Bcl-2 inhibition of neural death: decreased generation of reactive oxygen species, *Science*, 262, 74, 1993.
35. Hinkle, P. C., Butow, R. A., Racker, E., and Chance, B., Partial resolution of the enzymes catalyzing oxidative phosphorylation. XV. Reverse electron transfer in the flavin-cytochrome b region of the respiratory chain of beef heart submitochondrial particles, *J. Biol. Chem.*, 242, 5169, 1967.
36. Dawson, T. L., Gores, G. J., Nieminen, A. L., Herman, B., and Lemasters J. J., Mitochondria as a source of a reactive oxygen species during reductive stress in rat hepatocytes, *Am. J. Physiol.*, 264, C961, 1993.
37. Reynolds, I. J. and Hastings, T. G., Glutamate induces the production of reactive oxygen species in cultured forebrain neurons following NMDA receptor activation, *J. Neurosci.*, 15, 3318, 1995.
38. García-Ruiz, C., Collel, A., Morales, A., Kaplowitz, N., and Fernández-Checa, J.C., Role of oxidative stress generated from the mitochondrial electron transport chain and mitochondrial glutathione status in loss of mitochondrial function and activation of transcription factor-kB: studies with isolated mitochondria and rat hepatocytes, *Mol. Pharmacol.*, 48, 825, 1995.
39. Bondy, S. C. and Marwah, S., Stimulation of free radical production by fatty acids: relation to sterification and to degree of unsaturation, *FEBS Letts.*, 375, 53, 1995.
40. Soliman, E. F., Slikker, W., and Ali, S. F., Manganese-induced oxidative stress as measured by a fluorescent probe: an *in vitro* study, *Neurosci. Res. Comms.*, 17, 185, 1995.
41. Herrero, A. and Barja, G., ADP regulation of mitochondrial free radical production is different with Complex I- or Complex II-linked substrates: implications for the exercise paradox and brain hypermetabolism, *J. Bioenerg. Biomembr.*, 29, 241, 1997.
42. Herrero, A. and Barja, G., Sites and mechanisms responsible for the low rate of free radical production of heart mitochondria in the long-lived pigeon, *Mech. Aging Dev.*, 98, 95, 1997.
43. Sohal, R. S., Svensson, I., and Brunk, U.T., Hydrogen peroxide production by liver mitochondria in different species, *Mech. Aging Dev.*, 53, 209, 1990.
44. Ku, H. H., Brunk, U. T., and Sohal, R. S. Relationship between mitochondrial superoxide and hydrogen peroxide production and longevity of mammalian species, *Free Rad. Biol. Med.*, 15, 621, 1993.
45. Poderoso J. J., Careras, M. C., Lisderio, C., Riobó, N., Schöpfer, F., and Boveris, A., Nitric oxide inhibits electron transfer and increases superoxide radical production in rat heart mitochondria and submitochondrial particles, *Arch. Biochem. Biophys.*, 328, 85, 1996.
46. Ruch, W., Cooper, P. H., and Baggiolini, M., Assay of H_2O_2 production by macrophages and neutrophils with homovanillic acid and horse-radish peroxidase, *J. Immunol. Meth.*, 63, 347, 1983.
47. Lai, J. C. K. and Clark, J. B., Preparation of synaptic and nonsynaptic mitochondria from mammalian brain, *Methods in Enzymol.*, 55, 51, 1979.
48. Kettle, A. J., Carr, A. C., and Winterbourn, C. C., Assays using horseradish peroxidase and phenolic substrates require superoxide dismutase for accurate determination of hydrogen peroxide production by neutrophils, *Free Rad. Biol. Med.*, 17, 161, 1994.
49. Radi, R., Sims, S., Cassina, A., and Turrens, J. F., Roles of catalase and cytochrome c in hydroperoxide-dependent lipid peroxidation and chemiluminescence in rat heart and kidney mitochondria, *Free Rad. Biol. Med.*, 15, 653, 1993.
50. Hatefi, Y., The mitochondrial electron transport and oxidative phosphorylation system, *Ann. Rev. Biochem.*, 54, 1015, 1985.

24

Methods for the Study of Immune Cells in Aging

Gabriel Fernandes

CONTENTS

24.1 Introduction

The immune system undergoes change with age [1–6]. Numerous techniques are being developed to measure functional changes occurring in T cells, B cells, and accessory cells [7,8]. It is well established that age-related changes primarily occur in T cells [9–11]. However, T cells are now classified as naive and memory T cells. Memory T cells are further differentiated as Th-0, Th-1, and Th-2 cells. This shift generally occurs after encountering various antigens *in vivo* (microbial, viral, or non-self). Naive T cells, when cultured *in vitro*, undergo phenotypic changes after adding various activators [12–15]. T cell responses involve the generation of effector T cells that either recognize foreign antigens or interact with and destroy target cells displaying foreign antigens, and also initiate a delayed type hypersensitivity (DTH) response by attracting other immune cells to ward-off the infection.

The defective immune system arises with aging primarily from the gradual shift in the ratio of CD4$^+$ and CD8$^+$ cells, as well as from the age-related phenotypic changes occurring within these T cells, producing Th-1 and Th-2 like cytokines [16-19]. Th-1 lymphocytes are found to produce IL-2, IFN-γ, and TNF-β, whereas Th-2 lymphocytes are known to secrete IL-4, IL-5, IL-6, IL-9, IL-10, and IL-13, whereas Th-0 cells produce both kinds of cytokines (Th-1/Th-2) [20]. Aging is known to decrease IL-2 in humans as well as in both mice and rats, whereas IFN-γ, IL-3, IL-4, and IL-10 are found to rise in aged mice [21–23]. Although the loss of IL-2 is well established for aging rats, changes in Th-1/Th-2 cytokines are still not fully known. The field of immunology in the study of the aging process is becoming much more active due to the availability of various cell surface reagents and cDNA probes to investigate various cytokine genes and surface receptors involved in signal transduction pathways. The use of most recently developed immunological and molecular approaches will help us to understand both the intrinsic and extrinsic changes occurring in immune cells, particularly in the regulation of the host defense mechanism that controls infection and maintains optimum immune function during aging [6].

Various well-established techniques are being developed to study cellular and humoral immune functions in young and old animals [7,8]. These techniques can establish age-related functional changes in T cells and antigen-presenting cells (APC). To understand the complete, adaptive immune response, studies investigating the interaction of T and B cells in the production of antibodies are equally important [24–27]. Furthermore, B cell function is partially supported by other cell types such as macrophages and dendritic cells [28]. The latter cells are found to produce IL-12, which is recently known to influence the induction and functional activity of Th-1 cells, whereas IL-4 is found to promote Th-2 phenotypic changes with age [29,30].

Currently, changes in the Th-1/Th-2 subset ratio is viewed to cause decreased life span due to the rise in various age-associated diseases caused by a viral or bacterial infection, malignancy, and a rise in autoimmune disorders, including the development of renal disease [5,31–35]. Attempts are underway to study the cause of immune system failures as well as various intervention approaches to restore the functional activity of immune cells. These interventions include the use of monoclonal antibodies, various cytokines or anti-cytokine antibody therapies, and hormonal and gene therapies, and the manipulation of the immune system by diet or calorie restriction [36–46]. Also, numerous cross-sectional studies are undertaken to measure functional changes in isolated immune cells obtained from the blood of humans and from lymphoid organs such as thymus, spleen, and lymph nodes. Furthermore, besides peritoneal macrophages, bone marrow cells are also compared to study the age-related changes occurring in the hematopoietic cells [47,48].

24.2   Immunological Methods to Study Lymphocyte Function

24.2.1   Isolation of Lymphocytes

The first step in studying lymphocytes, is to isolate them from the blood or from the lymphoid tissues so that their functional activity can be analyzed *in vitro*. Both spleen and thymus or lymph nodes are collected aseptically in sterile PBS or RPM1 media. Tissues are minced with scissors, pressing against glass slides or petri dishes containing media. After removing clumps, the suspension is centrifuged 10 min at 1000 RPM; discard supernatant and resuspend in 10 ml with complete media containing 5% FCS; centrifuged again, and resuspeneded in a desired volume for counting viable and dead cells. Adjust cells accordingly, generally 5×10^6 cells per 1 ml.

In many experiments it is necessary to remove red blood cells (RBC) from spleen cells This can be carried out by using a commercially available ACK lysing buffer or can be prepared as follows:

ACK Lysing Buffer

   8.29 g NH_4Cl (0.15 M)

   1.0 g $KHCO_3$ (1.0 mM)

   37.2 mg Na_2EDTA (0.1 mM)

   Add 800 ml H_2O and adjust pH to 7.2–7.4 with 1 HCl

   Add H_2O to 1 liter

   Filter sterilize the buffer through a 0.2 mm filter and store at room temperature

An alternative method is to remove both dead cells and RBC using a one-step gradient method. The Ficoll and Hypaque procedure is routinely used for both blood and spleen cells [7]. This procedure is based on density differences between live cells (lower density) and dead cells (higher density), and has the advantage of separating RBC from live cells. Dead cells and RBC pass through a high-density solution (providing the desired low velocity). Sodium diatrizoate brings the solution to the appropriate density and osmotic strength during centrifugation, thereby, live cells remain floating on top of the high-density material.

The Ficoll and Hypaque Procedure

Resuspend cells in complete RPM1 media and adjust to 50 to 100×10^{-6} cells in 1 or 2 ml. Add 3 ml high-density solution in a 12 ml tube, then add slowly to the top cell suspension and centrifuge 15 min at 2000 rpm at room temperature. Slowly obtain a floating band of live cells from the top of the high-density solution, and avoiding the high-density solution, transfer cells using a pipet into a tube, add complete media and centrifuge 10 min at 1000 rpm. Pour off supernatant and resuspend cell pellet, repeat washing twice, and resuspend in complete RPM1 media with FCS, then count and adjust cells to a desired concentration [7].

24.2.2   Measurement of Lymphocyte Response to Mitogens

Once a mixture of T and B cells are obtained, a whole mixture of cells can be used for proliferative assays involving both T and B cells by using a 96-well microplate and by adding polyclonal mitogens (PHA, Con-A), which can induce T-cell proliferation, or pokeweed and LPS as mitogens for B-cell proliferation because they induce mitosis in the lymphocytes of many different specificities of clonal expansion. Although the polyclonal mitogens may not act directly on the antigen-specific receptors of lymphocytes (unlike superantigens), they seem to trigger most of the T-cell proliferation in humans and animals [7,8].

Mitogen responses are measured by the amount of ³H-TdR uptake into cellular DNA over the course of time (24, 28, 72, 96 hrs). Although viewed by researchers as a somewhat crude measurement of DNA synthesis, this type of study is still used by many to evaluate age-associated functional changes in T and B cells [49].

24.2.3   Enrichment of T or B Cells

For numerous immune function studies, the isolation of pure T or B cells are required to find whether functional defects occur with age in T cells, B cells, or macrophages. Enrichment of T cells can be carried out by using nylon wool or sephadex G-10 columns, to which B cells and macrophages readily attach and allow remaining T cells to pass through [50,51]. Alternatively, appropriate monoclonal antibodies against B cells and other adherent cells can be used with low toxic guinea pig or rabbit complement to remove APC cells and to obtain enriched T cells. Whereas, to obtain APC or B cells, monoclonal antibodies against T cells, such as Thy-1 or anti-CD3 or a mixture of anti-CD4 and anti-CD8, etc., could be used to lyse T cells [52,53]. Furthermore, the selective elimination of T cells or enrichment of T cell subsets, such as CD4+, CD8+ or B cells can also be carried out by following the panning technique and using antibodies against each of these T cell subsets to study defined-cell subset function [7,8,16,17,54]. Care should be taken to enrich T cells without using the same surface antibodies against the cells to be studied. Thus, negative enrichment is generally suitable to prevent non-specific activation of T cells by cell surface bound antibodies that may modify the functional activity of immune cells.

24.2.4   Measurement of T Cell Subsets Effector Function

To understand the more precise functional loss of T or B cells, particularly a defined role for CD4+ or CD8+ T cells, enrichment of subsets can also be carried out using a fluorescence-activated cell sorter (FACS) instrument [54]. This instrument is used to study the functional properties of individual T cell or B cell subsets by using monoclonal antibodies that bind to cell-surface proteins. As specific proteins are expressed exclusively on each subset, such as CD4+ or CD8+, label the cells with specific fluorescent monoclonal antibodies to tag individual cells, the mixed population of blood, spleen, and lymph node cells. When a cell mixture passes through a laser beam, if the droplet contains a cell, that cell will scatter the laser light and the information is then displayed in the form of a histogram. The cell sorter procedure allows one to isolate labeled or unlabeled cells that can be separated by the deflection of stream of cells that are selectively collected in a tube, then placed in micro-well plates for various functional assays.

In general, there are several approaches to studying CD4+ and CD8+ cell function using various cell activators to investigate T and B cell interaction or the cytotoxic function of CD8+ T cells, etc. For example, CD8+ T cell function is usually analyzed using a bioassay, the killing of a target cell such as 51 Cr labeled tumor cells by use of activated T or NK cells. This 51 Cr release technique is simple and carried out using a 96-well microtiter plate by adding various target to effector T cell ratios in an 4 hr, 51 Cr release assay. Live cells will take up, but do not spontaneously release radioactivity, whereas if labeled target cells are killed, the radioactive cromate is released, which can be harvested and counted by using a Gamma counter. In a similar fashion, proliferating target cells, such as tumor cells, can be labeled with ^{3}H-thymidine, which is incorporated into the replicating DNA. Again, when a cytotoxic T cell attaches to the labeled tumor cells, the DNA of the target cells is rapidly broken down and released into the supernatants, which can then be measured either as released or retained ^{3}H-TdR in the intact target cell.

Both these assays are found useful in the measurement of age-associated changes in cytotoxic T cell function in young and old animals. A decreased target cell killing is a sign of immunodeficiency, which may be one of the causes for increased malignancy during aging. Furthermore, one can augment the cytotoxic function of effector cells by culturing cells in the presence of added IL-2, which increases cytotoxic function in T cells in young animals; but, a decreased cytotoxic activity could occur by a failure of the cell to utilize IL-2 for further proliferation due to the loss of IL-2 receptors in aged T cells. Furthermore, cell function generally involves the activation of either Th-1 or Th-2 subsets bearing the specific antigen that stimulates the production of various Th-1 or Th-2 like cytokines, which are released by the T cell when it recognizes the specific antigen used, such as Con-A, anti-CD3 or co-stimulatory molecules (anti-CD28), etc., to stimulate the cells *in vitro*. Thus, T cell function, particularly CD4$^+$ T cells, can be studied by measuring the type and amount of proteins or cytokines released by culturing the cells at various time points *in vitro*.

24.2.5   Measurement of B Cell Function

It is well established that CD4$^+$ T cells induce B cells to secrete specific antibody. By utilizing young B cells, either T cells from young or old animals are mixed with B cells and activated by adding LPS to induce humoral immune response *in vitro*. The humoral immune response is mediated by antibody molecules that are secreted by plasma cells. Generally, helper CD4$^+$ T cells stimulate, first, the proliferation, then the differentiation of antigen-binding B cells. The specific interaction of specific antigen binding B cells (TD, TI-1, or TI-2 antigen) to T cells leads to the expression of the B cell stimulatory molecules, CD40 ligand (CD40L) on the helper T cell surface. This interaction activates secretion of the B cell stimulatory cytokines (e.g., IL-4, IL-5, and IL-6), which drives the proliferation and differentiation of the B cell as antibody secreting plasma cells [7,8].

Indeed, cytokines released by helper T cells induce isotype switching, which may be found impaired in old B cells. However, all mature B cells have surface IgM and IgD antigens. When activated, they initially secrete IgM. Through isotype switching, B cells can give rise to a progeny of various isotypes from IgG or IgE to IgG2a, IgG2b, or IgG3, etc. T cell dependent (TD) antigens are generally a diphtheria toxin or purified protein derivative (PPD), whereas T cell independent (T1) antigens are bacterial lypopolysaccharides, Brucella abortus. The TI-2 antigens include Pneumococeal polysaccharide and Salmonella polymerized flagellin, etc. These can help to establish defects to a recall antigen or a defective memory function in aging B cells. The loss of B cell activity can be caused by the loss of high-affinity IgG or IgA antibody response during aging. Thus, various immunological assays *in vitro* involving both T cells and B cells are extremely useful to the understanding of the immune cell's role in controlling infection during aging. However, similar studies, when undertaken in an intact animal, could provide additional information by collecting both serum and immune cells to establish the functional loss particularly to study the effect of obesity and/or various metabolic disorders during aging.

24.3   Measurement of Cytokine Production

24.3.1   Limiting Dilution Assay

Besides studying the general interaction of T and B cells, there are several well-defined assays in use to measure individual T and B cell function, such as limiting dilution (LD) assays. These assays measure the frequency of particular antigen or its activation by polyclonal mitogens, such as PHA, Con-A, SEB, etc., for which the LD assay works well [55,56]. This assay makes use of Poisson distribution, a statistical function that describes how functionally active cells are distributed at random in various age groups. For instance, T cells from young and old animals, including humans, are distributed in microwell plates, with or without varied numbers of test cells, along with specific activators to measure response to antigen, such as proliferation or cytokine production. Allow several days for proliferation and differentiation to take place in an *in vitro* culture system. The frequency of wells where there is no response is plotted against the number of cells initially added (0, 1, 10, 100, etc.) to the well. If cells of one type, typically antigen-specific T cells, are the sole limiting factor for obtaining a response, then a straight line is obtained from Poisson distribution. On average, one antigen-specific cell per well is measured when the frequency of negative wells is 37%. Thus, the frequency of antigen-specific cells in a population equals the reciprocal of the number of cells added to the well when 37% of the wells are negative. Thus, the LD assay actually measures the frequency of precursors in a population based on the number of effectors resulting from the original stimulated lymphocyte, rather than the activity of the original lymphocyte. This technique is found useful in measuring the function of naive versus memory T cells or CD4$^+$ versus CD8$^+$ cells. Besides measuring the proliferation, one can also apply this technique to measure more precisely the age-associated changes in cytokine production in PBL of humans or spleen cells.

24.3.2   ELISPOT Assay

The cytokine released by individual cells makes a distinct spot, which therefore is known as the enzyme-linked immunospot (ELISPOT) assay. This assay can also be used to detect specific antibody secretion by B cells by using antigen-coated surfaces to trap specific antibody, which are labeled with anti-immunoglobulins and used to detect the bound antibody [7,57,58]. Cells at various dilutions can also be plated into nitrocellulose-backed wells that have been precoated with a cytokine-specific monoclonal antibody to measure cytokine detection. After stimulation with a suitable activator for a number of days, cells are washed and alkaline phosphatase or horseradish peroxidase labeled anti-cytokine antibody, (recognizing a different epitope) is added. The bound cytokine is detected via an enzymatic colorimetric reaction. Each positive spot denotes a single cytokine producing cell. This technique is more sensitive and precise than the ELISA assays and can be combined with LD assay to establish the age-associated change in the frequency of cytokine producing cell. Although, the LD assay, ELISPOT, and ELISA assays have unique abilities to quantitate functional changes, each has their drawbacks and care should be taken to prevent artifacts during the assay. Furthermore, these assays are considered to depict immunogenic, rather than bioactive, forms of cytokines.

24.3.3   Sandwich ELISA Assay for Detection and Quantitation of Cytokines

This is a commonly used assay with numerous manufacturers offering several ready made kits to measure various cytokines in human and mice, including a few kits for rats. The following cytokine reagents are readily available for both mice and humans (IL-1a, IL-2, IL-3, IL-4, IL-5, IL-6, IL-7, IL-9, IL-10, IL-12(p40), IL-12 (p70), IL-13, IL-15, GM-CSF, IFN-γ, MCP-1, TNF-α, TNF-β, etc.). In this assay, the cytokine is characterized by its ability to bridge two monoclonal antibodies reacting with different epitopes on the cytokine molecule. The use of bulk purified T cells can use various activators, such as Con-A, and anti-CD3, co-stimulatory molecules such as anti-CD28, and various cytokines, such as IL-2, IL-4 and IL-12, to establish the Th-1 and Th-2 or Th-0 pattern of cytokine-secreted protein in the supernatants or mRNA levels by collecting the cell pellets in young and old animals or humans.

A detailed protocol of this commonly used assay is given below [7, 58, 59]:

Sandwich ELISA Protocol [11]

1. Dilute purified anti-cytokine capture antibody to 1-4 µg/ml in binding buffer. Add 50 µl diluted antibody to the wells of an enhanced protein binding ELISA plate (e.g., Nunc Maxisorb; Cat. #2442404).

2. Seal plate to prevent evaporation. Incubate overnight at 4°C.

3. Wash ≥4 times with PBS/Tween. For each wash, fill wells with approximately 200 µl PBS/Tween. Invert plate and pound it on absorbent material.

4. Block non-specific binding by adding 200 µl blocking buffer per well.

5. Seal plate and incubate at room temperature for 30 min.

6. Wash ≥ 3 times with PBS/Tween.

7. Add standards and samples diluted in blocking buffer/Tween at 100 µl per well.

8. Seal plate and incubate for 2 to 4 hrs at room temperature or overnight at 4°C.

9. Wash ≥ 4 times with PBS/Tween.

10. Dilute biotinylated anti-cytokine detecting antibody to 0.25 to 2 µl/ml in blocking buffer/Tween. Add 100 µl per well.

11. Seal plate and incubate at room temperature for 30 min.

12. Wash ≥ 6 times with PBS/Tween

13. Dilute avidin- or streptavidin-enzyme conjugate to a pre-titered optimal concentration in blocking buffer/Tween. Add 100 µl per well.

14. Seal plate and incubate at room temperature for 30 min.

15. Wash ≥ 8 times with PBS/Tween

16. Thaw ABTS substrate solution within 20 min of use. Add 10 µl of 30% H_2O_2 per 11 ml of substrate and vortex. Immediately dispense 100 µl into each well. Incubate at room temperature (5 to 80 min) for color development. Color reaction can be stopped by adding 50 µl of 1% SDS (optional).

17. Read OD at 405 nm.

Solutions

1. Binding Buffer: 0.1 M Na_2HPO_4, adjust pH to 9.0 with HCl.

2. PBS Solution: 80.0 g NaCl, 11.6 g Na_2HPO_4, 2.0 g KH_2PO_4, 2.0 g KCl; q.s. to 10 L; pH 7.0.

3. PBS/Tween: Add 0.5 ml of Tween-20 to 1L PBS.

4. Blocking Buffer: Add 10% fetal bovine serum (FBS), 10% newborn calf serum (NBCS) or 1% BSA (immunoassay grade) to PBS. If serum is used as blocking protein, filter blocking buffer before use.

5. Blocking Buffer/Tween: Add .5 ml Tween 10 to 1 L blocking buffer.

6. Substrate Buffer: Add 150 mg 2,2'-Azino-bis-(3-ethybenzthiazoline-6-sulfonic acid) (e.g., Sigma Cat. #A18888) to 500 ml of 0.1 M anhydrous citric acid (e.g., Fisher, Cat. #A-940) in ddH_2O; pH to 4.35 with NaOH. Aliquot 11 ml per vial and store at –20°C. Add 10 µl 30% H_2O_2 prior to use.

7. 30% H_2O_2: Store in aliquot at –20°C.

24.4 Methods to Measure Cytokine mRNA and Proteins in Lymphocytes

24.4.1   Semi-Quantitative Polymerase Chain Reaction (PCR) Technique

During the culturing of T cells, to obtain secreted cytokines, the activated cell pellet is used for purification of RNA for amplification, either from bulk cultures or tissue homogenates. The PCR method is very sensitive, allowing selective amplication of a particular stretch of cDNA that can be detected by direct hybridization of a probe to the DNA or by directly incorporating hot label into the PCR product. Furthermore, quantitative PCR techniques have made it possible to quantitate relative levels of the messages [7,60]. However, because there can be discrepancies between mRNA levels and secreted protein levels, the PCR method and cytokine production measurements are extremely useful when comparing age-related changes in T cell function. Of course, a measurement of only the cytokine could produce misleading results, as activated T cells not only produce cytokines, but also are able to utilize protein for further differentiation, thereby giving a false measurement of the actual level found. Thus, simultaneous measurements of mRNA and protein including the use of Western blotting technique may provide information of whether or not cytokines are utilized by activated T cells. To overcome the limitations of both PCR and cytokine assays, and to permit a precise quantitation of which T cell subsets are able to generate each cytokine, new assays are underway to study single cells by staining for intracellular cytokines. The following RT-PCR technique, derived from methods established in our laboratory [60, Frederick Ernani's Ph.D. dissertation, 1997] are found to be very sensitive to measuring cytokine expression in the immune cells of various animal tissues.

24.4.2   Semi-Quantitative RT-PCR Methods

Reverse Transcription

1. Prepare total RNA from cells or tissue by the Trizol® (Gibco BRL, Gaithersburg, MD) method as recommended by the manufacturer. Briefly, homogenize up to 100 mg of tissue or 1×10^6 cells in 1 ml of Trizol ®. This should be followed by chloroform extraction, isopropanol precipitation, and an ethanol wash according to the manufacturer's suggestion. Dissolve the RNA pellet in DEPC-treated water such that the concentration of the RNA is not diluted more than 1.2 µg/µl. The following technique refers to the Superscript II cDNA synthesis kit (Gibco BRL, Gaithersburg, MD).

2. Quantify the amount of RNA by aspirating 2 µl of the diluted RNA, then dilute in 1 ml of DEPC-treated water. Read the absorbance on a UV spectrophotometer at 260 nm. The concentration of the RNA can be determined in µg/µl by the following relationship: 1 O.D. 260 RNA = 40 µg/ml.

3. Load 10 µg of total RNA per tube. To each tube add 2 µl of oligo-dT and 2 µl of random primers (Gibco BRL, Gaithersburg, MD). Bring the tube volume up to 12 µl with DEPC-water. Heat the tubes up to 70°C for 10 min to denature the RNA. Immediately chill on ice.

4. To each tube, add 4 µl of 5X buffer, 2 µl of dTT, and 1 µl of dNTP. Mix well and spin down. Heat to the appropriate temperature for cDNA synthesis (range of 37°C to 54°C) for 2 minutes.

5. Leaving the tubes in the heating source (such as a thermal cycler), add 1 µl of Superscript II reverse transcriptase enzyme. Incubate for 1 hr at the same temperature as noted in step 4. Once the hour is over, dilute the cDNA with 80 µl autoclaved water. Keep frozen until used for PCR.

PCR

1. Create a master mix composed of 10 X Taq buffer, $MgCl_2$, dGTP, dCTP, dTTP, and dATP (reduced in concentration by 20% with respect to the other nucleotides), and autoclaved water. This mix can then be used as the base for any PCR. The concentration of the $MgCl_2$ can be changed to optimize each PCR reaction. The volume of the master mix should be 44.5 µl per 50 µl reaction.

2. For each PCR reaction add 37.5 µl of master mix. Add 1 µl of the forward and reverse primers per reaction to the master mix.

3. Add 39.5 µl of the master mix plus primer mix to each reaction tube.

4. Add 5 µl of the diluted cDNA per tube.

5. For each reaction, add 5 µl of master mix to a tube you label, "Hot Label." It is a good idea to increase the number of reactions worth of label by 10% to account for pipetting errors. Then add 1 µCi of 33-P dATP per reaction to the same tube. Add 0.5 µl of Taq enzyme per reaction to the labeled tube. Mix well.

6. Add 5.5 μl of "Hot Label" mix to each reaction.

7. If your thermal cycler does not have a heated lid, add mineral oil to each tube to prevent evaporation.

8. Cycle for the appropriate number of cycles. In order to make results quantitative, determine the range of logarithmic amplification. Use the number of cycles that falls well within this range.

9. To compare one cDNA to another, amplify the cDNAs using primers for the gene of interest and a house-keeping gene such as GAPDH or HPRT. Care should be taken when choosing a house-keeping gene because, under some circumstances, these genes can be regulated.

10. Resolve the radiolabeled PCR products using non-denaturing non-reducing PAGE (5% works well for a large range or product sizes). Dry the gel onto 3MM paper.

11. Expose the dried gel to a Phosphorimager screen (Molecular Dynamics, Sunnyvale, CA) or X-ray film. It is much easier to calculate the amount of product using a Phosphorimager; however, a scanning laser densitometer can measure the amount of signal using X-ray film. Another option is to run the products on an agarose gel. The products can be cut from the gel and counted in a liquid scintillation counter. However, this does not give the same resolution as the PAGE gel.

12. To compare one gene's expression in multiple tissues/animals, use the following formula: % Reduction = (Gene X signal tissue 1/HPRT signal tissue 1)/(Gene X signal tissue 2/HPRT signal tissue 2) × 100%. Tissue 1 must have the higher level of expression of Gene X.

24.4.3 Immunofluorescent Staining of Intracellular Cytokines for Flow Cytometric Analysis

At present various anti-cytokine monoclonal antibodies are found useful for staining intracellular cytokines in tissue and cell suspension by the use of multiparameter flow cytometric analyses for identification of individual cytokine-producing cells within unseparated cell populations [61,62]. For example, cell samples without removal of RBC can be stained for surface antigens and intracellular cytokines. The protocol, which follows in the next section, can be used to identify the phenotype and frequency of cell types defined by membrane antigens and intracellular cytokines. Staining for an individual cell surface antigen and two cytoplasmic cytokines has been used to identify and enumerate cell types which express cytokines in a restricted (e.g., Th-1 versus Th-2-like cells) or unrestricted (e.g., Th-0-like cells) manner. In addition to enabling highly specific and sensitive measurements of several parameters for individual cells simultaneously, this method has the capacity for rapid analysis of large numbers of cells which are required for making statistically significant measurements.

Staining of intracellular cytokines depends on the identification of cytokine-specific monoclonal antibodies compatible with a fixation-permeabilization procedure. Optimal intracellular cytokine staining has been reported using a combination of fixation with paraformaldehyde and subsequent permeabilization of cell membranes with the detergent saponin. Paraformaldehyde fixation allows preservation by detergent. Membrane permeabilization by saponin is critical for allowing the cytokine-specific monoclonal antibody to penetrate the cell membrane, cytosol, and membranes of the endoplasmic reticulum and Golgi apparatus.

Critical parameters for cytokine staining include the following: cell type and activation protocol, the time of cell harvest following activation, the inclusion of a protein transport inhibitor during cell activation, and the choice of anti-cytokine antibody.

24.4.3.1 Staining Procedure of Intracellular Cytokines

The polyclonal activators used in these models include, PMA, calcium ionophore/ionomycin, Con-A, PHA, SEB, anti-CD3 and anti-CD28.

Cell activation with PMA alone has been reported to cause a transient loss of $CD4^+$ expression from the surface of mouse T cells. Cell activation with PMA and calcium ionophore, together, has been reported to cause a greater and more sustained decrease in $CD4^+$ expression, and a decrease in $CD8^+$ expression in mouse thymocytes and mouse and human peripheral T lymphocytes. This assay will provide key information regarding the age-related changes in cytokine secreting T cells. We have recently adapted this technique to measure intracellular cytokine secreting cells in the peripheral blood obtained via tail or retroorbital bleeding carried out periodically without sacrificing the mice or rats [63,64].

At present, many investigators use either monensin or brefeldin A in cell culture systems. These agents block intracellular transport processes resulting in the accumulation of most cytokine proteins in the Golgi complex, thereby enhancing the ability to detect cytokine producing cells. Because these agents have a dose- and time-dependent cytotoxic effect, exposure must be limited. Monensin (1 to 3 µM) or brefeldin A (1 to 5 µg/ml) has been included in *in vitro* cultures for 4 to 6 hours prior to cell harvest. Following *in vivo* activation, freshly harvested *ex vivo* cells have been cultured in brefeldin A to enhance the cytokine staining signal. Investigators should be aware of possible effects of transport inhibitors on expression levels of cell surface markers. Brefeldin A has been found to cause decreased levels of $CD4^+$ staining. The following detailed PharMingen protocol is given below [58]:

Protocol

Positive Staining Controls: The technical data sheets for PharMingen's fluorochrome-conjugated anti-cytokine antibodies provide specific examples of *in vitro* culture systems, which can induce detectable frequencies of cytokine-producing cells at specific time points. Particularly important parameters for cell activation protocols include the use of protein transport inhibitors and the examination of multiple time points in the identification of optimal conditions for generating cells that can serve as positive controls.

Negative Staining Controls: The use of one of the following three controls is suggested in the discrimination of specific staining from artifactual staining.

1. Ligand blocking control: Important for demonstrating the specificity of staining. This control consists of the preincubation of the antibody in excess of its target cytokine. Any observed background is due to non-Fab-mediated staining.

2. Unconjugated antibody blocking control: Preincubation of fixed and remeabilized cells are washed with an excess of "cold" unconjugated anti-cytokine antibody before staining the cells with the fluorochrome-conjugated form of the same anti-cytokine antibody.

3. Isotype-matched control immunoglobulin: Stain the target cells with an isotype matched immunoglobulin of irrelevant specificity at the same concentration as the antibody of interest.

Multi-Color Staining for Intracellular Cytokines and Cell Surface Antigens

Harvest Cells: Viable cell populations may be stained in plastic tubes or microwell plates. Cells should be protected from light throughout staining and storage.

Block Fc Receptors: Reagents that block Fc receptors may be useful for reducing nonspecific staining.

1. In the mouse system, purified 2.4G2 antibody directed against FcgII/III receptors, can be used to block nonspecific staining by fluorochrome conjugated antibodies that is mediated by Fc receptors. To block mouse Fc receptors with Fc Block, preincubate cell suspension with 1 μg Fc Block/10^6 cells in 100 μl for 5 min at 4°C, then proceed to staining with the cell-surface antibody of interest in the presence of Fc block.

2. Fc receptors on human cells may be pre-blocked with an excess of irrelevant purified Ig from the species in which the antibody of interest was generated.

Stain Cell Surface Antigens:

1. Stain 10^6 cells in 100 μl of staining buffer with ≤ 0.5 μg of a fluorochrome-conjugated monoclonal antibody specific for a cell surface antigen such as CD3+, CD4+, CD8+, CD14+, or CD19+ (30 min, 4°C). Multi-color staining of different cell surface antigens can be done at this time for setting flow cytometric compensations.

2. Wash cells twice with staining buffer and pellet by centrifugation (250 × g).

Fix Cells: Thoroughly resuspend and fix cell with 100 μl of fixation buffer for 20 min at 4°C. Cells may be kept overnight in fixation buffer at 4°C.

Permeabilize Cells: Wash cells twice in Permeabilization Buffer and pellet.

Stain Intracellular Cytokines: Thoroughly resuspend fixed cells in 100 μl of permeabilization buffer containing a previously determined optimized concentration of a fluorochrome-conjugated anti-cytokine antibody or appropriate negative control. Incubate at 4°C for 30 min in the dark.

Staining Controls:

1. Ligand blocking control: Preblock anti-cytokine antibody with recombinant cytokine.
   a. Preincubate fluorochrome-labeled antibodies with cytokine diluted to the appropriate concentration in 100 μl permeabilization buffer at 4°C for 30 min.
   b. Resuspend fixed and permeabilized cells in 100 μl preblocked labeled anti-cytokine antibody and incubate 30 min at 4°C.

2. Unconjugated antibody control:
   a. Resuspend fixed and permeabilized cells in 50 μl permeabilization buffer containing unconjugated anti-cytokine antibody diluted to the appropriate concentration, and incubate 30 min at 4°C.
   b. After incubation, add fluorochrome labeled anti-cytokine antibody at optimal concentration in 50 μl permeabilization buffer and incubate 30 min at 4°C.

3. Isotype control:
   a. Resuspend cell pellet in 100 μl of permeabilization buffer containing a concentration of the isotype control antibody equal to that of the anti-cytokine antibody (<0.5 μg/106 cells). Wash cells twice in Staining Buffer and pellet.

Analysis
Thoroughly resuspend cells in Staining Buffer and analyze by flow cytometry. Set PMT voltage and compensation choosing cell surface staining controls. Set quadrant markers based on blocking controls.
Solutions

Staining Buffer
Dulbecco's PBS (DPBS) without Mg^{2+} or Ca^{2+}

1% heat-inactivated FCS

0.1% (w/v) sodium azide

adjust buffer pH to 7.4–7.6, filter (0.2 μm pore membrane), and store at 4°C.

Fixation Buffer
4% (w/v) paraformaldehyde

Add the paraformaldehyde to DPBS and warm in 50°C water bath (in fume hood) until the paraformaldehyde dissolves (1–3 hr). Adjust buffer pH to 7.4–7.6 and store at 4°C protected from light.

Permeabilization Buffer
Dulbecco's PBS (without Mg^{2+} or Ca^{2+})

1% heat-inactivated FCS

0.1% (w/v) sodium azide

0.1% (w/v) saponin

Adjust buffer pH to 7.4–7.6 and filter

2mM Monensin (stock solution)
Dissolve monensin in 200 proof ethanol. This may require some agitation.

Add NaOH to 1 mM final concentration.

Brefeldin A (stock solution)
Dissolve in 100% DMSO.

24.5 Flow Cytometry Techniques

24.5.1 Enumeration of Lymphocyte Population by Antibodies Specific for Cell-Surface Molecules

Immune cells have various functional activities that change with age and are usually identified and distinguished from each other on the basis of their differential expression of cell-surface proteins, which are detected by using specific monoclonal cellular determinants (CD) antibodies (7,54). A flow cytometer allows for the most precise identification and changes in blood and various lymphoid organ cells. We currently use a desktop FACScan instrument (Becton Dickinson) with CELLQuest software supported by a Macintosh Power PC. When cells are labeled with a single fluorescent antibody, the data is usually displayed in the form of a histogram of fluorescence intensity versus cell number. If two or more antibodies are used, each coupled to different fluorescent dyes, then the data are usually displayed in the form of a two-dimensional scatter diagram or as a contour plot, where the fluorescent of one dye-labeled antibody is plotted against a second. The result of a cell population labeling with no antibody can be further subdivided by its labeling with second antibody. This kind of analysis facilitates in identifying an age-related loss or rise of various lymphocyte subsets in various organs. Further and very precise quantitative data can be obtained by staining cells with various monoclonal antibodies against T cells and their subsets, $CD4^+$ and $CD8^+$, B cells, macrophages, and various co-receptors or co-stimulatory molecules. The FACS analysis not only provides the percentages for each lymphocyte subset but also provides data on age-related changes in the number of molecules or the antigen density on the cell surface of each type of immune cells. The fluorescent technique followed to enumerate the lymphocytes is given in detail below [54].

24.5.1.1 *Immunofluorescent Staining of Leukocytes for Flow Cytometry*

Procedure

1. Harvest cells from tissue, preparing a single cell suspension. Red blood cells may be removed by lysis or density gradient. Wash cells once in cold ash buffer (PBS/0.1% NaN_3/0.1% fetal bovine serum); centrifuge at 350 × g for 5 min. Resuspend cell pellet to a concentration of 2×10^7 cells/ml (i.e., 10^6 cells per 50 μl).

2. Dilute primary mAbs (e.g., unconjugated, biotinylated, or fluorochrome-conjugated mAbs) to predetermined optimal concentrations (see Staining Tips) in wash buffer and deliver to the wells of a U-bottom microtiter plate in volume of 50 μl.

3. Deliver 10^6 cells in 50 μl to each well already containing 50 μl of mAb (or 50 μl wash buffer for negative controls). Mix by gently vortexing or tapping.

4. Incubate at 4°C for 20 to 40 min in the dark.

5. Wash twice with 200 μl wash buffer (or three times if a biotin-conjugated primary antibody is used). After each centrifugation, 350 × g for 5 min, aspirate wells or flick plate to remove supernatant. Vortex gently or tap plate to loosen pellet prior to adding next wash or diluted secondary reagent.

6. If a second-step reagent is needed, resuspend cell pellet in 100 μl of appropriate secondary reagent (e.g., fluorochrome-conjugated avidin, streptavidin, anti-Ig allotype, anti-Ig isotype, polyclonal anti-Ig). For example, dilute antibody to ~1 μg per 100 μl in wash buffer and add this to each well containing the loosened cell pellet.

7. Incubate at 4°C for 20 to 40 min in the dark.

8. Wash twice with 200 μl wash buffer, as in Step 5. Use 100 μl wash buffer to transfer cell pellets to 0.4 ml aliquots of wash buffer (final concentration ~10^6 cells in 0.5 ml) in tubes appropriate for flow cytometer. Acquire sample data on flow cytometer as soon as possible after staining.

24.5.2   Measurement of Oxidative Response in Immune Cells

It is well established that an age-associated rise in free-radicals and a decreased antioxidant enzyme activity to inhibit reactive oxygen species (ROS) may contribute to an accelerated aging process. The oxidative response of macrophages or other immune cells can now be measured by using flow cytometry for identification of subpopulation of phagocytes residing in peritoneal cavity, or alveolar macrophages can be evaluated after phagocytosis of fluorescent latex beads. Further use of dichloroflurosein diaacetate (DCFH) or hydroethidine (HE) are nonfluorescent precursors that can be oxidized intracellularly by macrophages, or by T and B cells to fluoroscent DCF and ethidiene, respectively. Although DCFH is specific for intracellular H_2O_2, the red fluorescence of oxidized HE, allows for the combination with fluorescinated particles for phagocytosis [65,66].

24.5.2.1   *Flow Cytometry of Intracellular Oxidation*

Incubate desired cell suspensions at 37°C in tubes on a shaking platform for 10 min with DCFH (10μl) or HE (0.30 μg/ml). After loading, activators such as PMA or phagocytic particles or control buffer are added. At different time points, analyze cells using by a flow cytometer collecting 5–10000 events. Red fluorescence and green fluorescence can be recorded using a desktop FACScan. This technique can be applied to various subpopulation measurements using upto 3- or 4-color analyses of age-associated changes in the oxidative response of various immune cells.

24.5.3   Measurement of Apoptosis

Recently, new information describes the new molecules responsible for regulating cellular homeostasis and programmed cell death (PCD). Either an increased loss or an excess accumulation of immune cells by failure of PCD could cause immune deficiency by activation or decreased rate of programmed cell death [67,68,44]. The regulation of immunological homeostasis is found controlled by various surface molecules, including Fas, FasL, TGFβ, Nurr 77, p53, etc. Furthermore, apoptosis is prevented by increased expression of Bcl-2/Bcl/X protein, etc. Therefore, the use of the PCD assay has provided further significant insight into understanding the breakdown of the immune system, particularly during the rise of autoimmune disorders and aging.

FACS analysis can be used to assay both cell surface expression of lymphocyte subsets along with Annexin-V to measure apoptosis [69]. The use of Annexin-V provides information regarding the rate of apoptosis and necrosis within a defined T cell subset. Furthermore, with Fas, FasL, and Bcl-2, antibodies can be used to measure surface protein expression. Generally, one million cells are fixed with 0.25% paraformaldehyde and ethanol followed by incubation with hamster anti-mouse Fas or Bcl-2 (2mg/1 × 10⁶ cells) for 1 hr at 4°C in RPMI-1640 containing 1% BSA and 0.1% NaN₃. After several washes, a fluorescein isothiocyanate (FITC) conjugated rat mAb to hamster IgG is added. The second fluorochrome-labeled antibody is conjugated with phycoerythin (PE). The second labeled antibody is directed at CD3⁺, CD4⁺, CD8⁺, or CD19 and can also be utilized to locate the presence of Fas or Bcl-2 on lymphocyte subsets. Controls will include cells incubated with one of the labeled antibodies only, no antibody, and a non-relevant FITC or PE conjugated antibody.

24.5.3.1   *Flow Cytometric Detection of Apoptosis by TdT end Labeling (TUNEL/PI Assay)*

This assay employs terminal deoxynucleotidyl transferase (TdT) to tag 3'OH ends resulting from apoptotic fragmentation [70]. A direct, commercially available, single-step procedure for labeling DNA strand breaks with BODIPY-conjugated deoxynucleotides is carried out. Cells are labeled with BODIPY-dUTP (B-dUTP) and then counterstained with PI for bivariate flow cytometric measurements of apoptosis with relationship to cell cycle position and ploidy by DNA content. 1.0×10^6 cells are washed with phosphate buffered saline (PBS) and resuspended in 50 μL of deoxynucleotdiyl transferase (TdT) reaction buffer containing 10 μL of 5× concentrated buffer solution (1M potassium cacodylarte; 125mM Tris-HCl, pH 6.6; 1.26 mg/ml bovine serum albumin (BSA), 5 μL of 25 mM cobalt chloride, 0.5 μL (12.5 units) of TdT (Boehringer Mannheim), 0.25 μL of B-d-UTP, and distilled water to 50 μL. Cells are again incubated in the reaction buffer for 60 min at 37°C, rinsed twice in 15 mM EDTA pH 8.0, and once in 0.1% triton X-100 in PBS. Cells are then suspended in 2.5μg/ml PI in PBS with 0.1% DNAse-free RNAse (Sigma Chemical Co., St. Louis, MO). A minimum of 25 to 30K cell events can be collected.

24.6 Summary

Advances in immunological techniques and the availability of various monoclonal antibodies allow researchers to separate and define young and old, human or animal cells and permits a more precise functional measurement of immune cells. Recent techniques such as the LD assay, ELISPOT, and the sandwich ELISA technique allow for measurements on the production of various cytokines in bulk and/or at the individual cell level. Furthermore, intracellular cytokine staining techniques permit the identification of Th-0, TH-1, and Th-2 secreting cells in a well-defined lymphocyte subpopulation. Additional techniques, such as RT-PCR, can establish the mRNA levels in lymphocytes. Also, techniques to measure apoptosis and oxidative response can facilitate in establishing the extent of immunodeficiency and free radical generation, which may contribute to the loss of immune function and rise in malignancy and autoimmune disorders.

Acknowledgments

This work is supported in part by National Institutes of Health grants R01-AG10531, AG14541, AG13693, and P01 DE10863. The author wishes to thank PharMingen Company for providing technical advice and procedures. Thanks are also due to Corinne Price for her editorial assistance and Olga German in the preparation of this manuscript.

References

1. Yunis, E. J., Handwerger, B. S, Hallgren, H. M., Good, R. A., Fernandes, G. Aging and immunity. In: Cohen S; Ward PA (eds): *Mechanisms of Immunopathology.* New York, NY, John Wiley & Sons, 91–106, 1979.

2. Makinodan, T., Kay, M. M. B. Age influence on the immune system. *Advances in Immunology,* 29, 287, 1980.

3. Klinman, N. Antibody-specific immunoregulation and the immunodeficiency of aging. *Journal of Experimental Medicine*, 154, 547–551, 1981.

4. Thoman, M. L, Weigle, W. O. The cellular and subcellular bases of immunosenscence. Advances in Immunology, 46, 221, 1989.

5. Miller, R. A. Aging and immune function. *International Review of Cytology,* 124, 187, 1991.

6. Hodes, R. J. Fauci, A. S. (Eds). *Report of the Task Force on Immunology and Aging*, National Institute of Aging, National Institute of Allergy and Infectious Diseases, U. S. Department of Health and Human Services, National Institutes of Health, 1996.

7. Ausubel, F. M., Brent, R., Kingston, R. E., Moore, D. D., Siedman, J. G., Smith, J. A., Struhl, K. (Eds). *Current Protocols in Immunology*, John Wiley & Sons, New York, NY, 1991.

8. Janeway, C. A. and Travers, P. (Eds). *Immunobiology: The Immune System in Health and Disease.* Garland Publishing, Inc., New York, NY, 1994.

9. Murasko, D. M., Nelson, B. J., Silver, R., Matour, D. Immunologic response in an elderly population with a mean age of 85. *American Journal of Medicine*, 81, 612, 1986.

10. Weksler, M. E., Schwab, R., Huetz, F., Kim, Y. T. Coutinho, A. Cellular basis for the age-associated increase in autoimmune reactions. *International Immunology*, 2, 329, 1990.

11. Miller, R. A. Age-associated decline in precursor frequency for different T cell-mediated reactions, with preservation of helper or cytotoxic effect per precursor cell. *Journal of Immunology*, 132, 63, 1984.

12. Seder, R., Paul, W. Acquisition of lymphokine producing phenotype by CD4$^+$ T cells. *Annual Review of Immunology*, 12, 635, 1994.

13. Romagnani, S. Lymphokine production by human T cells in disease states. *Annual Review of Immunology*, 12, 227, 1994.

14. Mosmann, T. R., Sad, S. The expanding universe of T-cell subsets: Th1, Th2 and more. *Immunology Today*, 17, 138, 1996.

15. Carter, L., Dutton, R. Type 1 and Type 2: a fundamental dichotomy for all T-cell subsets. *Current Opinion of Immunology*, 8, 336, 1996.

16. Ernst, D. N., Hobbs, M. V., Torbett, B. E., Glasebrook, A. L., Rehse, M. A., Bottomly, K, Hayakawa, K., Hardy, R. R., Weigle, W. O. Differences in the expression profiles of CD45RB, Pgp-1, and 3G11 membrane antigens and in the patterns of lymphokine secretion by splenic CD4$^+$ T cells from young and aged mice. *Journal of Immunology*, 142, 1413, 1989.

17. Hobbs, M. V., Ernst, D. N., Torbett, B. E., Glasebrook, A. L., Rehse, M. A., McQuitty, D. N., Thoman, M. L., Bottomy, K., Rothermel, A. L., Noonan, D. J. Cell proliferation and cytokine production by CD4$^+$ cells from old mice. *Journal of Cellular Biochemistry*, 46, 312, 1991.

18. Kubo, M., Cinader, B. Polymorphism of age-related changes in interleukin (IL) production: differential changes of T helper subpopulations, synthesizing IL1, IL3 and IL4. *European Journal of Immunology*, 20, 1289, 1990.

19. Mosmann, T. R., Coffman, R. L. TH1 and TH2 cells: different patterns of lymphokine secretion lead to different functional properties. *Annual Review of Immunology*, 7, 145, 1989.

20. Constant, S. L., Bottomly, K. Induction of TH1 and TH2 CD4+ T cell responses: The alternative approaches (eds) Paul, W. E., Fathman, C. G., Metzger, H. In *Annual Review of Immunology*, Palo Alto, CA, 15, 297, 1997.

21. Weigle, W. O. The effect of aging on cytokine release and associated immune functions. In: *Immunological Problems in the Aged.* Immunology and Allergy Clinics of North America, edited by H. J. Zeitz. Philadelphia: W. B. Saunders Co., 1993, p. 551.

22. Ben-Yehuda, A, Weksler, M. S. Host resistance and the immune system. Clinics in Geriatric Medicine, 8, 701–711, 1992.

23. Ernst, D. N., Weigle, W. O., Noonan, D. J., McQuitty, D. N., Hobbs, M. V. The age-associated increase in IFN-g synthesis by mouse CD8+ T cells correlates with shifts in the frequencies of cell subsets defined by membrane CD44, CD45RB, 3G11, and MEL-14 expression. *Journal of Immunology*, 151, 575–587, 1993.

24. Fernandes, G., Handwerger, B. S., Yunis, E. J., Brown, D. M. Immune response in the mutant diabetic C57BL/Ks-db+ mouse. Discrepancies between *in vitro* and *in vivo* immunological assays. *Journal of Clinical Investigation*, 61, 243, 1978.

25. Goidi, E. A., Innes, J. B., Weksler, M. E. Immunological studies of aging. II. Loss of IgG and high avidity plaque-forming cells and increased suppressor cell activity in aging mice. *Journal of Experimental Medicine*, 144, 1037, 1976.

26. Gupta, S. Membrane signal transduction in T cells in aging humans. *Annals of New York Academy of Science*, 568, 277, 1989.

27. Klinman, N. R. Antibody-specific immunoregulation and the immunodeficiency of aging. *Journal of Experimental Medicine*, 154, 547, 1981.
28. Macatonia, S. E., Hsieh, C-S, Murphy, K. M., O'Garra, A. Dendritic cells and macrophages are required for Th1 development of CD4$^+$ T cells from $\alpha\beta$ TCR transgenic mice: IL-12 substitution for macrophages to stimulated IFN-γ production is IFN-γ-dependent. *International Immunology,* 5, 1119, 1993.
29. Dobber, R., Tielemans, M, Weerd, H. D., Nagelkerken L. Mel14+ CD4+ T cells from aged mice display functional and phenotypic characteristics of memory cells. *International Immunology,* 6, 1227–1234, 1994.
30. Bradley, L. M., Croft, M., Swain, S. L. T-cell memory: new perspectives. *Immunology Today,* 14, 197–199, 1993.
31. Bender, G. S., Johnson, M. P., Small, P. A. Influenza in senescent mice: impaired cytotoxic T-lymphocyte activity is correlated with prolonged infection. *Immunology,* 72, 514–519, 1991.
32. Engwerda, C. R., Handwerger, B. S., Fox, B. S. Aged T cells are hyporesponsive to constimulation mediated by CD28. *Journal of Immunology,* 152, 3740–3747, 1994.
33. Patel, P. J. Aging and antimicrobial immunity. Impaired production of mediator T cells as a basis for the decreased resistance of senescent mice to Listeriosis. *Journal of Experimental Medicine,* 154, 821, 1981b.
34. Meydani, S. N., Meydani, M., Verdon, C. P., Shapiro, A. A., Blumberg, J. B., Hayes, K. C. Vitamin E supplementation suppresses prostaglandin E2 synthesis and enhances the immune response of aged mice. *Mechanisms of Aging Development,* 34, 191, 1986.
35. Ershler, W. B. The influence of an aging immune system on cancer incidence and progression. *Journal of Gerontological Biological Science,* 48, B3, 1993.
36. Kelley, K. W., Brief, S., Westly, H. J., Novakofski, J., Bechtel, P. J., Simon, J., Walker, E. B. GH3 pituitary adenoma cells can reverse thymic aging in rats. *Proceedings of the National Academy of Sciences, U.S.A.,* 83, 5663, 1986.
37. Degelau, J. J., O'Leary, J. J., Hallgren, H. M. Relationship of memory and naive T cell subsets to diminished influenza vaccination response in nursing home elderly. *Aging: Immunology and Infectious Disease,* 5, 27–41, 1994.
38. Gabriel, H., Schmitt, B., Kindermann, W. Age-related increase of CD45RO$^+$ lymphocytes in physically active adults. *European Journal of Immunology,* 23, 2704–2706, 1993.
39. Nicholson, L. B., Greer, J. M., Sobel, R. A., Lees, M. B., Kuchroo, V. K. An altered peptide ligand mediates immune deviation and prevents autoimmune encephalomyelitis. *Immunity,* 3, 397, 1995.
40. Bretscher, P., Wei, G., Menon, J. Establishment of stable cell mediated immunity that makes 'susceptible' mice resistant to Leishmania major. *Science,* 257, 539, 1992.
41. Wang, L-F, Lin, J-Y, Hsieh, K-H, Lin, R-H. Epicutaneous exposure of protein antigen induces a predominant Th2-like response with high IgE production in mice. *Journal of Immunology,* 156, 4079, 1996.
42. Fernandes G. Nutritional factors: Modulating effects on immune function and aging. *Pharmacology Reviews,* 36:123S–129S, 1984.
43. Fernandes, G. Effects of calorie restriction and omega-3 fatty acids on autoimmunity and aging. *Nutrition Reviews* 53:S72–S79, 1995.
44. Troyer, D. A., Fernandes, G. Nutrition and Apoptosis. *Nutrition Research* 16:1959–1988, 1996.
45. Fernandes, G., Chandrasekar, B., Troyer, D. A., Venkatraman, J. T., Good, R. A. Dietary lipids and energy intake affects mammary tumor incidence and gene expression in MMTV/v-Ha-*ras*. *Proceedings of the Natlional Academy of Sciences, U.S.A.* 92:6494–6498, 1995.
46. Effros, R. B., Walford, R. L., Weindruch, R., Mitcheltree, C. Influences of dietary restriction on immunity to influenza in aged mice. *Journal of Gerontological Biology Sciences,* 46, B142–B147, 1991.
47. Harrison, D. E., Jordan, C. T., Zhong, R. K., Astle, C. M. Primitive hemapoietic stem cells: direct assay of most productive populations by competitive repopulation with simple binomial, correlation and covariance calculations. *Experimental Hematology,* 21, 206–219, 1993.
48. Weissman, I. L. Developmental switches in the immune system. *Cell,* 76, 207, 1994.

49. Strong, D. M., Ahmed, A. A., Thurman, G. B., Sell, K. W. *In vitro* stimulation of murine spleen cells using a microculture system and a multiple automated sample harvester. *Journal of Immunological Methods*, 2, 279, 1973.

50. Ly, I. A., Mishell, R. I. Separation of mouse spleen cells by passage through columns of sephadex G-10. *Journal of Immunological Methods*, 5, 239, 1974.

51. Julius, M.H., Simpson, E., Herzeberg, L. A. A rapid method for the isolation of functional thymus-derived murine lymphocytes. *European Journal of Immunology* 3: 645, 1973.

52. Rosenwasser, L. J. and Rosenthal, A. S. Adherent cell function in murine T lymphocte antigen recognition. I. A macrophage-dependent T cell proliferation assay in the mouse. *Journal of Immunology*, 120, 1991, 1978.

53. Sprent, J., Schaefer, M., Lo, D., Korngold, R. Functions of purified L3T3 and LyT-2+ cells *in vitro* and *in vivo. Immunology Reviews*, 91, 195, 1986.

54. Darzynkiewicz, Z., Robinson, J. P., and Crissman, H. A. Flow Cytometry, in *Methods of Cell Biology*. Academic Press, San Diego, CA. 1994.

55. Lefkovits, I., Waldman, H. (eds) *Limiting Dilution Analysis of Cells in the Immune System*. Cambridge University Press. Cambridge and New York, 1979.

56. Miller, R. A. Quantitation of functional T cells by limiting dilution. In *Current Protocols in Immunology*. Coligan, J. E., Kruisbeek, A. M., Margulies, D. H., Shevach, E. M. Struber, W. (Eds) Green Publishing Associates and Wiley-Interscience, New York, 3, 15–1, 1991.

57. Klinman, D. M. ELISPOT assay to detect cytokine-secreting murine and human cells, in *Current Protocols in Immunology*. Coligan, J. E., Kruisbeek, A. M. Margulies, D. H., Shevach, E. M., Struber, W. (Eds) Green Publishing Associates and Wiley-Interscience, New York, NY, 1994, 6, 19.9.

58. PharMingen Research Products Catalog, 1996–1997.

59. Fernandes, G., Venkatraman, J.T., Chandrasekar, B., Tomar, V., Zhao, W. Increased transforming growth factor β and decreased oncogene expression by omega-3 fatty acids in the spleens delays the onset of autoimmune disease in B/W mice. *Journal of Immunology*, 152, 5979–5987, 1994.

60. Melby, P. C., Darnell, B. J., Tyron, V. V. Quantitiative measurement of human cytokine gene expression by polymerase chain reaction. *Journal of Immunological Methods*, 15, 235, 1993.

61. Jung, T., Schauer, U., Heusser, C., Neumann, C., Rieger, C. Detection of intracellular cytokines by flow cytometry. *Journal of Immunology Methods*, 159, 197, 1993

62. Carter, L. L., Swain, S. L. Single cell analyses of cytokine production. *Current Opinion in Immunology*, 9, 177–182, 1997.

63. Fernandes, G., Zhao W. Effect of diet and age on Th-1 and Th-2 like lymphocyte subsets in the peripheral blood of C57BL/6 and Balb/C aging mice. *Journal of Allergy Clinical Immunology*, 99, (1017), S248, 1997.

64. Zhao, W., Fernandes, G. A simple immunophenotyping technique for detection of intracellular cytokines in mouse peripheral blood by flow cytometry. *Journal of Allergy and Clinical Immunology*, 99 (1205)S295, 1997.

65. Boissy, R. E., Trinkle, L. S., Nordlund, J. J. Separation of pigmented and albino melanocytes and the concomitant evaluation of endogenous peroxide content using flow cytometry. *Cytometry*, 10, 779, 1989.

66. Kobzik, L., Goldleski, J. L., Brain J. D. Selective down-regulation of a alveolar macrophage oxidation response to opsoninindependent phagocytosis. *Journal of Immunology*, 144, 4312, 1990.

67. Cohen, J. J., Duke, R. C., Fadok, V. A., Sellins, K. S. Apoptosis and programmed cell death in immunity. *Annual Review of Immunology*, 10, 267–2671, 1992.

68. Luan, X., Zhao, W., Chandrasekar, B., Fernandes, G. Calorie restriction reverses lymphocyte subset phenotype and increases apoptosis in MRL/*lpr* mice. *Immunology Letters*, 47, 181, 1995.

69. Vermes, I. A novel assay for apoptosis - flow cytometric detection of phosphatidylserine expression on early apoptotic cells using fluorescein labeled Annexin V. *Journal of Immunology Methods*, 184, 39, 1995.

70. Li, X., Traganos, F., Malamed, M. R., Darzynkiewicz, Z. Single step procedure for labeling DNA strand breaks with fluorescein- or BODIPY-conjugated deoxynucleotides: Detection of apoptosis and bromodeoxy uridine incorporation. *Cytometry*, 20, 1975, 1995.

Section G

Techniques for Assessing Age-Related Oxidative Modification

25

Lipid Peroxidation

Mitsuyoshi Matsuo and Takao Kaneko

CONTENTS

25.1 Introduction

Lipid peroxidation is oxygenation in which molecular oxygen is added to unsaturated lipids. Organic compounds are never instantaneously oxidized in air at room temperature, although most oxygenations are thermodynamically exergonic and are expected to proceed spontaneously. Molecular oxygen in air is in the ground state and also in a triplet state ($^3O_2$). Thus, it has the characteristics of the diradical; ground-state molecular oxygen does not react as a normal double bond, but as a diradical. To maintain spin conservation during the reaction, molecular oxygen must either react with another molecule bearing unpaired electrons or give rise to a triplet-state product. This imposes a kinetic restriction on its reactions, since stable triplet states are unusual. Electron transfer requires that the first electron be placed in a partially filled antibonding π-orbital. Again, this creates a barrier to electron-transfer reactions. As a result, molecular oxygen has a high potential for but also a kinetic barrier to oxidizing common biological materials.

Triplet molecular oxygen is physically or chemically excited to give singlet molecular oxygen ($^1O_2$) with an energy of 94 kJ/mol ($^1\Delta_g$) or 157 kJ/mol ($^1\Sigma_g^+$). Unsaturated lipids are peroxidized with $^1\Delta_g O_2$ to give lipid hydroperoxides (hereafter, $^1O_2$ represents $^1\Delta_g O_2$), although $^1\Sigma_g^+ O_2$ is not involved in chemical reactions because of its extremely short life.

For lipid peroxidation, either molecular oxygen or an unsaturated lipid needs to be activated. For example, an unsaturated lipid is activated to the carbon-centered unsaturated lipid radical for lipid peroxidation by an autoxidation mechanism, and molecular oxygen is activated to $^1O_2$ for lipid peroxidation by a singlet oxygen oxygenation mechanism. In addition, typical enzymatic lipid peroxidation catalyzed by lipoxygenase is found to proceed by a reaction mechanism similar to an autoxidation mechanism.

Lipid peroxidation takes place *in vivo*, as well as *in vitro*. The products of lipid peroxidation are lipid peroxides and, in addition, the primary products are lipid hydroperoxides. It should be noted that *in vivo* lipid peroxidation influences biological systems not only through the resulting lipid peroxides, but also through their decomposition products. The decomposition products, so-called secondary products, of lipid peroxidation include a variety of compounds such as hydrocarbons, epoxides, aldehydes, ketones, and carboxylic acids. In particular, (*E*)-4-hydroxy-2-nonenal is known to be a highly toxic compound. The indices of *in vivo* lipid peroxidation, therefore, should be related to the formation and decomposition of lipid peroxides.

25.2 Chemistry of Lipid Peroxidation

25.2.1 Formation of Lipid Peroxides

25.2.1.1 *Autoxidation*

The autoxidation mechanism for the formation of lipid hydroperoxides has been well established [1]. Lipid peroxidation is a radical chain reaction consisting of the following processes:

Initiation $$RH + I\bullet \rightarrow R\bullet + IH \tag{25.1}$$

Propagation $$R\bullet + O_2 \rightarrow ROO\bullet \tag{25.2}$$

$$ROO\bullet + RH \rightarrow ROOH + R\bullet \tag{25.3}$$

Termination $$R\bullet + R\bullet \rightarrow RR \tag{25.4}$$

$$R\bullet + ROO\bullet \rightarrow ROOR \tag{25.5}$$

$$ROO\bullet + ROO\bullet \rightarrow ROOR + O_2 \tag{25.6}$$

where RH represents an unsaturated lipid, I• a radical derived from an initiator, R• a carbon-centered lipid radical, ROO• the lipid peroxyl radical, ROOH lipid hydroperoxide, RR a lipid dimer, and ROOR lipid peroxide.

The first process is initiation, the second propagation, and the third termination. In initiation, a hydrogen atom is abstracted from a lipid molecule to yield the carbon-centered lipid radical (Eq. 25.1). In propagation, molecular oxygen is added to the radical to give the lipid peroxyl radical (Eq. 25.2), which abstracts a hydrogen atom from another lipid molecule to produce both lipid hydroperoxide and the carbon-centered lipid radical (Eq. 25.3). To the resulting carbon-centered lipid radical is added another molecular oxygen (Eq. 25.2). Here, a chain reaction starts. The chain reaction can continue until termination reactions occur (Eqs. 25.4, 25.5, and 25.6).

The autoxidation of linoleic acid and its methyl ester has been most extensively examined [2]. The autoxidation mechanism of linoleic acid is shown in Figure 25.1. The primary product of its autoxidation is hydroperoxyoctadecadienoic acid, which consists of (10E,12E)-9-, (10E,12Z)-9-, (9E,11E)-13-, and (9Z,11E)-13-hydroperoxy isomers. The ratio of the (E,E)-isomers to the (E,Z)-isomers is variable depending on the reaction conditions, e.g., reaction temperature and hydrogen donor concentration.

The autoxidation of triene and tetraene fatty acids, e.g., linolenic acid and arachidonic acid, leads to a much more complex mixture of products than that formed in the autoxidation of diene fatty acids, e.g., linoleic acid. [2] As shown in Figure 25.2, arachidonic acid has three carbon atoms (C_7, C_{10}, and C_{13}) flanked by two double bonds, so-called bisallylic methylene groups, from which hydrogen atoms are abstracted. Its autoxidation gives rise to 5-, 8-, 9-, 11-, 12-, and 15-hydroperoxyeicosatetraenoic acids. Reflecting the difference in the reactivity of hydrogen abstraction from these bisallylic methylene groups, the yields of these hydroperoxides are different. In addition, peroxyl radical cyclization takes place in these autoxidation pathways. Figure 25.2 shows the mechanism of peroxyl radical cyclization in the autoxidation of arachidonic acid.

The autoxidation chemistry is complex. Not only the autoxidation of unsaturated fatty acids but also that of unsaturated lipids may yield hydroperoxides, endoperoxides, epoxides, etc.

25.2.1.2 *Singlet Oxygen Oxygenation*

Unsaturated lipids are peroxidized by $^1O_2$ via two types of addition reaction, an ene reaction and cycloaddition, as shown in Figure 25.3 [3].

Hydroperoxides are formed from both the autoxidation and singlet oxygen oxygenation of unsaturated lipids. These two reactions can be differentiated, because the double bond of each substrate migrates to the next carbon atom during the formation of hydroperoxide

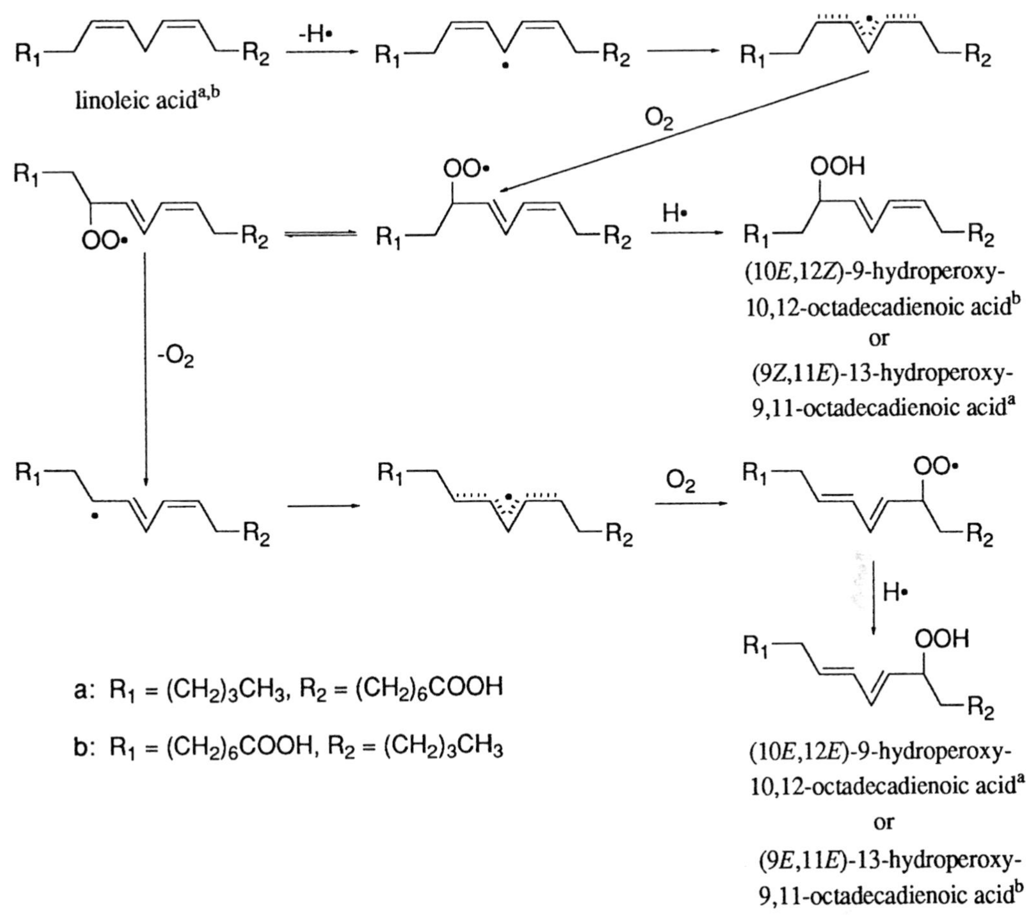

FIGURE 25.1
The autoxidation mechanism of linoleic acid.

due to singlet oxygen oxygenation. For this differentiation, the cooxidation of cholesterol can be utilized. When hydroperoxides are formed from unsaturated lipids in the presence of cholesterol, the formation of cholesterol 5α-hydroperoxide provides evidence for singlet oxygen oxygenation, while the formation of cholesterol 7α- and 7β-hydroperoxides is indicative of autoxidation (Figure 25.4).

25.2.1.3 Enzymatic and Enzyme-Assisted Oxygenations

Lipoxygenase and cyclooxygenase catalyze the controlled peroxidation of their fatty acid substrates to give hydroperoxides and/or endoperoxides, which are regiospecific and stereospecific and have important biological functions. Lipoxygenase refers to a group of enzymes that convert fatty acids with a (1Z,4Z)-pentadiene structure to hydroperoxides, and occurs in animals (e.g., platelets and leukocytes), plants (e.g., soybeans and tomatoes), and microorganisms [4,5]. Cyclooxygenase is a well-known prostaglandin-synthesizing enzyme that catalyzes the conversion of unsaturated fatty acids with 20 carbon atoms, e.g., arachidonic acid, to prostaglandin Hs via prostaglandin Gs [6]. Prostaglandin G is a hydroperoxyendoperoxide (Figure 25.2) and prostaglandin H is a hydroxyendoperoxide. The mechanism of these enzymatic oxygenations is thought to be an autoxidation mechanism.

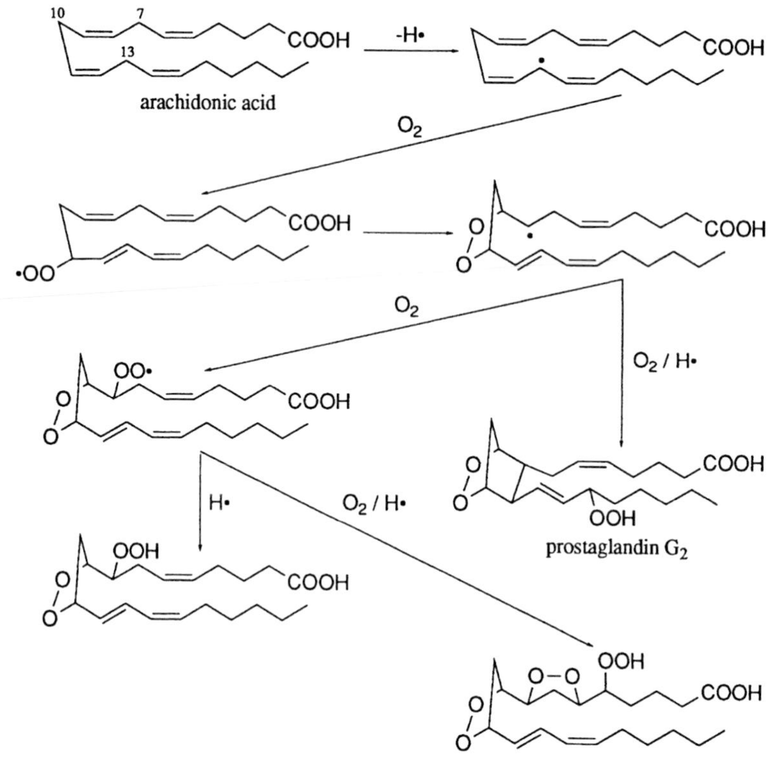

FIGURE 25.2
The mechanism of peroxyl radical cyclization in the autoxidation of arachidonic acid.

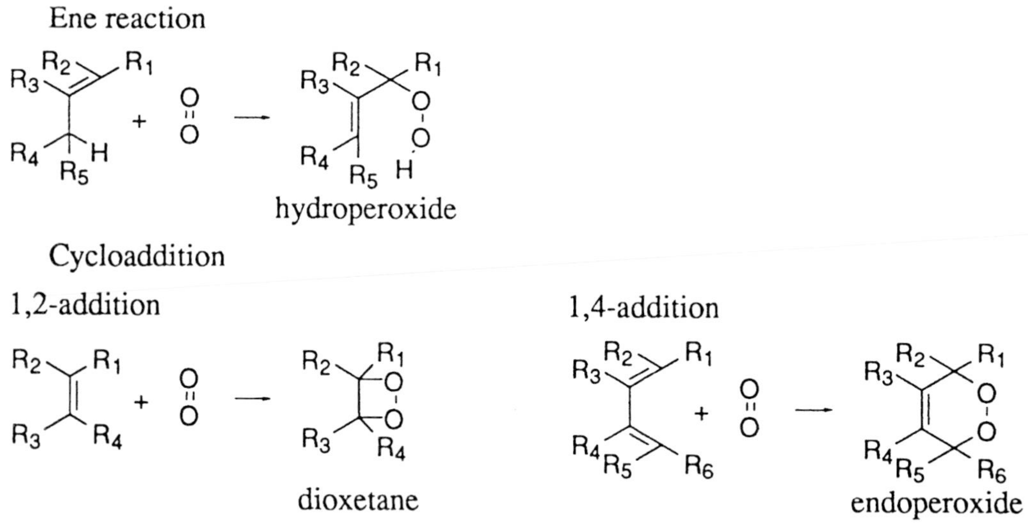

FIGURE 25.3
The mechanism of singlet oxygen oxygenation.

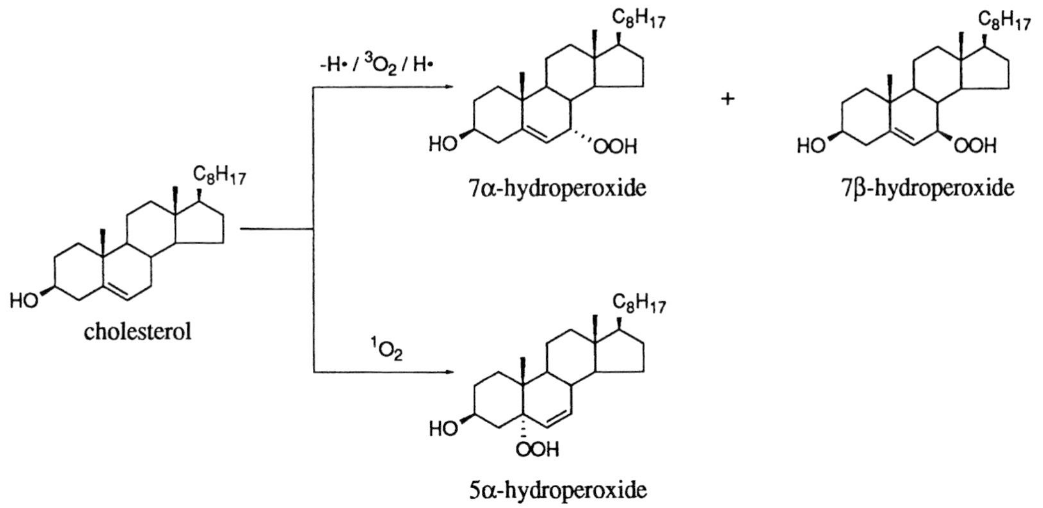

FIGURE 25.4
Products from the autoxidation and singlet oxygen oxygenation of cholesterol.

Lipid peroxidation can be enhanced by the xanthine–xanthine oxidase system which generates the superoxide radical (O_2^-). Usually, however, the superoxide radical acts to reduce iron(III) to iron(II), which stimulates peroxidation through the formation of the hydroxyl radical from the superoxide-dependent Fenton reaction, the so-called metal-catalyzed Haber-Weiss reaction, as shown below [7].

$$O_2^- + Fe^{3+} \rightarrow O_2 + Fe^{2+}$$

$$Fe^{3+} + H_2O_2 \rightarrow Fe^{3+} + HO^- + HO\bullet$$

Microsome fractions from animal tissues undergo lipid peroxidation in the presence of NADPH and Fe(III) chelated with ADP, EDTA, or some chelator. In microsomes, NADPH-cytochrome P-450 reductase, as well as reduced cytochrome P-450, is expected to reduce Fe(III) to Fe(II) [7].

The above lipid peroxidations cannot proceed without xanthine oxidase, NADPH-cytochrome P-450 reductase, or cytochrome P-450. These enzymes, however, act only to reduce iron(III) to iron(II), not to directly peroxidize unsaturated lipids. Thus, these lipid peroxidations are enzyme-assisted oxygenations.

25.2.2 Isomerization of Lipid Peroxides

Each hydroperoxy group of unsaturated lipid hydroperoxides migrates from the bonding carbon atom to another carbon atom in solution in the presence of molecular oxygen. For example, an isomer of linoleic acid hydroperoxide is isomerized to other isomers [8]. This isomerization is considered to proceed by a radical mechanism, because it is inhibited by anitoxidants and because the oxygen atoms of the hydroperoxy groups are exchangeable for the oxygen atoms of atmospheric molecular oxygen. Under aerobic conditions, any of (10E,12E)-9-, (10E,12Z)-9-, (9E,11E)-13-, and (9Z,11E)-13-hydroperoxy-octadecadienoic acids can be isomerized to give all of these isomers. Under argon, however, the (*E,E*)-isomers cannot be isomerized to the (*E,Z*)-isomers. Further, the isomerization is solvent-dependent. The isomerization rate is fast in benzene and very slow in di-*n*-propyl ether.

25.2.3 Decomposition of Lipid Peroxides

25.2.3.1 *Chemical Decomposition*

Peroxides are rather labile compounds, because the dissociation energy of the oxygen–oxygen bonds in their peroxy groups is exceptionally low: i.e., activation energies of 130 to 170 kJ/mol. Being practically stable without any interaction at room or body temperatures (e.g., *in vacuo* at 37°C), they decompose easily at high temperatures, at high concentrations, and/or in the presence of metal ions. For example, the half life of benzoyl peroxide is about 1 hour in an inert solvent at 95°C.

On pyrolysis, unsaturated fatty acid hydroperoxides give rise to complex decomposition products, including hydrocarbons, aldehydes, epoxides, caboxylic acids, dimers, polymers, and so forth [9]. The formation of hydrocarbons and aldehydes is explained on the basis of the mechanisms shown in Figure 25.5.

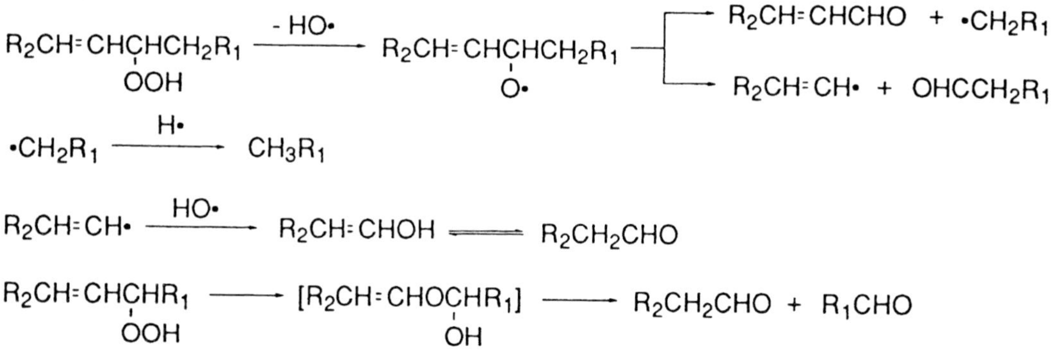

$$R_2CH=CHCHCH_2R_1 \quad \overset{-HO\bullet}{\longrightarrow} \quad R_2CH=CHCHCH_2R_1 \quad \begin{cases} R_2CH=CHCHO + \bullet CH_2R_1 \\ R_2CH=CH\bullet + OHCCH_2R_1 \end{cases}$$
$$\underset{OOH}{} \qquad \qquad \underset{O\bullet}{}$$

$$\bullet CH_2R_1 \quad \overset{H\bullet}{\longrightarrow} \quad CH_3R_1$$

$$R_2CH=CH\bullet \quad \overset{HO\bullet}{\longrightarrow} \quad R_2CH=CHOH \quad \rightleftharpoons \quad R_2CH_2CHO$$

$$R_2CH=CHCHR_1 \quad \longrightarrow \quad [R_2CH=CHOCHR_1] \quad \longrightarrow \quad R_2CH_2CHO + R_1CHO$$
$$\underset{OOH}{} \qquad \qquad \underset{OH}{}$$

FIGURE 25.5
The mechanism of pyrolysis of lipid peroxides.

On autoxidation, linoleic acid yields its 9- and 13-hydroperoxides, which decompose to hydrocarbons, aldehydes, and others. Figure 25.6 shows the high performance liquid chromatogram of the 2,4-dinitrophenylhydrazone derivatives of aldehydes from the autoxidation of linoleic acid hydroperoxides.

Peroxides also decompose under the catalysis of acids or bases. Acids cause the cleavage of oxygen–oxygen and oxygen–carbon bonds in hydroperoxides as shown in Figure 25.7. As will be discussed in Section 25.6.2.1., one of the indices of lipid peroxidation is the thiobarbituric acid (TBA) value. This value is thought to be related mainly to the amount of malondialdehyde (MDA), which is a rearranged product from lipid hydroperoxide. A possible mechanism for the formation of MDA is shown in Figure 25.8 [10].

Lipid peroxides decompose rapidly in the presence of free and chelated transition metal ions, including iron and copper ions, to give a variety of decomposition products [7]. For example, an iron(II) complex can react with lipid hydroperoxide (ROOH), as well as hydrogen peroxide. This causes the cleavage of an oxygen-oxygen bond to form the alkoxyl radical (RO•). The presence of a reducing agent, e.g., ascorbic acid, often enhances lipid peroxidation, because it regenerates the resulting iron(III) to iron(II).

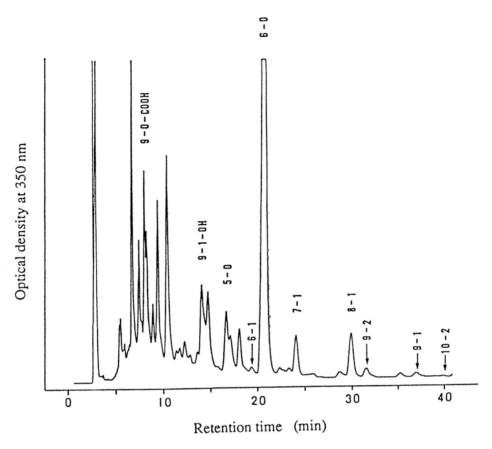

FIGURE 25.6

A high performance liquid chromatogram of the 2,4-dinitrophenylhydrazone derivatives of aldehydes from linoleic acid hydroperoxide. **Abbreviations** in the figure indicate the following aldehydes: 5-0, pentanal; 6-0 hexanal; 9-0-COOH, 9-oxononanoic acid; 6-1, 2-hexenal; 7-1, 2-heptenal; 8-1, 2-octenal; 9-1, 2-nonenal; 9-1-OH, 4-hydroxy-2-nonenal; 9-2, 2,4-nonadienal; 10-2 2,4-decadienal. The HPLC conditions were as follows: column, stainless steel, i.d. 4.6 mm, l. 25 cm; statonary phase, Nuclosil-ODS-5 (particle size 5μm); mobile phase, aceto-nitrile/water (70 : 30 to 80 : 20 v/v); gradient, 0.75%/min; flow rate, 3.5 ml/mim; column temperature, ambient; monitoring, optical density at 350 nm.

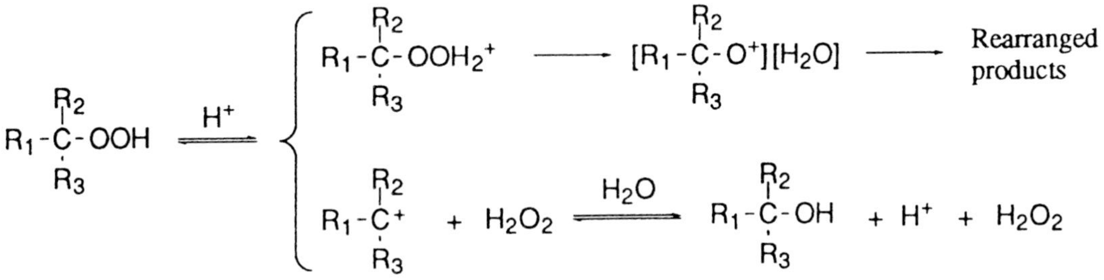

FIGURE 25.7

The mechanism of acid-catalyzed decomposition of lipid peroxides.

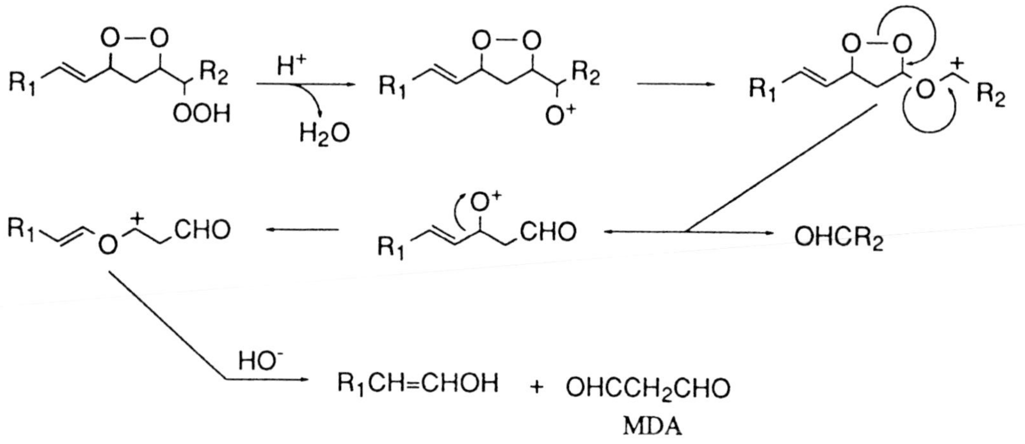

FIGURE 25.8
A possible mechanism for the formation of malondialdehyde.

$$ROOH + Fe^{2+} \text{ complex} \rightarrow RO\bullet + HO^- + Fe^{3+} \text{ complex}$$

In addition, an iron(III) complex can oxidize lipid hydroperoxide to the peroxyl radical ($ROO\bullet$).

$$ROOH + Fe^{3+} \text{ complex} \rightarrow ROO\bullet + H^+ + Fe^{2+} \text{ complex}$$

This means that the presence of iron ions enables the formation of both the alkoxyl and peroxyl radicals, although the reactions of iron(III) with hydroperoxides are much slower than those of iron(II).

Hydroperoxides are reduced by the following reducing agents to yield the corresponding alcohols: iodide, tin(II), sodium sulfite, sodium sulfide, phosphine, lithium aluminum hydride, sodium borohydride, hydrogen/metal catalyst, sodium alcoholate, zinc/acetic acid, aluminum isopropoxide, leuco methylene blue, hydrazine, amine, etc. [11]. As will be discussed in Section 25.5.3.1, the peroxide value, one of the indices of lipid peroxidation, is determined by the quantification of iodine formed from the reaction of iodide with peroxide. Triphenyl phosphine and sodium borohydride are recommended as reducing agents for the quantitative reduction of hydroperoxides to the corresponding alcohols.

25.2.3.2 Enzymatic Decomposition

Lipid peroxides decompose in the presence of some enzymes. Lipoxygenase catalyzes not only the peroxidation of unsaturated fatty acids, but also the decomposition of their hydroperoxides [12]. In the presence of soybean lipoxygenase I, linoleic acid hydroperoxide and arachidonic acid hydroperoxide are converted to the corresponding hydroxyepoxides and lipoxins, respectively.

Hydroperoxides are reduced by glutathione peroxidase or glutathione S-transferase to the corresponding alcohols as follows [13].

$$ROOH + 2\,GSH \rightarrow ROH + H_2O + GSSG$$

In this reaction, reduced glutathione (GSH) acts as a reducing agent, and is oxidized to oxidized glutathione (GSSG). In addition, hydrogen peroxide is reduced to water by glutathione peroxidase in the presence of GSH.

In plants, there is an isomerase that converts linoleic acid hydroperoxide to the corresponding α- or γ-ketoalcohol, and a lyase that converts the hydroperoxide to vinyl ether and aldehyde. [12] The odoriferous components of green leaves are (Z)-3-hexenol and (E)-2-hexenal, which may be derived from (Z)-3-hexenal, a decomposition product of linolenic acid hydroperoxide.

25.3 Accelerating and Retarding Factors for Lipid Peroxidation

25.3.1 Peroxidizability of Unsaturated Lipids

Substrates of lipid peroxidation are unsaturated lipids and molecular oxygen. Under anaerobic conditions, lipid peroxidation cannot occur. The rate-determining step of autoxidation is hydrogen abstraction from unsaturated lipids by the lipid peroxyl radical during propagation (eq. 25.3): i.e., the peroxyl radical is the chain carrier of lipid peroxidation. The rate of hydrogen abstraction is intimately related to the structure of unsaturated lipids. Practically, only hydrogen atoms at the bisallylic methylene groups are abstracted. Thus, the peroxidizability of unsaturated lipids depends on the number of the hydrogen atoms in the lipid molecule, although it is theoretically defined as the ratio $[k_p/(2k_t)^{1/2}]$ of the propagation rate constant (k_p) to the square root of double the termination rate constant (k_t). It has been found that on autoxidation of methyl oleate, methyl linoleate, and methyl linolenate at 30°C, the rate constants of hydrogen abstraction on an available hydrogen basis are 0.22, 31, and 59 $M^{-1}s^{-1}$, respectively, and the $k_p/(2k_t)^{1/2}s$ are 8.9×10^{-4}, 2.1×10^{-2}, and 3.9×10^{-2} $M^{-1/2}s^{-1/2}$, respectively [14].

25.3.2 Hydrogen Abstraction for the Initiation of Lipid Peroxidation

Hydrogen abstraction from a lipid molecule is necessary for the initiation of lipid peroxidation by autoxidation mechanisms, although it is unnecessary for singlet oxygen oxygenation. Usually, hydrogen abstraction is induced by radicals generated from the self decomposition of initiators, electron transfer of transition metals, photolysis, radiolysis, some enzyme reactions, and so forth. Initiators, such as 2,2'-azobis(2-amidinopropane) dihydrochloride and 2,2'-azobis(2,4-dimethylvaleronitrile), are agents that generate radicals from self-decomposition at ambient temperature. On the other hand, lipoxygenase and cyclooxygenase themselves appear to be able to abstract hydrogen atoms.

Active oxygen species, such as the superoxide radical, hydrogen peroxide, and the hydroxyl radical, are lipid peroxide-stimulating agents. Of course, the hydroxyl radical is a very reactive radical and acts as a hydrogen-abstracting agent. As shown before, the superoxide radical and hydrogen peroxide can be converted to the hydroxyl radical. Thus, antioxidant enzymes prevent the initiation of lipid peroxidation, since they scavenge the superoxide radical and hydrogen peroxide. Superoxide dismutase catalyzes the dismutation of the superoxide radical to molecular oxygen and hydrogen peroxide. Catalase catalyzes the dismutation of hydrogen peroxide to molecular oxygen and water. Further, glutathione peroxidase catalyzes the reduction of hydrogen peroxide and hydroperoxide to water and the corresponding alcohol, respectively.

25.3.3   Antioxidants

In biological systems, there are both antioxidants and reducing agents [15]. The former include α-tocopherol (vitamin E), β-carotene, ubiquinone, uric acid, and so forth. The latter include GSH, ascorbic acid, and so forth. Most are also radical scavengers and/or singlet oxygen quenchers. As shown above, GSH is very important as the substrate of glutathione peroxidase and glutathione S-transferase in the reduction of hydroperoxide and hydrogen peroxide. Thus, these agents may retard lipid peroxidation (see Section 25.6.1.2.2. and Figure 25.9).

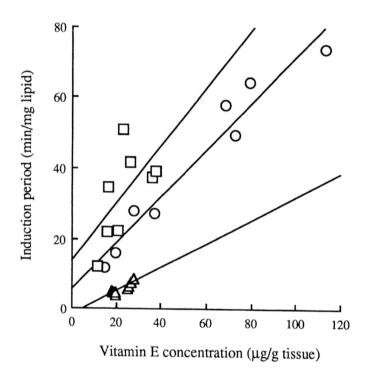

FIGURE 25.9

Relationship between the induction period of conjugated-diene formation and vitamin E concentration in tissues. The values for cerebrums, livers, and lungs are indicated by triangles, circles, and squares, respectively. (From Matsuo, M., Gomi, F., and Dooley, M. M., *Mech. Aging Dev.*, 64, 273, 1992. With permission.)

Typical lipid peroxidation-inhibiting antioxidants are the so-called chain breaking antioxidants, like α-tocopherol. If α-tocopherol (TOH) is added to a lipid peroxidation system (Eqs. 25.1–25.3), then the resulting lipidperoxyl radical (ROO•) abstracts a hydrogen atom from α-tocopherol to give lipid hydroperoxide (ROOH) and the α-tocopheroxyl radical (TO•). The α-tocopheroxyl radical does not abstract a hydrogen atom from any lipid molecule and is converted to nonradical products. Thus, the chain of lipid peroxidation is broken.

$$ROO\bullet + TOH \rightarrow ROOH + TO\bullet$$

$$ROO\bullet + TO\bullet \rightarrow \text{epoxyperoxide, nonradical products}$$

25.4 Preparation of Lipid Peroxides

Lipid peroxides are prepared nonenzymatically by both autoxidation and singlet oxygen oxygenation of unsaturated lipids and enzymatically by their lipoxygenase and cyclooxygenase reactions. Autoxidation is carried out by simple procedures and can be applied to the large scale preparation of lipid peroxides. However, it is accompanied by the formation of many secondary products and its rate is often slow. For singlet oxygen oxygenation, $^1O_2$ is commonly generated from $^3O_2$ by photosensitization as shown below:

$$^1Sens + h\nu \rightarrow {}^1Sens^* \rightarrow {}^3Sens^*$$

$$^3Sens^* + {}^3O_2 \rightarrow {}^1Sens + {}^1O_2$$

In the so-called Type II reaction, a triplet sensitizer ($^3Sens^*$) formed from a photochemically ($h\nu$) excited singlet sensitizer ($^1Sens^*$) interacts with $^3O_2$ to produce $^1O_2$. Dyes, including methylene blue, rose bengal, and protoporphylin, are used as sensitizers (^1Sens). Photosensitized oxygenation yields only hydroperoxides, but they are a mixture of nonconjugated and conjugated diene hydroperoxides. The reaction rate is relatively fast. Under appropriate conditions, the lipoxygenase reaction can give a specific hydroperoxide isomer in high yield and at a rapid rate, although the substrates are limited. It must be borne in mind that lipid hydroperoxides are labile at room temperature and their isomer composition depends on the preparative methods used.

Recently, chemoenzymatic methods for the regioselective, stereoselective synthesis of phospholipid hydroperoxides and triacylglycerol hydroperoxides has been developed [16, 17]. For example, by the combination of lipoxygenase and lipase reactions with organic reactions, phosphatidylcholines with a linoleic acid hydroperoxide residue or an arachidonic acid hydroperoxide residue can be prepared without contamination by regio- or stereo-chemical isomers. Such pure lipid hydroperoxides are very valuable because they have never been obtained by other methods.

25.4.1 Preparative Procedures

25.4.1.1. Autoxidation of Linoleic Acid

The autoxidation of linoleic acid is performed according to the method of Kaneko et al. [18]. Oxygen gas is bubbled at a flow rate of 100 ml/min in the dark at 40°C for 3 days into 0.1 M linoleic acid in benzene with a trace amount of cobalt(II) acetate. The reaction mixture is applied to a silica gel column and eluted stepwise with mixtures of n-hexane and diethyl ether (90 : 10, 80 : 20, 60 : 40, 40 : 60, 20 : 80, 0 : 100, v/v). Fractions are monitored by thin layer chromatography (TLC) on fluorescent indicator-containing Silica Gel 60 plates using a mixture of n-hexane, 2-propyl alcohol, and acetic acid (92 : 8 : 0.2, v/v/v) as the developing solvent. Appropriate fractions are combined and the solvent is evaporated under reduced pressure. The residue obtained is further fractionated by preparative high performance liquid chromatography (HPLC). The chromatographic conditions are as follows: column, stainless steel, i.d. 20 mm, l. 250 mm; stationary phase, Pegasil silica (particle size 5 μm); eluent, a mixture of *n*-hexane, ethanol, and acetic acid (97.5 : 2.45 : 0.05, v/v); flow rate, 10 ml/min; column temperature, anbient; monitoring, optical density at 234 nm.

25.4.1.2 Photosensitized Oxygenation of Methyl Linoleate

The photosensitized oxygenation of methyl linoleate is performed according to the method of Frankel et al. [19]. Oxygen gas is bubbled at 0°C into a methylene chloride solution of peroxide-free methyl linoleate in the presence of 10 mg methylene blue per 1 g methyl linoleate in an open tube exposed to a 1000-W, air-cooled tungsten light source through a 25-mm layer of water for the filtration of infrared radiation. The reaction is followed by TLC. The reaction products are separated on a silicic acid column using a mixture of diethyl ether and *n*-hexane as the eluent. Selected fractions are further separated by HPLC under the following conditions: column, stainless steel, i.d. 9.4 mm, l. 500 mm; stationary phase, Particil (particle size 10 μm); eluent, a mixture of *n*-hexane, methylene chloride, and ethyl acetate (6 : 4 : 1, v/v/v); flow rate, 1 ml/min. The purified products are characterized spectroscopically (see below).

25.4.1.3 Lipoxygenase-Catalyzed Oxygenation of Linoleic Acid

The lipoxygenase-catalyzed oxygenation of linoleic acid is performed according to the method of Kaneko and Matsuo [8]. A mixture of 2.8 g linoleic acid (10 mmol) and 10.5 mg soybean lipoxygenase I (1.53×10^5 units/mg) is incubated in 400 ml of 0.1 M ammonium chloride-ammonium hydroxide buffer, pH 9.2, at 5°C for 6 hours. Oxygen gas is bubbled through the solution during incubation. After acidification to pH 2–3 with 2N hydrochloric acid, the reaction mixture is extracted twice with diethyl ether. The ether layer is washed with water to be neutralized, and dried over anhydrous sodium sulfate. The solvent is evaporated under reduced pressure, and the residue obtained is chromatographed on a SilicAR CC-7 column by means of stepwise elution with 10, 20, and 30% diethyl ether in *n*-hexane to give crude (9Z,11E)-13-hydroperoxy-9,11-octadecadienoic acid. The crude hydroperoxide is purified by preparative HPLC as described in Section 25.4.1.1.

25.4.1.4 Chemoenzymatic Synthesis of Phosphatidylcholine Hydroperoxide

Hydroperoxylinoleoylphosphatidylcholine, 1-stearoyl-2-[13'-(S)-hydroperoxy-(9'Z,11'E)-octadecadienoyl]-*sn*-phosphatidylcholine (PCLOOH), is synthesized regioselectively and stereoselectively by chemoenzymatic methods [16]. Optically active 1-stearoyl-2-O-benzyl-*sn*-glycerol is prepared via the lipase-catalyzed enantioselective stearoylation of 2-O-benzylglycerol. The stearoylbenzylglycerol obtained is converted to 1-stearoyl-*sn*-phosphatidylcholine by treatment with phosphorus oxychloride and choline *p*-toluenesulphonate, followed by removal of the benzyl group. Using lipoxygenase, linoleic acid is converted to its hydroperoxide, 13-(S)-hydroperoxy-(9Z,11E)-octadecadienoic acid, the hydroperoxyl group of which is protected by peracetalization. The phosphatidylcholine is esterized with the perketal of linoleic acid hydroperoxide in the presence of dicyclohexylcarbodiimide and dimethylaminopyridine to give the perketal of PCLOOH. Deprotection of the perketal in a mixture of acetic acid, tetrahydrofuran, and water yields PCLOOH, which is purified by reverse-phase column chromatography. Its yield is 10% based on linoleic acid.

25.4.2 Separation of Lipid Peroxides from Biological Material

From tissue homogenates, total lipids are extracted with a mixture of chloroform and methanol (2 : 1, v/v) containing butylated hydroxytoluene (BHT), an antioxidant. After successive washing with the same solvent, the chloroform layers are combined. The chloroform extract is dried over anhydrous sodium sulfate, and the solvent is evaporated under reduced pressure or a nitrogen stream. The total lipid content is gravimetrically determined. Lipid peroxides are separated by chromatography, mainly HPLC, as described in Section 25.4.1.

25.4.3 Purity Assay

Since lipid peroxides are rather labile compounds, the purity of purified samples as well as crude samples should be assayed before use. Purity is determined by TLC, HPLC, iodometry, and ultraviolet (UV) spectroscopy. Impure lipid peroxide is expected to give more than one spot or a tailing spot by TLC and also more than one peak by HPLC. Its peroxide value may be less than 95% of the theoretical value (see Section 25.5.3.1). When lipid peroxide is prepared by autoxidation, purity can be estimated on the basis of UV absorption due to conjugated dienes and their molecular extinction coefficients (see Section 25.5.1.1).

25.4.4 Preservation

Generally speaking, lipid peroxides should be treated under anaerobic conditions at the lowest temperature possible. In addition, exposure of lipid peroxides to transition metal ions must be avoided. Lipid peroxides should be used within several days when dissolved in purified organic solvents such as *n*-hexane and ethanol and stored at temperatures below −80°C. As discussed in Section 25.2.2, linoleic acid hydroperoxides are rapidly isomerized in nonpolar solvents. Di-*n*-propyl ether is one of the best solvents for the storage of isomer-free hydroperoxides. For long-term storage, it is recommended that purified lipid peroxides be placed *in vacuo* or under argon-saturated solvent in an ampule at −196°C.

25.5 Analysis of Lipid Peroxides

Analyses of lipid peroxides are methodologically classified into spectral, chemical, and enzymatic analyses. In spectral analysis, pure lipid peroxides are directly quantified and their structures are determined. However, the sensitivity of spectral analysis is often insufficient for the measurement of lipid peroxides in biological material. By chemical analysis, pure lipid peroxides are quantified, and, in addition, the overall level of mixed lipid peroxides can be determined. Chemical analysis is a kind of indirect assay that depends on either oxidizing ability or the formation of decomposition products, so is not specific for each lipid peroxide. Enzymatic analysis permits the sensitive, specific assay of lipid peroxides. However, the cyclooxygenase-mediated assay discussed in Section 25.5.3.3 is specific for the peroxy groups of peroxides and not specific for each lipid peroxide itself, because it is based on the initiation rate of lipid peroxidation proportional to the number of the peroxy groups.

25.5.1   Spectral Analysis

25.5.1.1   *Ultraviolet Spectroscopy*

Ultraviolet Absorption

Conjugated dienes.   The autoxidation of unsaturated lipids with a (1Z,4Z)-pentadiene structure is accompanied by the formation of a conjugated diene structure (Figures 25.1 and 25.2), which absorbs UV light in a wavelength range of 233 ~ 236 nm (in ethanol) with molecular extinction coefficients (ε) of about $3 \times 10^4 \ M^{-1}cm^{-1}$ due to the π to π^* transition. This absorbance is used as an index of lipid peroxidation as described in Section 25.6.1.2. By the time course measurement of absorption, lipid peroxidation, in particular the early stage, can be quantitatively followed.

Detection limit.   Unsaturated lipid hydroperoxides can be analyzed with a detection limit of about 1.5 nmol/mg lipids.

Carbonyl group effect.   When the terminal carbon atom of a conjugated diene is bonded to a carbonyl group, the maximum absorption shifts to a longer wavelength of ~278 nm (in ethanol).

Conjugated polyenes.   As a chain of conjugated polyenes is lengthened, the wavelength of absorption becomes progressively greater and the molecular extinction coefficient also increases. For example, the absorption wavelengths of conjugated trienes and tetraenes are 268 and 315 nm, respectively.

Disadvantages

Nonconjugated dienes.   Nonconjugated dienes does not absorb UV light. Thus, unsaturated lipid hydroperoxides formed by singlet oxygen oxygenation can not be analyzed, because singlet oxygen attacks (1Z,4Z)-pentadiene structures to form both nonconjugated and conjugated dienes [20].

End absorption.   The measurement of conjugated dienes in peroxidized lipids is obstructed by the so-called end absorption, because unsaturated lipids themselves absorb UV light at slightly lower wavelengths than conjugated dienes. A difference spectrum between control and peroxidized samples gives a reasonable absorbance of the conjugated dienes. The absorption differences are slight, when only small amounts are to be measured. Presumably, this may be overcome by using second derivative spectra in the wavelength range [21]. In a second derivative spectrum, the absorption band of conjugated dienes appears as a sharp absorption minimum and is clearly separated from the end absorption.

Carbonyl compounds.   It should be noted that the decomposition of lipid peroxides gives several carbonyl compounds with UV absorption.

Biological materials.   It is difficult, further, for the measurement to be used directly for biological materials, because many of other biological compounds, such as proteins and nucleic acids, absorb strongly in the same wavelength range and their cooccurence produces a high background that interferes with the measurement of conjugated dienes. Thus, the fractionation of biological materials and/or the extraction of unsaturated lipids is necessary for measurement.

25.5.1.2 *Nuclear Magnetic Resonance Spectroscopy*

^{13}C Nuclear magnetic resonance spectroscopy. Nuclear magnetic resonance spectroscopy (NMR), in particular ^{1}H and ^{13}C NMR, is one of the most sophisticated techniques for the structure determination of organic compounds. The regio- and stereo-chemical isomers of unsaturated fatty acid hydroperoxides are analyzed by ^{13}C NMR. [22]

Analysis of methyl linoleate hydroperoxides. A mixture of methyl linoleate hydroperoxides from autoxidation is reduced with sodium borohydride to the corresponding dienols, because complication of the NMR spectra by the decomposition of hydroperoxides should be avoided. The dienols are separated from unreacted methyl linoleate by preparative TLC on a silica gel plate and further purified by preparative HPLC on a silica gel column. The assignment of ^{13}C NMR signals in their two (*E,E*)-isomers and two (*E,Z*)-isomers obtained can be made on the basis of both the effect of lanthanide shift reagents on these signals and the NMR data of other unsaturated fatty acids. This means methyl linoleate hydroperoxides are also composed of their two (*E,E*)-isomers and two (*E,Z*)-isomers. Further, the ^{13}C NMR spectrum of a reaction mixture from the autoxidation of methyl linoleate shows 16 olefinic carbon signals due to the conjugated diene structures of these four isomers. The ratio between the isomers is estimated from the integrals of intensities of these signals.

25.5.1.3 *Electron Spin Resonance Spectroscopy*

Detection of free radicals. Electron spin resonance (ESR) spectroscopy, which is also referred to as electron paramagnetic resonance (EPR) spectroscopy, is a technique by which the presence of unpaired electrons, namely free radicals, can be detected. Lipid peroxides cannot be directly detected by this technique. Since lipid peroxidation is a free radical chain-reaction, however, it is expected to be followed by ESR spectroscopy.

Spin-trapping. Usually, reactive radicals are difficult to observe by ordinary ESR spectroscopy because of their short lifetime. Spin-trapping methods have been developed for the detection of reactive radicals. Additions of radicals to spin traps, such as nitroso compounds and nitrons, often produce nitoxide radicals, so-called spin adducts, with long lifetimes. The spin-trapping method is a technique for the observation of long-lived spin adducts derived from short-lived radicals by ESR spectroscopy. Radicals formed in lipid peroxidation are detected by this method [23].

25.5.2 Chromatographic Analysis

25.5.2.1 *Thin Layer Chromatography*

Thin layer chromatography is one of the simplest methods for the detection of lipid peroxides.

Separation. Methyl linoleate hydroperoxide is separated on a silica gel plate developed with a mixture of *n*-hexane, diethyl ether, and acetic acid [24]. Phospholipid hydroperoxides are also separated from unreacted phospholipids by reverse phase TLC [25].

Detection.   The hydroperoxides are detected either under UV light irradiation or by the spraying of 5% potassium iodide in ethyl cellosolve. On heating of the silica gel plate sprayed with the detection reagent, hydroperoxides are colored yellow [24].

25.5.2.2 *High Performance Liquid Chromatography*

At present, HPLC is the best method for the detection, separation, and purification of lipid peroxides. The hydroperoxides of free and esterfied unsaturated fatty acids, phospholipids, and triacylglycerols are analyzed. The amount and/or concentration of lipid peroxides measured by HPLC is used as an index of lipid peroxidation as described in Section 25.6.1.1.

> **Separation.**   In the early stages of the autoxidation of unsaturated lipids, primarily lipid hydroperoxides are formed and quantitatively accumulated. These are separated by HPLC. Although the (E,E)- and (E,Z)-conjugated diene isomers of methyl linoleate hydroperoxides are only incompletely separated, the corresponding alcohols are perfectly separated [26]. Free unsaturated fatty acid hydroperoxides and the corresponding alcohols are separated without tailing when they are eluted with a small amount of acetic acid.

> **Detection by ultraviolet absorption.**   Each hydroperoxide purified is quantified on the basis of the measurement of optical density at 235 nm arising from its conjugated diene group as described in Section 25.5.1.1. However, this method depends not on a hydroperoxy group but on a conjugated diene group, and hence it cannot be used for the analysis of lipid hydroperoxides with no conjugated diene group, such as oleic acid hydroperoxide or cholesterol hydroperoxide.

> **Detection by chemiluminescence.**   Detectors for HPLC based on chemiluminescence (CL) and electrolytic reduction, which are directly related to the reactivity of hydroperoxy groups, have been developed [27,28]. HPLC with CL detection (CL-HPLC) depends on light emission from the hydroperoxide-induced oxidation of luminol or isoluminol. The hydroperoxides of phosphatidylcholine and phosphatidylethanolamine are quantified with a post-column CL-HPLC system. When a mixture of luminol and cytochrome *c* is used as a hydroperoxide-specific chemiluminescence reagent, the detection limit of these hydroperoxides is 10 pmol.

25.5.2.3 *Gas Chromatography–Mass Spectroscopy*

Gas chromatography-mass spectroscopy (GC-MS) and mass spectroscopy (MS) alone are utilized for the analysis of lipid peroxides [29]. There are GC-MS data for hydroperoxides and the corresponding alcohols of unsaturated fatty acids, their methyl esters, and phospholipids. Because the sensitivity of MS is extremely high, GC-MS, and MS are useful for the analysis of small amounts of sample.

> **Derivatization.**   However, the hydroperoxides themselves cannot be directly analyzed by these methods, since they decompose in a mass spectrometer. Thus, their derivatization is necessary for analysis. The following derivatizations are used: the methylation of carboxyl groups with diazomethane, the reduction of hydroperoxy groups to hydroxy groups with sodium borohydride or triphenylphosphine, the catalytic hydrogenation of double bonds, and the trimethylsilylation (TMS) or *tert*-butyldimethylsilylation (TBDMS) of hydroxyl groups.

The TBDMS derivatives have several advantages in that they are more stable than TMS derivatives and do not undergo hydrolysis in the presence of water, and their molecular weights may be measured because their molecular ion peaks are likely to appear on the mass spectra [30]. Lipid peroxides with a hydroperoxyfatty acid moiety, such as phospholipid hydroperoxides and triacylglycerol hydroperoxides, are hydrolyzed to free hydroperoxyfatty acids, the derivatives of which are analyzed.

25.5.3 Chemical and Enzymatic Analyses

25.5.3.1 *Iodometry*

Iodometry is an old method for the measurement of lipid peroxides. Many iodometric modifications have been reported [31,32]. The values of lipid peroxides estimated by iodometry are referred to as the peroxide value (POV).

> **Triiodide method.** The triiodide method is based on the oxidation of the iodide ion (I^-) to iodine (I_2) by lipid peroxides and on the spectroscopic estimation of triiodide anions (I_3^-) to which the iodine ion is quantitatively converted in the presence of excess iodide ions. It is performed with continuous monitoring of optical density at 358 nm until the reaction is completed.
>
> **Detection limit.** The detection limit of hydroperoxides is approximately 0.2 ~ 1.0 nmol. This sensitivity is as good or better than that of most of hydroperoxide assays. Thus, the assay is useful for the detection of hydroperoxides in biological systems.
>
> **Disadvantages.** The major disadvantage, however, is that molecular oxygen interferes so that the assay should be performed under anaerobic conditions. Several experimental approaches to this problem have been reported. For example, hydroperoxides are anaerobically reduced by iodide ions, unreacted iodide ions are protected against further oxidation by complexation with cadmium ions, and then the optical density of the reaction mixture at 353 nm is measured [32]. The assay may be affected by factors that disturb the equilibrium among I^-, I_2, and I_3^-. It is necessary, further, to take into account that transition metals, including iron ions and copper ions, react with both hydroperoxides and iodine.

25.5.3.2 *Heme*

Heme and the heme group of proteins, such as cytochrome *c*, decompose lipid peroxide to give reactive species including the alkoxyl, peroxyl, and hydroxyl radicals. These reactive species react with a chemiluminescence agent, e.g., luminol or isoluminol, under appropriate conditions to cause light emission, i.e., chemiluminescence. As shown in Section 25.5.2.2, this has been applied to CL-HPLC.

> **Detection limit.** Lipid peroxide can be quantified by the measurement of chemiluminescence intensity with a detection limit in the order of picomoles.

25.5.3.3 *Cyclooxygenase*

Cyclooxygenase catalyzes the formation of prostaglandin Gs as described in Section 25.2.1.3. The initiation of this reaction requires a very small amount of lipid hydroperoxide or hydrogen peroxide. Thus, it can be utilized as a sensitive, specific assay for these peroxides [32]. Each cyclooxygenase-catalyzed reaction is followed by molecular oxygen uptake and the period of time (lag time) required for the achievement of an optimal reaction rate is determined. The lag time approximates a linear function of the amount of peroxide added.

Detection limit.   This assay is useful in the quantification of fatty acid hydroperoxide in amounts ranging from 50 ~ 200 pmol.

Disadvantage.   Cyclooxygenase, however, may be activated by substances other than hydroperoxides. The effect of nonhydroperoxide activators can be estimated by the measurement of molecular oxygen uptake in samples in which hydroperoxides have been reduced by glutathione peroxidase in the presence of GSH.

25.5.3.4 *Glutathione Peroxidase*

As discussed in Section 25.2.3.2, glutathione peroxidase catalyzes the reduction of hydroperoxide and hydrogen peroxide using GSH as reducing agent. In a system containing peroxide, glutathione peroxidase, GSH, glutathione reductase, and NADPH, the consumption rate of NADPH depends on the concentration of peroxide. Here glutathione reductase acts to regenerate GSH from GSSG at the expense of NADPH to $NADP^+$. After hydrogen peroxide has been removed through preincubation with catalase, hydroperoxide can be assayed by the monitoring of the decrease in the optical density of NADPH at 340 nm [31].

Detection limit.   The detection limit of hydroperoxide concentration is expected to be 3 μM.

25.6 Indices of Lipid Peroxidation

25.6.1   Indices of Lipid Peroxide Formation

25.6.1.1 *Chromatographic Separation and Quantification of Lipid Peroxides*

Principle.   As discussed in Section 25.5.2.2, lipid peroxides can be separated and quantified by CL-HPLC. Thus, the amount and/or concentration of lipid peroxides measured is used as an index of lipid peroxidation.

Advantages and disadvantages.   Individual lipid peroxides can be analyzed qualitatively and quantitatively. A rather expensive CL-HPLC system is needed and the procedures are somewhat troublesome.

25.6.1.1.1   *Experimental Procedures for the Measurement of Phosphatidylcholine Hydroperoxide in Tissue Homogenates*

The measurement of phosphatidylcholine (PCOOH) in tissue homogenates is performed according to the method of Miyazawa et al. [33].

Lipid Extraction

Total lipid is extracted from rat tissues. To 200 mg tissues, 2 ml of 0.15 M sodium chloride containing 0.002% butylated hydroxytoluene as an antioxidant is added, and the mixture is homogenized in a Potter-type glass-Teflon homogenizer under ice cooling. To the homogenates obtained, 5 ml of a mixture of chloroform and methanol (2 : 1, v/v) is added and mixed vigorously for 1 min. The resulting mixture is centrifuged at 3,000 rpm for 10 min. The chloroform layer is separated and dried over anhydrous sodium sulfate. The chloroform solution is concentrated in a rotary evaporator and dried under a nitrogen stream. The total lipid obtained is weighed and diluted with 200 µl of the mixture of chloroform and methanol. A 20 µl portion of the solution is assayed by CL-HPLC.

Chemiluminescence-High Performance Liquid Chromatography

Apparatus and chromatographic conditions. Phosphatidylcholine in tissue lipids is measured with a high performance liquid chromatograph equipped with a nomal phase column and both UV and CL detectors. The chromatographic conditions are as follows: column, stainless steel, i.d. 4.6 mm, l. 250 mm; stationary phase, JASCO Finepak SIL (particle size 5 µm); mobile phase, chloroform/methanol (1 : 9, v/v); flow rate, 1.1 ml/min; column temperature, 30°C; monitoring, optical density at 234 nm.

Chemiluminescence measurement. After it passes through the UV detector, the eluate is mixed with a CL reagent in a post-column mixing joint. The CL reagent is a mixture of 10 µg/ml cytochrome c and 1 µg/ml luminol in 50 mM borate buffer at pH 9.3. The flow rate of the CL reagent is 1.0 ml/min. The generated CL is measured with the CL detector. A calibration curve is made using authentic PCOOH prepared from the photosensitized oxygenation of egg yolk PCOOH. The concentration of hydroperoxide is expressed as pmol of hydroperoxy groups.

25.6.1.1.2 Experimental Data of Phosphatidylcholine Hydroperoxide in Tissue Homogenates

Brain and liver phosphatidylcholine hydroperoxide contents in young and old rats. Brain and liver PCOOH contents in male and female rats at 1 and 18 months of age were measured [34]. The PCOOH was detected as a single sharp peak by HPLC. The contents of brain PCOOH from young and old male rats were 131 ± 46 and 573 ± 144 pmol/g brain, respectively, and those from young and old female rats 163 ± 41 and 575 ± 73 pmol/g brain, respectively. The contents of liver PCOOH from young and old male rats were 274 ± 74 and 2558 ± 348 pmol/g liver, respectively, and those from young and old female rats 409 ± 26 and 1917 ± 432 pmol/g liver, respectively. Each value represents a mean value with the standard deviation. Brain and liver PCOOH contents increase significantly in both male and female rats with advancing age. No age dependency, however, is observed for heart and lung PCOOH contents.

25.6.1.2 Conjugated Diene Formation

Principle. The autoxidation of unsaturated lipids having a (1Z,4Z)-pentadiene structure gives conjugated diene hydroperoxides with UV absorptions in a wavelength range of 233 ~ 236 nm (see Sections 2.1.1. and 5.1.1.). This absorbance is used as an index of lipid peroxidation [35].

Advantages and disadvantages. The overall level of lipid peroxidation is measured. The procedures are simple and the running cost is very low. A UV spectrometer is needed. The disadventages of this method are disccussed in Section 25.5.1.1.

25.6.1.2.1 Experimental Procedures for the Measurement of Conjugated Diene Formation in Lipid Extracts from Tissue Homogenates

Total lipids are extracted from tissue homogenates according the method of Burton et al. [36] and the measurement of conjugated diene formation in lipid extracts from tissue homogenates is performed according to the method of Matsuo et al. [37].

Lipid extraction.   Two milliliters of ethanol and 1 ml of 80 mM sodium dodecyl sulfate are added to 1 ml of 10% tissue homogenates in a 15-ml centrifuge tube with a ground-glass stopper. The mixture is vigorously mixed in the stoppered tube with a vortex mixer for 1 min, and 2 ml of *n*-heptane is added. It is vigorously mixed again and then centrifuged at 3000 rpm for 5 min. After phase separation, the heptane layer is transferred to a 10-ml round-bottomed flask and the heptane is evaporated under reduced pressure. The residue is dried *in vacuo* and weighed.

Autoxidation.   The residue obtained is dissolved in *n*-heptane with or without 1.34 mM 2,2′-azobis(2,4-dimethylvaleronitrile) brought to 3 ml in a quartz cuvette with a ground-glass stopper. Autoxidation is carried out in the stoppered cuvette under air at 30°C for an appropriate period of time.

Measurment of conjugated dienes.   The reaction is followed by the monitoring of the optical density of the heptane solution at 233 nm. The induction period (lag time) and initial rate of conjugated-diene formation in lipid extracts are estimated on the basis of changes in optical density.

25.6.1.2.2 Experimental Data of Conjugated Diene Formation in Lipid Extracts from Tissue Homogenates

Table 25.1 shows the induction periods and rates of initiator-induced conjugated-diene formation in lipid extracts from tissue homogenates of normal and vitamin E-deficient rats at different ages [37].

Initiator-induced conjugated-diene formation.   No conjugated-diene formation in lipid extracts from cerebral, liver, or lung homogenates was observed, when the extracts were incubated in *n*-heptane without any initiator. On the other hand, the optical density of each heptane solution at 233 nm increased with an increase in incubation time after an induction period, when the extracts were incubated in *n*-heptane with a fat-soluble initiator 2,2′-azobis(2,4-dimethylvaleronitrile).

Induction period of conjugated diene formation vs. vitamin E concentration.   The length of the induction period of lipid extracts from liver and lung homogenates increased or tended to increase with advancing age. The age-related extension of the induction period is due to an increase in vitamin E concentration. The induction periods for lipid extracts from vitamin E-deficient cerebral, liver, or lung homogenates are shorter than those from the corresponding normal tissue homogenates. As shown in Figure 25.9, the length of the induction period is proportional to the vitamin E concentration [37,38].

TABLE 25.1

Induction Periods and Rates of Initiator-Induced Conjugated-Diene Formation in Lipid
Extracts from Tissue Homogenates

Tissues	Age (months)	Conjugated-Diene Formation[a]			
		Induction period (min/mg lipid)	Rate (nmol diene/min/mg lipid)	Induction period (min/mg lipid)	Rate (nmol diene/min/mg lipid)
		Normal vitamin E status		Deficient vitamin E status	
Cerebrum	8	6.4 ± 1.9	123 ± 37	4.1 ± 1.0	112 ± 37
	26	8.3 ± 2.5	79.7 ± 5.4	4.7 ± 0.7	112 ± 46
Liver	8	58.2 ± 5.4	120 ± 13	11.7 ± 0.6	144 ± 41
	26	64.4 ± 17.1	144 ± 39	28.1 ± 17.3	165 ± 71
Lung	8	39.2 ± 6.3	95.2 ± 51.4	12.0 ± 4.0	147 ± 47
	26	41.9 ± 15.4	96.5 ± 22.7	22.4 ± 0.9	92.7 ± 23

[a] Each value is the mean value for 3 rats with the standard deviation.

Note: Reactions were carried out in the presence of 1.35 mM 2,2'-azobis(2,4-dimethylvaleronitrile) under air at 30°C.

Data are partially modified from Matsuo, M., Gomi, F., and Dooley, M. M., *Mech. Aging Dev.*, 64, 273, 1992. With permission.

Conjugated-diene formation rate.   There are no age-related changes in the rates of conjugated-diene formation observed after the induction periods. The conjugated-diene formation rate of lipid extracts from liver and lung homogenates tends to be enhanced by vitamin E deficiency.

25.6.1.3   Oxygen Absorption

Principle.   As shown in Section 25.2.1.1, the propagation of lipid peroxidation includes an oxygen absorption step for the formation of the lipid peroxyl radical. Further, oxygen absorption may occur in the subsequent decomposition reactions of lipid peroxidation. Lipid peroxidation necessarily involves oxygen absorption. Thus, the rate of oxygen absorption can be used as an overall index of lipid peroxidation. It is estimated on the basis of the time course measurement of the concentration of molecular oxygen in gas or liquid phase.

Methods.   A standard method is the measurement of dissolved oxygen concentration by a Clark-type oxygen electrode. Since this will not be discussed here, its users should refer to the literature [39]. Another sophisticated modern method is the measurement of oxygen concentration in gas phase with a pressure transducer.

Advantages and disadvantages.   The overall level of lipid peroxidation is measured. The procedures are simple and the running cost is low. An apparatus equipped with a pressure transducer or a Clark-type oxygen electrode is needed.

*25.6.1.3.1 Experimental Procedures for the Measurement of Oxygen Absorption
          in Tissue Homogenates*

Apparatus.   The measurement of oxygen absorption in tissue homogenates is
carried out with a pressure transducer [37]. Pressure change in a sealed,
flat-bottomed 20 ~ 30 ml flask with a stirring bar is measured using a pressure-
sensing apparatus composed of a pressure transducer, a DC amplifier, and an
XY recorder. The rate of oxygen absorption is estimated from the pressure
change. The flask is airtightly connected with a tube of small diameter to the
pressure transducer, which is held watertight in a stainless steel holder. These
are laid under water in a water bath. The ambient temperature must be maintained constant, because the pressure change is very sensitive.

Measurement.   In the flask, 2.5% tissue homogenates, freshly prepared from 50
mg tissue and 0.25% Triton X-100, are placed with or without 200 mM 2,2'-
azobis(2-amidinopropane) dihydrochloride, a water-soluble initiator. The mixture is gently stirred at 37°C and the pressure change is measured.

Detection limit.   Detection limits for this method are less than 50 pmol oxy-
gen/flask and less than 1 pmol/min/mg tissue oxygen absorption.

Gas analysis.   Before and after measurement, the composition of the gas in the
flask is analyzed by gas chromatography under the following conditions: column, stainless steel, i.d. 6 mm, l. 2.1 m, WG-100; solid phase, modified molecular sieve; carrier gas, helium; flow rate, 35 ml/min; monitoring, thermal
conductivity. By gas analysis, it is confirmed that the reduction in the pressure
inside the sealed flask has resulted from oxygen absorption into the tissue
homogenates.

25.6.1.3.2 Experimental Data of Oxygen Absorption in Tissue Homogenates

Table 25.2 shows the rates and induction periods of oxygen absorption of tissue homogenates from young and old rats in the presence or absence of 2,2'-azobis(2-amidionopro-
pane) dihydrochloride [37].

Initiator-independent oxygen absorption.   When tissue homogenates were incubated in the absence of any initiator at 37°C under air, oxygen was somewhat
absorbed into liver homogenates, very slightly into cerebral homogenates, and
not at all into lung homogenates. This oxygen absorption took place without
any induction period. The oxygen absorption rates of liver homogenates decreased gradually and became very low after one hour of measurement. After
this, the low rates were maintained for at least a few hours. Since the initial
oxygen absorption decreased in a time-dependent manner and was reduced to
about one-third by the addition of 1 mM potassium cyanide to the homogenates,
some enzymes, such as oxygenases, are presumed to be responsible. The low
oxygen absorption rates of cerebral homogenates remained unchanged for a
couple of hours.

Initiator-dependent oxygen absorption.   When tissue homogenates are incubated
with the water-soluble initiator, the oxygen absorptions of cerebral, liver, and
lung homogenates are greatly enhanced. The induction period for cerebral homogenates is shorter than those for liver and lung homogenates. There are no
age-related changes in the oxygen absorption rate or induction period.

TABLE 25.2
Rates and Induction Periods of Oxygen Absorption of Tissue Homogenates in the Presence or
Absence of an Initiator

Tissues	Age (months)	Oxygen absorption rate[a,b] (nmol oxygen/min/g tissue)			Induction period[a] (min/g tissue)
		No. additives	1 mM KCN	200 mM AAPH[c]	200 mM AAPH[c]
Cerebrum	8	5.0 ± 11.0	0	935 ± 44	97 ± 18
	26	3.7 ± 4.5	0	904 ± 22	103 ± 15
Liver	8	117 ± 49	42 ± 47	755 ± 31	190 ± 17
	26	130 ± 48	42 ± 47	757 ± 50	208 ± 32
Lung	8	0		479 ± 64	230 ± 43
	26	0		392 ± 69	210 ± 37

a Each value is the mean for 3 rats with the standard deviation. Reactions were carried out under air at 37°C.
b Initial oxygen absorption rates were measured unless otherwise mentioned.
c 2,2'-Azobis(2-amidipopropane) dihydrochloride.

Data are partially modified from Matsuo, M., Gomi, F., and Dooley, M. M., *Mech. Aging Dev.*, 64, 273, 1992.
With permission.

25.6.2   Indices of Lipid Peroxide Decomposition

25.6.2.1   The Thiobarbituric Acid Value

Principle.   The thiobarbituric acid (TBA) value is an oldest and most frequently
used index of lipid peroxidation [40]. It is based on the optical density of the
pink color produced when a sample and TBA are heated under acidic conditions
at about 95°C. [41, 42] As shown in Section 25.2.3.1, unsaturated lipid hydroperoxides decompose on pyrolysis and acid-catalyzed decomposition to give aldehydes including malondialdehyde. Under the above reaction conditions,
malondialdehyde reacts with two molecules of TBA to form a colored condensation product as shown in Figure 25.10.

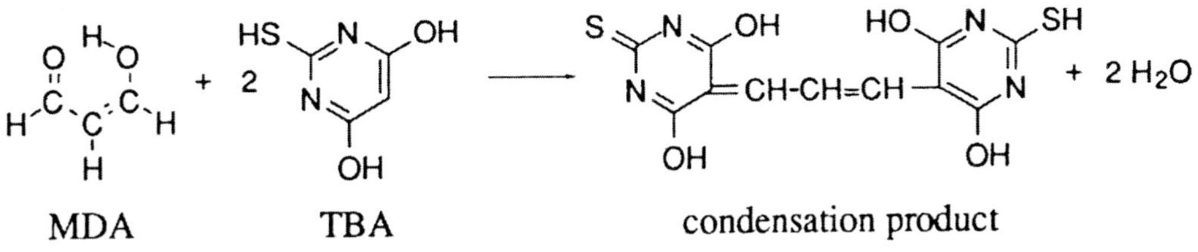

MDA   TBA   condensation product

FIGURE 25.10
The formation of the colored condensation product from malondialdehyde and thiobarbituric acid.

Usually, the condensation product is extracted with *n*-butanol, since it is readily extractable
into such organic solvents. The optical density of the *n*-butanol extract is measured at 535
nm (ε ca. 1.5×10^5 $M^{-1}cm^{-1}$). MDA prepared from the hydrolysis of its derivative, 1,1,3,3-
tetramethoxypropane or 1,1,3,3-tetraethoxypropane, immediately before use is used for

calibration because of the instability of MDA. Thus, the TBA value is expressed as MDA equivalent.

Preincubation of samples. The TBA value is estimated for unincubated 10% tissue homogenates (hereafter, referred to as TBAV$_u$) and for incubated 10% tissue homogenates (TBAV$_i$). For incubation, tissue homogenates are shaken at 37°C for 3 hours under air or an oxygen atmosphere. TBAV$_u$ is considered to reflect the results of the reaction of TBA with so-called TBA-reacting materials present initially in tissue homogenates and those that arise during the course of the procedures to measure TBA values. Thus, the difference between TBAV$_i$ and the corresponding TBAV$_u$ is expected to reflect, at least roughly, the formation of TBA-reacting materials during incubation.

Advantages. The TBA value is extremely sensitive. The procedures are simple and the running cost is very low.

Disadvantages. The value has considerable disadvantages as follows. It is based on the formation of secondary lipid peroxidation products including MAD and other aldehydes to form pink chromogens. There is the possibility that during the reaction, TBA-reacting materials may be produced not only from unsaturated lipids and their peroxides but also from other biological substances such as ribose, bile pigments, and glycoproteins.

Fluorometry. The pink chromogen for the TBA value can be fluorometrically measured at excitation and emission wavelenghts of 515 and 553 nm, respectively. In addition, as will be discussed in Section 25.6.2.3, the reaction of an amino compound with MDA gives a fluorescent substance. Using amino compounds such as 4,4-sulfonyldianiline and ethyl *p*-aminobenzoate, MDA concomitant with lipid peroxides can be fluorometrically measured as 1,4-dimethyl-1,4-dihydro-pyridine-3,5-dicarbaldehyde [43].

25.6.2.1.1   Experimental Procedures for the Measurement of the Thiobarbituric Acid Value of Tissue Homogenates

The TBA value is measured according to the method of Uchiyama and Mihara [44].

Reactions. In a 20 ml centrifuge tube with a screw cap, 0.2 ml of either fresh 10% homogenates or 10% homogenates shaken at 37°C for 3 hours under air is added to a mixture of 1 ml of 0.5% TBA and 3 ml of 1% phosphoric acid. The mixture is vigorously mixed in the screw-capped tube with a vortex mixer for 1 min, heated at about 95°C for 45 minutes, and then cooled to room temperature.

Chromogen extraction. After the addition of 4 ml of *n*-butanol, the mixture is vigorously mixed again and centrifuged at 3000 rpm for 10 min.

Measurement. The butanol layer is transferred to a glass cuvette and the optical densities are measured at 532 and 520 nm. The difference between the optical densities at 532 and 520 mn is used as the raw TBA value.

Calibration. The raw value is calibrated on the basis of a calibration curve made using hydrolyzed 1,1,3,3-tetramethoxypropane and the TBA value is expressed as MDA equivalent.

25.6.2.1.2   Experimental Data of the Thiobarbituric Acid Value of Tissue Homogenates

Table 25.3 shows the TBAV$_u$'s of tissue homogenates from young and old rats and the TBAV$_i$'s of tissue homogenates incubated under air [37].

TABLE 25.3

Thiobarbituric Acid Values of Tissue Homogenates

Tissues	Age (months)	TBA values[a] (nmol MDA[b] eq/g tissue)	
		Unincubated	Incubated[c]
Cerebrum	6	192 ± 17	665 ± 77
	30	170 ± 4	485 ± 47
Liver	6	175 ± 11	191 ± 20
	30	184 ± 31	177 ± 16
Lung	6	137 ± 10	227 ± 108
	30	136 ± 4	189 ± 34

[a] Each value is the mean for 3 rats with the standard deviation.
[b] Malondialdehyde equivalents.
[c] Before the measurement of TBA values, tissue homogenates were incubated under air at 37°C for 3 hours.

Data are partially modified from Matsuo, M., Gomi, F., and Dooley, M. M., *Mech. Aging Dev.*, 64, 273, 1992. With permission.

Tissue-dependent difference. Some tissue-dependent differences in the TBA values were found. The $TBAV_u$'s of lung homogenates were lower than those of cerebral and liver homogenates. The $TBAV_i$'s of cerebral homogenates were higher than those of liver and lung homogenates. The $TBAV_i$'s of liver and lung homogenates were similar to the $TBAV_u$'s of the corresponding tissue homogenates, whereas the $TBAV_i$'s of cerebral homogenates were more than three times as high as the $TBAV_u$'s.

Age-dependent decrease. No age-related changes in $TBAV_u$ were observed. Interestingly, the $TBAV_i$ of cerebral homogenates from young rats was higher than that from old rats. The $TBAV_i$'s of liver and lung homogenates from young rats also tended to be higher than those from old rats.

25.6.2.2 Exhaled Hydrocarbons

Principle. On pyrolysis and metal-induced decomposition, unsaturated lipid hydroperoxides give rise to hydrocarbons via the corresponding alkoxyl and carbon-centered radicals (see Section 25.2.3.1.). For example, linoleic acid is peroxidized to 13-hydroperoxy-9,11-octadecadienoic acid, which is converted to the 9,11-octadecadienic acid 13-oxyl radical. The β-scission of the radical yields the 1-pentyl radical and on hydrogen abstraction, *n*-pentane is formed. In fact, mammals, including humans, exhale such volatile hydrocarbons by breathing [45]. Further, the exhalation rates of hydrocarbons increase in animals treated with carbon tetrachloride, a lipid peroxidation-inducing chemical, and decrease in animals given vitamin E. The hydrocarbons can be quantified by gas chromatography.

Advantages. The overall level of *in vivo* lipid peroxidation is measured. The measurement of the exhalation rate of volatile hydrocarbons can be noninvasively and longitudinally made using whole animals.

Disadvantages. A complex hydrocarbon-collecting system needs to be assembled. When the volatile hydrocarbons are used as indices of *in vivo* lipid peroxidation, the following points should be borne in mind. They are minor products of lipid peroxidation, and their yields may be affected by several reaction factors, including the availability of transition metal ions and oxygen concentration. In addition, they may be produced by bacteria always present in the gut and on the skin. Nowadays, air is contaminated with hydrocarbons in the exhaust gases of cars and industrial plants, and this produces a high background of exhaled hydrocarbons, so that the air breathed by animals or human subjects must be purified. Good skill in training and handling rats and mice is necessary.

25.6.2.2.1 Experimental Procedures for the Measurement of Exhaled Hydrocarbons from Rats

Hydrocarbon-Collecting System

Figure 25.11 shows a diagram of a hydrocarbon-collecting system [46]. A rat is placed in a glass chamber compartmentalized into head and body portions with plastic and silicon rubber collars positioned around the neck. The apparatus is designed so that the breathing gas does not mix with intestinal gas in the chamber. For removal of hydrocarbon contaminants, purified air is repurified with a hydrocarbon-removing apparatus. In the repurified air, no *n*-pentane and only very small amounts of ethane are found. The repurified air is supplied into the head portion at a flow rate of 120 ml/min and into the body portion at a flow rate of 1000 ml/min, independently. This air flow system is an open system with a device preventing room air from flowing backward. Air pressure in the system is adjusted in such a way that a slight amount of air continues to leak out from the device.

Sampling

For removal of inhaled room air, the lungs of the rat are washed out for 10 minutes with repurified air into the head portion. After the air flow into the body portion is stopped, the

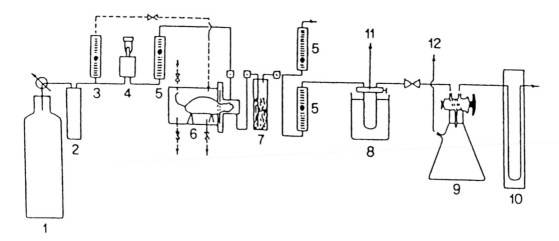

FIGURE 25.11
A hydrocarbon-collecting system:
Numbers indicate the following parts: 1. air cylinder, 2. hydrocarbon-removing apparatus, 3. flowmeter for air purge, 4. flow-controlling valve, 5. flowmeter, 6. glass chamber, 7. desiccator, 8. hydrocarbon trap with bath, 9. vacuum flask, 10. manometer, 11. gas chromatograph, 12. vacuum pump. (From Matsuo, M., Gomi, F., Kuramoto, K., and Sagai, M., *J. Gerontol.*, 48, B133, 1993. With permission.)

repurified air in the head portion carries exhaled hydrocarbons to a sampling loop trap with a six-way, nut-type valve for the opening and closing of the trap inlet and outlet. The outlet is connected to a vacuum flask. The trap is filled with activated alumina. The trap is immersed in a liquid nitrogen-ethanol bath at about −130°C. On being drawn from the outlet into the vacuum flask under a slightly reduced pressure, 300 ml of the air containing exhaled hydrocarbons is passed through the trap. After the hydrocarbons are collected in the trap, the outlet is closed. The liquid nitrogen-ethanol bath is taken off the trap. Immediately after a carrier gas of nitrogen is sent via the trap into a column of a gas chromatograph by the switching of the six-way valve, the trap is immersed in a hot water bath. The carrier gas carries the hydrocarbons released in the trap into the column.

Gas Chromatography

Gas chromatographic analysis is conducted on a gas chromatograph equipped with a six-way, gas-sample valve and a hydrogen-flame ionization detector.

> **Chromatographic conditions.** The conditions are as follows: column, stainless steel, i.d. 3 mm, l. 3.5 m; solid phase, 80-100 mesh activated alumina; carrier gas, nitrogen; flow rate, 40 ml/min; detector and injector temperatures, 260 and 160°C, respectively; column temperature, programmed from 60 to 280°C at a rate of 16°C/min for 13.8 min and then is held at 280°C for 6.2 min; monitoring, hydrogen-flame ionization. [45]

> **Calibration.** Using 1 ml of a standard gas containing 0.95 ppm ethane and 0.92 ppm n-pentane, peaks on gas chromatograms are assigned to these hydrocarbons and peak intensities are calibrated. The peak intensities are measured with a reporting integrator.

Calculation

According to the following equation, the exhalation rates of hydrocarbons are calculated and the exhalation rate of ethane is calibrated by means of the background value of ethane concentration in the repurified air (8.89 fmol/ml).

$$\left\{ \frac{[\text{an amount of the hydrocarbon trapped (pmol)}]}{[\text{a volume of air passed through a sampling loop trap (ml)}]} - B(\text{pmol/ml}) \right\}$$

$$\times \left\{ \frac{[\text{a flow rate of air into a chamber (ml/min)}]}{[\text{body weight (100 g)}]} \right\}$$

where B is 0.00889 for ethane and 0 for n-pentane.

25.6.2.2.2 Experimental Data of the Exhalation Rate of Hydrocarbons from Rats

> **Gas chromatogram.** Figure 25.12 shows a gas chromatogram of exhaled hydrocarbons from rats.

> **Effect of aging.** Figure 25.13 shows the exhalation rates of ethane and n-pentane from *ad libitum*-fed and food-restricted rats of different ages [46]. The exhalation rate of ethane or pentane from old, *ad libitum*-fed rats is higher than that of the corresponding hydrocarbon from young, *ad libitum*-fed rats.

Effect of food restriction.   For food-restricted rats, the collection of exhaled hydrocarbons was made on the days that they were fed. There was no difference between
the exhalation rates of each hydrocarbon from *ad libitum*-fed and food-restricted
rats aged 6 ~ 9 or 22 months. It was found, however, that the exhalation rate of
n-pentane from 28 ~ 30 month-old, food-restricted rats was significantly lower
than that from 28 ~ 30 month-old, *ad libitum*-fed rats, while the exhalation rate of
ethane from 28 ~ 30 month-old, food-restricted rats tended to be lower than that
from 28 ~ 30 month-old, *ad libitum*-fed rats. The exhalation rate of *n*-pentane from
the old, food-restricted rats is about one fourth of that from the old, *ad libitum*-fed
rats. In particular, it is noteworthy that the exhalation rate of *n*-pentane from old,
food-restricted rats is equivalent to that from young, *ad libitum*-fed rats.

25.6.2.3   Fluorescence

Fluorescent products from lipid peroxides.   It has been found that the decomposition of lipid peroxides forms fluorescent products. For example, MDA undergoes Schiff base formation with two molecules of amino compounds to give a
fluorescent aminoiminopropene derivative [47]. Further, three molecules of
MDA can be cyclized with one molecule of amino compounds to give fluorescent
1,4-disubstituted-1,4-dihydropyridine-3,5-dicarbaldehyde [48].

Lipofuscin.   Lipofuscin is a yellowish-brown, single membrane-bound granule in
cells that is known to accumulate in some tissues as a function of age [49]. It is
often referred to as age pigment and is considered to be a histological index of
aging in animal tissues. Its physicochemical characteristic is the emission of
golden yellow fluorescence, so-called autofluorescence, when excited with UV

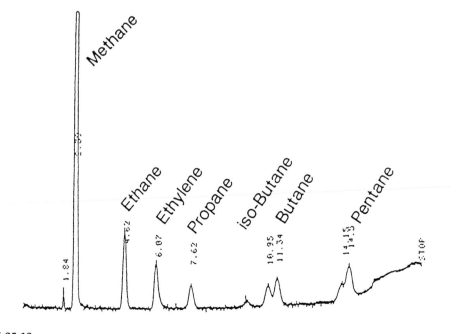

FIGURE 25.12
A gas chromatogram of exhaled hydrocarbons from an *ad libitum*-fed, 28-month-old rat. Numbers in the figure
are the retention times of exhaled hydrocarbons in minutes.

light. Its excitation and emission maxima are in the range of 360 ~ 395 nm and 430 ~ 460 nm, respectively.

Chloroform-soluble fluorescent substances. It has been reported, further, that tissue extracts prepared with a mixture of chloroform and methanol show fluorescence similar to that of lipofuscin, and that the concentration of fluorescent substances in tissues increases as a function of age. Both the fluorescent substances of lipofuscin and the chloroform-soluble fluorescent substances of tissues are expected to be related to the decomposition products of lipid peroxides, although they are not yet characterized chemically. Taken together, the fluorescence of tissue extracts is thought to be an index of *in vivo* lipid peroxidation.

Advantages and disadvantage. The procedures are simple and the running cost is low. A spectrofluorometer is needed. It must be borne in mind that the chemical basis of this method has not yet been established.

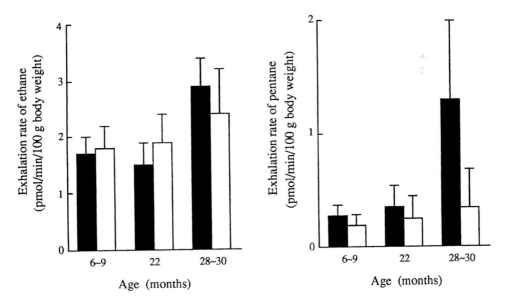

FIGURE 25.13
Exhalation rates of ethane and *n*-pentane from *ad libitum*-fed and food-restricted rats. Closed and open columns represent mean values of *ad libitum*-fed and food-restricted rats, respectively, with the standard deviations. (From Matsuo, M., Gomi, F., Kuramoto, K., and Sagai, M., *J. Gerontol.*, 48, B133, 1993. With permission.)

25.6.2.3.1 Experimental Procedures for the Measurement of Fluorescent Substances from Tissue Homogenates

Fluorescent substances are measured according to the method of Fletcher, Dillard, and Tappel [50].

Extraction of florescent substances. A Potter-type glass-Teflon homogenizer is used for the extraction of florescent substances. A 0.2 g sample of tissues is placed in the homogenizer, and 4 ml of a mixture of chloroform and methanol (2 : 1, v/v) is added. The mixture is homogenized for 1 min at 45°C, or for 1.5 min in the case of fibrous tissues such as the heart and skeletal muscle. The mixture is vigorously mixed on a vortex mixer, transferred to a centrifuge tube, and centrifuged at 3000 rpm for 5 min.

Measurement. A 1-ml aliquot of the chloroform layer is mixed with 0.1 ml of methanol in a quartz cuvette. Fluorescent intensity is measured at excitation and emission maxima of 365 and 435 nm, respectively, with a spectrofluorometer.

Calibration. For the calibration of fluorescence intensity and wavelength, 1 μg quinine bisulfate in 1 ml of 0.1 M sulfuric acid is used as a standard. The results are expressed as relative fluorescent units/g wet tissue.

Interference. Retinol is considered to be a fluorescent contaminant sometimes found in chloroform extracts of tissues. It can be removed by exposure of the extracts to high-intensity UV light for 30 sec.

25.6.2.3.2   *Experimental Data of Fluorescent Substances from Tissue Homogenates*

Fluorescent substances of testes and hearts in young and old mice.   For mice at 3, 10, and 23 months of age, relative fluorescence units in 0.2 g of testes were 29 ± 9, 81 ± 39, and 243 ± 10, respectively, and those of 0.2 g of hearts were 27 ± 9, 32 ± 9, and 241 ± 14, respectively, when the above quinine sulfate standard measured 1950 relative fluorescence units [51]. Fluorescent substances in chloroform-methanol extracts of mouse tissues increase as a function of age.

25.6.2.4   *Chemiluminescence*

Principle.   It has been observed that spontaneous photon emission (low-level chemiluminescence) arises from biological systems [52]. Usually, CL results from the decay of excited states in chemical reactions and is thought to be involved in free radical processes. On the other hand, CL is accompanied by tissue lipid peroxidation and depends on the presence of molecular oxygen. This seems to be reasonable, since molecular oxygen is required for the propagation of lipid peroxidation. In CL of tissues, light is found to be emitted at a wavelength range of 400 ~ 700 nm, which appears to be due to the formation of singlet oxygen and/or excited carbonyl species. Presumably, the CL is based on the decomposition of lipid peroxides, because the self-reaction of the peroxyl radicals, which can be derived from lipid peroxide, gives rise to singlet oxygen or excited carbonyl species by the Russell mechanism, and because the reaction of singlet oxygen with a double bond gives rise to dioxetane, as shown in Section 25.2.1.2, which decomposes to excited carbonyl species decaying with light emission.

Chemiluminescence enhancement.   CL is enhanced in rat liver microsomes with hydroperoxides and in livers of rats administered lipid peroxides orally, and correlates reasonably with MDA production in *tert*-butyl hydroperoxide-perfused rat livers.

Advantages and disadvantage.   The procedures are simple and the running cost is low. A relatively expensive CL measurement system is needed. The light emission mechanism of this method has not yet been established.

25.6.2.4.1   *Experimental Procedures for the Measurement of Low-Level Chemiluminescence from Fruit Flies and their Homogenates*

Low-level CL in fruit flies *Drosophila melanogester* is measured by the method of Sato et al. [53].

Apparatus. For CL measurement, a synchronous single photon counting apparatus equipped with a glass sample cell and a photomultiplier tube is used. The photomultiplier tube is kept at $-20°C$ for the reduction of background counts using a thermoelectronic cooler. CL spectra are recorded with a filter-type spectrometer.

Whole animals. Fruit flies are maintained in an environmentally controlled box in the dark at 60% humidity at 25 or 30°C for 5 days after emergence. Three hundred living flies are cooled to 4°C and placed in a glass sample cell 30 mm in diameter and 10 mm in height.

Homogenates. After another 100 flies are cooled to 4°C, homogenates are prepared in 4 ml ice-cold phosphate buffered saline using a Potter-type glass-Teflon homogenizer. The homogenates are placed in a glass sample cell 45 mm in diameter and 15 mm in height.

Measurement. CL spectra are obtained at 25 or 30°C for 20 min. The response range of these CL spectra is 430 ~ 830 nm. The CL intensity of the whole bodies and homogenates is expressed in terms of average counts per minute for a 20-minute measurement and corrected for background counts.

25.6.2.4.2   *Experimental Data of Low-Level Chemiluminescence from Fruit Flies and their Homogenates*

Low-level chemiluminescence from whole fruit flies and their homogenates. The CL intensity of 100 fruit flies at 30°C was 857 ± 23 cpm and 2.1-times greater than that at 25°C. [53] The CL intensity of homogenates from 100 fruit flies at 30°C was 2,906 ± 139 cpm and 1.7 times greater than that at 25°C. Each value represents a mean value with the standard error. Further, their oxygen consumption, mobility, and phosphatidylcholine hydroperoxide concentration increased at 30°C as compared with 25°C. Since their mean life span is 63 and 29 days at 25 and 30°C, respectively, it turns out that aging is accelerated at higher ambient temperatures. Thus, in aging-accelerated flies, the metabolic rate and lipid peroxidation may be enhanced.

25.7   Concluding Remarks

At present, many indices of lipid peroxidation are used in biomedical studies. In this chapter, only indices of lipid peroxidation that have so far been used in aging research are discussed. The indices fall into two main categories: one the index of the overall level of lipid peroxidation, and the other the indices of the formation of individual lipid peroxides. Except for the chromatographic quantification of lipid peroxides by CL-HPLC, all the indices discussed here reflect overall levels of lipid peroxidation. In biological systems, labile lipid peroxides are present in great variety and only very small amounts. Chromatography is the best technique by which these lipid peroxides can be separated from each other and quantitatively analyzed. Thus, the chromatographic quantification of lipid peroxides can be used as an index of the formation of individual lipid peroxides, although the resolution and sensitivity are still insufficient for the analysis of lipid peroxides in biological materials. It should be noted that each of the above indices means something different, although they are based on either the formation or decomposition of lipid peroxides. Whatever index is chosen, what it means and how related it is to lipid peroxidation must be understood. Whenever possible, two or more different indices should be used.

For aging research, cross-sectional studies have several considerable disadvantages. For example, the cross-sectional approach suffers from the following phenomena: average value changes in aging parameters from one age group to the next do not necessarily reflect the change that occurs in one individual with the passage of time, the effects of environmental changes can not be separated from the effects of age, and aging parameters are affected by the effects of selective mortality. In contrast, longitudinal studies are thought to be the best way. The exhalation rate of hydrocarbons from animals is of great value as an index of lipid peroxidation that can be longitudinally measured throughout the life-span, since the rate can be noninvasively measured without causing any damage to them. For longitudinal aging research, new noninvasive methods to measure lipid peroxidation need to be extensively developed.

Acknowledgment

The authors thank Dr. Margaret M. Dooley Ohto for her help in preparing this manuscript.

References

1. Gardner, H. W., Oxygen radical chemistry of polyunsaturated fatty acids, *Free Radical Biol. Med.*, 7, 65, 1989.
2. Porter, N. A., Chemistry of lipid peroxidation, *Methods Enzymol.*, 105, 274, 1984.
3. Wasserman, H. H. and Murray, R. W., Eds., *Singlet Oxygen*, Academic Press, New York, 1979.
4. Vliegenthart, J. F. G. and Veldink, G. A., Lipoxygenase, in *Free Radicals in Biology* V, Pryor, W. A. Ed., Academic Press, New York, 1982, 29.
5. Ingraham, L. L. and Meyer, D. L., *Biochemistry of Dioxygen*, Plenum, New York, 1985, pp. 211–219.
6. Gale, P. H. and Egan, R. W., Prostaglandin endoperoxide synthase-catalyzed oxidation reaction, in *Free Radicals in Biology* VI, Pryor, W. A. Ed., Academic Press, Orlando, 1984, 1.
7. Halliwell, B. and Gutteridge, J. M. C., Role of free radicals and catalytic metal ions in human disease: an overview, *Methods Enzymol.*, 186, 1, 1990.
8. Kaneko, T. and Matsuo, M., Isomerization of linoleic acid hydroperoxides under argon and under degassed conditions, *Chem. Pharm. Bull.*, 32, 332, 1984.
9. Frankel, E. N., Volatile lipid peroxidation products, *Prog. Lipid Res.*, 22, 1, 1982.
10. Frankel, E. N., and Neff, W. E., Formation of malonaldehyde from lipid oxidation products, *Biochim. Biophys. Acta*, 754, 264, 1983.
11. Hiatt, R., Hydroperoxides, in *Organic Peroxides* II, Swern, D., Ed., Wiley-Interscience, New York, 1971, pp. 49–52.
12. Gardner, H. W., Recent investigations into the lipoxygenase pathway of plants, *Biochim. Biophys. Acta*, 1084, 221, 1991.
13. Flohé, L., Glutathione peroxidase brought into focus, in *Free Radicals in Biology* V, Pryor, W. A. Ed., Academic Press, New York, 1982, 223.
14. Howard, J. A. and Ingold, K. U., Absolute rate constants for hydrocarbon autoxidation. VI. Alkyl aromatic and olefinic hydrocarbons, *Can. J. Chem.*, 45,793, 1967.
15. Niki, E., Antioxidants in relation to lipid peroxidation, *Chem. Phys. Lipids*, 44, 227, 1987.

16. Baba, N., Yoneda, K., Tahara, S., Iwasa, J., Kaneko, T., and Matsuo, M., A regioselective, stereoselective synthesis of a diacylglycerophosphocholine hydroperoxide by use of lipoxygenase and lipase, *J. Chem. Soc. Chem. Commun.*, 1281, 1990.
17. Baba, N., Hirota, N., Tahara, S., Nakajima, S., Iwasa, J., Kaneko, T., and Matsuo, M., Chemoenzymatic syntheses of triacylglyceride hydroperoxides, *Biosci. Biotech. Biochem.*, 56, 1694, 1992.
18. Kaneko, T., Nakano, S., Kaji, K., and Matsuo, M., Cytotoxicities of autoxidized polyunsaturated fatty acids toward cultured human umbilical vein endothelial cells, *Chem. Pharm. Bull.*, 36, 4203, 1988.
19. Frankel, E. N., Neff, W. E., Selke, E., and Weisleder, D., Photosensitized oxidation of methyl linoleate: Secondary and volatile thermal decomposition products, *Lipids*, 17, 11, 1982.
20. Terao, J. and Matsushita, S., Products formed by photosensitized oxidation of unsaturated fatty acid esters, *J. Am. Oil Chem. Soc.*, 54, 234, 1977.
21. Corongiu, F. P. and Milia, A., An improved and simple method for determining diene conjugation in autoxidized polyunsaturated fatty acids, *Chem.-Biol. Interactions*, 44, 289, 1983.
22. Frankel, E. N., Neff, W. E., and Weisleder, D., Determination of methyl linoleate hydroperoxides by ^{13}C nuclear magnetic resonance spectroscopy, *Methods Enzymol.*, 186, 380, 1990.
23. Janzen, E. G., A critical review of spin trapping in biological systems, in *Free Radical in Biology IV*, Pryor, W. A., Ed., Academic Press, New York, 1980, 116.
24. Oette, K., Identification of some lipid peroxides by thin layer chromatography, *J. Lipid Res.*, 6, 449, 1965.
25. Terao, J., Asano, I., and Matsushita, S., Preparation of hydroperoxy and hydroxy derivatives of rat liver phosphatidylcholine and phosphatidylethanolamine, *Lipids*, 20, 312, 1985.
26. Porter, N. A., Weber, B. A., Weenen, H., and Khan, J. A., Autoxidation of polyunsaturated lipids. Factors controlling the stereochemistry of product hydroperoxides, *J. Am. Chem. Soc.*, 102, 5597, 1980.
27. Miyazawa, T., Yasuda, K., and Fujimoto, K., Chemiluminescence-high performance liquid chromatography of phosphatidylcholine hydroperoxide, *Anal. Lett.*, 20, 915, 1987.
28. Yamamoto, Y., Frei, B., and Ames, B. N., Assay of lipid hydroperoxides using high-performance liquid chromatography with isoluminol chemiluminescence detection, *Methods Enzymol.*, 186, 371, 1990.
29. van Kuijk, F. J. G. M., Thomas, D. W., Stephens, R. J., and Dratz, E. A., Gas chromatography-mass spectrometry assays for lipid peroxides, *Methods Enzymol.*, 186, 388, 1990.
30. Corey, E. J. and Venkateswarlu, A., Protection of hydroxy group as tert-butyldimethylsilyl derivatives, *J. Am. Chem. Soc.*, 94, 6190, 1972.
31. Pryor, W. A. and Castle, L., Chemical methods for the detection of lipid hydroperoxides, *Methods Enzymol.*, 105, 293, 1984.
32. Kulmacz, R. J., Miller, Jr., J. F., Pendleton, R. B., and Lands, W. E. M., Cyclooxygenase initiation assay for hydroperoxides, *Methods Enzymol.*, 186, 431–438, 1990.
33. Miyazawa, T., Suzuki, T., Fujimoto, K., and Kaneda, T., Phospholipid hydroperoxide accumulation in liver of rats intoxicated with carbon tetrachloride and its inhibition by dietary α-tocopherol, *J. Biochem.*, 107, 689, 1990.
34. Miyazawa, T., Suzuki, T., and Fujimoto, K., Age-dependent accumulation of phosphatidylcholine hydroperoxide in the brain and liver of the rat, *Lipids*, 28, 789, 1993.
35. Recknagel, R. O. and Glende, Jr., E. A., Spectrophotometric detection of lipid conjugated dienes, *Methods Enzymol.*, 105, 331, 1984.
36. Burton, G. W., Webb, A., and Ingold, K. U., A mild, rapid and efficient method of lipid extraction for use in determining vitamin E/lipid ratios, *Lipids*, 20, 29, 1985.
37. Matsuo, M., Gomi, F., and Dooley, M. M., Age-related alterations in antioxidant capacity and lipid peroxidation in brain, liver, and lung homogenates of normal and vitamin E-deficient rats, *Mech. Aging Dev.*, 64, 273, 1992.
38. Gomi, F., Dooley, M.M., and Matsuo, M., Effects of oxygen inhalation on the antioxidant capacity of lungs, livers, and brains in normal and vitamin E-deficient rats at various ages, *J. Nutr. Sci. Vitaminol.*, 41, 141, 1995.
39. Green, M. J. and Hill, H. A. O., Chemistry of dioxygen, *Methods Enzymol.*, 105, 3, 1984.

40. Bird, B. P. and Draper, H. H., Comparative studies on different methods of malonaldehyde determination, *Methods Enzymol.*, 105, 299, 1984.
41. Esterbauer, H. and Cheeseman, K. H., Determination of aldehydic lipid peroxidation products: malonaldehyde and 4-hydroxynonenal, *Methods Enzymol.*, 186, 407, 1990.
42. Draper, H. H. and Handley, M., Malondialdehyde determination as index of lipid peroxidation, *Methods Enzymol.*, 186, 421, 1990.
43. Kikugawa, K., Kato, T., and Iwata, A., Determination of malonaldehyde in oxidized lipids by the Hantzsch fluorometric method, *Anal. Biochem.*, 174, 512, 1988.
44. Uchiyama, M. and Mihara, M., Determination of malondialdehyde precursor(s) in tissues by thiobarbituric acid test, *Anal. Biochem.*, 86, 271, 1978.
45. Sagai, M., and Tappel, A. L., Lipid peroxidation induced by some halomethanes as measured by *in vivo* pentane production in the rat, *Toxicol. Appl. Pharmacol.*, 49, 283, 1979.
46. Matsuo, M., Gomi, F., Kuramoto, K., and Sagai, M., Food restriction suppresses an age-dependent increase in the exhalation rate of pentane from rats, *J. Gerontol.*, 48, B133, 1993.
47. Chio, K. S. and Tappel, A. L., Synthesis and characterization of the fluorescent products derived from malonaldehyde and amino acids, *Biochemistry*, 8, 2821, 1969.
48. Kikugawa, K. and Ido, Y., Studies on peroxidized lipids. V. Formation and characterization of 1,4-dihydropyrodine-3,5-dicarbaldehydes as model of fluorescent compounds in lipofuscin, *Lipids*, 19, 600, 1984.
49. Sohal, R. S., Assay of lipofuscin/ceroid pigment *in vivo* during aging, *Methods Enzymol.*, 105, 484, 1984.
50. Fletcher, B. L., Dillard, C. J., and Tappel, A. L., Measurement of fluorescent lipid peroxidation products in biological systems and tissues, *Anal. Biochem.*, 52, 1, 1973.
51. Tappel, A. L., Fletcher, B. L., and Deamer, D., Effect of antioxidants and nutrients on lipid peroxidation fluorescent products and aging parameters in the mouse, *J. Gerontol.*, 28, 415, 1973.
52. Murphy, M. E. and Sies, H., Visible-range low-level chemiluminescence in biological systems, *Methods Enzymol.*, 186, 595, 1990.
53. Sato, T., Miyazawa, T., Kobayashi, M., Furukawa, H., and Inaba, H., Low-level chemiluminescence and life span of *Drosophila melanogaster*, *Gerontology*, 38, 50, 1992.

26

Measurement of Oxidative DNA Damage using the Technique of Gas Chromatography-Mass Spectrometry

Miral Dizdaroglu

CONTENTS

26.1 Introduction

Oxidative DNA damage produced by free radicals or other DNA-damaging agents has been implicated to play a role in mutagenesis, carcinogenesis, and aging [1]. Oxygen-derived species such as superoxide radical (O_2^-) and H_2O_2 are generated in all aerobic cells [1]. These species can also be generated in cells by exogenous sources (e.g., ionizing radiation) and may cause damage to biological molecules including DNA [1,2]. The toxicity of

O_2^- and H_2O_2, however, is thought to result from their metal ion-catalyzed conversion into the highly reactive hydroxyl radical ($^.$OH) [1]. The reactions of $^.$OH with DNA may modify heterocyclic bases and the sugar moiety in DNA, and cause formation of DNA-protein crosslinks in chromatin [3–5]. This type of damage to DNA, also called oxidative DNA damage, is subject to cellular repair processes by DNA repair enzymes [6,7]. DNA lesions can be cleaved out, and the DNA is repaired *in vivo*. Failure of DNA repair can have detrimental biological consequences to organisms [7]. Chemical characterization and quantification of DNA lesions may contribute to the understanding of the role of oxidative DNA damage in carcinogenesis, mutagenesis, and aging.

26.2 Background

Hydroxyl radicals react with the DNA bases by adding to the double bonds of these molecules [2–4]. Hydroxyl radicals also abstract an H atom from the methyl group of thymine and from the C–H bonds of the sugar moiety [2–4]. These addition and abstraction reactions lead to the formation of carbon- or nitrogen-centered radicals. If oxygen is present in the close proximity, base or sugar radicals react with oxygen at diffusion-controlled rates to give peroxyl radicals. Subsequent reactions of thus-formed base or sugar radicals lead to a variety of final products from each of the bases or the sugar moiety, and to DNA strand breaks [2–4]. Modified bases and some sugar products remain attached to the polynucleotide chain of DNA, whereas some other sugar products and intact bases are released from DNA. In addition, base radicals formed on DNA react with aromatic amino acids of neighboring proteins in chromatin or a base radical on DNA reacts with an amino acid radical on a protein [2,4,5]. Such reactions give rise to covalent DNA–protein crosslinks in cells [4,5]. Taken together, free radical reactions may generate a multitude of lesions in DNA including modified bases and sugars, strand breaks, and DNA–protein crosslinks.

In this chapter, the measurement of oxidative DNA damage by the technique of gas chromatography-mass spectrometry (GC/MS) is described.

26.3 Measurement of DNA Damage by GC/MS

26.3.1 Materials and Methods

Reagents and enzymes used in this methodology are available commercially from a number of suppliers [8].

26.3.1.1 *Reference Compounds*

Isobarbituric acid (5-hydroxyuracil), 5,6-dihydrothymine, 5,6-dihydrouracil, isodialuric acid (5,6-dihydroxyuracil), alloxan, 5-(hydroxymethyl)uracil, 4,6-diamino-5-formamidopyrimidine, isoguanine (2-hydroxyadenine), and xanthine-1,3-$^{15}N_2$ are available from Sigma Chemical Company (St. Louis, MO). 8-Hydroxyguanine (7,8-dihydro-8-oxoguanine) and 8-hydroxy-2'-deoxyguanosine are available from Schweizerhall Inc. (Piscataway, NJ) and Cayman Chemical Co. (Ann Arbor, MI), respectively. Reference compounds, and their stable isotope-labeled analogs, which are dealt with in this chapter, are available from Cambridge Isotope Laboratories (Andover, MA) or, on a custom-synthesis basis, from the Chemical Synthesis and Analysis Laboratory of Program Resources Inc./Dyncorp, National Cancer Institute-FCRD (Frederick, MD). The following stable isotope-containing analogs of modified DNA bases have become available recently: 5,6-dihydrothymine-1,3-$^{15}N_2$-2-^{13}C, 5,6-dihydrouracil-1,3-$^{15}N_2$-2-^{13}C, 5-hydroxy-5-methylhydantoin-1,3-$^{15}N_2$-2-^{13}C, alloxan-1,3-$^{15}N_2$-2,4-$^{13}C_2$, 5-hydroxyhydantoin-1,3-$^{15}N_2$-2,4-$^{13}C_2$, 5-hydroxyuracil-1,3-$^{15}N_2$-2-^{13}C, 5-hydroxy-6-hydrothymine-1,3-$^{15}N_2$-2-^{13}C-$^2H_2$, 5-(hydroxymethyl)uracil-2,4-$^{13}C_2$-α,α-$^2H_2$, 5-hydroxycytosine-1,3-$^{15}N_2$-2-^{13}C, *cis*-thymine glycol-$\alpha,\alpha,\alpha,6$-$^2H_4$, 5,6-dihydroxyuracil-1,3-$^{15}N_2$-2-^{13}C (isodialuric acid-1,3-$^{15}N_2$-2-^{13}C), 4,6-diamino-5-formamidopyrimidine-1,3-$^{15}N_2$-2-^{13}C-(5-formamido-^{15}N,^2H), 8-hydroxyadenine-1,3,7-$^{15}N_3$-2,8-$^{13}C_2$, 2,6-diamino-4-hydroxy-5-formamidopyrimidine-1,3-$^{15}N_2$-(5-amino-^{15}N)-2-^{13}C, and 8-hydroxyguanine-1,3-$^{15}N_2$-(2-amino-^{15}N)-2-^{13}C. 2′-Deoxyguanosine-$^{15}N_5$ is available from Cambridge Isotope Laboratories (Andover, MA). Recently, the syntheses of some labeled modified DNA bases have also been reported elsewhere [9–14].

26.3.1.2 *Hydrolysis*

For analysis by GC/MS, DNA or chromatin samples are hydrolyzed and subsequently derivatized. Hydrolysis is achieved either by treatment with acid or enzymatically.

26.3.1.2.1 *Acidic Hydrolysis*

Treatment of DNA with acids cleaves the glycosidic bonds between bases and sugar moieties in DNA, releasing intact and modified bases. Formic acid has been used for this purpose [15,16]. The use of other acids such as HF has also been reported [10]. For analysis of DNA–protein crosslinks, chromatin is hydrolyzed by the standard method of protein hydrolysis using 6 M HCl, which cleaves peptide bonds in proteins as well as glycosidic bonds in DNA to release DNA base–amino acid crosslinks.

Recently, it was found that most of the modified bases are stable under conditions of formic acid hydrolysis, and only a few undergo partial destruction, depending on the concentration of formic acid [16,17]. Furthermore, formic acid causes no significant formation of modified bases under the conditions used [13,16]. Of the concentrations tested, formic acid at a concentration of 60% has been found to be optimal for DNA hydrolysis [16].

The following procedure is used for DNA hydrolysis.

Procedure

DNA samples (10 to 50 μg) are treated with 0.5 ml of formic acid (60%) in evacuated and sealed tubes at 140°C for 30 min. Samples are then transferred into glass vials, frozen in liquid nitrogen and lyophilized overnight (approx. 15 h). If chromatin is to be hydrolyzed for detection of modified bases, the same hydrolysis procedure is followed. Formic acid hydrolysis causes deamination and dehydration of cytosine-derived products as follows: cytosine glycol yields a mixture of 5-hydroxycytosine and 5-hydroxyuracil, the former by dehydration and the latter by dehydration and deamination [18]. 5,6-Dihydrocytosine, 5-hydroxy-6-hydrocytosine and 5,6-dihydroxycytosine deaminate to give 5,6-dihydrouracil, 5-hydroxy-6-hydrouracil and 5,6-dihydroxyuracil, respectively [19]. Furthermore, alloxan, which is a product of cytosine, undergoes decarboxylation to give 5-hydroxyhydantoin [20].

For detection of DNA–protein crosslinks, chromatin samples containing 100 μg DNA are hydrolyzed with 0.5 ml of 6 M HCl in evacuated and sealed tubes at 120°C for 6 h. Subsequently, samples are transferred into glass vials, frozen in liquid nitrogen and lyophilized.

26.3.1.2.2   Enzymatic Hydrolysis

DNA can be hydrolyzed to nucleosides according to the following procedure: DNA samples (100 mg) are incubated in 0.5 ml of 10 mM Tris-HCl buffer, pH 8.5 (containing 2 mM MgCl$_2$) with deoxyribonuclease I (100 units), spleen exonuclease (0.01 unit), snake venom exonuclease (0.5 units) and alkaline phosphatase (10 units) at 37° for 24 h [21]. Samples are transferred into glass vials, frozen in liquid nitrogen, and lyophilized overnight.

Successful analysis by GC/MS of modified purine 2'-deoxynucleosides has been reported [21,22]. Moreover, enzymatic hydrolysis permits the GC/MS analysis of 8,5'-cyclopurine-2'-deoxynucleosides, which are not released from DNA by acidic hydrolysis [22]. On the other hand, the GC/MS analysis of the products of 2'-deoxycytidine renders more difficult, because of their poor gas chromatographic properties [21]. Deamination of 2'-deoxyadenosine products may also occur during hydrolysis due contaminating deaminase activity in the enzymes. Removal of excess salt from the hydrolysates and removal of deaminases from the enzymes may prevent problems associated with analysis of 2'-deoxycytidine and 2'-deoxyadenosine products [23].

26.3.1.3   Derivatization

DNA bases, nucleosides and DNA base–amino acid crosslinks must be converted into their volatile derivatives, because they are not sufficiently volatile for GC/MS analysis. Trimethylsilylation has been the derivatization mode of choice for this purpose in the past [24]. During trimethylsilylation, H atoms of the functional groups such as NH$_2$ and OH groups of DNA bases or amino acids are substituted by trimethylsilyl (Me$_3$Si) groups to give corresponding trimethylsilyl derivatives.

The following procedure can be used for trimethylsilylation.

Procedure

A mixture (0.1 ml) of bis(trimethylsilyl)trifluoroacetamide (containing 1% trimethylchlorosilane) and acetonitrile (4:1, by vol.) is added to lyophilized hydrolyzates of DNA or chromatin. Dry nitrogen is purged on the samples. Then, vials are sealed so that samples will be under nitrogen atmosphere during heating. For sealing of the vials, caps with polytetrafluoroethylene-coated septa are used. The amounts of the derivatization reagents can be modified according to the amount of DNA in samples. Sealed samples are heated at 120°C for 30 min. After derivatization, samples are cooled to room temperature. Without any further treatment, an aliquot (e.g., 1 µl) of each derivatized sample is injected into the injection port of the gas chromatograph. The transfer of samples to special vials may be required if an automatic sampler is used for injection. Such vials should also be sealed under nitrogen atmosphere after transferring the derivatized samples.

26.3.1.4   Instrumentation

A gas chromatograph equipped with a capillary inlet system and interfaced to a mass spectrometer is used. In the present technology, a computer controls various functions of these instruments, and stores and analyzes the data. Quadrupole mass spectrometers are well suited for analysis of components at low concentrations of a complex mixture. For this purpose, the selected-ion monitoring mode of mass spectral analysis is used, with which will be dealt in the forthcoming sections of this chapter. For separation by GC of derivatized hydrolyzates of DNA or chromatin, fused-silica capillary columns are used. Such columns provide high inertness, excellent separation efficiency and measurement of high sensitivity.

Column length may vary depending on the type of analysis. Generally, columns of 12.5 m length (0.2 mm internal diameter, 0.33 mm film thickness) are used for analysis of derivatized bases and nucleosides. A shorter column (e.g., 8 m) may be used for analysis of derivatized DNA base–amino acid crosslinks. Helium (ultra-high purity) serves as the carrier gas. The split injection is the preferred mode of injection if one wishes to avoid overloading the column. The split ratio (i.e., ratio of the carrier gas flow through the splitter vent to carrier gas flow through the column) may be adjusted according to the concentration of samples. An amount of ≈0.05 to 0.4 μg of hydrolyzed DNA on the GC column after splitting of injected sample is generally sufficient. The injection port of the gas chromatograph and the GC/MS interface are kept at 250°C. The temperature of the ion source of the mass spectrometer may be varied if the instrument used permits it. The recommended temperature is approx. 250°C. The injection port of the gas chromatograph contains a glass liner, which should be filled with silanized glass wool to allow homogenous vaporization of injected samples. Compounds eluting from the GC column are introduced into the ion source of the mass spectrometer and are ionized by bombardment with electrons at 70 eV of energy. Ions produced are analyzed by the quadrupole mass analyzer and mass spectra of the analytes are obtained. The electron-ionization (EI) mode is the generally used mode of ionization for analysis of DNA samples and is best suited for selected-ion monitoring.

26.3.2 Analysis of DNA Modifications

26.3.2.1 Analysis of Modified Bases

Gas chromatography on a fused silica capillary column permits separation of Me_3Si derivatives of a large number of modified bases from one another and from four intact DNA bases in a single analysis [15]. EI mass spectra of Me_3Si derivatives of modified DNA bases provide considerable structural detail for unequivocal identification. As those of the intact bases [25], these mass spectra contain an intense molecular ion ($M^{+\cdot}$ ion), an intense $(M - 15)^+$ ion, which results from the loss of a methyl radical from the $M^{+\cdot}$ ion, and other characteristic ions [15]. In some cases, an intense $(M - 1)^+$ ion resulting from loss of an H atom from the $M^{+\cdot}$ ion is also produced [15,25]. As an example, Figure 26.1 illustrates the mass spectrum of the Me_3Si derivative of 4,6-diamino-5-formamidopyrimidine, which is a product of adenine in DNA [3,4]. Ions at m/z 369, 368 and 354 represent $M^{+\cdot}$, $(M - 1)^+$ and $(M - 15)^+$ ions, respectively [15]. The most prominent ion at *m/z* 280 is produced by loss of $\cdot OSiMe_3$ (89 Da) from $M^{+\cdot}$ ion [i.e., $(M - 89)^+$ ion]. The intense ion at *m/z* 73 represents the Me_3Si group and is commonly observed in the mass spectra of Me_3Si derivatives [25]. It serves no diagnostic purpose. EI mass spectra of other modified DNA bases can be found elsewhere [15,26].

26.3.2.2 Analysis of Nucleosides

In the EI mass spectra of Me_3Si derivatives of modified 2'-deoxynucleosides, the most prominent ions are the (base + H)$^+$ ion [$(B + 1)^+$ ion] and the $(base + 1 - 15)^+$ ion [21]. On the other hand, $M^{+\cdot}$ and $(M - 15)^+$ ions possess low intensity. For example, in the mass spectra of Me_3Si derivatives of 8-hydroxy-2'-deoxynucleosides of purines, the $(B + 1)^+$ ion appears as the most prominent ion due to stabilization through an electron-donating substituent at the C-8 of the purine ring. The EI mass spectra of Me_3Si derivatives of 8,5'-cyclopurine 2'-deoxynucleosides are characterized by prominent ions containing the base plus portions of the sugar moiety [22]. The $M^{+\cdot}$ ion is also prominent because of its stabilization by the increased number of rings [22,27].

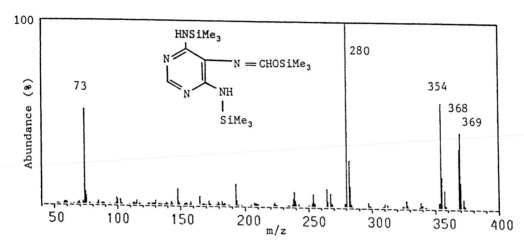

FIGURE 26.1
Electron-ionization mass spectrum of the Me₃Si derivative of 4,6-diamino-5-formamidopyrimidine (From Ref. 15).

26.3.2.3 Analysis of DNA Base–Amino Acid Crosslinks

Mass spectra of Me₃Si derivatives of DNA base–amino acid crosslinks contain ions resulting from typical fragmentations of base and amino acid moieties [26,28,29]. $M^{+\cdot}$ and $(M – 15)^+$ ions of low intensity are also present. For example, the cleavage of the bond between α- and β-carbons of the tyrosine moiety accompanied by an H atom transfer $[(M – 218 + 1)^+$ ion] leads to the most prominent ion (m/z 448) in the mass spectrum of the Me₃Si derivative of the thymine–tyrosine crosslink [29]. Resonance stabilization through the aromatic ring accounts for the high abundance of this ion. This typical cleavage also causes the formation of an intense ion of mass 218 Da when the charge is retained on the α-carbon without an H atom transfer. In the case of DNA base–aliphatic amino acid crosslinks, such fragmentations also occur [28].

26.3.2.4 Structures of DNA Modifications

Figures 26.2 and 26.3 illustrate the structures of modified DNA bases and nucleosides, and some DNA base–amino acid crosslinks that are amenable to measurement by GC/MS. These compounds are formed in DNA or chromatin by free radicals or other damaging agents that cause oxidative damage [4]. Formation of other DNA base–amino acid crosslinks in chromatin has also been reported [28,30].

26.4 Measurement of DNA Damage at Low Analyte Concentrations

26.4.1 Identification Using Selected-Ion Monitoring

When using GC/MS, the identification of components of a complex mixture at low concentrations is performed by the use of selected-ion monitoring (SIM) [31]. The previous knowledge of the mass spectra and the retention times of the analytes is required. In this mode, only a few characteristic ions of an analyte are monitored during the time period, in which

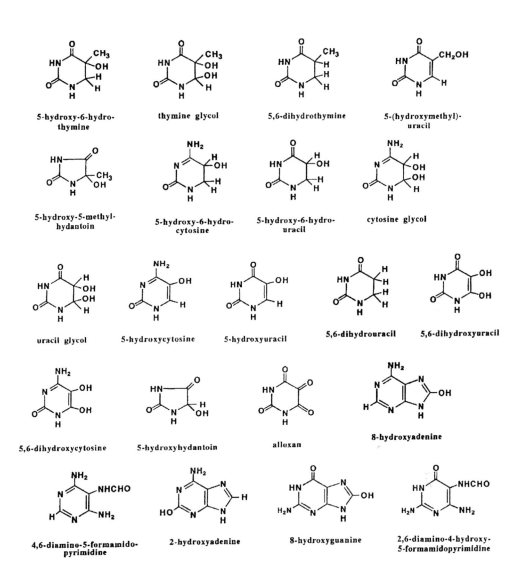

FIGURE 26.2
Structures of modified DNA bases that can be measured by GC/MS. Uracil derivatives illustrated here result from deamination of corresponding cytosine derivatives in DNA (see the text) (From Ref. 43).

the analyte elutes from the GC column. If the analyte is present in the mixture, signals of the monitored ions with their known intensities will line up at its expected retention time. In addition to ions and their relative intensities, the retention time of an analyte plays a crucial role in reliable identification. This is because gas chromatography on fused silica capillary columns permits measurement of retention times with great accuracy and precision.

26.4.2 Quantification

Accurate quantification of components of a complex mixture at low concentrations can also be achieved by GC/MS-SIM [31]. This is carried out by adding an aliquot of a suitable

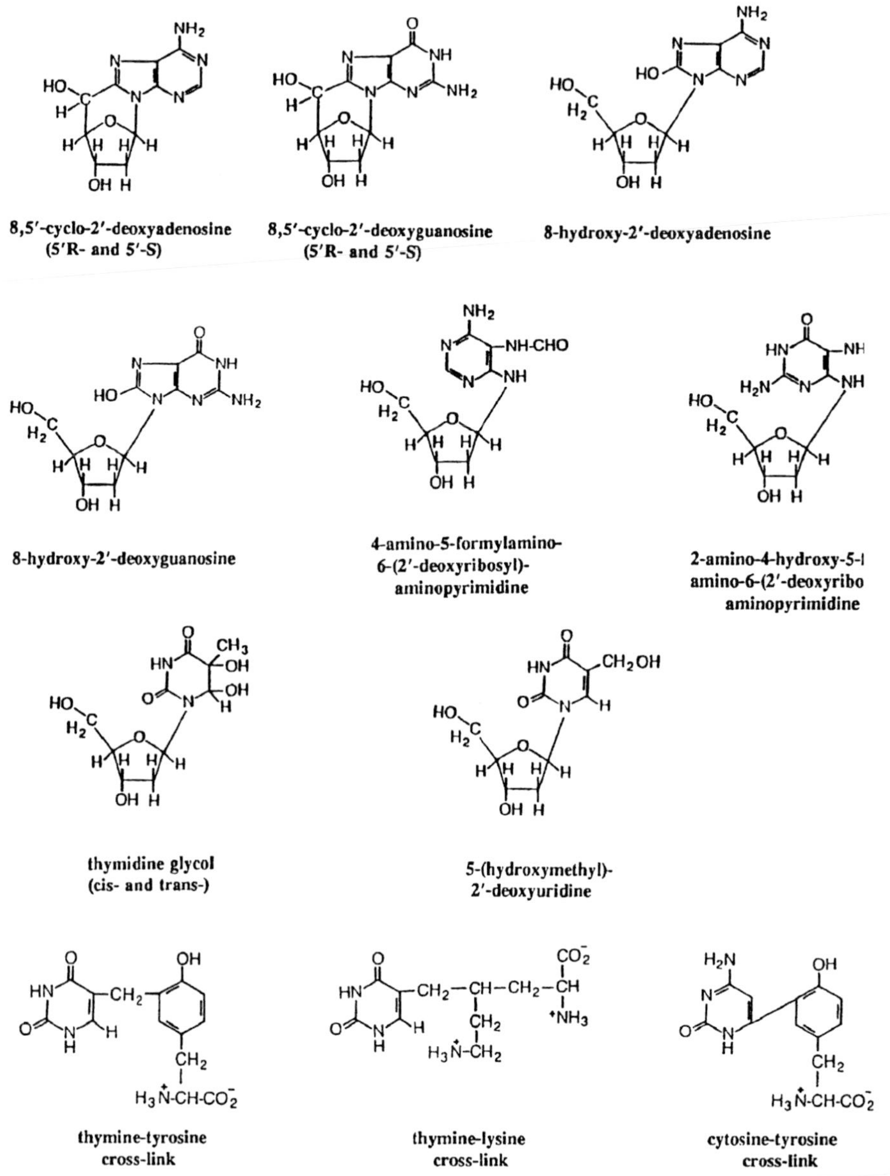

FIGURE 26.3

Structures of nucleosides and DNA base-amino acid crosslinks that can be measured by GC/MS (From Ref. 43).

internal standard to aliquots of DNA samples at an early stage such as prior hydrolysis. For this purpose, a stable-isotope labeled analog of an analyte is used [31]. The use of labeled analogs as internal standards permits compensation for possible losses of the analyte during sample preparation and GC/MS analysis, because the same physical and chemical properties of the analyte and its labeled analog. This procedure is called isotope-dilution mass spectrometry [31]. Recently, stable-isotope labeled analogs of modified DNA bases have become available (see Section 26.3.1).

Mass spectral fragmentation patterns of stable isotope-labeled analogs are similar to those of corresponding unlabeled compounds. Generally, ^{13}C- and ^{15}N-containing analogs are used. Compounds containing ^{2}H may also be used. However, labeled analogs with several ^{2}H atoms may have different gas chromatographic properties from those of their unlabeled counterparts such as different elution times. On the other hand, ^{13}C- and ^{15}N-labeled analogs co-elute with corresponding unlabeled compounds, indicating no isotope effect on elution behavior [20]. Masses of most ions in the mass spectra of labeled analogs are shifted to higher masses according to their isotope contents [20,31]. For example, in the mass spectrum of the Me$_3$Si derivative of 4,6-diamino-5-formamidopyrimidine-^{15}N$_3$,^{13}C,^{2}H, the characteristic ions appear at m/z 374 (M$^{+\cdot}$ ion), 373 [(M – 1)$^+$ ion], 359 [(M – 15)$^+$ ion] and 285 [(M – 89)$^+$ ion], representing a mass shift of 5 Da from the same ions in the mass spectrum of the Me$_3$Si derivative of 4,6-diamino-5-formamidopyrimidine [20] (Figure 26.4).

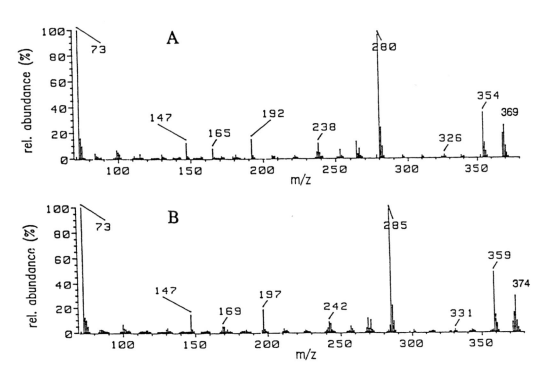

FIGURE 26.4
(A) EI-mass spectrum of the Me$_3$Si derivative of 4,6-diamino-5-formamidopyrimidine; (B) EI-mass spectrum of the Me$_3$Si derivative 4,6-diamino-5-formamidopyrimidine-^{15}N$_3$,^{13}C,^{2}H. (From Ref. 20).

Prior to analysis of DNA samples, calibration plots are obtained by analyzing mixtures of the analyte and its stable isotope-labeled analog with known quantities [31]. A number of prominent characteristic ions are monitored during GC/MS-SIM analysis. The ratios of ion currents at selected masses are plotted as a function of the ratios of the molar amounts of the analyte and its analog. A linear relationship of the ratio of ion currents to the ratio of quantities should be obtained. Subsequently, known amounts of labeled analogs are added to DNA samples and characteristic ions of an analyte and its labeled analog are recorded during the GC/MS-SIM analysis of derivatized DNA hydrolyzates. The quantity of an analyte is calculated using the areas of the ion-current profiles of the monitored ions and the corresponding calibration plots. For assessment of the DNA amount in samples, guanine can be quantified by using labeled guanine as an internal standard [11]. The synthesis of labeled guanine has been reported [11]. 2'-Deoxyguanosine-^{15}N$_5$, which is commercially available (see Materials and Methods), can be used instead of labeled guanine, because, unlike guanine, it is readily soluble in water. An aliquot of 2'-deoxyguanosine-^{15}N$_5$ should be added to DNA samples prior to acidic hydrolysis.

26.5 Selectivity and Sensitivity

The GC/MS-SIM technique provides high sensitivity and selectivity. The highest sensitivity for an analyte can achieved by monitoring the most abundant characteristic ions in its mass spectrum. A few characteristic ions of the analyte and its labeled analog should be monitored at the corresponding retention time for a reliable identification and quantification. For DNA components, sensitivities in the range of approximately 1 to 5 fmol per compound applied to the GC column, or in the range of 1 to 3 residues in 10^6 DNA bases can be achieved. The level of sensitivity of measurement may depend on the instrument, column type and other factors. Mass spectrometers equipped with a high energy dynode electron multiplier may provide somewhat higher sensitivity than those without it. Aliquots of hydrolyzates containing 0.05 to 0.4 µg of DNA injected onto the GC column after splitting of the sample are generally sufficient for reliable analysis.

26.6 Limitations and Comments

The GC/MS technique can be used for volatile compounds only, or for those that can be made volatile by derivatization. Me$_3$Si derivatives of DNA components reviewed here possess excellent properties for a successful GC analysis. Some modified DNA bases are acid-labile. Cytosine glycol is converted quantitatively into 5-hydroxycytosine by dehydration and into 5-hydroxyuracil by deamination and dehydration [18,19]. 5-Hydroxy-6-hydrocytosine and 5,6-dihydrocytosine deaminate to give corresponding uracil derivatives [19]. Alloxan, which is a cytosine-derived product in DNA, decarboxylates to yield 5-hydroxy-hydantoin [20]. The prominent product of guanine, 2,6-diamino-4-hydroxy-5-formamidodpyrimidine may be destroyed during acidic hydrolysis unless hydrolysis is carried out under vacuum in evacuated and sealed tubes. The stability of modified bases can be checked by GC/MS analysis before and after acidic treatment under the conditions of DNA hydrolysis. Problems associated with loss of an analyte during analysis are minimized by the use of its labeled analog as an internal standard, as was discussed earlier in the text.

Recently, the oxidation of guanine in DNA hydrolyzates during trimethylsilylation to give 8-hydroxyguanine has been reported [11,12]. It has been shown that oxygen present in the derivatization mixture causes artifactual formation of 8-hydroxyguanine from guanine. Furthermore, the derivatization at room temperature gave lower values for 8-hydroxyguanine than at 140°C [11]. This, however, does not consider the possibility of incomplete derivatization at room temperature. Values found by derivatization at room temperature were similar to those found by measurement using HPLC with electrochemical (EC) detection. Values obtained by derivatization at 140°C were similar to those reported using GC/MS [16]. Since the background values for 8-hydroxyguanine in DNA obtained by these two techniques differ from each other, it was concluded that derivatization at higher temperatures causes artifactual formation of 8-hydroxyguanine from guanine in DNA hydrolyzates [11,12]. However, these authors did not follow the same procedures that had been introduced previously, when the analysis of 8-hydroxyguanine by GC/MS was originally described. For example, hydrolysis was done originally under vacuum in evacuated and sealed tubes, and hydrolyzates were lyophilized, whereas the aforementioned work reported hydrolysis under argon without vacuum and removal of acid under a stream of argon. These differences in the methodology may have led to differences in the values obtained. Interestingly, other researchers did measurements using other techniques and no derivatization reported background values of 8-hydroxyguanine to be similar to those reported originally using GC/MS [32–35]. For example, Herbert et al. used guanase digestion to remove guanine from formic acid-hydrolyzates of DNA and then applied HPLC-EC to measure 8-hydroxyguanine [33]. The values obtained for untreated calf thymus DNA or γ-irradiated DNA were similar to those originally reported using GC/MS. A subsequent work by Kaur and Halliwell reported similar results using HPLC [34]. It is clear that oxygen causes formation of 8-hydroxyguanine from guanine in DNA hydrolyzates during derivatization. However, if derivatization is done under strictly nitrogen atmosphere without oxygen, oxidation of guanine may be avoided.

In recent years, the measurement of 8-hydroxyguanine as its nucleoside 8-hydroxy-2'-deoxyguanosine in *in vitro* and *in vivo* systems has been performed using HPLC–EC after hydrolysis of DNA with a mixture of enzymes [36]. The values reported were generally lower than those reported by GC/MS [37]. The underestimation of 8-hydroxyguanine by this technique has been discussed [37]. One possibility is that the enzymes used to hydrolyze DNA may not completely hydrolyze DNA. There is no clear evidence in the literature that the enzymes used completely remove 8-hydroxy-2'-deoxyguanosine from DNA. On the other hand, there is evidence that modified bases greatly affect the efficiency of exonucleases and endonucleases [38,39]. Thus, 8-hydroxyguanine has a severe inhibitive effect on diesterase action [39]. In the same context, formic acid does not cause formation of 8-hydroxyguanine from guanine [13,16]. Recently, Frenkel et al. showed that acidic pH used for nuclease P1 digestion may hydrolyze 8-hydroxy-2'-deoxyguanosine to give 8-hydroxyguanine, causing the loss of 8-hydroxy-2'-deoxyguanosine detected by HPLC–EC [40]. Using a more complex HPLC method, they also found background values of 8-hydroxyguanine in murine epidermal cells to be closer to those measured by GC/MS than those measured by HPLC–EC.

Recently, an artifactual formation of 5-hydroxycytosine and 8-hydroxyadenine during derivatization has been reported [14]. Similar to 8-hydroxyguanine values, the background values of these compounds were found to be lower than those reported originally using GC/MS. Again, these authors did not follow the same procedures as previous work using GC/MS. Moreover, in their conclusions, they did not take into consideration the background values of these two compounds, which had been reported previously in a large number of publications. The values found by measurement using GC/MS of 5-hydroxycytosine and 8-hydroxyadenine levels in various animal and human tissues and in cultured cells were similar to those reported by these authors after their prepurification of DNA samples to remove cytosine and adenine. It is beyond the scope of this article to review and cite all those papers. They can readily be found in the literature pertaining oxidative DNA damage *in vitro* and *in vivo*.

26.7 Applications

The GC/MS technique has been widely applied to the measurement of DNA damage in various *in vitro* and *in vivo* systems. Extensive reviews of these applications can be found elsewhere [4,41–43]. An interesting example of these applications was the measurement of DNA base damage *in vivo* in mouse forebrain after ischemia-reperfusion. A five-fold increase of mutations had been found in the reporter *lacI* gene in cortical DNA of transgenic mice after 30 min of forebrain ischemia and 8 h of reperfusion [44]. Mutations remained elevated at 24 h reperfusion, but DNA damage has not been determined. By the use of the GC/MS technique, a number of DNA lesions has been measured in mouse forebrain following 30 min of ischemia and up to 6 h of reperfusion. The levels of four premutagenic modified bases 8-hydroxyguanine, 5-hydroxycytosine, 2,6-diamino-4-hydroxy-5-formamidopyrimidine and 8-hydroxyadenine increased two- to fourfold during 10–20 min of reperfusion. These and other observations of this work were consistent with the notion that forebrain ischemia-reperfusion induces oxidative stress, which leads to DNA damage and is mutagenic to brain cells.

In recent years, the GC/MS technique has been applied to the measurement of DNA repair. Substrate specificities of a number of DNA repair enzymes called DNA glycosylases have been examined [19,45–48]. Since the GC/MS technique can identify and quantify a multitude of DNA base lesions in the same DNA sample at the same time, it can determine which lesions are excised and which lesions are not excised from DNA by a given DNA repair enzyme under the same conditions. This approach also facilitates the measurement of kinetics of excision [48]. Furthermore, repair of various DNA lesions has been measured in cultured human cells by the use of GC/MS [49].

26.8 Conclusions

Extensive research in recent years has demonstrated that the GC/MS technique is well suited for the measurement of DNA damage including base damage, sugar damage, and DNA–protein crosslinks in various *in vitro* and *in vivo* systems. This technique offers the sensitivity, selectivity, speed and versatility to solve a wide range of important measurement problems. It also permits studies of the substrate specificities of DNA repair enzymes as well as enzymatic repair of DNA damage in living cells. It appears that the GC/MS technique will also find a major role in studies of oxidative DNA damage in research on aging.

References

1. Halliwell, B. and Gutteridge, J. M. C., *Free Radicals in Biology and Medicine*, Second Edition, Clarendon Press, Oxford, 1989.
2. von Sonntag, C., *The Chemical Basis of Radiation Biology*, Taylor & Francis, London, 1987, 116 and 221.
3. Téoule, R., Radiation-induced DNA damage and its repair, *Int. J. Radiat. Biol.*, 51, 573, 1987.
4. Dizdaroglu, M., Oxidative damage to DNA in mammalian chromatin, *Mutat. Res.*, 275, 331, 1992.
5. Oleinick, N. L., Chiu, S., Ramakrishnan, N. and Xue, L., The formation, identification, and significance of DNA–protein cross-links in mammalian cells. *Br. J. Cancer*, 55 (Suppl. VIII), 135, 1987.

6. Demple, B. and Harrison, L., Repair of oxidative damage to DNA: enzymology and biology, *Ann. Rev. Biochem.*, 63, 915, 1994.

7. Wallace, S. S., DNA damages processed by base excision repair: biological consequences, *Int. J. Radiat. Biol.*, 66, 579, 1994.

8. Certain commercial equipment or materials are identified in this paper in order to adequately specify the experimental procedure. Such identification does not imply recommendation or endorsement by the National Institute of Standards and Technology, nor does it imply that the materials or equipment identified are necessarily the best available for the purpose.

9. Djuric, Z., Luongo, D. A. and Harper, D. A., Quantitation of 5-(hydroxymethyl)uracil in DNA by gas chromatography with mass spectral detection, *Chem. Res. Toxicol.* 4, 687, 1991.

10. Wagner, J. R., Analysis of oxidative cytosine products in DNA exposed to ionizing radiation, *J. Chim. Phys.*, 91, 1280, 1994.

11. Hamberg, M. and Zhang, L.-Y., Quantitative determination of 8-hydroxyguanine and guanine by isotope dilution mass spectrometry, *Anal. Biochem.*, 229, 336, 1995.

12. Ravanat, J.-L., Turesky, R. J., Gremaud, E., Trudel, L. J. and Stadler, R. H., Determination of 8-oxoguanine in DNA by gas chromatography-mass spectrometry and HPLC-electrochemical detection: overestimation of the background level of the oxidized base by the gas chromatography-mass spectrometry assay, *Chem. Res. Toxicol.*, 8, 1039, 1995.

13. Douki, T., Delatour, T., Paganon, F. and Cadet, J., Measurement of oxidative damage at pyrimidine bases in γ-irradiated DNA, *Chem. Res. Toxicol.*, 9, 1145, 1996.

14. Douki, T., Delatour, T., Bianchini, F. and Cadet, J., Observation and prevention of an artefactual formation of oxidized DNA bases and nucleosides in the GC-EIMS method, *Carcinogenesis*, 17, 347, 1996.

15. Dizdaroglu, M., Application of capillary gas chromatography-mass spectrometry to chemical characterization of radiation-induced base damage in DNA: implications for assessing DNA repair processes, *Anal. Biochem.*, 144, 593, 1985.

16. Nackerdien, Z., Olinski, R. and Dizdaroglu, M., DNA base damage in chromatin of γ-irradiated cultured human cells, *Free Rad. Res. Comms.*, 16, 259, 1992.

17. Fuciarelli, A. F., Wegher, B. J., Gajewski, E., Dizdaroglu, M. and Blakely, W. F., Quantitative measurement of radiation-induced base products in DNA using gas chromatography-mass spectrometry, *Radiat. Res.*, 119, 219, 1989.

18. Dizdaroglu, M., Holwitt, E., Hagan, M. P. and Blakely, W. F., Formation of cytosine glycol and 5,6-dihydroxycytosine in deoxyribonucleic acid on treatment with osmium tetroxide, *Biochem. J.* 235, 531,1986.

19. Dizdaroglu, M., Laval, J. and Boiteux, S., Substrate specificity of *Escherichia coli* endonuclease III: excision of thymine- and cytosine-derived lesions in DNA produced by radiation-generated free radicals. *Biochemistry*, 32, 12105, 1993.

20. Dizdaroglu, M., Quantitative determination of oxidative base damage in DNA by stable-isotope dilution mass spectrometry, *FEBS Lett.*, 315, 1, 1993.

21. Dizdaroglu, M., Characterization of free radical-induced damage to DNA by the combined use of enzymatic hydrolysis and gas chromatography-mass spectrometry, *J. Chromatogr.*, 367, 357, 1986.

22. Dirksen, M.-L., Blakely, W. F., Holwitt, E. and Dizdaroglu, M.,Effect of DNA conformation on the hydroxyl radical-induced formation of 8,5′-cyclopurine 2′-deoxyribonucleoside residues in DNA, *Int. J. Radiat. Biol.*, 54, 195, 1988.

23. Crain, P. F., Preparation and enzymatic hydrolysis of DNA and RNA for mass spectrometry, *Methods Enzymol.*, 193, 782, 1990.

24. Schram, K. H., Preparation of trimethylsilyl derivatives of nucleic acid components for analysis by mass spectrometry, *Methods Enzymol.*, 193, 791, 1990.

25. White V, E., Krueger, P. M. and McCloskey, J. A., Mass spectra of trimethylsilyl derivatives of pyrimidine and purine bases, *J. Org. Chem.* 37, 430, 1972.

26. Dizdaroglu, M., The use of capillary gas chromatography-mass spectrometry for identification of radiation-induced DNA base damage and DNA-amino acid cross-links, *J. Chromatogr.*, 295, 103, 1984.

27. McLafferty, F. W., *Interpretation of Mass Spectra*, Univ. Sci. Books, Mill Valley, CA, 1980.

28. Gajewski, E., Fuciarelli, A. F. and Dizdaroglu, M., Structure of hydroxyl radical-induced DNA–protein crosslinks in calf thymus nucleohistone *in vitro*. *Int. J. Radiat. Biol.*, 54, 445, 1988.

29. Dizdaroglu, M., Gajewski, E., Reddy, P. and Margolis, S. A., Structure of a hydroxyl radical-induced DNA–protein crosslink involving thymine and tyrosine in nucleohistone. *Biochemistry*, 28, 3625, 1989.

30. Gajewski, E., and Dizdaroglu, M., Hydroxyl radical–induced crosslinking of cytosine and tyrosine in nucleohistone. *Biochemistry*, 29, 977, 1990.

31. Watson, J. T., Selected-ion measurements, *Methods Enzymol.*, 193, 86, 1990.

32. Finnegan, M. T., Herbert, K. E., Evans, M. D., Farooq, S., Farmer, P., Podmore, I. D. and Lunec, J., Development of an assay to measure 8-oxoguanine using HPLC with electrochemical detection, *Biochem. Soc. Trans.*, 23, 431S, 1995.

33. Herbert, K. E., Evans, M. D., Finnegan, M. T., Farooq, S., Mistry, N., Podmore, I. D., Farmer, P. and Lunec, J., A novel HPLC procedure for the analysis of 8-oxoguanine in DNA, *Free Radic. Biol. Med.*, 20, 467, 1996.

34. Kaur, H. and Halliwell, B., Measurement of oxidized and methylated DNA bases by HPLC with electrochemical detection, *Biochem. J.*, 318, 21, 1996.

35. Devanaboyina, U.-S. and Gupta, R. C., Sensitive detection of 8-hydroxy-2'-deoxyguanosine in DNA by ^{32}P-postlabeling assay and the basal levels in rat tissues, *Carcinogenesis*, 17, 917, 1996.

36. Floyd, R. A., Watson, J. J., Wong, P. K., Altmiller, D. H. and Rickard, R. C., Hydroxyl free radical adduct of deoxyguanosine: sensitive detection and mechanism of formation, *Free Radic. Res. Comms.*, 1, 163, 1986.

37. Halliwell, B. and Dizdaroglu, M., The measurement of oxidative damage to DNA by HPLC and GC/MS techniques, *Free Radic. Res. Comms.*, 16, 75, 1992.

38. Dizdaroglu, M., Hermes, W., Schulte-Frohlinde, D. and von Sonntag, C., Enzymatic digestion of DNA γ-irradiated in aqueous solution. Separation of the digests by ion-exchange chromatography, *Int. J. Radiat. Biol.*, 33, 563, 1978.

39. Maccubin, A., Evans, M., Paul, C. R., Budzinski, E. E., Przybyszewski, J and Box, H. C., Enzymatyic excision of radiation-induced lesions from DNA model compounds, *Radiat. Res.*, 126, 21, 1991.

40. Frenkel, K., Zhong, Wei, H., Karkoszka, J., Patel, U., Rashid, K., Georgescu, M. and Solomon, J. J., Quantitative high-performance liquid chromatography analysis of DNA oxidized *in vitro* and *in vivo*, *Anal. Biochem.*, 196, 126, 1991.

41. Dizdaroglu, M., Chemical determination of free radical-induced damage to DNA. *Free Radic. Biol. Med.*, 10, 225, 1991.

42. Dizdaroglu, M., Measurement of radiation-induced damage to DNA at the molecular level, *Int. J. Radiat. Biol.*, 61, 175, 1992.

43. Dizdaroglu, M., Chemical determination of oxidative DNA damage by gas chromatography-mass spectrometry. *Methods Enzymol.*, 234, 3, 1994.

44. Liu, P. K., Hsu, C. Y., Dizdaroglu, M., Floyd, R. A., Kow, Y. W., Karakaya, A., Rabow, L. E. and Cui, J.-K., Damage, repair, and mutagenesis in nuclear genes after mouse forebrain ischemia-reperfusion, *J. Neurosc.*, 16, 6795, 1996.

45. Boiteux, S., Gajewski, E., Laval, J. and Dizdaroglu, M., Substrate specificity of the *Escherichia coli* Fpg protein (formamidopyrimidine-DNA glycosylase): excision of purine lesions in DNA produced by ionizing radiation or photosensitization, *Biochemistry*, 31, 106, 1992.

46. Zastawny, T. H., Doetsch, P. W. and Dizdaroglu, M., A novel activity of uracil DNA N-glycosylase: excision of isodialuric acid (5,6-dihydroxyuracil) from DNA, a major product of oxidative DNA damage, *FEBS Lett.*, 364, 255, 1995.

47. Dizdaroglu, M., Karakaya, A., Jaruga, P., Slupphaug, G. and Krokan, H. E., Novel activities of human uracil DNA N-glycosylase for cytosine-derived products of oxidative DNA damage, *Nucl. Acids Res.*, 24, 418, 1996.

48. Karakaya, A., Jaruga, P., Bohr, V., Grollman, A. P. and Dizdaroglu, M., Kinetics of Excision of Purine Lesions from DNA by *Escherichia coli* Fpg Protein, *Nucl. Acids Res.*, 25, 474, 1997.

49. Jaruga, P. and Dizdaroglu, M., Repair of products of oxidative DNA base damage in human cells, *Nucl. Acids Res.*, 24, 1389, 1996.

27

A Transgenic Mouse Model for Studying Mutations in vivo

Jan Vijg, Martijn E.T. Dollé, Michael E.T.I. Boerrigter, and Jan A. Gossen

CONTENTS

27.1 Introduction

Aging in higher organisms has been considered as the consequence of accumulated DNA mutations, causing the death of cells or rendering cells ineffective. This theory, which was postulated almost four decades ago [1–3], has never been experimentally tested due to the lack of suitable methods for quantitation and characterization of somatic mutational events in different organs and tissues. Indeed, the experimental data that have emerged either involve indirect methods or determinations on a very limited number of mitotically active cell types, mainly T-lymphocytes. Originally, background radiation was considered as the main cause of spontaneous mutagenesis. More recently it has become clear that endogenous DNA damage is much more frequent [4]. Major sources of endogenous DNA damage are the various normal and abnormal cellular processes that generate free radicals [4,5]. Error prone processing of DNA lesions during replication and repair results in mutations, that is, a variety of DNA sequence changes including point mutations, deletions, insertions and transpositions. Also large genome rearrangements, i.e., deletions, translocations should be considered as mutations.

Although detailed information about the types of mutation induced by free radicals is still lacking there is evidence that genome rearrangements are a major component of the spectrum. In fact, such mutations were among the first demonstrated to increase with age. Curtis and co-workers [6] looked at mouse liver parenchymal cell metaphase plates after partial hepatectomy and found considerably higher numbers of cells with abnormal chromosomes in old as compared to young animals (i.e., from about 10% of the cells in 4 to 5 month old mice to 75% in mice older than 12 months). Later, such large structural changes in DNA, i.e., aneuploidy, translocations, dicentrics) were observed to increase with donor age in white blood cells of human individuals, i.e., from about 2 to 4% of the cells being chromosomally aberrated in young individuals to about 6 times higher in the elderly [reviewed in Ref. 7]. Initially, the use of classical cytogenetic techniques only allowed to analyze small numbers of subjects and small numbers of cells. The use of more advanced methods, such as chromosome painting [8] and PCR assays for specific translocations [9], have amply confirmed the increase in cytogenetic damage with age.

With the development of tests based on selectable endogenous target genes, e.g., the hypoxanthine phosphoribosyl transferase (HPRT) locus [10], it became possible to assess the mutant frequency at these loci among T cells from human and animal donors. The results obtained with these assays suggest that mutant frequencies in humans go up with age from about 2×10^{-6} in young individuals to about 1×10^{-5} in middle aged and old individuals [11]. In mice the mutant frequency appeared to be somewhat higher, that is, from about 5×10^{-6} in young animals to about 3×10^{-5} in middle aged mice [12]. However, in both mice and men these values could be underestimates, due to the loss of mutants *in vivo* or *in vitro*. Indeed, results from Grist et al. [13], who assayed the HLA locus (using immunoselection for mutationally lost HLA antigen), indicate 2 to 3 times higher mutant frequencies. Values higher than HPRT were also found for the glycophorin A assay and the discrepancy has been explained in terms of the inability of the HPRT test to detect mitotic recombination events (HPRT is X-linked) and a relatively strong *in vivo* selection against mutants [13]. Indeed, HPRT mutant frequencies have been found to decrease with time following exposure to mutagenic agents, such as ethyl nitrosourea (ENU) and radiation, and has been considered as less suitable for long term monitoring [14].

In mice subjected to caloric restriction, the only intervention demonstrated to increase lifespan [15], HPRT mutant frequencies were found to increase with age at a significantly slower rate than in the ad libitum fed animals [12]. This suggests that the level of accumulated somatic mutations reflects biological rather than chronological age. More recently, the HPRT test was used on tubular epithelial cells of kidney tissue from 2- to 94-year old human donors. The mutation frequencies that were found were much higher than the above mentioned values for blood lymphocytes and were also found to increase with age from about 5×10^{-5} to about 2.5×10^{-4}; the increase with age appeared to be exponential rather than linear [16]. The higher mutant frequency in the kidney cells could reflect a relatively slow turnover as compared to T cells.

Thus, both cytogenetic analysis and the use of selectable marker genes like HPRT have provided clear evidence that mutations, large and small, occur *in vivo* and appear to accumulate with age, as predicted by the theory. Disadvantages of these methods include their limitation to actively proliferating cell types and their restriction to either very large microscopically visible alterations or mutational alterations in one single selectable gene. They are also labor intensive and subject to large variation. Moreover, especially for the HPRT test there appears to be *in vivo* and *in vitro* selection against mutant cells (see above). Ideally, to properly test the somatic mutation accumulation theory for different organs and tissues, including postmitotic organs, one needs a neutral marker at various places all over the genome that allows to rapidly and accurately probe for mutational changes in every organ and tissue as a function of age.

27.2 Background: Use of Transgenic Mice for Mutation Analysis *in vivo*

Cytogenetic analysis and the use of selectable marker genes have been very useful for confirming age-related mutation accumulation in, mainly, the blood forming tissues. With the advent of transgenic animal technology a completely different approach became possible. Some time ago we started to use this technology to generate transgenic mouse lines harboring lacZ reporter genes integrated in one or more chromosomes as part of a bacteriophage lambda vector that could be recovered and inspected in *E. coli* for mutations [17]. With this model, lacZ reporter gene copies could be rescued by *in vitro* packaging and plating, using host-restriction negative *E. coli* host cells [18]. Initially, mutant lacZ genes were detected as colourless plaques among the wild-type blue plaques. Later we developed a positive-selection system based on an *E. coli* host with an inactivated galE gene. Upon receiving a wild-type lacZ gene such cells, grown in the presence of the lactose analog p-gal, produce UDP-galactose which is highly toxic when it can not be converted into UDP glucose [19]. Thus, on p-gal containing medium only those cells infected with a mutant lacZ gene can give rise to a plaque [20]. This greatly decreased the cost and time involved in the assay.

An alternative model based on the same principle, but with the lacI gene as the reporter, was generated by Short and co-workers [21]. In this model mutational inactivation of the gene encoding the lacI repressor leads to de-repression of the lacZ gene, resulting in β-gal expression and a blue plaque among colourless wild-type plaques. The lacI model offers the advantage of a smaller sized mutational target (about 1000 bp instead of 3000) for which an extensive database of mutations in *E. coli* was already available [22]. Also for this model positive selection systems have been developed but are not yet widely used, probably because of the high background [22]. Indeed, one of the problems of the lacI system in studying spontaneous mutation frequencies is that a considerable part of the spectrum consists of mutations originating in *E. coli* [23].

Both the lacZ and lacI bacteriophage lambda systems have now been extensively used for studying mutations *in vivo*, mainly for genetic toxicological purposes [24, 25]. Comparisons have been made in one and the same animal between the transgene locus and the Dlb-1 [26] or HPRT [27] locus. (The Dlb-1 locus assay detects the loss of the binding site for the lectin *Dolichos biflorus* on the cell surface of the epithelial cells of the small intestine in Dlb-1ᵃ/Dlb-1ᵇ heterozygous mice.) It was found that transgene and natural gene react essentially the same to treatment with a mutagen. However, in blood lymphocytes a much lower background value was reported for the HPRT as compared to the (lacI) transgene

[27]. Explanations for this could be the loss of HPRT mutations due to selection *in vivo* or *in vitro* (see above), or the much higher methylation level of the transgene (which is not expressed). Deamination of 5′-methylcytosine is likely to be a major mechanism of spontaneous mutagenesis at such sites [24].

Another difference between the reporter transgene and an endogenous gene (in this case Dlb-1; Ref. 26) is the much lower sensitivity of the former for X-irradiation. In part, this can be explained by the fact that large deletion mutations will go undetected as a consequence of the minimum vector size required for efficient packaging of bacteriophage lambda vectors (i.e., between 42 and 52 kb). In addition, genome rearrangements involving the mouse flanking regions and therefore one of the cos sites will go undetected. It is not clear, however, why so few deletion mutations of moderate size, i.e., between 50 and 3,000 basepairs, have been detected with these models. Clearly, such deletion events are expected to occur, as they have been observed at the HPRT locus and other selectable genes [10]. One can speculate that the large amount of prokaryotic DNA in the lambda models will inhibit the activity of those DNA processing systems that in mammals are involved in DNA rearrangements. Since mutations, induced in the genome ad random, can be expected to have a higher functional impact when they involve large structural alterations than point mutations, it is especially for this reason that we decided to develop an alternative system based on integrated plasmid rather than bacteriophage lambda vectors.

27.3 System Description

Like in the bacteriophage lambda model, also in the plasmid model the vector with the reporter gene lacZ is integrated in multiple copies in a head-to-tail organization (see ref. 28 and also Figure 27.1). The reason that initially plasmid vectors have not been used as the vector of choice for transgenic mutation models is the notoriously low transformation efficiencies obtained with plasmids excised from their integrated state in the mammalian genome. Efficient intramolecular ligation of excised plasmids requires extensive dilution and the presence of the genomic DNA will also negatively influence the transformation efficiency. To solve this problem, methods were developed to recover the plasmids from genomic mouse DNA using magnetic beads coupled to the lac repressor protein, which selectively binds to the operator sequence in front of the lacZ gene [29,30]. After ligation, plasmids are introduced into galE⁻ *E. coli* host cells by means of electroporation. The galE positive selection is identical to that for bacteriophage lambda; only mutant lacZ plasmids give rise to a colony [31]. The rescue and mutant frequency determination procedure is schematically depicted in Figure 27.1.

This recently developed plasmid system offers two major advantages over the bacteriophage lambda models. First, its rescue efficiency is very high and, second, it allows the detection of a broad range of mutations including large deletions. The rescue efficiency of the system is mainly determined by the transformation efficiency of the *E. coli* host (galE⁻, ΔlacZ, host restriction negative). With this strain, transformation efficiencies of up to 5×10^{10} transformants per μg of vector DNA can be obtained and since the capacity of the magnetic beads to bind plasmids is not limiting, over a million copies can routinely be rescued in one single experiment. This number of transformants can still be loaded on one single 9-cm diameter plate without exhausting the ampicillin [30]. Such a large number of recovered reporter genes allows highly accurate mutant frequency determinations at a much lower costs than with the lambda systems.

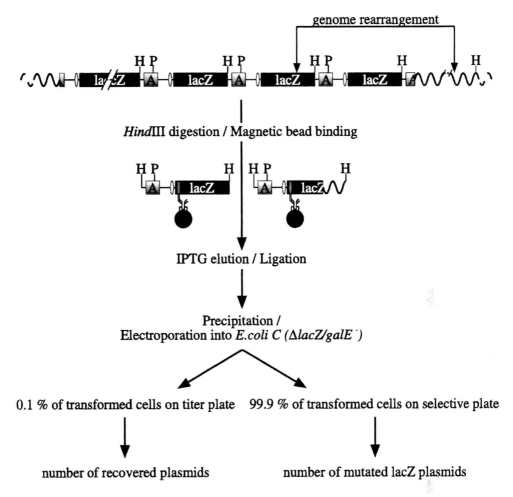

FIGURE 27.1

Schematic representation of the recovery of plasmids from their integrated state as a multiple-copy cluster in the mouse genome. The recovered and ligated plasmids are subsequently used to transform *E. coli*. A very small part is then plated on X-gal plates to determine the titer and the rest is plated on p-gal plates to select for the mutants. Genome rearrangements (e.g., deletions, translocations) with one breakpoint in the lacZ gene and the other in the mouse flanking genomic DNA can be recovered as mutant plasmids with a piece of mouse genomic DNA, i.e., from the breakpoint in the mouse flanking sequence to the first HindIII site. After PstI excision only internal plasmid copies can be recovered and, hence, no genome rearrangements involving the mouse flanking sequence can be detected. H = Hind III; P = PstI

In addition to its higher rescue efficiency, the characteristics of plasmid rescue should allow to detect a broad range of mutational events, including internal deletions as well as deletions with one breakpoint in the lacZ mutational target gene and the second breakpoint in the 3′ region flanking the concatemer. This is illustrated in Figure 27.1, showing excision of the plasmid from the mouse chromosomal DNA by using the restriction enzyme Hind III. Point mutations and deletion mutations in internal copies of the plasmid cluster are recovered on the p-gal selective plates, as long as the origin of replication, the operator sequence and the ampicillin resistance gene are not affected. Large deletions that involve the 3′ mouse flanking region can also be recovered. In such cases a fragment will be recovered that includes a mouse sequence, from the breakpoint in the flanking region to the nearest Hind III restriction site. As a consequence, these recovered plasmids should hybridize with labeled mouse total genomic DNA, which appeared to be the case [32,30; see also Figure 27.2].

27.4 Procedures

27.4.1 Transgenic Animals

The lacZ-plasmid transgenic mice currently available are C57Bl/6 transgenic mice harboring approximately 20 copies per haploid genome of the plasmid pUR288 containing the complete 3096-bp *E. coli* lacZ gene, integrated at different chromosomal sites: chromosomes 3, 4, and 11. One transgenic line was obtained with both the chromosome 3 and 4 vector clusters in a homozygous state. Preliminary results indicate that this transgenic line,

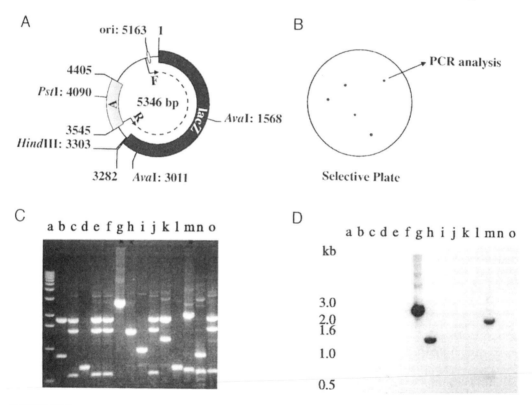

FIGURE 27.2

Mutant characterization. **(A)** Map of the pUR288 plasmid with the lacZ gene (1-3282), the PCR primer binding sites (F[orward] and R[eversed]) and the Ava I sites. Also the origin of replication, the ampicillin resistance gene with the Pst I site and the HindIII site are indicated. **(B)** Mutant colonies are taken from the selective plate and subjected to long-distance PCR using primers flanking the lacZ reporter. **(C)** PCR product is digested with AvaI and subjected to gel electrophoresis. After ethidium bromide staining, point mutants show a wild-type banding pattern (three bands of 1.8, 1.4 and 0.57 bp). Any deviation from this pattern indicates a size-change mutant (lanes b,d,g,h,i,k,l,m and n). **(D)** Hybridization of the pattern transferred to a nylon membrane with radiolabeled non-transgenic mouse genomic DNA reveals the mutants with a captured mouse sequence (lanes g, h, and m), i.e, potentially large genome rearrangements.

termed line 60, is not affected by the transgene; life span studies up until 30 months showed no apparent effects of the transgene on general health, fecundity, etc. Thus far, most experiments were carried out with this line. All animals were maintained in the animal facilities of the Beth Israel Hospital. Line 60 is now also available through The Jackson Laboratory (Bar Harbor, ME).

27.4.2   *E. Coli* Strain

The *E. coli* host strain used in this assay is *E. coli* C, ΔlacZ, galE⁻. It is kanamycin-resistant and sensitive to galactose.

27.4.3   **Reagents**

1. **5× Binding Buffer:**   50 mM Tris.HCL (made from a 1 M stock solution, pH 7.5), 5 mM EDTA, 50 mM $MgCl_2$, 25% glycerol (31.5 g/100 ml). Adjust pH to 6.8 with HCL and filter-sterilize the solution.

2. **IPTG Stock Solution:**   25 mg IPTG (isopropyl β-D-thiogalactopyranoside; Sigma) in ultrapure water. Filter-sterilize and store at –20°C.

3. **IPTG-Elution Buffer:**   10 mM Tris.HCL (made from a 1 M stock solution, pH 7.5), 1 mM EDTA, 125 mM NaCl. Filter-sterilize the solution.

4. **ATP Solution:**   10 mM ATP (Adenosine 5'-triphosphate; Sigma) in ultrapure water. Filter-sterilize and store at –80°C.

5. **HindIII Restriction Enzyme:**   New England Biolabs, 20 U/μl.

6. **Restriction Endonuclease NEBuffer #2:**   New England Biolabs.

7. **Magnetic Beads:**   Dynabeads M-450 sheep anti-mouse IgG (Dynal).

8. **Anti-β-galactosidase Monoclonal Antibody**   (Promega).

9. **LacI/LacZ Fusion Protein**   (available from Dynal, Norway, on request).

10. **Phosphate-Buffered Saline:**   Gibco BRL (without $MgCl_2$).

11. **T4 DNA Ligase:**   T4 DNA ligase (Gibco BRL, 1 U/μl) is diluted 10× in 1× T4 DNA ligase buffer (Gibco BRL).

12. **Glycogen:**   20 μg glycogen (Boehringer) per μl ultrapure water.

13. **Sodium Acetate:**   3M sodium acetate, pH 4.9, in ultrapure water.

14. **SOB Medium:**   20 g bacto-tryptone (Difco), 5 g bacto yeast extract (Difco), 0.5 g NaCl, 10 ml 250 mM KCl (add after everything else is dissolved) per liter distilled water. Adjust pH to 7.0 with 5 N NaOH and autoclave for 20 min (liquid cycle). Then add $MgCl_2$ to a final concentration of 5 mM ($MgCl_2$ should be autoclaved separately for 20 min).

15. **LB Medium:**   20 g LB Broth Base (Gibco BRL). Autoclave for 20 min (liquid cycle).

16. **LB Topagar:**   6.125 g/l LB Broth Base (Gibco BRL) and 6.125 g/l Antibiotic Medium 2 (Difco).

17. **Kanamycin:**   use as a 50 mg/ml solution, Sigma.

18. **Ampicillin:**   Sigma, make a stock solution of 50 mg/ml in ultrapure water.

19. **X-gal:** 5-bromo-4-chloro-3-indolyl-β-D galactoside; use as a 50 mg/ml solution, Promega.
20. **P-gal:** phenyl β-D-galactoside; use as powder, Sigma.
21. **Tetrazolium:** 2,3,5-triphenyl-2H-tetrazolium chloride; use as powder, Aldrich.
22. **Lysis Buffer:** 10 mM Tris-HCl, pH 8.0; 10 mM EDTA; 150 mM NaCl.

Note: It is important that all reagents are absolutely clean and contain no traces of plasmids (the aim of the assay is to detect lacZ minus plasmids at very-low frequency).

27.4.4 Preparation of LACI-LACZ Magnetic Beads

One ml of magnetic bead slurry is washed with 1 ml PBS, using a magnetic particle concentrator (Dynal) and resuspended in 850 μl PBS. Then 150 μl anti-β-galactosidase monoclonal antibody is added and the mixture is incubated for 1 h at 37°C while rotating. After incubation, the beads are washed 3 times with 1 ml PBS, resuspended in 950 μl PBS plus 50 μl lacI-lacZ fusion protein. The mixture is incubated for 2 h at 37°C while rotating. Subsequently, the beads are washed 3 times with 1 ml PBS and resuspended in 1 ml PBS. The lacI-lacZ beads can be stored at 4°C for up to 1 year.

27.4.5 Tissue Collection and DNA Extraction

1. Following sacrifice by CO_2-inhalation and decapitation, remove all tissues, place in 1.5-ml Eppendorf vials and freeze on dry ice. Maintain samples at –80°C until used.

2. Homogenize frozen tissues in 9 ml of lysis buffer (10 mM Tris-HCl, pH 8.0; 10 mM EDTA; 150 mM NaCl) in a 50-ml centrifuge tube with Brinkmann homogenizer. Add 1 ml of 10% SDS (final concentration: 1%), 120 μl of 10 mg/ml RNase A (final concentration: 120 μg/ml) and 200 μl of 25 mg/ml proteinase K (final concentration: 0.5 mg/ml) and digest the tissues overnight at 50°C in a hybridization oven while rotating.

3. The following morning add 1 vol phenol:chloroform:isoamyl alcohol (25:24:1, w/v) and gently shake the samples on a rocker for 30 min. Centrifuge the samples at 4000 × g for 20 min. Transfer the aqueous fraction to a new tube. Avoid transferring anything from the aqueous/organic interface. Repeat this step two to three times until a clean aqueous face is obtained.

4. Add 1/5 vol of 8M KAc and mix the samples gently. Add 1 vol chloroform and gently shake the samples on a rocker for 30 minutes. Centrifuge the samples at 4000 × g for 20 min.

5. Transfer the aqueous fraction to a clean tube. Avoid transferring anything from the aqueous/organic interface. Precipitate the DNA by adding 2 vol of ethanol and gently mixing. Spool out the precipitate with a sealed glass Pasteur pipet, swish in 70% ethanol to wash, allow to air-dry for a few minutes, then dissolve in 100-1000 μl 10 mM Tris-HCl, pH 8.0; 1 mM EDTA.

27.4.6 Preparation of Electrocompetent Cells

1. Add 50 µl of the lacZ⁻ galE⁻ *E. coli* C strain glycerol stock and 5 µl kanamycin to 10 ml LB medium in a 50-ml Falcon tube. Grow the cells overnight in an Innova 4000 incubator shaker (New Brunswick Scientific) at 37°C at 250 rpm.

2. Add 1.5 ml of the overnight culture to 500 ml LB medium in a 1-l Erlenmeyer. Grow the cells to an OD_{600} of 0.45. Distribute the cell suspension over six 500-ml tubes (Nalgene) and place on ice for 30 min.

3. Centrifuge for 15 min at 4000 × g in a Beckman R3C3 centrifuge. Resuspend the pellets in a total of one liter ice-cold ultrapure distilled water and centrifuge again for 15 min at 4000 g. Resuspend again in one liter ice cold ultrapure distilled water and repeat one more time (do not combine tubes).

4. Resuspend each pellet in 30 ml ice cold 10% glycerol and combine in two 50-ml tubes (Falcon). Centrifuge for 20 min at 4000 g, resuspend each pellet in 2.5 ml ice cold 10% glycerol and combine the contents of the 2 tubes. Measure the OD_{600} and adjust to 57 with ice cold 10% glycerol, i.e., 10 µl cell suspension in 3 ml LB medium should have an OD_{600} of 0.19.

5. Distribute the suspension in portions of 250 µl in eppendorf vials and freeze directly in a dry ice/ethanol bath. The electrocompetent cells can be kept at –80°C for several months.

27.4.7 Magnetic Bead Rescue of LACZ Plasmid from Mouse Genomic DNA

1. Gently resuspend on the edge of a vortex 60 µl of pelleted lacI-lacZ magnetic beads in a pre-made mixture of 15 µl 5× binding buffer, 2 µl Hind III (40 U), 10 to 50 µg genomic DNA (in 58 µl). Incubate for 1 h at 37°C while rotating.

2. After incubation, wash the beads three times with 250 µl 1× binding buffer (vortex gently) and resuspend in 75 µl IPTG-elution buffer plus 5 µl IPTG stock solution (vortex gently). Add 20 µl NEBuffer #2 and 100 µl ultra-pure water (vortex gently). Incubate for 30 min at 37°C while rotating.

3. Incubate at 65°C for 20 min and allow to cool down to room temperature. Spin the drops down and add 2 µl ATP solution (final concentration 0.1 mM) and 1 µl 0.1× T4 DNA ligase (total amount: 0.1 U). Gently vortex and incubate for 1 h at room temperature.

4. Resuspend the beads and pellet them. Then, transfer the supernatant to a clean tube. Precipitate the DNA for 1 h at –80°C, after adding 1.5 µl glycogen (30 µg), 0.1 volume sodium acetate (22 µl; vortex) and 2.5 volumes 95% ethanol (560 µl; mix). Centrifuge for 30 min in an eppendorf centrifuge at full speed, remove the ethanol, wash once with 250 µl 70% ethanol (vortex) and centrifuge 5 min in the same centrifuge at full speed. Remove all ethanol (use a pipet with a fine tip to remove the last traces) and allow the DNA pellet to dry for 10 to 15 min. Resuspend the DNA in 5 µl ultrapure water. The electrocompetent cells can now be added.

27.4.8 Electroporation, Plating, and Mutant Counting

1. Thaw electrocompetent cells on ice. Once thawed, directly add 60 µl cell suspension to the 5 µl DNA solution. Place the tube on the magnetic particle concentrator (to prevent carryover into the cuvette of magnetic beads possibly still left with the DNA).

2. Transfer the cells to a pre-chilled electroporation cuvette (0.1 cm electrode gap). Electroporate at 1.8 kV with 25 µF (Gene Pulser, BioRad) and 200 Ω (Pulse Controller, BioRad). Immediately add 1 ml ice-cold SOB medium. Transfer to a culture tube containing 1 ml SOB. Incubate for 30 min at 37°C while shaking (225 rpm). The time constant of the electroporation event should be 4.4 to 4.8.

3. Add 2 µl of the transformed cells (1:1000) to 2 ml SOB medium and combine with 13 ml LB topagar, containing ampicillin (end concentration 150 µg/ml), kanamycin (end concentration: 25 µg/ml), X-gal (end concentration: 75 µg/ml and tetrazolium (end concentration: 75 ng/ml). Plate into 9-cm petri dish. This is the titer plate.

4. Add 13 ml LB topagar to the rest of the transformed cells. In addition to ampicillin, kanamycin and tetrazolium, add p-gal in an end concentration of 0.3% (add p-gal directly as a powder to the topagar before adding the topagar to the transformed cells). X-gal is not necessary, but can be added also. Plate also in 9-cm petri dish. This is the selective plate.

5. Both titer and selective plate(s) are grown overnight at 37°C, for 14 to 16 h. The titer plate indicates the rescue efficiency. A typical yield is about 50,000 colonies (plasmid copies) per µg genomic DNA. Mutant frequencies are determined as the ratio of the number of colonies on the selective plates (visible as sharp dark-red points) versus the number of colonies on the titer plates (dark-red points with a much larger blue halo) times the dilution factor (1000 in this case). Mutant counting can best be done on a light table.

27.4.9 Mutant Characterization

Mutants can be characterized most conveniently by PCR. Alternatively, each mutant colony can be grown and plasmid DNA prepared by using miniprep kits. Below, only the protocol for PCR analysis is given. The entire procedure is also illustrated in Figure 27.2.

1. Transfer a single colony directly from a top agar plate or 1 µl from a mutant culture in LB to a thin-wall PCR tube and add 10 µl sterile ddH$_2$O. Incubate mixture at 94°C for 10 minutes into a thermocycler (with heated lid).

2. Mix the following components using the TaKaRa Ex Taq polymerase kit (Oncor) and primers pUR5101-F (5'-TCG CCA CCT CTG ACT TGA-3') and pUR3578-R (5'-GAA TAA GGG CGA CAC GGA-3'):

H$_2$O	24.5 µl
10× Ex Taq Buffer	5.0 µl
dNTP mix (2.5 mM each)	8.0 µl
20 µM pUR5101-F	1.0 µl
20 µM pUR3578-R	1.0 µl

     5 U/µl Ex Taq polymerase   0.5 µl

     Total   40.0 µl

3. Add the total 40 µl mixture to the 10 µl lysed cells in the PCR tube and run the following program :

   a. 94°C for 2 min, then

   b. 30 cycles of:94°C for 30 sec; 55°C for 45 sec; 70°C for 3 min.

   c. followed by a final extension at 70°C for 7 min and by cooling to 4°C

4. Digest 5 µl of the PCR product with 5 U AvaI and size-separate the fragments on a 1% agarose gel to determine product size. The PCR amplified wild-type construct has a total length of 3823 bp and the AvaI restricted fragments are sized at approximately 1.8, 1.4, and 0.57 kb, respectively.

27.5 Conclusion

27.5.1 Genome Rearrangements in Aging

The plasmid-based lacZ transgenic mouse system for *in vivo* mutation analysis described in this chapter should allow to experimentally test the somatic mutation theory of aging. Moreover, its capability to quantitate and characterize a broad range of mutations, provides the opportunity to specifically address the relationship between genome rearrangements and aging. In contrast to large DNA deletion events, point mutational changes occurring at low frequencies are unlikely to have phenotypic effects. On the other hand, relatively few genome rearrangements, involving many thousands of basepairs, can be expected to have major adverse effects on cell viability and functioning. Large deletions, including chromosomal aberrations, have been demonstrated to be preferentially induced by metabolically derived oxygen radicals [33], which are generally considered as a major cause of aging [34]. The possibility that large deletion mutations are more important than point mutations in causing age-related deterioration and death is also in keeping with the observed high levels of DNA rearrangements in cells from Werner's syndrome patients 35]. The gene defect in this autosomal recessive disorder with clinical symptoms resembling premature aging has recently been cloned and appeared to involve a member of the RecQ family of DNA helicases, that is, proteins that can be involved in various forms of DNA processing [36].

27.5.2 Validity of the Plasmid Model in Detecting Genome Rearrangements

Results thus far obtained with the plasmid model indicate mutational spectra in organs and tissues that consist for about 50% of size-change mutants and 50% no-change mutants, depending on the organ or tissue. The latter are assumed to represent point mutations or very small deletions or insertions, which have been confirmed by nucleotide sequencing. A major question regarding the validity of such data involved the potential contribution of the rescue procedure itself and/or the *E. coli* host to the mutation spectra obtained. This question has been extensively addressed by Dollé et al. [30] and led to the conclusion that virtually all mutants found are derived from the mouse. Extensive control experiments with plasmids grown in *E. coli* and subjected to the same rescue procedure did not provide evidence for artefactual mutations. Mutant frequencies of such plasmids were generally not higher than about 1×10^{-5}. However, since most of these mutations have probably arisen during the minipreparation procedure itself, it is not inconceivable that the background mutation frequency of the system is much lower than that. This is not unexpected since only mutations that would arise in *E. coli* during the first round of (plasmid) replication can be expected to allow its survival. In this respect, it is important to note that mutations in lacZ that do not completely inactivate the β-galactosidase activity can still be recovered in this system. Such so-called color mutants are easily recognizable by their light blue color on X-gal. Their mutant nature was confirmed by nucleotide sequencing [M. Dollé, unpublished]. So, theoretically a mixture of mutant and wild-type plasmids in the same bacterium will survive and scored as a mutant. To check for this possibility, we have grown a number of such color mutants in their *E. coli* host and re-introduced minipreparations of these plasmids in new *E. coli* hosts. The results indicated only colonies of the same light blue color. This strongly suggests that these color mutants arise in the mouse and not in *E. coli*.

Specifically with respect to the size-change mutants harboring a mouse sequence, i.e., the potentially large genome rearrangements, it is theoretically possible that these represent artifacts of HindIII digestion and ligation. To check for such a possibility, in some experiments all size-change mutants were re-digested with HindIII, which yielded one fragment only. This suggest that during the rescue procedure ligation was perfect. Some of the "mouse sequence mutants" were further characterized by nucleotide sequencing and physical mapping. Preliminary data confirm that such mutants involve large deletion or translocation events extending in the mouse flanking sequence as predicted.

27.5.3 Relevance of the Model for Age-Related Mutation Accumulation

In order to serve as a faithful marker for mutation accumulation in the aging genome, the integrated reporter gene should be representative for the genome overall and not deviate significantly from endogenous loci with respect to mutation susceptibility and spectrum. As discussed above, the observed mutation rates at the lacI and lacZ transgene loci are probably not too different from most selectable marker loci studied in T lymphocytes. Although Skopek et al. [27] reported a dramatic 10-fold difference between the lacI transgene and the endogenous HPRT gene, the latter may not be representative for most genes [13]. Moreover, endogenous genes are also likely to differ with respect to their mutation susceptibility and spectrum.

Regarding the lacZ transgenes, in our experience, most transgenic lines with the vector at different places in the genome have approximately the same mutant frequency. However, there are exceptions. In one transgenic line, with the lacZ-containing bacteriophage lambda vector integrated near the pseudoautosomal region of the X-chromosome, spontaneous somatic and germ line mutant frequencies were up to hundred times higher than in all other lines tested [37]. Genetic instability for this chromosomal region in the mouse was also found by others [38]. More recently, an about 20-fold higher spontaneous mutant frequency was reported for a bacteriophage lambda-based transgene on chromosome 7 [39]. These findings indicate the presence of unstable sites in the genome. However, the possibility that in some cases the transgenic integration actually creates a newly unstable locus can not be excluded. Although more research on this subject is necessary, taken together it seems reasonable to assume that most transgenic reporter gene lines represent the natural background mutation rate.

Thus far, the results obtained with the plasmid model on mutant frequencies and spectra in young versus old mice indicate a linear age-related increase in mutations in liver. In brain, mutant frequencies do not change with age [40]. Interestingly, a relatively large fraction of the mutation spectrum in liver consists of potentially large genome rearrangements, that is, size-change mutants harboring a mouse sequence (see above). In brain, this fraction is significantly smaller, i.e., about 1 to 3% as compared to up to 10% in liver. Moreover, these genome rearrangement events in the liver appeared to increase exponentially with age after about 24 months.

The key question with respect to the interpretation of such data is the functional relationship of the observed mutant frequencies and spectra in the reporter gene with actual pathophysiological changes during aging. In this respect, two strategies can be envisaged. The first involves the possibility of direct correlation. Indeed, the observed higher level of genome rearrangements and steeper age-related increase in mutant frequencies in liver correlates with the more severe pattern of multiple pathology in this organ as compared to brain [41,42]. More extensive correlative studies involving multiple organs and tissues are presently in progress.

A second possibility involves the generation of double and triple transgenic mice against the lacZ-plasmid background. For example, a transgenic mouse line overexpressing SOD and catalase can be expected to have lower background mutant frequencies and a longer life span. (The latter is suggested by results from Orr and Sohal with transgenic *Drosophila* overexpressing these two genes [43].) Also other candidate longevity assurance genes can be manipulated to test if life extension can be accomplished via a reduction in the mutation rate. Examples are to transgenetically upgrade various DNA transaction processes, e.g., DNA replication, DNA repair. The consistent demonstration that genes that control life span also control the spontaneous mutation rate would be evidence for a major role of somatic mutagenesis in the etiology of aging.

Acknowledgments

This work was supported by NIH grants AG10829, AG13314, amd CA75653.

References

1. Failla, G., The aging process and carcinogenesis, *Ann. N.Y. Acad. Sci.* 71, 1124–1135, 1958.
2. Szilard, L., On the nature of the aging process, *Proc. Natl. Acad. Sci. U.S.A.* 45, 35–45, 1959.
3. Strehler, B.L., Deletional mutations are the basic cause of aging: historical perspectives, *Mutation Res.* 338, 3–17, 1995.
4. Curtis, H. and Crowley, C., Chromosome aberrations in liver cells in relation to the somatic mutation theory of aging, *Radiat. Res.* 19, 337–344, 1963.
5. Ames, B.N., Shigenaga, M.K., and Hagen, T.M., Oxidants, antioxidants, and the degenerative diseases of aging. *Proc. Natl. Acad. Sci. U.S.A.* 90, 7915–7922, 1993.
6. Harman, D., Aging: A theory based on free radical and radiation chemistry, *J. Gerontol.* 11, 298–300, 1956.
7. Vijg, J. and Gossen, J.A., Somatic mutations and cellular aging. *Comp. Biochem. Physiol.* 104B, 429–437, 1993.

8. Ramsey, M.J., Moore II, D.H., Briner, J.F., Lee, D.A., Olsen, L.A., Senft, J.R., and Tucker, J.D., The effects of age and lifestyle factors on the accumulation of cytogenetic damage as measured by chromosome painting. *Mutation Res.* 338, 95–106, 1995.

9. Liu, Y., Hernandez, T., Shibata D., and Cortopassi, G., BCL2 translocation frequency rises with age in humans. *Proc. Natl. Acad. Sci. U.S.A.* 91, 8910–8914, 1994.

10. Albertini, R.J., Nicklas, J.A., O'Neill, J.P., and Robison, S.H., *In vivo* somatic mutations in humans: measurement and analysis, *Annu. Rev. Genet.*, 24, 305–326, 1990.

11. I.M. Jones, C.B. Thomas, B. Tucker, C.L. Thompson, P. Pleshanov, I. Vorobtsova, D.H. Moore II, Impact of age and environment on somatic mutation at the hprt gene of T lymphocytes in humans. *Mutation Res.* 338, 129–139, 1995.

12. Dempsey, J.L., Pfeiffer, M., and Morley, A.A., Effect of dietary restriction on *in vivo* somatic mutation in mice. *Mutat. Res.* 291, 141–145, 1993.

13. Grist, S.A., McCarron, M., Kutlaca, A., Turner, D.R., and Morley, A.A., *In vivo* human somatic mutation: frequency and spectrum with age. *Mutation Res.* 266, 189–196, 1992.

14. da Cruz, A.D., Curry, J., Curado, M.P., and Glickman, B., Monitoring hprt mutant frequency over time in T-lymphocytes of people accidentally exposed to high doses of ionizing radiation. *Environ. Mol. Mutagen.*, 27, 165–175, 1996.

15. Weindruch, R., Caloric restriction and aging. *Sci. Am.* 274, 46–52, 1996.

16. Martin, G.M., Ogburn, C.E., Colgin, L.M., Gown, A.M., Edland, S.D., and Monnat Jr, R.J., Somatic mutations are frequent and increase with age in human kidney epithelial cells. *Hum Mol. Genet.* 5, 215–221, 1996.

17. Gossen, J.A., de Leeuw, W.J.F., Tan, C.H.T., Lohman, P.H.M., Berends, F., Knook,D.L., Zwarthoff, E.C., and Vijg, J., Efficient rescue of integrated shuttle vectors from transgenic mice: a model for studying gene mutations *in vivo*. *Proc. Natl. Acad. Sci. U.S.A.* 86, 7971–7975, 1989.

18. Gossen, J.A. and Vijg, J., E. coli C: a convenient host strain for rescue of highly methylated DNA. *Nucleic Acids Res.* 16, 9343, 1988.

19. Malamy, M.H., Frameshift mutations in the lactose operon of E. coli. *Cold Spring Harbor Symp. Quant. Biol.* 31, 189, 1966.

20. Gossen, J.A. and Vijg, J., A selective system for LacZ⁻ phage using a galactose-sensitive E. coli host. *Biotechniques* 14, 326–330, 1993.

21. Kohler, S.W., Provost, G.S., Fieck, A., Kretz, P.L., Bullock, W.O., Sorge, J.A., Putman D.L., and Short, J.M., Spectra of spontaneous and mutagen-induced mutations in the LacI gene in transgenic mice, *Proc. Natl. Acad. Sci. U.S.A.* 88, 7958–7962, 1991.

22. de Boer, J.G., Erfle, H.L., Walsh, D., Holcroft, J., and Glickman, B.W., The use of lacI transgenic mice in genetic toxicology. In: G.P. Pfeifer (Ed.), *Technologies for Detection of DNA Damage and Mutations*, Part II, Plenum Press, New York, pp. 411–429, 1996.

23. Stuart, G.R., Gorelick, N.J., Andrews, J.L., de Boer, J.G., and Glickman, B.W., The genetic analysis of lacI mutations in sectored plaques from Big Blue transgenic mice. *Env. Mol. Mutagen.*, 28, 385–392, 1996.

24. Gossen, J.A. and Vijg, J., Transgenic mice as model systems for studying gene mutations *in vivo*. *Trends in Genetics* 9, 27–31, 1993.

25. Gorelick, N.J., Overview of mutation assays in transgenic mice for routine testing. *Environ. Mol. Mutagen.*, 25, 218–230, 1995.

26. Tao, K.S., Urlando C., and Heddle, J.A., Comparison of somatic mutation in a transgenic versus host locus, *Proc. Natl. Acad. Sci. U.S.A.* 90, 10681–85, 1993.

27. Skopek, T.R., Kort, K.L., and Marino, D.R., Relative sensitivity of the endogenous hprt gene and lacI transgene in ENU-treated Big Blue B6C3F1 mice. *Environ. Mol. Mutagen.* 26, 9–15, 1995.

28. Boerrigter, M.E.T.I., Dollé, M.E.T., Martus, H-J., Gossen, J.A., and Vijg, J., Plasmid-based transgenic mouse model for studying *in vivo* mutations. *Nature*, 377, 657–659, 1995.

29. Gossen, J.A., de Leeuw, W.J.F., Molijn A.C., and Vijg, J., Plasmid rescue from transgenic mouse DNA using LacI repressor protein conjugated to magnetic beads, *BioTechniques* 14, 624–629, 1993.

30. Dollé, M., Martus, H-J., Gossen, J.A., Boerrigter M.E.T.I., and Vijg, J., Evaluation of a plasmid-based transgenic mouse model for detecting *in vivo* mutations. *Mutagenesis* 11, 111–118, 1996.

31. Gossen, J.A., Molijn, A.C., Douglas, G.R., and Vijg, J., Application of galactose-sensitive *E.coli* strains as selective hosts for LacZ⁻ plasmids, *Nucleic Acids Res.* 20, 3254, 1992.

32. Gossen, J.A., Martus, H-J., Wei, J.Y., and Vijg, J., Spontaneous and X-ray-induced deletion mutations in a lacZ plasmid-based transgenic mouse model. *Mutat Res.* 331, 89–97, 1995.

33. Gille, J.J.P., van Berkel, C.G.M., and Joenje, H., Mutagenicity of metabolic oxygen radicals in mammalian cell cultures. *Carcinogenesis* 15, 2695–2699, 1994.

34. Martin, G.M., Austad S.N., and Johnson, T.E., Genetic analysis of aging: role of oxidative damage and environmental stresses. *Nature Genet.* 13, 25–34, 1996.

35. Fukuchi, K., Martin, G.M., Monnat Jr., R.J., Mutator phenotype of Werner syndrome is characterized by extensive deletions. *Proc. Natl. Acad. Sci. U.S.A.* 86, 5893–5897, 1989.

36. Yu, C-E., Oshima, J., Fu, Y-H., Wijsman, E.M., Hisama, F., Alisch et al., R., Positional cloning of the Werner's syndrome gene. *Science* 272, 258–262, 1996.

37. Gossen, J.A., de Leeuw, W.J.F., Verwest, A., Lohman, P.H.M., and Vijg, J., High somatic mutation frequencies in a LacZ transgene integrated on the mouse X-chromosome, *Mutation Res.*, 250, 423–429, 1991.

38. Kipling, D., Salido, E.C., Shapiro, and L.J., Cooke, H.J., High frequency *de novo* alterations in the long-range genomic structure of the mouse pseudoautosomal region, *Nature Genet.* 13, 78–82, 1996.

39. Leach, E.G., Gunther, E.J., Yeasky, T.M., Gibson, L.H., Yang-Feng, T.L., and Glazer, P.M., Frequent spontaneous deletions at a shuttle vector locus in transgenic mice. *Mutagenesis* 11, 49–56, 1996.

40. Dollé, M.E.T., Giese, H., Hopkins, C. L., Martus, H.-J., Hausdorff, J. M., and Vijg, J., Rapid accumulation of genome rearrangements in liver but not in brain of old mice. *Nature Genet.* 17, 431–434. 1997.

41. Bronson, R. T., Rate of occurrence of lesions in 20 inbred and hybrid genotypes of rats and mice sacrificed at 6-month intervals during the first years of life, in *Genetics of Aging II*, Harrison D. E., Ed., Telford Press, Caldwell, N.J., 1990, pp. 280–358.

42. Bronson, R. T. and Lipman, R. D., Reduction in rate of occurrence of age-related lesions in dietary restricted laboratory mice, *Growth, Development & Aging*, 55, 169–184, 1991.

43. Orr, W.C. and Sohal, R.C., Extension of life span by overexpression of superoxide dismutase and catalase in *Drosophila melanogaster. Science* 263, 1128–1130, 1994.

28

Protein Oxidation

Pamela E. Starke-Reed

CONTENTS

28.1 Introduction

Proteins play an important role in cellular function and therefore any alterations to them may significantly affect the ability of the organism, cell, tissue, or organ to function. The initial reports of conformational changes or oxidative modifications of proteins has led to a great number of studies examining protein modification during oxidative stress and aging. Modified unstable proteins accumulate in aging and interfere with normal homeostasis. Age-related changes in enzymes and other proteins include alterations in catalytic activities (increases and decreases) [1], altered heat stability [2,3], and altered folding characteristics [4]. The accumulation of these altered proteins may account for age-related problems and pathologies such as vision impairment (senile cataract), ischemic-reperfusion injury, failure of the immune system, decreased ability to heal wounds, and loss of cognitive function (Alzheimer's disease). The question remains as to why altered proteins accumulate in aging cells.

Major factors in determining the rates of degradation of various proteins are post-translational covalent modifications that make them more susceptible to proteolytic attack. The following modifications have been shown to mark the proteins for proteolytic degradation [5]: conjugation of ubiquitin to the proteolytic ε-amino group of lysine residues; phosphorylation of serine or threonine hydroxyl groups; oxidation of sulfhydryl groups; oxidation of non-heme iron clusters; deamidation of asparagine and glutamine residues, and the metal-catalyzed oxidation of amino acid side chains to derivatives. The age-related accumulation of altered proteins and enzymes reflects the balance between the rate of modification and elimination (degradation) of modified forms. With increasing age either the rate of modification increases or the rate of degradation decreases, or more likely it is a combination of both. It stands to reason that the age-related decrease in the levels of key metabolic regulatory enzymes and the accumulation of cellular protein as catalytically inactive, modified protein would seriously compromise some physiological functions.

28.2 Background

Although the free radical theory of aging was proposed over 40 years ago [6], much of the research has focused on the oxidation of nucleic acids (NAs) and lipids. During the past fifteen years, research has begun to focus on the alteration of proteins during aging. In the early 1980s, the Stadtman Laboratory demonstrated that many of the proteins susceptible to oxidative inactivation were also known to lose catalytic activity during aging [7]. The Stadtman Laboratory can be credited for initiating the extensive work on protein oxidation found in the literature today. Their work demonstrated a site-specific metal-catalyzed oxidative modification which provided the first consistently reliable marker of protein oxidation, which is chiefly responsible for the significant growth in the field of protein oxidation in physiological systems. Many of these studies have focused specifically on protein modification during the aging process. This laboratory has shown that specific amino acids are susceptible to oxidative modification by a number of enzymic and non-enzymic mixed function oxidation (MFO) systems [7]. Amici et al. [8] demonstrated that MCO reactions preferentially target histidine, arginine, lysine, methionine and cysteine residues, converting histidine to aspartate/asparagine residues, arginine to glutamic semialdehyde residues, lysine to 2-aminoadipic semialdehyde derivatives and cysteine to disulfide derivatives.

The original finding that oxidative modification of proteins leads to the formation of carbonyl derivatives of certain amino acid side chains provided one of the first biomarkers for determining levels of protein oxidation. Several studies have shown an age-dependent accumulation of oxidized proteins accompanied by the decrease in specific activities of key metabolic enzymes in many models of aging. Protein carbonyl content has been shown to increase with age in: (1) rat liver, brain, skeletal muscle [9–11], (2) human erythrocytes, fibroblasts [12] and brain [13], and (3) fly (cytosolic and mitochondrial fractions) [14]. Cultured fibroblasts from individuals with premature aging syndromes show a significant increase in levels of oxidized proteins when compared to age-matched controls [12]. The increase in protein carbonyl levels was associated with decreases in cellular enzyme activity. The oxidation of specific amino acids leads to the generation of carbonyl derivatives, the level of which can be used as a measure of oxygen radical-induced damage. Section 28.3 below will discuss the various methods for carbonyl determination.

It should be mentioned here that metal-catalyzed oxidation of amino acids is not the only process that leads to the formation of carbonyl derivatives. The interaction of proteins with reducing sugars or their oxidation products will also lead to the formation of carbonyl groups [15]; however, the contribution of glycation/glycoxidation products which have been identified (isopentosidine, N-carboxymethyl lysine) appear to be very low compared to the total carbonyl content found in proteins from old animals [16]. Glycation products are just briefly mentioned here because they are fully covered in Chapter 29 in this volume.

28.3 Protein Carbonyl Measurements

Building on the fact that oxidation of some amino acid side chains leads to the formation of protein carbonyl groups, several assays were developed to measure these products. The first reliable and consistent method to be utilized for carbonyl determination was the tritiated borohydride assay. Other methods were later developed whereby the carbonyl groups on proteins were derivatized with 2,4-dinitrophenylhydrazine and the resulting hydrazones were quantitated. The various methods are described below. All of these assays have been published in various journals, but have been brought together here to provide a complete reference for investigations into protein oxidation. Due to space limitation, this chapter will not cover in explicit detail all of the methods; for more in-depth detail see the referenced papers.

These methods can all be used for both purified as well as crude extract proteins. Crude extracts usually present more difficulties and therefore may require additional steps to obtain reproducible results. Because oxidatively modified proteins are highly susceptible to proteolytic degradation, it is essential to include a protease inhibitor cocktail in the homogenization or sonication buffer. The cocktail which has worked the best contains: 0.5 µg/ml leupeptin, 0.5 µg/ml aprotinin, 0.7 µg/ml pepstatin, and 40 µg/ml phenylmethylsulfonyl fluoride and 1.1 mM ethylenediamine tetraacetic acid (EDTA) in a 10 mM HEPES buffer, pH 7.4 [17].

Lipids can contribute carbonyl groups so it is necessary to achieve very good separation of the protein and lipid phases of the crude extracts. This is accomplished by centrifugation of the sample at 28,000 g for 30 minutes at 0°C. Proteins isolated from crude extracts may be contaminated with nucleic acids (NAs) which react with derivatizing reagents giving artifactual measures and unreliable results in all methods except for the HPLC method. Elimination of NAs can be accomplished by treating the samples with 1% streptomycin sulfate in 50 mM HEPES buffer (pH 7.2) for 15 min, followed by centrifugation at 11,000 g for 10 min to precipitate the NAs [18]. The presence of NAs can be monitored spectrophotometrically by measuring the absorbance at 260 nm. Elimination of the NAs is confirmed when the ratio of absorbance at 280 nm to 260 nm is greater than 1. Additional precipitation steps with 1% streptomycin sulfate may be required to achieve the appropriate ratio. Once the samples are free of NAs they can them be asayed for carbonyl content determination by several methods depending upon the type and amount of sample analyzed, and the laboratory equipment available.

28.3.1 Tritiated Borohydride Method

The first sensitive and reliable method to measure protein carbonyl groups is the reaction with tritiated borohydride. In this reaction all carbonyl groups such as aldehydes, keytones, and ketoacids are reduced to alcohols. This method also reduces carbonyl groups formed via Schiff bases with the ε-amino groups of lysine or the α-amino groups of the amino terminus and Schiff bases. This method as described below has been taken from the paper of Levine, et al. [19].

Reagents Needed:

100 mM sodium hydroxide

1 M Tris, with 10 mM EDTA, adjusted to pH 8.5 with HCL

10% (w/v) trichloroacetic acid (TCA)

6 M guanidine, with 20 mM potassium phosphate adjusted to pH 2.3 with trifluoroacetic acid

1 M sodium borohydride (NaBH$_4$) in 100 mM NaOH

100 mM sodium borotritide in NaOH with specific activity at 100 mCi/mol

Procedure

1. Suspend protein in 50 μl water
2. Remove contaminating NAs from sample
3. Add 6 μl of 1 M Tris-HCL, 10 mM EDTA, and 14 μl of 100 μM NaB^{3}H$_4$
4. Incubate at 37°C for 30 min
5. Add 10% TCA and let stand for 5 min in the hood to precipitate the protein
6. Centrifuge at 11,000 × g for 3 minutes
7. Draw off supernatant and wash pellet 2× with 1 ml of 10% TCA
8. Resuspend pellet at 37°C in 0.6 to 0.8 ml of guanidine solution for 15 min
9. Count the radioactivity in a liquid scintillation counter and determine protein recovery by standard direct spectrophotometric or chemical methods (i.e., Lowry)

This was the method of choice for measuring carbonyl groups in samples containing compounds such as myoglobin and hemoglobin which have absorbance peaks near the peak of the hydrazone derivatized product. Such compounds would interfere with measurements in the appropriate range and therefore can not be measured spectrophotometrically. A potential drawback to this method is the requirement for working with radioactive substances which not all laboratories are equipped to do.

28.3.2   2,4-DNPH Extraction Method

To avoid working with radioactivity, a newer method was developed based on the fact that carbonyl groups readily react with hydrazines to form stable hydrazones [19]. More protein (0.5 to 1 mg) is needed for this assay which may present difficulties depending upon the source of the sample. As mentioned above, proteins such as myoglobin or hemoglobin can not be measured this way due to peak absorbance interference, but most cellular proteins have been successfully evaluated with the derivatization methodology. The results obtained with the hydrazine methods are identical with those of the borohydride method.

Reagents Needed:

2 M 2,4-dinitrophenylhydrazine in 2 M HCl (2,4-DNPH)

20% (w/v) TCA

10% (w/v) TCA

Ethanol:ethyl acetate, 1:1 (v/v) (make fresh for each assay)

6 M guanidine hydrochloride solution with 20 mM potassium phosphate, adjusted to pH 2.3 with trifluroacetic acid (TFA).

Procedure

1. Remove NAs from crude extract samples.
2. Place 0.5 to 1.0 mg protein in a 12 ml glass tube, volume should be in the range of 300 to 500 μl. Set up parallel sample for blank if using crude extract.
3. Pipet 20% TCA in the same volume used for the protein giving a final concentration of 10% TCA. This step is not necessary if working with pure protein.
4. Precipitate the protein by mild centrifugation in a table-top centrifuge for 1 min.
5. Draw off the TCA solution and resuspend the protein sample in 500 μl of the 10 mM 2,4-DNPH solution, and add 500 μl of 2 M HCL for the blank if using a crude extract. Each crude protein sample must also have a blank run in parallel using 2 M HCL rather than the 2,4-DNPH.
6. Allow to stand at room temperature for 15 to 30 min.
7. Add 500 μl of 20% TCA and centrifuge in a table-top centrifuge for 3 min.
8. Draw-off and discard the supernatant, wash the pellet with the ethanol: ethyl acetate solution, let stand for 10 min and then centrifuge for 3 min.
9. Repeat step 8 two additional times.
10. Resuspend the precipitate in 0.5 ml 6 M guanidine solution.
11. Place samples in 1.5 ml plastic centrifuge tubes and centrifuge in microcentrifuge for 3 min to remove any insoluble debris.
12. Obtain spectrum, read derivitized sample against it's blank, or against water for pure protein. If possible read all blanks before reading the derivatized samples.
13. Calculate carbonyl content by using the maximum absorbance in the range of 360 to 390 nm using the molar absorption coefficient of 22,000 $M^{-1} cm^{-1}$.
14. Determine protein recovery by spectrophotometric method.

28.3.3 2,4-DNPH Derivatization Filtration Method

The second derivatization method is similar to the extraction methods, except, the derivatized protein is recovered by filtration [19]. It has been noted that proteins from crude extracts tend to aggregate during the centrifugation method causing problems with resuspension. This can severely limit the amount of protein contributing to the absorbance at 360 to 390 nm, giving artificially low values for the carbonyl level determination. It may also trap non-specifically bound reagent in the protein pellet, giving rise to artificially high absorbance in the 360 to 390 nm range. However, the extraction method is preferred when derivitizing pure proteins because their precipitates tend to be washed through the filter paper.

Additional Reagents Needed:

Whatman 3 MM filter paper disks
small vacuum manifold
small (10 ml) beakers
automatic mechanical rotating tray

Procedure

The procedure for the protein derivatization is the same as above with the following additions and changes:

1. During the derivatization procedure, the filter papers need to be treated as follows: place all filter papers (one for each sample, plus one for each blank and one for the spectrophotometric blank) in 6 M guanidine solution, rotating gently for 30 min. Wash the filter disks two times with de-ionized, distilled water and then place them in 10% TCA for 15 min.
2. Follow the extraction derivatization procedure as for above up to and including step 6.
3. Add 0.5 ml 20% TCA to each tube and immediately place on ice.
4. Place filter paper on the manifold under low vacuum. Wash it three times with 5 ml of ethanol:ethyl acetate solution. Then wash with 2 ml of 10% TCA. This will be used as the blank for the spectrophotometer.
5. Transfer precipitated protein to center of the filter disk. Wash it three times with 5 ml of ethanol:ethyl acetate solution. Wash with 2 ml of 10% TCA.
6. Place filter disk in beaker with 2 ml of 6 M guanidine solution, cover and gently rotate for 1.5 to 2 h at room temperature.
7. Carefully remove the guanidine solution and follow the extraction derivitization procedure above again starting at step 11.

This method is more tedious than the extraction procedure, but with crude extracts it gives more consistent results. For both of the spectrophotometric methods the use of a diode array spectrophotometer makes the determination much easier. The 360 to 390 nm peak can be read simultaneously with the 276-nm peak for protein determination. It is essential to determine the protein recovery to ensure protein was not lost during the processing of the sample.

28.3.4   2,4-DNPH HPLC Assay

Methods building on hydrazine labeling of carbonyl derivatives have been developed recently which provide more efficient ways of detecting oxidized proteins [20]. The first method utilizes high performance liquid chromatography (HPLC) gel filtration, and the second one employs a Western-blotting technique for sensitive and specific determinations of protein carbonyl groups. The new HPLC method is much more sensitive and specific and therefore requires less protein/sample. With the earlier derivatization methods 0.5 to 1.0 mg of protein was required; now, reproducible and reliable determination can be made using as little as 10 μg of protein containing 1 mol carbonyl/mol protein or 100 μg of protein containing 0.1 mol carbonyl/mol protein. This allows carbonyl determination to be made on single cell culture dishes or in small samples from biopsy and autopsy materials. The protein samples are separated by molecular weight which gives a more specific indication of the particular proteins which are oxidatively modified. Another advantage of this method, as with stand alone spectrophotometer method, is that most HPLCs today have a multichannel detector so that both the hydrazone formation and protein recovery determination can be done simultaneously.

Reagents Needed:
Buffer for Gel Filtration and Derivatization Blank

6 M guanidine HCL

0.5 M potassium phosphate buffer

Final pH is 2.5, adjusted with 10 M KOH

Make up 1 liter; stock solution is stable for months

Derivatization Solution

10 mM 2,4 DNPH

6.0 M guanidine

0.5 M potassium phosphate, use same as described above for derivitization blank.

Column: the best column found so far for this assay is the Zorbax GF450[8] using a flow rate of 2 ml/min.

Procedure

1. Divide sample into two equal amounts and place in 1.5 ml plastic tubes with tops; one for derivatization and one for the blank. Sample volume should be no greater than 75 μl.

2. Add the derivatizing buffer to the treatment sample or the blank buffer for the blank sample to the tubes in a volume 3 times that of the sample.

3. Let tubes stand for 15 to 30 minutes at room temperature.

4. Centrifuge the samples at 11,000 × g for 3 min in a microcentrifuge to remove insoluble debris.

5. Load sample into the column and monitor the chromatograms at 370 and 276 nm for hydrazone and protein levels, respectively. It is best if the detector can follow both wavelengths simultaneously; however, if only one can be monitored use the 370 nm and then determine protein in a separate assay. Using the specified column, the sample will begin to elute after 3 min and the reagent after 6 min.

6. Calculate the carbonyl content. This is done by the method of Levine et al. [20] using the following calculations:

$$\text{mol carbonyl/mol protein} = \frac{\left(\varepsilon_{\text{protein}276}\right)\left(\text{Area}_{370}\right)}{22{,}000\left(\text{Area}_{276} - 0.43\ \text{Area}_{370}\right)} \tag{28.1}$$

Use 50,000 for the molar absorptivity of the protein which is a good estimate of the molar absorptivity of a protein of average amino acid composition and moleculat weight of 50,000. Use the exact molar absorptivity of the protein if it is known. The result is a ratio. One can determine the mass of a peak by:

$$\text{mol} = \frac{\text{area} \times \text{flow}}{\varepsilon_M \times \text{path length}} \tag{28.2}$$

Given a flow rate of 2 ml/min and an $\varepsilon_{\text{M370nm}}$ of 22,000, the calculation for carbonyl is:

$$\text{pmol carbonyl} = 2.53\ \text{area}_{270\,\text{nm}} \tag{28.3}$$

For protein determination with $\varepsilon_{\text{M276nm}}$ of 50,000 the equation would be

$$\text{pmol protein} = 1.11\left(\text{Area}_{276\,\text{nm}} - 0.43\ \text{Area}_{370\,\text{nm}}\right) \tag{28.4}$$

with the area at 276 nm being corrected for any contribution of the hydrazine. For a molecular weight of 55,000 then

$$\text{ng protein} = 55.5\left(\text{Area}_{276\,\text{nm}} - 0.43\ \text{Area}_{370\,\text{nm}}\right) \tag{28.5}$$

This method can not be done with low-pressure systems such as the FPLC due to the high back pressure caused by the viscous guanidine solution. Some attempts have been successful using gravity gel filtration on single-use columns (PD-10 from Pharmacia), however, the results are not as consistent as with the HPLC method. Following derivatization and separation on this gravity column, the carbonyl is quantitated on the spectrophotometer as described above.

28.3.5 SDS Derivatization Methods for HPLC and Immunoblotting

Due to the fact that guanidine is very rough on the HPLC columns, an alternative method has been developed whereby proteins are derivatized in sodium dodecyl sulfate (SDS) for analysis on both HPLC and gel electrophoresis [20]. Although guanidine has been the denaturing reagent of choice in the previous methods, changing to SDS provides several advantages over guanidine: better separation of reagent from lower molecular weight components, lower back pressure with SDS, and elimination of salt corrosion providing a much longer column life.

28.3.5.1 HPLC Assay

Reagents Needed:
Gel Filtration Buffer

200 mM sodium phosphate, pH 6.5

1% SDS

20 mM 2,4-DNPH in 10% (v/v) trifluoroacetic acid (TFA)

10% (v/v) TFA (use for blank determination)

12% SDS

2 M Tris — free base not HCL salt

30% glycerol

Procedure

1. Prepare sample as before. Remove contaminating NAs. Sample should be 50 mM or less potassium because it is less soluble than the sodium salt.

2. Split sample into two equal portions, one for derivatization, one for the blank.

3. Add 1 vol of 12% SDS

4. Add 2 vol of 2,4-DNPH solution with mixing. For blank add two volumes of 10% TFA without 2,4-DNPH.

5. Set at room temperature for 15–30 min.

6. Add 1.5 sample vol of 2 M Tris/30% glycerol to neutralize the sample. Color will change from light yellow to orange.

7. For HPLC analysis, load sample on column and monitor the chromatograms as above with the guanidine system with the alteration that the hydrazones need to be monitored at 360 nm rather than 370 nm.

28.3.5.2 *Immunoblotting Assay*

Additional Reagents Needed:

Gel electrophoresis apparatus

Western-blotting apparatus

Antibodies to 2,4-dinitrophenyl (DNP) moiety, commercially available from Sigma (St. Louis, MO) in both monoclonal and polyclonal forms.

Labeled secondary anti-anti 2,4 DNP antibody.

Procedure

1. Derivatize samples exactly as above for the SDS HPLC procedure up to and including Step 6.

2. Load samples onto SDS-polyacrylamide gel and electrophorese according to the method of Laemmli [21].

3. Following electrophoresis, transfer gel to nitrocellulose and carry out standard Western-blotting technique.

4. Analyze by either colorometric or chemiluminescent techniques depending on label used.

With this method the sample can be split for analysis with both HPLC and SDS gel electrophoresis. Alternatively, samples from the HPLC gel filtration may also be loaded onto the SDS gel for Western blotting analysis of particular HPLC fractions. It may be helpful to include standard samples of known carbonyl content to help determine development times for the unknown samples.

28.3.6 Fluorescein Thiosemicarbazide Gel Electrophoresis

An additional method for a qualitative rather than a quantitative measurement of protein carbonyls is the reaction with fluorescein thiosemicarbazide for gel electrophoresis [19]. Fluorescein thiosemicarbazide reacts with carbonyl groups giving rise to fluorescent thiosemicarbazones. Labeled proteins are separated on a lithium dodecyl sulfate gel electrophoresis allowing the visualization of protein bands containing oxidized proteins. This method is of limited use for quantitation due to the instability of some of the derivatives and the variability of the quantum yield; however, it does facilitate visualization of proteins containing oxidized side chains and an indication of the molecular weights of the proteins which are oxidized. Another potential drawback of the method is that the derivative is stable at 0°C, but not room temperature, so electrophoresis must be done in the cold room. For this reason, lithium rather than the sodium dodecyl sulfate is used for electrophoresis.

Reagents Needed:

0.25% fluorescein thiosemicarbazide in dimethylformamide or dimethyl sulfoxide

1:1 ethanol:ethyl acetate (v/v)

20% TCA (w/v)

10% TCA (w/v)

Concentrated HCl

1 M Tris, adjusted to pH 8.5 with HCL

8 M Urea

Gel electrophoresis apparatus

Wratten 47B and #12 gelatin filters

Light source and Polaroid camera system

Procedure

1. Remove contaminating NAs.
2. Place 1 to 2 mg protein sample into 12 or 15 ml conical glass tube. Cover tubes with foil to protect reagents from light.
3. Add 20% cold TCA (equal volume of protein sample) to the tubes to precipitate protein.
4. Wash precipitate two times with 10% TCA
5. Suspend pellet in 1 ml fluorescein semicarbazide plus 1 drop of concentrated HCl. Flush tubes with nitrogen or argon gas to remove oxygen. Cover tube to keep oxygen out during the reaction

6. Keep sample at room temperature for 30 minutes, stirring frequently.

7. Add 20% TCA to the tubes and centrifuge in table-top centrifuge for 1 min to precipitate protein. Remove supernatant and invert tubes to completely drain the reagent. From this point on keep samples cold.

8. Add 50 μl cold 1 M Tris, pH 8.5 to neutralize the sample. Add 1 ml 8 M urea.

9. Carry out electrophoresis by the method of Laemmli [21] with the following changes: electrophoresis must be run in the cold room; lithium dodecyl sulfate is used in place of sodium dodecyl sulfate because it has better solubility at colder temperatures. The fluorescent derivative is more stable at a lower pH so the gel buffer should be pH 8.5 rather than the usual pH 8.8.

10. Load sample into gel and run electrophoresis as usual.

11. Immediately place finished gel between two glass plates and illuminate at 45° angle with a bright light through a Wratten 47B gelatin filter. Photograph the gel with the Polaroid camera with the #12 filter placed in front of the lens. Exposure times may vary and should be determined for each gel. Start with a 30 sec exposure and then optimize time depending on initial results. Longer exposure times may be needed for very light bands and shorter times for dark bands. It is necessary to use the camera to see the fluorescent bands because some may not be viable to the naked eye.

12. Stain the gel for protein determination following the photography step.

28.4 Dityrosine Formation Measurement

The free radical generation of dityrosine has been known for over 30 years [22], and recently, the effect on cellular proteins has been examined. Free radicals react with tyrosine to generate a tyrosyl radical which then through radical isomerization followed by diradical reaction and enolation lead to the formation of the stable endproduct dityrosine. Protein dityrosine content has been shown to increase in conditions associated with oxidative stress including: aging [23], UV radiation [24], γ-radiation [25], incubation with peroxides [26–29] and exposure to oxygen free radicals [30–32]. Two methods have been described for quantitation of dityrosine content in proteins. The first described by Giuluvi and Davies [33] uses the HPLC to measure ditryosine released from proteins following proteolytic digestion. The second method is a highly sensitive assay for both o-tyrosine and o,o'-dityrosine, which uses a combination of stable isotope gas chromatography and mass spectrometry (GC/MS) [34].

28.4.1 HPLC Following Proteolytic Digestion

It is best to synthesize a dityrosine standard for comparison of samples. This is accomplished by oxidizing L-tyrosine using a peroxidase such as horseradish peroxidase with the periodic addition of hydrogen peroxide. For a specific method of dityrosine standard formation see Giulive and Davies [33]. For the dityrosine measurement in proteins the following assay can be used.

Reagents Needed:

Protease (10 µl/ml proteinase K or pronase)

1.6 M perchloric acid

2 N KOH

Bondapak C_{18} Column (5 µm, 4.6 × 30 cm)

Solvent A: methanol:water (25:57, v/v), with 0.1% (v/v) TFA

Solvent B: acetonitrile:water (80:20, v/v), with 0.1% (v/v) TFA

HPLC equipped with a fluorodetector

Procedure

1. Digest samples for 1 to 4 h with the proteinase or pronase.
2. Precipitate with cold perchloric acid (PCA:sample; 3:1, v/v)
3. Chill samples on ice for 10 min, followed by centrifugation at 3000 g for 1 min.
4. Add 2 N KOH to neutralize samples.
5. Place in ice bath for 1 h, followed by centrifugation at 3000 g for 15 min. Remove the supernatant and discard the precipitate.
6. Load supernatant onto the HPLC column.
7. Elute with solvent A for 10 min, flow rate 0.8 ml/min.
8. Solvent B is run for 10 to 20 min with a gradient from 0 to 4%.
9. After 20 min and on run solvent B at 100%.
10. Detect with fluorodetector at $\lambda_{ex} = 315$ and $\lambda_{em} = 325$. For tyrosine use $\lambda_{ex} = 284$ and $\lambda_{em} = 325$.

28.4.2   Stable Isotope GC/MS

In an attempt to identify the tyrosine oxidation products generated by several *in vivo* systems, the laboratory of Dr. Jay Heinecke has successfully used an additional method for measuring the formation of o-tyrosine and o,o-dityrosine. This method identifies dityrosine content in amino acid hydrolysate by its fluorescence excitation and emission spectra, and by GC/MS. The formation is dependent upon both L-tyrosine and H_2O_2. Dityrosine formation can be easily measured by fluorescence measurement because the biphenolic crosslink formed is stable to acid hydrolysis and yields an intensely fluorescent product. Dityrosine formation may be used, in additional to carbonyl levels, as a marker of protein oxidation. Stable isotope dilution is used to quantify protein oxidation products by mass spectrometry. With this method a compound substituted with a stable isotope (^2H, ^{13}C, ^{18}O) is used to generate a standard. Intense signals are generated from the derivatives of halogenated esters which can be quantified in subnanogram quantities using negative-ion chemical ionization mass spectrometry [35]. Deuterated tyrosine is prepared according to the method of Heinecke et al. [36] from p-HO-C_6D_4-CH2CH[NH2]COOH using horseradish peroxidase and then isolated by ion exchange chromatography.

Reagents Needed:
Antioxidant Buffer

> 100 μM diethylenetriaminepentaacetic acid (DTPA)
>
> 1 mM butylated hydroxytoluene
>
> 1% ethanol
>
> 50 mM sodium phosphatebuffer, pH 7.4 passed over chelex 100 column to remove transition metal ions
>
> 10 mM 3-aminotriazole
>
> 10% TCA (w/v)

Buffer B

> 100 μM DTPA, pH 7.4., passed over chelex column to remove transition metals ions
>
> 6 N HCl with 1% benzoic acid and 1% phenol (w/v)
>
> o-[^{13}C]tyrosine (see Ref. 35)
>
> o,o-[^{13}C]dityrosine (see Ref. 35)
>
> argon gas
>
> C-18 solid-phase extraction column (3 ml)
>
> 0.1% TFA
>
> 25% methanol (v/v)
>
> vacuum
>
> HCl/*n*-propanol (1:3, v/v)
>
> nitrogen gas
>
> heptafluorobutyric anhydride/ethyl acetate (1:3, v/v)
>
> GC with a 12 m DB-1 capillary column (0.20 mm id, 0.33 microm film thickness) interfaced with a MS with extended mass range.

Procedure

1. Place sample in ice-cold antioxidant buffer (1:10, sample wt/v) add 10 mM 3-aminotriazole and homogenize.
2. Suspend 5 to 10 mg tissue sample in 1 ml buffer B and dialyze against buffer B at 4°C for 24 h.
3. Remove sample lipids by incubation of sample with 3 ml methanol and 7 ml of water-washed diethyl ether on ice for 10 min.
4. Precipitate protein by centrifugation for 10 min at 500 g.
5. Extract the pellet once with ice-cold water-washed diethyl ether and immediately subject to amino acid hydrolysis.
6. Dry sample under vacuum.
7. Add isotopically labeled standards and resuspend sample in 0.5 ml 6 N HCl with 1% benzoic acid and 1% phenol.

8. Hydrolyze sample at 110°C under argon for 24 hours.

9. Wash C 18 column sequentially with 6 ml of 50 m*M* sodium phosphate buffer, pH 7.4, 0.1 mM DTPA and 6 ml 0.1% TFA.

10. Add small amount of 10% TCA and pass over column, elute with 2 ml of 25% methanol.

11. Dry eluted amino acid under vacuum for derivatization. Usual recovery for oxidation products is greater than 80%.

12. Add 200 µl HCl/*n*-propanol and heat at 65°C or 1 h to convert amino acids to carboxylic acid esters.

13. Evaporate excess reagent under N_2. Add 50 µl heptafluorobutyric anhydride/ethyl acetate. Heat samples for 15 min at 65°C.

14. Analyze amino acids in a gas chromatograph equipped with a 12 m DB-1 column interfaced with a mass spectrometer with extended mass range. Set injection and detector temperature at 250°C.

15. Obtain full-scan mass spectra and selected ion monitoring of the N-propyl, heptafluorobutyric derivatives of the samples and the ^{13}C-labeled amino acids in the negative-ion chemical ionization mode with methane as the gas regent.

16. Use base ion of each amino acid for quantitation.

17. Monitor the ratio of ion currents of the two most abundant ions in each amino acid and its interval standard to ensure that interfering ions were not co-eluted with the analyte. Sensitivity limits for all amino acids is usually < 1 nmol (signal to noise >10).

28.5 Protein Oxidation Secondary to Lipid Peroxidation

Proteins can also become oxidized by the byproducts of lipid peroxidation. Lipid peroxidation has been shown to increase with aging so its effect on proteins should also be considered when looking at protein oxidation. Lipid peroxidation results in the formation of α,β-unsaturated aldehydes with the major product being 4-hydroxy-2-alkenals resulting from such peroxidations. 4-hydroxy-2-alkenals will react with sulfhydryl groups on proteins to form thioether adducts via the Michael-type addition. 4-hydroxynonenol has been shown to be the most abundant aldehyde formed via lipid peroxidation reactions. Recent work by Szweda et al. [37] has demonstrated the α,β-double bond of 4-hydroxynonenol (4-HNE) reacts through a Michael addition with the ε-amino groups of lysine thereby forming secondary amino groups containing an aldehyde group. It also reacts through a Michael addition with the imidazole group of histidine residues in proteins. LDL oxidative modification by 4-hydroxynonenol was shown to be associated with the loss of lysine and histidine residues [38]. Because such lipid peroxidation products can oxidize proteins it is necessary to detect these products when examining protein oxidation in the aging process. The Stadtman Laboratory has developed a method to detect 4-hydroxynonenal-lysine and 4-hydroxynonenal-histidine adduct in proteins [39].

Reagents Needed:
Standards Preparation Reagents

> N-acetylhistidine for standard 4-HNE-histidine adduct
>
> N-acetyllysine for standard 4-HNE-lysine adduct
>
> 2 mM 4-hydroxynonenal
>
> 50 mM sodium phosphate buffer, pH 7.2
>
> 0.05% TFA
>
> HPLC Solvent A: water
>
> HPLC solvent B: acetonitrile
>
> TSK-Gel ODS-80 reverse-phase HPLC column (0.46 × 25 cm)

^{3}H Labeled 4-HNE Modified Proteins

> 20% TCA
>
> 8 M guanidine-HCl/13 mM EDTA/133 mM Tris pH 7.2 solution
>
> 0.1 M EDTA/1 N NaOH/1 M NaB^{3}H$_4$
>
> 1 N HCL
>
> PD-10 column

Acid Hydrolysis of Labeled Proteins

> 0.1 mg 4-HNE-modified proteins
>
> 1–10 nmols purified 4-HNE-N-acetylhistidine or 4-HNE-N-acetyllysine
>
> 10 mM EDTA
>
> 1 N NaOH
>
> 0.1 M NaBH$_4$
>
> 6 N HCl
>
> nitrogen gas
>
> 50 mM sodium phosphate buffer, pH 8.0, with 1 mM EDTA.

Preparation of standard 4-HNE-acetylhistidine and 4-HNE-N-acetyllysine adducts

1. Add 50 mg *N*-acetylhistidine or *N*-acetyllysine to 5 to 10 mM 4-HNE in 2 ml of sodium phosphate buffer, pH 7.2.

2. Incubate for 20 h at 37°C.

3. Isolate products by reverse phase HPLC using a linear gradient of 0.05% TFA in solvent A – solvent B (time = 0, 100% A, 20 min, 0% A), using a flow rate of 1 ml/min.

4. With this method expect 4-HNE-*N*-acetyllysine to elute at 10.8 minutes and 4-HNE-*N*-acetylhistidine to elute at 11.2 min.

Preparation of 4-HNE modified proteins

1. Add 1 mg protein to 2 mM 4-HNE in 1 ml of 50 mM sodium phosphate buffer, pH 7.2, and incubate for 2 h at 37°C.
2. Add 400 µl of protein sample to an equal volume of 20% TCA.
3. Precipitate protein by centrifugation at 11,000 g for 3 min at room temperature.
4. Discard supernatant and resuspend pellet with the 8 M guanidine-HCL solution.
5. Place 40 µl of 0.1 M EDTA, 40 µl 1 N NaOH and 40 µl NaB^3H_4 in a 1.5 mo sarstedt tube fitted with an O ring and cap.
6. Add resuspended protein sample and incubate capped for 1 h at 37°C.
7. Terminate reaction by adding 1 N HCl and then load on a PD-10 column equilibrated with 6 M guanidine-HCl.
8. Collect eluate in 500 µl fractions and determine radioactivity by liquid scintillation and protein recovery of the sample by spectrophotometric methods.

Procedure for Acid hydrolysis of 4-HNE-modified proteins:

1. Place 0.1 mg of 4-HNE modified protein or 1 to 10 nmoles of purified 4-HNE-N-acetylhistidine or 4-HNE-N-acetyllysine in hydrolysis vial.
2. Add 10 µl each of 10 mM EDTA, 1 N NaOH and 0.1 M $NaBH_4$ to the vial and incubate at 37°C for 1 h.
3. Terminate reaction by the addition of 30 µl of 1 N HCl.
4. Concentrate in vacuum centrifuge.
5. Add 200 µl 6 N HCl and flush top of vial with nitrogen for 1 min.
6. Cap vial and heat at 110°C for 20 h.
7. Concentrate in vacuum centrifuge.
8. Resuspend sample in the 50 mM sodium phosphate buffer, pH 8.0.
9. Preform amino acid analysis according to the method of Uchida and Stadtman [39].

This method separates the 4-HNE-histidine and 4-HNE-lysine derivatives from all other animo acids. For all the proteins which have been tested so far by this method including insulin, glyceraldehyde-3-phosphate dehydrogenase, bovine serum albumin and LDL, the number of histidine residues which disappeared could almost always be accounted for by the number of 4-HNE-histidine derivatives identified. Unfortunately there is not as consistent a correlation with lysine; the number of lysine residues disappearing is usually greater than the 4-HNE-lysine derivative detected.

28.6 Conclusion

Although this chapter has shown several well developed methods by which to measure protein oxidation, it should be noted that these methods measure only certain types of oxidation products and probably indicate only a small fraction of the total protein oxidative modification during aging. The formation of carbonyl derivatives on amino acid side chains, dityrosine formation and 4-HNE reactions products are only a few of the modifications that may be produced during oxidative stress or aging. Further work needs to be pursued to evaluate other types of oxidative damage to proteins. These measures give us a very good start to indicate that an oxidative stress had occurred, but to fully understand the consequences oxidative stress may have during the aging process we need to be able to detect all types of protein oxidation reactions no matter how subtle they may be.

Another point which needs mentioning is that protein oxidation has generally become accepted as a measure of oxidative damage within the cell or tissue. Studies have shown that oxidative modification of proteins may play a role in normal regulation and turnover of proteins and enzymes [3], therefore the level of protein oxidation should be correlated with a decrease in function or another negative consequence. Previous studies have linked the accumulation of oxidized proteins with negative consequences, such as decreases in catalytic activity of enzymes [3], cross-linking of proteins [40], and decreases in cognitive ability [41].

Acknowledgments

I would like to thank Dr. Earl R. Stadtman and Dr. Rodney L. Levine for all their contributions, without which the preparation of this chapter would not have been possible.

References

1. Rothstein, M., Recent developments in the age-related alteration of enzymes: a review, *Mech. Aging. Devel.*, 6, 241, 1977.
2. Gershon, H. and Gershon, D., Detection of inactive enzyme molecules in aging organisms, *Nature*, 227, 1214, 1970.
3. Oliver, C. N., Levine, R. L. and Stadtman, E. R., A role of mixed-function oxidation reactions in the accumulation of altered enzyme forms during aging, *J. Am. Geriat. Soc.*, 35, 947, 1987.
4. Zhou, J. Q. and Gafni, A., Exposure of rat muscle phosphoglycerate kinase to a non-enzymatic MFO system generates the old forms of the enzyme, *J. Geront.*, 46, B217, 1991.
5. Stadtman, E.R., Biochemical Markers of Aging, *Experimental Gerontology*, 23, 327,1988.
6. Harman D., Aging: a theory based on free radical and radiation chemistry, *J .Gerontol.*, 11, 298, 1956.
7. Fucci, L., Oliver, C. N., Coon, M. J., and Stadtman, E.R. Inactivation of key metabolic enzymes by mixed-function oxidation reactions: possible implications in protein turnover and aging, *Proc. Natl. Acad. Sci. U.S.A.*, 80, 1521, 1983.
8. Amici, A., Levine, R. L., Tsai, L., and Stadtman, E. R., Conversion of amino acid residues in proteins and amino acid homopolymers to carbonyl derivatives by metal-catalyzed oxidation reactions, *J. Biol. Chem.*, 264, 3341, 1989.
9. Starke-Reed, P. E. and Oliver, C. N., Protein oxidation and proteolysis during aging and oxidative stress, *Archive Biochem. Biophys.*, 275, 559, 1989.
10. Starke-Reed, P. E. and Oliver, C. N., Unpublished results.
11. Witt, E.H., Abraham, R.Z., Viguie, C.A., Starke-Reed, P. E., and Packer, L., Exercise, oxidative damage and effects of antioxidant manipulation, *J. Nutr.*, 122, 758, 1992.

12. Oliver, C. N., Ahn, B.-W., Moerman, E. J., Goldstein, S., and Stadtman, E., Age-related changes in oxidized proteins, *J. Biol. Chem.*, 262, 5488, 1987.

13. Smith, C. D., Carney, J. M., Starke-Reed, P. E., Oliver, C. N., Stadtman, E. R., Floyd, R. A., and Markesbery, W. R., Excess brain protein oxidation and enzyme dysfunction in normal aging and Alzheimer's disease, *Proc. Natl. Acad. Sci. U.S.A.*, 88, 10540, 1991.

14. Sohal, R. S., Agarwal, S., Dubey, A., and Orr, W. C., Protein oxidative damage is associated with life expectancy of house flies, *Proc. Natl. Acad. Sci. U.S.A.*, 90, 7255, 1993.

15. Wolf, P.S. and Dean, R.T., Glucose autoxidation and protein modification, *Biochem. J.*, 245, 243.

16. Monnier, V.M., Nonenzymatic glycosylation, the Maillard reaction and the aging process, *J. Gerontol. Biol. Sci.*, 45, B105, 1990.

17. Starke-Reed, P. E. and Oliver, C. N., Protein oxidation and proteolysis during aging and oxidative stress. *Arch. Biochem. Biophys.*, 275, 559, 1989.

18. Ahn, B., Rhee, S. G., and Stadtman, E. R., Use of fluorescein hydrazide and fluorescein thiosemicarbazide reagents for the fluorometric determination of protein carbonyl groups and for the detection of oxidized protein on polyacrylamide gels, *Anal. Biochem.*, 161 (2), 245, 1987.

19. Levine, L., Garland, D., Oliver,.N., Amici, A., Climent, I., Lenz, A.-G., Ahn, B.-W., Shaltiel, S., and Stadtman, E. R., Determination of carbonyl content in oxidatively modified proteins, *Methods Enz.*, 186, 464, 1990.

20. Levine, R. L., Williams, J. A., Stadtman, E. R., and Shacter, E., Carbonyl assay for determination of oxidatively modified proteins. *Methods Enz.*, 233, 237, 1994.

21. Laemmli, U. K., Cleavage of structural proteins during the assembly of the head of bacteriophage T4, *Nature*, 277 (259), 680, 1970.

22. Gross, A. J. and Sizer, I. W., The oxidation of tyramine, tyrosine, and related compounds by peroxidase, *J. Biol. Chem.*, 234, 1611, 1959.

23. Garcia-Castineiras, S., Dillon, J., and Specter, A., Detection of bityrosine in cataractous human lens protein, *Science*, 199, 897, 1978.

24. Lehrer, S. S. and Fasman, G. D., Ultraviolet irradiation effects in poly-L-tyrosine and model compounds. Identification of bityrosine as a photoproduct, *Biochemistry*, 6 (3), 757, 1967.

25. Boguta, G. and Dancewicz, A.M., Radiation-induced dimerization of tyrosine and glycyltyrosine in aqueous solutions, *Int. J. Radiat. Biol. Relat. Stud. Phys. Chem. Med.*, 39 (2), 163, 1981.

26. Sizer, J. W., Oxidation of proteins by tyrosine and peroxidase, *Adv. Enzymol.*, 14, 129, 1953.

27. Wagley, P. F., Sizer, I. W., Diamond, L. K., and Allen, F. H., The inactivation of RH-antibodies by peroxidase, *J. Immunol.*, 64, 85, 1950.

28. Aeshbach, R., Amado, R., and Neukum, H., Formation of dityrosine cross-links in proteins by oxidation of tyrosine residues, *Biochim. Biophys. Acta.*, 439 (2), 292, 1976.

29. Foerder, C. A. and Shapiro, B. M., Release of ovoperoxidase from sea urchin eggs hardens the fertilization membrane with tyrosine crosslinks, *Proc. Natl. Acad. Sci. U.S.A.*, 74 (10), 4214, 1977.

30. Davies, K. J. A., Protein damage and degradation by oxygen radicals I general aspects, *J. Biol. Chem.*, 262, 9895, 1987.

31. Tew, D. and Ortiz de Montellano, P. R., The myoglobin protein radical. Coupling of Tyr-103 to Tyr-151 in the H_2O_2-mediated cross-linking of sperm whale myoglobin, *J. Biol. Chem.*, 263 (33), 17880, 1988.

32. Guilivi, C. and Davies, K. J. A., Dityrosine and tyrosine oxidation products are endogenous markers for the selective proteolysis of oxidatively modified red blood cell hemoglobin by (the 19 S) proteasome, *J. Biol Chem.*, 268 (12), 8752, 1993.

33. Giulivi, C. and Davies, K. J. A., Dityrosine: a marker for oxidatively modified proteins and selective proteolysis, *Methods Enzymol.*, 333, 363, 1994.

34. Leeuwenburgh, C., Hardy, M. M., Hazen, S. L., Wagner, P., Oh-ishi, S., Steinbrecher, U. P., and Heinecke, J. W., Reactive nitrogen intermediates promote low density lipoprotein oxidation in human atherosclerotic intima, *J. Biol. Chem.*, 272: 1433, 1997.

35. Turk, J., Stump, W.T., Wolf, B.A., Easom, R.A., and McDaniel, M.L., Quantitative stereochemical analysis of subnanogram amounts of 12-hydroxy(5,8,10,14)-eicosatetraenoic acid by sequential chiral phase liquid chromatography and stable isotope dilution mass spectrometry, *Anal Biochem.*, 174, 580, 1988.

36. Heinecke, J.W., Li, W., Daehnke, III, H.L., and Goldstein, J.A., Dityrosine, a specific marker of oxidation is synthesized by the myeloperoxidase-hydrogen peroxide system of human neutrophils and macrophages, *J. Bio. Chem.*, 268, 4069, 1993.
37. Szweda, L.I., Uchida, L., Tsai, L.,and Stadtman, E.R., Inactivation of glucose-6-phosphate dehydrogenase by 4-hydroxynonenal: selective modification of an active site lysine, *J. Biol. Chem.*, 268, 3342, 1993.
38. Uchida, K. and Stadtman, E.R., Modification of residues in proteins by reaction with 4-hydroxynonenal, *Proc. Natl. Acad. Sci. U.S.A.* 89, 4544, 1992.
39. Uchida, U. and Stadtman, E.R., Quantitation of 4-hydroxynonenal protein adducts. *Methods Enzymol.* 233, 371,1994.
40. Friguet, B, Stadtman, E.R., and Szweda, L.I., Modification of glucose-6-phosphate dehydrogenase by 4-hydroxy-2-nonenal:formation of crossed-linked protein which inhibits the multicatalytic protease, *J. Biol. Chem.*, 269, 21639, 1994.
41. Carney,J.M., Starke-Reed, P.E., Oliver, C.N., Landon, R.W., Cheng, M.S., Wu, J.F., and Floyd, R.A., Age-related increase in brain protein oxidation, decrease in enzyme activity, and loss in temporal and spacial memory by chronic administration of the spin-trapping compound n-tert-butyl-phenylnitrone, *Proc. Natl. Acad. Sci. U.S.A.*, 88, 3633, 1991.

29

Glycation, Glycoxidation, and Other Maillard Reaction Products

Vincent M. Monnier, John F. Fogarty, Camille S. Monnier, and David R. Sell

CONTENTS

29.1 Introduction

Proteins that are exposed to reducing sugars undergo postsynthetic modification by the Maillard reaction. Sugar adducts, UV active, and fluorescent products are formed which result from sugar dehydration, oxidation, fragmentation, and rearrangement reactions. *In vitro*, the reaction is extremely complex leading to hundreds of compounds [1]. *In vivo*, however, many of the compounds observed *in vitro* are unlikely to form because the activation energy used by food chemists studying the reaction is much higher. Furthermore, many of the reactive intermediates can be metabolized enzymatically, thus limiting the extent of protein modification. Yet, more and more products of the advanced Maillard reaction are being discovered *in vivo*.

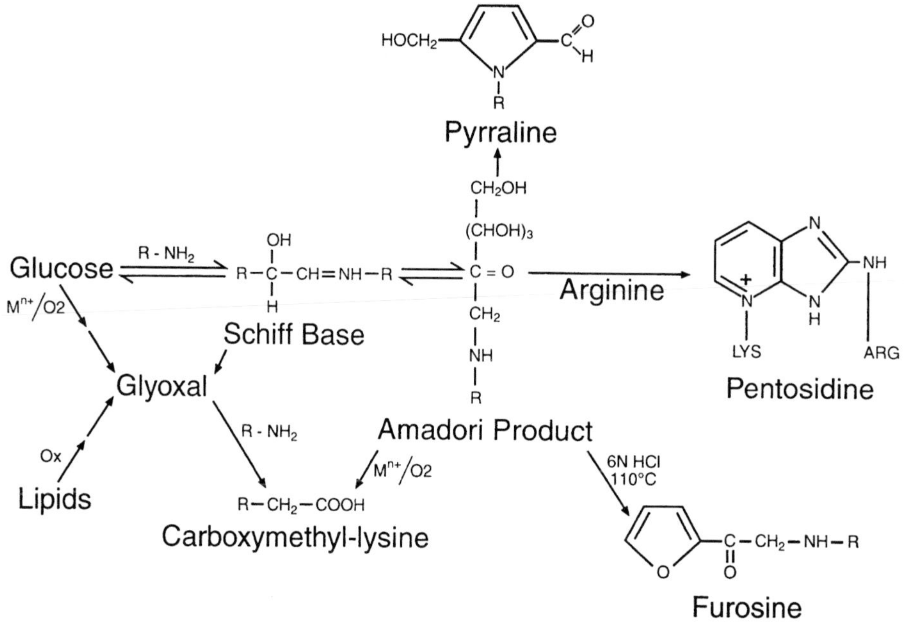

FIGURE 29.1
Selected pathways of the Maillard reaction with structures discussed in this chapter.

Figure 29.1 depicts a simplified scheme of the Maillard reaction. The reaction is initiated by the nonenzymatic glycation of primary amino groups, i.e., the condensation reaction of reducing sugars (e.g., glucose) with protein amino groups to form a Schiff base adduct which undergoes further rearrangement to the more stable Amadori product. Schiff bases can be quantitated indirectly following displacement of the labile bound sugar using acetic acid and quantitation of the sugar released [2], or by using tritiated sodium cyanoborohydride and isolation of the glucitolyl-amino acid residues by boronate affinity chromatography [3]. If, however, sodium borohydride instead of cyanoborohydride is used, total glucose adducts, i.e., Amadori products and Schiff base adducts will be measured. In general, the amount of Schiff base adducts does not exceed 10% of the Amadori product [4]. Although Schiff bases are thought to play an important role in the fragmentation of sugars by the Namiki pathway [5,6], there is currently no rationale for quantitating them in biological tissues.

The Amadori product of glucose is the single major product of the Maillard reaction *in vivo*. Its levels are strongly dependent on glycemia and protein turnover rate [7] (Table 29.1). Quantitation of the Amadori product in hemoglobin provides information on mean cumulative glycemia over the preceding 5 to 6 weeks and such assay is routine in the clinical setting. At equivalent glycemic levels as e.g., in various tissues within the same animal, Amadori products reflect protein turnover rate [8,9]. It is important to bear in mind that not all sites on a protein can form Amadori products, and that certain sites are highly favored, such as lys-525 in albumin [10]. Both *in vivo* and *in vitro*, levels of Amadori products are in a steady state which is strongly dependent on mean glycemia. Thus, levels of glycated lysine residues in crystallins do not increase with age in the human lens [11], and glycation increases only slightly with age in human skin collagen [12]. In rodents, however, there is a marked increased in Amadori products in the first 4 to 6 months, both in collagen [13,14] and lens crystallins [15]. The reason for this increase in rodents but not in humans is unclear.

TABLE 29.1
Levels of Selected Maillard Reaction Products in Mammalian Tissues*

Maillard Compound	Specimen	Levels	Method	Ref.
Fructose-lysine (Amadori product)	Albumin (human) (diabetic)	250– 500 pmol/mg 400–1,500 pmol/mg	RIA	105
	Albumin (human) (diabetic)	3–7 nmol/mg 7–24 nmol/mg	ELISA	106
	Plasma protein (human) Peritoneal fluid	1.1 ± 1.6 nmol/mg^1 1.2 ± 2.1 nmol/mg^1	Furosine HPLC	30
	LDL (human)	0.76–0.80 mmol/mol lys	Furosine SIMS-GC/MS	66
	Lens crystallins (human)	0.5–2.0 mmol/mol lys (no change with age)	Furosine SIMS-GC/MS	28
	(dog) normal diabetic (moderate) diabetic (severe)	100 ± 20 pmol/mg^1 510 ± 620 pmol/mg^1 2250 ± 680 pmol/mg^1	Furosine HPLC	107
	Skin collagen (human) normal diabetic (type 1)	2.5–5.0 mmol/mol lys 5–25 mmol/mol lys (very small age-change)	Furosine SIMS-GC/MS	24
	(human) normal diabetic (type 1) diabetic (intense therapy)	400 ± 100 pmol/mg^1 920 ± 250 pmol/mg^1 600 ± 75 pmol/mg^1	Furosine HPLC	67; see also 108
	(rat, Brown-Norway) normal diet. restr.	1.7–2.1 mmol/mol lys 1.2–1.4 mmol/mol lys	Furosine SIMS-GC/MS	26

TABLE 29.1 (continued)

Levels of Selected Maillard Reaction Products in Mammalian Tissues*

Maillard Compound	Specimen	Levels	Method	Ref.
Fructose-lysine (continued)	(rat, Fischer 344)			
	normal	290–450 pmol/mg[1]	Furosine	14
	diet restr.	290–350 pmol/mg[1]	HPLC	
		(from 5–25mos)		
	(mouse, C57BL)			
	normal	100–500 pmol/mg[1]	Furosine	14
	diet. restr.	100–350 pmol/mg[1]	HPLC	
		(from 2–25 mos)		
Carboxymethyllysine	Serum (human)			
	Normal	73.2 ± 16.9 pmol/mg	HPLC	65
	Hemodialysis F8[2]	308.8 ± 94.0 pmol/mg		
	Hemodialysis F80[2]	275.5 ± 79.1 pmol/mg		
	Peritoneal dialysis	284.5 ± 98.2 pmol/mg		
	LDL (human)	0.034–0.06 mmol/mol lys	SIMS-GC/MS	66
	LDL$_{ox}$ (human)[3]	0.67 –10.9 mmol/mol lys		
	Skin (human)			
	normal	~1.5 mmol/mol lys (0–80 yrs)	SIMS-GC/MS	24
	diabetes (type 1)	0.5–2.0 mmol/mol lys (20–80 yrs)		
	(rat)			
	normal	0.07–0.13 mmol/mol lys	SIMS-GC/MS	26
	diet restr.	0.08–0.10 mmol/mol lys		
		increase from 0–30 mos[4]		
	Lens crystallins (human)	~1.0–8.0 mmol/mol lys, increase from 1-100 yrs[4]	SIMS-GC/MS	28
	Implanted tendons for 28 days into:		HPLC	109
	normal rat	7 ± 2 pmol/mg		
	diabetic	35 ± 7 pmol/mg		

Carboxyhydroxymethyl-lysine	Skin (human) normal	see original ref.	Furosine SIMS-GC/MS	27
	diabetic	0.5–4.0 mmol/mol lys	Furosine SIMS-GC/MS	108
Pentosidine	Plasma protein (human) normal diabetic ESRD[5]	0.3–1.6 pmol/mg 0.5–4.8 pmol/mg 5–55 pmol/mg	HPLC	110
	Hemolysate (human) diabetic ESRD[5]	0.09–0.21 pmol/mg 0.05–0.22 pmol/mg 0.2 –1.5 pmol/mg	HPLC	110
	Urine (free pentosidine)	3–7 uM (control) 5–20 uM (diabetic)	HPLC	111
	Urine (total pentosidine) Control: Diabetic:	4.2 ± 1.4 umol/mol creat. 8.8 ± 4.3 umol/mol creat.	HPLC	111
	Peritoneal fluid protein	7.7 ± 2.7 pmol/mg prot.	HPLC	30
	Skin collagen[6] Shrew Rat (ad libit.) Rat (calor.restr.) Cow Pig Rhesus monkey Squirrel monkey Human	0–8 pmol/mg (0–3 yrs) 0–3 pmol/mg (0–24 mos) 0–2 pmol/mg (0–24 mos) 0–30 pmol/mg (0–14 yrs) 0–20 pmol/mg (0–14 yrs) 2–12 pmol/mg (0–25 yrs) 0–18 pmol/mg (0–25 yrs) 0–100 pmol/mg (0–100 yrs)	HPLC	31
	Skin (rat) normal dietary restr	2.0–5.0 umol/mol lys 1.8–2.7 umol/mol lys (from 10–30 mos)		26
	Cartilage (human)	0–80 mmol/mol collagen (from 0–80 yrs)	HPLC	112

TABLE 29.1 (continued)

Levels of Selected Maillard Reaction Products in Mammalian Tissues*

Maillard Compound	Specimen	Levels	Method	Ref.
Pentosidine (continued)	Dura mater	25–225 pmol/mg from (0–80 yrs)	HPLC	23
	Lens crystallins (hum.) water soluble:	0–0.6 pmol/mg	HPLC	25
	water insoluble:	1.0–1.5 pmol/mg		
	brunescent lens:	2–5-fold increased		
	diabetic (human)	no increase	HPLC	25
		increase (1.5–2.5 fold)	HPLC	113
	(dog, water insol.) normal	1.2 ± 0.3 pmol/mg	HPLC	107
	diabetic (moderate)	1.3 ± 2.7 pmol/mg		
	diabetic (severe)	15.5 ± 16.1 pmol/mg		
	Rat tail tendon 6 mos	1.24 ± 0.22 pmol/mg	HPLC	31
	18 mos	2.48 ± 0.60 pmol/mg		
Pyrraline	Albumin-rich fract.[7] (human) normal	20–40 pmol/mg	ELISA	93
	diabetic	30–60 pmol/mg		
	Plasma (human)[7] normal	115 ± 36 uM	ELISA	92
	diabetic	211 ± 103 uM		
	Plasma (rat)[7] normal	194 ± 79 uM	ELISA	92
	diabetic	627 ± 189 uM		
	Plasma (human) normal	12.8 ± 5.6 pmol/mg	HPLC	104
	diabetic	21.6 ± 9.6 pmol/mg		

	HPLC	100
Lens crystallins (human 50-70yrs)		
normal	30.9 ± 10.2 pmol/mg	
cataract	48.4 ± 12.6 pmol/mg	
diabetic	28.4 ± 15.3 pmol/mg	

* Adapted from the literature.

1 Furosine values were not corrected for a 70% loss during acid hydrolysis.

2 F8 and F80 refers to dialysis membrane type, whereby F80 membrane indicates a larger pore size than F8.

3 LDL were oxidized with 5 uM Cu(II) for 24 hr.

4 Indicates age-related increase from the lowest to the highest value.

5 ESRD = end stage renal disease.

6 Curvilinear increase with age found for all mammalian species examined.

7 The discrepancy between the levels in albumin-rich fractions vs plasma proteins stems from the way the assay was calibrated. Plasma values are probably 100 times too high.

One of the major reasons for studying the Maillard reaction in aging is that its advanced products, also called advanced glycation end products (AGEs), are stable products which accumulate in aging tissues. The accumulation of AGEs in extracellular proteins has been associated with the pathological complications of aging, diabetes and end stage renal disease. Particular disorders associated with AGE formation are cataracts [16,17], retinopathy [18], nephropathy [19], vascular disease [20] and Alzheimer's disease [21] to name just a few. It is postulated that the accumulation of AGE crosslinks in long-lived proteins of skin, tendon, aorta, and other collagen-rich tissues leads to a decrease in solubility and elasticity, and increased resistance to enzyme digestion.

The term glycoxidation is used to delineate AGEs that are generated through oxidative fragmentation of Amadori products [22]. The most widely studied glycoxidation products, pentosidine and carboxymethyllysine, have been structurally characterized and are known to accumulate in lens crystallins and many collagen-rich tissues with age [23–28]. Under conditions of oxidative or metabolic stress, as in diabetes and uremia, physiological levels of these two products can accumulate faster and at higher levels than in healthy individuals [29,30]. Of importance for gerontological studies is that caloric restriction in rodents has been found to diminish the formation rate of both CML and pentosidine [26,31], and that the formation of the latter has been found to occur at a rate inversely related with maximal lifespan in several mammalian species [31].

Much debate presently exists over the primary source of AGEs *in vivo*. Whereas the Amadori compound is expected to account for much of the glycoxidation products formed in skin collagen, conditions associated with oxidative stress and increased levels of transition metals as e.g., in uremia [32] may activate alternative pathways of AGE formation. These include glucose autooxidation (e.g., glyoxal and arabinose) [33–35], ascorbate catabolites and short-chain carbohydrates (e.g., ascorbate, pentoses, and tetroses) [36,37].

Finally, it now appears that AGE products may also originate from pathways unrelated to the Maillard reaction with sugars, such as the oxidation of lipids [38], the oxidation of serine by myeloperoxidase [39], or through physiological metabolic pathways as in the case of methylglyoxal and 3-deoxyglucosone [40,41]. Interestingly, protein glycation may itself catalyze lipid peroxidation [42] and may thus contribute toward atherosclerosis [43].

29.2 Glycated Proteins

29.2.1 General Methods

The choice of an assay for quantitation of glycation in proteins depends on the type of information one wishes to obtain. If the question is what fraction of total protein has been glycated (Amadori product), such fraction can be selectively isolated using a *boronate affinity column* which specifically binds native or borohydride reduced glycated proteins (or amino acids) [44,45], essentially regardless of the number and type of residues that have been modified in the molecule. The data are then expressed in terms of mol% of modified molecules, whereby the actual number of Amadori products may be equal to or larger than 1 mol/mol. For example, it is now known that albumin from diabetic plasma can contain up to 14 molecules of Amadori products per molecule of albumin, as determined by MALDI mass-spectrometry [46]. However, this does not mean that 100% of circulating albumin molecules have been modified. In clinical assays glycated hemoglobin is expressed as mol%, whereby the glycated fraction can be quantitated not only by boronate affinity chromatography, but also by ion exchange chromatography, electrophoresis or immunological methods [47]. Boronate affinity minicolumns are available commercially (e.g., Pierce Glyco-gel II cat.#20201) as a kit assay with detailed instructions provided by the manufacturer.

Protein glycation can also be determined *colorimetrically* using either treatment with oxalic acid to release furfuraldehyde which is then reacted with thiobarbituric acid to give a compound which absorbs at 443 nm^2, or by reaction with nitro blue tetrazolium, a method which is used routinely for quantitation of "fructosamine" in clinical specimens [48–50]. Both methods work well with proteins and other substrates that have been glycated *in vitro*. However, both assays are prone to interference by other reducing agents (e.g., ascorbate) present in biological specimens. Immunological methods for quantitation of glycated proteins have been developed. To our knowledge, however, no one has succeeded in obtaining antibodies which react specifically with the Amadori product. In contrast, excellent antibodies have been raised against the borohydride reduced glucitolyl-lysine adduct in glycated proteins which can be used in ELISAs and for immunohistochemistry [51–54].

Glycated albumin in human and rodent specimens can now be quantitated immunologically with a commercially available monoclonal antibody [55]. The antibody apparently does not recognize the Amadori product per se, but a conformational change in the protein that is specific for glycated albumin (unpublished). A kit can be purchased from Exocell, Inc. (Philadelphia, PA).

29.2.2   Nitroblue Tetrazolium Assay

Principle: The nitroblue tetrazolium (NBT) assay was introduced by Baker et al.in 1982 [50] and is based on the ability of Amadori products to reduce NTB to the purple monoformazan dye under alkaline conditions. Detailed studies on the mechanism of the reaction have been published leading to the conclusion that superoxide dismutase in biological specimens may interfere with the assay, and that the presence of non-ionic detergents may be helpful in decreasing the interference.

Sample Preparation: The NBT assay can be used with synthetic and biological specimens. Reducing agents, such as ascorbate may interfere.

Procedure: Adapted from Baker et al. [48]: To 1.8 ml of a reaction mixture containing 250 μmol NBT in 0.1 mol/L sodium carbonate buffer (pH 10.3) with 0.1 mmol/L diethylenetriamine pentaacetic acid (DTPA) and 0.1 mmol/L phytic acid, add 200 μl of test specimen or fructosamine standard solution (0, 200, 400, 600, and 800 μmol/L (1-deoxy-1-morpholino fructose, Sigma D-6149) dissolved in the carbonate buffer containing 40 g/l BSA and 22g/l Triton-X-100 or Brij 35. Incubate at 37°C for 20 min and read absorbance at 530 nm.

29.2.3   Furosine Assay by HPLC

Principle: Protein glycation can be quantitated in absolute terms with the furosine assay. Furosine is the the acid conversion product of ε-fructosyl-lysine [56] (Figure 29.1) Lysine residues are together with the N-terminal amino acid, the only possible glycation site in a protein. Protein hydrolysis yields about 30% of the fructose-lysine adduct in the form of furosine (ε-*N*-(2-furoylmethyl)-L-lysine) and the yield depends on a number of factors [56,57]. Furosine assays are considered to be the most sensitive and specific indicators of protein glycation.

Sample Preparation.   The procedure below is applicable to furosine, pentosidine, and carboxymethyllysine (see Section 29.3). Because advanced Maillard reaction products are present in low levels compared to the total pool of amino acids, it is important to thoroughly deaerate specimens prior to acid hydrolysis in order to prevent artefacts which cochromatograph in the HPLC chromatogram.

Collagen-Rich Samples

The following description represents one method to analyze collagen tissue samples by HPLC, but can generally be applied to other tissues of similar nature.

Samples need to be cleaned and fat tissue should be removed. If specimens are to be stored frozen for a length of time, storage should be done under nitrogen or argon to minimize lipid peroxidation which is likely to enhance the yield of CML. *Important:* If specimens need to be sonicated, sonication should be done under nitrogen or argon. Addition of 10 mM aminoguanidine HCl (prepared by addition of equimolar concentration of HCl to aminoguanidine bicarbonate) helps diminishing artifactual CML production from lipid peroxidation (unpublished).

1. Similar weights of fresh tissues should be preferentially be used from each sample. This can be accomplished by blotting washed tissue samples on absorbent paper and weighing out at least 10 mg wet weight. Samples should then be minced into fine pieces.
2. Delipidate by suspending minced portions in 5 ml chloroform:methanol (2:1, v:v) and shake with rotary shaker at 4°C for 18 hours.
3. Centrifuge at 2,500 rpm and discard supernatant.
4. Rehydrate the fluffy pellet in 50% methanol for several hours.
5. Centrifuge at 2,500 rpm and discard methanol.
6. Suspend pellet in metal-free water and transfer into labeled teflon-lined screw-cap test tubes and discard water.
7. Resuspend pellet in 2 ml of 6 N hydrochloric acid.

Generally, protein concentration should be not more than 3mg/ml HCl. To limit oxidative formation of glycoxidation products during acid hydrolysis the HCl should be thoroughly deaerated by bubbling nitrogen for 30 min., and sample tubes should be flushed with nitrogen or argon gas after adding the acid to the sample tube. Immediately after flushing, tubes should be tightly sealed with teflon-lined caps and heated at 110°C for 18 h. It is recommended that tubes be briefly shaken by hand 60 min after onset of incubation so that the partially solubilized pellet is finely distributed. Evaporate the acid to dryness with e.g., a Savant Rotary Evaporator (Speedvac AS160) and reconstitute in 500 µl of the HPLC buffer A (described below). An aliquot of 50 µl should be saved for determination of collagen concentration by the hydroxyproline colorimetric assay of Stegeman and Stalder as described by Sell and Monnier [23].

Plasma Protein

Precipitate 5 mg of protein (or about 100µl) with 10% cold trichloroacetic acid (TCA) on ice. Centrifuge at 2,500 rpm and discard the acid. Wash pellets twice with 5% TCA. Delipidate the pellet as above and resuspend in 2 ml of 6 N hydrochloric acid. To limit oxidative formation of glycoxidation products during acid hydrolysis use thoroughly deaerated HCl and flush samples with nitrogen or argon gas for 10 min. after adding the acid to the sample tube. Process as above.

HPLC Procedure

The quantitation of furosine is carried out using a method modified from Resmini et al. [58,59]. Furosine standard is commercially available (Neosystem Laboratoire, code SC494, SNPE North America, Princeton, NJ, 609-987-9424). Equal amount of protein (ca. 200 µg) from each sample is injected onto a reverse phase C8 column (Alltech no. specfuro, Sedriano, MI). Buffer $\underline{A}$ consists of 0.4% acetic acid (v/v). Buffer $\underline{B}$ consists of 0.4% acetic acid (v/v) and 0.27% Kcl (w/v). Initially, buffer $\underline{A}$ is run isocratically for 12.5 minutes, then a gradient from 0% $\underline{B}$ to 10% $\underline{B}$ for 7 minutes, and finally isocratic at 10% $\underline{B}$ for a further 15.5 minutes, for a total of 35 minutes at a flow rate of 1.2 ml/min. The elution of furosine is monitored using an absorbance detector at 280 nm. Furosine elutes at ≈28 min.

29.2.4   Furosine Assay by GC/MS

Principle: Furosine together with carboxymethyllysine can be assayed in protein hydrolysates in a combined procedure utilizing trifluoroacetyl derivatives according to the method developed by Dunn et al. [27]. Furosine is detected using selected ion monitoring at m/z = 110 (see also Wells-Knecht, 1995 [60])

Sample Preparation: A portion of fructoselysine (3 to 4%) may convert to carboxymethyllysine during acid hydrolysis. Therefore, a separate set of samples (3 mg) should be used for CML measurement. The samples are prepared by first reducing with 500 µl of 0.1 M NaBH$_4$ in 1 M NaOH for 12 hours at room temperature.This procedure stabilizes the Amadori product. After addition of 6 N HCl to release excess NaBH$_4$, samples are evaporated.

Remove excess borate by adding 1 ml of acetic acid in methanol (1:10) for 30 sec at 65°C and dry under nitrogen stream. Samples are then acid hydrolyzed as described previously. Dried hydrolysates are derivatized for analysis by GC/MS. Resuspend in 5 ml of 1 N methanolic HCl and heat at 65°C for 30 min. Evaporate solvent with nitrogen stream at room temperature. Resuspend samples in 500 µl of dry methylene chloride. Add 1 ml of trifluoroacetic anhydride, mix, and incubate at room temperature for 1 hour. Remove solvent in the presence of a nitrogen stream. Resuspend sample in 150 µl of methylene chloride.

Procedure: Inject 2 µl of solution for GC/MS analysis. Use a 30m Rtx-5 (95% methyl, 5% phenyl) capillary column (Restek, Bellefonte, PA) with the following temperature program: 70°C initially for 2 min; increase to 260°C at a rate of 5 deg/min followed by 290°C at 15 deg/min; maintain for 4 min at 290°C. Carboxymethyllysine retention time is ca. 30 min. Furosine retention is ca. 37 min. For an alternative procedure using TMS derivative, see Glomb and Monnier [6].

29.2.5   Borohydride Reduction and Labeling of Amadori Products with ³H-NaBH$_4$

Principle: Borohydride reduction can be used for two purposes: first, for labeling with tritium in position 2 of the product, e.g., 1-deoxyglucitolyl-lysine, and second, to render the Amadori product inactive toward oxidizing agents and to diminished the number of fluorescent and other contaminants formed during acid hydrolysis. It is also useful to reduce the Amadori product for ELISAs in which NaOH is being used to enhanced the immunore-activity [61].

Sample Preparation: Borohydride reduction can be done with any protein. Insoluble col-lagen should be minced finely and resuspended by sonication in PBS containing 1mM DTPA (chelating agent) under N_2. Solid specimens of Amadori products can also be reduced.

Procedure: The widely quoted procedure of Bookchin and Gallop [62] consisting of reduction for 10 min at room temperature and 50 min on ice has been found in our labora-tory insufficient for complete reduction of glycated proteins.

For Proteins

Prepare a protein solution (2 to 10 mg/ml) in PBS 2X. To a fresh solution of 500mM $NaBH_4$ in 1 mM NaOH, add 100 to 900 μl of protein solution (final pH should not exceed 7.4 oth-erwise protein fragmentation may occur!). Iincubate for 4 hrs and dialyze against PBS to remove unreacted reagents, and then against the appropriate buffer as required by the sub-sequent procedure (water if hydrolysis; column equilibration buffer if boronate affinity). Precipitation by slow addition (frothing!) of 150 μl of cold 50% TCA to the solution on ice is also possible. Addition of 50 μl of amyl alcohol before TCA is helpful to prevent frothing.

For Non-Dialyzable Products

Same procedure except that the $NaBH_4$ concentration should be at least in 50–100M excess compared to Amadori products. The reaction is stopped by adding 50 μl of 6N HCl per ml of solution. If HPLC is the next step, there is generally no need to remove the salt. For GC/MS, removal of excess borate salt which interferes with the derivatization procedure and clogs-up the column is necessary. See description under "Furosine by GC/MS."

29.3 Glycoxidation Products

Carboxymethyllysine

Carboxymethyllysine (CML) was first described by Ahmed et al. [63] as a metal catalyzed Amadori degradation product. It is an acid stable, colorless, non-UV active molecule which, like pentosidine, can form from any reducing sugar and ascorbate, but also through lipid peroxidation products(see above). It has been found in lens, collagen [24,27,64] and plasma proteins [65,66]. CML and pentosidine levels in skin are highly correlated [26,29,67]. Because CML, in contrast to pentosidine, can also stem from lipid peroxidation, measurment of both may be needed to obtain information on the *in vivo* source of CML as for example in arteries affected by atherosclerosis [68]. It is important to stress that CML is, as of this date, quantitatively the single most important AGE product known to form *in vivo*.

Pentosidine

The AGE pentosidine is a fluorescent imidazo (4,5b) pyridinium ring crosslink between lysine and arginine residues that was first isolated from human dura mater [23]. It's exci-tation/emission maxima are at 335/385nm. It has since been shown to be a ubiquitous compound in human tissues that increases with age in long-lived proteins of skin, dura mater, and cartilage [69]. Pentosidine represents about 1% of tissue crosslinks, but it has been shown to serve as a potentially reliable biological marker of protein aging, glycoxidative stress, and age-related diseases. Pentosidine levels correlates with severity of complications in insulin-dependent diabetic subjects [29,70,71]. Pentosidine is stable to acid hydrolysis and can thus be assayed in protein hydrolysates. At high alkaline pH it slowly degrades (unpublished observation).

Fluorescence at 370/440 nm

Protein-bound fluorescence at 370/440 nm was first used as indirect evidence that nonenzymatic browning had occured in aging human lenses and collagen [16,72]. It is usually expressed in arbritrary units. This type of fluorescence forms readily in most proteins incubated with reducing sugars, but oxidizing conditions are required for it to form from glucose [73]. *In vivo*, however, similar fluoresence can originate from lipid peroxidation [74] and other oxidizing processes [22] and is therefore not specific. Specific molecules with similar fluorescence have been found in lens (LM-1) [75], and more recently in reactions involving glucose and lysine [76–79].

29.3.1 Carboxymethylysine Assay by HPLC with Post-Column Detection

Principle.   CML is detected in the acid hydrolysate of proteins. Since it is not UV active it must be detected with a post-column detection system such as o-phthalaldehyde (OPA). In our laboratory we use a sequential dual column system for biological specimens, whereas a single column system is sufficent for quantitation of CML in model reactions *in vitro*. Samples are injected into a C-18 reverse phase column, and the fractions containing CML are collected, dried and reinjected into a C-18 column with different solvent system. This procedure is somewhat cumbersome but in view of the fact the OPA method is highly sensitive and therefore also prone to contaminants, much more reliable results are obtained with the dual than single column method.

Sample Preparation: Acid hydrolysates prepared as described previously are needed. Delipidation is important since lipid peroxidation may create substantial amounts of CML rapidly. In those cases in which delipidation is not feasible or when lipid oxidation cannot be avoided, addition of 10 mM aminoguanidine hydrochloride helps decrease artifactual CML formation.

HPLC Procedure (Dual-Column Procedure)

First, ascertain the retention time of CML by injection of standard onto a reverse phase C18 column (Vydac No. 218TP104, Hesperia, CA). Standards of CML can be made from N-α-acetyllysine and glyoxylic acid under reductive alkylation and purification according to Knecht et al. [80]. Buffer $\underline{A}$ consists of water with heptafluorobutyric acid (0.01 M; Sigma H-no. 7133). Buffer $\underline{B}$ consists 60% acetonitrile (v/v, Fisher no. A994SK-4) and heptafluorobutyric acid (0.01 M). Buffer B is run isocratically at a flow rate of 1ml/min. Detection of CML is by post-column derivatization with o-phthalaldehyde in the presence of 2-mercaptoethanol at excitation wavelength 340 nm/emission wavelength 455 nm on a spectrofluorometer (method described below). Standard CML elutes at ≈12 min. After determination of CML retention, equal amounts of protein (ca. 200 µg) from each sample is injected without post-column derivatization, and collection time is for six min (≈ minute 9 to 15). The collected solvent is evaporated and each pellet is resuspended in 200 μl buffer <u>A</u> for CML detection (described below). Equal amounts (50 μl) are re-injected onto a reverse phase C18 column (Vydac No. 218TP54, Hesperia, CA). Buffer <u>A</u> consists of 5% 1-propanol (v/v; Sigma no. 29,328-8), sodium dodecyl sulfate (SDS, 3g/L; Fluka no. 71725) and monobasic sodium phosphate (1g/L; Mallinckrodt no. 7892). Buffer <u>B</u> consists of 60% 1-propanol (v/v) with the same amount of SDS and monobasic sodium phosphate as buffer <u>A</u>. Solvent flow is a linear gradient from 15% to 22% buffer <u>B</u> over 30 min at a flow rate of 1ml/min. Elution of CML is detected by post-column derivatization with o-phthaldialdehyde (OPA) in the presence of 2-mercaptoethanol at excitation wavelength 340 nm/emission wavelength 455 nm on a spectrofluorometer. CML elutes at ca. 28 min.

The post-column buffer consists of boric acid 24.73g/L (Aldrich no. 33,906-7) at pH 9.7 adjusted with KCl (Aldrich no. 30,656-8), OPA 800mg/L (Aldrich no. P3,940-0), 2-mercaptoethanol 2ml/L (Aldrich no. M370-1) and 1g Brij35/l (Aldrich no. 85,536-6). The mixture should be filtered before the addition of Brij. The OPA buffer should be run at a flow of 0.5 ml/minute with the post-column eluent using a mixing tee (Upchurch no. U466) and mixing column (Supelco no. 5-8319) packed with glass beads (Supelco no.5-9201).

29.3.2 Carboxymethylysine Assay by GC/MS

Principle: The method is the same as for furosine by GC/MS, except that the samples are reduced with $NaBH_4$ to prevent artifactual CML formation during hydrolysis. Utilization of selected ion monitoring methodology and electron impact ionization mode allows to quantitate both CML and furosine in the same run, whereby CML trifluoroacetyl derivative is assayed as ion at m/z = 392 [60].

Sample preparation: See "Furosine by GC/MS," Section 29.2.4.

Procedure: See "Furosine by GC/MS," Section 29.2.4.

29.3.3 Pentosidine Assay by HPLC

Principle and Synthesis of Pentosidine Standard: Pentosidine is a minor compound of the advanced Maillard reaction, and therefore it is present in small quantities both *in vivo* and *in vitro*. There are currently no commercial sources of pentosidine. Pentosidine can be prepared as described by Grandhee and Monnier [81]. Higher yields are obtained when the reaction is initiated from ribated N-α-acetyl lysine. The latter is then reacted with N-acetyl arginine to form N-acetylpentosidine which is deacetylated by heating in 6N HCL overnight (yield 3 to 10%).

Pentosidine cannot be derivatized and assayed by GC/MS because of its polarity. A fluorescent HPLC detector is needed because it is present *in vivo* in quantities not detectable with a UV detector. Tissues need to be acid hydrolyzed for release of pentosidine.

Synthesis of Ribated N-α-acetyl Lysine

It can be relatively easily prepared by boiling under reflux while stirring N-α-acetyl-lysine (1M) with D-ribose (1.5M) in 250 ml methanol for 2 hrs. The compounds will slowly go into solution. Add very small amounts of water if necessary. Evaporate to dryness and dilute in 500ml of 0.2M pyridine-formate at pH 3.25 (check pH at the end). Save 1 ml for thin layer chromatography. Load onto a Dowex-50W column in the H+-form and equilibrate with the loading buffer. The pH of the effluent must be the same as that of the loading buffer! Monitor the effluent by TLC with cellulose plates (EM Science #5552) in methanol:pyridine:acetic acid: H_2O = 40:20:6:10) detecting with both ninhydrin (500 mg ninhydrin in 203 ml ethanol and 40 ml acetic acid) and triphenyl tetrazolium chloride (TTC, 4g in 100 ml and 104 ml NaOH 1 M). Collect and pool the fractions that are both ninhydrin and TTC positive, evaporate under reduced pressure to dryness, repeat twice by adding methanol to remove excess pyridine. A brown, oily residue is obtained. The Amadori product is then obtained as a solid by diluting in 250 ml methanol and dropwise addition of this solution to an Erlenmeyer flask containing 2000 ml cold ethyl acetate while stirring. A fluffy, white-creamy precipitate is obtained. Let stir overnight. The product is highly hygroscopic. Do not try to filter! Decant most of the ethyl acetate. A solid powder can be obtained by transfering the slurry into polypropylene 50 ml culture tubes and evaporating the excess solvent with N_2. The residual solid can be crushed into powder under an atmosphere of nitrogen. Yield 50% on the average.

Synthesis of N-α-Acetyl Pentosidine and Standard Pentosidine

Incubation of ribated lysine with N-α-acetyl-arginine (1:3) in phosphate buffer leads to the formation of a single major fluorescent compound which can be detected by HPLC (excitation 335 nm, emission 385 nm) using a C-18 reverse phase column as a peak eluting at 17 min using the HPLC program for pentosidine (see below). The peak is collected and acid hydrolyzed to yield the deacetylated pentosidine peak which elutes at 30 min. Using repeated injections and collection, sufficient pentosidine can be obtained for a UV spectrum from which pentosidine concentration can be calculated assuming a molar extinction coefficient of 4522 Mol^{-1} cm^{-1} L (in 0.1N HCl) [23].

HPLC Procedure

For the detection of pentosidine equal amounts of protein (ca. 200 µg collagen, 5 mg plasma) from each sample is injected onto a reverse phase C18 column (Vydac No. 218TP104, Hesperia, CA). Buffer A consists of HPLC water with heptafluorobutyric acid (0.01 M; Sigma H-# 7133). Buffer B consists of 60% acetonitrile (v/v, Fisher no. A994SK-4) and heptafluorobutyric acid (0.01 M). Initially, the mobile phase is at 16% B; the program consists of a linear gradient to 28% B for 35 minutes at a flow rate of 1ml/min. Elution of pentosidine is detected by fluorescence at excitation wavelength 335 nm/emission wavelength 385. Pentosidine elutes at ≈30 min.

Acid hydrolysis creates fluorescent artifacts which may interfere with pentosidine determination in certain samples (e.g., for plasma, and when pentosidine concentration is very small). Therefore, in order to quantitate levels accurately, a dual-column system has been developed consisting sequentially of reverse-phase and ion-exchange separation (Odetti et al. 1992). First, pentosidine is collected from each sample using the column, solvents and program described above. Generally, eluent is collected three minutes before and after the standard retention time. The collected solvent is evaporated and each pellet is resuspended in 250 µl of cation exchange buffer A (described below).

Injection of the collected material is on a sulfopropyl cation-exchange column (Protein-Pak, SP 5PW. 7.5 × 75 mm, Waters). Buffer A consists of 0.02 M sodium acetate (pH 4.47). Buffer B consists of 0.02 M sodium acetate (pH 4.47) and 0.3 M NaCl. Initially, the mobile phase is at 0% B; the program consists of a curved gradient to 20% B for 40 minutes at a flow rate of 1ml/min. Elution of pentosidine is detected by fluorescence at excitation wavelength 335 nm/emission wavelength 385. Pentosidine elutes at ≈25 min.

29.4 Immunological Assays of Glycoxidation Products and AGEs

29.4.1   General Considerations

Immunoreactive AGEs can be defined as any protein that has been incubated for a length of time (e.g., 30 days) with high glucose concentration (typically 1M) in high phosphate buffer (e.g., 0.2M) at pH 7.4 without chelating agents and used for immunization. First developed by Nakayama and Makita et al. [82–84] but also by Horiuchi [85,86], such antisera or monoclonal antibodies are able to crossreact with proteins from various biological sources, such as plasma proteins from animals or individuals with diabetes or end stage renal disease, or in aging human lenses and skin [53,68,83–89].

Subsequent studies in two different laboratories showed unequivocally that such anti-AGE antibodies recognized CML as the major determinant [38,90] whereby other comparatively minor epitopes are present as well [90]. From this perspective, it is likely that anti-AGE antibodies will be equivalent to anti-CML antibodies for most ELISAs, but immunochemical differences may emerge in situ.

In contrast to antibodies that have been generated to the native (e.g., anti-glycoalbumin [55]) or the sugar-incubated protein, antibodies to specific advanced Maillard reaction products can be generated against the hapten-modified protein as in the case of anti-pyrraline and anti-pentosidine antibodies [91–93]. In these instances, the synthetic AGE is covalently linked to the carrier using the dicyclohexyl carbodiimide reagent.

One general problem emerges with the design of antibodies to Maillard reaction modified proteins which has to do with the affinity of the antibody for the AGE-protein versus the affinity of the antibody for the free compound. For example, anti-AGE antibodies raised against glucose -modifed protein have a high affinity against AGE-protein or CML-protein synthesized by reductive alkylation with glyoxylic acid but have a 1000-fold lower affinity against free CML [38], thus leading to the false negative result that CML is not the epitope recognized by the antibody [84].Thus, such antibody is not suitable for determination of free CML. In contrast, Schleicher et al. had only a 50-fold loss in affinity for free CML when they used KLH-CML as immunogen (see below).

The situation, however, is quite the opposite in the case of anti-pyrraline and anti-pentosidine antibodies. Because the hapten is chemically coupled to the protein, the effect of the spacer arm leads to an antibody with high affinity for the free molecule but low affinity against the protein modified *in vivo* or during Maillard reaction *in vitro* due to steric hindrance. ELISAs based on such antibodies provide an accurate measure of the Maillard compound if the latter is released from the protein by hydrolysis, either enzymatically, or by acid for pentosidine and base for pyrraline, respectively [91].

29.4.2   Immunoreactive "AGE" ELISA

Principle: Rabbits are immunized against AGE-RNAse and the antiserum is tested against AGE-BSA as coating agent [84]. AGE-KLH can be used as immunogen as described by Reddy [38] but crossreactivity with normal human serum albumin may occur [94]. AGE protein standards are not universally defined and therefore only relative AGE values are obtained. Makita et al. defined one AGE unit as the amount of antibody-reactive material that is equivalent to 1 μg of BSA standard which is prepared by incubation of albumin (50mg/ml) with 0.5M glucose in 0.2M Na/PO$_4$ buffer. *Note:* do not use chelating agents and anaerobic conditions except for the explicit purpose of suppressing CML formation which is the major epitope in most antisera raised against glucose-incubated proteins!

Sample Preparation: In view of the fact that CML may form from lipids during sample preparation (homogenization, sonication), serum should be partially delipidated by centrifugation at 100,000 g for 30 min (lipids float on top) prior to storage under nitrogen or argon at –80°C. All specimens should be delipidated in chloroform : methanol 3:1 prior to protein extraction. Insoluble specimens can be partially solubilized by short-term treatment with trypsin or collagenase. Avoid pronase or proteinase K as these might release free CML which has much lower affinity for the antibody.

Procedure for Competitive ELISA (adapted from Makita et al. [84])

Immunize New Zealand White rabbits by injecting in multiple sites intradermally 10–100 µg AGE-protein emulsified in completed Freund adjuvans, and boost after 10 days using AGE protein in incomplete Freund adjuvans. Save preimmune serum. Thereafter, boost weekly until satisfactory titer has been obtained (e.g., 1: 2000 or more for half maximal binding to AGE-coated plates). Use AGE-BSA (10 ug/ml PBS) as coating agent. Add 100 ul/well to a 96-well Nunc microtiter plate (Nunc Immunoplate, GBCO, Grand Island, NY), incubate for 6 hrs at 37°C in moist chamber. Wash three times with 150 ul of a solution containing PBS, 0.05% Tween 20, 1 mM sodium azide (PBS-Tween). Block by incubation for 60 min.under shaking with 200 ul of PBS containing 2% goat serum, 0.1% BSA and 1 ml NaN_3. If high background levels are noticed, use chicken ovalbumin instead. After washing with Tween-PBS, add 50 ul of a solution containing the antiserum (e.g., 1:1000 dilution, or as determined in preliminary experiments) mixed with the test substance. This solution is conveniently prepared by mixing a 100 ul of antiserum 2× concentrated with the 2× concentrated test solution. Apply to microtiter plate in duplicate, incubate for 2 hrs at room temperature under shaking and wash. Develop plates with an anti-rabbit alkaline phosphatase conjugated antibody. Stop the reaction by adding 50 ul of 2.5N NaOH. Read absorbance at 420 nm. *Note:* for optimal results absorbance rate of increase without inhibitor should not reach more than 1.0 absorbance units in less than 30 min. If the absorbance develops too fast, the concentration of the coating agent and/or antibody titer needs to be decreased. Results are expressed as B/B_0 calculated as experimental OD- background OD (no antibody)/total OD (no competitor). A calibration curve is constructed. Good experimental data points should fall on the slope. One unit of AGE-BSA is arbitrarily defined as the amount of AGE-BSA which inhibits 50% of absorbance.

29.4.3 Carboxymethyllysine by ELISA

Principle: CML is the single major advanced glycation compound which has been detected so far in reactions involving glucose (and most other reducing sugars) and proteins at physiological pH in phosphate buffer.Therefore, it is not surprising that it has been found to be the major epitope recognized in "AGE proteins" [38,90]. For this reason it is relatively easy to generate antibodies to CML modified proteins. One such method has been described by Schleicher [95,96]. The immunogen consists of KLH that is modified with glyoxylic acid under reducing conditions.

Sample Preparation: Specimens need to be in soluble form. Thus, the method is not applicable for quantitation of CML in insoluble collagen-rich tissue fractions. However, CML can be released by acid hydrolysis. In that case, antibody affinity decreases from 50–100-fold [95].

Procedure: Modified according to Schleicher [95].

Immunogen
10 mg KLH (BSA or HSA as coating agents) are dissolved in 2 ml 0.3 M sodium tetraborate pH 9.2 at 4°C and 1 mg $NaCNBH_3$ is added. 12 µmoles glyoxylic acid are added every 6 min. up to 60 min and the reaction is carried out for another 30 min.

Purification of Anti-CML Antibodies
The rabbit antiserum is passed over a column of CML-BSA coupled to Sepharose lysine and the antibody is eluted with 0.2 M glycine buffer (pH 2.5).

ELISA
The plates are coated with 100 µl of CML-BSA, incubated for 2hrs at room temperature, washed with Tween/PBS (0.01% Tween 20 in PBS containing 0.1% BSA). Blocking is performed with 200 µl of 2.7% crotein (Boehringer, Mannheim, Germany) for 1 hr and washed. The plates are incubated for 2 hr with 100 µl of the antibody solution (antiseum 1:1,000, or purified antibody 1:80) that was preincubated with and without competitor for 2 hrs at room temperature. After three washings, 100 µl of goat anti-rabbit IgG conjugated with peroxidase (1:8000) is added to each well and incubated for 1 hr. After five washings, the plate is developed with phenylenediamine substrate for 30 min in the dark and read at 450 nm with a microplate reader.

Calibration
There is no single accepted method for calibration of a CML ELISA. Schleicher et al. [96] used a BSA:CML standard thought to contain CML in a ratio 10:1. However, serum concentrations of CML determined by this ELISA were 50–100 times lower than by HPLC as determined in our laboratory, again illustrating the inherent problem of standardizing ELISAs for protein modification by Maillard reaction products. It is recommended that home-made standards be defined in their absolute content of CML and that correction factors be utilized as far as possible.

29.4.4   Pentosidine by ELISA

Principle: Rabbit polyclonal anti-pentosidine antibodies were developed by Taneda and Monnier [91]. Purified N-acetylpentosidine was coupled to KLH and used for immunization.The purified IgG fraction is used in a competitive enzyme-linked immunosorbent assay using pentosidine coupled to bovine serum albumin as coating agent. The reader is referred to the original procedure which is described in great detail.

Sample Preparation: See original procedure. Sample preparation is critical since pentosidine is present in low levels, especially in plasma proteins. The assay works best with acid hydrolyzed material prepared as described above for "Furosine by HPLC." Protein K or pronase E digested material can also be utilized. However, commercial specimens may naturally contain some pentosidine which may interfere with the assay. Pentosidine content in these batches should be assayed by HPLC and levels should not exceed the working range of the assay, i.e., final concentration 1 to 10 pmol per 100 µl/well.

Procedure: The reader is referred to original procedure. The assay is standardized with free pentosidine synthesized as described under "pentosidine by HPLC."

29.4.5 Pyrraline by ELISA

Principle: Pyrraline, also called pyrrole lysine (ε-(2-formyl-5-hydroxymethyl-pyrrol-1-yl)-L-norleucine) is a non-oxidative product of the advanced Maillard reaction. It has been detected in proteins exposed to glucose or Amadori products *in vitro* and, in elevated levels, in plasma from diabetic individuals, in urine from patients with end-stage renal disease and in human lenses [92,93,97–101]. Pyrraline was detected in lens Descemet membrane and lens capsule [102]. Its precise relationship with tissue aging is unknown. In one study, pyrraline could not be detected by ELISA in any samples [103]. However, its presence in biological specimens has been ascertained by GC/MS [101]. The ELISA described below is based on an anti-pyrraline monoclonal antibody that was raised against hapten-coupled caproyl pyrraline to KLH. The antibody has high affinity towards free pyrraline and pyrraline in histological tissue sections, and a lower affinity toward soluble glycated protein containing pyrraline together with other Maillard products.

Sample Preparation: Best results are obtained with solubilized, defatted proteins. Acid hydrolysis destroys pyrraline, but alkaline hydrolysis (dissolved 3 to 5 mg protein in 0.8 ml containing Ba(OH)$_2$ • 8 H$_2$0) followed by neutralization with 4N sulfuric acid is effective at releasing protein-bound pyrraline [100,104]. Digestion with pronase E (0.5% w/w) 3 times during 24 hours is effective. Appropriate blank with enzyme-only is needed.

Procedure: The procedure for generation of poly- and monoclonal anti-pyrraline antibodies in mice has been described in detail by Miyata [92].

Acknowledgments

This work was supported in part by grants AG 05601 and EY 07099 from the National Insitute of Aging and the National Eye Institute, respectively.

References

1. Ledl F, Schleicher E. New Aspects of the Maillard reaction in foods and in the human body, *Angew. Chem.*, 29, 565, 1990.
2. Fluckiger R, Gallop PM. Measurement of nonenzymatic protein glycosylation, *Meth. Enzymol.*, 106, 77, 1984.
3. Vlassara H, Brownlee M, Cerami A. Nonenzymatic glycosylation of peripheral nerve protein in diabetes mellitus, *Proc. Natl. Acad. Sci. (USA)*, 78, 5190, 1981.
4. Mortensen HB, Christophersen C. Glycosylation of human hemoglobin A kinetics and mechanisms studies by isoelectric focussing, *Biochim. Biophys. Acta*, 707, 154, 1982.
5. Namiki M, Hayashi T. A new mechanism of the Maillard reaction involving sugar fragmentation and free radical formation. In: Waller GR, Feather MS, eds. The Maillard Reaction in Foods and Nurtition. ACS Symposium Series. Vol. 215. Washington D.C.: American Chemical Society, 1983:21.
6. Glomb MA, Monnier VM. Mechanism of protein modification by glyoxal and glycolaldehyde reactive intermediates of the Maillard reaction, *J. Biol. Chem.*, 270, 10017, 1995.

7. Baynes JW, Thorpe SR, Murtiashaw MH. Nonenzymatic glucosylation of lysine residues in albumin. *Meth. Enzymol.* Vol. 106: Academic Press, Inc, New York, 1984:88.

8. Schleicher E, Wieland OH. Specific quantitation by HPLC of protein (lysine) bound glucose in human serum albumin and other glycosylated proteins, *J. Clin. Chem. Clin. Biochem.*, 19, 81, 1981.

9. Baynes JW, Watkins NG, Fisher CI, et al. The Amadori product on protein: structure and reactions. In: Baynes JW, Monnier VM, eds. The Maillard Reaction in aging, diabetes and nutrition. *Prog. Clin. Biol. Res.* Vol. 304. New York: Alan R. Liss, Inc., 1989:43.

10. Iberg N, Fluckiger R. Nonenzymatic glycosylation of albumin *in vivo*, *J. Biol. Chem.*, 261, 13442, 1986.

11. Patrick JS, Thorpe SR, Baynes JW. Nonenzymatic glycosylation of protein does not increase with age in normal human lenses, *J. Gerontol.*, 45, B18, 1990.

12. Schnider SL, Kohn RR. Glucosylation of human collagen in aging and diabetes mellitus, *J. Clin. Invest.*, 66, 1179, 1980.

13. Oimomi M, Kitamura Y, Nishimoto S, Matsumoto S, Hatanaka H, Baba S. Age-related acceleration of glycation of tissue proteins in rats, *J. Gerontol.*, 41, 695, 1986.

14. Sell DR. Aging promotes the increase of early glycation Amadori product as assessed by furosine levels in rodent skin collagen. The relationship to dietary restriction and glycoxidation, *Mech. Aging. Dev.*, in press, 1997.

15. Perry RE, Swamy MS, Abraham EC. Progressive changes in lens crystallin glycation and high-molecular-weight aggregate fromation leading to cataract development in streptozotocin-diabetic rats, *Exp. Eye Res.*, 44, 269, 1987.

16. Monnier VM, Cerami A. Nonenzymatic browning *in vivo*: Possible process for aging long-lived proteins, *Science*, 211, 491, 1981.

17. Abraham E, Tsai C, Abraham A, Swamy M. Formation of early and advanced glycation products of lens crystallins with erythrose, ribose and glucose. In: Finot PA, Aeschbacher HU, Hurrell RF, Liardon R, eds. The Maillard Reaction in Food Processing, Human Nutrition and Physiology. Basel: Birkhauser Verlag, 1990:437.

18. Hammes H-P, Brownlee M, Edelstein D, Saleck M, Martin S, Federlin K. Aminoguanidine inhibits the development of accelerated diabetic retinopathy in the spontaneous hypertensive rat, *Diabetologia*, 37, 32, 1994.

19. Doi T, Vlassara H, Kirstein M, Yamada Y, Striker GE, Striker LJ. Receptor-specific increase in extracellular matrix production in mouse mesangial cells by advanced glycosylation end products is mediated via platelet-derived growth factor, *Proc. Natl. Acad. Sci. (USA)*, 89, 2873, 1992.

20. Schmidt AM, Mora R, Cao R, et al. The endothelial cell binding site for advanced glycation end products consists of a complex: an integral membrane protein and a lactoferrin-like polypeptide, *J. Biol. Chem.*, 269, 9882, 1994.

21. Smith MA, Taneda S, Richey PL, et al. Advanced Maillard reaction end products are associated with Alzheimer disease pathology, *Proc. Natl. Acad. Sci. (USA)*, 91, 5710, 1994.

22. Baynes JW. Perspectives in Diabetes: Role of oxidative stress in development of complications in diabetes, *Diabetes*, 40, 405, 1991.

23. Sell DR, Monnier VM. Structure elucidation of a senescence cross-link from human extracellular matrix, *J. Biol. Chem.*, 264, 21597, 1989.

24. Dyer DG, Dunn JA, Thorpe SR, et al. Accumulation of Maillard reaction products in skin collagen in diabetes and aging, *J. Clin. Invest.*, 91, 2463, 1993.

25. Nagaraj RH, Sell DR, Prabhakaram M, Ortwerth BJ, Monnier VM. High correlation between pentosidine protein crosslinks and pigmentation implicates ascorbate oxidation in human lens senescence and cataractogenesis, *Proc. Natl. Acad. Sci (USA)*, 88, 10257, 1991.

26. Cefalu WT, Bell-Farrow, A.D., Wang, Z.Q, Sonntag, W.E, Fu, M.X, Baynes, J.W, Thorpe, S.R. Caloric restriction decreases age-dependent accumulation of the glycoxidation products Ne-(carboxymethyl)lysine and pentosidine, *J. Gerontol.*, 50A, B337, 1995.

27. Dunn JA, McCance DR, Thorpe SR, Lyons TJ, Baynes JW. Age-dependent accumulation of Ne-(carboxymethyl)lysine and Ne-(carboxymethyl)hydroxylysine in human skin collagen, *Biochemistry*, 30, 1205, 1991.

28. Dunn JA, Patrick JS, Thorpe SR, Baynes JW. Oxidation of glycated proteins: Age-dependent accumulation of N-(Carboxymethyl)lysine in lens proteins, *Biochemistry*, 28, 9464, 1989.

29. McCance DR, Dyer DG, Dunn JA, et al. Maillard reaction products and their relation to complications in insulin-dependent diabetes mellitus, *J. Clin. Invest.*, 91, 2470, 1993.

30. Friedlander MA, Wu YC, Elgawish A, Monnier VM. Early and advanced glycosylation end products. Kinetics of formation and clearance in peritoneal dialysis, *J. Clin. Inv.*, 97, 728, 1996.

31. Sell DR, Lane MA, Johnson WA, et al. Longevity and the genetic determination of collagen glycoxidation kinetics in mammalian senescence, *Proc. Natl. Acad. Sci. (USA)*, 93, 485, 1996.

32. Allfrey AC. Role of iron and oxygen radicals in the progression of chronic renal failure, *Am. J. Kid. Dis.*, 23, 183, 1994.

33. Zyzak DV, Richardson JM, Thorpe SR, Baynes JW. Formation of reactive intermediates from Amadori compounds under physiological conditions, *Arch. Biochem. Biophys.*, 316, 547, 1995.

34. Wells-Knecht KJ, Zyzack DV, Litchfield JE, Thorpe SR, Baynes JW. Mechanism of autoxidative glycosylation: identification of glyoxal and arabinose as intermediates in the autoxidative modification of proteins by glucose, *Biochemistry*, 34, 3702, 1995.

35. Wolff SP, Dean RT. Glucose autoxidation and protein modification: The potential role of 'autoxidative glycosylation' in diabetes, *Biochem. J.*, 245, 250, 1987.

36. Swamy MS, Tsai C, Abraham A, Abraham EC. Glycation mediated lens crystallin aggregation and cross-linking by various sugars and sugar phosphates *in vitro*, *Exp. Eye Res.*, 56, 177, 1993.

37. Ortwerth BJ, Feather MS, Olesen PR. The precipitation and cross-linking of lens crystallins by ascorbic acid, *Exp. Eye Res.*, 47, 155, 1988.

38. Reddy S, Bichler J, Wells-Knecht KJ, Thorpe SR, Baynes JW. Ne-(Carboxymethyl)lysine is a dominant advanced glycation end product (AGE) antigen in tissue proteins, *Biochemistry*, 34, 10872, 1995.

39. Anderson MM, Hazen SL, Hsu FF, Heinecke JW. Human neutrophils employ the myeloperoxidase-hydrogen peroxide-chloride system to convert hydroxy-amino acids into glycolaldehyde, 2-hydroxypropanal and acrolein: A mechanism for the generation of highly reactive a-hydroxy and a,b-unsaturated aldehydes by phagocytes at sites of inflammation, *J. Clin. Inv.*, 99, 397, 1997.

40. Thornalley PJ. The Glyoxalase System in Health and Disease, *Mol. Aspects. Med.*, 14, 289, 1993.

41. Szwergold BS, Kappler F, Brown TR. Identification of fructose 3-phosphate in the lens of diabetic rats, *Science*, 247, 451, 1990.

42. Hicks M, Delbridge L, Yue DK, Reeve TS. Catalysis of lipid peroxidation by glucose and glycosylated collagen, *Biochem. Biophys. Res. Comm.*, 151, 649, 1988.

43. Mullarkey CJ, Edelstein D, Brownlee M. Free radical generation by early glycation products: A mechanism for accelerated atherogenesis in diabetes, *Biochem. Biophys. Res. Commun.*, 173, 932, 1990.

44. Vlassara H, Brownlee M, Cerami A. Nonenzymatic glycosylation of peripheral nerve protein in diabetes mellitus, *Proc. Natl. Acad. Sci. (USA)*, 78, 5190, 1981.

45. Gould BJ, Hall PM. m-Aminophenylboronate affinity ligands distinguish between nonenzymically glycosylated proteins and glycoproteins, *Clin.Chim. Acta*, 163, 225, 1987.

46. Lapolla A, Fedele D, Seraglia R, et al. A new effective method for the evaluation of glycated intact plasma proteins in diabetic subjects, *Diabetologia*, 38, 1076, 1995.

47. Miedema K, Casparie T. Glycosylated glycohemoglobins: biochemical evaluation and clinical utility, *Ann. Clin. Biochem.*, 21, 2, 1984.

48. Baker JR, Zyzak DV, Thorpe SR, Baynes JW. Mechanism of fructosamine assay: Evidence against role of superoxide as intermediate in nitroblue tetrazolium reduction, *Clin. Chem.*, 39, 2460, 1993.

49. Baker JR, Zyzak DV, Thorpe SR, Baynes JW. Chemistry of the fructosamine assay: D-Glucosone is the product of oxidation of Amadori compounds, *Clin. Chem.*, 40, 1950, 1994.

50. Johnson RN, Metcalf PA, Baker JR. Fructosamine: a new approach to the estimation of serum glycosylprotein. An index of diabetic control, *Clin. Chim. Acta*, 127, 87, 1982.

51. Kato M, Nakayama H, Makita Z, et al. Radioimmunoassay for non-enzymatically glycated serum proteins, *Horm. Metabol. Res*, 21, 245, 1989.

52. Myint T, Hoshi S, Ookawara T, Miyazawa N, Suzuki K, Taniguchi N. Immunological detection of glycated proteins in normal and streptozotocin-induced diabetic rats using anti hexitol-lysine IgG, *Biochim. Biophys. Acta*, 1272, 73, 1995.

53. Nakamura Y, Horii Y, Nishino T, et al. Immunohistochemical localization of advanced glycosylation endproducts in coronary atheroma and cardiac tissue in diabetes mellitus, *Amer. J. Path.*, 143, 1649, 1993.

54. Kelly SB, Olerud JE, Witztum JL, Cutriss LK, Gown AM, Odland GF. A method for localizing the early products of nonenzymatic glycosylation in fixed tissue, *J. Invest. Derm.*, 93, 327, 1989.

55. Cohen MP, Hud E. Production and characterization of monoclonal anitbodies against human glycoalbumin, *J. Immunol. Meth.*, 117, 121, 1989.

56. Finot PA, Bricout J, Viani R, Mauron J. Identification of a new lysine derivative obtained upon acid hydrolysis of heated milk, *Experientia*, 24, 1097, 1968.

57. Bujard E, Finot PA, Madelaine R, Van Kiet AL, Isely DeA. Mesure de la disponibilite et du blocage de la lysine dans les laits industriels, *Ann. Nutr. Alim.*, 32, 291, 1978.

58. Resmini P, Pellegrino L, Battelli G. Accurate quantification of furosine in milk and dairy products by a direct HPLC method, *Ital. J. Food Sci.*, 173, 1990.

59. Wu YC, Monnier VM, Friedlander MA. Reliable determination of furosine in human serum and dialysate proteins by high-performance liquid chromatography, *J. Chrom.*, B667, 328, 1995.

60. Wells-Knecht MC, Thorpe SR, Baynes JW. Pathways of formation of glycoxidation products during glycation of collagen, *Biochemistry*, 34, 15134, 1995.

61. Niwa T, Katsuzaki T, Miyazaki S, et al. Immunohistochemical detection of imidazolone, a novel advanced glycation end product, in kidneys and aortas of diabetic patients, *J. Clin. Invest.*, 99, 1272, 1997.

62. Bookchin RM, Gallop PM. Structure of hemoglobin A1c: Nature of the N-terminal beta chain blocking group, *Biochem. Biophys. Res. Comm.*, 32, 86, 1968.

63. Ahmed MU, Thorpe SR, Baynes JW. Identification of Ne-(carboxymethyl)lysine as a degradation product of fructoselysine in glycated protein, *J. Biol. Chem.*, 261, 4889, 1986.

64. Slight SH, Prabhakaram M, Shin DB, Feather MS, Ortwerth BJ. The extent of N e-(carboxymethyl)lysine formation in lens proteins and polylysine by autoxidation products of ascorbic acid, *Biochim. Biophys. Acta*, 1117, 199, 1992.

65. Friedlander MA, Randall CP, Baumgartner GP, Wu YC, Deoreo PB. Formation of protein-bound and clearance of free AGEs in end stage renal failure, *Amer. J. Kid. Dis.*, submitted, 1997.

66. Fu MX, Requena JR, Jenkins AJ, Lyons TJ, Baynes JW, Thorpe SR. The advanced glycation end product, Ne-(carboxymethyl)lysine, is a product of both lipid peroxidation and glycoxidation reactions, *J. Biol. Chem.*, 271, 9982, 1996.

67. Monnier VM, Kenny D, Fogarty JF, et al. The relationship between collagen glycation, glycoxidation and the complications of diabetes in IDDM, *in press*, 1997.

68. Kume S, Takeya M, Mori T, et al. Immunohistochemical and ultrastructural detection of advanced glycation end products in atherosclerotic lesions of human aorta with a novel specific monoclonal antibody, *Amer. J. Pathol.*, 147, 654, 1995.

69. Sell DR, Monnier VM. Aging of long-lived proteins: collagen, elastin, proteoglycans and lens crystallins. In: Masoro E, ed. *Handbook of Physiology: Physiology of Aging*. Vol. 14, 1994:235.

70. Beisswenger PJ, Moore LL, Brinck-Johnson T, Curphey TJ. Increased collagen-linked pentosidine levels and advanced glycosylation end products in early diabetic nephropathy, *J. Clin. Invest*, 92, 212, 1993.

71. Sell DR, Lapolla A, Odetti P, Fogarthy J, Monnier VM. Pentosidine formation in skin correlates with severity of complications in individuals with long-standing IDDM, *Diabetes*, 41, 1286, 1992.

72. Monnier VM, Kohn RR, Cerami A. Accelerated age-related browning of human collagen in diabetes mellitus, *Proc Natl Acad*, 81, 583, 1984.

73. Fu M-X, Wells-Knecht KJ, Blackledge JA, Lyons TJ, Thorpe SR, Baynes JW. Glycation, glycoxidation and crosslinking of collagen by glucose: Kinetics, mechanisms and inhibition of late stages of the Maillard reaction, *Diabetes*, 43, 1, 1993.

74. Odetti P, Pronzato MA, Noberasco G, et al. Relationships between glycation and oxidation related fluorescences in rat collagen during aging, *Lab. Invest.*, 70, 61, 1994.

75. Nagaraj RH, Monnier VM. Isolation and Characterization of a blue fluorophore from human eye lens crystallins; *In vitro* formation from Maillard reaction with ascorbate and ribose, *Biochimica et Biophysica Acta.*, 1116, 34, 1992.

76. Ienaga K, Nakamura K, Hochi T, et al. Crosslines, fluorophores in the age-related cross-linked proteins, *Contrib. Nephrol.*, 112, 42, 1995.

77. Ienaga K, Karita H, Hochi T, et al. Crossline-like structures accumulates as fluorescent advanced glycation end products in renal tissue of rats with diabetic nephropathy, *Proc. Japan Acad.*, 72B, 79, 1996.

78. Nakamura K, Hasegawa T, Fukunaga Y, Ienaga K. Crosslines A and B as candidates for the flurophores in age-and diabetes-related cross-linked proteins, and their diacetates produced by Maillard reaction of alpha-N-acetyl-L-lysine with D-glucose, *Chem. Soc. Chem. Comm.*, 992, 1992.

79. Nakamura K, Nakazawa Y, Ienaga K. Acid-stable fluorescent advanced glycation end products: Vesperlysines A, B, and C are formed as crosslinked products in the Maillard reaction bewteen lysine or proteins with glucose, *Biochem. Biophys. Res. Comm.*, 232, 227, 1997.

80. Knecht KF, Dunn JA, McFarland KF, et al. Effect of diabetes and aging on carboxymethyllysine levels in human urine, *Diabetes*, 40, 190, 1991.

81. Grandhee SK, Monnier VM. Mechanism of formation of the Maillard protein cross-link pentosidine, *J. Biol. Chem.*, 266, 11649, 1991.

82. Nakayama H, Kato M, Makita Z, et al. A radioimmunoassay for an advanced glycosylation endproduct, *J. Immunol. Meth.*, 112, 57, 1988.

83. Makita ZMD, Randoff SPD, Rayfield EJMD, et al. Advanced glycosylation end products in patients with diabetic nephropathy, *New Engl. J. Med.*, 836, 1991.

84. Makita Z, Vlassara H, Cerami A, Bucala R. Immunochemical detection of advanced glycosylation end products *in vivo*, *J. Biol. Chem.*, 267, 5133, 1992.

85. Horiuchi S, Araki N, Morino Y. Immunochemical approach to characterize advanced glycation end products of the Maillard reaction, *J. Biol. Chem.*, 266, 7329, 1991.

86. Araki N, Ueno N, Chakrabarti B, Morino Y, Horiuchi S. Immunochemical evidence for the presence of advanced glycation end products in human lens proteins and its positive correlation with aging, *J. Biol. Chem.*, 267, 10211, 1992.

87. Miyata T, Oda O, Inagi R, et al. Beta2-microglobulin modified with advanced glycation end products is a major component of hemodialysis-associated amyloidosis, *J. Clin. Invest*, 92, 1243, 1993.

88. Meng J, Sakata N, Takebayashi S, et al. Advanced glycation end products of the Maillard reaction in aortic pepsin-insoluble and pepsin-soluble collagen from diabetic rats, *Diabetes*, 45, 1037, 1996.

89. Beisswenger P, Makita Z, Curphey TJ, et al. Formation of immunochemical advanced glyco-sylation end products precedes and correlates with early manifestations of renal and retinal disease in diabetes, *Diabetes*, 44, 824, 1995.

90. Ikeda K, Higashi T, Sano H, et al. Ne-(carboxymethyl)lysine protein adduct is a major immu-nological epitope in proteins modified with advanced glycation end products of the Maillard reaction, *Biochemistry*, 35, 8075, 1996.

91. Taneda S, Monnier VM. ELISA of pentosidine, an advanced Maillard reaction product, in biological specimens, *Clin. Chem.*, 1984.

92. Miyata S, Monnier VM. Immunochemical detection of advanced glycosylation end prod-ucts in diabetic tissues using monoclonal antibody to pyrraline, *J. Clin. Invest.*, 89, 1102, 1992.

93. Hayase F, Nagaraj RH, Miyata S, Njoroge FG, Monnier VM. Aging of proteins: immunological detection of a glucose-derived pyrrole formed during Maillard reaction *in vivo*, *J. Biol. Chem.*, 264, 3758, 1989.

94. Friedlander MA. Personal Communication

95. Schleicher ED, Wagner E, Nerlich AG. Increased accumulation of the glycosylation product Ne-(carboxymethyl)lysine in human tissues in diabetes and aging, *J. Clin. Inv.*, 99, 457, 1997.

96. Gempel KE, Wagner EM, Schleicher ED. Production and characterization of antibodies against carboxymethyllysine-modified proteins. In: Labuza TP, Reineccius GA, Monnier VM, O'Brien J, Baynes JW, eds. *Maillard Reactions in Chemistry, Food and Health*. Cambridge, UK: The Royal Society of Chemistry, 1994:393.

97. Hayase F, Kato H. Maillard reaction products from D-glucose and butylamine, *Agric. Biol. Chem.*, 49, 467, 1985.

98. Henle T, Klostermeyer H. Determination of protein-bound 2-amino-6-(2-formyl-1-pyrrolyl)-hexanoic acid ('pyrraline') by ion exchange chromatography and photodiode array detection, *Z. Lebensm. Unters. Forsch.*, 196, 1, 1993.

99. Nagaraj RH, Portero-Otin M, Monnier VM. Pyrraline ether crosslinks as a basis for protein crosslinking by the advanced Maillard reaction in aging and diabetes, *Arch. Biochem. Biophys.*, 325, 152, 1996.

100. Nagaraj RH, Sady C. The presence of a glucose-derived Maillard product in the human lens, *FEBS lett.*, 382, 234, 1996.

101. Odani H, Shinzato T, Matsumoto Y, et al. First evidence for accumulation of protein-bound and protein-free pyrraline in human plasma by mass spectrometry, *Biochem. Biophys. Res. Comm.*, 224, 237, 1996.

102. Marion MS, Carlson EC. Immunoelectron microscopic analyses of Maillard reaction products in bovine anterior lens capsule and Descemet's membrane, *Biochim. Biophys. Acta*, 1191, 33, 1994.

103. Smith PR, Somani HH, Thornalley PJ, Benn J, Sonksen PH. Evidence against the formation of 2-amino-6-(2-formy-5-hydroxymethyl-pyrrol-l-yl)-hexanoic acid ("pyrraline") as an early-stage product or advanced glycation end product in non-enzymic protein glycation, *Clin. Sci.*, 84, 87, 1993.

104. Portero-Otin M, Nagaraj RH, Monnier VM. Chromatographic evidence for pyrraline formation during protein glycation *in vitro* and *in vivo*, *Biochim. Biophys. Acta*, 1247, 74, 1995.

105. Nakayama H, Taneda S, Manda N, et al. Radioimmunoassay for nonenzymatically glycated protein in human serum, *Clin. Chim. Acta*, 158, 293, 1986.

106. Nakayama H, Makita Z, Kato M, et al. Quantitative enzyme-linked immunosorbent assay (ELISA) for non-enzymatically glycated serum protein, *J. Immunol. Meth.*, 99, 95, 1987.

107. Nagaraj RH, Kern TS, Sell DR, Fogarty J, Engerman RL, Monnier VM. Evidence of a glycemic threshold for the formation of pentosidine in diabetic dog lens but not in collagen, *Diabetes*, 45, 587, 1996.

108. Lyons TJ, Bailie KE, Dyer DG, Dunn JA, Baynes JW. Decrease in skin collagen glycation with improved glycemic control in patients with insulin-dependent diabetes mellitus, *J. Clin. Invest.*, 87, 1910, 1991.

109. Elgawish A, Glomb M, Friedlander MA, Monnier VM. Involvement of hydrogen peroxide in collagen crosslinking by high glucose *in vitro* and *in vivo*, *J. Biol. Chem.*, 271, 12964, 1996.
110. Odetti P, Fogarty J, Sell DR, Monnier VM. Chromatographic quantitation of plasma and erythrocyte pentosidine in diabetic and uremic subjects, *Diabetes*, 41, 153, 1992.
111. Takahashi M, Ohishi T, Aoshima H, et al. The Maillard protein cross-link pentosidine in urine from diabetic patients, *Diabetologia*, 36, 664, 1993.
112. Uchiyama A, Ohishi T, Takahashi M, et al. Fluorophores for aging human articular cartilage, *J. Biochem.*, 110, 714, 1991.
113. Lyons TJ, Silbestri G, Dunn JA, Dyer DG, Baynes JW. Role of glycation in modification of lens crystallins in diabetic and nondiabetic senile cataracts, *Diabetes*, 40, 1010, 1991.

Index

A

T - #1069 - 101024 - C0 - 254/178/33 [35] - CB - 9780849331121 - Gloss Lamination